Lecture Notes in Computer Science 13431

More information about this series at https://link.springer.com/bookseries/558

Linwei Wang · Qi Dou · P. Thomas Fletcher ·
Stefanie Speidel · Shuo Li (Eds.)

Medical Image Computing and Computer Assisted Intervention – MICCAI 2022

25th International Conference
Singapore, September 18–22, 2022
Proceedings, Part I

Springer

Editors
Linwei Wang
Rochester Institute of Technology
Rochester, NY, USA

P. Thomas Fletcher
University of Virginia
Charlottesville, VA, USA

Shuo Li
Case Western Reserve University
Cleveland, OH, USA

Qi Dou
Chinese University of Hong Kong
Hong Kong, Hong Kong

Stefanie Speidel
National Center for Tumor Diseases
(NCT/UCC)
Dresden, Germany

ISSN 0302-9743 ISSN 1611-3349 (electronic)
Lecture Notes in Computer Science
ISBN 978-3-031-16430-9 ISBN 978-3-031-16431-6 (eBook)
https://doi.org/10.1007/978-3-031-16431-6

This Springer imprint is published by the registered company Springer Nature Switzerland AG
The registered company address is: Gewerbestrasse 11, 6330 Cham, Switzerland

Preface

We are pleased to present the proceedings of the 25th International Conference on Medical Image Computing and Computer-Assisted Intervention (MICCAI) which – after two difficult years of virtual conferences – was held in a hybrid fashion at the Resort World Convention Centre in Singapore, September 18–22, 2022. The conference also featured 36 workshops, 11 tutorials, and 38 challenges held on September 18 and September 22. The conference was also co-located with the 2nd Conference on Clinical Translation on Medical Image Computing and Computer-Assisted Intervention (CLINICCAI) on September 20.

MICCAI 2022 had an approximately 14% increase in submissions and accepted papers compared with MICCAI 2021. These papers, which comprise eight volumes of Lecture Notes in Computer Science (LNCS) proceedings, were selected after a thorough double-blind peer-review process. Following the example set by the previous program chairs of past MICCAI conferences, we employed Microsoft's Conference Managing Toolkit (CMT) for paper submissions and double-blind peer-reviews, and the Toronto Paper Matching System (TPMS) to assist with automatic paper assignment to area chairs and reviewers.

From 2811 original intentions to submit, 1865 full submissions were received and 1831 submissions reviewed. Of these, 67% were considered as pure Medical Image Computing (MIC), 7% as pure Computer-Assisted Interventions (CAI), and 26% as both MIC and CAI. The MICCAI 2022 Program Committee (PC) comprised 107 area chairs, with 52 from the Americas, 33 from Europe, and 22 from the Asia-Pacific or Middle East regions. We maintained gender balance with 37% women scientists on the PC.

Each area chair was assigned 16–18 manuscripts, for each of which they were asked to suggest up to 15 suggested potential reviewers. Subsequently, over 1320 invited reviewers were asked to bid for the papers for which they had been suggested. Final reviewer allocations via CMT took account of PC suggestions, reviewer bidding, and TPMS scores, finally allocating 4–6 papers per reviewer. Based on the double-blinded reviews, area chairs' recommendations, and program chairs' global adjustments, 249 papers (14%) were provisionally accepted, 901 papers (49%) were provisionally rejected, and 675 papers (37%) proceeded into the rebuttal stage.

During the rebuttal phase, two additional area chairs were assigned to each rebuttal paper using CMT and TPMS scores. After the authors' rebuttals were submitted, all reviewers of the rebuttal papers were invited to assess the rebuttal, participate in a double-blinded discussion with fellow reviewers and area chairs, and finalize their rating (with the opportunity to revise their rating as appropriate). The three area chairs then independently provided their recommendations to accept or reject the paper, considering the manuscript, the reviews, and the rebuttal. The final decision of acceptance was based on majority voting of the area chair recommendations. The program chairs reviewed all decisions and provided their inputs in extreme cases where a large divergence existed between the area chairs and reviewers in their recommendations. This process resulted in the acceptance of a total of 574 papers, reaching an overall acceptance rate of 31% for MICCAI 2022.

In our additional effort to ensure review quality, two Reviewer Tutorials and two Area Chair Orientations were held in early March, virtually in different time zones, to introduce the reviewers and area chairs to the MICCAI 2022 review process and the best practice for high-quality reviews. Two additional Area Chair meetings were held virtually in July to inform the area chairs of the outcome of the review process and to collect feedback for future conferences.

For the MICCAI 2022 proceedings, 574 accepted papers were organized in eight volumes as follows:

- Part I, LNCS Volume 13431: Brain Development and Atlases, DWI and Tractography, Functional Brain Networks, Neuroimaging, Heart and Lung Imaging, and Dermatology
- Part II, LNCS Volume 13432: Computational (Integrative) Pathology, Computational Anatomy and Physiology, Ophthalmology, and Fetal Imaging
- Part III, LNCS Volume 13433: Breast Imaging, Colonoscopy, and Computer Aided Diagnosis
- Part IV, LNCS Volume 13434: Microscopic Image Analysis, Positron Emission Tomography, Ultrasound Imaging, Video Data Analysis, and Image Segmentation I
- Part V, LNCS Volume 13435: Image Segmentation II and Integration of Imaging with Non-imaging Biomarkers
- Part VI, LNCS Volume 13436: Image Registration and Image Reconstruction
- Part VII, LNCS Volume 13437: Image-Guided Interventions and Surgery, Outcome and Disease Prediction, Surgical Data Science, Surgical Planning and Simulation, and Machine Learning – Domain Adaptation and Generalization
- Part VIII, LNCS Volume 13438: Machine Learning – Weakly-supervised Learning, Machine Learning – Model Interpretation, Machine Learning – Uncertainty, and Machine Learning Theory and Methodologies

We would like to thank everyone who contributed to the success of MICCAI 2022 and the quality of its proceedings. These include the MICCAI Society for support and feedback, and our sponsors for their financial support and presence onsite. We especially express our gratitude to the MICCAI Submission System Manager Kitty Wong for her thorough support throughout the paper submission, review, program planning, and proceeding preparation process – the Program Committee simply would not have be able to function without her. We are also grateful for the dedication and support of all of the organizers of the workshops, tutorials, and challenges, Jianming Liang, Wufeng Xue, Jun Cheng, Qian Tao, Xi Chen, Islem Rekik, Sophia Bano, Andrea Lara, Yunliang Cai, Pingkun Yan, Pallavi Tiwari, Ingerid Reinertsen, Gongning Luo, without whom the exciting peripheral events would have not been feasible. Behind the scenes, the MICCAI secretariat personnel, Janette Wallace and Johanne Langford, kept a close eye on logistics and budgets, while Mehmet Eldegez and his team from Dekon Congress & Tourism, MICCAI 2022's Professional Conference Organization, managed the website and local organization. We are especially grateful to all members of the Program Committee for their diligent work in the reviewer assignments and final paper selection, as well as the reviewers for their support during the entire process. Finally, and most importantly, we thank all authors, co-authors, students/postdocs, and supervisors, for submitting and presenting their high-quality work which made MICCAI 2022 a successful event.

We look forward to seeing you in Vancouver, Canada at MICCAI 2023!

September 2022

Linwei Wang
Qi Dou
P. Thomas Fletcher
Stefanie Speidel
Shuo Li

Organization

General Chair

Shuo Li	Case Western Reserve University, USA

Program Committee Chairs

Linwei Wang	Rochester Institute of Technology, USA
Qi Dou	The Chinese University of Hong Kong, China
P. Thomas Fletcher	University of Virginia, USA
Stefanie Speidel	National Center for Tumor Diseases Dresden, Germany

Workshop Team

Wufeng Xue	Shenzhen University, China
Jun Cheng	Agency for Science, Technology and Research, Singapore
Qian Tao	Delft University of Technology, the Netherlands
Xi Chen	Stern School of Business, NYU, USA

Challenges Team

Pingkun Yan	Rensselaer Polytechnic Institute, USA
Pallavi Tiwari	Case Western Reserve University, USA
Ingerid Reinertsen	SINTEF Digital and NTNU, Trondheim, Norway
Gongning Luo	Harbin Institute of Technology, China

Tutorial Team

Islem Rekik	Istanbul Technical University, Turkey
Sophia Bano	University College London, UK
Andrea Lara	Universidad Industrial de Santander, Colombia
Yunliang Cai	Humana, USA

Clinical Day Chairs

Jason Chan	The Chinese University of Hong Kong, China
Heike I. Grabsch	University of Leeds, UK and Maastricht University, the Netherlands
Nicolas Padoy	University of Strasbourg & Institute of Image-Guided Surgery, IHU Strasbourg, France

Young Investigators and Early Career Development Program Chairs

Marius Linguraru	Children's National Institute, USA
Antonio Porras	University of Colorado Anschutz Medical Campus, USA
Nicole Rieke	NVIDIA, Deutschland
Daniel Racoceanu	Sorbonne University, France

Social Media Chairs

Chenchu Xu	Anhui University, China
Dong Zhang	University of British Columbia, Canada

Student Board Liaison

Camila Bustillo	Technische Universität Darmstadt, Germany
Vanessa Gonzalez Duque	Ecole centrale de Nantes, France

Submission Platform Manager

Kitty Wong	The MICCAI Society, Canada

Virtual Platform Manager

John Baxter	INSERM, Université de Rennes 1, France

Program Committee

Ehsan Adeli	Stanford University, USA
Pablo Arbelaez	Universidad de los Andes, Colombia
John Ashburner	University College London, UK
Ulas Bagci	Northwestern University, USA
Sophia Bano	University College London, UK
Adrien Bartoli	Université Clermont Auvergne, France
Kayhan Batmanghelich	University of Pittsburgh, USA

Hrvoje Bogunovic	Medical University of Vienna, Austria
Ester Bonmati	University College London, UK
Esther Bron	Erasmus MC, the Netherlands
Gustavo Carneiro	University of Adelaide, Australia
Hao Chen	Hong Kong University of Science and Technology, China
Jun Cheng	Agency for Science, Technology and Research, Singapore
Li Cheng	University of Alberta, Canada
Adrian Dalca	Massachusetts Institute of Technology, USA
Jose Dolz	ETS Montreal, Canada
Shireen Elhabian	University of Utah, USA
Sandy Engelhardt	University Hospital Heidelberg, Germany
Ruogu Fang	University of Florida, USA
Aasa Feragen	Technical University of Denmark, Denmark
Moti Freiman	Technion - Israel Institute of Technology, Israel
Huazhu Fu	Agency for Science, Technology and Research, Singapore
Mingchen Gao	University at Buffalo, SUNY, USA
Zhifan Gao	Sun Yat-sen University, China
Stamatia Giannarou	Imperial College London, UK
Alberto Gomez	King's College London, UK
Ilker Hacihaliloglu	University of British Columbia, Canada
Adam Harrison	PAII Inc., USA
Mattias Heinrich	University of Lübeck, Germany
Yipeng Hu	University College London, UK
Junzhou Huang	University of Texas at Arlington, USA
Sharon Xiaolei Huang	Pennsylvania State University, USA
Yuankai Huo	Vanderbilt University, USA
Jayender Jagadeesan	Brigham and Women's Hospital, USA
Won-Ki Jeong	Korea University, Korea
Xi Jiang	University of Electronic Science and Technology of China, China
Anand Joshi	University of Southern California, USA
Shantanu Joshi	University of California, Los Angeles, USA
Bernhard Kainz	Imperial College London, UK
Marta Kersten-Oertel	Concordia University, Canada
Fahmi Khalifa	Mansoura University, Egypt
Seong Tae Kim	Kyung Hee University, Korea
Minjeong Kim	University of North Carolina at Greensboro, USA
Baiying Lei	Shenzhen University, China
Gang Li	University of North Carolina at Chapel Hill, USA

Xiaoxiao Li	University of British Columbia, Canada
Jianming Liang	Arizona State University, USA
Herve Lombaert	ETS Montreal, Canada
Marco Lorenzi	Inria Sophia Antipolis, France
Le Lu	Alibaba USA Inc., USA
Klaus Maier-Hein	German Cancer Research Center (DKFZ), Germany
Anne Martel	Sunnybrook Research Institute, Canada
Diana Mateus	Centrale Nantes, France
Mehdi Moradi	IBM Research, USA
Hien Nguyen	University of Houston, USA
Mads Nielsen	University of Copenhagen, Denmark
Ilkay Oksuz	Istanbul Technical University, Turkey
Tingying Peng	Helmholtz Zentrum Muenchen, Germany
Caroline Petitjean	Université de Rouen, France
Gemma Piella	Universitat Pompeu Fabra, Spain
Chen Qin	University of Edinburgh, UK
Hedyeh Rafii-Tari	Auris Health Inc., USA
Tammy Riklin Raviv	Ben-Gurion University of the Negev, Israel
Hassan Rivaz	Concordia University, Canada
Michal Rosen-Zvi	IBM Research, Israel
Su Ruan	University of Rouen, France
Thomas Schultz	University of Bonn, Germany
Sharmishtaa Seshamani	Allen Institute, USA
Feng Shi	United Imaging Intelligence, China
Yonggang Shi	University of Southern California, USA
Yang Song	University of New South Wales, Australia
Rachel Sparks	King's College London, UK
Carole Sudre	University College London, UK
Tanveer Syeda-Mahmood	IBM Research, USA
Qian Tao	Delft University of Technology, the Netherlands
Tolga Tasdizen	University of Utah, USA
Pallavi Tiwari	Case Western Reserve University, USA
Mathias Unberath	Johns Hopkins University, USA
Martin Urschler	University of Auckland, New Zealand
Maria Vakalopoulou	University of Paris Saclay, France
Harini Veeraraghavan	Memorial Sloan Kettering Cancer Center, USA
Satish Viswanath	Case Western Reserve University, USA
Christian Wachinger	Technical University of Munich, Germany
Hua Wang	Colorado School of Mines, USA
Hongzhi Wang	IBM Research, USA
Ken C. L. Wong	IBM Almaden Research Center, USA

Fuyong Xing	University of Colorado Denver, USA
Ziyue Xu	NVIDIA, USA
Yanwu Xu	Baidu Inc., China
Pingkun Yan	Rensselaer Polytechnic Institute, USA
Guang Yang	Imperial College London, UK
Jianhua Yao	Tencent, China
Zhaozheng Yin	Stony Brook University, USA
Lequan Yu	University of Hong Kong, China
Yixuan Yuan	City University of Hong Kong, China
Ling Zhang	Alibaba Group, USA
Miaomiao Zhang	University of Virginia, USA
Ya Zhang	Shanghai Jiao Tong University, China
Rongchang Zhao	Central South University, China
Yitian Zhao	Chinese Academy of Sciences, China
Yefeng Zheng	Tencent Jarvis Lab, China
Guoyan Zheng	Shanghai Jiao Tong University, China
Luping Zhou	University of Sydney, Australia
Yuyin Zhou	Stanford University, USA
Dajiang Zhu	University of Texas at Arlington, USA
Lilla Zöllei	Massachusetts General Hospital, USA
Maria A. Zuluaga	EURECOM, France

Reviewers

Alireza Akhondi-asl
Fernando Arambula
Nicolas Boutry
Qilei Chen
Zhihao Chen
Javid Dadashkarimi
Marleen De Bruijne
Mohammad Eslami
Sayan Ghosal
Estibaliz Gómez-de-Mariscal
Charles Hatt
Yongxiang Huang
Samra Irshad
Anithapriya Krishnan
Rodney LaLonde
Jie Liu
Jinyang Liu
Qing Lyu
Hassan Mohy-ud-Din
Manas Nag
Tianye Niu
Seokhwan Oh
Theodoros Pissas
Harish RaviPrakash
Maria Sainz de Cea
Hai Su
Wenjun Tan
Fatmatulzehra Uslu
Fons van der Sommen
Gijs van Tulder
Dong Wei
Pengcheng Xi
Chen Yang
Kun Yuan
Hang Zhang
Wei Zhang
Yuyao Zhang
Tengda Zhao

Yingying Zhu
Yuemin Zhu
Alaa Eldin Abdelaal
Amir Abdi
Mazdak Abulnaga
Burak Acar
Iman Aganj
Priya Aggarwal
Ola Ahmad
Seyed-Ahmad Ahmadi
Euijoon Ahn
Faranak Akbarifar
Cem Akbaş
Saad Ullah Akram
Tajwar Aleef
Daniel Alexander
Hazrat Ali
Sharib Ali
Max Allan
Pablo Alvarez
Vincent Andrearczyk
Elsa Angelini
Sameer Antani
Michela Antonelli
Ignacio Arganda-Carreras
Mohammad Ali Armin
Josep Arnal
Md Ashikuzzaman
Mehdi Astaraki
Marc Aubreville
Chloé Audigier
Angelica Aviles-Rivero
Ruqayya Awan
Suyash Awate
Qinle Ba
Morteza Babaie
Meritxell Bach Cuadra
Hyeon-Min Bae
Junjie Bai
Wenjia Bai
Ujjwal Baid
Pradeep Bajracharya
Yaël Balbastre
Abhirup Banerjee
Sreya Banerjee
Shunxing Bao
Adrian Barbu
Sumana Basu
Deepti Bathula
Christian Baumgartner
John Baxter
Sharareh Bayat
Bahareh Behboodi
Hamid Behnam
Sutanu Bera
Christos Bergeles
Jose Bernal
Gabriel Bernardino
Alaa Bessadok
Riddhish Bhalodia
Indrani Bhattacharya
Chitresh Bhushan
Lei Bi
Qi Bi
Gui-Bin Bian
Alexander Bigalke
Ricardo Bigolin Lanfredi
Benjamin Billot
Ryoma Bise
Sangeeta Biswas
Stefano B. Blumberg
Sebastian Bodenstedt
Bhushan Borotikar
Ilaria Boscolo Galazzo
Behzad Bozorgtabar
Nadia Brancati
Katharina Breininger
Rupert Brooks
Tom Brosch
Mikael Brudfors
Qirong Bu
Ninon Burgos
Nikolay Burlutskiy
Michał Byra
Ryan Cabeen
Mariano Cabezas
Hongmin Cai
Jinzheng Cai
Weidong Cai
Sema Candemir

Qing Cao
Weiguo Cao
Yankun Cao
Aaron Carass
Ruben Cardenes
M. Jorge Cardoso
Owen Carmichael
Alessandro Casella
Matthieu Chabanas
Ahmad Chaddad
Jayasree Chakraborty
Sylvie Chambon
Yi Hao Chan
Ming-Ching Chang
Peng Chang
Violeta Chang
Sudhanya Chatterjee
Christos Chatzichristos
Antong Chen
Chao Chen
Chen Chen
Cheng Chen
Dongdong Chen
Fang Chen
Geng Chen
Hanbo Chen
Jianan Chen
Jianxu Chen
Jie Chen
Junxiang Chen
Junying Chen
Junyu Chen
Lei Chen
Li Chen
Liangjun Chen
Liyun Chen
Min Chen
Pingjun Chen
Qiang Chen
Runnan Chen
Shuai Chen
Xi Chen
Xiaoran Chen
Xin Chen
Xinjian Chen
Xuejin Chen
Yuanyuan Chen
Zhaolin Chen
Zhen Chen
Zhineng Chen
Zhixiang Chen
Erkang Cheng
Jianhong Cheng
Jun Cheng
Philip Chikontwe
Min-Kook Choi
Gary Christensen
Argyrios Christodoulidis
Stergios Christodoulidis
Albert Chung
Özgün Çiçek
Matthew Clarkson
Dana Cobzas
Jaume Coll-Font
Toby Collins
Olivier Commowick
Runmin Cong
Yulai Cong
Pierre-Henri Conze
Timothy Cootes
Teresa Correia
Pierrick Coupé
Hadrien Courtecuisse
Jeffrey Craley
Alessandro Crimi
Can Cui
Hejie Cui
Hui Cui
Zhiming Cui
Kathleen Curran
Claire Cury
Tobias Czempiel
Vedrana Dahl
Tareen Dawood
Laura Daza
Charles Delahunt
Herve Delingette
Ugur Demir
Liang-Jian Deng
Ruining Deng

Yang Deng
Cem Deniz
Felix Denzinger
Adrien Depeursinge
Hrishikesh Deshpande
Christian Desrosiers
Neel Dey
Anuja Dharmaratne
Li Ding
Xinghao Ding
Zhipeng Ding
Ines Domingues
Juan Pedro Dominguez-Morales
Mengjin Dong
Nanqing Dong
Sven Dorkenwald
Haoran Dou
Simon Drouin
Karen Drukker
Niharika D'Souza
Guodong Du
Lei Du
Dingna Duan
Hongyi Duanmu
Nicolas Duchateau
James Duncan
Nicha Dvornek
Dmitry V. Dylov
Oleh Dzyubachyk
Jan Egger
Alma Eguizabal
Gudmundur Einarsson
Ahmet Ekin
Ahmed Elazab
Ahmed Elnakib
Amr Elsawy
Mohamed Elsharkawy
Ertunc Erdil
Marius Erdt
Floris Ernst
Boris Escalante-Ramírez
Hooman Esfandiari
Nazila Esmaeili
Marco Esposito
Théo Estienne
Christian Ewert
Deng-Ping Fan
Xin Fan
Yonghui Fan
Yubo Fan
Chaowei Fang
Huihui Fang
Xi Fang
Yingying Fang
Zhenghan Fang
Mohsen Farzi
Hamid Fehri
Lina Felsner
Jianjiang Feng
Jun Feng
Ruibin Feng
Yuan Feng
Zishun Feng
Aaron Fenster
Henrique Fernandes
Ricardo Ferrari
Lukas Fischer
Antonio Foncubierta-Rodríguez
Nils Daniel Forkert
Wolfgang Freysinger
Bianca Freytag
Xueyang Fu
Yunguan Fu
Gareth Funka-Lea
Pedro Furtado
Ryo Furukawa
Laurent Gajny
Francesca Galassi
Adrian Galdran
Jiangzhang Gan
Yu Gan
Melanie Ganz
Dongxu Gao
Linlin Gao
Riqiang Gao
Siyuan Gao
Yunhe Gao
Zeyu Gao
Gautam Gare
Bao Ge

Rongjun Ge
Sairam Geethanath
Shiv Gehlot
Yasmeen George
Nils Gessert
Olivier Gevaert
Ramtin Gharleghi
Sandesh Ghimire
Andrea Giovannini
Gabriel Girard
Rémi Giraud
Ben Glocker
Ehsan Golkar
Arnold Gomez
Ricardo Gonzales
Camila Gonzalez
Cristina González
German Gonzalez
Sharath Gopal
Karthik Gopinath
Pietro Gori
Michael Götz
Shuiping Gou
Maged Goubran
Sobhan Goudarzi
Alejandro Granados
Mara Graziani
Yun Gu
Zaiwang Gu
Hao Guan
Dazhou Guo
Hengtao Guo
Jixiang Guo
Jun Guo
Pengfei Guo
Xiaoqing Guo
Yi Guo
Yuyu Guo
Vikash Gupta
Prashnna Gyawali
Stathis Hadjidemetriou
Fatemeh Haghighi
Justin Haldar
Mohammad Hamghalam
Kamal Hammouda
Bing Han
Liang Han
Seungjae Han
Xiaoguang Han
Zhongyi Han
Jonny Hancox
Lasse Hansen
Huaying Hao
Jinkui Hao
Xiaoke Hao
Mohammad Minhazul Haq
Nandinee Haq
Rabia Haq
Michael Hardisty
Nobuhiko Hata
Ali Hatamizadeh
Andreas Hauptmann
Huiguang He
Nanjun He
Shenghua He
Yuting He
Tobias Heimann
Stefan Heldmann
Sobhan Hemati
Alessa Hering
Monica Hernandez
Estefania Hernandez-Martin
Carlos Hernandez-Matas
Javier Herrera-Vega
Kilian Hett
David Ho
Yi Hong
Yoonmi Hong
Mohammad Reza Hosseinzadeh Taher
Benjamin Hou
Wentai Hou
William Hsu
Dan Hu
Rongyao Hu
Xiaoling Hu
Xintao Hu
Yan Hu
Ling Huang
Sharon Xiaolei Huang
Xiaoyang Huang

Yangsibo Huang
Yi-Jie Huang
Yijin Huang
Yixing Huang
Yue Huang
Zhi Huang
Ziyi Huang
Arnaud Huaulmé
Jiayu Huo
Raabid Hussain
Sarfaraz Hussein
Khoi Huynh
Seong Jae Hwang
Ilknur Icke
Kay Igwe
Abdullah Al Zubaer Imran
Ismail Irmakci
Benjamin Irving
Mohammad Shafkat Islam
Koichi Ito
Hayato Itoh
Yuji Iwahori
Mohammad Jafari
Andras Jakab
Amir Jamaludin
Mirek Janatka
Vincent Jaouen
Uditha Jarayathne
Ronnachai Jaroensri
Golara Javadi
Rohit Jena
Rachid Jennane
Todd Jensen
Debesh Jha
Ge-Peng Ji
Yuanfeng Ji
Zhanghexuan Ji
Haozhe Jia
Meirui Jiang
Tingting Jiang
Xiajun Jiang
Xiang Jiang
Zekun Jiang
Jianbo Jiao
Jieqing Jiao
Zhicheng Jiao
Chen Jin
Dakai Jin
Qiangguo Jin
Taisong Jin
Yueming Jin
Baoyu Jing
Bin Jing
Yaqub Jonmohamadi
Lie Ju
Yohan Jun
Alain Jungo
Manjunath K N
Abdolrahim Kadkhodamohammadi
Ali Kafaei Zad Tehrani
Dagmar Kainmueller
Siva Teja Kakileti
John Kalafut
Konstantinos Kamnitsas
Michael C. Kampffmeyer
Qingbo Kang
Neerav Karani
Turkay Kart
Satyananda Kashyap
Alexander Katzmann
Anees Kazi
Hengjin Ke
Hamza Kebiri
Erwan Kerrien
Hoel Kervadec
Farzad Khalvati
Bishesh Khanal
Pulkit Khandelwal
Maksim Kholiavchenko
Ron Kikinis
Daeseung Kim
Jae-Hun Kim
Jaeil Kim
Jinman Kim
Won Hwa Kim
Andrew King
Atilla Kiraly
Yoshiro Kitamura
Stefan Klein
Tobias Klinder

Lisa Koch
Satoshi Kondo
Bin Kong
Fanwei Kong
Ender Konukoglu
Aishik Konwer
Bongjin Koo
Ivica Kopriva
Kivanc Kose
Anna Kreshuk
Frithjof Kruggel
Thomas Kuestner
David Kügler
Hugo Kuijf
Arjan Kuijper
Kuldeep Kumar
Manuela Kunz
Holger Kunze
Tahsin Kurc
Anvar Kurmukov
Yoshihiro Kuroda
Jin Tae Kwak
Francesco La Rosa
Aymen Laadhari
Dmitrii Lachinov
Alain Lalande
Bennett Landman
Axel Largent
Carole Lartizien
Max-Heinrich Laves
Ho Hin Lee
Hyekyoung Lee
Jong Taek Lee
Jong-Hwan Lee
Soochahn Lee
Wen Hui Lei
Yiming Lei
Rogers Jeffrey Leo John
Juan Leon
Bo Li
Bowen Li
Chen Li
Hongming Li
Hongwei Li
Jian Li
Jianning Li
Jiayun Li
Jieyu Li
Junhua Li
Kang Li
Lei Li
Mengzhang Li
Qing Li
Quanzheng Li
Shaohua Li
Shulong Li
Weijian Li
Weikai Li
Wenyuan Li
Xiang Li
Xingyu Li
Xiu Li
Yang Li
Yuexiang Li
Yunxiang Li
Zeju Li
Zhang Li
Zhiyuan Li
Zhjin Li
Zi Li
Chunfeng Lian
Sheng Lian
Libin Liang
Peixian Liang
Yuan Liang
Haofu Liao
Hongen Liao
Ruizhi Liao
Wei Liao
Xiangyun Liao
Gilbert Lim
Hongxiang Lin
Jianyu Lin
Li Lin
Tiancheng Lin
Yiqun Lin
Zudi Lin
Claudia Lindner
Bin Liu
Bo Liu

Chuanbin Liu
Daochang Liu
Dong Liu
Dongnan Liu
Fenglin Liu
Han Liu
Hao Liu
Haozhe Liu
Hong Liu
Huafeng Liu
Huiye Liu
Jianfei Liu
Jiang Liu
Jingya Liu
Kefei Liu
Lihao Liu
Mengting Liu
Peirong Liu
Peng Liu
Qin Liu
Qun Liu
Shenghua Liu
Shuangjun Liu
Sidong Liu
Tianrui Liu
Xiao Liu
Xingtong Liu
Xinwen Liu
Xinyang Liu
Xinyu Liu
Yan Liu
Yanbei Liu
Yi Liu
Yikang Liu
Yong Liu
Yue Liu
Yuhang Liu
Zewen Liu
Zhe Liu
Andrea Loddo
Nicolas Loménie
Yonghao Long
Zhongjie Long
Daniel Lopes
Bin Lou
Nicolas Loy Rodas
Charles Lu
Huanxiang Lu
Xing Lu
Yao Lu
Yuhang Lu
Gongning Luo
Jie Luo
Jiebo Luo
Luyang Luo
Ma Luo
Xiangde Luo
Cuong Ly
Ilwoo Lyu
Yanjun Lyu
Yuanyuan Lyu
Sharath M S
Chunwei Ma
Hehuan Ma
Junbo Ma
Wenao Ma
Yuhui Ma
Anderson Maciel
S. Sara Mahdavi
Mohammed Mahmoud
Andreas Maier
Michail Mamalakis
Ilja Manakov
Brett Marinelli
Yassine Marrakchi
Fabio Martinez
Martin Maška
Tejas Sudharshan Mathai
Dimitrios Mavroeidis
Pau Medrano-Gracia
Raghav Mehta
Felix Meissen
Qingjie Meng
Yanda Meng
Martin Menten
Alexandre Merasli
Stijn Michielse
Leo Milecki
Fausto Milletari
Zhe Min

Tadashi Miyamoto
Sara Moccia
Omid Mohareri
Tony C. W. Mok
Rodrigo Moreno
Kensaku Mori
Lia Morra
Aliasghar Mortazi
Hamed Mozaffari
Pritam Mukherjee
Anirban Mukhopadhyay
Henning Müller
Balamurali Murugesan
Tinashe Mutsvangwa
Andriy Myronenko
Saad Nadeem
Ahmed Naglah
Usman Naseem
Vishwesh Nath
Rodrigo Nava
Nassir Navab
Peter Neher
Amin Nejatbakhsh
Dominik Neumann
Duy Nguyen Ho Minh
Dong Ni
Haomiao Ni
Hannes Nickisch
Jingxin Nie
Aditya Nigam
Lipeng Ning
Xia Ning
Sijie Niu
Jack Noble
Jorge Novo
Chinedu Nwoye
Mohammad Obeid
Masahiro Oda
Steffen Oeltze-Jafra
Ayşe Oktay
Hugo Oliveira
Sara Oliveira
Arnau Oliver
Emanuele Olivetti
Jimena Olveres
Doruk Oner
John Onofrey
Felipe Orihuela-Espina
Marcos Ortega
Yoshito Otake
Sebastian Otálora
Cheng Ouyang
Jiahong Ouyang
Xi Ouyang
Utku Ozbulak
Michal Ozery-Flato
Danielle Pace
José Blas Pagador Carrasco
Daniel Pak
Jin Pan
Siyuan Pan
Yongsheng Pan
Pankaj Pandey
Prashant Pandey
Egor Panfilov
Joao Papa
Bartlomiej Papiez
Nripesh Parajuli
Hyunjin Park
Sanghyun Park
Akash Parvatikar
Magdalini Paschali
Diego Patiño Cortés
Mayank Patwari
Angshuman Paul
Yuchen Pei
Yuru Pei
Chengtao Peng
Jialin Peng
Wei Peng
Yifan Peng
Matteo Pennisi
Antonio Pepe
Oscar Perdomo
Sérgio Pereira
Jose-Antonio Pérez-Carrasco
Fernando Pérez-García
Jorge Perez-Gonzalez
Matthias Perkonigg
Mehran Pesteie

Jorg Peters
Terry Peters
Eike Petersen
Jens Petersen
Micha Pfeiffer
Dzung Pham
Hieu Pham
Ashish Phophalia
Tomasz Pieciak
Antonio Pinheiro
Kilian Pohl
Sebastian Pölsterl
Iulia A. Popescu
Alison Pouch
Prateek Prasanna
Raphael Prevost
Juan Prieto
Federica Proietto Salanitri
Sergi Pujades
Kumaradevan Punithakumar
Haikun Qi
Huan Qi
Buyue Qian
Yan Qiang
Yuchuan Qiao
Zhi Qiao
Fangbo Qin
Wenjian Qin
Yanguo Qin
Yulei Qin
Hui Qu
Kha Gia Quach
Tran Minh Quan
Sandro Queirós
Prashanth R.
Mehdi Rahim
Jagath Rajapakse
Kashif Rajpoot
Dhanesh Ramachandram
Xuming Ran
Hatem Rashwan
Daniele Ravì
Keerthi Sravan Ravi
Surreerat Reaungamornrat
Samuel Remedios
Yudan Ren
Mauricio Reyes
Constantino Reyes-Aldasoro
Hadrien Reynaud
David Richmond
Anne-Marie Rickmann
Laurent Risser
Leticia Rittner
Dominik Rivoir
Emma Robinson
Jessica Rodgers
Rafael Rodrigues
Robert Rohling
Lukasz Roszkowiak
Holger Roth
Karsten Roth
José Rouco
Daniel Rueckert
Danny Ruijters
Mirabela Rusu
Ario Sadafi
Shaheer Ullah Saeed
Monjoy Saha
Pranjal Sahu
Olivier Salvado
Ricardo Sanchez-Matilla
Robin Sandkuehler
Gianmarco Santini
Anil Kumar Sao
Duygu Sarikaya
Olivier Saut
Fabio Scarpa
Nico Scherf
Markus Schirmer
Alexander Schlaefer
Jerome Schmid
Julia Schnabel
Andreas Schuh
Christina Schwarz-Gsaxner
Martin Schweiger
Michaël Sdika
Suman Sedai
Matthias Seibold
Raghavendra Selvan
Sourya Sengupta

Carmen Serrano
Ahmed Shaffie
Keyur Shah
Rutwik Shah
Ahmed Shahin
Mohammad Abuzar Shaikh
S. Shailja
Shayan Shams
Hongming Shan
Xinxin Shan
Mostafa Sharifzadeh
Anuja Sharma
Harshita Sharma
Gregory Sharp
Li Shen
Liyue Shen
Mali Shen
Mingren Shen
Yiqing Shen
Ziyi Shen
Luyao Shi
Xiaoshuang Shi
Yiyu Shi
Hoo-Chang Shin
Boris Shirokikh
Suprosanna Shit
Suzanne Shontz
Yucheng Shu
Alberto Signoroni
Carlos Silva
Wilson Silva
Margarida Silveira
Vivek Singh
Sumedha Singla
Ayushi Sinha
Elena Sizikova
Rajath Soans
Hessam Sokooti
Hong Song
Weinan Song
Youyi Song
Aristeidis Sotiras
Bella Specktor
William Speier
Ziga Spiclin
Jon Sporring
Anuroop Sriram
Vinkle Srivastav
Lawrence Staib
Johannes Stegmaier
Joshua Stough
Danail Stoyanov
Justin Strait
Iain Styles
Ruisheng Su
Vaishnavi Subramanian
Gérard Subsol
Yao Sui
Heung-Il Suk
Shipra Suman
Jian Sun
Li Sun
Liyan Sun
Wenqing Sun
Yue Sun
Vaanathi Sundaresan
Kyung Sung
Yannick Suter
Raphael Sznitman
Eleonora Tagliabue
Roger Tam
Chaowei Tan
Hao Tang
Sheng Tang
Thomas Tang
Youbao Tang
Yucheng Tang
Zihao Tang
Rong Tao
Elias Tappeiner
Mickael Tardy
Giacomo Tarroni
Paul Thienphrapa
Stephen Thompson
Yu Tian
Aleksei Tiulpin
Tal Tlusty
Maryam Toloubidokhti
Jocelyne Troccaz
Roger Trullo

Chialing Tsai
Sudhakar Tummala
Régis Vaillant
Jeya Maria Jose Valanarasu
Juan Miguel Valverde
Thomas Varsavsky
Francisco Vasconcelos
Serge Vasylechko
S. Swaroop Vedula
Roberto Vega
Gonzalo Vegas Sanchez-Ferrero
Gopalkrishna Veni
Archana Venkataraman
Athanasios Vlontzos
Ingmar Voigt
Eugene Vorontsov
Xiaohua Wan
Bo Wang
Changmiao Wang
Chunliang Wang
Clinton Wang
Dadong Wang
Fan Wang
Guotai Wang
Haifeng Wang
Hong Wang
Hongkai Wang
Hongyu Wang
Hu Wang
Juan Wang
Junyan Wang
Ke Wang
Li Wang
Liansheng Wang
Manning Wang
Nizhuan Wang
Qiuli Wang
Renzhen Wang
Rongguang Wang
Ruixuan Wang
Runze Wang
Shujun Wang
Shuo Wang
Shuqiang Wang
Tianchen Wang
Tongxin Wang
Wenzhe Wang
Xi Wang
Xiangdong Wang
Xiaosong Wang
Yalin Wang
Yan Wang
Yi Wang
Yixin Wang
Zeyi Wang
Zuhui Wang
Jonathan Weber
Donglai Wei
Dongming Wei
Lifang Wei
Wolfgang Wein
Michael Wels
Cédric Wemmert
Matthias Wilms
Adam Wittek
Marek Wodzinski
Julia Wolleb
Jonghye Woo
Chongruo Wu
Chunpeng Wu
Ji Wu
Jianfeng Wu
Jie Ying Wu
Jiong Wu
Junde Wu
Pengxiang Wu
Xia Wu
Xiyin Wu
Yawen Wu
Ye Wu
Yicheng Wu
Zhengwang Wu
Tobias Wuerfl
James Xia
Siyu Xia
Yingda Xia
Lei Xiang
Tiange Xiang
Deqiang Xiao
Yiming Xiao

Hongtao Xie
Jianyang Xie
Lingxi Xie
Long Xie
Weidi Xie
Yiting Xie
Yutong Xie
Fangxu Xing
Jiarui Xing
Xiaohan Xing
Chenchu Xu
Hai Xu
Hongming Xu
Jiaqi Xu
Junshen Xu
Kele Xu
Min Xu
Minfeng Xu
Moucheng Xu
Qinwei Xu
Rui Xu
Xiaowei Xu
Xinxing Xu
Xuanang Xu
Yanwu Xu
Yanyu Xu
Yongchao Xu
Zhe Xu
Zhenghua Xu
Zhoubing Xu
Kai Xuan
Cheng Xue
Jie Xue
Wufeng Xue
Yuan Xue
Faridah Yahya
Chaochao Yan
Jiangpeng Yan
Ke Yan
Ming Yan
Qingsen Yan
Yuguang Yan
Zengqiang Yan
Baoyao Yang
Changchun Yang
Chao-Han Huck Yang
Dong Yang
Fan Yang
Feng Yang
Fengting Yang
Ge Yang
Guanyu Yang
Hao-Hsiang Yang
Heran Yang
Hongxu Yang
Huijuan Yang
Jiawei Yang
Jinyu Yang
Lin Yang
Peng Yang
Pengshuai Yang
Xiaohui Yang
Xin Yang
Yan Yang
Yifan Yang
Yujiu Yang
Zhicheng Yang
Jiangchao Yao
Jiawen Yao
Li Yao
Linlin Yao
Qingsong Yao
Chuyang Ye
Dong Hye Ye
Huihui Ye
Menglong Ye
Youngjin Yoo
Chenyu You
Haichao Yu
Hanchao Yu
Jinhua Yu
Ke Yu
Qi Yu
Renping Yu
Thomas Yu
Xiaowei Yu
Zhen Yu
Pengyu Yuan
Paul Yushkevich
Ghada Zamzmi

Ramy Zeineldin
Dong Zeng
Rui Zeng
Zhiwei Zhai
Kun Zhan
Bokai Zhang
Chaoyi Zhang
Daoqiang Zhang
Fa Zhang
Fan Zhang
Hao Zhang
Jianpeng Zhang
Jiawei Zhang
Jingqing Zhang
Jingyang Zhang
Jiong Zhang
Jun Zhang
Ke Zhang
Lefei Zhang
Lei Zhang
Lichi Zhang
Lu Zhang
Ning Zhang
Pengfei Zhang
Qiang Zhang
Rongzhao Zhang
Ruipeng Zhang
Ruisi Zhang
Shengping Zhang
Shihao Zhang
Tianyang Zhang
Tong Zhang
Tuo Zhang
Wen Zhang
Xiaoran Zhang
Xin Zhang
Yanfu Zhang
Yao Zhang
Yi Zhang
Yongqin Zhang
You Zhang
Youshan Zhang
Yu Zhang
Yubo Zhang
Yue Zhang
Yulun Zhang
Yundong Zhang
Yunyan Zhang
Yuxin Zhang
Zheng Zhang
Zhicheng Zhang
Can Zhao
Changchen Zhao
Fenqiang Zhao
He Zhao
Jianfeng Zhao
Jun Zhao
Li Zhao
Liang Zhao
Lin Zhao
Qingyu Zhao
Shen Zhao
Shijie Zhao
Tianyi Zhao
Wei Zhao
Xiaole Zhao
Xuandong Zhao
Yang Zhao
Yue Zhao
Zixu Zhao
Ziyuan Zhao
Xingjian Zhen
Haiyong Zheng
Hao Zheng
Kang Zheng
Qinghe Zheng
Shenhai Zheng
Yalin Zheng
Yinqiang Zheng
Yushan Zheng
Tao Zhong
Zichun Zhong
Bo Zhou
Haoyin Zhou
Hong-Yu Zhou
Huiyu Zhou
Kang Zhou
Qin Zhou
S. Kevin Zhou
Sihang Zhou

Tao Zhou
Tianfei Zhou
Wei Zhou
Xiao-Hu Zhou
Xiao-Yun Zhou
Yanning Zhou
Yaxuan Zhou
Youjia Zhou
Yukun Zhou
Zhiguo Zhou
Zongwei Zhou
Dongxiao Zhu
Haidong Zhu
Hancan Zhu
Lei Zhu
Qikui Zhu
Xiaofeng Zhu
Xinliang Zhu
Zhonghang Zhu
Zhuotun Zhu
Veronika Zimmer
David Zimmerer
Weiwei Zong
Yukai Zou
Lianrui Zuo
Gerald Zwettler
Reyer Zwiggelaar

Outstanding Area Chairs

Ester Bonmati	University College London, UK
Tolga Tasdizen	University of Utah, USA
Yanwu Xu	Baidu Inc., China

Outstanding Reviewers

Seyed-Ahmad Ahmadi	NVIDIA, Germany
Katharina Breininger	Friedrich-Alexander-Universität Erlangen-Nürnberg, Germany
Mariano Cabezas	University of Sydney, Australia
Nicha Dvornek	Yale University, USA
Adrian Galdran	Universitat Pompeu Fabra, Spain
Alexander Katzmann	Siemens Healthineers, Germany
Tony C. W. Mok	Hong Kong University of Science and Technology, China
Sérgio Pereira	Lunit Inc., Korea
David Richmond	Genentech, USA
Dominik Rivoir	National Center for Tumor Diseases (NCT) Dresden, Germany
Fons van der Sommen	Eindhoven University of Technology, the Netherlands
Yushan Zheng	Beihang University, China

Honorable Mentions (Reviewers)

Chloé Audigier	Siemens Healthineers, Switzerland
Qinle Ba	Roche, USA

Meritxell Bach Cuadra	University of Lausanne, Switzerland
Gabriel Bernardino	CREATIS, Université Lyon 1, France
Benjamin Billot	University College London, UK
Tom Brosch	Philips Research Hamburg, Germany
Ruben Cardenes	Ultivue, Germany
Owen Carmichael	Pennington Biomedical Research Center, USA
Li Chen	University of Washington, USA
Xinjian Chen	Soochow University, Taiwan
Philip Chikontwe	Daegu Gyeongbuk Institute of Science and Technology, Korea
Argyrios Christodoulidis	Centre for Research and Technology Hellas/Information Technologies Institute, Greece
Albert Chung	Hong Kong University of Science and Technology, China
Pierre-Henri Conze	IMT Atlantique, France
Jeffrey Craley	Johns Hopkins University, USA
Felix Denzinger	Friedrich-Alexander University Erlangen-Nürnberg, Germany
Adrien Depeursinge	HES-SO Valais-Wallis, Switzerland
Neel Dey	New York University, USA
Guodong Du	Xiamen University, China
Nicolas Duchateau	CREATIS, Université Lyon 1, France
Dmitry V. Dylov	Skolkovo Institute of Science and Technology, Russia
Hooman Esfandiari	University of Zurich, Switzerland
Deng-Ping Fan	ETH Zurich, Switzerland
Chaowei Fang	Xidian University, China
Nils Daniel Forkert	Department of Radiology & Hotchkiss Brain Institute, University of Calgary, Canada
Nils Gessert	Hamburg University of Technology, Germany
Karthik Gopinath	ETS Montreal, Canada
Mara Graziani	IBM Research, Switzerland
Liang Han	Stony Brook University, USA
Nandinee Haq	Hitachi, Canada
Ali Hatamizadeh	NVIDIA Corporation, USA
Samra Irshad	Swinburne University of Technology, Australia
Hayato Itoh	Nagoya University, Japan
Meirui Jiang	The Chinese University of Hong Kong, China
Baoyu Jing	University of Illinois at Urbana-Champaign, USA
Manjunath K N	Manipal Institute of Technology, India
Ali Kafaei Zad Tehrani	Concordia University, Canada
Konstantinos Kamnitsas	Imperial College London, UK

Pulkit Khandelwal	University of Pennsylvania, USA
Andrew King	King's College London, UK
Stefan Klein	Erasmus MC, the Netherlands
Ender Konukoglu	ETH Zurich, Switzerland
Ivica Kopriva	Rudjer Boskovich Institute, Croatia
David Kügler	German Center for Neurodegenerative Diseases, Germany
Manuela Kunz	National Research Council Canada, Canada
Gilbert Lim	National University of Singapore, Singapore
Tiancheng Lin	Shanghai Jiao Tong University, China
Bin Lou	Siemens Healthineers, USA
Hehuan Ma	University of Texas at Arlington, USA
Ilja Manakov	ImFusion, Germany
Felix Meissen	Technische Universität München, Germany
Martin Menten	Imperial College London, UK
Leo Milecki	CentraleSupelec, France
Lia Morra	Politecnico di Torino, Italy
Dominik Neumann	Siemens Healthineers, Germany
Chinedu Nwoye	University of Strasbourg, France
Masahiro Oda	Nagoya University, Japan
Sebastian Otálora	Bern University Hospital, Switzerland
Michal Ozery-Flato	IBM Research, Israel
Egor Panfilov	University of Oulu, Finland
Bartlomiej Papiez	University of Oxford, UK
Nripesh Parajuli	Caption Health, USA
Sanghyun Park	DGIST, Korea
Terry Peters	Robarts Research Institute, Canada
Theodoros Pissas	University College London, UK
Raphael Prevost	ImFusion, Germany
Yulei Qin	Tencent, China
Emma Robinson	King's College London, UK
Robert Rohling	University of British Columbia, Canada
José Rouco	University of A Coruña, Spain
Jerome Schmid	HES-SO University of Applied Sciences and Arts Western Switzerland, Switzerland
Christina Schwarz-Gsaxner	Graz University of Technology, Austria
Liyue Shen	Stanford University, USA
Luyao Shi	IBM Research, USA
Vivek Singh	Siemens Healthineers, USA
Weinan Song	UCLA, USA
Aristeidis Sotiras	Washington University in St. Louis, USA
Danail Stoyanov	University College London, UK

Ruisheng Su	Erasmus MC, the Netherlands
Liyan Sun	Xiamen University, China
Raphael Sznitman	University of Bern, Switzerland
Elias Tappeiner	UMIT - Private University for Health Sciences, Medical Informatics and Technology, Austria
Mickael Tardy	Hera-MI, France
Juan Miguel Valverde	University of Eastern Finland, Finland
Eugene Vorontsov	Polytechnique Montreal, Canada
Bo Wang	CtrsVision, USA
Tongxin Wang	Meta Platforms, Inc., USA
Yan Wang	Sichuan University, China
Yixin Wang	University of Chinese Academy of Sciences, China
Jie Ying Wu	Johns Hopkins University, USA
Lei Xiang	Subtle Medical Inc, USA
Jiaqi Xu	The Chinese University of Hong Kong, China
Zhoubing Xu	Siemens Healthineers, USA
Ke Yan	Alibaba DAMO Academy, China
Baoyao Yang	School of Computers, Guangdong University of Technology, China
Changchun Yang	Delft University of Technology, the Netherlands
Yujiu Yang	Tsinghua University, China
Youngjin Yoo	Siemens Healthineers, USA
Ning Zhang	Bloomberg, USA
Jianfeng Zhao	Western University, Canada
Tao Zhou	Nanjing University of Science and Technology, China
Veronika Zimmer	Technical University Munich, Germany

Mentorship Program (Mentors)

Ulas Bagci	Northwestern University, USA
Kayhan Batmanghelich	University of Pittsburgh, USA
Hrvoje Bogunovic	Medical University of Vienna, Austria
Ninon Burgos	CNRS - Paris Brain Institute, France
Hao Chen	Hong Kong University of Science and Technology, China
Jun Cheng	Institute for Infocomm Research, Singapore
Li Cheng	University of Alberta, Canada
Aasa Feragen	Technical University of Denmark, Denmark
Zhifan Gao	Sun Yat-sen University, China
Stamatia Giannarou	Imperial College London, UK
Sharon Huang	Pennsylvania State University, USA

Anand Joshi	University of Southern California, USA
Bernhard Kainz	Friedrich-Alexander-Universität Erlangen-Nürnberg, Germany and Imperial College London, UK
Baiying Lei	Shenzhen University, China
Karim Lekadir	Universitat de Barcelona, Spain
Xiaoxiao Li	University of British Columbia, Canada
Jianming Liang	Arizona State University, USA
Marius George Linguraru	Children's National Hospital, George Washington University, USA
Anne Martel	University of Toronto, Canada
Antonio Porras	University of Colorado Anschutz Medical Campus, USA
Chen Qin	University of Edinburgh, UK
Julia Schnabel	Helmholtz Munich, TU Munich, Germany and King's College London, UK
Yang Song	University of New South Wales, Australia
Tanveer Syeda-Mahmood	IBM Research - Almaden Labs, USA
Pallavi Tiwari	University of Wisconsin Madison, USA
Mathias Unberath	Johns Hopkins University, USA
Maria Vakalopoulou	CentraleSupelec, France
Harini Veeraraghavan	Memorial Sloan Kettering Cancer Center, USA
Satish Viswanath	Case Western Reserve University, USA
Guang Yang	Imperial College London, UK
Lequan Yu	University of Hong Kong, China
Miaomiao Zhang	University of Virginia, USA
Rongchang Zhao	Central South University, China
Luping Zhou	University of Sydney, Australia
Lilla Zollei	Massachusetts General Hospital, Harvard Medical School, USA
Maria A. Zuluaga	EURECOM, France

Contents – Part I

Heart and Lung Imaging

Brain Development and Atlases

Progression Models for Imaging Data with Longitudinal Variational Auto Encoders

Benoît Sauty(✉) and Stanley Durrleman

Inria, Sorbonne Université, Institut du Cerveau - Paris Brain Institute - ICM, Inserm, CNRS, AP-HP, Paris, France
benoit.sautydechalan@icm.institute.org

Abstract. Disease progression models are crucial to understanding degenerative diseases. Mixed-effects models have been consistently used to model clinical assessments or biomarkers extracted from medical images, allowing missing data imputation and prediction at any timepoint. However, such progression models have seldom been used for entire medical images. In this work, a Variational Auto Encoder is coupled with a temporal linear mixed-effect model to learn a latent representation of the data such that individual trajectories follow straight lines over time and are characterised by a few interpretable parameters. A Monte Carlo estimator is devised to iteratively optimize the networks and the statistical model. We apply this method on a synthetic data set to illustrate the disentanglement between time dependant changes and inter-subjects variability, as well as the predictive capabilities of the method. We then apply it to 3D MRI and FDG-PET data from the Alzheimer's Disease Neuroimaging Initiative (ADNI) to recover well documented patterns of structural and metabolic alterations of the brain.

Keywords: Variational auto encoders · Mixed-effects models · Disease progression models · Alzheimer's Disease

1 Introduction

Estimating progression models from the analysis of time dependent data is a challenging task that helps to uncover latent dynamics. For the study of neurodegenerative diseases, longitudinal databases have been assembled where a set of biomarkers (medical images, cognitive scores and covariates) are gathered for individuals across time. Understanding their temporal evolution is of crucial importance for early diagnosis and drug trials design, especially the imaging biomarkers that can reveal a silent prodromal phase.

In this context, several approaches have been proposed for the progression of scalar measurements such as clinical scores or volumes of brain structures [12,18,29] or series of measurements across brain regions forming a network [4,21]. These approaches require the prior segmentation and extraction of the

L. Wang et al. (Eds.): MICCAI 2022, LNCS 13431, pp. 3–13, 2022.
https://doi.org/10.1007/978-3-031-16431-6_1

measurements from the images. Providing progression models for high dimensional structured data without prior processing is still a challenging task. The difficulty is to provide a low dimensional representation of the data, where each patient's trajectory admits a continuous parametrization over time. It should allow sampling at any time point, be resilient to irregularly spaced instances and disentangle temporal alterations from the inter-patients variability. 0

1.1 Related Work

When dealing with high dimensional data, it is often assumed that the data can be encoded into a low dimensional manifold where the distribution of the data is simple. Deep Generative Neural Networks such as Variational Auto Encoders (VAE) [19] allow finding such embeddings. Several approaches have explored longitudinal modeling for images within the context of dimensionality reduction.

Recurrent Neural Networks (RNN) provide a straightforward way to extract information from sequential data. Convolutional networks with a recurrent structure have been used for diagnosis prediction using MRI [11] or PET [23] scans in Alzheimer's Disease (AD). The main caveat of these approaches is that the recurrent structure is highly sensible to the temporal spacing between instances which is troublesome in the context of disease modeling, where visits are often missing of irregularly spaced.

Mixed-effects models provide an explicit description of the progression of each patient, allowing to sample at any timepoint. Through a time reparametrization, all patients are aligned on a common pathological timeline, and individual trajectories are parametrised as small variations (random effects) around a reference trajectory (fixed effects) that can be seen as the average scenario. Now considered a standard tool in longitudinal modeling [21,28,29], mixed-effects models have yet been scarcely used for images within the context of dimension reduction. In [24], a RNN outputs the parameters of a mixed-effect model that describes patients' trajectories as straight lines in the latent space of a VAE across time.

Self supervised methods have proposed to alleviate the need for labels, in our case the age of the patients at each visit. In [9], the encoder of a VAE learns a latent time variable and a latent spatial variable to disentangle the temporal progression from the patient's intrinsic characteristics. Similarly, in [30], the encoder is penalized with a cosine loss that imposes one direction in the latent space that corresponds to an equivalent of time. Both these methods allow the model to learn a temporal progression that does not rely on the clinical age of the patients, offering potential for unlabeled data, at the cost of interpretability of the abstract timeline and the ability to sample at any given timepoint.

Longitudinal VAEs architectures have been proposed in order to endow the latent space with a temporal structure. Namely, Gaussian Process VAEs (GPVAE) [7] introduced a more general prior for the posterior distribution in the latent space, in the form of a Gaussian Process (GP) that depends on the age of the patients [13] as well as a series of covariates [2,26]. This approach poses challenges as to the choice of parametrization for the Gaussian Process, and does not provide an expected trajectory for each patient.

Diffeomorphic methods provide progression models for images. The main approaches are based off of the geodesic regression framework [3,25] and allow learning a deformation map that models the effect of time on the images for a given subject. While providing high resolution predictions, these methods show limited predictive abilities further in time when compared to mixed-effects models, that aggregate information from all the subjects at different stages of the disease [6].

1.2 Contributions

In this context, we propose to endow the latent space of a VAE with a linear mixed-effect longitudinal model. While in [24], the networks predict the random effects from visits grouped by patients, we propose to enrich a regular VAE that maps each individual visits to a latent representation, with an additional longitudinal latent model that describes the progression of said representations over time. A novel Monte Carlo Markov Chain (MCMC) procedure to jointly estimate the VAE and the structure of its representation manifold is proposed. To sum up the contributions, we:

1. use the **entire 3D scan** without segmentation or parcellation to study relations across brain regions in an unsupervised manner,
2. proceed to **dimension reduction** using a convolutional VAE with the added constraint that latent representations must comply with the structure of a generative statistical model of the trajectories,
3. provide a **progression model** that disentangles temporal changes from changes due to inter-patients variability, and allows sampling patients' trajectories at any timepoint, to infer missing data or predict future progression,
4. demonstrate this method on a synthetic data set and on both **MRI** and **PET scans** from the Alzheimer's Disease Neuroimaging Initiative (ADNI), recovering known patterns in normal or pathological brain aging.

2 Methodology

2.1 Representation Learning with VAEs

Auto Encoders are a standard tool for non-linear dimensionality reduction, comprised of an *encoder* network q_ϕ that maps high dimensional data $x \in \mathcal{X}$ to $z \in \mathcal{Z}$, in a smaller space refered to as the *latent space*, and a *decoder* network $p_\theta : z \in \mathcal{Z} \mapsto \hat{x} \in \mathcal{X}$. VAEs [19] offer a more regularized way to populate the latent space. Both encoder and decoder networks output variational distributions $q_\phi(z|x)$ and $p_\theta(x|z)$, chosen to be multivariate Gaussian distributions. Adding a prior $q(z)$, usually the unit Gaussian distribution $\mathcal{N}(0, \mathrm{I})$, on $\mathcal{Z}$ allows to derive a tractable Evidence Lower BOund for the log-likelihood $ELBO = \mathcal{L}_{recon} + \beta\mathcal{L}_{KL}$ where $\mathcal{L}_{recon}$ is the ℓ_2 reconstruction error, $\mathcal{L}_{KL}$ is the Kullback-Leibler (KL) divergence between the approximate posterior and the prior on the latent space and β balances reconstruction error and latent space regularity [16].

2.2 Longitudinal Statistical Model

In this section, we propose a temporal latent variables model that encodes disease progression in the low-dimensional space $\mathcal{Z}$. Given a family of observations from N patients $\{x_{i,j}\}_{1\leqslant i\leqslant N}$, each observed at ages $t_{i,j}$ for $1 \leqslant j \leqslant n_i$ visits, and their latent representations $\{z_{i,j}\}$, we define a statistical generative model with

$$z_{i,j} = p_0 + \left[e^{\xi_i}(t_{i,j} - \tau_i)\right] v_0 + w_i + \varepsilon_{i,j}$$

where e^{ξ_i} and τ_i, respectively the *acceleration factor* and the *onset age* of patient i, allow an affine time warp aligning all patients on a common pathological timeline, and $w_i \in \mathcal{Z}$ is the *space shift* that encodes inter-subjects variability, such as morphological variations across regions that are independent from the progression. These parameters position the individual trajectory with respect to the typical progression that is estimated at the population level. These three parameters form the **random effects** of the model ψ_r. Vectors w_i and v_0 need to be orthogonal in order to uniquely identify temporal and spatial variability. We choose the Gaussian priors for the noise $\varepsilon_{i,j} \sim \mathcal{N}(0, \sigma_\varepsilon^2)$ and random effects $\tau_i \sim \mathcal{N}(t_0, \sigma_\tau^2)$, $\xi_i \sim \mathcal{N}(0, \sigma_\xi^2)$ and $w_i \sim \mathcal{N}(0, \mathrm{I})$. The parameters $p_0 \in \mathcal{Z}$, $v_0 \in \mathcal{Z}$, $t_0 \in \mathbb{R}$ are respectively a reference *position*, *velocity* and *time* and describe the average trajectory. Together with the variances $\sigma_\varepsilon, \sigma_\tau, \sigma_\xi$, they form the **fixed-effects** of the model ψ_f. We note $\psi = (\psi_r, \psi_f)$.

2.3 Longitudinal VAE

We combine dimension reduction using a regular β-VAE and the aforementioned statistical model to add a temporal structure to the latent space. To do so, we consider a composite loss that accounts for both the VAE loss and the goodness-of-fit of the mixed-effect model:

$$\mathcal{L} = \mathcal{L}_{recon} + \beta\mathcal{L}_{KL} + \gamma\mathcal{L}_{align} \text{ where } \begin{cases} \mathcal{L}_{recon} = \sum_{i,j} ||x_{i,j} - \hat{x}_{i,j}||^2 \\ \mathcal{L}_{KL} \quad = \sum_{i,j} KL(q_\phi(z|x_{i,j})||\mathcal{N}(0,\mathrm{I})) \\ \mathcal{L}_{align} = \sum_{i,j} ||z_{i,j} - \eta_\psi^i(t_{i,j})||^2 \end{cases}$$

where $z_{i,j}$ and $\hat{x}_{i,j}$ are the modes of $q_\phi(x_{i,j})$ and $p_\theta(z_{i,j})$, and $\eta_\psi^i(t_{i,j}) = p_0 + \left[e^{\xi_i}(t_{i,j} - \tau_i)\right] v_0 + w_i$ is the expected position of the latent representation according to the longitudinal model and γ balances the penalty for not aligning latent representations with the linear model. Since the loss is invariant to rotation in $\mathcal{Z}$, we set $p_0 = 0$ and $v_0 = (1, 0, \cdots, 0)$.

Since $\mathcal{L}_{align}$ is a ℓ_2 loss in the latent space, it can be seen as the log-likelihood of a Gaussian prior $z_{i,j} \sim \mathcal{N}(\eta_\psi^i(t_{i;j}), \mathrm{I})$ in the latent space, which defines an elementary Gaussian Process, and which supports the addition of GP priors in the latent space of VAEs to model longitudinal data [13,26]. Besides, $\mathcal{X}$ can be seen as a Riemannian manifold, the metric of which is given by the pushforward of the Euclidean metric of $\mathcal{Z}$ through the decoder, such that trajectories in $\mathcal{X}$ are geodesics, in accordance with the Riemannian modeling of longitudinal data

[5,14,24,28,29]. The metric on $\mathcal{X}$ thus allows to recover non linear dynamics, as is often the case for biomarkers. Our approach thus bridges the gap between the deep learning approach to longitudinal data and a natural generalization of well studied disease progression models to images.

Network Implementation and Estimation. Both the encoder and decoder are chosen to be vanilla convolutional Neural Networks (4 layers of Convolution with stride/BatchNorm/ReLU and transposition for decoder) with a dense layer towards the latent space, as described in Fig. 1. The implementation is in PyTorch and available at https://github.com/bsauty/longitudinal-VAEs.

Algorithm 1: Monte Carlo estimation of the Longitudinal VAE

Input : Longitudinal visits $\{(x_{i,j}, t_{i,j})\}$ and hyperparameters β and γ.
Output: Estimation of (ϕ, θ) for the VAE and ψ for the temporal model.
Init : Initialize (ϕ, θ) as a regular β-VAE.
Set $k = 0$ and $z^0 = q_\phi(x)$;
while *not converged* **do**
- ***Simulation***
  - Draw candidates $\psi_r^c \sim p(.|\psi_f^k)$; `// Sampling with prior` $p(.|\psi_f^k)$
  - $\forall i, j$ compute $\eta^i_{\psi_k}(t_{i,j})$; `// Expected latent trajectories`
  - Compute likelihood ratio $\omega = \min\left(1, \frac{q(\psi_r^c|z^k, \psi_f^k)}{q(\psi_r^k|z^k, \psi_f^k)}\right)$
  - **if** $u \sim \mathcal{U}(0,1) > \omega$ **then** $\psi_r^{k+1} \leftarrow \psi_r^c$ **else** $\psi_r^{k+1} \leftarrow \psi_r^k$
- ***Approximation*** Compute sufficient statistics S^k for ψ_r^k
- ***Maximisation*** $\psi_f^{k+1} \leftarrow \psi_f^*(S^k)$
- ***VAE optimization*** Run one epoch using $\mathcal{L}$ with the target latent representation $\eta^i_{\psi^k}$ for $\mathcal{L}_{align}$ and update $z^{k+1} \leftarrow q_\phi(x)$

end

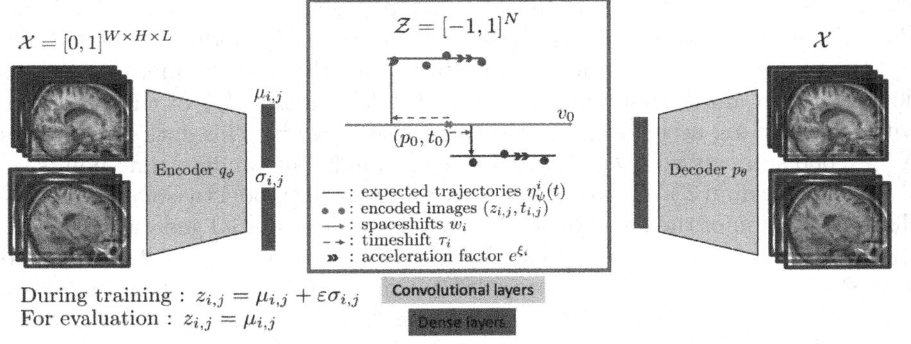

Fig. 1. Images $\{x_{i,j}\}$ are encoded into $\mathcal{Z}$ such that the $\{z_{i,j}\}$ are close to the estimated latent trajectories. Individual trajectories (straight lines) are parametrized with w_i,τ_i and e^{ξ_i} as variations around the reference trajectory (orange arrow). (Color figure online)

The difficulty lies in the joint estimation of (θ, ϕ) and ψ, which are co-dependant since $\mathcal{L}_{align}$ depends on η^i_ψ and ψ depends on the encoded representation $z = q_\phi(x)$. The longitudinal statistical model is part of a family of geometric models that have been studied in [20,29]. Given $\{z_{i,j}\}$, we can proceed to a Maximum a Posteriori estimation of ψ with the MCMC-SAEM procedure in which the estimation step of an EM algorithm is replaced by a stochastic approximation. See [1,22] for details. Given the target trajectories $\{\eta^i_\psi\}$, the weights from both networks of the VAE is optimized through backpropagation using $\mathcal{L}$ with an optimizer with randomized batches. Both estimation schemes are iterative so we designed a Monte Carlo estimator for (ϕ, θ, ψ), presented in Algorithm 1, alternating between both schemes.

Once calibrated with a training set, we freeze the VAE parameters (ϕ, θ) and fixed effects ψ_f, and learn the individual parameters ψ_r, via gradient descent of the likelihood, to personalize the model for new subjects.

Hyperparameters are $\dim(\mathcal{Z})$, which should be small enough to allow the mixed-effect model to be interpretable but big enough to reach good reconstruction; β, which should minimize overfitting while not impairing reconstruction quality; and γ, which should also not be too big to avoid loosing contextual information in $\mathcal{Z}$. These parameters were set using grid search. Lastly, the MCMC-SAEM is computationaly inexpensive compared to backpropagation so memory footprint and runtime are similar to training a regular VAE. All training was performed with a Quadro RTX4000 8Go GPU.

3 Experiments and Results

3.1 Results on Synthetic Experiments

We first validated our approach on a synthetic data set of images of silhouettes of dimension 64×64 [10] . Over time, the silhouette raises its left arm. Different silhouettes are generated by varying the relative position of the three other limbs, all of them raising their left arm in time. The motion is modulated by varying the time stamp at which the motion starts and the pace of motion. This is done using an affine reparametrization of the time stamp $t_{i,j}$ of the silhouette with Gaussian log-acceleration factor ξ_i and onset age τ_i. This data set contains $N = 1,000$ subjects with $n = 10$ visits each, sampled at random time-points.

We choose $\dim(\mathcal{Z}) = 4$ to evaluate the ability of our model to isolate temporal changes (motion of the left arm) from the independent spatial generative factors (the position of the other 3 limbs). Results are displayed in Fig. 2. The 5-fold reconstruction mean squared errors (MSE) (times 10^{-3}) for train/test images are $7.88 \pm .22/7.93 \pm .29$, showing little over-fitting. Prediction error for missing data, when trained on half-pruned data set, is $8.1 \pm .78$ which shows great extrapolation capabilities. A thorough benchmark of six former approaches on this data set was provided in [9] displaying similar MSE to ours. Although a couple of approaches [9,30] also disentangle time from space, ours is the only one to yield the true generative factors, with the direction of progression encoding the motion of the left arm and the 3 spatial directions orthogonal to it encoding legs spreading, legs translation, and right arm position respectively.

3.2 Results on 3D MRI and PET Scans

We then applied the method to 3D T1w MRI and FDG-PET scans from the public ADNI database (http://adni.loni.usc.edu). For MRI, we selected two cohorts: patients with a confirmed AD diagnosis at one visit at least (N = 783 patients for a total of N_{tot} = 3,685 images) and Cognitively Normal (CN) patients at all visits for modeling normal aging (N = 886 and N_{tot} = 3,205). We considered PET data for AD patients only (N = 570 and N_{tot} = 1,463). Images are registered using the T1-linear and PET-linear pipelines of the Clinica software [27] and resampled to $80 \times 96 \times 80$ resolution.

We set $\dim(\mathcal{Z}) = 16$ for both modalities, as it is the smallest dimension that captured the reported dynamics with satisfying resolution. For MRI, error (10^{-3}) for train/test reconstruction and imputation on half-pruned data set for the AD model are $14.15 \pm .12/15.33 \pm .23$ and $18.65 \pm .76$, again showing little over-fit and good prediction abilities. In Fig. 3, the reference trajectory for AD patients reveals the structural alterations that are typical of AD progression. The control trajectory displays alterations more in line with normal aging.

We tested differences in the mean of individual parameters between subgroups using Mann-Whitney U test within a 5-fold cross-validation. AD average onset age occured earlier for women than for men: 72.2±.4 vs. 73.7±.6 years, $p < 3.10^{-7}{\scriptstyle \pm 5.10^{-8}}$). APOE-$\varepsilon$4 mutation carriers experience also earlier onset

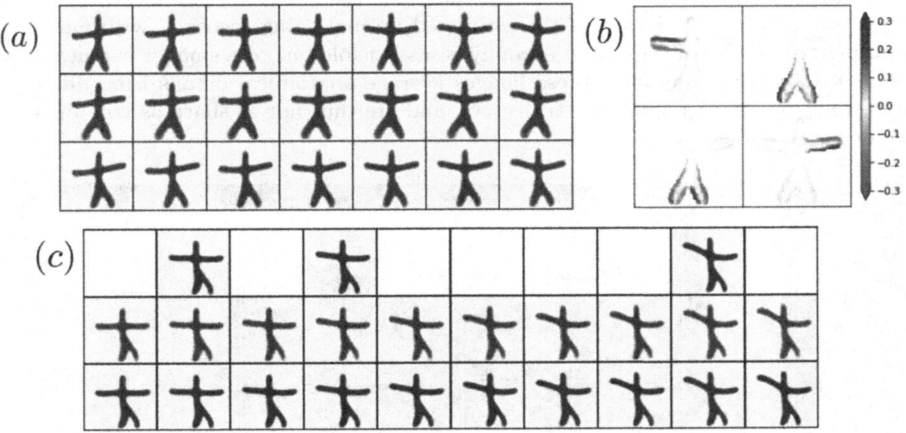

Fig. 2. Synthetic experiment. (a) the average trajectory over time (left to right) on first row, followed by its translation in the directions $w_1 = (0, 1, 0, 0)$ and $w_2 = (0, 0, 1, 0)$ in the latent space (second and third row). (b) the gradient of the image at $p_0 = 0$ in the 4 directions of the latent space v_0, w_1, w_2, w_3. (c) data of a test subject (first row), its reconstructed trajectory (second row) and the ground truth (third row)

than non-carriers (71.8±.2 vs 73.1±.4, $p < 3.3.10^{-2}{\pm 6.10^{-3}}$) and greater pace of progression (.1±3.10^{-2} vs −.08±2.10^{-2}, $p < 1.4.10^{-4}{\pm 6.10^{-3}}$). The normal aging model shows an earlier onset for men (71.2±.4 vs 73.7±.6, $p < 3.10^{-10}{\pm 6.10^{-11}}$). These results are in line with the current knowledge in AD progression [15,17] and normal aging [8]. For PET scans, the 5-fold train/test reconstruction MSE (10^{-2}) are 4.71 ± .32/5.10 ± .23 (Fig. 4).

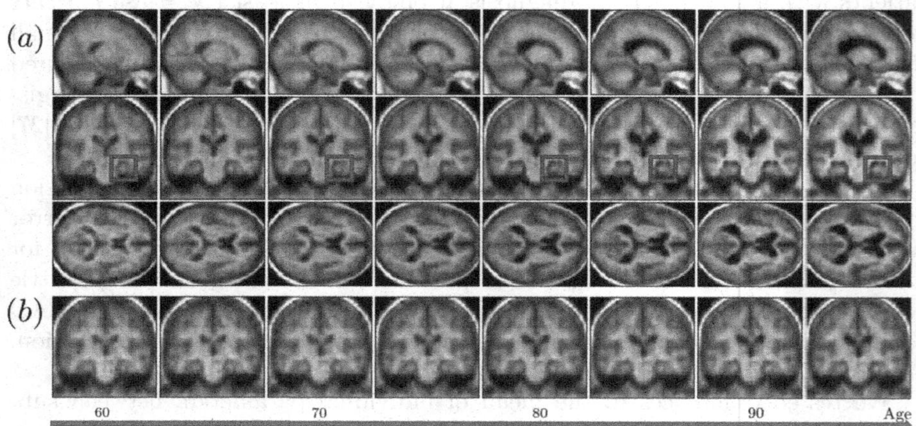

Fig. 3. (a) Sagittal, coronal and axial views of the population trajectory over pathological time (left to right) for the AD cohort. Enlargment of the ventricles and atrophy of the cortex and the hippocampus are visible. Red squares around the hippocampus are positioned at the 5 stages of the Schelten' scale used in the clinics to evaluate AD progression. (b) Coronal view of the estimated normal aging scenario, with matched reparameterized age distribution. Atrophy is also visible but to a smaller extent. As is common in atlasing methods, these images average anatomical details from different subjects to provide a population trajectory, and are thus not as sharp as true images. (Color figure online)

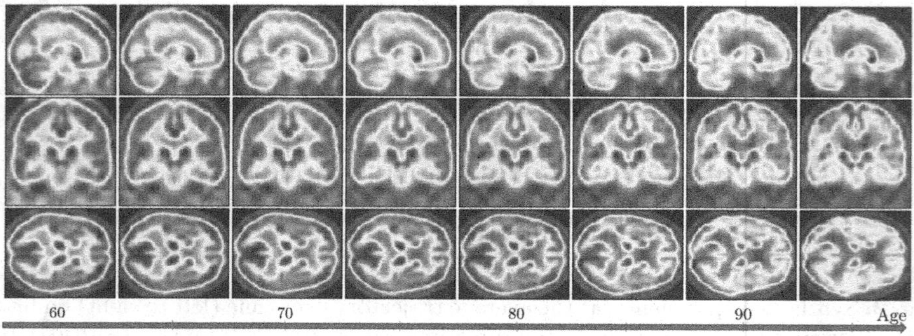

Fig. 4. Sagittal, coronal and axial views of the average trajectory for FDG-PET scans, showing decreased level of metabolism across brain regions.

4 Conclusion

We proposed a generative Variational Autoencoder architecture that maps longitudinal data to a low dimensional Euclidean space, in which a linear spatio-temporal structure is learned to accurately disentangle the effects of time and inter-patient variability, while providing interpretable individual parameters (onset age and acceleration factor). This is the first approach to provide a progression model for 3D MRI or PET scans and it relies on vanilla deep learning architectures that only require the tuning of the loss balance. We showed that it bridges the gap between former approaches to handle longitudinal images, namely GP-VAEs, and Riemannian disease progression models. The method applied to MRI and PET data retrieves known patterns of normal and pathological brain aging but without the need to extract specific biomarkers. It does not only save time but also makes the approach independent of prior choice of biomarkers.

Current work focuses on linking this progression model of brain alterations with cognitive decline, and exploring disease sub-types in the latent space.

Acknowledgments. This work was funded in part by grant number 678304 (ERC) and 826421 (TVB-Cloud) from H2020 programme, and ANR-10-IAIHU-06 (IHU ICM), ANR-19-P3IA-0001 (PRAIRIE) and ANR-19-JPW2-000 (E-DADS) from ANR.

References

1. Allassonnière, S., Kuhn, E., Trouvé, A.: Construction of Bayesian deformable models via a stochastic approximation algorithm: a convergence study. Bernoulli **16**(3), 641–678 (2010)
2. Ashman, M., So, J., Tebbutt, W., Fortuin, V., Pearce, M., Turner, R.E.: Sparse gaussian process variational autoencoders. arXiv preprint arXiv:2010.10177 (2020)
3. Banerjee, M., Chakraborty, R., Ofori, E., Okun, M.S., Viallancourt, D.E., Vemuri, B.C.: A nonlinear regression technique for manifold valued data with applications to medical image analysis. In: Proceedings of the IEEE Conference On Computer Vision and Pattern Recognition, pp. 4424–4432 (2016)
4. Bauer, S., May, C., Dionysiou, D., Stamatakos, G., Buchler, P., Reyes, M.: Multiscale modeling for image analysis of brain tumor studies. IEEE Trans. Biomed. Eng. **59**(1), 25–29 (2011)
5. Bône, A., Colliot, O., Durrleman, S.: Learning distributions of shape trajectories from longitudinal datasets: a hierarchical model on a manifold of diffeomorphisms. In: CVPR 2018 - Computer Vision and Pattern Recognition 2018, Salt Lake City, United States, June 2018. https://hal.archives-ouvertes.fr/hal-01744538
6. Bône, A., et al.: Prediction of the progression of subcortical brain structures in Alzheimer's disease from baseline. In: Cardoso, M.J., et al. (eds.) GRAIL/MFCA/MICGen -2017. LNCS, vol. 10551, pp. 101–113. Springer, Cham (2017). https://doi.org/10.1007/978-3-319-67675-3_10
7. Casale, F.P., Dalca, A.V., Saglietti, L., Listgarten, J., Fusi, N.: Gaussian process prior variational autoencoders. arXiv preprint arXiv:1810.11738 (2018)
8. Coffey, C.E., et al.: Sex differences in brain aging: a quantitative magnetic resonance imaging study. Arch. Neurol. **55**(2), 169–179 (1998)

9. Couronné, R., Vernhet, P., Durrleman, S.: Longitudinal self-supervision to disentangle inter-patient variability from disease progression. In: de Bruijne, M., et al. (eds.) MICCAI 2021. LNCS, vol. 12902, pp. 231–241. Springer, Cham (2021). https://doi.org/10.1007/978-3-030-87196-3_22
10. Couronné, R., Vernhet, P.: Starmen longitudinal (2021). https://doi.org/10.5281/zenodo.5081988
11. Cui, R., Liu, M., et al.: RNN-based longitudinal analysis for diagnosis of Alzheimer's disease. Comput. Med. Imaging Graph. **73**, 1–10 (2019)
12. Fonteijn, H.M., et al.: An event-based model for disease progression and its application in familial Alzheimer's disease and Huntington's disease. NeuroImage **60**(3), 1880–1889 (2012)
13. Fortuin, V., Baranchuk, D., Rätsch, G., Mandt, S.: GP-VAE: deep probabilistic time series imputation. In: International Conference On Artificial Intelligence and Statistics, pp. 1651–1661. PMLR (2020)
14. Gruffaz, S., Poulet, P.E., Maheux, E., Jedynak, B., Durrleman, S.: Learning Riemannian metric for disease progression modeling. Adv. Neural Inf. Process. Syst. **34**, 23780–23792 (2021)
15. Gurvich, C., Hoy, K., Thomas, N., Kulkarni, J.: Sex differences and the influence of sex hormones on cognition through adulthood and the aging process. Brain Sci. **8**(9), 163 (2018)
16. Higgins, I., et al.: BETA-VAE: Learning basic visual concepts with a constrained variational framework (2016)
17. Jack, C.R., et al.: Age, sex, and apoe $\varepsilon 4$ effects on memory, brain structure, and β-amyloid across the adult life span. JAMA Neurol. **72**(5), 511–519 (2015)
18. Jedynak, B.M., et al.: A computational neurodegenerative disease progression score: method and results with the Alzheimer's disease neuroimaging initiative cohort. Neuroimage **63**(3), 1478–1486 (2012)
19. Kingma, D.P., Welling, M.: Auto-encoding variational bayes. arXiv preprint arXiv:1312.6114 (2013)
20. Koval, I., et al.: Statistical learning of spatiotemporal patterns from longitudinal manifold-valued networks. In: Descoteaux, M., Maier-Hein, L., Franz, A., Jannin, P., Collins, D.L., Duchesne, S. (eds.) MICCAI 2017. LNCS, vol. 10433, pp. 451–459. Springer, Cham (2017). https://doi.org/10.1007/978-3-319-66182-7_52
21. Koval, I., et al.: AD Course Map charts Alzheimer's disease progression. Sc. Rep. **11**(1), -1-6 (2021). https://doi.org/10.1038/s41598-021-87434-1, https://hal.inria.fr/hal-01964821
22. Kuhn, E., Lavielle, M.: Coupling a stochastic approximation version of EM with an MCMC procedure. ESAIM: Probabil. Statist **8**, 115–131 (2004)
23. Liu, M., Cheng, D., Yan, W., Initiative, A.D.N., et al.: Classification of Alzheimer's disease by combination of convolutional and recurrent neural networks using FDG-pet images. Front. Neuroinform. **12**, 35 (2018)
24. Louis, M., Couronné, R., Koval, I., Charlier, B., Durrleman, S.: Riemannian geometry learning for disease progression modelling. In: Chung, A., Gee, J., Yushkevich, P., Bao, S. (eds.) Information Processing in Medical Imaging. IPMI 2019. LNCS, vol. 11492, pp. 542–553. Springer, Cham (2019)
25. Niethammer, M., Huang, Y., Vialard, F.-X.: Geodesic regression for image time-series. In: Fichtinger, G., Martel, A., Peters, T. (eds.) MICCAI 2011. LNCS, vol. 6892, pp. 655–662. Springer, Heidelberg (2011). https://doi.org/10.1007/978-3-642-23629-7_80

26. Ramchandran, S., Tikhonov, G., Kujanpää, K., Koskinen, M., Lähdesmäki, H.: Longitudinal variational autoencoder. In: International Conference on Artificial Intelligence and Statistics, pp. 3898–3906. PMLR (2021)
27. Routier, A., et al.: Clinica: an open-source software platform for reproducible clinical neuroscience studies. Front. Neuroinform. **15**, 689675 (2021)
28. Sauty, B., Durrleman, S.: Riemannian metric learning for progression modeling of longitudinal datasets. In: ISBI 2022-International Symposium on Biomedical Imaging (2022)
29. Schiratti, J.B., Allassonniere, S., Colliot, O., Durrleman, S.: Learning spatiotemporal trajectories from manifold-valued longitudinal data. In: Neural Information Processing Systems, vol. 28, Advances in Neural Information Processing Systems, Montréal, Canada, December 2015. https://hal.archives-ouvertes.fr/hal-01163373
30. Zhao, Q., Liu, Z., Adeli, E., Pohl, K.M.: Longitudinal self-supervised learning. Med. Image Anal. **71**, 102051 (2021)

Boundary-Enhanced Self-supervised Learning for Brain Structure Segmentation

Feng Chang[1], Chaoyi Wu[1], Yanfeng Wang[1,2], Ya Zhang[1,2(✉)], Xin Chen[3], and Qi Tian[4]

[1] Cooperative Medianet Innovation Center, Shanghai Jiao Tong University, Shanghai, China
ya_zhang@sjtu.edu.cn

[2] Shanghai AI Laboratory, Shanghai, China

[3] Huawei Inc., Shenzhen, China

[4] Huawei Cloud & AI, Shenzhen, China

Abstract. To alleviate the demand for a large amount of annotated data by deep learning methods, this paper explores self-supervised learning (SSL) for brain structure segmentation. Most SSL methods treat all pixels equally, failing to emphasize the boundaries that are important clues for segmentation. We propose Boundary-Enhanced Self-Supervised Learning (BE-SSL), leveraging supervoxel segmentation and registration as two related proxy tasks. The former task enables capture boundary information by reconstructing distance transform map transformed from supervoxels. The latter task further enhances the boundary with semantics by aligning tissues and organs in registration. Experiments on CANDI and LPBA40 datasets have demonstrated that our method outperforms current SOTA methods by 0.89% and 0.47%, respectively. Our code is available at https://github.com/changfeng3168/BE-SSL.

Keywords: Self-supervised learning · Registration · Supervoxels · Brain structure segmentation

1 Introduction

Brain structure segmentation is a fundamental task in medical image analysis of many brain diseases, such as computer-aided diagnosis and treatment planning. In the last decade, deep learning-based methods have made remarkable progress in brain image analysis, such as brain tumor segmentation [7] and brain disorder diagnosis [16]. However, the requirement for large-scale annotated images for training has greatly limited their application to a wider range of clinical tasks, considering the prohibitive cost of medical data annotation.

Self-supervised learning (SSL) [4,5,9,14,17–20], proposed as a new supervision paradigm that learns representations without explicit supervision, has recently received considerable attention in medical imaging community.

The existing SSL methods generally fall into two categories: discriminative methods and reconstruction methods [10]. For medical image analysis, typical

L. Wang et al. (Eds.): MICCAI 2022, LNCS 13431, pp. 14–23, 2022.
https://doi.org/10.1007/978-3-031-16431-6_2

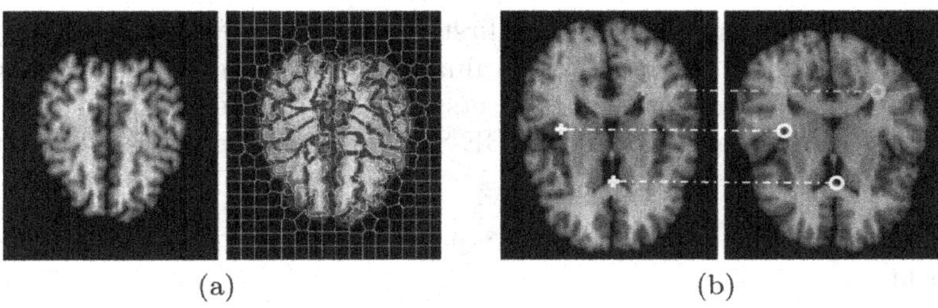

Fig. 1. (a) The boundaries generated by the supervoxel algorithm (yellow) and ground truth (blue). (b) Given a specific anatomical location in the left figure, one can roughly find the corresponding location in the right, even without clinical experience. (Color figure online)

discriminative SSL methods include rotating [3], RubikCube [20] and contrastive learning [14], mostly borrowing ideas from approaches for natural images. Reconstruction methods, such as Genesis [19] and PCRL [18], instead, learn latent representations by reconstructing original inputs from corrupted inputs. The transform-restoration process is expected to better capture low-level features, and as a result, reconstruction methods generally outperform discriminative methods in medical imaging tasks [6,19].

The reconstruction-based SSL methods, despite their promise in medical image segmentation, share a common drawback, *i.e.,* failing to emphasize the edges and boundaries that represent important clues for segmentation, with the reconstruction loss treating all pixels equally. To combat the above drawback, we here explore how to implicitly emphasize the representation of boundaries. We first resort to supervoxel, a semantic-insensitive over-segmentation of volumetric medical images, which is expected to capture boundaries of interest without supervision. As shown in Fig. 1(a), the boundaries from supervoxels cover the majority of ground truth boundaries. Brain structure involves a large number of semantic classes (*e.g.,* 28 classes for CANDI [11]). To further capture the semantics of different structure classes, we attempt to leverage image registration as a proxy task for self-supervision. Different from supervoxels, the alignment of the matching tissues and organs during image registration necessarily involves semantic-aware boundary information.

Based on the above speculation, this paper proposes a Boundary-Enhanced Self-Supervised Learning (BE-SSL) method for brain structure segmentation, by leveraging supervoxel segmentation and registration as two related proxy tasks. Specifically, a two-branch network with a shared encoder and independent decoders is introduced (Fig. 2). The supervoxel branch learns to generate over-segmented supervoxels, while the registration branch learns to transform the input volume to a given atlas. The two branches are forced to share the same encoder, attempting to make the encoder capture the boundary and layout information of the images. At the test time, considering the supervoxel branch is more closely related to segmentation, its decoder is adopted for further finetuning. To the best of our knowledge, we are the first to introduce registration as a proxy task for self-supervised learning. We evaluate the proposed BE-SSL method on two brain structure segmentation datasets: CANDI [11] and LPBA40 [13]. The experimental results have shown that BE-SSL outperforms state-of-the-art self-supervised learning methods.

2 Method

2.1 Overview of the Framework

Figure 2 presents the overall framework of BE-SSL, consisting of two branches, the supervoxel branch and the registration branch, with one shared encoder E. The supervoxel branch, with the encoder E and the decoder D_{sv}, learns to segment the input volumes into supervoxels. The registration branch, with the encoder E and the decoder D_{reg}, learns to register input volumes to the selected atlas.

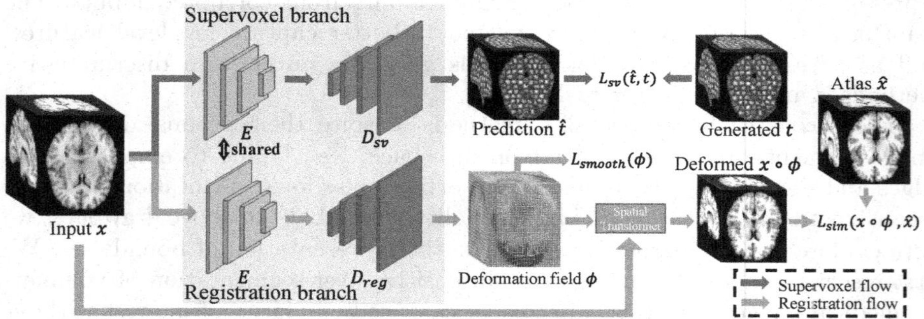

Fig. 2. The overall framework of BE-SSL.

2.2 Supervoxel Branch

Supervoxels [1], as semantic-insensitive over-segmentation of volumetric medical images, provide information of the potential boundaries based on intensity gradients. To turn the segmentation labels of supervoxels into boundary labels, a straightforward way is to use the gradients to transform supervoxels into boundaries. However, the boundaries generated in this way are expected to be quite thin and narrow. We thus propose to transform the supervoxels into a form of distance transform map (DTM) [12], which not only presents whether a voxel is on the boundary but also provides its distance to the nearest boundary.

Figure 3 shows the flowchart to transform supervoxels s into distance transform map t. For each voxel in the supervoxel of interest, its corresponding value in the distance transform map is calculated as:

$$t(p) = \begin{cases} \inf_{q \in \partial G} \|p - q\|_2, & p \in G_{in} \\ 0, & \text{others} \end{cases}, \tag{1}$$

where p and q are coordinates of voxels, G is the supervoxel of interest, G_{in} and ∂G refer to the coordinates sets inside and on boundaries of G, respectively, and $||p - q||_2$ is the Euclidean distance between the two coordinates p and q.

To make the network learn to embed the boundary information, we then treat the distance transform map as the target for reconstruction. An advantage of reconstructing the distance transform map is that by formulating it as a voxel-wise regression problem, every voxel contributes to providing the boundary information. With the distance transform map t as supervision, the supervoxel branch is optimized by minimizing the Mean Square Error (MSE) loss. Denote the prediction of the network as $\hat{t}$, the space of input volumes as Ω. The MSE loss is formulated as follow:

$$\mathcal{L}_{sp}(\hat{t}, t) = \frac{1}{|\Omega|} \sum_{p \in \Omega} [\hat{t}(p) - t(p)]^2. \tag{2}$$

2.3 Registration Branch

Registration is a classical task in medical image analysis and can be operated without annotations. The insight of registration on the brain images is to match the boundaries of different brain structures. Thus, registration can learn the boundary information more related to the downstream brain structure classes compared with the supervoxel branch. We adopt registration as a proxy task to enhance the semantic boundary information.

Denote x and $\hat{x}$ as an input image and the atlas. The self-supervised registration progress is described as follows. Taking x as input, the registration network computes a deformation field ϕ. Each value in ϕ represents an offset distance, and the symbol $\circ$ refers to the deformation operator for x, which consists of voxel shifting and interpolation. We can adopt ϕ to x through the spatial transformer module and get the deformed image $x \circ \phi$. The detail of the spatial transformer

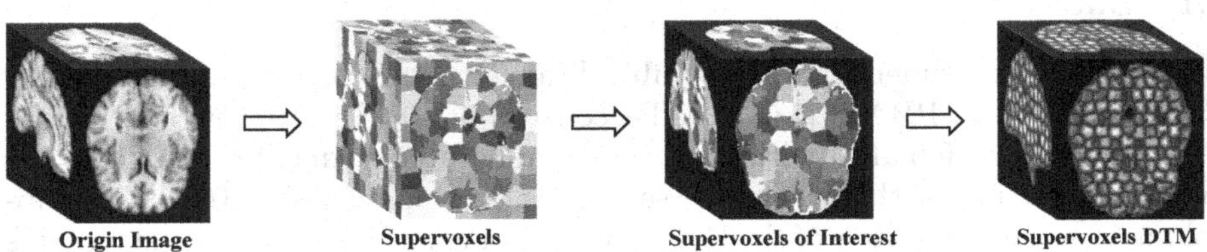

Fig. 3. Flowchart of supervoxel distance transform map generation.

module can be found in [8]. We follow [2] and employ two loss functions for registration: $\mathcal{L}_{sim}$ that penalizes differences in appearance of $x \circ \phi$ and $\hat{x}$, and $\mathcal{L}_{smooth}$ that penalizes local spatial variations in ϕ.

$$\begin{aligned}\mathcal{L}_{reg} &= \mathcal{L}_{sim}(x \circ \phi, \hat{x}) + \lambda \mathcal{L}_{smooth}(\phi) \\ &= \frac{1}{|\Omega|} \sum_{p \in \Omega} [(x \circ \phi)(p) - \hat{x}(p)]^2 + \lambda \sum_{p \in \phi} ||\nabla \phi(p)||^2,\end{aligned} \tag{3}$$

where λ controls the strength of regularization, $\mathcal{L}_{sim}$ is the Mean Squared Error. We set $\lambda = 1$ as [2] suggests and use a gradient loss to constrain the gradient of the deformation field ϕ.

Atlas Selection. The atlas $\hat{x}$ would better be chosen by experts. If lacking it, we can also select it from the training set. For example, given a set of brain images $\mathcal{X} = \{x_1, x_2, x_3, ..., x_n\}$, we tend to select the image most close to the mean of the set as the atlas $\hat{x}$:

$$\hat{x} = \min_{x_i \in \mathcal{X}} \mathcal{L}_{dis}(x_i, \frac{1}{n} \sum_{j=1}^{n} x_j) \tag{4}$$

where x_i is the i-th image in $\mathcal{X}$, $\frac{1}{n} \sum_{j=1}^{n} x_j$ represents the mean of the set, and MSE is used to measure the distance $\mathcal{L}_{dis}$ between images.

2.4 Overall Loss Function

To train the two-branch network simultaneously in an end-to-end fashion, the final objective function for optimizing the model is formulated as:

$$\min_{E, D_{sv}, D_{reg}} \quad \mathcal{L}_{sv} + \gamma \mathcal{L}_{reg}, \tag{5}$$

where γ is a hyper-parameter to balance two branches. In our experiments, we set $\gamma = 1$ empirically.

3 Experiment

3.1 Datasets

We carry out experiments with two public brain structure segmentation datasets: CANDI [11] and LPBA40 [13]. CANDI consists of 103 T1-weighted MRI scans (57 males and 46 females) with fine-grained annotated segmentation labels. We follow the setting of the previous research [15], cropping $160 \times 160 \times 128$ sub-volumes around the center of the brain and choosing 28 classes of brain structures as our segmentation objects. LPBA40 is a cohort of 40 T1-weighted MRI images collected from 40 healthy, normal volunteers with 56 classes of brain structures labeled. The final Dice score is calculated by averaging the Dice scores of all structures.

For each dataset, we randomly split it into 60%, 15%, 25% for training, validation, and testing, respectively. At the training stage, the whole training set *without* any annotation mask is used to train the BE-SSL network in a self-supervised fashion. At the fine-tuning stage, a portion of the training set *with* the corresponding annotation masks is provided to fine-tune the network. To show the performance of our method with a different amount of labeled data, we gradually increase the proportion of labeled data for fine-tuning, from 10%, 30%, 50%, 70% to 100%. Note that test images for target tasks are never exposed to model training or fine-tuning.

3.2 Experiments Setup and Implementation Details

Backbones. Both the supervoxel branch and the registration branch adopt 3D U-Net architecture as [2], with the encoder shared by the two branches.

Supervoxels. For the supervoxel branch, SLIC [1] is adopted to obtain the 'ground-truth' supervoxels of volumetric images. To alleviate the impact of extraneous parts in supervoxels, the supervoxels in the background area are ruled out to get supervoxels of interest.

Training. Models are trained using an Adam optimizer with a learning rate of 3×10^{-4} and a weight decay of 3×10^{-5}. ReduceLROnPlateau scheduler is employed to adjust the learning rate according to the validation loss.

Fine-tuning. The supervoxel branch is employed for fine-tuning, considering the close relation of supervoxel and segmentation. We apply data augmentation involving scaling, Gaussian noise, Gaussian blur, brightness, contrast and gamma as proposed in nnU-Net [7]. Similarly, an adam optimizer is applied with a learning rate of 3×10^{-4} and a weight decay of 3×10^{-5}. ReduceLROnPlateau scheduler is used to control the learning rate.

For a fair comparison, we control all variables the same except for model initialization to compare with other SOTA self-supervised learning methods and report the mean and standard deviation of five-fold cross-validation segmentation results in the fine-tuning stage. All experiments are conducted using PyTorch framework with a GeForce RTX 3090 GPU.

4 Experimental Results

Comparison with SOTA: We compare our method with several state-of-the-art SSL methods [14,18–20]. Table 1 and Table 2 present the results for CANDI and LPBA40, respectively. For CANDI, with 10%, 30%, 50%, 70%, and 100% annotations, our method improves the Dice by 2.62%, 1.40%, 1.67%, 1.52%, and 1.54%, respectively, compared to trained from scratch. Similar results are shown for LPBA40. While our method shows more improvement over train-from-scratch in the case of limited annotations, same as existing SSL methods, it is worth

Table 1. Comparison of Dice(%) with SOTA methods on CANDI dataset

Method	Proportion					Avg
	10%	30%	50%	70%	100%	
Scratch	$80.57_{\pm 0.39}$	$83.58_{\pm 0.39}$	$83.90_{\pm 0.43}$	$84.49_{\pm 0.50}$	$84.70_{\pm 0.64}$	83.45
3D-Rot [14]	$80.71_{\pm 0.38}$	$83.48_{\pm 0.07}$	$84.00_{\pm 0.28}$	$84.59_{\pm 0.30}$	$85.06_{\pm 0.18}$	83.57
3D-Jig [14]	$80.23_{\pm 0.28}$	$83.52_{\pm 0.22}$	$84.40_{\pm 0.19}$	$84.85_{\pm 0.15}$	$85.30_{\pm 0.21}$	83.66
3D-Cpc [14]	$80.10_{\pm 0.42}$	$83.51_{\pm 0.18}$	$84.26_{\pm 0.17}$	$84.50_{\pm 0.17}$	$85.03_{\pm 0.18}$	83.48
PCRL [18]	$80.20_{\pm 0.73}$	$82.75_{\pm 0.60}$	$84.11_{\pm 0.58}$	$84.70_{\pm 0.13}$	$84.96_{\pm 0.23}$	83.34
Genesis [19]	$80.14_{\pm 0.99}$	$83.14_{\pm 0.73}$	$84.14_{\pm 0.39}$	$83.99_{\pm 0.52}$	$84.79_{\pm 0.20}$	83.24
RubikCube [20]	$81.45_{\pm 0.82}$	$84.29_{\pm 0.31}$	$84.76_{\pm 0.28}$	$85.19_{\pm 0.24}$	$85.84_{\pm 0.17}$	84.31
Reg	$81.51_{\pm 0.73}$	$84.10_{\pm 0.25}$	$84.88_{\pm 0.12}$	$85.32_{\pm 0.19}$	$85.59_{\pm 0.15}$	84.28
SV	$82.57_{\pm 0.57}$	$84.79_{\pm 0.15}$	$85.35_{\pm 0.07}$	$85.54_{\pm 0.32}$	$86.04_{\pm 0.17}$	84.86
BE-SSL	$\mathbf{83.19}_{\pm 0.27}$	$\mathbf{84.98}_{\pm 0.33}$	$\mathbf{85.57}_{\pm 0.27}$	$\mathbf{86.01}_{\pm 0.16}$	$\mathbf{86.24}_{\pm 0.28}$	**85.20**

Table 2. Comparison of Dice(%) with SOTA methods on LPBA40 dataset

Method	Proportion					Avg
	10%	30%	50%	70%	100%	
Scratch	$71.92_{\pm 0.18}$	$76.09_{\pm 0.45}$	$77.35_{\pm 0.09}$	$77.81_{\pm 0.23}$	$78.33_{\pm 0.25}$	76.30
3D-Rot [14]	$71.24_{\pm 0.47}$	$76.36_{\pm 0.25}$	$77.60_{\pm 0.21}$	$77.99_{\pm 0.08}$	$78.38_{\pm 0.18}$	76.31
3D-Jig [14]	$71.66_{\pm 0.32}$	$76.56_{\pm 0.23}$	$77.62_{\pm 0.34}$	$78.05_{\pm 0.23}$	$78.58_{\pm 0.17}$	76.49
3D-Cpc [14]	$70.69_{\pm 0.31}$	$75.67_{\pm 0.39}$	$77.05_{\pm 0.25}$	$77.50_{\pm 0.21}$	$78.11_{\pm 0.14}$	75.81
PCRL [18]	$71.40_{\pm 0.41}$	$76.17_{\pm 0.27}$	$77.64_{\pm 0.17}$	$77.76_{\pm 0.35}$	$78.40_{\pm 0.17}$	76.27
Genesis [19]	$71.83_{\pm 0.51}$	$76.24_{\pm 0.34}$	$77.47_{\pm 0.36}$	$77.98_{\pm 0.13}$	$78.41_{\pm 0.17}$	76.39
RubikCube [20]	$72.13_{\pm 0.35}$	$77.04_{\pm 0.34}$	$78.16_{\pm 0.16}$	$78.65_{\pm 0.10}$	$79.07_{\pm 0.10}$	77.01
Reg	$72.23_{\pm 0.12}$	$76.89_{\pm 0.35}$	$77.91_{\pm 0.30}$	$78.14_{\pm 0.23}$	$78.78_{\pm 0.26}$	76.80
SV	$73.19_{\pm 0.20}$	$77.42_{\pm 0.15}$	$78.51_{\pm 0.19}$	$78.79_{\pm 0.07}$	$79.27_{\pm 0.10}$	77.43
BE-SSL	$\mathbf{73.23}_{\pm 0.25}$	$\mathbf{77.46}_{\pm 0.14}$	$\mathbf{78.60}_{\pm 0.16}$	$\mathbf{78.81}_{\pm 0.11}$	$\mathbf{79.30}_{\pm 0.11}$	**77.48**

noting that our method achieves clear improvement even with 100% annotations, which is encouraging because the gains of many existing SSL methods diminish quickly with increased annotations, suggesting that our method can help the network to better capture boundary. Compared with the best performed SSL method, RubikCube [20], our method achieves a gain of 0.89% and 0.47% on average, respectively, for CANDI and LPBA40, suggesting that our method is a general SSL training method for structure segmentation tasks.

Ablation Study: To show the effectiveness of different parts in our method, we independently evaluate the impact of the supervoxel branch and the registration branch. The results are reported in as 'SV' and 'Reg' respectively in Table 1 and Table 2. As shown in the Tables, both branches can improve the performance compared to train-from-scratch. The supervoxel branch and the registration branch lead to a gain of 1.41% and 0.83% in Dice, respectively for CANDI, with similar results on the LPBA40, demonstrating the effectiveness of the two proxy tasks. Combining these two proxy tasks further improves the

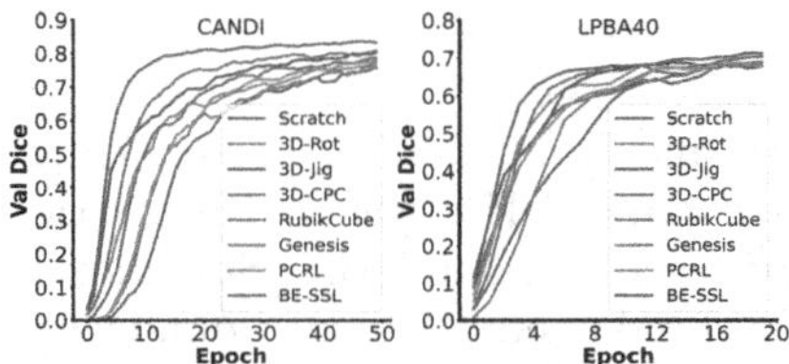

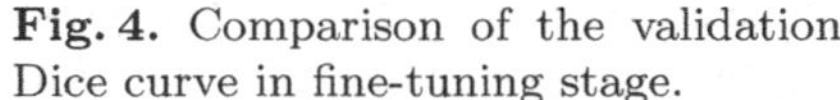

Fig. 4. Comparison of the validation Dice curve in fine-tuning stage.

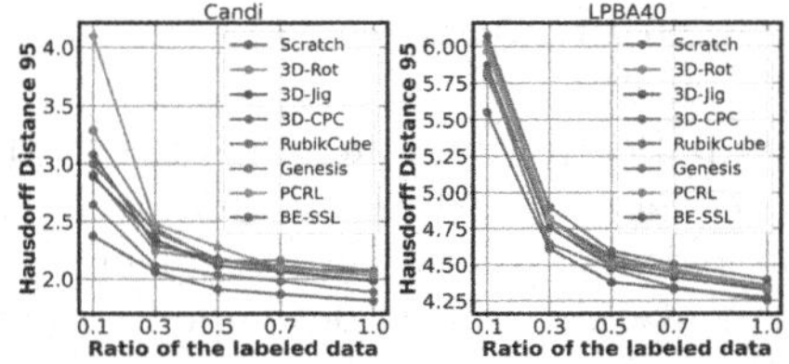

Fig. 5. Comparison with SOTA methods using HD95(↓) metric.

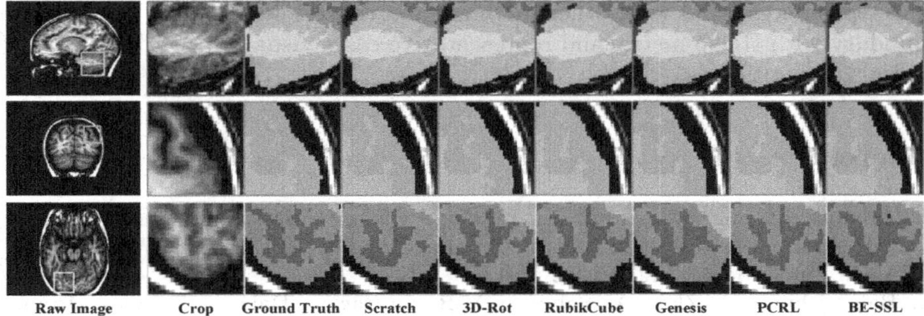

Fig. 6. Visual analysis of segmentation results when finetuning on CANDI dataset with 10% labeled data.

performance of segmentation, increasing the Dice score to 85.20%, indicating the two tasks providing complementary representation, one enhancing the fundamental boundaries and the other enhancing the semantic boundaries.

Analysis of Convergence Speed. Figure 4 shows the validation Dice curves of BE-SSL compared with other methods in the initial convergence process in the case of fine-tuning with 10% labeled data. It indicates that the proposed BE-SSL can speed up the convergence and boost performance.

Analysis of Boundary Enhancement. To investigate the effectiveness of BE-SSL on boundary enhancement, 95th percentile Hausdorff distance (HD95), a metric focusing on segmentation boundary, is adopted. As shown in Fig. 5, compared to SOTA methods, BE-SSL achieves the lowest HD95 value in most cases, demonstrating its effectiveness in capturing boundaries.

Figure 6 provides the comparison of visual segmentation results of the proposed BE-SSL and other methods. BE-SSL can obtain better segmentation boundaries when dealing with regions tricky to segment, such as the cerebral gyrus.

5 Conclusion

This paper proposes a self-supervised learning method BE-SSL, harmonizing two novel pretext tasks, Supervoxel Segmentation and Self-supervised Registration learning, to boost the segmentation performance on boundaries. Supervoxel Segmentation task is developed to predict the fundamental visual boundaries of the input volumes. Registration learning is employed to boost the ability by enhancing the semantic boundary of the target brain structure classes. The proposed BE-SSL outperforms several SOTA self-supervised learning methods on two brain structure segmentation datasets. Additionally, we analyze its convergence speed compared with other methods and explain its capacity to enhance boundary with HD95 metric and representative cases visualization, demonstrating our assumption.

References

1. Achanta, R., Shaji, A., Smith, K., Lucchi, A., Fua, P., Süsstrunk, S.: SLIC superpixels compared to state-of-the-art superpixel methods. IEEE Trans. Pattern Anal. Mach. Intell. **34**(11), 2274–2282 (2012)
2. Balakrishnan, G., Zhao, A., Sabuncu, M.R., Guttag, J., Dalca, A.V.: VoxelMorph: a learning framework for deformable medical image registration. IEEE Trans. Med. Imaging **38**(8), 1788–1800 (2019)
3. Chen, L., Bentley, P., Mori, K., Misawa, K., Fujiwara, M., Rueckert, D.: Self-supervised learning for medical image analysis using image context restoration. Med. Image Anal. **58**, 101539 (2019)
4. Chen, X., He, K.: Exploring simple Siamese representation learning. In: Proceedings of the IEEE/CVF Conference on Computer Vision and Pattern Recognition, pp. 15750–15758 (2021)
5. Doersch, C., Gupta, A., Efros, A.A.: Unsupervised visual representation learning by context prediction. In: Proceedings of the IEEE International Conference on Computer Vision, pp. 1422–1430 (2015)
6. Feng, R., Zhou, Z., Gotway, M.B., Liang, J.: Parts2Whole: self-supervised contrastive learning via reconstruction. In: Albarqouni, S., et al. (eds.) DART/DCL -2020. LNCS, vol. 12444, pp. 85–95. Springer, Cham (2020). https://doi.org/10.1007/978-3-030-60548-3_9
7. Isensee, F., Jaeger, P.F., Kohl, S.A., Petersen, J., Maier-Hein, K.H.: NNU-Net: a self-configuring method for deep learning-based biomedical image segmentation. Nat. Methods **18**(2), 203–211 (2021)
8. Jaderberg, M., et al.: Spatial transformer networks. Adv. Neural Inf. Process. Syst. **28**, 1–9 (2015)
9. Jena, R., Singla, S., Batmanghelich, K.: Self-supervised vessel enhancement using flow-based consistencies. In: de Bruijne, M., et al. (eds.) MICCAI 2021. LNCS, vol. 12902, pp. 242–251. Springer, Cham (2021). https://doi.org/10.1007/978-3-030-87196-3_23
10. Jin, Y., Buntine, W., Petitjean, F., Webb, G.I.: Discriminative, generative and self-supervised approaches for target-agnostic learning. arXiv preprint arXiv:2011.06428 (2020)

11. Kennedy, D.N., Haselgrove, C., Hodge, S.M., Rane, P.S., Makris, N., Frazier, J.A.: CANDIShare: a resource for pediatric neuroimaging data (2012)
12. Ma, J., et al.: How distance transform maps boost segmentation CNNs: an empirical study. In: Medical Imaging with Deep Learning, pp. 479–492. PMLR (2020)
13. Shattuck, D.W., et al.: Construction of a 3d probabilistic atlas of human cortical structures. Neuroimage **39**(3), 1064–1080 (2008)
14. Taleb, A., et al.: 3d self-supervised methods for medical imaging. Adv. Neural Inf. Process. Syst. **33**, 18158–18172 (2020)
15. Wang, S., et al.: LT-Net: label transfer by learning reversible voxel-wise correspondence for one-shot medical image segmentation. In: Proceedings of the IEEE/CVF Conference on Computer Vision and Pattern Recognition, pp. 9162–9171 (2020)
16. Zhang, L., Wang, M., Liu, M., Zhang, D.: A survey on deep learning for neuroimaging-based brain disorder analysis. Front. Neurosci. **14**, 779 (2020)
17. Zhang, X., Feng, S., Zhou, Y., Zhang, Y., Wang, Y.: SAR: scale-aware restoration learning for 3D tumor segmentation. In: de Bruijne, M., et al. (eds.) MICCAI 2021. LNCS, vol. 12902, pp. 124–133. Springer, Cham (2021). https://doi.org/10.1007/978-3-030-87196-3_12
18. Zhou, H.Y., Lu, C., Yang, S., Han, X., Yu, Y.: Preservational learning improves self-supervised medical image models by reconstructing diverse contexts. In: Proceedings of the IEEE/CVF International Conference on Computer Vision, pp. 3499–3509 (2021)
19. Zhou, Z., Sodha, V., Pang, J., Gotway, M.B., Liang, J.: Models genesis. Med. Image Anal. **67**, 101840 (2021)
20. Zhuang, X., Li, Y., Hu, Y., Ma, K., Yang, Y., Zheng, Y.: self-supervised feature learning for 3D medical images by playing a Rubik's cube. In: Shen, D., et al. (eds.) MICCAI 2019. LNCS, vol. 11767, pp. 420–428. Springer, Cham (2019). https://doi.org/10.1007/978-3-030-32251-9_46

Domain-Prior-Induced Structural MRI Adaptation for Clinical Progression Prediction of Subjective Cognitive Decline

Minhui Yu[1], Hao Guan[1], Yuqi Fang[1], Ling Yue[2(✉)], and Mingxia Liu[1(✉)]

[1] Department of Radiology and BRIC, University of North Carolina at Chapel Hill, Chapel Hill, NC 27599, USA
mxliu@med.unc.edu

[2] Department of Geriatric Psychiatry, Shanghai Mental Health Center, Shanghai Jiao Tong University School of Medicine, Shanghai 200030, China
bellinthemoon@hotmail.com

Abstract. Growing evidence shows that subjective cognitive decline (SCD) among elderly individuals is the possible pre-clinical stage of Alzheimer's disease (AD). To prevent the potential disease conversion, it is critical to investigate biomarkers for SCD progression. Previous learning-based methods employ T1-weighted magnetic resonance imaging (MRI) data to aid the future progression prediction of SCD, but often fail to build reliable models due to the insufficient number of subjects and imbalanced sample classes. A few studies suggest building a model on a large-scale AD-related dataset and then applying it to another dataset for SCD progression via transfer learning. Unfortunately, they usually ignore significant data distribution gaps between different centers/domains. With the prior knowledge that SCD is at increased risk of underlying AD pathology, we propose a domain-prior-induced structural MRI adaptation (DSMA) method for SCD progression prediction by mitigating the distribution gap between SCD and AD groups. The proposed DSMA method consists of two parallel *feature encoders* for MRI feature learning in the labeled source domain and unlabeled target domain, an *attention block* to locate potential disease-associated brain regions, and a *feature adaptation module* based on maximum mean discrepancy (MMD) for cross-domain feature alignment. Experimental results on the public ADNI dataset and an SCD dataset demonstrate the superiority of our method over several state-of-the-arts.

Keywords: Subjective cognitive decline · MRI · Domain adaptation

Supplementary Information The online version contains supplementary material available at https://doi.org/10.1007/978-3-031-16431-6_3.

L. Wang et al. (Eds.): MICCAI 2022, LNCS 13431, pp. 24–33, 2022.
https://doi.org/10.1007/978-3-031-16431-6_3

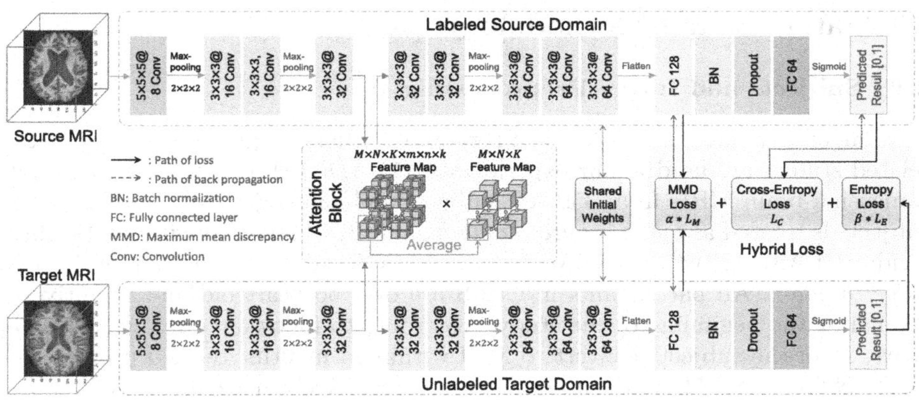

Fig. 1. Illustration of the proposed domain-prior-induced structural MRI adaptation (DSMA) framework for automated SCD progression prediction.

1 Introduction

Subjective cognitive decline (SCD) is the self-perception of persistently worsening cognitive capacity without an acute event or deficit in neuropsychological examination [1,2], and it is estimated that 1 in 4 cognitively normal individuals over the age of 60 would experience SCD [3]. Previous studies suggest that people with SCD have AD-related brain patterns, electromagnetic signatures, and even a higher mortality risk [4–6]. Some subjects with SCD progress to mild cognitive impairment (MCI), the prodromal stage of AD that displays irreversible neuropathological alteration, in a few years [1,7]. These subjects are considered potential populations for treatment trials. It is critical to predict the future progression of SCD for early intervention of MCI and AD. Previous studies develop machine learning methods for MRI-based SCD progression prediction but generally suffer from small sample sizes [8]. A few studies tackle the sample-size limitation by applying a model trained on large-scale AD and MCI related datasets to small-scale SCD datasets [9]. However, they ignore the inherent data distribution gap across different datasets/domains caused by different disease stages, imaging scanners, and/or scanning protocols [10].

To this end, we propose a domain-prior-induced structural MRI adaptation (DSMA) method for automated SCD progression prediction. As shown in Fig. 1, the DSMA is constructed of a labeled source domain and an unlabeled target domain, where two feature encoders are used for end-to-end MRI feature extraction and prediction. We employ a maximum mean discrepancy (MMD) based feature adaptation module for cross-domain feature alignment. An attention block is used to locate potential disease-associated brain regions. To our knowledge, this is among the first attempts that employ domain adaptation to minimize the distribution gap between AD-related datasets for SCD progression prediction.

2 Materials and Methodology

2.1 Subjects and Image Preprocessing

A total of 871 subjects with 3T T1-weighted MRIs acquired from two AD-related studies are used in our experiments: (1) ADNI [11] with predominantly Caucasian brain MRIs from 359 AD subjects and 436 cognitively normal (CN) subjects is treated as *labeled source domain*, and (2) Chinese Longitudinal Aging Study (CLAS) [12] with brain MRIs from 76 SCD subjects is used as *unlabeled target domain*. All participants in CLAS are over 60 years old and cognitively normal at the beginning of the study. According to follow-up outcomes after 7 years, 24 of the subjects convert to MCI within 84 months, termed progressive SCD (pSCD), and 52 SCD subjects remain cognitively normal, termed stable SCD (sSCD). In the CLAS study, T1-weighted 3D magnetization prepared rapid gradient echo (MPRAGE) structural MRIs are acquired through a 3T MRI scanner (Siemens, MAGNETOM VERIO, German) with the following scanning parameters: repetition time (TR) = 2,300 ms, echo time (TE) = 2.98 ms, flip angle of 9°, matrix of 240×256, field of view of 240×256 mm, slice thickness of 1.2 mm, and slice number of 176. That is, the scanners and scanning parameters involved in CLAS are different from ADNI[1], leading to significant heterogeneity between these two datasets. The demographic information of the studied subjects from CLAS and ADNI is listed in the *Supporting Information.*

All 3D MR images are preprocessed through standard procedures, containing skull stripping, intensity inhomogeneity correction, resampling to the resolution of $1 \times 1 \times 1\,\text{mm}^3$, and spatial normalization to the Montreal Neurological Institute (MNI) space. To remove uninformative background, we crop each MRI to $160 \times 180 \times 160$ and ensure that the entire brain remains within each volume.

2.2 Proposed Method

Feature Encoder. To extract local-to-global semantic features of MRI scans from both domains, we use a parallel convolution neural network (CNN) architecture with two branches of identical feature encoders that share initial weights. As shown in Fig. 1, the top branch is used for the labeled source domain and the bottom branch is used for the unlabeled target domain. Each branch is composed of 4 convolution blocks with increasing filter numbers. The stride is fixed at $1 \times 1 \times 1$. All convolution layers have the kernel size of $3 \times 3 \times 3$ except that the first one has kernels of size $5 \times 5 \times 5$. It is equivalent to two stacked $3 \times 3 \times 3$ layers in terms of effective receptive field but contains fewer parameters [13], thus the model complexity is reduced. After the convolution layers, we use a fully connected (FC) layer with 128 neurons, a batch normalization (BN), a dropout layer with a 0.3 dropout rate, and another FC layer with 64 neurons. The derived MRI features for the source domain and the target domain are then fed into a sigmoid function to predict category labels (*i.e.*, pSCD or sSCD).

[1] http://adni.loni.usc.edu/methods/documents/mri-protocols/.

Feature Adaptation. To effectively mitigate the cross-domain data distribution gap, we align the extracted MRI features derived from two feature encoders by minimizing an MMD loss. The FC layers located at the end of model can extract task-specific features and can be directly adapted, thus are considered desired layers to employ MMD-based feature alignment [14]. We select the first FC layer with 128 neurons to perform MMD-based feature adaptation. Specifically, the MMD-based loss is defined as:

$$L_M = \frac{1}{S^2}|| \sum\nolimits_{i=1}^{S} \theta_s(x_i^s) - \sum\nolimits_{j=1}^{S} \theta_t(y_j^t)||^2 \tag{1}$$

where S is the number of subjects in both domains per training batch, θ_s and θ_t denote parameters of two feature encoders of source and target domains, and x_i^s and y_j^t denote an input source sample and a target sample, respectively.

Attention Block. To locate the potential biomarker area and remove unrelated/redundant information, an attention block is incorporated into the proposed model to enhance those discriminative regions voxel-wisely. We put the attention block in the intermediate convolution layers because it contains relatively high-level attention while still keeping enough details. Specifically, we divide the activation map $\mathbf{Q} \in \mathbb{R}^{(M \times m) \times (N \times n) \times (K \times k)}$ into $M \times N \times K$ identical small volumes of size $m \times n \times k$, and then apply an average operation to each of them, and finally multiply the averages with each weight in the small volumes. The activation map $\mathbf{B} \in \mathbb{R}^{M \times N \times K}$ after the average operation is expressed as:

$$\mathbf{B}(I, J, H) = \frac{\sum\nolimits_{a=(I-1)\times m}^{I \times m} \sum\nolimits_{b=(J-1)\times n}^{J \times n} \sum\nolimits_{c=(H-1)\times k}^{H \times k} w_{a,b,c}}{m \times n \times k} \tag{2}$$

where w is the weights in the original activation map; I, J, H denote the location of each small volume; m, n, k denote the length of three dimensions of the small volumes; and a, b, c denote the location of each single weight within the small area. The feature maps $\mathbf{F} \in \mathbb{R}^{(M \times m) \times (N \times n) \times (K \times k)}$ is then formulated as:

$$\mathbf{F}(i, j, h) = \mathbf{B}\left(\left\lceil \frac{i}{m} \right\rceil, \left\lceil \frac{j}{n} \right\rceil, \left\lceil \frac{h}{k} \right\rceil\right) \times w_{i,j,h} \tag{3}$$

where i, j, h denote the coordinates of each weight in the original activation map, and the symbol $\lceil \cdot \rceil$ denotes ceiling function of a real number. The volume sizes in our case are $M = 7, N = 8, K = 7, m = 5, n = 5, k = 5$. In Eq. (3), F is the product of the original activation map and the rescaled activation map. For the former, each of the neurons can be viewed as a local feature. For the latter, each of the rescaled values can be viewed as a global feature for each of the blocks. This operation enhances the important areas while weakening the unrelated areas. To get the attention map, we average all the feature maps in a chosen layer, which represents the information the layer has learned, thus we can expect that it shows potential SCD-associated brain regions in MRI.

Hybrid Loss. To work with different intentions of each branch, we deploy three different loss functions and add them together to derive the overall loss of the proposed DSMA model. In the top branch with labeled source data, we use a cross-entropy loss for the supervised classification. Since this is a binary problem, we use a sigmoid function in the last layer to generate a score for each input. Thus, we can further express the cross-entropy loss as:

$$L_C = -\frac{1}{S}\sum_{i=1}^{S} l_i log(p_i) + (1 - l_i) log(1 - p_i) \tag{4}$$

where l_i denotes the ground-truth label of the i-th subject, and p_i denotes the estimated probability score of an input MRI belonging to a specific category (*e.g.*, pSCD or sSCD). In the bottom branch with unlabeled target data, we use an entropy loss to further constrain the model, which can be expressed as:

$$L_E = -\frac{1}{S}\sum_{i=1}^{S} p_i log(p_i) + (1 - p_i) log(1 - p_i) \tag{5}$$

where p_i is the estimated probability score for the i-th input MRI. The hybrid loss L_H of the proposed DSMA can be formulated as:

$$L_H = L_C + \alpha L_M + \beta L_E \tag{6}$$

where α and β are the hyperparameters to regularize L_M and L_E, respectively. The hybrid loss is used as the loss function in model training, while only the two losses L_M and L_C participate in the back-propagation. The entropy loss L_E is used to constrain the bottom branch model in Fig. 1 in an unsupervised manner.

Implementation. In the ***training*** phase, our method is executed via a two-stage optimization process. The 1st stage is to use labeled AD and CN source data to pretrain a classification model (see top branch of Fig. 1). This stage lets the model learn the source domain data to a satisfactory accuracy, and provides informative initial weights for next steps. We firstly use 90% of data to train and 10% data to validate the model to ensure the model is working, and then use all the data to boost the learning before next step. In the 2nd stage, we jointly train the top and the bottom branches in Fig. 1 by optimizing the proposed hybrid loss defined in Eq. (6) without using any label information of target data. Inspired by [1], we intuitively assume that the distribution of sSCD and pSCD is close to CN and AD, respectively. During the 2nd training stage, the model learns to map the samples of sSCD and pSCD towards CN and AD, respectively. In the ***test*** phase, the top branch model in Fig. 1 is used for inference, where target MRIs are used as test data. In this way, we expect the sSCD to be classified on the side of NC, and pSCD to be classified on the side of AD.

For model training, we use Adam optimizer [15] with a decaying learning rate that starts at 1×10^{-4}. The batch size is set as 5 at the 1st stage and 2 at the 2nd stage. The proposed DSMA method is implemented with TensorFlow 2.0 on a workstation equipped with an RTX 3090 GPU (24 GB memory). It takes around 0.04 s to predict the category label of an input 3D T1-weighted MRI scan, which means it is applicable in real-time SCD to MCI converting prediction.

Table 1. Results (%) of different methods in MRI-based SCD progression prediction (*i.e.*, pSCD vs. sSCD classification) with best results shown in boldface, as well as p-values via paired sample t-test between DSMA and each of the competing methods.

Method	AUC (%)	ACC (%)	BAC (%)	SEN (%)	SPE (%)	F1 (%)	$p < 0.05$
SVM	56.25 ± 0.00	55.26 ± 0.00	56.09 ± 0.00	58.33 ± 0.00	53.85 ± 0.00	45.16 ± 0.00	Yes
RF	50.07 ± 8.11	47.11 ± 4.19	47.66 ± 5.03	49.17 ± 12.19	46.15 ± 7.20	36.53 ± 6.83	Yes
VoxCNN	63.86 ± 1.32	57.89 ± 1.66	59.13 ± 1.93	62.50 ± 2.64	55.77 ± 1.22	48.39 ± 2.04	Yes
CORAL	63.97 ± 2.57	58.42 ± 3.49	59.74 ± 4.04	63.33 ± 5.53	56.15 ± 2.55	49.03 ± 4.28	Yes
ITL	64.63 ± 2.46	59.47 ± 1.29	60.96 ± 1.49	65.00 ± 2.04	56.92 ± 0.94	50.32 ± 1.58	Yes
DSMA (Ours)	$\mathbf{68.80 \pm 2.28}$	$\mathbf{62.63 \pm 3.87}$	$\mathbf{64.62 \pm 4.48}$	$\mathbf{70.00 \pm 6.12}$	$\mathbf{59.23 \pm 2.83}$	$\mathbf{54.19 \pm 4.74}$	-

3 Experiment

Experimental Settings. Six metrics are used to evaluate model performance: area under the ROC curve (AUC), sensitivity (SEN), specificity (SPE), accuracy (ACC), balanced accuracy (BAC), and F1-Score (F1). In the experiments, we use the MRI scans of AD and CN subjects from the ADNI dataset as labeled source data, and all MR images from the CLAS dataset as unlabeled target data.

Competing Methods. Five methods are used for comparison: (1) support vector machine (**SVM**) that uses a linear SVM (with $C = 1$) as the classifier and a 90-dimensional gray matter (GM) volume from 116 regions-of-interests (ROIs) (defined by the anatomical automatic labeling atlas) per MR image; (2) random forest (**RF**) that uses random forest (with 100 trees) as the classifier and 90-dimensional volumetric GM features of MR images; (3) **VoxCNN** [16], a state-of-the-art deep learning architecture for the classification of AD spectrum with 3D structural MRIs, which is performed in a transfer learning manner, trained on AD and CN subjects from ADNI and directly tested on the pSCD and sSCD subjects from CLAS; (4) **CORAL** [17], a correlation alignment method that performs domain adaptation by minimizing the covariance between the output of the parallel feature encoders for source and target domain (with the same feature encoders as DSMA); and (5) inductive transfer learning (**ITL**) [18], that is a transfer learning method with ADNI as training set and CLAS as test set. The ITL method uses the same feature decoder and attention block as our DSMA, and shares the same pretrained weights. That is, the ITL method can be regarded as a degenerated version of the DSMA, with no effort to reduce the distribution gap between two domains. For the fairness of comparison, all the hyperparameter settings of the competing methods are to the largest extent kept the same as our method in the experiments. Except for the SVM method, we repeat each method 5 times with random initialization.

Prediction Results. Results produced by different methods of SCD progression prediction are shown in Table 1. From Table 1, we have several observations. *First*, the deep learning methods (*i.e.*, VoxCNN, CORAL, ITL and DSMA)

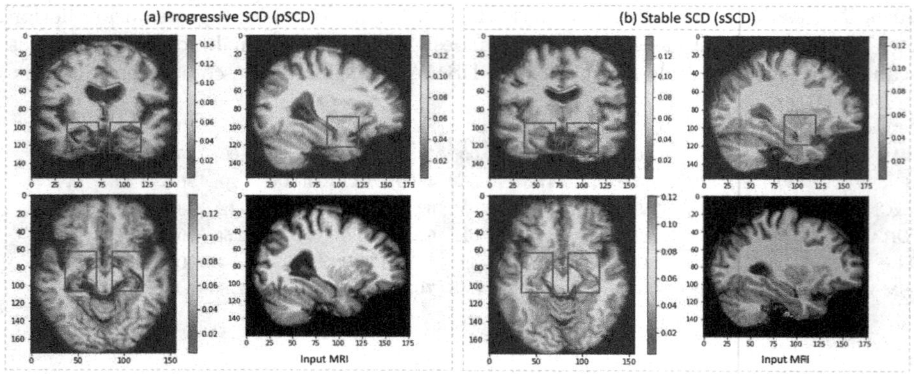

Fig. 2. Attention map identified by DSMA for subjects with (a) pSCD, and (b) sSCD in three views. The color bar shows the intensity of the discriminative power.

Table 2. Results (%) of our DSMA and its four variants in MRI-based SCD progression prediction in the form of mean ± standard deviation (best results shown in boldface).

Method	AUC (%)	ACC (%)	BAC (%)	SEN (%)	SPE (%)	F1 (%)	p<0.05
DSMA-AT	53.77 ± 6.48	54.21 ± 4.88	54.87 ± 5.65	56.67 ± 7.73	53.08 ± 3.57	43.87 ± 5.98	Yes
DSMA-IN	51.39 ± 4.47	50.00 ± 3.33	50.00 ± 3.85	50.00 ± 5.27	50.00 ± 2.43	38.71 ± 4.08	Yes
DSMA-MMD	64.42 ± 2.89	58.55 ± 2.19	59.90 ± 2.52	63.54 ± 3.45	56.25 ± 1.59	49.19 ± 2.67	Yes
DSMA-3L	59.38 ± 7.30	56.84 ± 5.16	57.92 ± 5.97	60.83 ± 8.16	55.00 ± 3.77	47.10 ± 6.32	Yes
DSMA (Ours)	**68.80 ± 2.28**	**62.63 ± 3.87**	**64.62 ± 4.48**	**70.00 ± 6.12**	**59.23 ± 2.83**	**54.19 ± 4.74**	–

are overall better than the traditional machine learning methods (*i.e.*, SVM and RF). The possible reason is that machine learning methods largely rely on hand-crafted MRI features while deep learning methods can learn classification-oriented MRI features in an end-to-end manner. *Second*, among four deep learning methods, our DSMA achieves the best results in terms of six metrics, with the p-values less than 0.05. To be noticed, we apply the same transfer learning strategy to VoxCNN and ITL, but ITL yields better performance than VoxCNN. This suggests that even without domain adaptation, our model structure used in ITL is superior to VoxCNN, possibly due to the attention mechanism used in ITL. *Third*, our DSMA generally outperforms the CORAL method, which may imply that MMD loss is more suitable for our case compared to CORAL loss. *Besides*, the superiority of our DSMA over ITL demonstrates the advantage of the proposed feature adaptation strategy for SCD progression prediction.

Identified Attention Map. The attention maps generated by DSMA model for subjects with pSCD and sSCD are shown in Fig. 2(a)–(b). It is derived by averaging all feature maps in the 2nd convolution layer of the feature encoder in Fig. 1. From Fig. 2(a)–(b), we can observe that the asymmetry changes in hippocampus, parahippocampal regions, and amygdala are detected by our DSMA

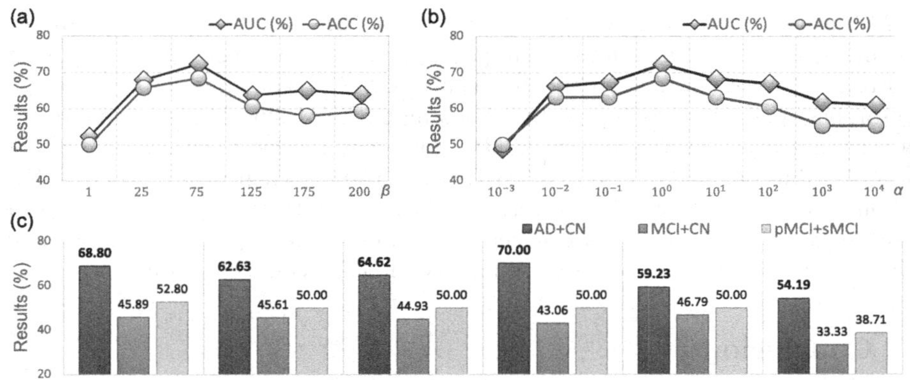

Fig. 3. Results of our DSMA in pSCD vs. sSCD classification with (a) different β and fixed $\alpha = 1$, (b) different α and fixed $\beta = 75$, and (c) different source domains.

between pSCD and sSCD subjects. These regions have been reported to be helpful for SCD/MCI identification in previous studies [19–21].

Ablation Study. We compare DSMA with its four variants, including (1) **DSMA-AT** without any attention mechanism, (2) **DSMA-IN** without model pretraining, (3) **DSMA-MMD** with only the MMD loss in back-propagation, and (4) **DSMA-3L** with all three losses in back-propagation. For the fair comparison, these four variants share a similar network architecture as DSMA, and the results in pSCD vs. sSCD classification are reported in Table 2. It can be observed from Table 2 that our DSMA yields significantly better performance than its four variants ($p < 0.05$). This further suggests the effectiveness of the proposed attention mechanism, model pretraining, and back-propagation strategy.

Hyperparameter Analysis. We study the influence of two hyperparameters (α and β) in Eq. (6) by varying their values within $[10^{-3}, 10^{4}]$ and [1, 200], respectively. Experimental results are reported in Fig. 3(a)–(b). This figure suggests that the performance drops dramatically when either of the two hyperparameters is very small, implying that both of the constraints are important for the proposed DSMA. With $\alpha = 1$, our method produces good results when $\beta \in \{25, 75\}$. With $\beta = 75$, the model yields the best performance when α is 10^0.

Influence of Source Domains. Besides AD+CN subjects that have been used as source data in the above experiments, there are other related domains in the AD spectrum, such as MCI that consists of 262 progressive MCI (pMCI) and 405 stable MCI (sMCI) subjects in ADNI. Thus, we also investigate the impact of different source domains. We compare our result with two other cases: (1)

DSMA using *MCI+CN* and (2) DSMA with *pMCI+sMCI* as source domains. Results achieved by three methods in pSCD vs. sSCD classification are reported in Fig. 3(c). This figure suggests that using AD+CN as source data yields the overall best performance, especially with a high SEN. Using MCI+NC as source data cannot produce satisfactory results. This may be due to the complexity of data distributions of MCI and NC MRIs. The results when using pMCI+sMCI as source domain are slightly better, but they are still not as good as AD+CN case, probably because the biomarker differences between pMCI and sMCI are relatively mild, which increase the difficulty in classifying the source data.

4 Conclusion

This paper presents a domain-prior-induced structural MRI adaptation (DSMA) method for automated progression prediction of SCD within 7 years, guided by the prior knowledge that SCD is at increased risk of underlying AD pathology. The DSMA contains two feature encoders for MRI feature learning in labeled source and unlabeled target domains, an attention block to locate disease-related brain regions, and a feature adaptation module to mitigate cross-domain MRI heterogeneity. Experimental results on 871 subjects with T1-weighted structural MRIs suggest the superiority of our method over several state-of-the-arts. Considering the potential complementary information of different data modalities, it is interesting to fuse imaging and non-imaging biological measurements to further boost the prediction performance, which will be the future work.

Acknowledgements. This work was partly supported by NIH grants (Nos. AG073297 and AG041721).

References

1. Jessen, F., et al.: A conceptual framework for research on subjective cognitive decline in preclinical Alzheimer's disease. Alzheimer's Dementia **10**(6), 844–852 (2014)
2. Wolfsgruber, S., et al.: Differential risk of incident Alzheimer's disease dementia in stable versus unstable patterns of subjective cognitive decline. J. Alzheimer's Dis. **54**(3), 1135–1146 (2016)
3. Röhr, S., et al.: Estimating prevalence of subjective cognitive decline in and across international cohort studies of aging: a COSMIC study. Alzheimer's Res. Ther. **12**(1), 1–14 (2020)
4. Peter, J., et al.: Gray matter atrophy pattern in elderly with subjective memory impairment. Alzheimer's Dementia **10**(1), 99–108 (2014)
5. Nakamura, A., et al.: Electromagnetic signatures of the preclinical and prodromal stages of Alzheimer's disease. Brain **141**(5), 1470–1485 (2018)
6. Luck, T., Roehr, S., Jessen, F., Villringer, A., Angermeyer, M.C., Riedel-Heller, S.G.: Mortality in individuals with subjective cognitive decline: results of the Leipzig longitudinal study of the aged (LEILA75+). J. Alzheimers Dis. **48**(s1), S33–S42 (2015)

7. Morris, J.C., et al.: Mild cognitive impairment represents early-stage Alzheimer disease. Arch. Neurol. **58**(3), 397–405 (2001)
8. Schultz, S.A., et al.: Subjective memory complaints, cortical thinning, and cognitive dysfunction in middle-age adults at risk of AD. Alzheimer's Dementia Diagn. Assess. Dis. Monit. **1**(1), 33–40 (2015)
9. Liu, Y., Yue, L., Xiao, S., Yang, W., Shen, D., Liu, M.: Assessing clinical progression from subjective cognitive decline to mild cognitive impairment with incomplete multi-modal neuroimages. Med. Image Anal. **75**, 102266 (2022)
10. Guan, H., Liu, M.: Domain adaptation for medical image analysis: a survey. IEEE Trans. Biomed. Eng. **69**(3), 1173–1185 (2021)
11. Jack Jr, C.R., et al.: The Alzheimer's disease neuroimaging initiative (ADNI): MRI methods. J. Magn. Reson. Imaging Official J. Int. Soc. Magn. Reson. Med. **27**(4), 685–691 (2008)
12. Xiao, S., et al.: The china longitudinal ageing study: overview of the demographic, psychosocial and cognitive data of the Shanghai sample. J. Ment. Health **25**(2), 131–136 (2016)
13. Simonyan, K., Zisserman, A.: Very deep convolutional networks for large-scale image recognition. arXiv preprint arXiv:1409.1556 (2014)
14. Tzeng, E., Hoffman, J., Zhang, N., Saenko, K., Darrell, T.: Deep domain confusion: maximizing for domain invariance. arXiv preprint arXiv:1412.3474 (2014)
15. Kingma, D.P., Ba, J.: Adam: a method for stochastic optimization. arXiv preprint arXiv:1412.6980 (2014)
16. Korolev, S., Safiullin, A., Belyaev, M., Dodonova, Y.: Residual and plain convolutional neural networks for 3D brain MRI classification. In: IEEE 14th International Symposium on Biomedical Imaging, pp. 835–838. IEEE (2017)
17. Sun, B., Feng, J., Saenko, K.: Return of frustratingly easy domain adaptation. In: Proceedings of the AAAI Conference on Artificial Intelligence, vol. 30 (2016)
18. Zhuang, F., et al.: A comprehensive survey on transfer learning. Proc. IEEE **109**(1), 43–76 (2020)
19. Karas, G., et al.: Global and local gray matter loss in mild cognitive impairment and Alzheimer's disease. Neuroimage **23**(2), 708–716 (2004)
20. Stoub, T.R., deToledo Morrell, L., Stebbins, G.T., Leurgans, S., Bennett, D.A., Shah, R.C.: Hippocampal disconnection contributes to memory dysfunction in individuals at risk for Alzheimer's disease. Proc. Natl. Acad. Sci. **103**(26), 10041–10045 (2006)
21. Dickerson, B.C., et al.: Medial temporal lobe function and structure in mild cognitive impairment. Ann. Neurol. **56**(1), 27–35 (2004)

3D Global Fourier Network for Alzheimer's Disease Diagnosis Using Structural MRI

Shengjie Zhang[1,2], Xiang Chen[1,2], Bohan Ren[3], Haibo Yang[1,2], Ziqi Yu[1,2], Xiao-Yong Zhang[1,2(✉)], and Yuan Zhou[4(✉)]

[1] Institute of Science and Techonology for Brain-inspired Intelligence, Fudan University, Shanghai, China

[2] Key Laboratory of Computational Neuroscience and Brain-Inspired Intelligence (Fudan University), Ministry of Education, Shanghai, China

xiaoyong_zhang@fudan.edu.cn

[3] Department of Cyber Science and Technology, Beihang University, Beijing, China

[4] School of Data Science, Fudan University, Shanghai, China

yuanzhou@fudan.edu.cn

Abstract. Deep learning models, such as convolutional neural networks and self-attention mechanisms, have been shown to be effective in computer-aided diagnosis (CAD) of Alzheimer's disease (AD) using structural magnetic resonance imaging (sMRI). Most of them use spatial convolutional filters to learn local information from the images. In this paper, we propose a 3D Global Fourier Network (GF-Net) to utilize global frequency information that captures long-range dependency in the spatial domain. The GF-Net contains three primary components: a 3D discrete Fourier transform, an element-wise multiplication between frequency domain features and learnable global filters, and a 3D inverse Fourier transform. The GF-Net is trained by a multi-instance learning strategy to identify discriminative features. Extensive experiments on two independent datasets (ADNI and AIBL) demonstrate that our proposed GF-Net outperforms several state-of-the-art methods in terms of accuracy and other metrics, and can also identify pathological regions of AD. The code is released at https://github.com/qbmizsj/GFNet.

Keywords: Alzheimer's disease · Global fourier network · Multi-instance learning · MRI

1 Introduction

Alzheimer's disease (AD) has become one of the most prevalent neurological disorders due to the increasing population of AD [1]. In AD patients, their cognitive function is progressively impaired, accompanied by irreversible neurological

Supplementary Information The online version contains supplementary material available at https://doi.org/10.1007/978-3-031-16431-6_4.

L. Wang et al. (Eds.): MICCAI 2022, LNCS 13431, pp. 34–43, 2022.
https://doi.org/10.1007/978-3-031-16431-6_4

damage. An early diagnosis of AD could provide valuable information in the subsequent treatment, thereby delaying the onset of late stage symptoms, such as amnesia. Such a diagnosis can be achieved by structural magnetic resonance imaging (sMRI) which can capture the morphological changes induced by brain atrophy in a non-invasive way [2].

Existing methods for computer-aided diagnosis (CAD) of AD using sMRI could be categorized into traditional machine learning (ML) and deep learning (DL) [3]. Traditional ML methods follow the pipeline of regions-of-interest (ROIs) extraction and feature classification [4–10]. However, manual ROI extraction or feature construction is laborious and time-consuming. DL methods, especially convolutional neural networks (CNN), extract image features automatically, hence may greatly improve the performance of AD diagnosis [11–19].

Currently, several DL methods have been proposed for this task. Some methods replace the machine learning classifiers with neural networks while keeping the pre-defined ROIs [17,18]. Some other methods train CNN in a multi-instance learning strategy [14]. Typically, these DL methods use spatial convolution to extract local features [15]. However, spatial convolution means using local receptive fields in vision, which ignores global connections between disease-related brain regions.

To address this problem, we propose a *global Fourier network* (GF-Net). The GF-Net divides an input image into several sub-regions (patches) and consists of a patch embedding operation and a sequence of *global Fourier blocks* (GF block). Each block contains a 3D discrete Fourier transform, an element-wise multiplication, and a 3D inverse Fourier transform. The element-wise multiplication is introduced to learn filters in the frequency domain to capture global information. The GF-Net is trained by a *multi-instance learning* (MIL) strategy, which randomly drops some patches (filling with zero) to generate additional instances for training. Our model was applied on two AD datasets and showed promising performance. To illustrate the effectiveness of the extracted features, we present two interpretability analyses, *saliency map* and *Shapley value*, to highlight the pathological regions.

2 Methods

An overview of our architecture is given in Fig. 1. It consists of a patch embedding operation and a series of global Fourier blocks (GF block). Let $\mathbf{x} \in \mathbb{R}^{H\times W\times D\times C}$ denote the input image, where H, W, D denote the height, width, and depth respectively, C denotes the number of channels (1 for sMRI). We first extract several 3D patches $\{\mathbf{x}_i \in \mathbb{R}^{P\times P\times P\times C} : i = 1,\dots,N\}$ from $\mathbf{x}$, where P is the patch size, $N = HWD/P^3$ is the number of patches. The patches are embedded in a E-dimensional space by linear transformation $\mathbf{z}_i = \mathbf{W}\text{vec}(\mathbf{x}_i)$, which is implemented by E 3D convolutions with a stride equal to P. Denote the resulting tensor by $\mathbf{z} \in \mathbb{R}^{\tilde{H}\times\tilde{W}\times\tilde{D}\times E}$ ($\tilde{H} = \frac{H}{P}$, $\tilde{W} = \frac{W}{P}$, $\tilde{D} = \frac{D}{P}$). This tensor will go through a sequence of GF blocks for prediction.

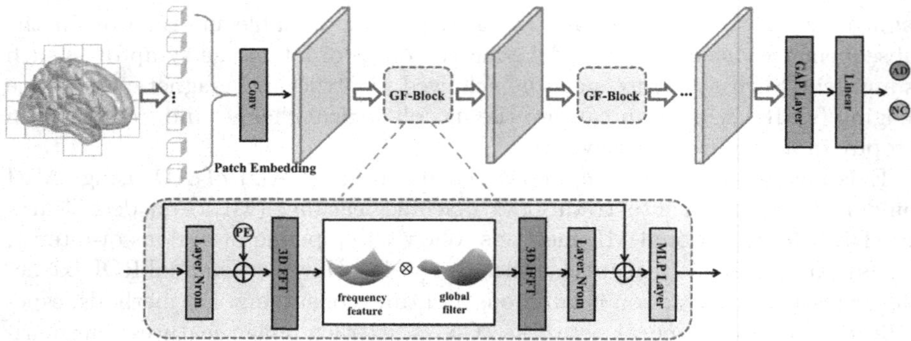

Fig. 1. The overall architecture of the GF-Net. It is a frequency attention model that consists of a sequence of blocks. Each block features a 3D discrete Fourier transform to convert features from the spatial domain to the frequency domain, an element-wise multiplication between frequency features and learnable global filters, and a 3D inverse Fourier transform to convert the features back to the spatial domain.

2.1 Global Fourier Block

A GF block adds $\mathbf{z}$ to its frequency filtered version before feeding it to a multi-layer perceptron (MLP) f_{MLP}:

$$f_{GF}(\mathbf{z}) = f_{MLP}(\mathbf{z} + f_{FF}(\mathbf{z})),$$

where $f_{FF} : \mathbb{R}^{\tilde{H} \times \tilde{W} \times \tilde{D} \times E} \rightarrow \mathbb{R}^{\tilde{H} \times \tilde{W} \times \tilde{D} \times E}$ denotes a channel-wise frequency filtering operation by Fourier transform. We first explain f_{FF} in details below.

Frequency Filtering by 3D Fourier Transform. Simply put, f_{FF} is composed of a layer normalization (LN), a positional embedding (PE), Fourier and inverse Fourier transform, and finally another LN (see Fig. 1). Given input $\mathbf{z}$, the LN operation normalizes it along the channels such that the values at each voxel has a zero mean and a standard deviation of 1. The PE operation adds a learnable vector to each voxel. After these two operations, denote the resulting tensor by $\tilde{\mathbf{z}}$ and its jth channel by $\tilde{\mathbf{z}}^j$. The 3D image $\tilde{\mathbf{z}}^j$ has a size $\tilde{H} \times \tilde{W} \times \tilde{D}$ and can be seen as a discretized version of a 3D function $\tilde{z}^j : \mathbb{R}^3 \rightarrow \mathbb{R}$. We use the 3D Fourier transform to capture the global information of this function:

$$\hat{z}^j(\xi) = \mathcal{F}(\tilde{z}^j)(\xi) = \int_{\mathbb{R}^3} \tilde{z}^j(\mathbf{y}) e^{-i2\pi \mathbf{y} \cdot \xi} d\mathbf{y},$$

where $\xi \in \mathbb{R}^3$. For a discretized image $\tilde{\mathbf{z}}^j$, a fast Fourier transform (FFT) implementation can be used to convert it to a tensor in complex values. Combining all these tensors, we have $\hat{\mathbf{z}} \in \mathbb{C}^{\tilde{H} \times \tilde{W} \times \tilde{D} \times E}$, which can be seen as features in the frequency domain. These features are multiplied element-wisely by a learnable filter $\mathbf{K} \in \mathbb{C}^{\tilde{H} \times \tilde{W} \times \tilde{D} \times E}$:

$$\hat{\mathbf{z}}' = \mathbf{K} \odot \hat{\mathbf{z}}$$

where $\odot$ is the element-wise multiplication (Hadamard product). Then, a 3D inverse Fourier transform is performed channel-wisely to convert the frequency domain features to the spatial domain. Denote the output by $\tilde{\mathbf{z}}' \in \mathbb{R}^{\tilde{H}\times\tilde{W}\times\tilde{D}\times E}$. Finally, another LN is applied to $\tilde{\mathbf{z}}'$ and the result is the output of f_{FF}.

Theoretically, multiplication in the frequency domain is equivalent to convolution in the spatial domain. We have some empirical results showing that as the filter size increases in spatial convolution, the performance improves in our framework, possibly due to better learned global information (see Fig. S1 in the supplementary material). However, the computational complexity of spatial convolution in the 3D domain increases rapidly as the filter size increases. Hence, filtering in the frequency domain becomes a more viable approach. Similar results have also been found in a Transformer-like framework [20].

Multi-layer Perceptron Layer. The MLP layer in the GF block reshapes $\mathbf{z} + f_{FF}(\mathbf{z})$ to a matrix of size $\tilde{H}\tilde{W}\tilde{D} \times E$ and linearly transforms its columns, followed by an activation function. The linear transformation is designed to keep the resulting size unchanged (i.e. $\tilde{H}\tilde{W}\tilde{D} \times E$) such that the resulting matrix can be reshaped back to the original size $\tilde{H} \times \tilde{W} \times \tilde{D} \times E$. Hence, the whole GF block preserves the tensor size.

The above GF block is repeated multiple times. The output of the last GF block is fed to a classifier layer for prediction.

2.2 Classifier Layer and Loss Function

The classifier layer contains a global average pooling (GAP) layer and a linear layer. The GAP layer calculates the average value along the channel dimension. Then the resulting 3D image is flattened and the linear layer is applied to the flattened vector to produce the logits. Finally, a standard cross entropy loss is used for training.

2.3 Multi-instance Learning Strategy

To avoid the over-fitting problem, we drop some patches randomly to generate multiple instances for training. Dropping patches is implemented by filling the patches with zero [21]. Hence, for each image, we obtain multiple instances that have the same size and share the same class label as the original image [22]. For example, given an image, we can randomly pick 30, 40, 50 patches for dropping, which results in 4 different instances that share the same label.

3 Experiments and Results

3.1 Dataset Description and Experimental Setup

The brain structural images (T1WI) acquired at 3.0T MRI systems, were provided by the Alzheimer's Disease Neuroimaging Initiative (ADNI) and the Australian Imaging, Biomarker & Lifestyle Flagship Study of Ageing (AIBL). We pre-processed the sMRI images by spatial normalization, intensity normalization, and background removal. The spatial normalization registered the images to the MNI152 template by affine transformation as implemented by the FSL package in Python. The intensity normalization transforms the intensities linearly such that they have a zero mean and a standard deviation of 1. Any voxel with intensity lower than -1 was set to -1 and those with intensity higher than 2.5 were truncated to 2.5. The background removal step stripped the skull according to the MNI152 template and set the voxels outside the skull to -1. Quality check of these steps was performed by visual inspection. After quality check, 417 images were left from ADNI (229 for NC, 188 for AD), and 380 images from AIBL (320 for NC, 62 for AD). All of them were acquired within 6 months from the date of diagnosis.

In the experiment, 60% of the ADNI data were used for network training, 20% for validation and the remaining 20% for testing. The entire AIBL data were kept for testing. The experiment was repeated 10 times with random training/validation/test split and the mean and standard deviation of the accuracy/sensitivity/specificity/F1-score were calculated.

3.2 Implementation Details

Our model was implemented in PyTorch and accelerated by 2 NVIDIA A-6000 GPUs and 4 NVIDIA V-100 GPUs. The Gaussian error linear unit (GELU) was used as the activation function in the MLP of the GF blocks. The patch size was set to $10 \times 10 \times 10$. The batch size was set to 10. The learning rate was set to 0.0001 and the maximal number of training epochs was set to 1000 with a weight decay of 0.9 for each 100 epochs. The number of GF blocks was set to 8. For each image, we dropped 1/16, 1/8, 1/4, 1/2 patches according to the patch indices randomly 10 times, resulting in a training set whose size is 40 times the original size.

3.3 Experimental Results on ADNI and AIBL

To show the effectiveness of the proposed GF-Net, we compared it with VBM [23], CNN3D [24], ResNet3D [25], FCN [14], and ViT3D [26]. VBM calculates gray matter (GM) densities and uses a linear SVM classifier. The other competing methods were run according to their online code. The results are shown in Table 1.

Table 1. Results for AD classification (AD vs. NC) on ADNI and AIBL. The results (in %) were calculated based on 10 random training/validation/test splits. The rows in italic have their numbers directly copied from the papers. The "# of Images" column has a format of (number of images for AD, number of images for NC). The best result in each category is boldfaced.

Dataset	Method	# of Images	Accuracy	Sensitivity	Specificity	F1-score	AUC
ADNI	VBM	(188,229)	82.5 ± 4.4	75.5 ± 4.7	88.9 ± 3.2	82.8 ± 5.7	80.1 ± 4.2
	CNN3D		84.8 ± 3.8	86.2 ± 3.3	84.5 ± 4.8	85.3 ± 4.2	84.6 ± 4.4
	FCN		78.6 ± 5.7	82.2 ± 4.3	76.5 ± 6.4	79.2 ± 6.8	78.2 ± 5.3
	ViT3D		85.5 ± 2.9	87.9 ± 3.6	86.8 ± 3.7	87.3 ± 3.6	85.7 ± 3.5
	ResNet3D		87.7 ± 3.5	90.2 ± 2.8	89.7 ± 3.0	90.0 ± 3.5	86.2 ± 4.0
	Salvatore et al. [5]	*(137,162)*	*76.0*	–	–	–	–
	Cuingnet et al. [6]	*(137,162)*	*88.6*	*81.0*	***95.0***	*87.4*	–
	Eskildsen et al. [7]	*(194,226)*	*86.7*	*80.4*	*92.0*	*85.8*	–
	Cao et al. [8]	*(192,229)*	*88.6*	*85.7*	*90.4*	*88.0*	–
	Lin et al. [19]	*(188,229)*	*88.8*	–	–	–	–
	Tong et al. [10]	*(198,231)*	*90.0*	*86.0*	*93.0*	*89.4*	–
	Li et al. [18]	*(199,229)*	*89.5*	*87.9*	*90.8*	*89.3*	–
	Qiu et al. [16]	*(188,229)*	*83.4*	*76.7*	*88.9*	*82.4*	–
	H-FCN [14]	*(389,400)*	*90.5*	*89.7*	*91.3*	*90.5*	–
	DA-Net [15]	*(389,400)*	*92.4*	*91.0*	*93.8*	***92.4***	–
	GF-Net	(188,229)	$\mathbf{94.1 \pm 2.8}$	$\mathbf{93.2 \pm 2.4}$	90.6 ± 2.6	91.8 ± 2.6	$\mathbf{93.5 \pm 2.7}$
AIBL	VBM	(62,320)	81.7 ± 4.4	75.5 ± 4.7	86.3 ± 4.1	80.4 ± 6.1	81.5 ± 4.2
	CNN3D		86.2 ± 2.9	70.2 ± 5.7	88.7 ± 3.2	78.3 ± 4.8	80.3 ± 5.5
	FCN		77.2 ± 5.5	74.4 ± 4.6	78.8 ± 5.1	76.5 ± 4.8	76.8 ± 5.1
	ViT3D		87.5 ± 2.6	88.2 ± 3.4	91.8 ± 1.9	89.9 ± 4.0	87.9 ± 4.1
	ResNet3D		88.0 ± 3.6	91.1 ± 2.4	83.7 ± 4.6	87.2 ± 3.5	87.8 ± 3.7
	GF-Net		$\mathbf{93.2 \pm 2.4}$	$\mathbf{93.3 \pm 2.6}$	$\mathbf{94.6 \pm 3.3}$	$\mathbf{94.0 \pm 2.7}$	$\mathbf{93.8 \pm 2.9}$

Several observations could be derived from Table 1: 1) The three DL methods—CNN3D, ViT3D, ResNet3D—yield better performance than the traditional VBM method, suggesting the superiority of DL algorithms on this task. 2) The proposed GF-Net outperforms the competing methods by a relatively large margin in all the metrics. We also calculated the p-values and the p-values were less than 0.05. 3) Our results are even better than almost all the results in the literature even though that our results are calculated based on 10 random splits.

Additional results on classifying progressive mild cognitive impairment (pMCI) and stable mild cognitive impairment (sMCI) are available in Table S1 in the supplementary material (AUC: 87.8 ± 3.6). Similar to classifying AD and NC, GF-Net achieved superior accuracy on classifying pMCI and sMCI.

3.4 Ablation Study

We performed two ablation studies. First, we investigated the influence of the number of GF blocks. Using the ADNI dataset, we selected the number of blocks from {4,6,8,10} and reported the results in Table 2. As shown in Table 2, most of the metrics increased when the number of blocks increased from 4 to 8. However, when the number of blocks reached 10, the performance declined slightly, which could be attributed to the complexity of the network that results in over-fitting. Nevertheless, the proposed network in this case still outperformed other state-of-the-art methods with a large margin, demonstrating its robustness.

Table 2. Ablation study (results in %) on the number of blocks on ADNI.

# of blocks	Accuracy	Sensitivity	Specificity	F1-score	AUC
4	93.3 ± 3.1	92.1 ± 2.7	91.6 ± 2.4	91.8 ± 2.7	91.5 ± 3.6
6	93.5 ± 3.5	91.7 ± 2.3	91.3 ± 2.2	91.5 ± 3.1	93.1 ± 3.0
8	94.1 ± 2.8	93.2 ± 2.4	90.6 ± 2.6	91.8 ± 2.6	93.5 ± 2.7
10	93.4 ± 2.7	91.4 ± 2.6	91.4 ± 2.7	91.4 ± 3.2	92.5 ± 3.5

In the previous experiment, the patch size is fixed as $10 \times 10 \times 10$. Now, we show the effect of different patch sizes. Table 3 shows the classification results when the patch size ranges in $10 \times 10 \times 10$, $15 \times 15 \times 15$, $20 \times 20 \times 20$, $25 \times 25 \times 25$. From Table 3, we can see that our GF-Net is also not sensitive to the change of patch size. The patch size of $10 \times 10 \times 10$ shows the best performance, while the rest sizes also outperform the current state-of-the-art methods.

Table 3. Ablation study (results in %) on the patch size on ADNI.

Patch size	Accuracy	Sensitivity	Specificity	F1-score	AUC
10 × 10 × 10	94.1 ± 2.8	93.2 ± 2.4	90.6 ± 2.6	91.8 ± 2.6	93.5 ± 2.7
15 × 15 × 15	93.7 ± 3.1	91.2 ± 3.7	90.3 ± 3.4	90.8 ± 3.6	92.3 ± 2.9
20 × 20 × 20	93.5 ± 3.4	90.7 ± 3.9	91.1 ± 2.9	90.9 ± 3.1	92.0 ± 3.1
25 × 25 × 25	92.4 ± 2.5	89.9 ± 3.5	90.2 ± 3.1	90.0 ± 2.7	91.4 ± 3.6

3.5 Interpretation by Saliency Map and Shapley Value

To investigate the discriminative regions for classifying AD and NC, we employed two interpretation analyses, saliency map and Shapley value. Figure 2 shows the corresponding saliency map, which calculates the derivative of the logit with respect to the input image. Figure 2 shows that most of the discriminative voxels (in red) are located in the brain regions closely associated with clinical diagnosis, such as hippocampus, amygdala, and thalamus [27,28].

We also use Shapley value explanation to quantitatively evaluate the importance of the extracted features, i.e. the tensor before the last linear layer. Specifically, the features from all the instances are reduced to 3 dimensions by PCA, called PC_1, PC_1, PC_3. They are combined with non-imaging features: mini-mental state examination (MMSE), age, APOE, and gender. The combined features are fed into XGboosting for classification. Finally, the Shapley values are calculated for all the features and shown in Fig. 3.

It is not surprising that MMSE has the widest spread since MMSE measures the mental state of the patients (the lower, the more severe), which can be considered as another ground truth to classify AD and NC. Furthermore, our imaging

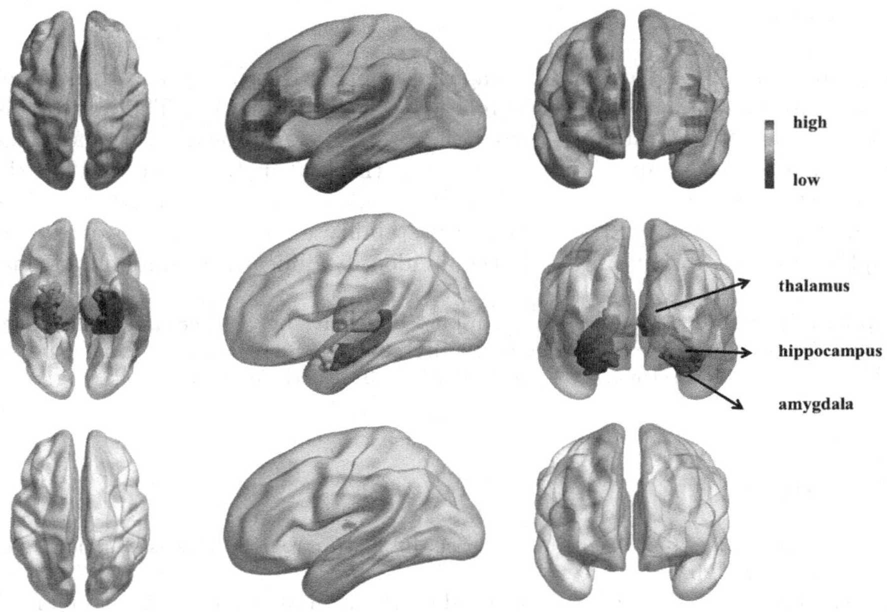

Fig. 2. Saliency map of the trained network on an example image. The three rows show the saliency map, clinically related reference regions, saliency map truncated to the reference regions respectively.

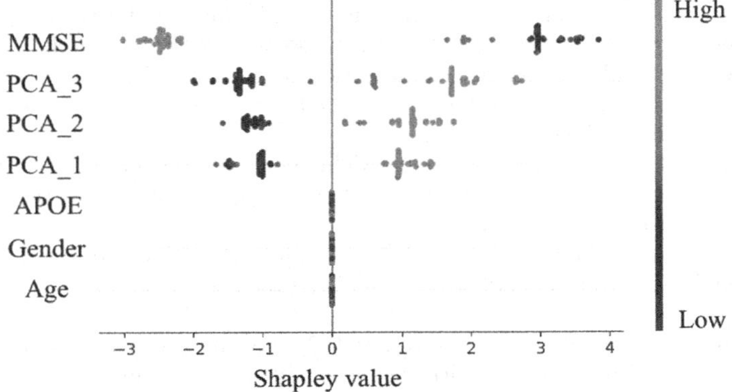

Fig. 3. Shapley values for analyzing importance scores of the image features extracted by GF-Net, when the image features are combined with the non-imaging features for classification. The spread of the Shapley values of a feature reflects its impact on the model output.

features have a spread close to MMSE, outperforming the other non-imaging features by a large margin. This quantitative interpretation demonstrates the effectiveness of our method again.

4 Conclusion

We propose a 3D GF-Net for AD diagnosis using sMRI. The GF-Net uses frequency filtering to capture disease-related global information. The network is trained by an MIL strategy to avoid the over-fitting problem. The classification results on ADNI and AIBL demonstrate that our method can significantly outperform other state-of-the-art methods for classifying AD and NC (also for classifying pMCI and sMCI). Two interpretability analyses, saliency map and Shapley value, show that our method could identify clinically meaningful regions.

Acknowledgments. This work was supported in part by Science and Technology Commission of Shanghai Municipality (20ZR1407800), Shanghai Municipal Science and Technology Major Project (No.2018SHZDZX01), and Fudan Univerisity startup fund (JIH2305006Y).

References

1. Knopman, D.S., et al.: Alzheimer disease. In: Nature reviews Disease Primers, vol. 7.1, pp. 1–21 (2021)
2. Damulina, A., et al.: Cross-sectional and longitudinal assessment of brain iron level in Alzheimer disease using 3-T MRI. Radiology **296**(3), 619–626 (2020)
3. Wen, J., et al.: Convolutional neural networks for classification of Alzheimer's disease: overview and reproducible evaluation. Med. Image Anal. **63**, 101694 (2020)
4. Zeng, N., et al.: A new switching-delayed-PSO-based optimized SVM algorithm for diagnosis of Alzheimer's disease. Neurocomputing **320**, 195–202 (2018)
5. Salvatore, C., et al.: Magnetic resonance imaging biomarkers for the early diagnosis of Alzheimer's disease: a machine learning approach. Front. Neurosci. **9**, 307 (2015)
6. Cuingnet, R., et al.: Automatic classification of patients with Alzheimer's disease from structural MRI: a comparison of ten methods using the ADNI database. Neuroimage **56**(2), 766–781 (2011)
7. Eskildsen, S.F., et al.: Prediction of Alzheimer's disease in subjects with mild cognitive impairment from the ADNI cohort using patterns of cortical thinning. Neuroimage **65**, 511–521 (2013)
8. Cao, P., et al.: Nonlinearity-aware based dimensionality reduction and over-sampling for AD/MCI classification from MRI measures. Comput. Biol. Med. **91**, 21–37 (2017)
9. Sørensen, L., Nielsen, M., Initiative, A.D.N., et al.: Ensemble support vector machine classification of dementia using structural MRI and mini-mental state examination. J. Neurosci. Methods **302**, 66–74 (2018)
10. Tong, T., et al.: Multiple instance learning for classification of dementia in brain MRI. Med. Image Anal. **18**(5), 808–818 (2014)
11. Khvostikov, A., et al.: 3D CNN-based classification using sMRI and MD-DTI images for Alzheimer disease studies. arXiv preprint arXiv:1801.05968 (2018)

12. Liu, M., et al.: Landmark-based deep multi-instance learning for brain disease diagnosis. Med. Image Anal. **43**, 157–168 (2018)
13. Zhou, B., et al.: Learning deep features for discriminative localization. In: Proceedings of the IEEE Conference on Computer Vision and Pattern Recognition, pp. 2921–2929 (2016)
14. Lian, C., et al.: Hierarchical fully convolutional network for joint atrophy localization and Alzheimer's disease diagnosis using structural MRI. IEEE Trans. Pattern Anal. Mach. Intell. **42**(4), 880–893 (2018)
15. Zhu, W., et al.: Dual attention multi-instance deep learning for Alzheimer's disease diagnosis with structural MRI. IEEE Trans. Med. Imaging **40**(9), 2354–2366 (2021)
16. Qiu, S., et al.: Development and validation of an interpretable deep learning framework for Alzheimer's disease classification. Brain **143**(6), 1920–1933 (2020)
17. Li, H., Habes, M., Fan, Y.: Deep ordinal ranking for multi-category diagnosis of Alzheimer's disease using hippocampal MRI data. arXiv preprint arXiv:1709.01599 (2017)
18. Li, F., Liu, M., Initiative, A.D.N., et al.: Alzheimer's disease diagnosis based on multiple cluster dense convolutional networks. Comput. Med. Imaging Graph. **70**, 101–110 (2018)
19. Lin, W., et al.: Convolutional neural networks-based MRI image analysis for the Alzheimer's disease prediction from mild cognitive impairment. Front. Neurosci. **12**, 777 (2018)
20. Rao, Y., et al.: Global filter networks for image classification. Adv. Neural. Inf. Process. Syst. **34**, 980–993 (2021)
21. He, K., et al.: Masked autoencoders are scalable vision learners (2021). arXiv: 2111.06377 [cs.CV]
22. Couture, H.D., Marron, J.S., Perou, C.M., Troester, M.A., Niethammer, M.: Multiple instance learning for heterogeneous images: training a CNN for histopathology. In: Frangi, A.F., Schnabel, J.A., Davatzikos, C., Alberola-López, C., Fichtinger, G. (eds.) MICCAI 2018. LNCS, vol. 11071, pp. 254–262. Springer, Cham (2018). https://doi.org/10.1007/978-3-030-00934-2_29
23. Ashburner, J., Friston, K.J.: Voxel-based morphometry-the methods. Neuroimage **11**(6), 805–821 (2000)
24. Kruthika, K.R. HD Maheshappa, Alzheimer's disease neuroimaging initiative. CBIR System using Capsule Networks and 3D CNN for Alzheimer's disease diagnosis. Inf. Med. Unlocked **14**, 59–68 (2019)
25. He, K., et al.: Deep residual learning for image recognition. In: Proceedings of the IEEE Conference on Computer Vision and Pattern Recognition, pp. 770–778 (2016)
26. Dosovitskiy, A., et al.: An image is worth 16x16 words: transformers for image recognition at scale. arXiv preprint arXiv:2010.11929 (2020)
27. Gerischer, L.M., et al.: Combining viscoelasticity, diffusivity and volume of the hippocampus for the diagnosis of Alzheimer's disease based on magnetic resonance imaging. NeuroImage Clin. **18**, 485–493 (2018)
28. Shao, W., et al.: Hypergraph based multi-task feature selection for multimodal classification of Alzheimer's disease. Comput. Med. Imaging Graph. **80**, 101663 (2020)

CASHformer: Cognition Aware SHape Transformer for Longitudinal Analysis

Ignacio Sarasua[1,2](✉), Sebastian Pölsterl[2], and Christian Wachinger[1,2]

[1] School of Medicine, Technical University of Munich, Munich, Germany
ignacio@ai-med.de

[2] Lab for Artificial Intelligence in Medical Imaging (AI-Med), KJP, LMU Klinikum, Munich, Germany

Abstract. Modeling temporal changes in subcortical structures is crucial for a better understanding of the progression of Alzheimer's disease (AD). Given their flexibility to adapt to heterogeneous sequence lengths, mesh-based transformer architectures have been proposed in the past for predicting hippocampus deformations across time. However, one of the main limitations of transformers is the large amount of trainable parameters, which makes the application on small datasets very challenging. In addition, current methods do not include relevant non-image information that can help to identify AD-related patterns in the progression. To this end, we introduce CASHformer, a transformer-based framework to model longitudinal shape trajectories in AD. CASHformer incorporates the idea of pre-trained transformers as universal compute engines that generalize across a wide range of tasks by freezing most layers during fine-tuning. This reduces the number of parameters by over 90% with respect to the original model and therefore enables the application of large models on small datasets without overfitting. In addition, CASHformer models cognitive decline to reveal AD atrophy patterns in the temporal sequence. Our results show that CASHformer reduces the reconstruction error by 73% compared to previously proposed methods. Moreover, the accuracy of detecting patients progressing to AD increases by 3% with imputing missing longitudinal shape data.

1 Introduction

Alzheimer's disease (AD) is a complex neurodegenerative disorder that is characterized by progressive atrophy in the brain [12]. A spatio-temporal model of neuroanatomical changes is instrumental for understanding atrophy patterns and predicting patient-specific trajectories. For inferring such a model, longitudinal neuroimaging data can be used, but they are usually highly irregular with non-uniform follow-up visits and dropouts. Transformers provide a flexible approach that can incorporate different sequence length inputs and incomplete time series.

Supplementary Information The online version contains supplementary material available at https://doi.org/10.1007/978-3-031-16431-6_5.

L. Wang et al. (Eds.): MICCAI 2022, LNCS 13431, pp. 44–54, 2022.
https://doi.org/10.1007/978-3-031-16431-6_5

Hence, they are well suited for modeling longitudinal neuroimaging data. In addition, recent work has combined geometric deep learning on anatomical meshes with transformers to obtain a model that is sensitive to small shape changes in the hippocampus [24].

However, one of the main limitations of transformers is their large number of parameters. Their huge success in Natural Language Processing (NLP) [3] and Computer Vision [4] is also based on the availability of large datasets in these domains. Medical datasets are typically much smaller, especially from longitudinal studies. This limits the application of transformers, particularly because deeper transformer networks are thought to be preferred [15]. At the same time, recent research established pre-trained transformers as *universal compute engines* [17] that generalize across domains, e.g., pre-training on NLP and fine-tuning on tasks like numerical computation and vision. Based on this seminal work, we investigate whether transformers – pre-trained on non-medical applications – are helpful for creating a spatio-temporal model of progression to AD.

Related research in generative shape modeling suggests that adding prior information (e.g., diagnosis) reduces the reconstruction error [9]. However, in longitudinal modeling, diagnosis might remain the same along a patient's trajectory. Fortunately, the AD Assessment Scale (ADAS; [19]) cognitive score presents a fine-grained measure of cognitive decline along the longitudinal sequence. Including such information in a model can help aligning inter-patient data based on disease progression, as observed in [18].

Given these considerations, we introduce CASHformer, a Cognition Aware SHape Transformer for longitudinal shape analysis. CASHformer generates embeddings from the input meshes of the hippocampus using a SpiralResNet [1]. The embeddings are input into a transformer network, which has been pre-trained on a large non-medical dataset. Following the Frozen Pre-Trained Transformer [17], only a few layers are fine-tuned, which allows us to train deeper transformer architectures, while keeping the number of trainable parameters small. For explicitly modeling cognitive decline, we introduce ADAS embeddings and an ADAS cost function that acts as regularizer. Our experiments demonstrate that CASHformer reduces the reconstruction error by 73% with respect to previously proposed methods and increases the AD-progressor classification accuracy by 3% with imputing missing shapes.

Related Work. Long short-term memory (LSTM) networks have been previously applied to detecting AD, either given a single scan [5,6], or a temporal sequence [10]. Convolutional neural networks (CNNs) have been applied to longitudinal modeling of brain images in [2,28]. These methods work on full brain images and cannot capture subtle changes in subcortical structures. In addition, they explicitly enforce linearity in latent representations within one patient. Our models take advantage of Multi-Head attention layers in transformers and positional embeddings to implicitly enforce this behavior. Given their capabilities of capturing subtle changes in brain structures, deep neural networks have been

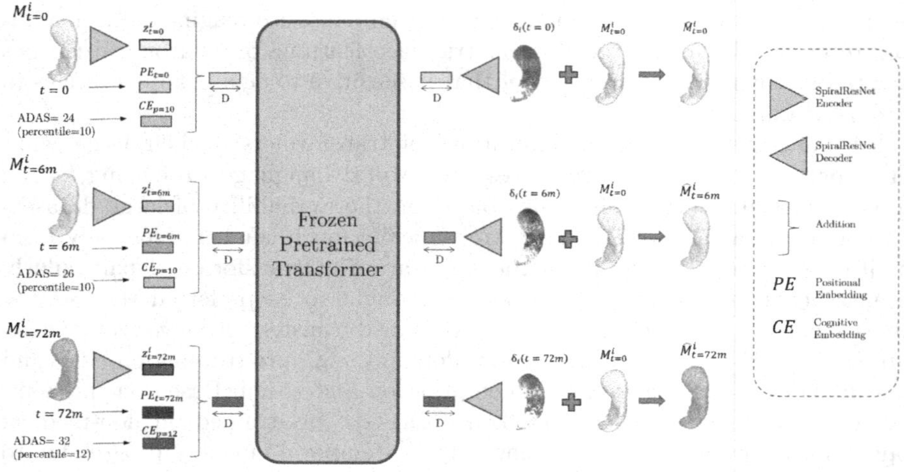

Fig. 1. Overview of the proposed CASHformer framework.

proposed to work on anatomical meshes [1,23]. Transformer networks [26] have disrupted the fields of Natural Language Processing (NLP) [3,13], computer vision [4], and medical image analysis [14,25,27]. Recently, they have also been applied to study anatomical structures like the hippocampus [24]. While the latter is closely related to ours, we propose to include cognitive performance as prior information, and a more efficient training strategy that allows us to train deeper transformer architectures with limited data.

2 Methods

Figure 1 illustrates our proposed CASHformer framework. Let $\{M_t^i \,|\, t = 0, ..., T_i\}$ be the set of hippocampi meshes of subject i from the time of enrollment $t = 0$ to the last visit $t = T_i$. The *SpiralResNet encoder* generates an embedding for every available mesh $\{z_t^i \,|\, t = 0, ..., T_i\}$ of dimension D. The latent representations are then modulated by a set of learnable positional encodings and *cognitive score embeddings*. The former keeps track of the ordering of the sequence [4], while the latter adds information about the cognitive evolution of the patient. These latent representations are fed into a *pre-trained transformer* and decoded by a *SpiralResNet decoder*, yieling the deformation fields $\{\delta_i(t) \,|\, t = 0, ..., T_i\}$ that are applied to a reference shape. The model is trained end-to-end using a combination of reconstruction loss and *cognitive aware loss*.

2.1 Mesh Network

Since meshes do not follow a grid-like structure, operations like convolution or pooling are not as straightforward as for images. Therefore, we need a method that can define convolutional and down/up-sampling operations for such structures. Let V_i be a vertex of a triangular mesh and $S(i,l)$ an ordered set consisting of l neighboring vertices of V_i, that are inside a spiral defined by choosing an arbitrary direction in counter-clockwise manner. The *spiral convolution operation* for layer k for features $\mathbf{x}_i$ associated to the i-th vertex is defined as $\mathbf{x}_i^{(k)} = \gamma^{(k)}(\|_{j\in S(i,l)}\,(\mathbf{x}_j^{(k-1)}))$, where γ denotes Multi-Layer Perceptrons (MLP) and $\|$ is the concatenation operation [8]. We work with a residual version [1], where multiple SpiralConvolution layers are concatenated with ELU and batch-norm layers, and residual connections are added (see Fig. S1 in the supplemental material). For down-sampling the mesh after each convolution block, vertex pairs with the lowest quadratic error are contracted [7]. For up-sampling the meshes, the barycentric coordinates of the closest triangle in the downsampled mesh are used [22] (supplemental Fig. S2). Given that all our shapes are registered to a template (see more in Sect. 3.1), we can precompute the coordinates of the down-sampled and up-sampled meshes once for the template shape and then apply it to every patient in the dataset.

2.2 Frozen Pre-trained Transformer

Transformers are formed by a set of encoder blocks, where each of them consists of a Multi-Head-Attention and an MLP, preceded by a Layer Normalization (LN) block [26]. In addition, residual connections are added after every block and GELU is used as non-linearity in the MLP block. The input is a sequence of embeddings, which in case of NLP are generated through tokenization methods that convert words into feature vectors like Byte Pair Encoding(BPE) [21]. Subsequently, transformers were applied to image recognition tasks, by passing patches of the input image through a convolutional neural network to generate the embeddings, which achieved state-of-the-art performance on ImageNet [4]. However, one of the main issues of these architectures is their large number of trainable parameters. While this feature allows transformers to perform great on large datasets, it limits their application for smaller ones.

Inspired by the Frozen Pre-Trained Transformer [17], we use a pre-trained model and only fine-tune the LN blocks. This not only reduces the amount of trainable parameters by 90%, but also allows us to apply larger transformer models (e.g., 24 blocks) to our smaller dataset [21]. To generate the embeddings that are the input to the transformer, we substitute the input layers of the transformers (i.e., CNN for ViT, and BPE tokenizer for GPT-2) by a SpiralResNet, described in Sect. 2.1.

2.3 Cognitive Embeddings and Cognitive Decline Aware Loss

The Alzheimer's Disease Assessment Scale (ADAS) cognitive score is a widely used measure to study the progression of cognitive impairment [19]. Therefore, we introduce a set of learnable Cognitive Embeddings (CE) that modulate the input latent representations coming from the mesh network (similar as the positional encoding ones [4]). The ADAS score ranges from 0 to 85. Since generating

85 embeddings would unnecessarily increase the number of trainable parameters, we quantize the whole range in 20 intervals based on the percentile they occupy within the training set (e.g., those shapes corresponding to ADAS score between percentile 10 and 15 are modulated using the same embedding). This results in a set of 20 embeddings of size D. The CE and mesh embedding are then summed to form a combined embedding (see Fig. 1).

Patients that experience acute cognitive decline also show larger deformation in their hippocampus [18]. We incorporate this clinical knowledge by introducing a regularization term, the Cognitive Decline Aware (CDA) loss. The CDA loss enforces those patients experiencing a larger cognitive decline to have a larger norm of their deformation field. To this end, we maximize the cosine similarity (CS) between the line describing the ADAS score evolution and the sequence of the deformation field's norms $\mathcal{L}_{\text{CDA}}(\delta(t), ADAS(t)) = \text{CS}(\mathbf{\Delta}_{\delta(t)}, \mathbf{\Delta}_{ADAS(t)})$, where $\mathbf{\Delta}_{\delta(t)} = \left[\|\delta(t=0)\|, \ldots, \|\delta(t=T)\|\right]^\top$, $\mathbf{\Delta}_{ADAS(t)} = m \cdot [0, ..., T]^\top$, and $m = (ADAS(T) - ADAS(0))/T$.

2.4 Training Procedure

As illustrated in Fig. 1, CASHformer predicts the mesh deformation field $\delta_i(t)$ with respect to a *reference mesh*, which is a person's baseline shape $M^i_{t=0}$.

Missing Shapes: Transformer networks use padding tokens to fill up the positions of missing inputs. These are ignored during training by adding key masks [26]. However, the model still predicts an output for those positions,which allows us to impute missing values during inference. For our application, we use the encoding of the reference shape, as produced by the mesh network, to generate the "missing value" embedding. As for the CE, we train an extra embedding to account for missing shapes.

Data Augmentation: Inspired by [3], we apply random masking and shuffling to the input embeddings. In particular, 35% of the shapes' embeddings are substituted by the reference shape's embedding and 15% are shuffled along the sequence. Contrary to the real missing shapes, these are not ignored during training; the self-attention masks do take them into consideration.

Loss Function: The model is trained using a combination of Mean Square Error (MSE) and the CDA loss:

$$\mathcal{L} = \textstyle\sum_{i=0}^{P} \sum_{t=0}^{T_i} ||M^i_{t=0} + \delta_i(t) - M^i_t||^2 - \lambda \mathcal{L}_{\text{CDA}}(\delta_i(t), ADAS_i(t)), \quad (1)$$

where λ (empirically set to 10^{-6}) controls the contribution of the CDA loss, and P is the number of patients in the set.

3 Experiments

3.1 Data Processing

Given its relevance in AD pathology [12], we use the shape of the left hippocampus for all our experiments. Our data is a subset of the Alzheimer's Disease

Neuroimaging Initiative (ADNI) data (adni.loni.usc.edu) [11]. Structural MRI scans are segmented with FIRST [20] from the FSL Software Library, which provides meshes for the segmented samples. FIRST segments the subcortical structures by registering the MRI scan to a reference template, creating voxel-wise correspondences (and therefore, also vertex-wise) between the template and every sample in the dataset. This template is used for all the pre-computations of the mesh network in Sect. 2.1. We limit the number of follow-ups to 8 (from baseline to $T_{max} = 72$ months). Since our focus is on modeling cognitive decline, we only include patients that have been diagnosed with MCI or AD. Our subset of the ADNI data consists of 845 patients split 70/10/20 (train/validation/test) following a data stratification strategy that accounts for age, sex and diagnosis, so they are represented in the same proportion in each set. The average number of follow-up scans per patient is 3.32 and only 3.6% of the patients have attended all the follow-up sessions.

3.2 Implementation Details

For the mesh encoder and decoder, we followed [1] autoencoder's architecture design (more details in supplemental Fig. S1). For the transformer, we evaluated three state-the-art pre-trained transformer architectures: GPT-2 [21] trained on the WebText dataset, ViT_{base}(12 blocks) and ViT_{large}(24 blocks) trained on ImageNet-21k and fine-tuned on ImageNet-1k[1]. As baselines, we compare to TransforMesh [24], the substitution of the transformer part (including the positional and cognitive embeddings) by LSTM networks[2], and training GPT-2 and ViT networks from scratch. Note that only TransforMesh has been used in mesh-based progression modeling before. More details about the architecture design can be found in supplemental Table S1.

3.3 Longitudinal Shape Modeling

We evaluate each model on three types of tasks: shape interpolation, shape extrapolation, and trajectory prediction. For each of these experiments, a set of input shapes, M_t^i, are removed for every patient. For a fairer comparison to the other methods, and to simulate a more realistic scenario where a patient missed the visit at time t, we do not input the ADAS score for that follow-up (i.e. we use the extra CE described in Sect. 2.4). The corrupted sequence is passed through the network and the Mean Absolute Error (MAE) is computed between M_t^i and $\hat{M}_t^i$, where $\hat{M}_t^i$ is the predicted mesh at visit t.

Interpolation: From each patient in the test set, the shape in the middle of the input sequence, $M_{t=\mu}^i$ with $\mu = \lfloor T_i/2 \rfloor$, is removed. The goal of this experiment is to evaluate the capabilities of the models to predict a missing shape using both past and previous information.

[1] https://github.com/rwightman/pytorch-image-models.
[2] https://pytorch.org/docs/stable/generated/torch.nn.LSTM.html.

Table 1. Top: median error and median absolute deviation for interpolation, extrapolation, and trajectory experiments are reported. Errors were multiplied by 10^3 to facilitate presentation. Bottom: evaluation of each of the contributions in CASHformer.

Method	Pre-train	Interpolation	Extrapolation	Trajectory	Parameters
Small TransforMesh [24]	✗	5.454 ± 0.289	5.573 ± 0.317	5.592 ± 0.294	$10.4M$
Base TransforMesh [24]	✗	6.352 ± 0.334	6.589 ± 0.381	6.560 ± 0.379	$38.7M$
LSTM(3 blocks)	✗	6.610 ± 0.060	6.610 ± 0.060	6.832 ± 0.341	$20.1M$
LSTM(12 blocks)	✗	5.263 ± 0.094	5.258 ± 0.088	5.653 ± 0.308	$62.6M$
GPT-2 [21]	✗	5.258 ± 0.124	5.267 ± 0.127	5.692 ± 0.351	$130M$
ViT_{base} [4]	✗	3.713 ± 0.076	3.713 ± 0.073	4.040 ± 0.262	$86.4M$
ViT_{large} [4]	✗	6.514 ± 0.290	6.610 ± 0.305	6.709 ± 0.286	$309M$
CASHformer w/ GPT-2	✓	6.111 ± 0.222	6.362 ± 0.324	6.353 ± 0.311	$5.99M$
CASHformer w/ GPT-2	✗	3.683 ± 0.223	3.953 ± 0.335	3.955 ± 0.324	$130M$
CASHformer w/ ViT_{base}	✓	3.200 ± 0.256	3.402 ± 0.348	3.418 ± 0.366	$5.99M$
CASHformer w/ ViT_{large}	✓	$\mathbf{1.472 \pm 0.257}$	$\mathbf{1.718 \pm 0.337}$	$\mathbf{1.801 \pm 0.391}$	$6.84M$
Base model in CASHformer	Pre-train	CE	CDA	Interpolation	Parameters
ViT_{large}	✗	✗	✗	6.514 ± 0.290	$309M$
ViT_{large}	✓	✗	✗	2.610 ± 0.149	$6.82M$
ViT_{large}	✓	✓	✗	1.623 ± 0.261	$6.84M$
ViT_{large}	✓	✓	✓	1.472 ± 0.257	$6.84M$

Extrapolation: From every patient in the test set, we remove from the sequence the shape $M^i_{t=T_i}$ and input the remaining shapes to the network. This experiment aims to measure the performance of each model to predict the mesh of the last available visit, based on all the previous visits.

Trajectory Prediction: The third experiment is similar to the extrapolation experiment, but it predicts shapes that are more distant in time. Therefore, we only input the shape belonging to the baseline scan $M^i_{t=0}$, and predict all the shapes that are at least 2 years apart ($24\text{m} \leq t \leq T_i$).

3.4 Results

Mesh Prediction. Table 1 reports the median error and the median absolute deviation across all patients and visits. Regarding the baseline methods, ViT_{base} performs best. The second best performing baseline is the deeper version of LSTM (with 12 blocks). Compared to its more shallow counterpart with 3 blocks, LSTM12 reduces the error by approximately 20%, but triples the number of parameters. We observe the opposite behaviour for TransforMesh, where the Small TransforMesh performs better than the Base TransforMesh. Note that the TransforMesh architectures use the regular SpiralNet++ as their mesh backbone, while the rest of the methods use SpiralResNet. For CASHformer, ViT_{large} leads to the best results with a reduction of the reconstruction error by 60% w.r.t. the best performing baseline (ViT_{base}) and 73% w.r.t. to previously proposed TransforMesh, while having 90% and 33% less trainable parameters, respectively. Interestingly, the improvement of CASHformer w/ ViT_{base} compared to ViT_{large} is much smaller. These results are in line with prior studies about the superiority of deeper transformer architectures [15]. The comparison between CASHformer

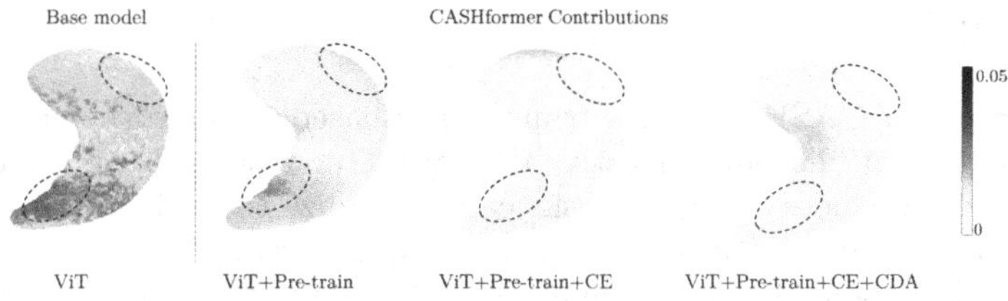

Fig. 2. Hippocampus reconstruction error for the contributions of CASHformer. The highlighted areas correspond to the medial part of the body in the subiculum and parasubiculum areas, and the lateral part of the body in the CA1 area.

with ViT and GPT-2 models shows that vision transformers are more suitable for our application. We believe that the distribution of the latent space in NLP models, which is generated from input embeddings following tokenization methods like BPE [21], is too different to our embeddings coming from a mesh network. For this reason, we also evaluated the effect of only using the CE and the CDA loss for GPT-2, without pre-training. We observe that adding these two features reduces the reconstruction error, w.r.t. using the original GPT-2, by 30%.

Given the superior performance of ViT_{large}, we study the effect of each contribution in Table 1 and Fig. 2. Fine-tuning the LN blocks reduces the reconstruction error by 60% with respect to the original model. Adding the CE yields a further reduction of the error by 38%, and adding the CDA loss by another 10%. Figure 2 shows the average reconstruction error along all patients for the interpolation experiment. We can observe that adding ADAS information reduces the reconstruction error in the medial part of the body in the subiculum and parasubiculum areas, and the lateral part of the body of the CA1 area (highlighted in the figure). These parts of the hippocampus have been found to suffer larger atrophy with progression of dementia [16]. Adding cognitive information drives more attention to those AD-related areas.

Trajectory Classification. As final experiment, we evaluate the differentiation between MCI subjects that remain stable and those that progress to AD within the study period. Once we classify based on the original shape sequence, and once we classify based on imputing the missing shapes with CASHformer w/ ViT_{large}. As classifier, we fine-tune a separate ViT_{base} model. The mean classification accuracy without imputation is 0.73 and with imputation is 0.76 (t-test $P = 0.003$). A boxplot of the results is in supplemental Fig. S3. The increase of the classification accuracy by 3% after imputation by CASHformer not only shows that our model can be used to boost performance, but it also confirms that it is able to learn meaningful representations that model the evolution of hippocampus atrophy in AD.

4 Conclusion

We have proposed CASHformer, a transformer-based framework for the longitudinal modeling of neurodegenerative diseases. Our results demonstrated that pre-trained transformers on vision tasks can generalize to medical tasks, despite the large domain gap, supporting their role as universal compute engines. We believe that this opens up new avenues for using large transformers models on medical tasks, despite scarcity of data. Moreover, our results illustrated the importance of including cognition data in the model through CE and the CDA loss to focus on AD-related shape changes. Finally, we showed the capability of CASHformer to impute missing shapes for improving the discrimination between MCI subjects that remain stable and progress to AD.

Acknowledgment. This research was partially supported by the Bavarian State Ministry of Science and the Arts and coordinated by the bidt, and the Federal Ministry of Education and Research in the call for Computational Life Sciences (DeepMentia, 031L0200A). We gratefully acknowledge the computational resources provided by the Leibniz Supercomputing Centre (www.lrz.de).

References

1. Azcona, E.A., et al.: Analyzing brain morphology in Alzheimer's disease using discriminative and generative spiral networks. bioRxiv (2021)
2. Couronné, R., Vernhet, P., Durrleman, S.: Longitudinal self-supervision to disentangle inter-patient variability from disease progression. In: de Bruijne, M., et al. (eds.) Longitudinal self-supervision to disentangle inter-patient variability from disease progression. LNCS, vol. 12902, pp. 231–241. Springer, Cham (2021). https://doi.org/10.1007/978-3-030-87196-3_22
3. Devlin, J., Chang, M.W., Lee, K., Toutanova, K.: BERT: pre-training of deep bidirectional trans-formers for language understanding. In: NAACL (2019)
4. Dosovitskiy, A., et al.: An image is worth 16x16 words: transformers for image recognition at scale. arXiv:2010.11929 (2020)
5. Dua, M., Makhija, D., Manasa, P., Mishra, P.: A CNN-RNN-LSTM based amalgamation for Alzheimer's disease detection. J. Med. Biol. Eng. **40**(5), 688–706 (2020)
6. Feng, C., et al.: Deep learning framework for Alzheimer's disease diagnosis via 3d-CNN and FSBI-LSTM. IEEE Access **7**, 63605–63618 (2019)
7. Garland, M., Heckbert, P.S.: Surface simplification using quadric error metrics. In: Proceedings of the 24th Annual Conference on Computer Graphics and Interactive Techniques, pp. 209–216 (1997)
8. Gong, S., Chen, L., Bronstein, M., Zafeiriou, S.: SpiralNet++: a fast and highly efficient mesh convolution operator. In: Proceedings of the IEEE/CVF International Conference on Computer Vision Workshops (2019)
9. Gutiérrez-Becker, B., Wachinger, C.: Learning a conditional generative model for anatomical shape analysis. In: Chung, A.C.S., Gee, J.C., Yushkevich, P.A., Bao, S. (eds.) Information Processing in Medical Imaging. LNCS, vol. 11492, pp. 505–516. Springer, Cham (2019). https://doi.org/10.1007/978-3-030-20351-1_39
10. Hong, X., et al.: Predicting Alzheimer's disease using LSTM. IEEE Access **7**, 80893–80901 (2019)

11. Jack, C.R., et al.: The Alzheimer's disease neuroimaging initiative (ADNI): MRI methods. J. Magn. Resonan. Imaging **27**(4), 685–691 (2008)
12. Jack, C.R., Holtzman, D.M.: Biomarker modeling of Alzheimer's disease. Neuron **80**(6), 1347–1358 (2013)
13. Lewis, M., et al.: BART: denoising sequence-to-sequence pre-training for natural language generation, translation, and comprehension]. In: Proceedings of the 58th Annual Meeting of the Association for Computational Linguistics, pp. 7871–7880 (2020)
14. Li, S., et al.: Few-shot domain adaptation with polymorphic transformers. In: de Bruijne, M., et al. (eds.) MICCAI 2021. LNCS, vol. 12902, pp. 330–340. Springer, Cham (2021). https://doi.org/10.1007/978-3-030-87196-3_31
15. Li, Z., et al.: Train large, then compress: rethinking model size for efficient training and inference of transformers. arXiv preprint arXiv:2002.11794 (2020)
16. Lindberg, O., et al.: Shape analysis of the hippocampus in Alzheimer's disease and subtypes of frontotemporal lobar degeneration. J. Alzheimer's Dis. JAD **30**(2), 355 (2012)
17. Lu, K., Grover, A., Abbeel, P., Mordatch, I.: Pretrained transformers as universal computation engines. arXiv preprint arXiv:2103.05247 (2021)
18. Mofrad, S.A., Lundervold, A.J., Vik, A., Lundervold, A.S.: Cognitive and MRI trajectories for prediction of Alzheimer's disease. Sci. Rep. **11**(1), 1–10 (2021)
19. Mohs, R.C., et al.: Development of cognitive instruments for use in clinical trials of antidementia drugs: additions to the Alzheimer's disease assessment scale that broaden its scope. Alzheimer disease and associated disorders (1997)
20. Patenaude, B., Smith, S.M., Kennedy, D.N., Jenkinson, M.: A Bayesian model of shape and appearance for subcortical brain segmentation. NeuroImage **56**(3), 907–922 (2011)
21. Radford, A., et al.: Language models are unsupervised multitask learners. OpenAI Blog **1**(8), 9 (2019)
22. Ranjan, A., Bolkart, T., Sanyal, S., Black, M.J.: Generating 3d faces using convolutional mesh autoencoders. In: Ferrari, V., Hebert, M., Sminchisescu, C., Weiss, Y. (eds.) ECCV 2018. LNCS, vol. 11207, pp. 725–741. Springer, Cham (2018). https://doi.org/10.1007/978-3-030-01219-9_43
23. Sarasua, I., Lee, J., Wachinger, C.: Geometric deep learning on anatomical meshes for the prediction of Alzheimer's disease. In: 2021 IEEE 18th International Symposium on Biomedical Imaging (ISBI), pp. 1356–1359. IEEE (2021)
24. Sarasua, I., Pölsterl, S., Wachinger, C.: TransforMesh: a transformer network for longitudinal modeling of anatomical meshes. In: Lian, C., Cao, X., Rekik, I., Xu, X., Yan, P. (eds.) MLMI 2021. LNCS, vol. 12966, pp. 209–218. Springer, Cham (2021). https://doi.org/10.1007/978-3-030-87589-3_22
25. Valanarasu, J.M.J., Oza, P., Hacihaliloglu, I., Patel, V.M.: Medical transformer: gated axial-attention for medical image segmentation. In: de Bruijne, M., et al. (eds.) Medical transformer: Gated axial-attention for medical image segmentation. LNCS, vol. 12901, pp. 36–46. Springer, Cham (2021). https://doi.org/10.1007/978-3-030-87193-2_4
26. Vaswani, A., et al.: Attention is all you need. In: Proceedings of the 31st International Conference on Neural Information Processing Systems, pp. 6000–6010 (2017)

27. Yu, S., et al.: MIL-VT: multiple instance learning enhanced vision transformer for fundus image classification. In: de Bruijne, M., et al. (eds.) MICCAI 2021. LNCS, vol. 12908, pp. 45–54. Springer, Cham (2021). https://doi.org/10.1007/978-3-030-87237-3_5
28. Zhao, Q., Liu, Z., Adeli, E., Pohl, K.M.: Longitudinal self-supervised learning. Med. Image Anal. **71**, 102051 (2021)

Interpretable Differential Diagnosis for Alzheimer's Disease and Frontotemporal Dementia

Huy-Dung Nguyen(✉), Michaël Clément, Boris Mansencal, and Pierrick Coupé

University Bordeaux, CNRS, Bordeaux INP, LaBRI, UMR 5800, 33400 Talence, France
huy-dung.nguyen@u-bordeaux.fr

Abstract. Alzheimer's disease and Frontotemporal dementia are two major types of dementia. Their accurate diagnosis and differentiation is crucial for determining specific intervention and treatment. However, differential diagnosis of these two types of dementia remains difficult at the early stage of disease due to similar patterns of clinical symptoms. Therefore, the automatic classification of multiple types of dementia has an important clinical value. So far, this challenge has not been actively explored. Recent development of deep learning in the field of medical image has demonstrated high performance for various classification tasks. In this paper, we propose to take advantage of two types of biomarkers: structure grading and structure atrophy. To this end, we propose first to train a large ensemble of 3D U-Nets to locally discriminate healthy versus dementia anatomical patterns. The result of these models is an interpretable 3D grading map capable of indicating abnormal brain regions. This map can also be exploited in various classification tasks using graph convolutional neural network. Finally, we propose to combine deep grading and atrophy-based classifications to improve dementia type discrimination. The proposed framework showed competitive performance compared to state-of-the-art methods for different tasks of disease detection and differential diagnosis.

Keywords: Deep grading · Differential diagnosis · Multi-disease classification · Alzheimer's disease · Frontotemporal dementia

1 Introduction

Alzheimer's disease (AD) and Frontotemporal dementia (FTD) are the first and third leading causes of early-onset dementia [3]. The detection of these diseases is critical for the development of novel therapies. Besides, people with FTD are often misdiagnosed with AD, although these diseases have different clinical diagnostic criteria [27,33], due to similar clinical symptoms such as a behavior or language disorder [1] and brain atrophy [29]. This is especially true for behavioral variant of FTD (bvFTD) which is the most common variant of FTD [33]. Indeed,

L. Wang et al. (Eds.): MICCAI 2022, LNCS 13431, pp. 55–65, 2022.
https://doi.org/10.1007/978-3-031-16431-6_6

many studies have shown that cognitive tests fail to accurately identify FTD from AD population [14,40]. This raises the need for an early and accurate differential diagnosis to determine specific intervention and slow down the disease progression. Consequently, a multi-class differential diagnosis method - able to differentiate AD, FTD and cognitively normal (CN) subjects - would be a highly valuable tool in clinical practice. Indeed, such tool can assist clinicians in making more informed decision in a general context.

The atrophy patterns of AD and FTD can be identified with the help of structural magnetic resonance imaging (sMRI) [9,28]. Moreover, the affected regions may be different between diseases [8]. Hence, it should be beneficial to use sMRI for disease detection and differential diagnosis. In the past, some methods have been proposed for this problem using volumetric and shape measurements obtained from sMRI [9,32]. However, the large majority of existing methods considered only binary classification problems (*e.g.*, AD *vs.* CN, FTD *vs.* CN, FTD *vs.* AD). For the multi-class differential diagnosis, only a few works have been proposed [5,13,22]. Moreover, the majority of existing approaches in this domain used traditional machine learning methods based on handcrafted features that may not fully exploit the image information. Therefore, Deep learning methods have been recently studied. However, the results of these approaches are usually not easily interpretable. This limits our understanding about the disease patterns.

In this paper, we propose a new method to perform specific-disease diagnosis (*i.e.*, AD *vs.* CN and FTD *vs.* CN) and differential diagnosis (*i.e.*, AD *vs.* FTD and AD *vs.* FTD *vs.* CN). Our purpose is to expand the knowledge about dementia sub-types and to offer an accurate diagnosis tool in a real clinical scenario. To this end, our contributions are twofold. First, we generate a 3D grading map reflecting the abnormality level of brain structures. This interpretable biomarker may assist clinicians in localizing the abnormal regions of brain, allowing a deeper understanding about multiple disease signatures. To do so, we extend the recently proposed Deep Grading (DG) framework [30] by training it with multiple types of dementia (*i.e.*, AD and FTD). Then, we classify these DG features using a graph convolutional neural network (GCN) [19] to better capture disease signatures. Second, we propose to ensemble the GCN decision with a support vector machine (SVM) using brain structure volumes to improve differential diagnosis accuracy. By combining structure grading and structure atrophy, the proposed framework offers state-of-the-art performance in both disease detection and differential diagnosis.

2 Materials and Method

2.1 Datasets

In this study, we used 2036 MRIs extracted from multiple open access databases: the Alzheimer's Disease Neuroimaging Initiative (ADNI) [15], the Open Access Series of Imaging Studies (OASIS) [20], the Australian Imaging, Biomarkers and Lifestyle (AIBL) [10], the Minimal Interval Resonance Imaging in Alzheimer's

Table 1. Number of participants used in our study.

Dataset	CN	AD	FTD
ADNI1	191	191	-
ADNI2	149	181	-
AIBL	233	47	-
OASIS3	658	97	-
MIRIAD	23	46	-
NIFD	136	-	74

Disease (MIRIAD) [23] and the Frontotemporal lobar Degeneration Neuroimaging Initiative (NIFD)[1]. All the baseline T1-weighted MRIs available in these databases were used. The NIFD dataset contains FTD patients and CN subjects while other datasets contain AD patients and CN subjects. In NIFD dataset, we only used the behavioral variant sub-type (bvFTD) which is the most prevalent form of FTD. We use all data available in ADNI1 (*i.e.*, 191 AD and 191 CN) and 90 subjects from NIFD (*i.e.*, 45 FTD, 45 CN) for training and validation. We use stratified splitting strategy to obtain 80% training and 20% validation data. To eliminate possible biases during training, the same number of patients and healthy people with no significant difference in age distribution were chosen. The other subjects (*i.e.*, 1199 CN, 371 AD and 29 FTD) were used only at the final evaluation. Table 1 describes the number of participants of each dataset used in this study.

2.2 Preprocessing

The preprocessing of the T1w MRI consisted of 5 steps: (1) denoising [25], (2) inhomogeneity correction [37], (3) affine registration into MNI152 space ($181 \times 217 \times 181$ voxels at $1\,\text{mm} \times 1\,\text{mm} \times 1\,\text{mm}$) [2], (4) intensity standardization [24] and (5) intracranial cavity (ICC) extraction [26]. Then, we used AssemblyNet[2] [7] with its default hyper-parameters to segment 133 brain structures (see Fig. 1). The brain structure segmentation was used to measure the structure volumes (*i.e.*, normalized volume in % of ICC) and aggregate information to compute the structure grading (see Sect. 2.3 and Fig. 1).

2.3 Method

Recently, a Deep Grading (DG) framework has been proposed for AD detection as an efficient and interpretable tool [30]. Here, we proposed to extend this approach to differential diagnosis and to combine it with atrophy-based features through an ensemble strategy (see Fig. 1).

[1] Available at https://ida.loni.usc.edu/.
[2] https://github.com/volBrain/AssemblyNet.

Grading-Based Classification: First, a preprocessed MRI was downsampled with a factor of 2 to reduce the computational cost. This was used to extract $k \times k \times k$ (*i.e.*, $k = 5$) overlapping patches of the same size (*i.e.*, $32 \times 48 \times 32$ voxels) uniformly distributed across the whole brain. We used $m = k \times k \times k$ (*i.e.*, $m = 125$) 3D U-Nets to grade these patches. Concretely, each of the m patch locations was analyzed by one specific U-Net. This U-Net was trained to predict a 3D grading map whose each voxel reflects the degree of similarity to normal or abnormal group. The obtained grading values were assembled to reconstitute a global grading map which was interpolated to the original input size. The training procedure of deep grading part was similar to [30]. However, we considered the brain changes of both AD and FTD as abnormal patterns instead of specific patterns of AD. Thus, for the ground-truth, we assign the value 1 (resp. -1) to all voxels inside a patch extracted from an AD/FTD patient (resp. CN subject). All voxels outside of ICC are set to 0. After that, we computed $s = 133$ average grading scores (one per brain structure) using a structural segmentation (obtained with AssemblyNet [7]). By doing this, each subject is encoded into an s-dimensional vector. Then, we defined a fully-connected graph of s nodes. Each node embeds structure grading score and subject's age. A GCN classifier with 3 layers of 32 channels was used to classify this graph. While training the GCN model, due to the imbalance nature of training set, we applied oversampling technique to balance classes in order to make the model more robust.

Atrophy-Based Classification: Parallel to the deep grading model, we used the volume of the 133 brain structures as an additional feature vector to represent atrophy information. To exploit these features, we trained a linear SVM model for the same classification problem. The data used in training the SVM model was the same as training the deep grading model. We used a grid search of three kernels (linear, polynomial, and gaussian) and 500 values of the hyper-parameter C in $[10^{-5}, 10^{5}]$ on the validation set for tuning hyper-parameters. During training, we used balanced weights (available in scikit-learn library [31]) to compensate for class imbalance.

Finally, we fused the probability vector of GCN and SVM by estimating their best linear combination on training data for the final decision. The training data used for the deep grading model and all the classifiers (*i.e.*, GCN, SVM or ensemble) came from the same subjects.

3 Experimental Results

3.1 Ablation Study for Binary Classification Tasks

Table 2 shows the results of our ablation study dedicated to binary classification tasks. First, we propose to perform dementia diagnosis (*i.e.*, AD and FTD *vs.* CN), AD diagnosis (*i.e.*, AD *vs.* CN), FTD diagnosis (*i.e.*, FTD *vs.* CN) and 2-class differential diagnosis (*i.e.*, AD *vs.* FTD).

Multiple Diseases Training for Specific Disease Classifications. In this part, we assessed the influence of mixing multiple diseases in training for specific

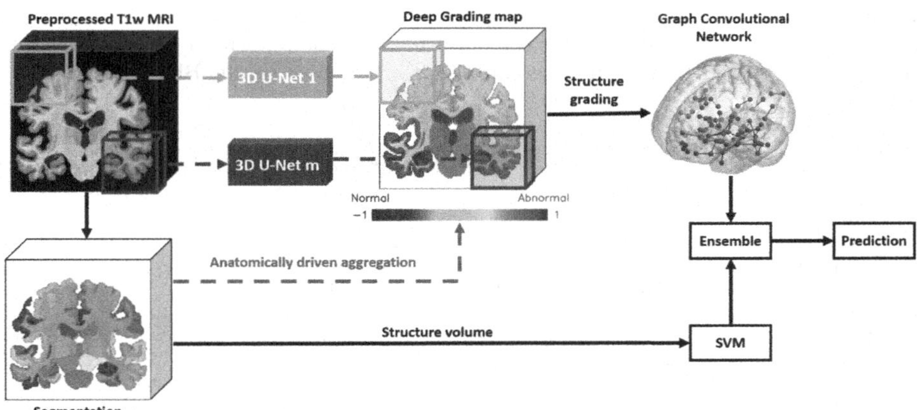

Fig. 1. An overview of the proposed method. The T1w image, its segmentation and the deep grading map are taken from an AD patient.

disease classifications. To do this, we trained deep grading model + classifier respectively with ADNI1 ($N = 382$), NIFD ($N = 90$) and ADNI1 + NIFD ($N = 472$) (see Table 2). For AD diagnosis, we observed that mixing AD and FTD (using ADNI1 + NIFD) during training can improve the model performance compared to training only on AD, CN subjects (see exp. 1 *vs.* 3; exp. 7 *vs.* 9). For FTD diagnosis, this training strategy yielded better results for all classifiers (see exp. 2 *vs.* 3; exp. 5 *vs.* 6; exp. 8 *vs.* 9). Overall, training on multiple diseases can improve, in most of cases, the performance compared to directly train the classifier for a specific task.

Combining Grading and Atrophy for Better Diagnosis. We observed that GCN based on DG features obtained the best results for almost all the binary diagnosis tasks compared to SVM with balanced accuracies (BACC) similar or higher than 88%. For 2-class differential diagnosis AD *vs.* FTD, SVM based on atrophy features outperformed GCN and obtained 89.5% in BACC. Consequently, we propose to combine both strategies in order to take advantage of the high capability of deep grading framework to detect pathologies and the good performance of atrophy-based classification to differentiate AD from FTD. As expected, the combined model (see exp. 9) yielded the best performance for all the considered classification tasks (see Table 2).

3.2 Performance for Multi-disease Classification

Table 3 presents the results obtained for the 3-class differential diagnosis (*i.e.*, AD *vs.* CN *vs.* FTD). We studied the performance of GCN using deep grading, SVM using brain structure volumes and the proposed combination. As previously, the proposed combination of deep grading and atrophy-based features obtained the best results. Indeed, this model yielded better performance for all of three metrics: accuracy (ACC), balanced accuracy (BACC) and area under curve (AUC).

Table 2. Ablation study of our method for binary classification tasks. We use the balanced accuracy (BACC) to assess the performance. The reported performances are the average of 10 repetitions and presented in %. Red: best result, Blue: second best result.

No.	Training set	Classifier	Dementia diagnosis Dem. *vs.* CN $N = 1554$	AD diagnosis AD *vs.* CN $N = 1525$	FTD diagnosis FTD *vs.* CN $N = 1183$	Differential diagnosis AD *vs.* FTD $N = 400$
1	ADNI1	GCN	-	89.6	-	-
2	NIFD	GCN	-	-	88.0	-
3	ADNI1+NIFD	GCN	**90.3**	**90.1**	**92.3**	74.6
4	ADNI1	SVM	-	86.3	-	-
5	NIFD	SVM	-	-	88.4	-
6	ADNI1+NIFD	SVM	86.7	86.2	91.7	**89.5**
7	ADNI1	Ensemble	-	89.9	-	-
8	NIFD	Ensemble	-	-	88.4	-
9	ADNI1+NIFD	Ensemble	90.5	90.3	93.3	89.7

Table 3. Performance of different models for the multiple disease classification. We denote "Sen." for sensitivity. The reported performances are the average of 10 repetitions and presented in %. The best performances are in red.

Classifier	ACC	BACC	AUC	CN Sen	AD Sen	FTD Sen
GCN	85.5	80.7	91.3	91.2	66.3	84.5
SVM	89.3	83.4	94.8	92.9	77.9	76.9
Ensemble	90.4	85.4	95.3	93.4	81.0	81.7

GCN obtained the best results only for FTD sensitivity. These ensembling results were then chosen to compare with current state-of-the-art methods.

3.3 Comparison with State-of-the-Art Methods

Most of existing methods are based on classification using machine learning of volume-based or surface-based features. Kim *et al.* used a linear discriminant analysis to classify cortical thickness (Cth) for dementia diagnosis and 2-class differential diagnosis (*i.e.*, AD *vs.* FTD) [18]. Möller *et al.* used an SVM to classify tissue density map for binary classification tasks [28]. In [41], Yu *et al.* computed volume-based scores and used a simple threshold for different classifications. Bron *et al.* used an SVM based on tissues density maps for many classification tasks. A few methods based on deep learning have been also proposed. Ma *et al.* used a generative adversarial network (GAN) to classify structure volumes and Cth for 3-class differential diagnosis [22]. In [13], Hu *et al.* used a Convolutional Neural Network (CNN) on intensities for multi-class classification.

Tables 4 and 5 respectively summarize the current performance of state-of-the-art methods proposed for binary classifications and 3-class classification. It should be noted that comparison has to be considered with caution. First, the data used for evaluation are not the same between methods. For instance, we use bvFTD patients (like [28]) while others mixed several variants [5,13,18,22]. The metrics can also be different, some studies (*i.e.*, [13]) used ACC which is known to be sub-optimal when classes are unbalanced while other used BACC.

Table 4. Comparison of our method with current state-of-the-art methods for binary classification tasks. Our reported performances are the average of 10 repetitions and presented in %. Red: best result. The balanced accuracy (BACC) is used to assess the model performance except for [13] that used accuracy (ACC).

Method	Dementia diagnosis Dem. *vs.* CN	AD diagnosis AD *vs.* CN	FTD diagnosis FTD *vs.* CN	Differential diagnosis AD *vs.* FTD
CNN on intensities [13]	-	77.2	68.0	81.3
LDA on Cth [18]	86.2	-	-	89.8
SVM on tissue map [28]	-	85.0	72.5	78.5
Threshold on volumes [41]	85.7	-	-	83.0
Our method	90.5	90.3	93.3	89.7

Table 5. Comparison of our method with current state-of-the-art methods for 3-class differential diagnosis AD *vs.* FTD *vs.* CN. Red: best result. We denote ACC for accuracy, BACC for balanced accuracy and AUC for area under curve.

Method	ACC	BACC	AUC
SVM on tissue maps [5]	-	75.0	90.0
GAN on Cth and volume [22]	88.3	85.3	-
CNN on intensities [13]	66.8	-	-
Our method	90.4	85.4	95.3

For binary classification problems (see Table 4), our method shows higher performance than other methods in specific disease diagnosis. For the 2-class differential diagnosis (AD *vs.* FTD) our result was comparable with the best one. For the 3-class differential diagnosis (see Table 5), our method presented better performance in all metrics compared to other studies. Furthermore, contrary to other deep learning based methods, we provide an interpretable grading map allowing to localize abnormal regions in brain (see Sect. 3.4). These results highlight the potential value of our method in clinical practice.

3.4 Interpretation of Deep Grading Map

To assess the interpretability of the deep grading map, we computed the average grading map for each group (*i.e.*, CN, AD and FTD). Figure 2 shows sagittal and coronal views of these average grading maps. For the group of heathy people (*i.e.*, CN), all brain regions were detected as normal as expected. For the AD group, the average grading map highlighted regions around hippocampus. This area is well-known to be affected by AD [36]. For FTD group, the abnormal regions were localized around the ventromedial frontal cortex. This region presents significant atrophy in FTD patients [34].

We further investigated the top 10 structures with highest grading score in AD and FTD group. For AD group, eight over ten structures highlighted by the average grading map were known to be related to AD: left ventral diencephalon [21], left hippocampus [11], left amygdala and left inferior lateral ventricle [6], left parahippocampal gyrus and left entorhinal area [17], left thalamus [16], left basal forebrain [38]. For FTD group, we also found several structures related to the disease: bilateral anterior cingulate gyrus [4], left medial frontal cortex and right middle frontal gyrus [35], bilateral frontal pole [39].

Moreover, the abnormality map of FTD appeared asymmetric between left and right hemisphere. This result is in line with our current knowledge on this disease [12] since bvFTD patients usually exhibit higher atrophy in the right hemisphere.

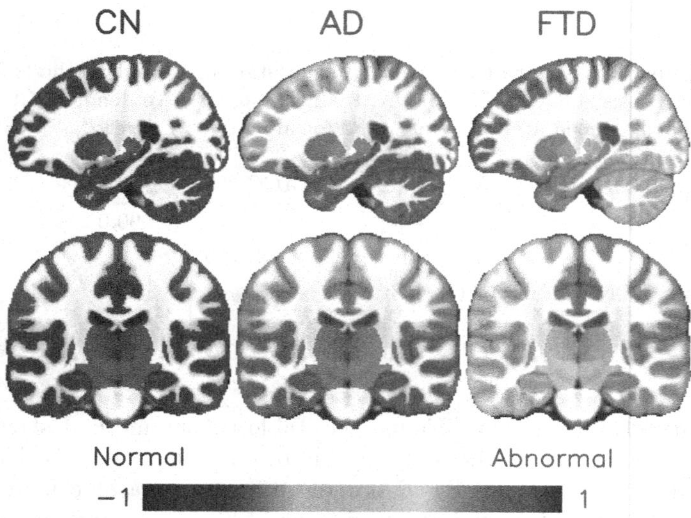

Fig. 2. Average grading map per group of subjects in the MNI space with neurological orientation (with the right of the patient at the right).

4 Conclusion

In this paper, we proposed a new method for specific disease diagnosis (*i.e.*, AD *vs.* CN, FTD *vs.* CN) and differential diagnosis (*i.e.*, AD *vs.* FTD, AD *vs.* FTD *vs.* CN). Our purpose was to expand the knowledge about different dementia sub-types and offer an accurate diagnosis tool in a real clinical scenario. To this end, we extended a recent deep grading framework for multiple classification problems. First, we showed that our training strategy using multiple diseases (*i.e.*, AD and FTD) can improve the performance of specific disease diagnosis. Second, we proposed to combine the high capacity in disease detection offered by structure grading with the high accuracy on differential diagnosis provided by structure atrophy. As a result, our ensembling models achieved state-of-the-art in performance for various classification tasks. Finally, the grading maps offer an easily interpretable tool to investigate dementia signatures over the entire brain. The structures highlighted by our grading map highly correlate with current physiopathological knowledge on AD and FTD. This presents a potential value of our method in clinical practice.

Acknowledgments. This work benefited from the support of the project DeepvolBrain of the French National Research Agency (ANR-18-CE45-0013). This study was achieved within the context of the Laboratory of Excellence TRAIL ANR-10-LABX-57 for the BigDataBrain project. Moreover, we thank the Investments for the future Program IdEx Bordeaux (ANR-10-IDEX-03-02 and RRI "IMPACT"), the French Ministry of Education and Research, and the CNRS for DeepMultiBrain project.

References

1. Alladi, S., et al.: Focal cortical presentations of Alzheimer's disease. Brain **130**, 2636–2645 (2007)
2. Avants, B.B., et al.: A reproducible evaluation of ANTs similarity metric performance in brain image registration. Neuroimage **54**, 2033–2044 (2011)
3. Bang, J., et al.: Frontotemporal dementia. The Lancet **386**, 1672–1682 (2015)
4. Brambati, S.M., et al.: A tensor based morphometry study of longitudinal gray matter contraction in FTD. Neuroimage **35**, 998–1003 (2007)
5. Bron, E.E., et al.: Multiparametric computer-aided differential diagnosis of Alzheimer's disease and frontotemporal dementia using structural and advanced MRI. Eur. Radiol. **27**, 3372–3382 (2017)
6. Coupé, P., et al.: Lifespan changes of the hum brain in Alzheimer's disease. Sci. Rep. **9**, 3998 (2019)
7. Coupé, P., et al.: AssemblyNet: a large ensemble of CNNs for 3D whole brain MRI segmentation. Neuroimage **219**, 117026 (2020)
8. Davatzikos, C., et al.: Individual patient diagnosis of AD and FTD via high-dimensional pattern classification of MRI. Neuroimage **41**, 1220–1227 (2008)
9. Du, A.T., et al.: Different regional patterns of cortical thinning in Alzheimer's disease and frontotemporal dementia. Brain **130**, 1159–1166 (2006)
10. Ellis, K.A., et al.: The Australi Imaging, Biomarkers and Lifestyle (AIBL) study of aging: methodology and baseline characteristics of 1112 individuals recruited for a longitudinal study of Alzheimer's disease. Int. Psychogeriatr. **21**, 672–687 (2009)
11. Frisoni, G.B., et al.: The clinical use of structural MRI in Alzheimer disease. Nat. Rev. Neurol. **6**, 67–77 (2010)
12. Harper, L., et al.: An algorithmic approach to structural imaging in dementia. J. Neurol. Neurosurg. Psychiatry **85**, 692–698 (2014)

13. Hu, J., et al.: Deep learning-based classification and voxel-based visualization of frontotemporal dementia and Alzheimer's disease. Front. Neurosci. **14**, 626154 (2021)
14. Hutchinson, A.D., et al.: Neuropsychological deficits in frontotemporal dementia and Alzheimer's disease: a meta-analytic review. J. Neurol. Neurosurg. Psychiatry **78**, 917–928 (2007)
15. Jack, C.R., et al.: The Alzheimer's Disease Neuroimaging Initiative (ADNI): MRI methods. J. Magn. Reson. Imaging **27**, 685–691 (2008)
16. de Jong, L.W., et al.: Strongly reduced volumes of putamen and thalamus in Alzheimer's disease: an MRI study. Brain **131**, 3277–3285 (2008)
17. Kesslak, J.P., et al.: Quantification of magnetic resonance scans for hippocampal and parahippocampal atrophy in Alzheimer's disease. Neurology **41**, 51–54 (1991)
18. Kim, J.P., et al.: Machine learning based hierarchical classification of frontotemporal dementia and Alzheimer's disease. NeuroImage Clin. **23**, 101811 (2019)
19. Kipf, T.N., et al.: Semi-supervised classification with graph convolutional networks. In: 5th International Conference on Learning Representations, ICLR 2017 (2017)
20. LaMontagne, P.J., et al.: OASIS-3: longitudinal neuroimaging, clinical, and cognitive dataset for normal aging and Alzheimer disease. Radiol. Imaging (2019)
21. Lebedeva, A.K., et al.: MRI-based classification models in prediction of mild cognitive impairment and dementia in late-life depression. Front. Aging Neurosci. **9**, 13 (2017)
22. Ma, D., et al.: Differential diagnosis of frontotemporal dementia, Alzheimer's disease, and normal aging using a multi-scale multi-type feature generative adversarial deep neural network on structural magnetic resonance images. Front. Neurosci. **14**, 853 (2020)
23. Malone, I.B., et al.: MIRIAD-public release of a multiple time point Alzheimer's MR imaging dataset. Neuroimage **70**, 33–36 (2013)
24. Manjón, J.V., et al.: Robust MRI brain tissue parameter estimation by multistage outlier rejection. Magn. Reson. Med. **59**, 866–873 (2008)
25. Manjón, J.V., et al.: Adaptive non-local means denoising of MR images with spatially varying noise levels. J. Magn. Reson. Imaging **31**, 192–203 (2010)
26. Manjón, J.V., et al.: Nonlocal intracranial cavity extraction. Int. J. Biomed. Imaging **2014**, 820205 (2014)
27. McKhann, G.M., et al.: The diagnosis of dementia due to Alzheimer's disease: recommendations from the National Institute on Aging-Alzheimer's Association workgroups on diagnostic guidelines for Alzheimer's disease. Alzheimer's Dement. **7**, 263–269 (2011)
28. Möller, C., et al.: Alzheimer disease and behavioral variant frontotemporal dementia: automatic classification based on cortical atrophy for single-subject diagnosis. Radiology **279**, 838–848 (2016)
29. Neary, D., et al.: Frontotemporal dementia. Lancet Neurol. **4**, 771–780 (2005)
30. Nguyen, H.D., et al.: Deep grading based on collective artificial intelligence for AD diagnosis and prognosis. In: Interpretability of Machine Intelligence in Medical Image Computing, and Topological Data Analysis and Its Applications for Medical Data, vol. 12929, pp. 24–33 (2021)
31. Pedregosa, F., et al.: Scikit-learn: machine learning in python. J. Mach. Learn. Res. **12**, 2825–2830 (2011)
32. Rabinovici, G., et al.: Distinct MRI atrophy patterns in autopsy-proven Alzheimer's disease and frontotemporal lobar degeneration. Am. J. Alzheimer's Dis. Other Dement. **22**, 474–488 (2008)

33. Rascovsky, K., et al.: Sensitivity of revised diagnostic criteria for the behavioural variant of frontotemporal dementia. Brain **134**, 2456–2477 (2011)
34. Rosen, H.J., et al.: Patterns of brain atrophy in frontotemporal dementia and semantic dementia. Neurology **58**, 198–208 (2002)
35. Rosen, H.J., et al.: Neuroanatomical correlates of behavioural disorders in dementia. Brain **128**, 2612–2625 (2005)
36. Schuff, N., et al.: MRI of hippocampal volume loss in early Alzheimer's disease in relation to ApoE genotype and biomarkers. Brain **132**, 1067–1077 (2009)
37. Tustison, N.J., et al.: N4ITK: improved N3 bias correction. IEEE Trans. Med. Imaging **29**, 1310–1320 (2010)
38. Whitehouse, P., et al.: Alzheimer's disease and senile dementia: loss of neurons in the basal forebrain. Science **215**, 1237–1239 (1982)
39. Wong, S., et al.: Contrasting prefrontal cortex contributions to episodic memory dysfunction in behavioural variant frontotemporal dementia and Alzheimer's disease. PLoS ONE **9**, e87778 (2014)
40. Yew, B., et al.: Lost and forgotten? Orientation versus memory in Alzheimer's disease and frontotemporal dementia. J. Alzheimer's Dis. JAD **33**, 473–481 (2013)
41. Yu, Q., et al.: An MRI-based strategy for differentiation of frontotemporal dementia and Alzheimer's disease. Alzheimer's Res. Therapy **13**, 23 (2021)

Is a PET All You Need? A Multi-modal Study for Alzheimer's Disease Using 3D CNNs

Marla Narazani[1(✉)], Ignacio Sarasua[1,2], Sebastian Pölsterl[2], Aldana Lizarraga[1], Igor Yakushev[1], and Christian Wachinger[1,2]

[1] School of Medicine, Technical University of Munich, Munich, Germany
marla.narazani@tum.de

[2] Lab for Artificial Intelligence in Medical Imaging (AI-Med), KJP, LMU Klinikum, Munich, Germany

Abstract. Alzheimer's Disease (AD) is the most common form of dementia and often difficult to diagnose due to the multifactorial etiology of dementia. Recent works on neuroimaging-based computer-aided diagnosis with deep neural networks (DNNs) showed that fusing structural magnetic resonance images (sMRI) and fluorodeoxyglucose positron emission tomography (FDG-PET) leads to improved accuracy in a study population of healthy controls and subjects with AD. However, this result conflicts with the established clinical knowledge that FDG-PET better captures AD-specific pathologies than sMRI. Therefore, we propose a framework for the systematic evaluation of multi-modal DNNs and critically re-evaluate single- and multi-modal DNNs based on FDG-PET and sMRI for binary healthy vs. AD, and three-way healthy/mild cognitive impairment/AD classification. Our experiments demonstrate that a single-modality network using FDG-PET performs better than MRI (accuracy 0.91 vs 0.87) and does not show improvement when combined. This conforms with the established clinical knowledge on AD biomarkers, but raises questions about the true benefit of multi-modal DNNs. We argue that future work on multi-modal fusion should systematically assess the contribution of individual modalities following our proposed evaluation framework. Finally, we encourage the community to go beyond healthy vs. AD classification and focus on differential diagnosis of dementia, where fusing multi-modal image information conforms with a clinical need.

1 Introduction

With life expectancies rising globally, dementia is becoming a growing concern for individuals and society. Dementia is characterized by a progressive cogni-

M. Narazani and I. Sarasua—These authors contributed equally to this work.

Supplementary Information The online version contains supplementary material available at https://doi.org/10.1007/978-3-031-16431-6_7.

L. Wang et al. (Eds.): MICCAI 2022, LNCS 13431, pp. 66–76, 2022.
https://doi.org/10.1007/978-3-031-16431-6_7

tive impairment that eventually requires individuals to be completely dependent upon caregivers. While this process cannot be reversed, recent efforts have focused on diagnosing subjects at an early stage to improve disease management [4]. A particular focus has been on Alzheimer's Disease (AD), given that it is the most common form of dementia and benefits from large data-sharing initiatives [21]. To date, a wide range of diagnostic tools are available for diagnosing AD: magnetic resonance imaging (MRI), positron emission tomography (PET), cerebrospinal fluid (CSF), demographics, cognitive tests, and genetic alterations [1]. Structural MRI (sMRI) captures regional atrophy of the brain, whereas FDG-PET measures the brain's glucose metabolism. FDG-PET plays a major role in the clinical diagnosis of AD. It can detect functional brain changes in AD early in the disease progression and can help to differentiate AD from other causes of dementia such as frontotemporal and Lewy body dementia [22]. In the memory clinic, MRI and FDG-PET are among the most common neuroimaging methods used [26], where FDG-PET is considered to have a higher diagnostic and prognostic accuracy [3,12].

Recently, studies on deep learning (DL) techniques have emerged that showed that distinguishing healthy controls from AD subjects becomes more accurate when learning from MRI *and* FDG-PET, rather than a single modality [24,29,30]. However, this scenario is very different from that in a memory clinic. In the clinic, the main objective is differential diagnosis to determine the type of dementia, whereas studies on DL merely considered a single type of dementia, namely AD [24,29,30]. When considering that both modalities assess neural degeneration, but AD-specific changes are better captured by FDG-PET than MRI [3,12], it seems surprising why combining MRI and FDG-PET with DL would be beneficial when AD is the only form of dementia that is being studied.

In this work, we critically re-evaluate single- and multi-modal DL models based on FDG-PET and structural MRI for classifying healthy vs. AD subjects. We study three different modes of multimodal fusion: early, middle, and late fusion. We evaluate each to investigate whether it truly benefits from multimodal data by performing ablation studies for which MRI and FDG-PET images are paired randomly. Contrary to previous work, our experiments show that FDG-PET alone is sufficient for AD diagnosis, which conforms with established clinical knowledge about biomarkers in AD. We argue that future work on multimodal fusion should follow our proposed evaluation framework to systematically assess the contribution of individual modalities.

Related Work. Most DL models for AD prediction are single-modal (see [7] for an overview). In [9], the authors propose a 2D convolutional neural network (CNN) using slices of sMRI volumes. However, recent work has shifted towards 3D CNN architectures for analyzing sMRI [2,8,14,18,19,23]. A sparse autoencoder is combined with a CNN in [23]. Korolev et al. [18] compare a 3D-VGG and 3D-Resnet architecture. Both [2] and [8] use a 3D CNN for whole brain MRIs. Regarding work related to FDG-PET, a 2D CNN has been used in [6,20], and a 3D CNN in [28]. Finally, several works combined sMRI and FDG-PET

[24,29,30]. In [24], the authors propose an early fusion approach by overlaying gray matter (GM) tissues from MRI with the FDG-PET scans and evaluate the effectiveness of their fusion strategy using a 3D CNN. In [30], a three-stage framework based on middle and late fusion using MRI, FDG-PET, and single nucleotide polymorphisms is proposed. The authors of [10] combine a 3D CNN and LSTM. Finally, in [15], an early and a late fusion approach are presented based on a 3D-VGG. The works on multi-modal fusion unanimously concluded that fusing sMRI and FDG-PET improves prediction accuracy over using a single modality, which conflicts with the established clinical knowledge that FDG-PET better captures AD-specific pathologies than sMRI [3,12].

2 Methods

To determine the contribution of each modality in a multi-modal DNN, we propose a systematic evaluation framework. First, we consider each modality in isolation by using a single branch 3D CNN. Next, we consider the joint contribution of multiple modalities using a 3D CNN with either early, late, or middle fusion (see Fig. 1). To assess whether multi-modal inputs are truly helpful, we perform ablation experiments where MRI and FDG-PET images are paired randomly. This allows us to quantify to importance of each modality.

2.1 CNN Architecture

We use a 3D ResNet as the base architecture for all models (more details in supplemental Fig. S1). It comprises 12 convolutional layers with kernel size 3^3 in total. We use four residual learning blocks consisting of two convolutional layers followed by batch normalization (BN) [16] and rectified linear unit (ReLU) activation. We half the spatial resolution of feature maps in the last three residual blocks by using a stride of 2. Finally, we perform global average pooling across the spatial dimensions of the feature maps and use two linear layers to output a log-probability. We use dropout in each residual block to reduce overfitting.

2.2 Fusion Strategies

We consider three strategies for fusing multi-modal data: early, late, and middle fusion (see Fig. 1). All three strategies follow the base CNN architecture described above. Next, we describe the fusion strategies in detail.

Early Fusion. In early fusion, raw modalities are combined directly before being passed to the network. Here, we follow the strategy proposed in [24]: gray matter maps are obtained via Voxel-Based Morphometry (VBM) and used to mask the FDG-PET intensities. In the resulting volume, the intensities of the FDG-PET are effectively weighted by the MRI intensities. The network is a single branch network that receives the combined MRI-FDG-PET volume as input.

Late Fusion. Late fusion is the most straight-forward approach to fuse multi-modal data. Rather than fusing the images, it fuses the latent representations of

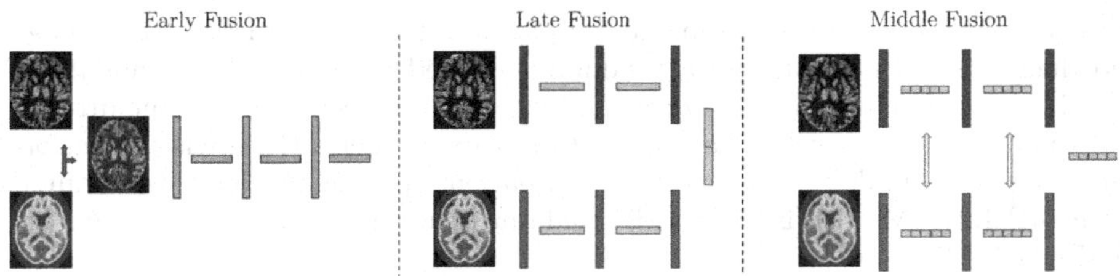

Fig. 1. Overview of the three fusion strategies. Early fusion combines the MRI and PET inputs in a single volume. Late fusion concatenates the latent representations coming from each independent network. Middle fusion exchanges channels of the intermediate feature maps along the network.

two separate networks. Here, we train two independent 3D ResNet branches, one for MRI and one for FDG-PET. The features obtained from each branch after global average pooling are then concatenated and passed through a Multi-layer perceptron (MLP) [128, 64, number of classes] to obtain a log-probability that accounts for both sources of information.

Middle Fusion. While early and late fusion are common in multi-modal analysis, we also explore an approach that fuses intermediate representations of modality-specific networks [27]. In this approach, modality-specific information are fused by dynamically exchanging feature maps between sub-networks of different modalities. This bi-directional exchange of information is self-guided by considering individual channel importance, which is measured by the magnitude of the BN scaling factor. This process is carried out under the ℓ_1 regularization to penalize exchanging all channels. To further encourage sharing of information, convolutional filter weights are shared across modalities. Note that BN layers are not shared in order to determine the channel importance for each individual modality. To the best of our knowledge, channel exchange has not been applied for multi-modal fusion for AD prediction before.

2.3 Evaluation Scheme

Our main objective is to rigorously evaluate whether MRI is truly relevant for diagnosing AD when FDG-PET is available too. For all of our experiments, we train the models using FDG-PET and MRI data from the same patient. During inference we define three different experiments based on the input data: (i) correct, (ii) random PET, and (iii) random MRI. We use balanced accuracy (BACC) to assess the predictive performance of models, because it is insensitive to the relative frequency of class labels [5].

Correct. This strategy follows the standard training and testing scheme. For each fusion strategy, we test the networks based on FDG-PET and MRI scans from the same patient. If both modalities would be relevant for AD diagnosis, we would expect this scenario to yield the highest predictive performance. It serves as a baseline for the remaining experiments.

Random MRI. In this experiment, we pair a patient's true FDG-PET image and diagnosis with an MRI of a randomly selected patient. If both modalities would be relevant for the final decision, we would expect a significant drop in performance with respect to the *Correct MRI* experiment. Otherwise, if performance remains similar, the contribution of patient-specific anatomy, as captured by the MRI, the MRI adds little additional information that is not available from the FDG-PET.

Random PET. This experiment is similar to the previous experiment, but this time we pair the correct MRI and diagnosis with a randomly selected FDG-PET from another patient. The conclusions we can derive from this experiment are the same as in the previous experiment, but focus on assessing the contribution of FDG-PET.

2.4 Data Processing and Training Strategy

We use pre-processed FDG-PET scans and T1-weighted MRI scans from the Alzheimer's disease neuroimaging initiative (ADNI; [17]) database. Full details about the pre-processing steps can be found at the ADNI website for FDG-PET[1] and for MRI[2]. Both scans were additionally processed using SPM[3] and CAT12 [13]. MRI scans were processed using the standard VBM pipeline in CAT12[4]. We use the gray matter (GM) tissue area of the brain as an input to the network. FDG-PET scans were normalized and registered to the MNI152 template [11] with 1.5 mm^3 voxel size. We performed min-max scaling to rescale the image intensity values to the range between 0 and 1. The final image size for both modalities is $113 \times 137 \times 113$.

Our dataset comprises 257 patients with AD, 370 healthy controls (CN), and 611 patients with mild cognitive impairment (MCI); see the supplemental Table S1 for additional information. We split the data into train/validation/test sets with sizes roughly in 65/15/20% of the full dataset. We perform cross-validation across 5 splits, based on a data stratification strategy that accounts for sex, age and diagnosis. We only include baseline visits scans so that only a single scan per patient is available. We train models for two tasks (i) binary classification of healthy controls (CN) vs. patients with AD, and (ii) three-way classification of CN vs. MCI vs. AD. All models are trained end-to-end using a cross-entropy loss and data augmentation during training (up to $8°$ angle rotation and 8 mm translation in each dimension). More information about the training setup can be found in the supplemental Table S2.

3 Results

Testing on Random PET or MRI. Table 1 reports the results for the experiments described in Sect. 2.3, for binary and three-way classification. We observe

1 http://adni.loni.usc.edu/methods/pet-analysis.
2 http://adni.loni.usc.edu/methods/mri-analysis.
3 https://www.fil.ion.ucl.ac.uk/spm/software/spm12.
4 http://www.neuro.uni-jena.de/cat12/CAT12-Manual.pdf.

Table 1. Overview of the evaluation scheme for correct data, random MRI or random PET. Numbers are mean balanced accuracy (BACC) and standard deviation across folds.

	Random MRI	Random PET	BACC 2-Class	BACC 3-Class
Early fusion	✗	✗	0.885 ± 0.041	0.573 ± 0.023
Early fusion	✗	✓	0.696 ± 0.026	0.414 ± 0.015
Early fusion	✓	✗	0.720 ± 0.015	0.470 ± 0.028
Middle fusion	✗	✗	0.893 ± 0.036	0.530 ± 0.034
Middle fusion	✗	✓	0.527 ± 0.020	0.366 ± 0.028
Middle fusion	✓	✗	0.890 ± 0.020	0.528 ± 0.025
Late fusion	✗	✗	0.896 ± 0.019	0.577 ± 0.029
Late fusion	✗	✓	0.597 ± 0.029	0.368 ± 0.027
Late fusion	✓	✗	0.786 ± 0.080	0.527 ± 0.038

that when testing on the correct pair of scans, all fusion approaches perform similarly for both tasks with two exceptions: Early Fusion achieves a mean BACC approximately 0.01 lower for binary classification, and Middle Fusion a BACC approximately 0.04 lower for three-class classification. Overall, we observe a significant drop in performance between these two tasks, which is expected given that MCI is not a true diagnosis, but a syndrome, which makes it highly heterogeneous, especially with limited amount of training data.

Interestingly, if we look at the results for the middle and late fusion models when testing on partially random data, we observe a much larger drop in performance when the FDG-PET is randomized; the accuracy is close to random chance. On the other hand, randomizing the MRI data has much lower impact on the overall performance. For binary classification the mean BACC drops around 0.11 for late fusion and merely 0.003 for middle fusion, which is much lower than for the random PET experiments: 0.299 and 0.366, respectively. For early fusion, results for both randomized experiments experience a significant drop compared to using the original data. This outcome is expected, since early fusion results in a single volume where the MRI acts as a mask to select regions from the FDG-PET. If the pair of images is from different patients, anatomies are not perfectly aligned and early fusion will remove important areas. Hence, the effect of randomizing the MRI or the FDG-PET leads to a similar loss in information and comparable drop in performance.

Training on Random MRI: The performance difference between randomizing the FDG-PET data vs. the MRI (see Table 1) suggests that both modalities do not have the same contribution to the models' final decision. We decided to further evaluate this hypothesis by defining an additional experiment: during training, the FDG-PET remains associated to a specific patient, but the MRI is exchanged with a random subject. Table 2 shows the results for two- and three-class. Note that results for the original data (Correct) are identical to those in Table 1. For binary classification with correct data, middle and late fusion outperform early fusion by at least 0.08 in mean BACC. Single modality PET yields the best performance on correct data. When using a random MRI, the BACC for early fusion decreases, but improves for late fusion, matching the BACC of the single modality PET. For three-classes with correct data, using PET and MRI data performs similarly with a 0.03 improvement for early and late fusion, while middle fusion decreases in performance by 0.01 compared to using only PET. For random MRI, we observe a strong improvement for middle fusion (0.168) and late fusion (0.081), while the accuracy for early fusion decreases to chance level.

Table 2. Training and testing on correct, and random MRI. Numbers are mean balanced accuracy and standard deviation across folds.

	CN vs. AD		CN vs. MCI vs. AD	
	Correct	Random MRI	Correct	Random MRI
PET only	0.905 ± 0.015	—	0.541 ± 0.034	—
MRI only	0.866 ± 0.029	—	0.536 ± 0.062	—
Early fusion	0.885 ± 0.041	0.729 ± 0.034	0.573 ± 0.023	0.365 ± 0.037
Middle fusion	0.893 ± 0.036	0.863 ± 0.026	0.530 ± 0.034	0.698 ± 0.087
Late fusion	0.896 ± 0.019	0.906 ± 0.022	0.577 ± 0.029	0.658 ± 0.015

Post-hoc Explanation via Relevance Maps: Relevance maps are a helpful way of assessing the decision-making process of a classification model. In this work, we use them to quantify how much individual modalities contribute to the final prediction of the network. We use Integrated Gradients (IG; [25]) because its axiomatic approach allows us to precisely quantify how much the MRI and FDG-PET of a multi-modal CNN contribute to a particular prediction. Given a patient's images and a baseline, which is defined by the user (in our case a black volume), IG computes voxel-wise contributions by integrating along the path from the baseline input to the real input. Since the sum of all voxel-wise IG scores equals the predicted log-probability, we can summarize the total contribution of the MRI and FDG-PET by summing over the IG scores for the respective modality. Figure 2 depicts the average absolute importance across 42 correctly classified AD patients by the late fusion model for CN vs. AD. This example clearly illustrates that the PET contributes significantly more to the overall predictions. Overall, the PET contributes 1.77 times more to a prediction than the MRI (sum of —IG— is 33.8 vs. 19.1), which confirms our results from above.

4 Discussion

We performed a thorough evaluation of the different methods across 5 splits. In our first set of experiments, we observed that when training on correct data but

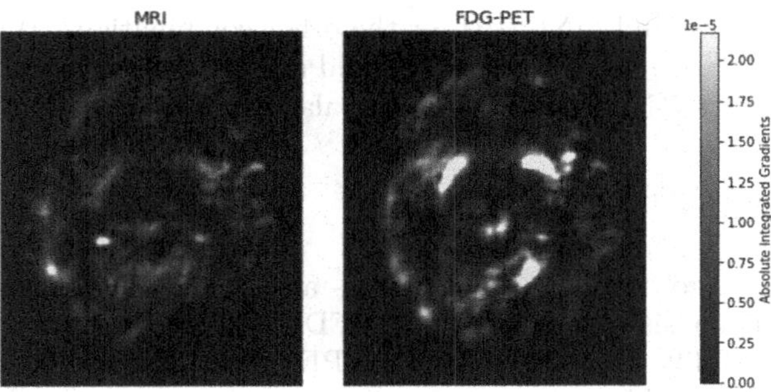

Fig. 2. Mean absolute integrated gradients across 42 correctly classified AD patients by the late fusion model. Illustrated is an axial slice located at the center of the volume.

introducing random FDG-PET or MRI data at test time, both the middle and late networks were more sensitive to changes of PET. While this is already a strong indicator of the bias of the neural network, our second set of experiments (Table 2) give us better insights on the reasons behind this phenomenon. First of all, the AD vs. CN classification experiments are consistent across Tables 1 and 2, which proves that the middle and late fusion networks rely mostly on FDG-PET. These results are supported by the relevance maps in Fig. 2.

For the three-class experiment, the BACC is below 60%, confirming the difficulty of the task. MCI subjects are a heterogeneous group that may also suffer from other types of dementia. Therefore, the amount of data required to train a predictive model for this task is much larger than in a two-class setting. For this challenging task, the usage of random MRIs led to a steep increase in accuracy for middle and late fusion. We believe that randomizing the MRI data serves as an augmentation mechanism during training. Given that in each epoch, the model sees a different pair of FDG-PET and MRI scans, this is likely making the networks more robust to alterations during inference.

Our results, while being aligned to previous medical findings, are in disagreement with previous literature that favored the fusion of MRI and FDG-PET for AD prediction. One reason for this difference could be that randomly exchanging image pairs during training leads to a larger effective training data size, which in turn allows the network to be more robust to changes in the data distribution during inference (similarly observed in Table 2 for the random MRI experiments). Additionally, by increasing the number of branches (e.g. two branches for the late fusion) the amount of trainable parameters is almost doubled, which allows the network to define more complex decision boundaries. This also makes the networks more prone to overfitting as observed in the three-class experiment when comparing late fusion on correct or random MRI. Finally, another potential reason is the importance of the PET pre-processing. For instance, [24] use a different pre-processing for the image fusion (for which they report high accuracy) and different input for the uni-modal and concatenation networks. GM is used in image fusion and MNI-MRI for the other combinations. When we compared the performance between processed and un-processed PET data, we noted a decrease of about 7% (t-test P=0.01) in balanced accuracy.

5 Conclusion

In this work, we rigorously evaluated single- and multi-modal deep neural networks for AD diagnosis based on MRI and FDG-PET. Our results demonstrate that a single-modality network using FDG-PET performs best for healthy/AD classification. While this is in contrast with previous work on deep learning for modality fusion, it does conform with the established clinical knowledge that FDG-PET better captures AD-specific patterns of neurodegeneration than MRI. We argue that recent work on multi-modal fusion, while technically sound, are largely disconnected from the established clinical knowledge about biomarkers in AD. We argue that future work on multi-modal fusion for AD diagnosis should take the existing clinical knowledge better into account, and systematically assess the contribution of individual modalities following our experimental setup. In the future, we plan to conduct experiments for MCI vs. NC, validate our hypotheses on different datasets and test other classification models.

Acknowledgment. This research was partially supported by the Bavarian State Ministry of Science and the Arts and coordinated by the bidt, and the Federal Ministry of Education and Research in the call for Computational Life Sciences (DeepMentia, 031L0200A). We gratefully acknowledge the computational resources provided by the Leibniz Supercomputing Centre (www.lrz.de).

References

1. Aisen, P.S., Cummings, J., Jack, C.R., Morris, J.C., Sperling, R.: On the path to 2025: understanding the Alzheimer's disease continuum. Alzheimers Res. Ther. **9**(1), 60 (2017)
2. Basaia, S., et al.: Automated classification of Alzheimer's disease and mild cognitive impairment using a single MRI and deep neural networks. NeuroImage Clin. **21**, 101645 (2019)
3. Bloudek, L.M., Spackman, D.E., Blankenburg, M., Sullivan, S.D.: Review and meta-analysis of biomarkers and diagnostic imaging in Alzheimer's disease. J. Alzheimer's Dis. **26**(4), 627–645 (2011)
4. Borson, S., et al.: Improving dementia care: the role of screening and detection of cognitive impairment. Alzheimer's Dement **9**(2), 151–159 (2013)
5. Brodersen, K.H., Ong, C.S., Stephan, K.E., Buhmann, J.M.: The balanced accuracy and its posterior distribution. In: 20th International Conference on Pattern Recognition, pp. 3121–3124 (2010)
6. Ding, Y., Sohn, J.H., Kawczynski, M.G., Trivedi, H., Harnish, R., et al.: A deep learning model to predict a diagnosis of Alzheimer disease by using 18F-FDG PET of the brain. Radiology **290**(2), 456–464 (2019)

7. Ebrahimighahnavieh, M.A., Luo, S., Chiong, R.: Deep learning to detect Alzheimer's disease from neuroimaging: a systematic literature review. Comput. Methods Programs Biomed. **187**, 105242 (2020)
8. Esmaeilzadeh, S., Belivanis, D.I., Pohl, K.M., Adeli, E.: End-to-end Alzheimer's disease diagnosis and biomarker identification. In: MLMI, pp. 337–345 (2018)
9. Farooq, A., Anwar, S., Awais, M., Rehman, S.: A deep CNN based multi-class classification of Alzheimer's disease using MRI. In: IST, pp. 1–6 (2017)
10. Feng, C., et al.: Deep learning framework for Alzheimer's disease diagnosis via 3D-CNN and FSBi-LSTM. IEEE Access **7**, 63605–63618 (2019)
11. Fonov, V., et al.: Unbiased average age-appropriate atlases for pediatric studies. Neuroimage **54**(1), 313–327 (2011)
12. Frisoni, G.B., et al.: Imaging markers for Alzheimer disease: which vs how. Neurology **81**(5), 487–500 (2013)
13. Gaser, C., Dahnke, R., et al.: Cat-a computational anatomy toolbox for the analysis of structural MRI data. HBM **2016**, 336–348 (2016)
14. Hosseini-Asl, E., Gimel'farb, G., El-Baz, A.: Alzheimer's disease diagnostics by a deeply supervised adaptable 3D convolutional network. arXiv preprint arXiv:1607.00556 (2016)
15. Huang, Y., et al.: Diagnosis of Alzheimer's disease via multi-modality 3D convolutional neural network. Front. Neurosci. **13**, 509 (2019)
16. Ioffe, S., Szegedy, C.: Batch normalization: accelerating deep network training by reducing internal covariate shift. In: ICML, pp. 448–456 (2015)
17. Jack, C.R., Jr., et al.: The Alzheimer's disease neuroimaging initiative (ADNI): MRI methods. J. Magn. Resonan. Imaging **27**(4), 685–691 (2008)
18. Korolev, S., Safiullin, A., Belyaev, M., Dodonova, Y.: Residual and plain convolutional neural networks for 3D brain MRI classification. In: ISBI, pp. 835–838 (2017)
19. Li, F., Cheng, D., Liu, M.: Alzheimer's disease classification based on combination of multi-model convolutional networks. In: IST, pp. 1–5 (2017)
20. Liu, M., Cheng, D., Yan, W., et al.: Classification of Alzheimer's disease by combination of convolutional and recurrent neural networks using FDG-PET images. Front. Neuroinform. **12**, 35 (2018)
21. Livingston, G., Sommerlad, A., Orgeta, V., Costafreda, S.G., Huntley, J., et al.: Dementia prevention, intervention, and care. The Lancet **390**(10113), 2673–2734 (2017)
22. Marcus, C., Mena, E., Subramaniam, R.M.: Brain PET in the diagnosis of Alzheimer's disease. Clin. Nucl. Med. **39**(10), e413 (2014)
23. Payan, A., Montana, G.: Predicting Alzheimer's disease: a neuroimaging study with 3D convolutional neural networks. arXiv preprint arXiv:1502.02506 (2015)
24. Song, J., Zheng, J., Li, P., Lu, X., Zhu, G., Shen, P.: An effective multimodal image fusion method using MRI and PET for Alzheimer's disease diagnosis. Front. Digit Health **3**, 19 (2021)
25. Sundararajan, M., Taly, A., Yan, Q.: Axiomatic attribution for deep networks. In: ICML, pp. 3319–3328 (2017)
26. Teipel, S., Kilimann, I., Thyrian, J.R., Kloppel, S., Hoffmann, W.: Potential role of neuroimaging markers for early diagnosis of dementia in primary care. Curr. Alzheimer Res. **15**(1), 18–27 (2017)
27. Wang, Y., Huang, W., Sun, F., Xu, T., Rong, Y., Huang, J.: Deep multimodal fusion by channel exchanging. NeurIPS **33**, 4835–4845 (2020)
28. Yee, E., et al.: Quantifying brain metabolism from FDG-PET images into a probability of Alzheimer's dementia score. Hum. Brain Mapp. **41**(1), 5–16 (2020)

29. Zhang, D., et al.: Multimodal classification of Alzheimer's disease and mild cognitive impairment. Neuroimage **55**(3), 856–867 (2011)
30. Zhou, T., Thung, K.H., Zhu, X., Shen, D.: Effective feature learning and fusion of multimodality data using stage-wise deep neural network for dementia diagnosis. Hum. Brain Mapp. **40**(3), 1001–1016 (2019)

Unsupervised Representation Learning of Cingulate Cortical Folding Patterns

Joël Chavas[1(✉)], Louise Guillon[1], Marco Pascucci[1], Benoît Dufumier[1,2], Denis Rivière[1], and Jean-François Mangin[1]

[1] Université Paris-Saclay, CEA, CNRS, NeuroSpin, Baobab, Gif-sur-Yvette, France
{joel.chavas,louise.guillon}@cea.fr
[2] LTCI, Télécom Paris, IPParis, Paris, France

Abstract. The human cerebral cortex is folded, making sulci and gyri over the whole cortical surface. Folding presents a very high inter-subject variability, and some neurodevelopmental disorders are correlated to local folding structures, named folding patterns. However, it is tough to characterize these patterns manually or semi-automatically using geometric distances. Here, we propose a new methodology to identify typical folding patterns. We focus on the cingulate region, known to have a clinical interest, using so-called skeletons (3D representation of folding patterns). We compare two models, $\beta - VAE$ and SimCLR, in an unsupervised setting to learn a relevant representation of these patterns. We add a decoder to SimCLR to be able to analyse latent space. Specifically, we leverage the data augmentations used in SimCLR to propose a novel kind of augmentations based on folding topology. We then apply a clustering on the latent space. Cluster folding averages, interpolation in the latent space and reconstructions reveal new pattern structures. This structured representation shows that unsupervised learning can help in the discovery of still unknown patterns. We will gain further insights into folding patterns by using new priors in the unsupervised algorithms and integrating other brain data modalities. Code and experiments are available at github.com/neurospin-projects/2021_jchavas_lguillon_deepcingulate.

Keywords: beta-VAE · SimCLR · Contrastive learning · Folding pattern · Cortex

1 Introduction

The human cortex is convoluted, made of folds, called gyri, separated by grooves, the sulci. Contrary to macaque, whose cortical folding follows a systematic

J. Chavas and L. Guillon—Contributed equally.

Supplementary Information The online version contains supplementary material available at https://doi.org/10.1007/978-3-031-16431-6_8.

L. Wang et al. (Eds.): MICCAI 2022, LNCS 13431, pp. 77–87, 2022.
https://doi.org/10.1007/978-3-031-16431-6_8

scheme, human cortex folding is highly variable, making it a fingerprint of each individual [27]. Although this diversity seems, first, intractable, neuroanatomists have succeeded in defining a partially reproducible scheme, which has led to the nomenclature of sulci used in neuroscience [21]. But each sulcus can have a large number of patterns, which hinders its reliable identification (Fig. 1B). Deep learning could be a real lever to deal with this tremendous inter-individual variability.

Shapes of the folding patterns are particularly interesting to study as they are "trait" features, they remain during lifespan, contrary to "state" features (e.g. sulci depth or width) that are not stable throughout life [3]. Some works tried to decipher folding patterns and identify the most common shapes. Historically, this was done visually [29], enabling to define central sulcus knob and the omega-shape of the mid-fusiform sulcus in particular [28,30]. However, manually finding relevant geometrical shapes is very hard due to the high diversity of folding patterns. Thus some studies tried to automate the characterization of folding patterns mainly based on geometric distances [8,20,25]. More recently, [24] trained neural network classifiers to map geometric shapes to folding patterns applied to the broken-H shape pattern in the orbitofrontal region.

However, characterizing the full diversity of folding patterns remains out of reach for these automatic geometric methods. Unsupervised deep learning methods is a natural next step: they have been used for detecting anomalies in folding shapes [12], but they have not been used yet to characterize the normal inter-individual variability of folding patterns.

Numerous approaches try to tackle unsupervised representation learning problems. On the one hand, auto-encoders (AE) are generative models that build a latent space comprising much fewer dimensions than the input, suggesting that the representations could be more easily understood, leading eventually to pattern discovery. For example, [12] showed that β-VAE are promising to detect anomalies of folding patterns.

On the other hand, self-supervised methods, particularly contrastive learning models, have proved to be very powerful. The foundation contrastive model, SimCLR [6], permits structuring the obtained latent space without using any labels. Its strength lies in the possibility to integrate prior information either by choosing the adapted random augmentations or by integrating into the loss function similarity information from other modalities.

Many works start to apply such framework to biomedical imaging. Thus, [26] proposed 3D versions of several self-supervised tasks on various objectives including brain tumor segmentation. Self-supervised methods offer the opportunity to leverage additional prior information from medical data. For instance, [10] applied contrastive learning to brain MRI and took advantage of available metadata such as age and sex. This accelerating research on nearby fields shows that it is the right moment to apply self-supervised learning to the folding pattern characterization problem.

This study aims to pave the way for unsupervised deep learning to systematize the identification of typical folding patterns across the cortex in the future. More specifically, we aim to compare two unsupervised deep learning models in the task of obtaining a latent space structured enough to bring out folding patterns. To achieve this goal, we developed a deep learning pipeline that focuses on the folding pattern of predefined regions. We tested the pipeline on the cingulate region, as it is sufficiently variable to justify the use of our methods, and it has a clinical interest for psychiatric disorders [2,22,31]. Then, we chose, adapted and compared two powerful and standard unsupervised methods, namely a contrastive learning model, SimCLR [6] to which we added a decoder, and a generative model, β-VAE [13]. Last, we proposed ways to analyze the results which are new and challenging with respect to classical deep learning literature as the input sample topologies are very different from classical 3D images.

Our contributions are three-fold. First, we implement an efficient topology-based augmentation for SimCLR. Second, we propose models reconstruction of cingulate patterns as alternative to local average folding pattern to explain the encoded features and last, a preprocessing pipeline to study efficiently folding patterns. Our work constitutes a first step to increase the knowledge on folding patterns that will permit to find biomarkers for brain disease.

2 Methods

2.1 Pre-processing

From brain MRI images, we used skeletons whose concept was first introduced in [19]. They consist in 3D images of the cortical folds obtained with BrainVISA/Morphologist preprocessing pipeline (https://brainvisa.info/web/). Skeletons' voxels are divided into background and folds (Fig. 1A). Fold voxels can hold several values depending on their topological meaning (fold bottom, fold junction, etc.). Using this input enables to focus on the folding geometry and eliminates some biases such as age or site.

We focus our study on the cingulate region of the right hemisphere (Fig. 1A). We learned a mask of the cingulate and paracingulate sulci over a database where the folds were manually labelled [1]. In short, labeled subjects were first affinely normalized to a standard brain referential (ICBMc2009); then, each subject voxel belonging to the sulci of interest increments a sulcus-specific mask. We combined and dilated these two resulting masks to get a simple Region of Interest (ROI). We then applied this final mask to skeleton images of any unlabeled brain. Our final input is a 2-mm resolution 3D crop of dimension $20 \times 40 \times 40$ (Fig. 1B) with integer values representing local topologies.

2.2 Learning Cingulate Region Representations

We compared two unsupervised deep learning models: an autoencoder-based model and a contrastive learning framework.

β-VAE. AE-based models are commonly used to learn representations and to model the inter-subject variability. With an encoder θ, they enable to project data from input space $\mathcal{X}$ onto a latent space $\mathcal{Z}$ comprising much fewer dimensions. The latent code is then reconstructed thanks to a decoder ϕ. β-VAE [13],

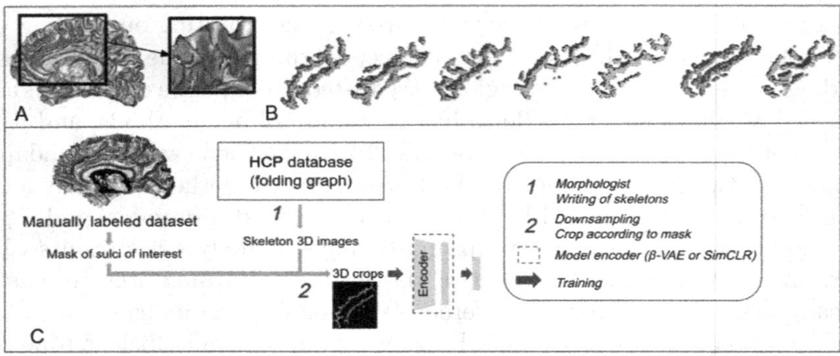

Fig. 1. *Framework to study skeletons in the cingulate region of the HCP dataset.* **A)** Sample crop of the cingulate region represented as buckets superimposed with the white matter mesh. Red voxels are *bottom voxels.* **B)** Samples of the studied crops, given as inputs to the unsupervised algorithms. **C)** Whole pipeline. We generate crops of the cingulate region based on a manually labeled dataset. We train both models (β-VAE and SimCLR), we then infer and perform downstream analysis of our two models. (Color figure online)

an extension of VAE [15], is particularly interesting as the latent space is constrained to follow a prior distribution and input data are encoded as a distribution. The objective function is a combination of the reconstruction error and the matching of two distributions using the Kullback-Liebler (KL) divergence. The two terms are weighted thanks to β, which enables to improve latent factors disentanglement [13]. β-VAE is trained to maximize:

$$\mathcal{L}(\theta, \phi; \mathbf{x}, \mathbf{z}, \beta) = \mathbb{E}_{q_\phi(\mathbf{z}|\mathbf{x})}[\log p_\theta(\mathbf{x}|\mathbf{z})] - \beta \mathcal{D}_{KL}(q_\phi(\mathbf{z}|\mathbf{x})||p(\mathbf{z})) \quad (1)$$

where $p(\mathbf{z})$ corresponds to the prior distribution (here, a reduced centered Gaussian distribution) and $q_\phi(\mathbf{z}|\mathbf{x})$, the posterior distribution. We ran the model on binarized skeletons.

SimCLR. SimCLR is an instance discrimination contrastive model. For each sample x of the batch of size N, we generate at each epoch two views x_i and x_j, whose model outputs are respectively z_i and z_j. The model trains to bring together views from the same image, that is it minimizes $\sum_{i=1}^{N} \ell_{i,j=pos(i)} + \sum_{j=1}^{N} \ell_{j,i=pos(i)}$, $\ell_{i,j}$ being the loss function for a positive pair of examples (τ is a temperature parameter):

$$\ell_{i,j} = -\log \frac{\exp(\text{sim}(z_i, z_j)/\tau)}{\sum_{k=1,k\neq i}^{2N} \exp(\text{sim}(z_i, z_k)/\tau)}, \quad (2)$$

View generations are the algorithms specific to our problem: they use the discrete topology of the fold skeleton. For each fold, the bottom line voxels can be distinguished from the inner part of the fold surface because they do not split the skeleton background into two different local connected components [19] (Figure 1A). Then, the bottom line tag permits to define a topology-based augmentation, which conserves the bottom lines in all views but remove the inner part of some folds. The first view combines random $[-10, 10]^{\circ}$ rotations over all axes and a 60% rolling cutout with only bottom lines kept inside the cutout volume. The second view combines random $[-10, 10]^{\circ}$ rotations over all axes followed by a 60% rolling cutout with the whole skeleton conserved inside the cutout whereas only bottom values are kept outside the cutout volume. All views are then binarized. This topology-based augmentation forces the model to learn the sheet-based structure of the fold-based skeleton.

To decode SimCLR latent code, we freeze SimCLR weights and train a decoder whose input layer is the representation space. The decoder backbone is the one of the $\beta - VAE$ decoder, and the decoder loss is the cross entropy reconstruction error.

2.3 Identifying Folding Patterns

Characterizing Folding Shapes. To identify folding patterns, data are encoded to the latent space of both models and reduced to a 2-dimensions space with t-SNE algorithm. The reduction to two dimensions enables to get more hints of the learned representations and to analyze subjects groups more easily. A clustering is then performed with hierarchical affinity propagation (AP) algorithm [11]. One advantage of AP is that the number of expected clusters does not have to be precised. However it may output a very large number of clusters, making it difficult to understand from an anatomical point of view. Hence, following the method used in [20], we applied the algorithm in an iterative way until a maximum number of five clusters is found. We stress out that the maximum of five clusters is an arbitrary number and that it has not a biological meaning beyond facilitating our understanding.

The analysis of the main anatomical characteristics of the clusters can be done either on the latent codes or on the input space based on cluster labels. The first method enables to understand the encoded characteristics in the latent dimensions. For $\beta - VAE$ we generated images corresponding to clusters' centroids from their latent codes which are next decoded. Then, we travelled between clusters through the latent space to analyse variations across dimensions. For SimCLR, we reconstructed latent representation of the nearest subject of each cluster centroid. The second method computes the local *per*-cluster averaging pattern in the input space [25].

3 Experiments and Results

Datasets. We use HCP database[1] in which MRI images were obtained with a Siemens Skyra Connectom scanner with isotropic resolution of 0.7 mm. We focused on the right hemisphere of 550 subjects. 80% of subjects were used for training and the remaining 20% were used for validation.

[1] https://www.humanconnectome.org/.

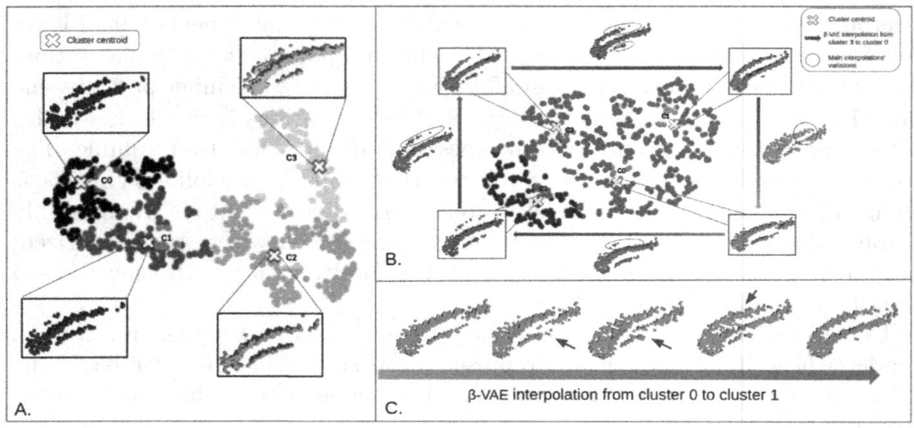

Fig. 2. *β-VAE and SimCLR latent spaces analysis.* (**A.**) t-SNE representation of SimCLR latent space. Insets are decoded latent codes of nearest neighbours for each cluster centroid. (**B.**) t-SNE representation of β-VAE latent space. Insets are decoded latent codes of cluster centroids and of interpolations between cluster centroids. Leftmost and rightmost patterns of (**C.**) are respectively the decoded latent code of cluster 0 and 1 centroids of the β-VAE model. Intermediate patterns of (**C.**) are obtained traveling through the latent space and then decoded. (Color figure online)

Model Implementation. Our β-VAE comprises fully convolutional encoder and decoder of symmetrical architectures with three convolutional blocks and two fully connected layers. The backbone of our SimCLR model is the DenseNet [14], followed by two fully connected projection overheads based on [9] benchmark on 3D MRI images. To adapt to our smaller input, we reduced the size of the DenseNet network down to two dense blocks. We call latent space, the representation space of the SimCLR model, which has a better representation quality than output space [6].

To find the best hyperparameters (size of the latent space for both models, β value for β-VAE and temperature τ for SimCLR), we performed a gridsearch where the best combination is chosen based on the loss value, the silhouette score on the latent space and the reconstruction abilities for β-VAE. We obtained $\beta = 2$ (tested range 1–8) and $\tau = 0.1$ (tested range 0.01–0.3), as well as a latent size of 4 (tested range 2–150) for both models, which enabled to balance between the model performance and the clustering quality. Training of 300 epochs lasted for approximately 1 h and 2 h for β-VAE and SimCLR respectively, on an Nvidia Quadro RTX5000 GPU.

Latent Space Structure. Figure 2A and B present clustering results. The silhouette score with AP on the latent space is 0.17 and 0.42, respectively for β-VAE and SimCLR. It becomes 0.43 and 0.44 when applied to the t-SNE space, indicating a tendency towards a clustered distribution with close clusters. This range of score is common when dealing with complex data such as neuroimaging modalities [16]. For both models, four clusters were identified but the organization of the latent space is different: β-VAE latent space seems to distinguish four groups of subjects, separated only with a thin boundary whereas SimCLR latent space is more structured and could be interpreted as a manifold, consistent with the biological reality of folding patterns.

Deciphering the Patterns. Both models were able to produce reconstructions that are compliant with the inputs, presenting a simplified version of the scene which enables to focus and bring out the most important features (Fig. 2A and B). For SimCLR, the black cluster seems to have a small paracingulate, which is a pattern described in the literature [20]. The brown and orange cluster seems to correspond to two parallel sulci without a callosal sulcus. Based on the latent space organization, similar to a manifold, it is interesting to analyse the reconstructions in terms of evolution. The curvature of the longest sulcus becomes more bent from the black to the yellow cluster.

In the case of $\beta - VAE$, cluster 0 (green) shows another pattern defined in the literature [20]: a split anterior cingulate sulcus. Cluster 1 (blue) presents a long paracingulate, while the pink has a shorter paracingulate. Lastly, the indigo presents a slight paracingulate. In Fig. 2C, interpolating from one cluster to another shows that the latent space is continuous and regular, and we can progressively see the change of patterns as indicated by the arrows. More detailed and complete interpolations are presented in supplementary materials.

It is interesting to link these decoder outputs to the cluster average of the folding pattern based on the input space (Fig. 3). For SimCLR model, the black cluster average could correspond to a simple anterior cingulate. Subjects of the brown cluster could have a sketch of the paracingulate sulcus, which increases in length in the orange average to present two long parallel sulci. Finally, the yellow average also includes a sketch of a sulcus parallel to the anterior cingulate, but in the left part of the ROI, where it is not usually called a paracingulate sulcus in anatomical literature. Both methods, cluster averages and decoders, represent something different: in the first case, it is the geometrically-aligned average of all subjects in a cluster; in the latter case, it is the reconstruction of one representative subject from the latent space. They can converge either to the same (orange and yellow cluster) or to an apparently different (black cluster) representation.

For $\beta - VAE$, when comparing with centroids' generation, we find a similar shape for the green average (cluster 0): cingulate split in two. Conversely, for the blue cluster, based only on the average pattern, we could interpret a simple cingulate, but in the light of the reconstructions, the swollen anterior part could represent a paracingulate. Cingulate and paracingulate could be merged in the average representation due to positional variations among subjects. For indigo and pink folding averages, only the highly variable paracingulate pits are kept contrary to the decoder reconstructions. Thus, only the decoder part will permit to get rid of the too complex inter-individual variability and to focus on the main shapes.

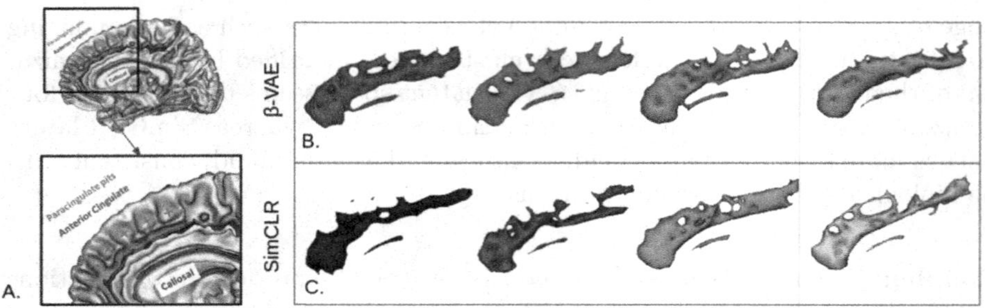

Fig. 3. *Representative patterns as cluster averages.* **A)** Description of typical folding structures in our ROI using the icbm152 average template. **B), C)** Local average sulci obtained for each cluster with β-VAE and SimCLR encodings respectively. Colors match cluster colors of Fig. 2. (Color figure online)

4 Discussion and Conclusion

Our work proposes several method contributions that can be useful for the community. We introduced topology-based augmentations in the SimCLR setting, which is directly applicable for studies working on skeletons or similar inputs [23]. It is all the more interesting as the augmentations used for contrastive views is still under investigation to understand what makes good views, especially in biomedical images. Moreover, we added a decoder to SimCLR and analyzed SimCLR and β-VAE reconstructions to recover folding patterns. Last, we proposed a preprocessing based on a mask, enabling to focus on the region of interest, while avoiding the disadvantages of parallelepipedic bounding boxes used in [12].

Our work also finds a structured latent space for the cingulate region with both models, β-VAE and SimCLR. The organization obtained with SimCLR seems more consistent with anatomical reality of folding patterns and can be linked to folding manifolds [18]. In return, the generative and regularization aspect of the $\beta - VAE$ is a real lever to understand the learned representations and ease the analysis of this complex region. Note that, if we chose a higher maximum number of clusters, it would have been tricky to analyse but it could be closer to reality. In addition, according to the distribution of SimCLR latent space, a finer clustering, with a higher granularity could be of interest.

To encourage a structured and well separated latent space, we wish to introduce in future cluster objectives in the learning phase both for generative models [7], and for contrastive models [4,5,17].

Another line of research will be to adapt our model further to the folding topology by developing other topology-based augmentations and by introducing other specific priors such as the geometry-based similarity measure between input samples [25].

Finally, we found both cluster averages and decoder outputs to be similar to known cingulate patterns that correlate with executive functions and psychiatric disorders [3]. This similarity makes us firmly believe that such latent space structures could correlate with medically relevant parameters. Our study is therefore a first step towards the systematization of the search for main region-specific patterns to then analyze their potential correlations with human cognition and disease.

Acknowledgments. We thank colleagues from the Deep learning journal club for thorough discussions. We thank Zaccharie Ramzi for helping set up rotation augmentations. This research received funding from the European Union's Horizon 2020 Research and Innovation Programme under Grant Agreement No. 945539 (HBP SGA3), from the FRM DIC20161236445, the ANR-19-CE45-0022-01 IFOPASUBA, the ANR-20-CHIA-0027-01 FOLDDICO. Data were provided in part by the Human Connectome Project funded by the NIH.

References

1. Borne, L., Rivière, D., Mancip, M., Mangin, J.F.: Automatic labeling of cortical sulci using patch- or CNN-based segmentation techniques combined with bottom-up geometric constraints. Med. Image Anal. **62**, 101651 (2020)
2. Borst, G., et al.: Folding of the anterior cingulate cortex partially explains inhibitory control during childhood: a longitudinal study. Dev. Cogn. Neurosci. **9**, 126–135 (2014)
3. Cachia, A., et al.: Longitudinal stability of the folding pattern of the anterior cingulate cortex during development. Dev. Cogn. Neurosci. **19**, 122–127 (2016)
4. Caron, M., Bojanowski, P., Joulin, A., Douze, M.: Deep clustering for unsupervised learning of visual features. In: Ferrari, V., Hebert, M., Sminchisescu, C., Weiss, Y. (eds.) Computer Vision – ECCV 2018. LNCS, vol. 11218, pp. 139–156. Springer, Cham (2018). https://doi.org/10.1007/978-3-030-01264-9_9
5. Caron, M., Misra, I., Mairal, J., Goyal, P., Bojanowski, P., Joulin, A.: Unsupervised learning of visual features by contrasting cluster assignments. In: Thirty-fourth Conference on Neural Information Processing Systems (NeurIPS2020), vol. 33, pp. 9912–9924 (2020)
6. Chen, T., Kornblith, S., Norouzi, M., Hinton, G.: A simple framework for contrastive learning of visual representations. In: Proceedings of the 37th International Conference on Machine Learning, pp. 1597–1607. PMLR (2020)
7. Danks, D., Yau, C.: BasisDeVAE: interpretable simultaneous dimensionality reduction and feature-level clustering with derivative-based variational autoencoders. In: Proceedings of the 38th International Conference on Machine Learning, pp. 2410–2420. PMLR (2021)
8. Duan, D., et al.: Exploring folding patterns of infant cerebral cortex based on multi-view curvature features: methods and applications. Neuroimage **185**, 575–592 (2019)
9. Dufumier, B., Gori, P., Battaglia, I., Victor, J., Grigis, A., Duchesnay, E.: Benchmarking CNN on 3D anatomical brain MRI: architectures, data augmentation and deep ensemble learning. arXiv:2106.01132 (2021)
10. Dufumier, B., et al.: Contrastive learning with continuous proxy meta-data for 3D MRI classification. In: de Bruijne, M., et al. (eds.) MICCAI 2021. LNCS, vol. 12902, pp. 58–68. Springer, Cham (2021). https://doi.org/10.1007/978-3-030-87196-3_6
11. Frey, B.J., Dueck, D.: Clustering by passing messages between data points. Science **315**(5814), 972–976 (2007)

12. Guillon, L., Cagna, B., Dufumier, B., Chavas, J., Rivière, D., Mangin, J.F.: Detection of abnormal folding patterns with unsupervised deep generative models. In: Abdulkadir, A., et al. (eds.) MLCN 2021. LNCS, vol. 13001, pp. 63–72. Springer, Cham (2021). https://doi.org/10.1007/978-3-030-87586-2_7
13. Higgins, I., et al.: beta-VAE: learning basic visual concepts with a constrained variational framework. In: 5th International Conference on Learning Representations, ICLR (2017)
14. Huang, G., Liu, Z., Van Der Maaten, L., Weinberger, K.Q.: Densely connected convolutional networks. In: 2017 IEEE Conference on Computer Vision and Pattern Recognition (CVPR), pp. 2261–2269 (2017)
15. Kingma, D.P., Welling, M.: Auto-encoding variational Bayes. In: 2nd International Conference on Learning Representations, ICLR (2014)
16. Lebenberg, J., et al.: Mapping the asynchrony of cortical maturation in the infant brain: a MRI multi-parametric clustering approach. Neuroimage **185**, 641–653 (2019)
17. Li, Y., Hu, P., Liu, Z., Peng, D., Zhou, J.T., Peng, X.: Graph contrastive clustering. Proc. AAAI Conf. Artif. Intell. **35**(10), 8547–8555 (2021)
18. Mangin, J.F., et al.: Spatial normalization of brain images and beyond. Med. Image Anal. **33**, 127–133 (2016)
19. Mangin, J.F., Frouin, V., Bloch, I., Régis, J., Lopez-Krahe, J.: From 3D magnetic resonance images to structural representations of the cortex topography using topology preserving deformations. J. Math. Imaging Vis. **5**(4), 297–318 (1995)
20. Meng, Y., Li, G., Wang, L., Lin, W., Gilmore, J.H., Shen, D.: Discovering cortical sulcal folding patterns in neonates using large-scale dataset. Hum. Brain Mapp. **39**(9), 3625–3635 (2018)
21. Ono, M., Kubik, S., Abarnathey, C.D.: Atlas of the Cerebral Sulci. Thieme-Stratton Corp, Stuttgart, New York, 1er édition edn. (1990)
22. Provost, J.B.L., et al.: Paracingulate sulcus morphology in men with early-onset schizophrenia. Br. J. Psychiatry **182**(3), 228–232 (2003)
23. Rao, H., et al.: A self-supervised gait encoding approach with locality-awareness for 3D skeleton based person re-identification. IEEE Trans. Pattern Anal. Mach. Intell. **43**(1) (2021). arXiv: 2009.03671
24. Roy, A., McMillen, T., Beiler, D.L., Snyder, W., Patti, M., Troiani, V.: A pipeline to characterize local cortical folds by mapping them to human-interpretable shapes. Tech. rep. (2020). bioarxiv:2020.11.25.388785
25. Sun, Z.Y., et al.: The effect of handedness on the shape of the central sulcus. Neuroimage **60**(1), 332–339 (2012)
26. Taleb, A., et al.: 3D self-supervised methods for medical imaging. In: Larochelle, H., Ranzato, M., Hadsell, R., Balcan, M.F., Lin, H. (eds.). In: 34th Conference on Neural Information Processing Systems (NeurIPS 2020), vol. 33, pp. 18158–18172 (2020)
27. Wachinger, C., Golland, P., Kremen, W., Fischl, B., Reuter, M.: BrainPrint: a discriminative characterization of brain morphology. Neuroimage **109**, 232–248 (2015)
28. Weiner, K.S., et al.: The mid-fusiform sulcus: a landmark identifying both cytoarchitectonic and functional divisions of human ventral temporal cortex. Neuroimage **84**, 453–465 (2014)
29. White, L.E., et al.: Structure of the human sensorimotor system. I: morphology and cytoarchitecture of the central sulcus. Cereb. Cortex **7**(1), 18–30 (Feb 1997)

30. Yousry, T.A., et al.: Localization of the motor hand area to a knob on the precentral gyrus. A new landmark. Brain **120**(1), 141–157 (1997)
31. Yücel, M., et al.: Morphology of the anterior cingulate cortex in young men at ultra-high risk of developing a psychotic illness. Br. J. Psychiatry **182**(6), 518–524 (2003)

Feature Robustness and Sex Differences in Medical Imaging: A Case Study in MRI-Based Alzheimer's Disease Detection

Eike Petersen[1(✉)], Aasa Feragen[1], Maria Luise da Costa Zemsch[1], Anders Henriksen[1], Oskar Eiler Wiese Christensen[1], Melanie Ganz[2,3], for the Alzheimer's Disease Neuroimaging Initiative

[1] Technical University of Denmark DTU Compute, Kgs. Lyngby, Denmark
{ewipe,afhar}@dtu.dk

[2] Department for Computer Science, University of Copenhagen, Copenhagen, Denmark

[3] Rigshospitalet, Neurobiology Research Unit, Copenhagen, Denmark
melanie.ganz@nru.dk

Abstract. Convolutional neural networks have enabled significant improvements in medical image-based diagnosis. It is, however, increasingly clear that these models are susceptible to performance degradation when facing spurious correlations and dataset shift, leading, e.g., to underperformance on underrepresented patient groups. In this paper, we compare two classification schemes on the ADNI MRI dataset: a simple logistic regression model using manually selected volumetric features, and a convolutional neural network trained on 3D MRI data. We assess the robustness of the trained models in the face of varying dataset splits, training set sex composition, and stage of disease. In contrast to earlier work in other imaging modalities, we do not observe a clear pattern of improved model performance for the majority group in the training dataset. Instead, while logistic regression is fully robust to dataset composition, we find that CNN performance is generally improved for both male and female subjects when including more female subjects in the training dataset. We hypothesize that this might be due to inherent differences in the pathology of the two sexes. Moreover, in our analysis, the logistic regression model outperforms the 3D CNN, emphasizing the utility of manual feature specification based on prior knowledge, and the need for more robust automatic feature selection.

Data used in preparation of this article was obtained from the Alzheimers Disease Neuroimaging Initiative (ADNI) database (http://www.adni-info.org/). The investigators within the ADNI contributed to the design and implementation of ADNI and/or provided data, but did not participate in analysis or writing of this report.

Supplementary Information The online version contains supplementary material available at https://doi.org/10.1007/978-3-031-16431-6_9.

L. Wang et al. (Eds.): MICCAI 2022, LNCS 13431, pp. 88–98, 2022.
https://doi.org/10.1007/978-3-031-16431-6_9

Keywords: Deep learning · MRI · Alzheimer's disease · Robustness

1 Introduction

In recent years, various groups have reported highly accurate detection of Alzheimer's disease (AD) and progressive mild cognitive impairment (pMCI) – which represents an earlier disease stage that continues to progress into AD or other types of dementia [17] – based on magnetic resonance imaging (MRI) volumes using convolutional neural networks (CNNs) [27]. Simultaneously, the potential brittleness of deep learning has become apparent due to issues like model underspecification [8], spurious correlations [12], and susceptibility to dataset shift [5]. Multiple reviews in the medical domain have shown that deep learning-based publications often suffer from inadequate model reporting and overly optimistic performance estimates [14,28]. Wen et al. [27] found in their systematic review that half of the studies reporting on MRI-based AD detection using deep learning potentially suffered from data leakage, likely leading to inflated performance estimates. Additionally, in clinical applications with low-dimensional input spaces, several systematic reviews have found no performance benefit of machine learning-based techniques over simple logistic regression [7,18], raising the question of whether simple models based on manually extracted, low-dimensional features may also perform competitively in medical imaging.

In a parallel development, the question of sex and gender-related performance disparities of machine learning models has lately received a lot of attention [19, 24]. Studies have shown that CNNs can accurately identify a patient's age, sex, and ethnicity from chest x-ray images [6,29], and that chest x-ray classifiers tend to underdiagnose underserved patient populations [24]. Larrazabal et al. [15] have analyzed the effect of training dataset sex imbalance on a chest x-ray classifier's performance for male and female subgroups, finding a consistent decrease in performance for the underrepresented sex. To the authors' knowledge, similar analyses have yet to be performed in the brain MRI classification setting, even though it is known that AD presents differently in males and females [17,22].

Training on a sex-imbalanced dataset and then evaluating model performance on the underrepresented group represents a particular type of *dataset shift* [23], and the close connection between model robustness (to dataset shift and other challenges) and algorithmic fairness has been emphasized recently [2]. To achieve robustness to dataset shifts, the *feature space* in which classification is performed plays a crucial role: to be robust against dataset shifts, estimation must be performed in a feature space that gives rise to a classifier that is optimal across different environments [3]. This feature space can either be manually crafted or automatically inferred, as is usually the case in deep learning [1].

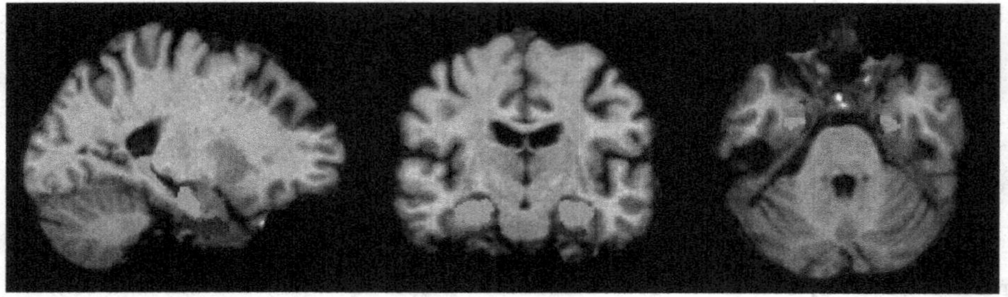

Fig. 1. Slices of an exemplary recording used in this study, skull-stripped and registered to a common space. Hippocampus and entorhinal cortex (yellow and red, segmented using FreeSurfer) are highlighted. Left to right: sagittal, coronal, and axial view. (Color figure online)

Table 1. Composition of the dataset used in this study (based on the ADNI dataset [13]), stratified by sex and field strength.

Diagnosis	Male	Female	1.5 T	3 T	Total
AD	181 (54.03%)	154 (45.97%)	175 (52.24%)	160 (47.76%)	335
HC	282 (43.52%)	366 (56.48%)	219 (33.80%)	547 (66.20%)	648
pMCI	149 (58.66%)	105 (41.34%)	184 (72.44%)	70 (27.56%)	254
sMCI	155 (59.62%)	105 (40.38%)	122 (46.92%)	138 (53.08%)	260

In this paper, we analyze the robustness of two different MRI-based classifiers to distribution shifts. Both classifiers are trained to detect Alzheimer's disease, based on different feature representations. The first is a simple logistic regression model using manually selected volumetric features, which are obtained using standard MRI processing tools [4,11]. The second is a CNN using full 3D MRI volumes. Both models are considered state-of-the-art for Alzheimer's disease classification. For analyzing the classifiers' robustness, we consider two separate types of distribution shifts: Firstly, we analyze the effect of differing training dataset sex and age compositions on the performance for male/female or young/old test subjects, similar to the analysis of Larrazabal et al. [15]. And secondly, like various other groups have done [27] (although not combined with the sex and age imbalance analyses we perform), we evaluate the performance of classifiers trained on subjects diagnosed with AD and healthy controls on a test set consisting of subjects with stable and progressive MCI.

2 Methods

2.1 Dataset

We use MRI volumes from the ADNI dataset [13] (700, 463, and 334 recordings from ADNI1, ADNI2, and ADNI3, respectively), including a single T1-weighted structural MRI volume (acquired on different MR scanners using a field strength of either 1.5 T or 3 T) per subject in our experiments. Like various other studies [27], we consider two groups of subjects, giving rise to two classification tasks: firstly, healthy control (HC) and Alzheimer's disease (AD) subjects and, secondly, stable and progressive mild cognitive impairment (sMCI, pMCI) subjects. Subjects were labeled as pMCI in our analysis if they were diagnosed with MCI at the time of the recording and then were diagnosed with AD at any point during the following five years. (This definition only encompasses (progressive) *amnestic* MCI, since non-amnestic MCI may develop into non-AD dementias [17].) Subjects were labeled as sMCI if they were diagnosed with MCI at the time of the recording, not diagnosed with AD at any point during the five-year follow-up period, and if there was at least one valid diagnosis from years three to five after the initial recording was made. The dataset is summarized in Table 1. The median age across the whole dataset was 73.0 years; the average age of the different subject cohorts was 74.99 ± 7.08 (males), 72.70 ± 6.94 (females), 75.13 ± 7.83 (AD/pMCI cases), 73.08 ± 6.58 (HC/sMCI cases), 75.70 ± 7.91 (male AD/pMCI cases), and 71.96 ± 6.46 (female AD/pMCI cases).

2.2 AD Classification Using Logistic Regression

Intracranial volume (ICV) was quantified as the combined volumes of gray- and white matter and cerebrospinal fluid [16], segmented with SPM12 [4]. Hippocampal volume (HCV) and entorhinal cortex volume (ECV) was extracted from structural MRI volumes using FreeSurfer v7.1.1 [11], see Fig. 1 for an example.

A simple logistic regression (LR) model of the form

$$P(\text{AD}) \approx q_{\text{AD}} = \sigma(\theta_1 \cdot \text{Age} + \theta_2 \cdot \text{ICV} + \theta_3 \cdot \text{HCV} + \theta_4 \cdot \text{ECV}), \quad (1)$$

with $\sigma(\cdot)$ denoting the sigmoid function, was fitted using stochastic gradient descent for 4000 epochs (initial learning rate 10^{-3}, automatic learning rate scheduling – if there is no improvement in the validation loss for 10 epochs, the learning rate is halved –, momentum 0.9, early stopping based on the validation loss, batch size 256). The loss function was binary cross-entropy with L_2 regularization (regularization constant 10^{-4}).

2.3 AD Classification Using CNNs

The subject-specific MRI images were skull-stripped and registered to a common space (MNI305) using FreeSurfer v7.1.1 [11], resulting in spatially normalized grayscale volumes of the dimensionality $256 \times 256 \times 256\,\text{mm}^3$. These were then cropped to include the whole brain ($186 \times 186 \times 191\,\text{mm}^3$). Dataset augmentation was performed using torchio [21]: During training, 80% of the training samples were transformed using one (randomly selected) of the following transformations: rotation, elastic deformation, flipping, blurring, addition of Gaussian noise, and addition of an MRI bias field, spike, ghosting, or motion artifact.

A convolutional neural network (CNN) was trained, using the preprocessed 3D MRI volumes as inputs and the subject's AD/HC label as the target. The model architecture was inspired by Tinauer et al. [25] and Wen et al. [27], and was (manually) selected based on validation set performance. The convolutional part of the model consists of eight consecutive convolution layers (all with sixteen channels), each followed by a rectified linear unit (ReLu). The first two layers use kernels of size $5 \times 5 \times 5$; the subsequent six layers use $3 \times 3 \times 3$ kernels. Every second layer uses a stride of two, thus progressively reducing image resolution. This is followed by a dropout layer ($p = 0.5$), a fully connected linear layer with 32 output channels, a ReLu, another dropout layer ($p = 0.5$), and the final, fully connected classification layer with sigmoid activation. The model has 337,000 trainable parameters, which are estimated using the Adam optimization algorithm (initial learning rate 10^{-4}, automatic learning rate scheduling based on the validation loss as described above, 200 epochs, batch size 6, early stopping based on the validation loss) to minimize binary cross-entropy (no regularization).

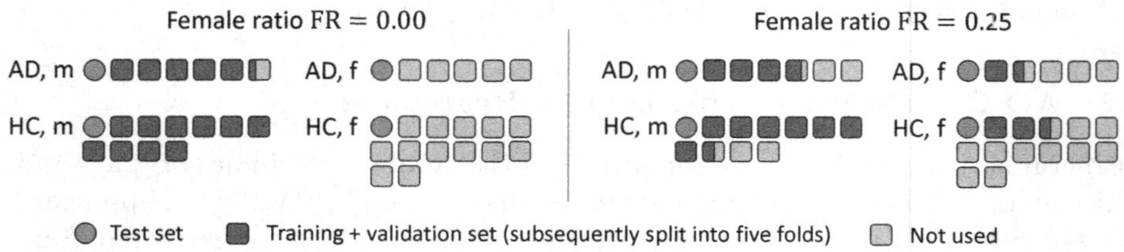

Fig. 2. Visualization of the AD/HC dataset and two exemplary splits. Each marker represents 25 subjects (rounded). For clarity, only one of the five test sets and only two of the five training and validation datasets for this test set (all with different male–female ratios $\mathrm{FR} = \frac{n_f}{n_f + n_m}$, with n_f and n_m the number of females and males in the combined training and validation dataset) are shown.

2.4 Performance and Robustness Evaluation

To evaluate the two models' overall performance and robustness, we trained them on 125 (LR) and 50 (CNN) different training datasets and evaluated them on six different test sets. First, five AD/HC test sets of size 100 were drawn from the full dataset with 25 samples each of male/female AD/HC subjects, with no subject overlap between the test sets. For each of the five test sets, five training and validation sets of combined size 379 were drawn from the remaining AD/HC subjects (without replacement), with five different male-to-female ratios (0% to 100% females in 25% increments) and a constant fraction of 34.1% AD cases, corresponding to the fraction of AD cases in the whole dataset. The gender-specific datasets were nested: female subjects used at a female ratio (FR) of 50% represent a subset of female subjects used at an FR of 75%, etc.

Each of the 25 combined training and validation sets was then split randomly into five folds following standard recommendations [26]. For the logistic regression, each fold was used once as the validation set, resulting in a total of five training–validation set combinations per test set (of which there are five) and sex ratio (of which there are five), yielding a total of 125 different training and validation datasets. For the CNNs, due to the computational effort, only the first two of the five fold combinations were used per sex ratio, thus yielding a total of 50 different training and validation datasets. Each model was then evaluated on both the respective AD/HC test set and the full sMCI/pMCI dataset (which had not been used for training), calculating performance metrics on both males and females. Figure 2 illustrates the full AD/HC dataset and two exemplary data splits.

The decision thresholds for all models were selected to maximize the geometric mean of sensitivity and specificity on the validation dataset [10]. All models were implemented and trained using PyTorch Lightning v.1.5.9 [9] and training one CNN took about twelve hours on a single NVIDIA Titan X GPU. When the performance of models trained on the same datasets was compared, statistical significance was assessed using Wilcoxon signed-rank tests.

An analogous analysis was also performed with respect to age groups instead of sex; its details and results can be found in the supplementary material.

3 Results

Figure 3 shows the distribution of the area under the curve (AUC) of the receiver-operating characteristic and the accuracy (ACC) achieved by the trained models on the different test sets. To assess the dependence of model performance for males and females on the training dataset sex composition, regression lines were fit and a t-test was performed to assess whether the slope was significantly different from zero. To account for multiple comparisons, a Bonferroni correction with a factor of eight was performed. The figure reports the corrected p-values.

As also found in previous studies [27], both models drop in performance on sMCI vs. pMCI classification compared to HC vs. AD classification, for which they were trained. The LR model achieves consistently high AUC and ACC across all training and validation sets, and performance is similar for male and female test subjects, although significantly better for females ($p_{\text{corr}} = 5.5 \times 10^{-4}$ and $p_{\text{corr}} = 5.9 \times 10^{-22}$ for AUC on AD/HC and pMCI/sMCI, respectively, with Bonferroni correction factor 2). For the LR model, all regression slopes were non-significant ($p_{\text{corr}} \gg 0.05$), indicating no significant effect of dataset composition on model performance. The CNN performed significantly worse compared to the LR ($p_{\text{corr}} = 2.8 \times 10^{-25}$ for AUC, $p_{\text{corr}} = 6.2 \times 10^{-12}$ for ACC, with Bonferroni correction factor 2), and it exhibited a stronger dependence of performance for male and female test subjects on the training dataset composition. Three statistically significant positive correlations were observed: between FR and AUC in male pMCI/sMCI subjects ($p_{\text{corr}} = 5.4 \times 10^{-3}$), and between FR and ACC in female ($p_{\text{corr}} = 2.1 \times 10^{-2}$) and male ($p_{\text{corr}} = 1.9 \times 10^{-2}$) pMCI/sMCI subjects.

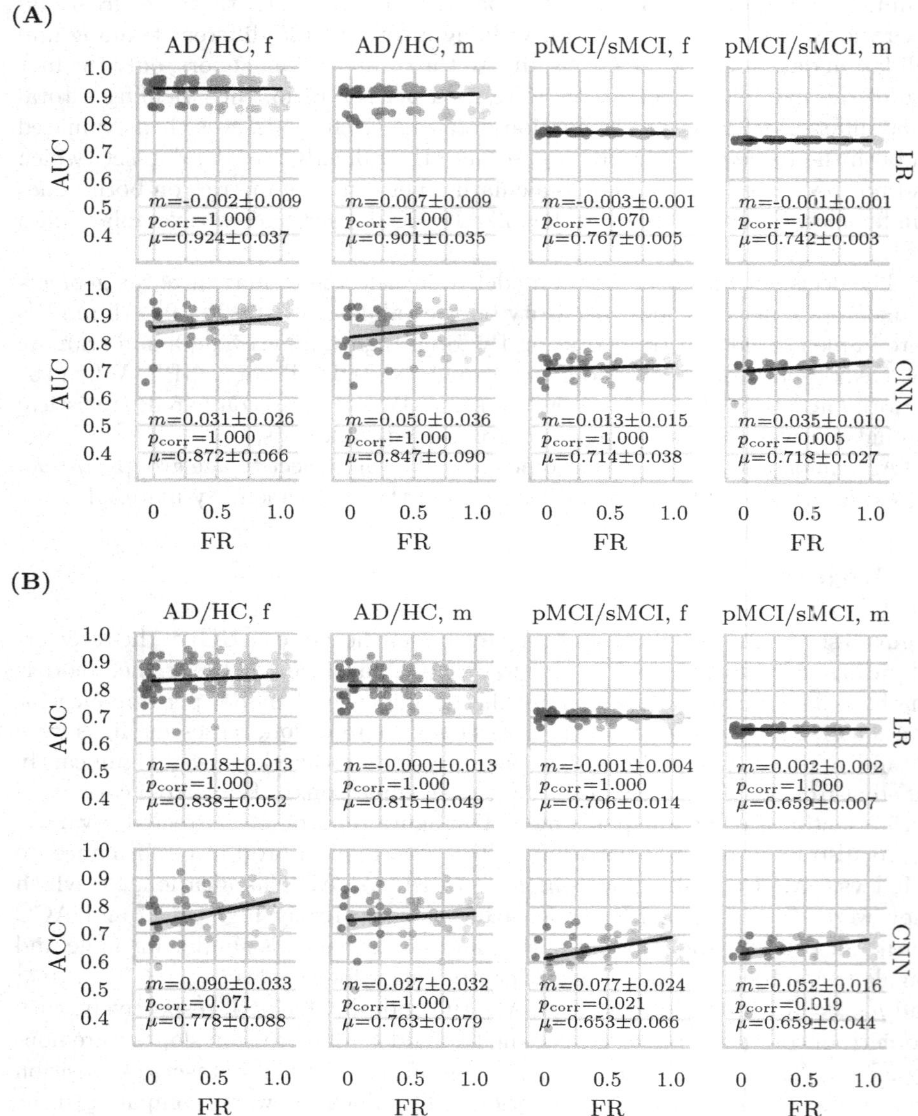

Fig. 3. Distribution of (A) the area under the curve (AUC) of the receiver-operating characteristic and (B) the accuracy achieved by the trained models (first row in both panels: LR, second row: CNN) on the different test sets. FR: ratio of female subjects in the training and validation set, m: slope of the regression line ($\pm$ standard deviation), Bonferroni corrected p-value null hypothesis: $m = 0$, μ: average AUC / ACC ($\pm$ standard deviation) across all sex ratios. Random jitter is added to all x coordinates to prevent excessive overlap – all points are sampled at FR $\in \{0, 0.25, 0.5, 0.75, 1.0\}$, as indicated by the five different colors. Note the truncation of all y-axes.

4 Discussion and Conclusion

In the present study, motivated by recent work demonstrating performance disparities in medical image classifiers between patient populations [15,24], we have analyzed the robustness of two state-of-the-art brain MRI classifiers and their associated feature representations to multiple types of distribution shift. Logistic regression using manually extracted volumetric features was compared to a standard 3D CNN classifier using the full 3D MRI volumes as inputs. Both models performed within the range reported for state-of-the-art CNN-based AD classification in the review of Wen et al. [27] for studies without suspected data leakage. The (valid) studies that report higher performance either use significantly more recordings or multiple modalities, and they only report the performance of a single training run. (Some of our runs performed significantly better than the mean accuracy reported above, see Fig. 3.) The LR model, using manually selected volumetric features, performed significantly better on average than the 3D CNN using the full MRI images. While this might change for larger dataset sizes, our sample size is comparable to typical sample sizes used in this domain.

We analyzed the effect of different ratios of male and female examples in the training dataset on the performance of the trained classifier for male and female test subjects. In our study, only the CNN's performance showed a dependence on training dataset composition, while the LR model was robust to this variation. This is not surprising, given that the LR model is based on a given representation (the extracted brain volumes and subject age), whereas the CNN performs representation learning and is, thus, prone to tailor the learned representation specifically to the training dataset. An important avenue for future research concerns the question of whether environment-invariant feature representations can be inferred automatically; an end towards which various methods have been proposed [3,20,30]. This would potentially allow for mitigating the effect of training dataset composition on male and female test subject performance. Generalization to different populations remains an unsolved problem in brain MRI AD analysis [27], with important consequences for the general robustness and fairness of the resulting systems. Finally, as a general trend, we observe that the influence of training dataset composition on accuracy seems to be larger than the influence on AUC, as indicated by larger slopes in Fig. 3. This appears plausible, given that the accuracy is influenced by one additional, dataset-dependent processing step compared to AUC: threshold selection.

Interestingly, we find that CNN performance (as measured by AUC and ACC) is improved in both men and women by including more women in the training dataset. This differs from the results of our age splitting experiment, where, while most relationships are not statistically significant, we observe performance on younger subjects to improve the more younger subjects are in the training dataset, whereas performance on older subjects tends to decline. These contrasting results underline the complexity of the relationship between dataset composition and subgroup performance. We hypothesize that the beneficial influence of female subjects in the training dataset on model performance on male test subjects may be explained by sex differences in pathological severity between AD study subjects. Several epidemiological studies have shown that neurodegeneration and clinical symptoms occur more rapidly for females once a diagnosis is suspected [22]. Moreover, women might present stronger cognitive decline than men with the same level of brain pathology [17], thus being diagnosed with AD at an earlier stage of disease progression than men. This might render female subjects more helpful as training examples to a CNN classifier compared to male subjects, and it would also explain why we generally observe better performance for females subjects. More research is certainly required to further investigate and explain these observations, however.

One potential limitation of our study concerns the preprocessing employed for both models. The employed segmentation steps are partially based on MRI atlases, which have been extracted from study databases. Thus, even in the cases in which the training dataset nominally contained no males or no females, this preprocessing step still incorporated some information about male and female subjects. The degree to which this influenced our analyses is challenging to quantify, but it represents a potentially confounding factor that might cause us to underestimate the effect of the training dataset composition. However, practical applications also typically employ atlas-based preprocessing, making our analysis closer to practical scenarios than if we had omitted these preprocessing steps.

Code & Data Availability. The full code for all implemented models, dataset splitting, and statistical analyses is available online in our GitHub repository: https://github.com/e-pet/adni-bias.

Acknowledgements. We thank Morten Rieger Hannemose for helpful comments on the manuscript and the statistical analysis. This research was supported by Danmarks Frie Forskningsfond (9131-00097B), the Novo Nordisk Foundation through the Center for Basic Machine Learning Research in Life Science (NNF20OC0062606) and the Pioneer Centre for AI, DNRF grant number P1. Data collection and sharing for this project was funded by the ADNI (National Institutes of Health Grant U01 AG024904). ADNI is funded by the National Institute on Aging, the National Institute of Biomedical Imaging and Bioengineering, and through generous contributions from private sector institutions. The Canadian Institutes of Health Research is providing funds to support ADNI clinical sites in Canada. Private sector contributions are facilitated by the Foundation for the National Institutes of Health (www.fnih.org). The grantee organization is the Northern California Institute for Research and Education, and the study is coordinated by the Alzheimer's Disease Cooperative Study at the University of California, San Diego. ADNI data are disseminated by the Laboratory for Neuro Imaging at the University of California, Los Angeles.

References

1. Abrol, A., et al.: Deep learning encodes robust discriminative neuroimaging representations to outperform standard machine learning. Nat. Commun. **12**(1), 1–7 (2021). https://doi.org/10.1038/s41467-020-20655-6

2. Adragna, R., Creager, E., Madras, D., Zemel, R.: Fairness and robustness in invariant learning: A case study in toxicity classification. In: NeurIPS Workshop on Algorithmic Fairness through the Lens of Causality and Interpretability (2020). https://arxiv.org/abs/2011.06485
3. Arjovsky, M., Bottou, L., Gulrajani, I., Lopez-Paz, D.: Invariant risk minimization. arXiv (2019). https://arxiv.org/abs/1907.02893
4. Ashburner, J.: SPM: a history. Neuroimage **62**(2), 791–800 (2012). https://doi.org/10.1016/j.neuroimage.2011.10.025
5. Azulay, A., Weiss, Y.: Why do deep convolutional networks generalize so poorly to small image transformations? J. Mach. Learn. Res. **20**(184), 1–25 (2019). http://jmlr.org/papers/v20/19-519.html
6. Banerjee, I., et al.: Reading race: AI recognises patient's racial identity in medical images. arXiv (2021). https://arxiv.org/abs/2107.10356
7. Cowling, T.E., Cromwell, D.A., Bellot, A., Sharples, L.D., van der Meulen, J.: Logistic regression and machine learning predicted patient mortality from large sets of diagnosis codes comparably. J. Clin. Epidemiol. **133**, 43–52 (2021). https://doi.org/10.1016/j.jclinepi.2020.12.018
8. D'Amour, A., et al.: Underspecification presents challenges for credibility in modern machine learning. CoRR (2020). https://arxiv.org/abs/2011.03395
9. Falcon, W.: The PyTorch Lightning team: PyTorch Lightning (version 1.5.9) (2019). https://www.pytorchlightning.ai
10. Fernández, A., García, S., Galar, M., Prati, R.C., Krawczyk, B., Herrera, F.: Performance measures. In: Learning from Imbalanced Data Sets, pp. 47–61. Springer, Cham (2018). https://doi.org/10.1007/978-3-319-98074-4_3
11. Fischl, B.: Freesurfer. Neuroimage **62**(2), 774–781 (2012). https://doi.org/10.1016/j.neuroimage.2012.01.021
12. Geirhos, R., et al.: Shortcut learning in deep neural networks. Nat. Mach. Intell. **2**(11), 665–673 (2020). https://doi.org/10.1038/s42256-020-00257-z
13. Jack, C.R., et al.: The Alzheimer's disease neuroimaging initiative (ADNI): MRI methods. J. Magn. Resonan. Imaging **27**(4), 685–691 (2008). https://doi.org/10.1002/jmri.21049
14. Jacobucci, R., Littlefield, A.K., Millner, A.J., Kleiman, E.M., Steinley, D.: Evidence of inflated prediction performance: a commentary on machine learning and suicide research. Clin. Psychol. Sci. **9**(1), 129–134 (2021). https://doi.org/10.1177/2167702620954216
15. Larrazabal, A.J., Nieto, N., Peterson, V., Milone, D.H., Ferrante, E.: Gender imbalance in medical imaging datasets produces biased classifiers for computer-aided diagnosis. Proc. Natl. Acad. Sci. **117**(23), 12592–12594 (2020). https://doi.org/10.1073/pnas.1919012117
16. Malone, I.B., et al.: Accurate automatic estimation of total intracranial volume: a nuisance variable with less nuisance. NeuroImage **104**, 366–372 (2015). https://doi.org/10.1016/j.neuroimage.2014.09.034
17. Mielke, M., Vemuri, P., Rocca, W.: Clinical epidemiology of Alzheimer's disease: assessing sex and gender differences. Clin. Epidemiol. **6**, 37 (2014). https://doi.org/10.2147/clep.s37929
18. Nusinovici, S., et al.: Logistic regression was as good as machine learning for predicting major chronic diseases. J. Clin. Epidemiol. **122**, 56–69 (2020). https://doi.org/10.1016/j.jclinepi.2020.03.002
19. Obermeyer, Z., Powers, B., Vogeli, C., Mullainathan, S.: Dissecting racial bias in an algorithm used to manage the health of populations. Science **366**(6464), 447–453 (2019). https://doi.org/10.1126/science.aax2342

20. Pawlowski, N., Castro, D.C., Glocker, B.: Deep structural causal models for tractable counterfactual inference. In: Advances in Neural Information Processing Systems, vol. 33, pp. 857–869. Curran Associates, Inc. (2020), https://proceedings.neurips.cc/paper/2020/file/0987b8b338d6c90bbedd8631bc499221-Paper.pdf
21. Pérez-García, F., Sparks, R., Ourselin, S.: TorchIO: a python library for efficient loading, preprocessing, augmentation and patch-based sampling of medical images in deep learning. Comput. Methods Programs Biomed. **208**, 106236 (2021). https://doi.org/10.1016/j.cmpb.2021.106236
22. Podcasy, J.L., Epperson, C.N.: Considering sex and gender in Alzheimer disease and other dementias. Dialogues Clin. Neurosc. **18**(4), 437–446 (2016). https://doi.org/10.31887/dcns.2016.18.4/cepperson
23. Quiñonero-Candela, J., Sugiyama, M., Lawrence, N.D., Schwaighofer, A.: Dataset Shift in Machine Learning. MIT Press, Cambridge (2009)
24. Seyyed-Kalantari, L., Zhang, H., McDermott, M.B.A., Chen, I.Y., Ghassemi, M.: Underdiagnosis bias of artificial intelligence algorithms applied to chest radiographs in under-served patient populations. Nat. Med. **27**(12), 2176–2182 (2021). https://doi.org/10.1038/s41591-021-01595-0
25. Tinauer, C., et al.: Interpretable brain disease classification and relevance-guided deep learning. medRxiv (2021). https://doi.org/10.1101/2021.09.09.21263013
26. Varoquaux, G., et al.: Assessing and tuning brain decoders: cross-validation, caveats, and guidelines. NeuroImage **145**, 166–179 (2017). https://doi.org/10.1016/j.neuroimage.2016.10.038
27. Wen, J., et al.: Convolutional neural networks for classification of Alzheimer's disease: overview and reproducible evaluation. Med. Image Anal. **63**, 101694 (2020). https://doi.org/10.1016/j.media.2020.101694
28. Wynants, L., et al.: Prediction models for diagnosis and prognosis of COVID-19: systematic review and critical appraisal. BMJ **369**, m1328 (2020). https://doi.org/10.1136/bmj.m1328
29. Yi, P.H., et al.: Radiology "forensics": determination of age and sex from chest radiographs using deep learning. Emerg. Radiol. **28**(5), 949–954 (2021). https://doi.org/10.1007/s10140-021-01953-y
30. Zhao, Q., Adeli, E., Pohl, K.M.: Training confounder-free deep learning models for medical applications. Nat. Commun. **11**(1), 1–9 (2020). https://doi.org/10.1038/s41467-020-19784-9

Extended Electrophysiological Source Imaging with Spatial Graph Filters

Feng Liu[1(✉)], Guihong Wan[2], Yevgeniy R. Semenov[2], and Patrick L. Purdon[2]

[1] School of Systems and Enterprises, Stevens Institute of Technology, Hoboken, NJ 07030, USA
fliu22@stevens.edu

[2] Massachusetts General Hospital, Harvard Medical School, Boston, MA 02114, USA

Abstract. Electrophysiological Source Imaging (ESI) refers to the process of localizing the brain source activation patterns given measured Electroencephalography (EEG) or Magnetoencephalography (MEG) signal from the scalp. Recent studies have focused on designing sophisticated neurophysiologically plausible regularizations or efficient estimation frameworks to solve the ESI problem, with the underlying assumption that brain source activation has some specific structures. Estimation of both source location and its extents is important in clinical applications. However, estimating the high dimensional extended location is challenging due to the highly coherent columns in the leadfield matrix, resulting in a reconstructed spiky spurious sources. In this work, we describe an efficient and accurate framework by exploiting the graph structure defined in the 3D mesh of the brain. Specifically, we decompose the graph signal representation in the source space into low-, medium-, and high-frequency subspaces, and project the source signal into the graph low-frequency subspace. We further introduce a low-rank representation with temporal graph regularization in the projected space to build the ESI framework, which can be efficiently solved. Experiments with simulated data and real world EEG data demonstrated the superiority of the proposed paradigm for estimating brain source extents.

Keywords: EEG/MEG source imaging · Spatial graph filter · Graph signal processing · Low-rank representation

1 Introduction

Complex firing neurons and interactions between neural circuits at different brain regions serve as the underpinning for our brain functionality. The brain physiological and cognitive behaviors generate electromagnetic and metabolic signals

F. Liu and G. Wan—Contributed equally to this work.

Supplementary Information The online version contains supplementary material available at https://doi.org/10.1007/978-3-031-16431-6_10.

L. Wang et al. (Eds.): MICCAI 2022, LNCS 13431, pp. 99–109, 2022.
https://doi.org/10.1007/978-3-031-16431-6_10

that can be measured with different brain imaging modalities, which can be further categorized into non-invasive modalities, such as Electroencephalogram (EEG) and Magnetoencephalogram (MEG) [23], functional magnetic resonance imaging (fMRI), positron emission tomography (PET), and single-photon emission computed tomography (SPECT) [11], and invasive measurement modalities, such as stereoelectroencephalography (sEEG) and electrocorticography (ECoG). The EEG/MEG measurement is a direct measurement of electrical firing patterns between neurons, while fMRI, as another important brain imaging modality, measures the blood-oxygen-level-dependent signal, which is a secondary metabolic signal. EEG/MEG has a unique advantage of high temporal resolution up to 1 ms compared to 1 s for fMRI. However, given that the dimension of source signal to be estimated is much higher than the number of EEG/MEG sensors, an ill-conditioned inverse problem must be solved to localize the activated brain source extents. This problem is referred to as the EEG/MEG source localization or EEG/MEG source imaging (ESI) problem [15,17,19,21].

Estimating the correct location as well as the *source extents* is important for detecting epilepsy zones [4]. Source extents refers to a focally extended area of activation, which presents the seizure onset focal region [25]. The spatial structure of source signal is commonly used to improve the accuracy of ESI, such as using total variation (TV) [3]. Defining TV regularization on the irregular 3D mesh can help algorithms render an extended source activation pattern. However, TV regularization may not always work well (as illustrated in Sect. 2.2), and defining TV regularization will incur a high dimensional transform matrix $V \in \mathbb{R}^{P \times N}$ (N is the number of sources and $P = \sum_{i=1}^{N} d_i$, with d_i representing the number of sources adjacent to source i in the 3D mesh) as each row of V characterizes the connectivity relationship between any of two sources defined in the 1st-order spatial gradient. The high dimensional transform matrix will significantly increase the computation complexity in the downstream optimization.

To address the aforementioned problems, instead of estimating the source signal in the original source space, we proposed to conduct ESI in a projected subspace spanned by low-frequency graph basis to (1) improve the accuracy of source extent estimation by eliminating the spatial high frequency spurious activation pattern; (2) reduce the number of latent variables for efficient estimation. We introduced a low-rank representation framework to estimate the parameters in the projected subspace for ESI. In summary, the main contributions are as follows: (I) For the first time, we delineated reasons that using total variation may not provide desirable solutions. (II) We proposed to approximate the source extent activation in a subspace composed of low band-pass graph filters instead of using total variation regularization. (III) To solve the ESI problem in low-frequency spatial graph subspace, we proposed a low-rank representation framework with graph regularization, and described an efficient algorithm to solve the optimization problem.

2 Problem and Motivation

In this section, we first introduce the ESI inverse problem, and how TV regularization is defined. Then, by providing an example, we show the possible reasons that the traditional hand crafted total variation is not working.

2.1 EEG/MEG Source Imaging Problem

EEG data are mostly generated by pyramidal cells in the gray matter with an orientation perpendicular to the cortex. The ESI forward model can be expressed as $Y = KS + E$, where $Y \in \mathbb{R}^{C \times T}$ is the EEG/MEG measurements, C is the number EEG/MEG channels, T is the number of time points, $K \in \mathbb{R}^{C \times N}$ is the *leadfield* matrix which characterizes the mapping from brain source space to EEG/MEG channel space, $S \in \mathbb{R}^{N \times T}$ represents the electrical potentials in N source locations for all the T time points, and E is the uncertainty/noise. The ESI inverse problem is to estimate S given the EEG/MEG measurements and the leadfield matrix. Since channel number C is much smaller than the number of sources N, estimating S becomes ill-posed and has infinitely number of solutions. In order to find a unique solution, different regularizations were introduced by using prior assumptions of the source solution. More specifically, S can be obtained by solving the following minimization problem:

$$\underset{S}{\operatorname{argmin}} \|Y - KS\|_F^2 + \lambda R(S), \tag{1}$$

where $\|\cdot\|_F$ is the Frobenius norm. The first term of Eq. (1) is called *data fitting* which tries to explain the observed EEG data, and the second term is called *regularization* term which is imposed to find a unique solution of Eq. (1) by using sparsity or other neurophysiology inspired regularizations. If $R(S)$ equals ℓ_2 norm, the problem is called minimum norm estimate (MNE) [9]; if $R(S)$ is defined using ℓ_1 norm, the problem becomes minimum current estimate (MCE) [26].

As the cortex is discretized with 3D meshes, simply employing ℓ_1 norm on S will result in an estimated source distributed across the cortex instead of an extended continuous area of activation. In order to encourage source extents estimation, Ding proposed to use a sparse constraint in the transformed domain by introducing TV [2]. Other studies (e.g., [16,24,27]) used the same TV definition. The TV was defined to be the ℓ_1 norm of the first order spatial gradient using a linear transform matrix $V \in \mathbb{R}^{P \times N}$, where N is the number of voxels/sources, P equals the sum of the degrees of all source nodes [2]. The framework with sparsity and total variation constraints are given as follows:

$$S = \underset{S}{\operatorname{argmin}} \|Y - KS\|_F^2 + \alpha\|S\|_{1,1} + \beta\|VS\|_{1,1} \tag{2}$$

where $\|\cdot\|_{1,1}$ represents ℓ_1 norm on both row and column of a matrix. For example, $\|A\|_{1,1}$ is defined as $\sum_j \sum_i |A_{ij}|$. The first term is the data fitting term, the second term is the sparsity term, and the third term is the total variation term. Ideally, the TV regularization promotes source extents estimation.

2.2 Why TV Sometimes is Not Working?

A simple yet realistic example: we generated the ground truth signal s_1 and EEG data y using the forward model defined in Eq. (1) plus Gaussian noise. The

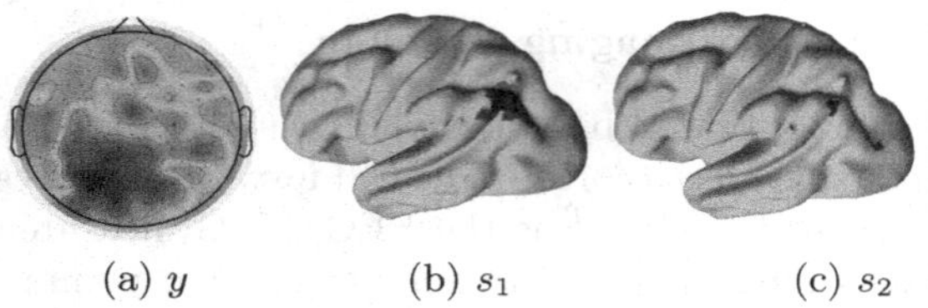

(a) y (b) s_1 (c) s_2

Fig. 1. Example that TV may fail to work. (a) The EEG topoplot which is represented as y; (b) the ground truth signal with total variance property; (c) the reconstructed source signal. For solution s_1 in (b), data fitting $\|y-Ks_1\|_2^2$: **0.620**, sparsity cost $\|s_1\|_1$: **8.926**, TV cost$\|Vs_1\|_1$: **22.602**; for solution s_2 in (c): data fitting $\|y-Ks_2\|_2^2$: **0.616** , sparsity cost $\|s_2\|_1$: **3.336**, TV cost $\|Vs_2\|_1$: **15.436**. Although solution s_2 in (c) is not preferred over s_1 in (b), its cost in all three terms of Eq.(2) is lower than s_1.

ground truth activation has a pattern with "extended" area of activation, which is preferred by using TV regularization. The leadfield matrix K is obtained from a realistic "New York Head" model [14]. The example is illustrated in Fig. 1. Instead of the ground truth signal s_1, using TV regularization yielded a solution s_2 that has a lower cost in *all three terms* of Eq. (2), which means that for any hyper-parameter settings of α and β, the solution s_2 can always provide a lower objective value in Eq. (2). The solution s_2 in Fig. 1 is undesirable since it has distributed unrealistic "spiky" activation across the cortex. The possible reasons can be summarized as follows: (a) high similarity between columns in the leadfield matrix; (b) unequal contribution to the magnitude of EEG. In order to address the problem, an important insight we want to leverage is that: *the "spiky" spurious activation lies in the high frequency space of the eigenvectors from the graph Laplacian matrix.* The above assumption motivates us to use low band-pass filters to reduce the level of spurious activation.

3 The Proposed Framework

3.1 Preliminary

Let $\mathcal{G} = \{\mathcal{V}, \mathcal{E}, W\}$ represents the weighted graph generated from 3D mesh, where $\mathcal{V} = \{v_1, \ldots, v_N\}$ is the set for all N nodes, $\mathcal{E}$: $\mathcal{V} \times \mathcal{V}$ denotes all edges, W is the weighted adjacency matrix, and W_{ij} is the $\{i, j\}$th element of W, representing the connection strength between node i and node j. A graph signal $s \in \mathbb{R}^N$ is defined on the graph nodes $\{v_1, \ldots, v_N\}$ with coefficient of $s_i \in \mathbb{R}$, for $i \in \{1, \ldots, N\}$. The Laplacian matrix L is defined as $L = D - W$, where D is a diagonal matrix with $D_{ii} = d_i = \sum_{j \neq i} W_{ij}$. The Laplacian matrix L is real symmetric, therefore its set of eigenvalues $0 = \lambda_1 \leq \lambda_2 \leq \cdots \leq \lambda_N$, and the associated eigenvectors $U = [u_1, u_2, \ldots, u_N]$ are corresponding to the graph's Fourier modes [13,20]. The graph Fourier transform $\hat{s}$ of a signal s is given as $\hat{s} = U^T s$. The inverse graph Fourier transform is defined as $s = U\hat{s}$.

By analogy to the continuous Fourier transfer, u_i is the basis associated with a frequency defined on the graph. Although previous literature mentioned that each u_i is associated with a "frequency" on the graph, its frequency is low when i is small (λ_i is small), and frequency is high when i is large (λ_i is large), there is no formal definition for a quantitative measurement of frequency on the graph in analogy to the time-series domain.

Definition 1: Normalized Graph Frequency (NGF) $f_{\mathcal{G}}$ is a function of u_i which is defined as:

$$f_{\mathcal{G}}(u_i) = \frac{z_{sf}(u_i)}{\mathrm{Tr}(L)},$$

where $\mathrm{Tr}(L)$ is the trace of L; z_{sf} represents the total number of sign flips of u_i between any two connected nodes on $\mathcal{G}$:

$$z_{sf}(u_i) = \sum_{m=1}^{N} \sum_{n \in \mathcal{N}(m)} \mathbb{I}(u_i(m)u_i(n) < 0)/2, \tag{3}$$

where $\mathcal{N}(m)$ denotes all the neighbors of node m, and $\mathbb{I}(\cdot)$ is the indicator function to check if the values of u_i at node m and n have a sign flip. The sign flip makes an analogy to counting the number of zero crossings of a basis signal within a given window for a time series data. We constructed the Laplacian matrix within first order neighbors, second order neighbors, and third order neighbors, respectively. The associated NGF spectrum is illustrated in Fig. S-1.1 in the supplementary material. Observe that the NGF value for the eigenvector increases as the eigenvalue increases. According to NGF values, we further separated all eigenvectors in U into low-, medium-, and high-frequency filters: $U = [U_L, U_M, U_H]$, with each being able to be served as low-pass filter, band-pass filter, and high-pass filter.

3.2 Objective Function

A set of graph filters with lower frequencies in U are used to reduce the dimension of solution space. Specifically, we project graph signals S onto a subspace spanned by $\hat{U}$ with low NGF values. The data fitting term can be written as $\|Y - K\hat{U}\hat{U}^T S\|_F^2$. Let $\hat{S} = \hat{U}^T S$ and $\hat{K} = K\hat{U}$, then the data fitting term can be rewritten as $\|Y - \hat{K}\hat{S}\|_F^2$. Solving S is equivalent to finding parameters in the subspace $\hat{U}$. Our goal is to estimate $\hat{S}$, and all regularizations will be added to $\hat{S}$ instead of S.

We argue that during a short period of task-related event-related potentials, the corresponding activated sources are sparse and remain activated across the time course, which makes the S to be a matrix with sparsity in the spatial direction [6]. Here, $\hat{S}$ is not necessarily sparse due to projection into the graph eigenvector space. Given S can usually be assumed as row sparse [6], it has a low rank property as well. Since $\mathrm{rank}(\hat{S}) = \mathrm{rank}(\hat{U}S) \leq \min(\mathrm{rank}(S), \mathrm{rank}(\hat{U}))$, we assume that the $\hat{S}$ has a low rank property. To address the ill-posedness of the ESI inverse problem, we pose a low-rank constraint and still use a sparse regularization on $\hat{S}$. For simplicity of notations, we use K to represent $\hat{K}$ and S to represent $\hat{S}$ in the remaining sections.

Low Rank Representation Framework: The low-rank representation (LRR) framework for the EEG inverse problem is introduced as follows:

$$\min_{S,E} \|S\|_* + \beta\|S\|_{1,1} + \lambda\|E\|_{1,1} \quad \text{s.t.} \quad Y = KS + E. \tag{4}$$

where β and λ are positive parameters to balance the rank function, sparsity cost, and reconstruction error. As ℓ_1 on the error term is more robust to outliers, we use ℓ_1 in the row-rank model [18]. Due to the discrete nature of the rank function, it is a common practice to use a surrogate nuclear norm $\|\cdot\|_*$ instead.

LRR Model with Graph Regularization: In order to promote temporal smoothness [16], we use the graph regularization in the temporal direction. It is defined to penalize the difference of two neighboring source signals: $R_t(S) = \sum_{i,j=1}^{N} \|s_i - s_j\|_2^2 M_{ij}$, where s_i is the i-th column of the matrix S, and a binary matrix M is designed as follows: $M_{ij} = 1$, if $s_i \in \mathcal{N}_k(s_j)$ or $s_j \in \mathcal{N}_k(s_i)$ and $M_{ij} = 0$, otherwise. Here $\mathcal{N}_k(s_i)$ is the set containing k temporally closest points to s_i. This formulation intends to force neighboring source signal to similar patterns. By defining D as a diagonal matrix whose entries are row sums of the symmetric matrix M and denoting $G = D - M$, $R_t(S)$ can be rewritten as: $R_t(S) = \sum_{i,j=1}^{N} ({s_i}^T s_i + {s_j}^T s_j - 2{s_i}^T s_j)M_{ij} = 2\text{Tr}(SGS^T)$. The final low-rank representation with temporal graph structures is described as:

$$\min_{S,E} \|S\|_* + \lambda\|E\|_{1,1} + \beta\|S\|_{1,1} + \alpha\text{Tr}(SGS^T) \quad \text{s.t.} \quad Y = KS + E, \tag{5}$$

where $\lambda, \beta, \alpha > 0$ are tuning parameters to balance the trade-off of different terms. Our proposed model is able to enforce row-sparse via low-rank prior, and temporal smoothness via temporal graph regularization.

The detailed derivation of Problem (5) can be found in the supplementary material, and the final algorithm based on alternating direction method of multipliers (ADMM) [1] is summarized in Algorithm 1 in the supplementary material.

4 Numerical Experiments

In this section, we conducted extensive numerical experiments to validate the effectiveness of the proposed method under different Signal Noise Ratios (SNRs) using both synthetic and real EEG data. The benchmark algorithms include MNE [9], MCE [26], sLORETA [22], $\ell_{2,1}$ [6] and TV-ADMM [16], where MCE and $\ell_{2,1}$ both employ sparse regularization in the spatial direction, and sLORETA and MNE are ℓ_2-based methods.

4.1 Simulation Experiments

We first performed experiments on synthetic EEG data with known ground truth (GT) activation patterns. We quantitatively evaluated the performance of each algorithm based on two metrics: *Localization error* (LE), which measures the geodesic distance between two extended source locations using the Dijkstra shortest path algorithm; *Area under curve* (AUC), which is particularly useful to characterize the overlap of two extended source locations. Better performance for localization is expected if LE is close to 0 and AUC is close to 1.

Forward Model: We used a real head model to calculate the leadfield matrix. The head model was calculated based on T1-MRI images from a 26-year old male subject scanned at the Massachusetts General Hospital, Boston, MA [12]. Brain tissue segmentation and source surface reconstruction were conducted using FreeSurfer [5]. We used a 128-channel BioSemi EEG cap layout, coregistered EEG channels with the head model using Brainstorm, and further validated on MNE-Python toolbox [7]. The source space contains 1026 sources in each hemisphere, resulting in a leadfield matrix K of size 128×2052. We used 600 eigenvectors of the graph Laplacian matrix L with lowest NGF values as the graph frequency filters.

Experimental Settings: To generate EEG data, we selected a random location in the source space, and activated the neighbors connecting to the selected source location on the 3D brain mesh. We used 3 neighborhood levels (1-, 2-, and 3-level neighborhood) to represent extended areas of different sizes (illustrated in Fig. S-1.2 in the supplementary material), and activated the whole "patch" of sources with source space EEG time series data generated using 5th-order autoregressive (AR) model [10]. Then we used the forward model to generate scalp EEG data under different SNR levels (30 dB, 20 dB and 10 dB). We set the length of EEG data in each experiment to be 1 s 100 Hz sampling rate. Grid search for parameter tuning on additional generated dataset was performed.

The performance comparison between the proposed method and benchmark algorithms on LE and AUC is summarized in Table 1. The boxplots of LE and AUC are presented in Fig. 3. As can be seen from Table 1, the proposed algorithm performs well compared to other benchmark algorithms under different settings of SNR and levels of activated neighbors. Given that we used low band-pass graph filters, the high frequency spurious noise is much less observed in our reconstructed signal, as illustrated in Fig. 2.

Table 1. Performance comparison.

	Level of Neighbors (LNs)	Source with one LN		Source with two LNs		Source with three LNs	
SNR	Method	LE	AUC	LE	AUC	LE	AUC
30dB	MNE	18.449 ± 9.202	0.947 ± 0.038	18.981 ± 8.011	0.929 ± 0.043	20.200 ± 8.478	0.917 ± 0.034
	MCE	18.454 ± 7.101	0.652 ± 0.086	22.616 ± 3.448	0.598 ± 0.039	26.448 ± 4.846	0.567 ± 0.025
	sLORETA	12.869 ± 6.088	0.938 ± 0.050	15.759 ± 6.421	0.920 ± 0.036	18.829 ± 5.762	0.900 ± 0.032
	$\ell_{2,1}$	10.674 ± 8.064	0.656 ± 0.104	16.262 ± 10.574	0.615 ± 0.083	19.661 ± 8.350	0.597 ± 0.079
	TV-ADMM	25.630 ± 8.816	0.939 ± 0.022	18.919 ± 4.646	0.966 ± 0.020	17.596 ± 5.933	**0.963 ± 0.021**
	Proposed	**10.141 ± 6.284**	**0.984 ± 0.013**	**7.793 ± 3.716**	**0.983 ± 0.013**	**10.667 ± 7.801**	**0.963 ± 0.035**
20dB	MNE	28.900 ± 10.125	0.907 ± 0.045	27.923 ± 10.361	0.888 ± 0.061	33.883 ± 7.910	0.860 ± 0.037
	MCE	36.185 ± 6.949	0.606 ± 0.067	36.545 ± 7.912	0.570 ± 0.035	43.895 ± 5.487	0.546 ± 0.022
	sLORETA	19.193 ± 5.247	0.911 ± 0.046	20.747 ± 6.317	0.893 ± 0.047	25.266 ± 5.124	0.863 ± 0.029
	$\ell_{2,1}$	**11.991 ± 7.670**	0.648 ± 0.088	14.289 ± 6.193	0.602 ± 0.058	19.066 ± 5.781	0.572 ± 0.046
	TV-ADMM	26.701 ± 8.555	0.935 ± 0.028	18.816 ± 7.534	0.968 ± 0.019	19.794 ± 6.615	**0.955 ± 0.023**
	Proposed	13.486 ± 7.470	**0.976 ± 0.018**	**12.057 ± 6.771**	**0.969 ± 0.027**	**15.086 ± 8.232**	0.945 ± 0.043
10dB	MNE	51.002 ± 14.358	0.835 ± 0.067	56.686 ± 10.784	0.801 ± 0.061	62.229 ± 9.318	0.772 ± 0.047
	MCE	63.853 ± 7.924	0.558 ± 0.050	67.478 ± 4.965	0.539 ± 0.027	71.246 ± 4.959	0.529 ± 0.021
	sLORETA	32.731 ± 7.333	0.854 ± 0.047	37.441 ± 6.587	0.817 ± 0.036	41.013 ± 6.176	0.784 ± 0.030
	$\ell_{2,1}$	**17.166 ± 9.277**	0.625 ± 0.072	**16.509 ± 5.698**	0.601 ± 0.051	24.517 ± 4.334	0.559 ± 0.037
	TV-ADMM	31.358 ± 12.804	0.932 ± 0.029	39.607 ± 11.501	0.934 ± 0.024	45.006 ± 11.522	**0.909 ± 0.030**
	Proposed	19.870 ± 11.218	**0.957 ± 0.031**	22.261 ± 9.494	**0.935 ± 0.036**	**23.504 ± 12.090**	**0.909 ± 0.063**

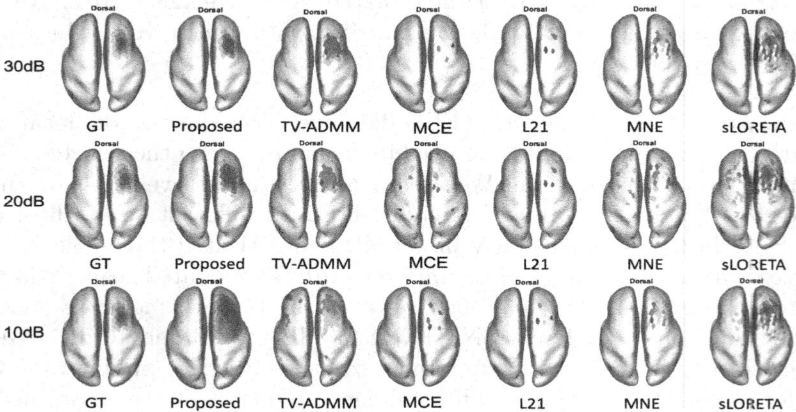

Fig. 2. Source reconstruction comparison under different SNR levels. The activated neighborhood level is 3.

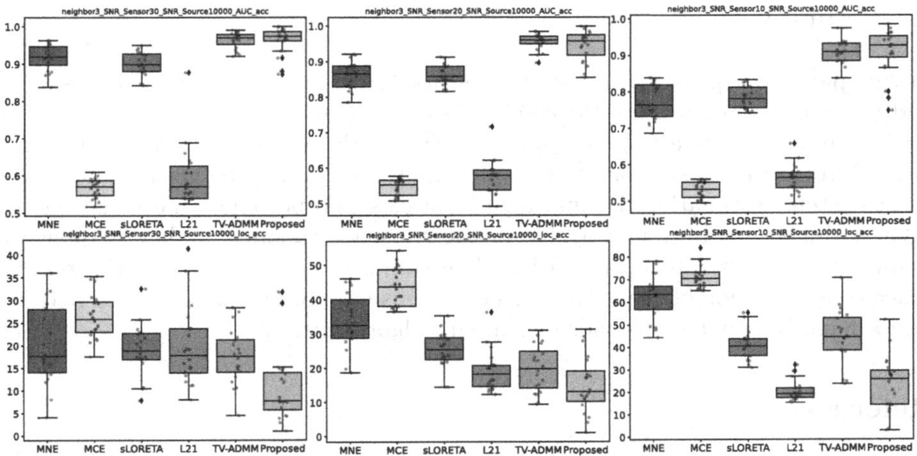

Fig. 3. Performance of all the algorithms based on AUC (upper) and LE (bottom) under different level of noises (from left to right: SNR = 30, 20, 10 dB).

4.2 Real Data Experiments

We analyzed a real dataset that is publicly accessible through the MNE-Python package [7]. The EEG/MEG data were collected when the subject was given auditory and visual stimuli. Data description can be found in [8]. The leadfield matrix was constructed using MNE-Python. There are 7498 sources distributed over the cortical surfaces. In this experiment, we focused on source localization on the stimuli of left auditory stimuli (LAS). There are 66 epochs for LAS. The reconstructed source at $t = 0.08$ s by different algorithms are presented in Fig. 4. The results show that our algorithm can locate the activation pattern in the auditory cortex. The other competing methods show highly diffuse or spurious activations, similar to the results from the synthetic data. MNE, sLORETA gave a overdiffuse solution, while MCE and $\ell_{2,1}$ provides spurious source reconstructions not only located in the auditory cortex.

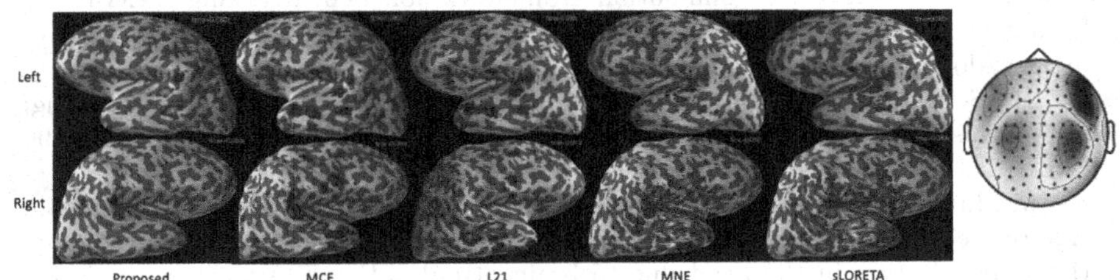

Fig. 4. Source activation patterns from MEG data at $t = 0.08$ s.

5 Conclusion

In this paper, we proposed a new framework based on graph signal processing and low rank representation for EEG/MEG extended source imaging. The proposed framework has a principle way of reducing the spurious source activation located in high frequency band of graph signal, thus improving the accuracy for source extents estimation. Estimating of source signal becomes an estimation of a lower dimensional latent variable in the subspace spanned by low frequency graph filters. A low rank optimization algorithm is used to reconstruct the graph space representation of the source signal. Extensive experiments showed a better performance compared to other benchmark algorithms.

References

1. Boyd, S., Vandenberghe, L.: Convex Optimization. Cambridge University Press (2004)
2. Ding, L.: Reconstructing cortical current density by exploring sparseness in the transform domain. Phys. Med. Biol. **54**(9), 2683–2697 (2009). https://doi.org/10.1088/0031-9155/54/9/006
3. Ding, L., He, B.: Sparse source imaging in electroencephalography with accurate field modeling. Hum. Brain Mapp. **29**(9), 1053–1067 (2008)
4. Ding, L., et al.: EEG source imaging: correlate source locations and extents with ECoG and surgical resections in epilepsy patients. J. Clin. Neurophysiol. **24**(2), 130–136 (2007)
5. Fischl, B.: Freesurfer. Neuroimage **62**(2), 774–781 (2012)
6. Gramfort, A., Kowalski, M., Hämäläinen, M.: Mixed-norm estimates for the M/EEG inverse problem using accelerated gradient methods. Phys. Med. Biol. **57**(7), 1937 (2012)
7. Gramfort, A., et al.: MNE software for processing MEG and EEG data. Neuroimage **86**, 446–460 (2014)
8. Gramfort, A., Strohmeier, D., Haueisen, J., Hämäläinen, M.S., Kowalski, M.: Time-frequency mixed-norm estimates: sparse M/EEG imaging with non-stationary source activations. Neuroimage **70**, 410–422 (2013)
9. Hämäläinen, M.S., Ilmoniemi, R.J.: Interpreting magnetic fields of the brain: minimum norm estimates. Med. Biol. Eng. Comput. **32**(1), 35–42 (1994)
10. Haufe, S., Ewald, A.: A simulation framework for benchmarking EEG-based brain connectivity estimation methodologies. Brain Topogr. **32**(4), 625–642 (2016). https://doi.org/10.1007/s10548-016-0498-y
11. He, B., Sohrabpour, A., Brown, E., Liu, Z.: Electrophysiological source imaging: a noninvasive window to brain dynamics. Annu. Rev. Biomed. Eng. **20**, 171–196 (2018)
12. He, M., Liu, F., Nummenmaa, A., Hämäläinen, M., Dickerson, B.C., Purdon, P.L.: Age-related EEG power reductions cannot be explained by changes of the conductivity distribution in the head due to brain atrophy. Front. Aging Neurosci. **13**, 632310 (2021)
13. Huang, W., Bolton, T.A., Medaglia, J.D., Bassett, D.S., Ribeiro, A., Van De Ville, D.: A graph signal processing perspective on functional brain imaging. Proc. IEEE **106**(5), 868–885 (2018)

14. Huang, Y., Parra, L.C., Haufe, S.: The New York head - a precise standardized volume conductor model for EEG source localization and tES targeting. Neuroimage **140**, 150–162 (2016)
15. Jiao, M., et al.: A graph fourier transform based bidirectional long short-term memory neural network for electrophysiological source imaging. Front. Neurosci. **16**, 867466 (2022)
16. Liu, F., Rosenberger, J., Lou, Y., Hosseini, R., Su, J., Wang, S.: Graph regularized EEG source imaging with in-class consistency and out-class discrimination. IEEE Trans. Big Data **3**(4), 378–391(2017)
17. Liu, F., Wang, L., Lou, Y., Li, R.C., Purdon, P.L.: Probabilistic structure learning for EEG/MEG source imaging with hierarchical graph priors. IEEE Trans. Med. Imaging **40**(1), 321–334 (2020)
18. Liu, G., Lin, Z., Yan, S., Sun, J., Yu, Y., Ma, Y.: Robust recovery of subspace structures by low-rank representation. IEEE Trans. Pattern Anal. Mach. Intell. **35**(1), 171–184 (2013)
19. Michel, C.M., Murray, M.M., Lantz, G., Gonzalez, S., Spinelli, L., de Peralta, R.G.: EEG source imaging. Clin. Neurophysiol. **115**(10), 2195–2222 (2004)
20. Ortega, A., Frossard, P., Kovačević, J., Moura, J.M., Vandergheynst, P.: Graph signal processing: overview, challenges, and applications. Proc. IEEE **106**(5), 808–828 (2018)
21. Ou, W., Hämäläinen, M.S., Golland, P.: A distributed spatio-temporal EEG/MEG inverse solver. Neuroimage **44**(3), 932–946 (2009). https://doi.org/10.1016/j.neuroimage.2008.05.063
22. Pascual-Marqui, R.D.: Standardized low-resolution brain electromagnetic tomography (sLORETA): technical details. Methods Find. Exp. Clin. Pharmacol. **24**(Suppl D), 5–12 (2002)
23. Phillips, C., Rugg, M.D., Friston, K.J.: Anatomically informed basis functions for EEG source localization: combining functional and anatomical constraints. Neuroimage **16**(3), 678–695 (2002)
24. Sohrabpour, A., Lu, Y., Worrell, G., He, B.: Imaging brain source extent from EEG/MEG by means of an iteratively reweighted edge sparsity minimization (IRES) strategy. Neuroimage **142**, 27–42 (2016)
25. Sohrabpour, A., He, B.: Exploring the extent of source imaging: recent advances in noninvasive electromagnetic brain imaging. Curr. Opin. Biomed. Eng. **18**, 100277 (2021)
26. Uutela, K., Hämäläinen, M., Somersalo, E.: Visualization of magnetoencephalographic data using minimum current estimates. Neuroimage **10**(2), 173–180 (1999)
27. Zhu, M., Zhang, W., et al.: Reconstructing spatially extended brain sources via enforcing multiple transform sparseness. Neuroimage **86**, 280–293 (2014)

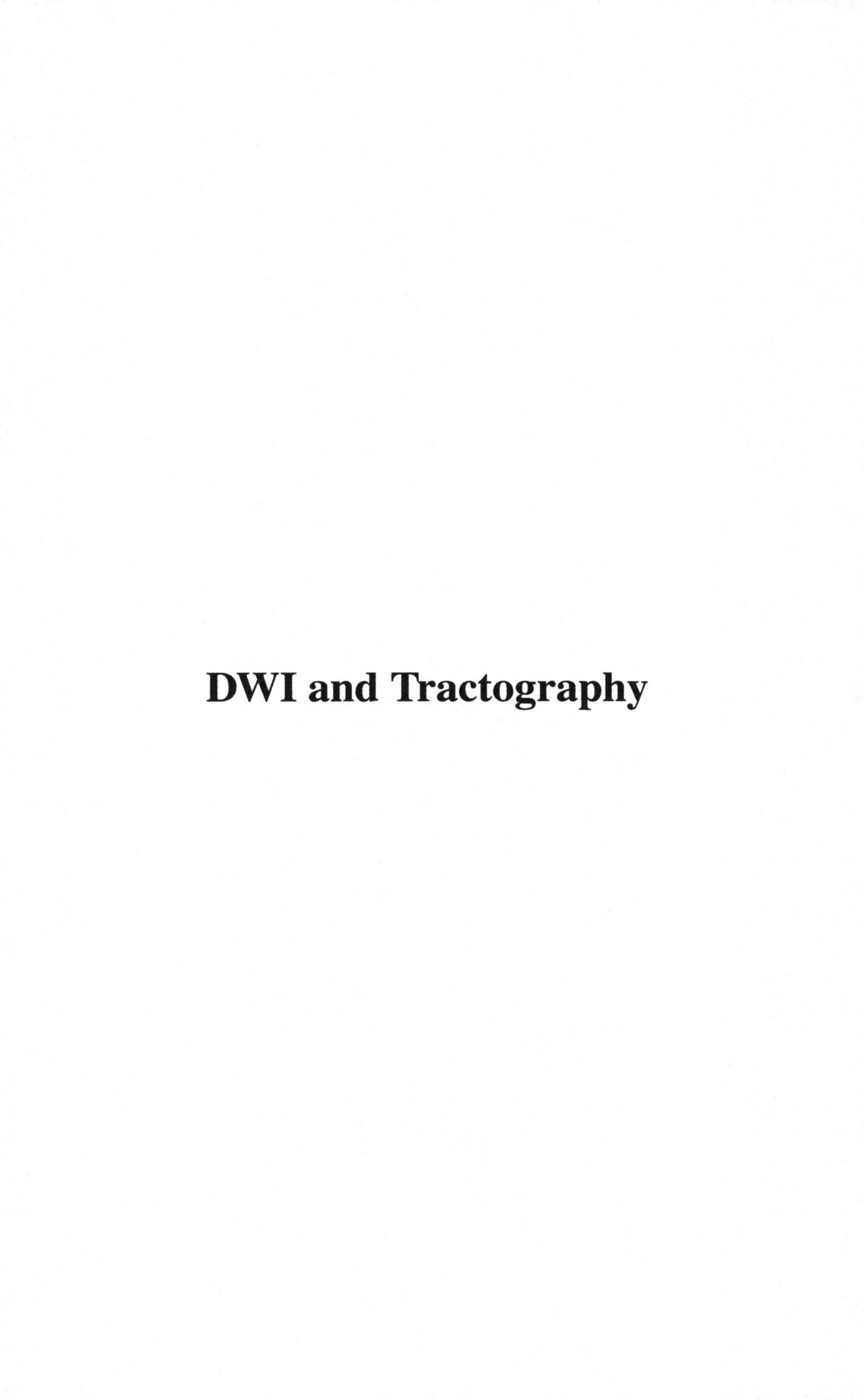

DWI and Tractography

Hybrid Graph Transformer for Tissue Microstructure Estimation with Undersampled Diffusion MRI Data

Geng Chen[1,2](✉), Haotian Jiang[3], Jiannan Liu[4], Jiquan Ma[4](✉), Hui Cui[5], Yong Xia[1], and Pew-Thian Yap[2]

[1] National Engineering Laboratory for Integrated Aero-Space-Ground-Ocean Big Data Application Technology, School of Computer Science and Engineering, Northwestern Polytechnical University, Xi'an, China
geng.chen@nwpu.edu.cn

[2] Department of Radiology and Biomedical Research Imaging Center (BRIC), University of North Carolina at Chapel Hill, Chapel Hill, NC, USA

[3] Department of Software Engineering, Heilongjiang University, Harbin, China

[4] Department of Computer Science and Technology, Heilongjiang University, Harbin, China
majiquan@hlju.edu.cn

[5] Department of Computer Science and Information Technology, La Trobe University, Melbourne, Australia

Abstract. Advanced contemporary diffusion models for tissue microstructure often require diffusion MRI (DMRI) data with sufficiently dense sampling in the diffusion wavevector space for reliable model fitting, which might not always be feasible in practice. A potential remedy to this problem is by using deep learning techniques to predict high-quality diffusion microstructural indices from sparsely sampled data. However, existing methods are either agnostic to the data geometry in the diffusion wavevector space (q-space) or limited to leveraging information from only local neighborhoods in the physical coordinate space (x-space). Here, we propose a hybrid graph transformer (HGT) to explicitly consider the q-space geometric structure with a graph neural network (GNN) and make full use of spatial information with a novel residual dense transformer (RDT). The RDT consists of multiple densely connected transformer layers and a residual connection to facilitate model training. Extensive experiments on the

G. Chen and H. Jiang—Contributed equally to this work. This work was supported in part by the Fundamental Research Funds for the Central Universities under Grant D5000220213, the Natural Science Foundation of Heilongjiang Province under Grant LH2021F046, the National Natural Science Foundation of China under Grant 62171377, and the Key Research and Development Program of Shaanxi Province under Grant 2022GY-084. P.-T. Yap was supported in part by the United States National Institutes of Health (NIH) through grant MH125479.

Supplementary Information The online version contains supplementary material available at https://doi.org/10.1007/978-3-031-16431-6_11.

L. Wang et al. (Eds.): MICCAI 2022, LNCS 13431, pp. 113–122, 2022.
https://doi.org/10.1007/978-3-031-16431-6_11

data from the Human Connectome Project (HCP) demonstrate that our method significantly improves the quality of microstructural estimations over existing state-of-the-art methods.

Keywords: Diffusion MRI · GNNs · Transformer · Microstructure imaging

1 Introduction

Diffusion magnetic resonance imaging (DMRI) provides a rich characterization of brain tissues by capturing the anisotropic motion of water molecules. However, microstructural models, such as diffusion kurtosis imaging (DKI) [11] and neurite orientation dispersion and density imaging (NODDI) [21], usually require DMRI data with sufficiently dense sampling in q-space. For example, 90 diffusion gradients are recommended for the NODDI model [21]. Increasing the number of diffusion gradients inevitably prolongs acquisition time and can therefore be impractical.

Deep learning techniques have been used to overcome this limitation [1,5, 9,10,12,13,18–20]. For instance, Golkov et al. [10] proposed a multiple layer perceptron (MLP) to learn the relationship between sparsely samples q-space data and high-quality microstructural index maps. Ye et al. [18] improved the accuracy of microstructural estimation with a deep ℓ_0 encoder. Graph neural networks (GNNs) have been introduced for microstructural estimation by graph-based q-space learning [5]. Gibbons et al. [9] simultaneously computed GFA and the NODDI parameters with CNNs by leveraging spatial information. DeepDTI [13] employs 10-layer 3D CNNs to predict high-fidelity diffusion tensor imaging (DTI) metrics from only one b_0 image and six diffusion-weighted images. Aliotta et al. [1] focused on a more challenging problem of performing DTI fitting with 3 diffusion-weighted images by using MLP and U-Net [8]. Despite being shown to be effective, existing methods suffer from two drawbacks. First, methods based on CNN/MLP are agnostic to the data structure in q-space, neglecting important information such as the angular relationships between data points. Second, methods based on GNNs are typically focused on voxel-wise estimation and neglect information across the x-space, particularly from distant locations, for improved performance.

To address these issues, we propose a hybrid graph transformer (HGT) for accurate tissue microstructure estimation with the joint x-q space information from undersampled DMRI data. HGT leverages joint x-q space learning with a hybrid architecture consisting of a GNN and a transformer [16]. Specifically, we first encode the geometric structure of q-space as a graph. We adopted a GNN called topology adaptive graph convolutional network (TAGCN) [7], which is highly efficient and requires less memory. The features generated by the GNN are reshaped into feature maps for further x-space learning with a residual dense transformer (RDT). Transformer, a relatively recent deep learning technique primarily based on a self-attention mechanism, is remarkably more effective than traditional CNNs in capturing long-distance dependencies. The RDT consists of a set of building blocks, each of which has a residual dense unit that is built with multiple densely connected transformer layers and a residual connection to facilitate training. To the best of our knowledge, this is the first work on leveraging transformer for tissue microstructure estimation. Evaluation with data from the Human Connectome Project (HCP)[1] [14] indicates that our method substantially improves the accuracy of microstructural estimation, particularly with data of fewer diffusion gradients.

2 Method

2.1 Overall Architecture

Our goal is to estimate tissue microstructure by jointly considering q-space information with a GNN and x-space information with a transformer (Fig. 1).

Let the input DMRI slice be $\mathbf{x} \in \mathbb{R}^{H\times W\times C}$, where H, W, and C represent height, width, and the number of gradient directions, respectively. We reshape $\mathbf{x}$ to $\hat{\mathbf{x}} \in \mathbb{R}^{HW\times C}$ by flattening two dimensions. Next, $\hat{\mathbf{x}}$ is fed to a TAGCN for q-space learning. After q-space learning, the features $\mathbf{x}_{\mathrm{q}} \in \mathbb{R}^{HW\times 4C}$ are reshaped to $\hat{\mathbf{x}}_{\mathrm{q}} \in \mathbb{R}^{H\times W\times 4C}$ for further x-space learning by the RDT, a novel transformer architecture with residual and dense connections. Finally, each microstructure index is obtained through a linear layer.

2.2 Learning in q-Space

Following [3,4,17], we encode the geometric structure of q-space data points as a graph $\mathcal{G}$, which is defined by a binary affinity matrix $\mathbf{A} = \{a_{i,j}\}$ with $a_{i,j}$ denoting the affinity weight between two q-space data points indexed by i and j. We set the corresponding affinity weight $a_{i,j} = 1$, if their angle $\theta_{i,j}$ is smaller than an angular threshold θ, and to zero otherwise. The diffusion signals of each voxel can be viewed as a function on the graph.

Our q-space learning module is based on TAGCN [7], which consists of a set of fixed-size learnable filters for efficient convolution on graphs. Let $\mathbf{x}_c^{(\ell)} \in \mathbb{R}^{N_\ell}$ be the c-th feature on all graph nodes at the ℓ-th hidden layer with N_ℓ denoting the number of graph nodes. The f-th feature $\mathbf{x}_f^{(\ell+1)}$ at the $(\ell+1)$-th layer is given by

$$\mathbf{x}_f^{(\ell+1)} = \sigma\left(\sum_{c=1}^{C_\ell} \mathbf{G}_{c,f}^{(\ell)} \mathbf{x}_c^{(\ell)} + b_f^{(\ell)} \mathbf{1}_{N_\ell}\right), \tag{1}$$

where $\sigma(\cdot)$ represents a ReLU activation function, $b_f^{(\ell)}$ denotes a learnable bias, $\mathbf{1}_{N_\ell}$ represents a vector of ones with dimension N_ℓ, and $\mathbf{G}_{c,f}^{(\ell)} = \sum_{k=1}^{K} \mathbf{g}_{c,f,k}^{(\ell)} \hat{\mathbf{A}}^k$ is a graph filter with $\mathbf{g}_{c,f,k}^{(\ell)}$ and $\hat{\mathbf{A}}$ denoting a vector of the graph filter polynomial coefficients and the normalized affinity matrix of the graph, respectively.

[1] https://db.humanconnectome.org/.

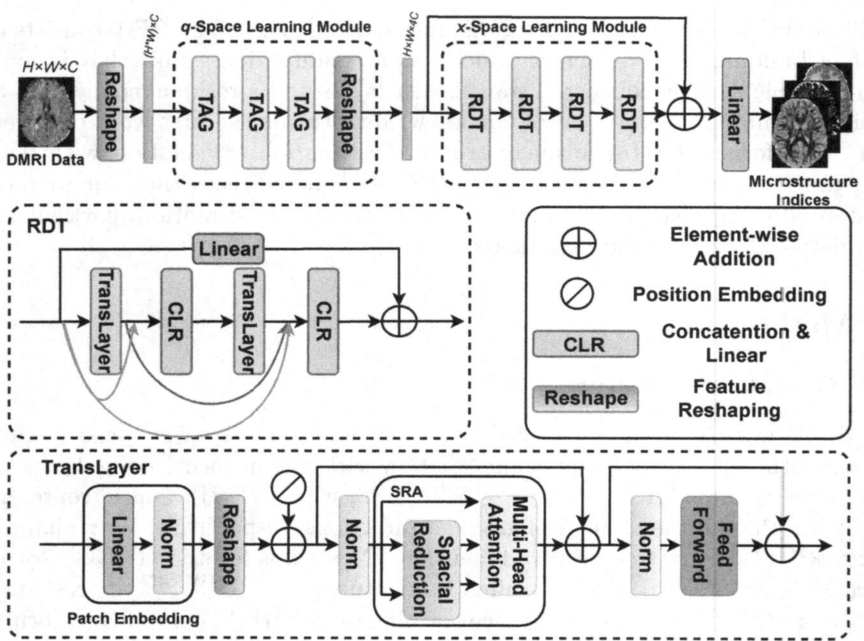

Fig. 1. Overview of HGT. The model is divided into two modules: q-space learning with a GNN and x-space learning with a transformer. RDT denotes residual dense transformer. TransLayer is short for the transformer layer. SRA denotes spatial-reduction attention.

In TAGCN [7], $\mathbf{G}_{c,f}^{(\ell)}\mathbf{x}_c^{(\ell)}$ is equivalent to the convolution on a graph with an arbitrary topology by using a set of filters with filter sizes ranging from 1 to K.

2.3 Learning in x-Space

Our x-space learning module, RDT, is based on a residual dense architecture combined with a transformer. As shown in Fig. 1, it consists of multiple cascaded residual dense transformer (RDT) blocks, each of which consists of densely connected transformer layers and a residual connection to facilitate training. The transformer layer is based on the pyramid vision transformer (PVT) [16], which improves multi-head attention (MHA) with spatial-reduction attention (SRA). The inputs of SRA are a query Q, a key K, and a value V. The advantage of SRA lies in its ability of reducing the spatial scale of K and V before the attention operation, which provides an ideal solution to resolve the problem of excessive memory and calculation of MHA. Mathematically, SRA is defined as

$$\mathtt{SRA}(Q, K, V) = \mathtt{concat}\left(\mathtt{head}_0, \ldots, \mathtt{head}_N\right) W^O, \tag{2}$$

where $W^O \in \mathbb{R}^{C\times C}$ are parameters of linear projections with C denoting the number of channels, $\mathtt{concat}(\cdot)$ is the concatenation operation, N is the number of heads, and $\texttt{head}_j$ represents the attention coefficients for each head and is defined as

$$\texttt{head}_j = \texttt{attention}\left(QW_j^Q, \texttt{SR}(K)W_j^K, \texttt{SR}(V)W_j^V\right), \tag{3}$$

where $W_j^Q \in \mathbb{R}^{C\times d_{\text{head}}}$, $W_j^K \in \mathbb{R}^{C\times d_{\text{head}}}$, and $W_j^V \in \mathbb{R}^{C\times d_{\text{head}}}$ are linear projection parameters with d_{head} denoting the dimension of each head, $\texttt{attention}(\cdot)$ is the attention operator as in [15], and $\texttt{SR}(\cdot)$ reduces the spatial dimensions of K and V:

$$\texttt{SR}(\mathbf{x}) = \texttt{norm}\left(\texttt{reshape}\left(\mathbf{x}, R\right) W^S\right), \tag{4}$$

where $\mathbf{x} \in \mathbb{R}^{(HW)\times C}$ is an input sequence, R is the reduction ratio, $\texttt{reshape}(\cdot)$ adjusts the dimension of an input sequence from $(HW) \times C$ to $\frac{HW}{R^2} \times \left(R^2C\right)$, $W^S \in \mathbb{R}^{\left(R^2C\right)\times C}$ is a linear projection to reduce the dimension of the input sequence to C, and $\texttt{norm}(\cdot)$ denotes layer normalization [2]. With the help of $\texttt{SR}(\cdot)$, the computation of attention coefficients is significantly reduced by R^2 times compared with the original MHA.

3 Experiments

3.1 Dataset

Our dataset consists of 21 subjects randomly selected from the HCP, which is partitioned to 10 subjects for training, 1 subject for validation, and 10 subjects for testing. We consider two undersampling patterns: (i) Two shells with 60 uniform gradients, i.e., 30 for $b = 1000$ s/mm^2, 30 for $b = 2000$ s/mm^2; (ii) One shell with 30 uniform gradients for $b = 1000$ s/mm^2. We train our network to predict NODDI-derived indices, including intra-cellular volume fraction (ICVF), isotropic volume fraction (ISOVF), and orientation dispersion index (ODI).

3.2 Implementation Details

Data Preparation: We first performed brain extraction and removed the non-brain regions from DMRI data. 2D slices were then extracted from the DMRI data of 10 training subjects to form our training dataset, consisting of 1,110 slices (110×10) in total. We set the angular threshold $\theta = 45°$ to construct the graph for q-space learning. Finally, the DMRI data were normalized by dividing each DW image with the average of all b_0 images.

Model Training: The network was implemented using PyTorch 1.7.1 and PyTorch-Geometric (PyG) 2.0.6[2]. We trained the network with an NVIDIA GeForce GTX 2080 GPU with 8GB RAM. The ADAM optimizer was adopted with an initial learning rate of 0.0005. The network was trained with an MSE loss function. The best model parameters were selected based on the validation set.

[2] https://github.com/pyg-team/pytorch_geometric.

Table 1. Quantitative evaluation using PSNR and SSIM for two-shell undersampling (60 gradient directions total for $b = 1000, 2000\,\text{s/mm}^2$). The best results are in **bold**. All: ICVF & ISOVF & OD.

Method	PSNR ↑				SSIM ↑			
	ICVF	ISOVF	OD	All	ICVF	ISOVF	OD	All
AMICO [6]	15.35	25.11	17.56	17.78	0.692	0.791	0.675	0.720
U-Net [8]	23.32	27.13	20.48	22.84	0.834	0.835	0.659	0.776
U-Net++ [22]	23.91	27.56	21.25	23.51	0.851	0.844	0.705	0.800
MLP [10]	23.66	26.80	22.17	23.80	0.832	0.810	0.763	0.801
GCNN [5]	23.78	27.42	22.47	24.10	0.838	0.842	0.776	0.819
MEDN [18]	23.90	27.39	22.56	24.18	0.839	0.837	0.782	0.820
MEDN+ [18]	24.69	27.59	23.01	24.70	0.869	0.849	0.798	0.838
HGT	**24.95**	**28.66**	**23.46**	**25.18**	**0.880**	**0.882**	**0.813**	**0.858**

Comparison Methods: The gold standard microstructure was computed via AMICO [6] using the full set of 270 diffusion gradients. AMICO [6], MLP [10], GCNN [5], MEDN [18], MEDN+ [18], U-Net [8], and U-Net++ [22] were used as comparison baselines. The MLP was implemented with a multi-layer feed-forward neural network with parameters described in [10]. The GCNN was implemented using PyG with a network architecture similar to [5]. We implemented MEDN and MEDN+ using the source code provided in [18]. Note that U-Net and U-Net++ were modified based on the code provided at GitHub[3] by replacing their loss functions with the MSE loss, consistent with our network.

3.3 Evaluation Results

For the two-shell case, Table 1 shows that HGT outperforms all competing methods for different microstructural indices (i.e., ICVF, ISOVF, and OD) and their combination (i.e., All) in terms of peak signal-to-noise ratio (PSNR) and structural similarity index measure (SSIM). Compared with MEDN+, HGT improves the PSNR by large margins of 1.1%, 3.9%, and 2.0% for ICVF, ISOVF, OD, respectively. Visual results (Fig. 2) indicate that HGT is closest to the gold standard with the smallest prediction errors among all methods. For the competing methods, MLP, MEDN, and GCNN ignore x-space information and perform predictions voxel-wise, while U-Net and U-Net++ ignore the geometric structure in q-space. Our results demonstrate that simultaneously considering q-space and x-space information substantially improves microstructural estimations.

The results for the single-shell case are shown in Table 2. Compared with MEDN+, HGT improves PSNR by 14.7%, 10.0% and 3.8% for ICVF, ISOVF, OD, respectively. It is worth noting that U-Net and U-Net++ perform better than the voxel-wise methods, i.e., MLP, GCNN, and MEDN because x-space

[3] https://github.com/4uiiurz1/pytorch-nested-unet.

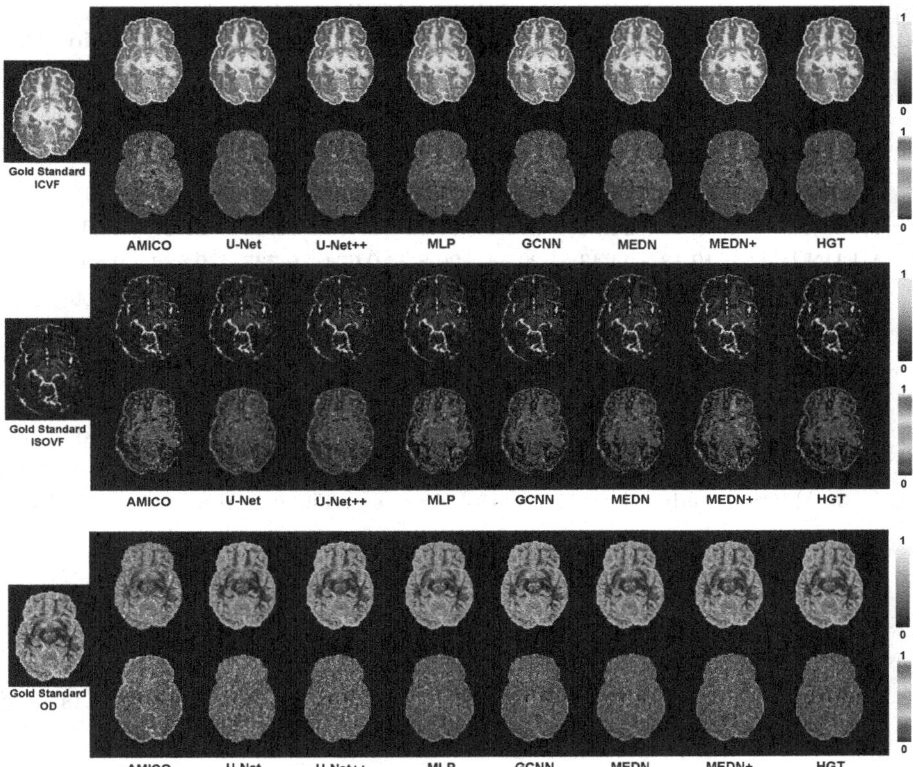

Fig. 2. Visual comparison of the prediction results for two-shell undersampling (60 gradients total for $b = 1000, 2000\,\mathrm{s/mm^2}$).

information plays an important role in the tissue microstructure estimation when q-space information is limited, i.e., data from less gradients are available. In addition, AMICO fails for the single-shell DMRI data. We show the visual results for the single-shell case as part of the supplementary materials. The results are consistent with Fig. 2, indicating that HGT significantly improves the quality of microstructure index maps.

3.4 Ablation Study

We perform an extensive ablation study to investigate the effectiveness of q-space learning module and x-space learning module. As shown in Table 3, the ablation study is completed under the condition of two-shell case with 60 gradients. The basic model (A) consists of multiple cascaded PVT layers without the q-space learning module and residual dense connections.

Effectiveness of q-space learning module: We first investigate the effectiveness of x-space learning module by creating ablated version (B), where the

Table 2. Quantitative evaluation using PSNR and SSIM for single-shell undersampling (30 gradient directions total for $b = 1000\,\text{s/mm}^2$). The best results are in **bold**. All: ICVF & ISOVF & OD.

Method	PSNR ↑				SSIM ↑			
	ICVF	ISOVF	OD	All	ICVF	ISOVF	OD	All
AMICO [6]	8.09	14.90	12.03	10.80	0.020	0.273	0.380	0.210
MLP [10]	19.15	24.07	20.31	20.72	0.670	0.706	0.667	0.681
MEDN [18]	19.18	24.43	20.54	20.87	0.674	0.737	0.684	0.698
GCNN [5]	19.27	24.48	20.55	20.92	0.676	0.742	0.681	0.700
MEDN+ [18]	19.14	24.67	21.30	21.14	0.684	0.754	0.717	0.719
U-Net [8]	19.95	25.11	20.41	21.28	0.727	0.787	0.651	0.722
U-Net++ [22]	19.94	25.49	20.87	21.51	0.750	0.798	0.688	0.745
HGT	**21.96**	**27.13**	**22.12**	**23.16**	**0.814**	**0.850**	**0.757**	**0.807**

Table 3. Ablation study shown for the two-shell case. The best results are in **bold**.

Model	q-Space	x-Space Learning Module			PSNR ↑			
	Learning Module	Transformer	Residual	Dense	ICVF	ISOVF	OD	All
(A)		✓			24.22	28.26	23.23	24.75
(B)	✓	✓			24.48	28.60	23.40	24.97
(C)	✓	✓	✓		24.82	28.61	23.33	25.06
(D)	✓	✓	✓	✓	**24.95**	**28.66**	**23.46**	**25.18**

q-space learning module is added to basic model (A). Model (B) outperforms (A) for the NODDI-derived indices, demonstrating that the q-space learning module improves performance by explicitly modeling the q-space geometric structure with graph representation.

Effectiveness of Residual Learning: We then investigate the effectiveness of residual learning by adding residual connections to (B), resulting in (C). The results indicate that residual learning improves the overall performance, especially for the ICVF index.

Effectiveness of Dense Connections: Finally, the full version of HGT, i.e., (D), outperforms (C) for all three indices, confirming the effectiveness of dense connections.

4 Conclusion

In this work, we have proposed a novel deep learning model, called HGT, for estimating tissue microstructure from undersampled DMRI data. Thanks to the hybrid architecture built with GNN and transformer, HGT is able to make full use of angular neighborhood information in q-space and spatial relationships in x-space to improve the estimation accuracy. Evaluations based on the HCP data demonstrate that HGT substantially improves the quality of microstructural index maps.

References

1. Aliotta, E., Nourzadeh, H., Patel, S.H.: Extracting diffusion tensor fractional anisotropy and mean diffusivity from 3-direction DWI scans using deep learning. Magn. Reson. Med. **85**(2), 845–854(2020)
2. Ba, J.L., Kiros, J.R., Hinton, G.E.: Layer normalization. arXiv preprint arXiv:1607.06450 (2016)
3. Chen, G., Dong, B., Zhang, Y., Lin, W., Shen, D., Yap, P.T.: Denoising of infant diffusion MRI data via graph framelet matching in x-q space. IEEE Trans. Med. Imaging **38**(12), 2838–2848 (2019)
4. Chen, G., Dong, B., Zhang, Y., Lin, W., Shen, D., Yap, P.T.: XQ-SR: joint x-q space super-resolution with application to infant diffusion MRI. Med. Image Anal. **57**, 44–55 (2019)
5. Chen, G., et al.: Estimating tissue microstructure with undersampled diffusion data via graph convolutional neural networks. In: Martel, A.L., et al. (eds.) MICCAI 2020. LNCS, vol. 12267, pp. 280–290. Springer, Cham (2020). https://doi.org/10.1007/978-3-030-59728-3_28
6. Daducci, A., Canales-Rodríguez, E.J., Zhang, H., Dyrby, T.B., Alexander, D.C., Thiran, J.P.: Accelerated microstructure imaging via convex optimization (AMICO) from diffusion MRI data. Neuroimage **105**, 32–44 (2015)
7. Du, J., Zhang, S., Wu, G., Moura, J.M., Kar, S.: Topology adaptive graph convolutional networks. arXiv preprint arXiv:1710.10370 (2017)
8. Falk, T., et al.: U-Net: deep learning for cell counting, detection, and morphometry. Nat. Methods **16**(1), 67–70 (2019)
9. Gibbons, E.K., et al.: Simultaneous NODDI and GFA parameter map generation from subsampled q-space imaging using deep learning. Magn. Reson. Med. **81**(4), 2399–2411 (2019)
10. Golkov, V., et al.: Q-space deep learning: twelve-fold shorter and model-free diffusion MRI scans. IEEE Trans. Med. Imaging **35**(5), 1344–1351 (2016)
11. Jensen, J.H., Helpern, J.A., Ramani, A., Lu, H., Kaczynski, K.: Diffusional kurtosis imaging: the quantification of non-gaussian water diffusion by means of magnetic resonance imaging. Magn. Reson. Med. **53**(6), 1432–1440 (2005)
12. Park, J., et al.: DIFFnet: diffusion parameter mapping network generalized for input diffusion gradient schemes and b-values. IEEE Trans. Med. Imaging **41**, 491–499 (2021)
13. Tian, Q., et al.: DeepDTI: high-fidelity six-direction diffusion tensor imaging using deep learning. Neuroimage **219**, 117017 (2020)
14. Van Essen, D.C., et al.: The WU-Minn human connectome project: an overview. Neuroimage **80**, 62–79 (2013)
15. Vaswani, A., et al.: Attention is all you need. In: Advances in Neural Information Processing Systems 30 (2017)
16. Wang, W., et al.: Pyramid vision transformer: a versatile backbone for dense prediction without convolutions. In: Proceedings of the IEEE/CVF International Conference on Computer Vision, pp. 568–578 (2021)

17. Yap, P.-T., Dong, B., Zhang, Y., Shen, D.: Tight graph framelets for sparse diffusion MRI q-space representation. In: Ourselin, S., Joskowicz, L., Sabuncu, M.R., Unal, G., Wells, W. (eds.) MICCAI 2016. LNCS, vol. 9902, pp. 561–569. Springer, Cham (2016). https://doi.org/10.1007/978-3-319-46726-9_65
18. Ye, C.: Estimation of tissue microstructure using a deep network inspired by a sparse reconstruction framework. In: International Conference on Information Processing in Medical Imaging, pp. 466–477. Springer (2017)
19. Ye, C., Li, X., Chen, J.: A deep network for tissue microstructure estimation using modified LSTM units. Med. Image Anal. **55**, 49–64 (2019)
20. Ye, C., Li, Y., Zeng, X.: An improved deep network for tissue microstructure estimation with uncertainty quantification. Med. Image Anal. **61**, 101650 (2020)
21. Zhang, H., Schneider, T., Wheeler-Kingshott, C.A., Alexander, D.C.: NODDI: practical in vivo neurite orientation dispersion and density imaging of the human brain. Neuroimage **61**(4), 1000–1016 (2012)
22. Zhou, Z., Rahman Siddiquee, M.M., Tajbakhsh, N., Liang, J.: UNet++: a nested U-Net architecture for medical image segmentation. In: Stoyanov, D., et al. (eds.) DLMIA/ML-CDS -2018. LNCS, vol. 11045, pp. 3–11. Springer, Cham (2018). https://doi.org/10.1007/978-3-030-00889-5_1

Atlas-Powered Deep Learning (ADL) - Application to Diffusion Weighted MRI

Davood Karimi(✉) and Ali Gholipour

Computational Radiology Laboratory of the Department of Radiology at Boston Children's Hospital, and Harvard Medical School, Boston, MA, USA
davood.karimi@childrens.harvard.edu

Abstract. Deep learning has a great potential for estimating biomarkers in diffusion weighted magnetic resonance imaging (dMRI). Atlases, on the other hand, are a unique tool for modeling the spatio-temporal variability of biomarkers. In this paper, we propose the first framework to exploit both deep learning and atlases for biomarker estimation in dMRI. Our framework relies on non-linear diffusion tensor registration to compute biomarker atlases and to estimate atlas reliability maps. We also use non-linear tensor registration to align the atlas to a subject and to estimate the error of this alignment. We use the biomarker atlas, atlas reliability map, and alignment error map, in addition to the dMRI signal, as inputs to a deep learning model for biomarker estimation. We use our framework to estimate fractional anisotropy and neurite orientation dispersion from down-sampled dMRI data on a test cohort of 70 newborn subjects. Results show that our method significantly outperforms standard estimation methods as well as recent deep learning techniques. Our method is also more robust to higher measurement down-sampling factors. Our study shows that the advantages of deep learning and atlases can be synergistically combined to achieve unprecedented biomarker estimation accuracy in dMRI.

Keywords: Deep learning · Atlas · Estimation · Diffusion MRI

1 Introduction

Diffusion weighted magnetic resonance imaging (dMRI) is the de-facto tool for probing the brain micro-structure in vivo. Biomarkers estimated with dMRI are used to study normal and abnormal brain development [1]. Estimation of these biomarkers from the dMRI signal entails solving inverse problems that range from non-linear least squares to complex non-convex optimization problems [11, 20]. Accurate estimation depends on dense high-quality measurements, which may be difficult or impossible to obtain in some applications. Deep learning (DL), has emerged as an alternative to classical optimization-based methods. DL models can learn the complex relation between the dMRI signal and the biomarkers from training data. Recent studies have used DL models to estimate

L. Wang et al. (Eds.): MICCAI 2022, LNCS 13431, pp. 123–132, 2022.
https://doi.org/10.1007/978-3-031-16431-6_12

diffusion tensor, multi-compartment models, and fiber orientation distribution, to name a few [8,9,15].

Statistical atlases are models of expected anatomy and anatomical variation. They are routinely used in medical image analysis to characterize normal anatomy and for identifying anatomical differences between normal and/or abnormal populations. dMRI atlases have been used to study brain development [17] and degeneration [12], and for other purposes [10,23].

However, no prior work has attempted to use DL and atlases for dMRI biomarker estimation in a unified framework. Such a framework makes much intuitive sense. While DL methods are effective in learning the mapping between the dMRI signal and biomarkers, atlases can supply additional useful information that is absent from the local diffusion signal. The information contained in the atlas can be particularly useful where the local signal is not adequate for accurate estimation, such as when the number of measurements is low or the signal is noisy. Studies have shown that atlases can improve the performance of DL methods for segmentation (e.g., [5]). There have also been efforts to use different sources of prior information with non-DL estimation methods such as Bayesian techniques in dMRI [25,29]. However, to the best of our knowledge, no prior work has used atlases within a DL framework for dMRI biomarker estimation.

On the other hand, designing such a framework is not trivial. An atlas only represents the average of a population and it lacks the variations among the individuals. Also, the match between an atlas and an individual brain is complex and varies between subjects as well as within the brain of a subject. It is not clear how these information can be incorporated into a machine learning framework.

In this work, we propose a framework that brings together DL models and atlases for dMRI biomarker estimation. Our framework offers methods to compute the reliability of the atlas and its degree of correspondence with a subject in a spatially-varying manner. We use our proposed framework to estimate fractional anisotropy and neurite orientation dispersion from down-sampled dMRI data and show that it is more accurate than several alternative techniques.

2 Materials and Methods

2.1 Data

We used 300 dMRI scans from the Developing Human Connectome Project (DHCP) dataset [3]. Each scan is from a different subject. We used 230 scans for model development and the remaining 70 scans for evaluation. The gestational age (GA) of the subjects ranged between 31 and 45 weeks. Analysis of dMRI data for this age range is challenged by high free water content, incomplete myelination, and low data quality [6].

2.2 Atlas Development

Because of rapid brain development in the neonatal period, a single atlas cannot represent the entire age range [22,28]. Therefore, we build atlases at one-week intervals. To build atlases for week 35, for example, we use subjects with GA between 34.5 and 35.5 weeks. For each GA, we build the atlases from 10 subjects, which our experience and prior works [17,22] show is a sufficient number.

We focus on two biomarkers: 1) Fractional anisotropy (FA) from the diffusion tensor model, which is a widely used biomarker, and 2) Orientation dispersion (OD) from the NODDI model [31], which is a more complex model than diffusion tensor and which may be a more specific biomarker in some applications [2,21].

Given the dMRI volumes $\{s_i\}_{i=1}^n$ for n subjects, we compute the biomarker(s) of interest separately for each subject. Regardless of which biomarker(s) are explored, we also always compute the diffusion tensor for each subject because we use the tensors for accurate spatial alignment. This is shown in Fig. 1, where we have used p and q to denote the biomarkers considered in this work and T to denote the diffusion tensor. Given the set $\{T_i\}_{i=1}^n$ of subject tensors, we compute a set of transformations $\{\Phi_i\}_{i=1}^n$ that align these tensors into a common atlas space and compute a mean tensor $\bar{T}$. This is done using an iterative approach that computes a series of rigid, affine, and diffeomorphic non-rigid registrations and updates $\bar{T}$ at every iteration, rather similar to common practice [17,22]. Specifically, we perform five iterations of rigid registration, followed by five iterations of affine registration, and finally ten iterations of non-rigid registration. The final registration transform Φ_i for subject tensor T_i is the composition of the final affine and non-rigid transforms. Tensor-based alignment, which is the central step in creating the atlases was performed using the method of [32,33].

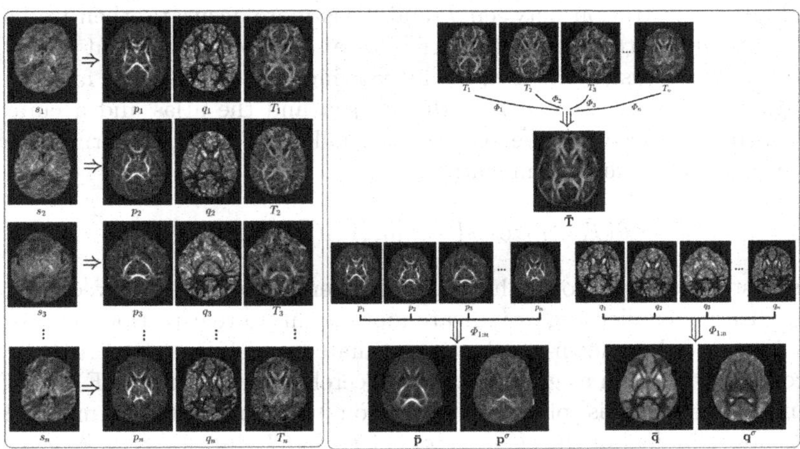

Fig. 1. Left: Biomarkers and diffusion tensor are computed for each subject. Right: tensor-to-tensor registration is used to compute the atlas and atlas confidence maps.

We use the transformations $\{\Phi_i\}_{i=1}^n$ to move the biomarker maps into the common atlas space. We compute the mean of the transformed biomarker maps as the biomarker atlas. We compute the standard deviation of the transformed biomarker maps as a measure of atlas confidence. Formally, for biomarker p:

$$\bar{p} = \text{mean}\big[\Phi_i(p_i)\big]_{i=1}^{n}, \quad p^{\sigma} = \text{std}\big[\Phi_i(p_i)\big]_{i=1}^{n} \tag{1}$$

where mean and standard deviation are computed in a voxel-wise manner across subjects. Larger values of p^{σ} indicate higher variability/disagreement between the biomarker maps used to create the atlas and, hence, *lower* confidence. Figure 1 shows example atlas and atlas confidence maps for FA and OD.

2.3 Biomarker Estimation for an Individual Subject

Given the dMRI volume for a subject, s_k, we compute the biomarker(s) for that subject via the following steps (shown for an example subject in Fig. 2).

Step 1: Atlas-to-Subject Alignment. To exploit the information encoded in an atlas, we need to accurately register it to the subject space. We use tensor-to-tensor registration for this alignment. Hence, we first compute the diffusion tensor, T_k, for the subject. We then compute affine+non-linear transforms that map the atlas tensor $\bar{T}$ to the subject tensor, T_k. We use Φ_k to denote the composition of these affine and non-linear transforms. We denote the atlas tensor transformed to the subject space with $\bar{T}_k = \Phi_k(\bar{T})$.

The registration between the atlas tensor and the subject tensor is never perfect. The accuracy of this registration is a potentially important piece of information because it indicates where the atlas information is more reliable. If at voxel i the registration between $\bar{T}$ and T_k is more accurate, then we have a higher incentive to trust the biomarker atlas at the location of that voxel. The accuracy of this registration is spatially varying and depends on factors such as the degree of similarity between the subject and the atlas and accuracy of the registration procedure. We propose the following practical formulation to estimate the error of this alignment:

$$\Phi_k^{\text{err}} = \theta\big(\bar{T}_k, T_k\big).\exp\Big(-\min\big[\text{FA}(\bar{T}_k), \text{FA}(T_k)\big]/\tau\Big). \tag{2}$$

The first term, θ, denotes the angle between the major eigenvectors of $\bar{T}_k$ and T_k. Clearly, smaller angles indicate more accurate registration. The second term is introduced to down-weight the estimated registration accuracy for less anisotropic tensors such as gray matter and cerebrospinal fluid (CSF). For CSF, for example, the tensor is spherical and the computed orientation of major eigen-vector is not reliable, but the eigenvectors of T_k and $\bar{T}_k$ may be very close to each other by chance, hence artificially making θ very small. By using the minimum of the FAs, if either $\bar{T}_k$ or T_k has a low anisotropy, the second term will have a larger value. We set $\tau = 0.2$, which we found empirically to work well.

Step 2: Estimation Using a DL Model. In our framework, in addition to the diffusion signal s_k, we also use the prior information encoded in the biomarker atlas as described above. Specifically, for estimating the biomarker p for subject

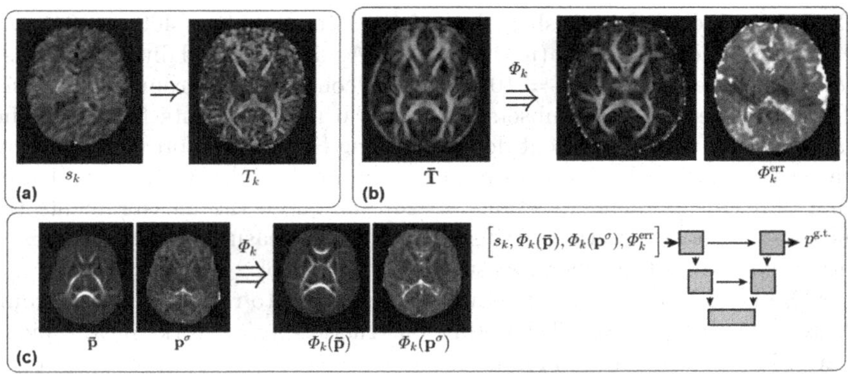

Fig. 2. Our method for estimating biomarkers using atlases and DL. (a) We compute the diffusion tensor T_k from the subject data. (b) We compute the tensor-to-tensor registration ϕ_k from the atlas to the subject, and the error of this registration Φ_k^{err} (Eq. 2). (c) We align the atlas and atlas confidence map to the subject using Φ_k. These information are concatenated to the dMRI volume to form the input to the DL model.

k, we have three additional pieces of information; 1) the biomarker atlas registered to the subject space $\Phi_k(\bar{p})$, 2) the biomarker atlas confidence registered to the subject space $\Phi_k(\bar{p})$, and 3) registration error Φ_k^{err}. Since these three pieces of information are spatially varying and aligned with s_k, we simply concatenate them to form the input to our DL model (Fig. 2c). For example, for estimating biomarker p for subject k, the input to the network is $\left[s_k, \Phi_k(\bar{\mathbf{p}}), \Phi_k(\mathbf{p}^\sigma), \Phi_k^{\text{err}}\right]$. The estimation target is the ground-truth, $p^{\text{g.t.}}$, computed as described below.

We used a U-Net as our DL architecture. The exact architecture is not critical and not the focus of this paper. We used patches of size 48^3 voxels that, given small neonatal brains, cover most of the brain. Use of patches is common practice and also acts as data augmentation. We set the number of feature maps in the first stage of the network to be 12, which was the largest possible on our GPU memory. For training, we sampled blocks from random locations in the training images. At test time, we used a sliding window with a stride of 16 voxels in each dimension to estimate the biomarker for an input dMRI volume of arbitrary size.

Compared Methods and Evaluation Strategy. For FA estimation, we compared our method with: 1) Constrained weighted linear least-squares (**CWLLS**) [19], which is the standard method; 2) **Deep-DTI** [26], which is a recent DL method based on CNNs. This method exploits the anatomical T2 image, in addition to the diffusion signal, for estimation. Hence, for Deep-DTI we also used the T2 image, which we registered to the dMRI volume. For OD estimation, we compared our method with: 1) **Dmipy** [7], which follows a standard optimization-based estimation approach, 2) Microstructure Estimation using a Deep Network (**MEDN+**) [30], which is a DL method that has been inspired by AMICO [4] and significantly outperforms AMICO too.

Each scan included multishell data: $b = 0$ (n=20), $b = 400$ (n=64), $b = 1000$ (n=88), and $b = 2600$ (n=128). For FA, as suggested in [13], we used all 88 measurements in the $b = 1000$ shell to compute the ground truth using CWLLS. We then selected subsets of 12 and 6 measurements from this shell for each subject, which represent down-sampling factors of approximately 7 and 15, respectively. To select 6 measurements, as in [14,26], we considered the 6 optimal diffusion gradient directions of [24] and chose the measurements that were closest to those directions. We selected the 12 measurements to be close to uniformly spread on the sphere, as suggested in [13,16].

For OD (and NODDI), we used all measurements to reconstruct the ground truth using Dmipy [7]. For DL-based reconstruction, prior works have typically used 20–60 measurements from more than one shell [8,30]. Here, we chose either 6 and 15 measurements from each of the $b = 1000$ and $b = 2600$ shells, for a total of 12 and 30 measurements, which represent downsampling factors of approximately 24 and 10, respectively. We selected these measurements to be close to uniformly spread on the sphere, using an approach similar to [16]. For a fair comparison, for both FA and OD we used the same down-sampled datasets for our method and for all competing techniques.

Implementation and Training. We used DTI-TK [32] to compute all registrations. In creating the atlases, we used MRtrix [27] to compute the diffusion tensors and FA, and we used Dmipy [7] to compute OD. We implemented all DL techniques in TensorFlow 1.14 under Python 3.7. We ran all algorithms on a Linux machine with 16 CPU cores and an NVIDIA GeForce GTX 1080 GPU.

We trained our model by minimizing the ℓ_2 norm between the predicted and ground truth biomarker using Adam [18], a batch size of 10, and an initial learning rate of 10^{-4} that was reduced by half every time the validation loss did not decrease after a training epoch. For the competing methods, we followed the training procedures recommended in the original papers.

3 Results and Discussion

3.1 Comparison with Other Techniques

Tables 1 and 2 show the reconstruction error for different methods for FA and OD, respectively, computed on the 70 independent test subjects. For both FA and OD, the DL methods were substantially more accurate than the standard methods (i.e., CWLLS and Dmipy). Our proposed method achieved lower errors than the other DL methods for both FA and OD at both down-sampling factors. We used paired t-tests to compare our method with the other methods. For both FA and OD and at both down-sampling factors, the estimation error for our method was significantly lower than any of the compared methods ($p < 0.001$). We observed similar improvements across the gestational age. Figure 3 shows example reconstruction results for different methods. Our method achieves lower errors than other methods across the brain for both FA and OD.

Table 1. FA estimation errors for the proposed method and compared methods.

No. of measurements	CWLLS	Deep-DTI	Proposed
$n = 6$	0.111 ± 0.014	0.048 ± 0.008	$\mathbf{0.040 \pm 0.005}$
$n = 12$	0.053 ± 0.007	0.044 ± 0.006	$\mathbf{0.039 \pm 0.005}$

Table 2. OD estimation errors for the proposed method and compared methods.

No. of measurements	Dmipy	MEDN+	Proposed
$n = 16$	0.138 ± 0.032	0.064 ± 0.028	$\mathbf{0.047 \pm 0.004}$
$n = 30$	0.096 ± 0.029	0.052 ± 0.030	$\mathbf{0.044 \pm 0.005}$

For both FA and OD, our proposed method showed a smaller increase in error as the down-sampling rate increased. Specifically, the FA estimation error for our method increased by 2.5% as the number of measurements was reduced from 12 to 6, compared with 9% for Deep-DTI. For OD, the estimation error for our method increased by 7% as the number of measurements was decreased from 30 to 16, compared with 23% for MEDN+.

The training time for our model was 10 h, compared with 10–60 hours for the other DL methods. To estimate FA or OD for a dMRI test volume, our model required 73 ± 20 seconds. Approximately 90% of this time was spent on computing the atlas-to-subject registration. The average computation time for the other DL methods ranged from 10 s to 30 min for MEDN+. The average computation times for CWLLS and Dmipy were, respectively, 20 s and 3.2 h. Nonetheless, computation time for dMRI analysis is not a critical factor since fast estimation is typically not a requirement.

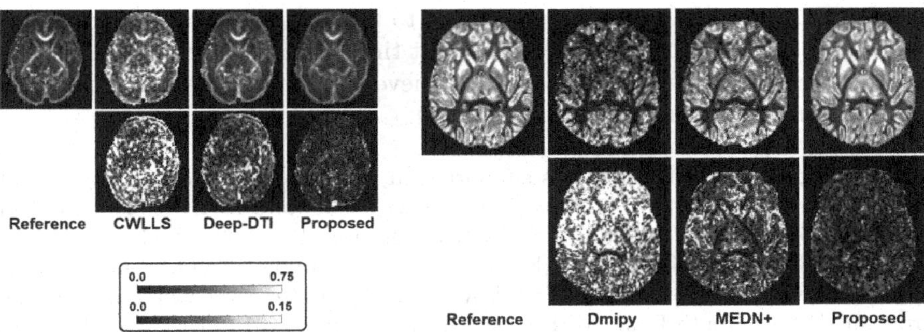

Fig. 3. Example FA (left) and OD (right) reconstructions by different methods. In each of the four examples, the bottom row shows maps of absolute estimation error.

3.2 Ablation Experiments

Table 3 shows the results of some ablation experiments. They show that the error of our method increases significantly when we discard the atlas information. When we only use the dMRI signal (last column in Table 3), the FA estimation error of our method is close to Deep-DTI (Table 1) and the OD estimation error of our method is slightly worse than CNN-NODDI (Table 2). Table 3 also shows that all three extra pieces of information contribute to the model accuracy, but the contribution of atlas and atlas confidence is larger than the contribution of alignment error. With one exception (FA between first and second column), all differences were significant (p< 0.01).

Table 3. Results of some ablation experiments. In the column headings, we use p to denote either FA or OD.

Input to the model	$[s_k, \Phi_k(\bar{\mathbf{p}}), \Phi_k(\mathbf{p}^\sigma), \Phi_k^{\text{err}}]$	$[s_k, \Phi_k(\bar{\mathbf{p}}), \Phi_k(\mathbf{p}^\sigma)]$	$[s_k, \Phi_k(\bar{\mathbf{p}})]$	s_k
FA, $n = 6$	$\mathbf{0.040 \pm 0.005}$	0.041 ± 0.005	0.044 ± 0.006	0.047 ± 0.006
OD, $n = 16$	$\mathbf{0.047 \pm 0.004}$	0.050 ± 0.004	0.055 ± 0.011	0.060 ± 0.010

4 Conclusions

Our results show that a framework to unify atlases and DL can lead to superior results in dMRI biomarker estimation. The success of this approach is, in part, because we build the atlas via tensor-based diffeomorphic registration, which enables accurate alignment of white matter structures. Although many DL models can learn spatial correlations, they are not capable of encoding the information obtained via tensor-based diffeomorphic deformation that we used in building the atlas. We cannot claim that the design of our framework is "optimal". For example, atlas-to-subject alignment may be improved by incorporating anatomical MRI information in addition to the diffusion tensor information. Nonetheless, our work has shown, for the first time, that spatio-temporal atlases can be used within a DL framework to achieve superior biomarker estimation accuracy from down-sampled data.

Acknowledgment. This study was supported in part by the National Institutes of Health (NIH) under grants R01EB031849, R01NS106030, and R01EB032366; and in part by the Office of the Director of the NIH under grant S10OD0250111.

The DHCP dataset is provided by the developing Human Connectome Project, KCL-Imperial-Oxford Consortium funded by the European Research Council under the European Union Seventh Framework Programme (FP/2007–2013)/ERC Grant Agreement no. [319456]. We thank the families who supported this trial.

References

1. Alexander, D.C., et al.: Imaging brain microstructure with diffusion MRI: practicality and applications. NMR Biomed. **32**(4), e3841 (2019)

2. Andica, C., et al.: Neurite orientation dispersion and density imaging reveals white matter microstructural alterations in adults with autism. Mol. Autism **12**(1), 1–14 (2021)
3. Bastiani, M., et al.: Automated processing pipeline for neonatal diffusion MRI in the developing human connectome project. Neuroimage **185**, 750–763 (2019)
4. Daducci, A., et al.: Accelerated microstructure imaging via convex optimization (AMICO) from diffusion MRI data. Neuroimage **105**, 32–44 (2015)
5. Diniz, J.O.B., et al.: Esophagus segmentation from planning CT images using an atlas-based deep learning approach. Comput. Methods Programs Biomed. **197**, 105685 (2020)
6. Dubois, J., et al.: The early development of brain white matter: a review of imaging studies in fetuses, newborns and infants. Neuroscience **276**, 48–71 (2014)
7. Fick, R.H., et al.: The Dmipy toolbox: diffusion MRI multi-compartment modeling and microstructure recovery made easy. Front. Neuroinform. **13**, 64 (2019)
8. Gibbons, E.K., et al.: Simultaneous NODDI and GFA parameter map generation from subsampled q-space imaging using deep learning. Magn. Reson. Med. **81**(4), 2399–2411 (2019). https://doi.org/10.1002/mrm.27568
9. Golkov, V., et al.: q-space deep learning: twelve-fold shorter and model-free diffusion MRI scans. IEEE Trans. Med. Imaging **35**(5), 1344–1351 (2016)
10. Hagler, D.J., Jr., et al.: Automated white-matter tractography using a probabilistic diffusion tensor atlas: application to temporal lobe epilepsy. Hum. Brain Mapp. **30**(5), 1535–1547 (2009)
11. Harms, R.L., et al.: Robust and fast nonlinear optimization of diffusion MRI microstructure models. Neuroimage **155**, 82–96 (2017)
12. Hasan, K.M., et al.: Serial atlas-based diffusion tensor imaging study of uncomplicated mild traumatic brain injury in adults. J. Neurotrauma **31**(5), 466–475 (2014)
13. Jones, D.K., et al.: Optimal strategies for measuring diffusion in anisotropic systems by magnetic resonance imaging. Magn. Reson. Med. **42**(3), 515–525 (1999)
14. Karimi, D., Gholipour, A.: Diffusion tensor estimation with transformer neural networks. arXiv preprint arXiv:2201.05701 (2022)
15. Karimi, D., et al.: Deep learning-based parameter estimation in fetal diffusion-weighted MRI. Neuroimage **243**, 118482 (2021)
16. Karimi, D., et al.: Learning to estimate the fiber orientation distribution function from diffusion-weighted MRI. Neuroimage **239**, 118316 (2021)
17. Khan, S., et al.: Fetal brain growth portrayed by a spatiotemporal diffusion tensor MRI atlas computed from in utero images. Neuroimage **185**, 593–608 (2019)
18. Kingma, D.P., Ba, J.: Adam: A method for stochastic optimization. In: Proceedings of the 3rd International Conference on Learning Representations, (ICLR) (2014)
19. Koay, C.G., et al.: A unifying theoretical and algorithmic framework for least squares methods of estimation in diffusion tensor imaging. J. Magn. Reson. **182**(1), 115–125 (2006)
20. Novikov, D.S., et al.: Quantifying brain microstructure with diffusion MRI: theory and parameter estimation. NMR Biomed. **32**(4), e3998 (2019)
21. Palacios, E.M., et al.: The evolution of white matter microstructural changes after mild traumatic brain injury: a longitudinal DTI and NODDI study. Sci. Adv. **6**(32), eaaz6892 (2020)
22. Pietsch, M., et al.: A framework for multi-component analysis of diffusion MRI data over the neonatal period. Neuroimage **186**, 321–337 (2019)
23. Saghafi, B., et al.: Spatio-angular consistent construction of neonatal diffusion MRI atlases. Hum. Brain Mapp. **38**(6), 3175–3189 (2017)

24. Skare, S., Hedehus, M., Moseley, M.E., Li, T.Q.: Condition number as a measure of noise performance of diffusion tensor data acquisition schemes with MRI. J. Magn. Reson. **147**(2), 340–352 (2000)
25. Taquet, M., Scherrer, B., Boumal, N., Peters, J.M., Macq, B., Warfield, S.K.: Improved fidelity of brain microstructure mapping from single-shell diffusion MRI. Med. Image Anal. **26**(1), 268–286 (2015)
26. Tian, Q., et al.: DeepDTI: high-fidelity six-direction diffusion tensor imaging using deep learning. Neuroimage **219**, 117017 (2020)
27. Tournier, J.D., et al.: MRtrix3: a fast, flexible and open software framework for medical image processing and visualisation. Neuroimage **202**, 116137 (2019)
28. Uus, A., et al.: Multi-channel 4D parametrized atlas of macro-and microstructural neonatal brain development. Front. Neurosci. **15**, 661704 (2021)
29. Veraart, J., Rajan, J., Peeters, R.R., Leemans, A., Sunaert, S., Sijbers, J.: Comprehensive framework for accurate diffusion MRI parameter estimation. Magn. Reson. Med. **70**(4), 972–984 (2013)
30. Ye, C.: Tissue microstructure estimation using a deep network inspired by a dictionary-based framework. Med. Image Anal. **42**, 288–299 (2017)
31. Zhang, H., Schneider, T., Wheeler-Kingshott, C.A., Alexander, D.C.: NODDI: practical in vivo neurite orientation dispersion and density imaging of the human brain. Neuroimage **61**(4), 1000–1016 (2012)
32. Zhang, H., Yushkevich, P.A., Alexander, D.C., Gee, J.C.: Deformable registration of diffusion tensor MR images with explicit orientation optimization. Med. Image Anal. **10**(5), 764–785 (2006)
33. Zhang, H., Yushkevich, P.A., Rueckert, D., Gee, J.C.: Unbiased white matter atlas construction using diffusion tensor images. In: Ayache, N., Ourselin, S., Maeder, A. (eds.) MICCAI 2007. LNCS, vol. 4792, pp. 211–218. Springer, Heidelberg (2007). https://doi.org/10.1007/978-3-540-75759-7_26

One-Shot Segmentation of Novel White Matter Tracts via Extensive Data Augmentation

Wan Liu[1], Qi Lu[1], Zhizheng Zhuo[2], Yaou Liu[2], and Chuyang Ye[1](✉)

[1] School of Integrated Circuits and Electronics, Beijing Institute of Technology, Beijing, China
chuyang.ye@bit.edu.cn

[2] Department of Radiology, Beijing Tiantan Hospital, Capital Medical University, Beijing, China

Abstract. Deep learning based methods have achieved state-of-the-art performance for automated *white matter* (WM) tract segmentation. In these methods, the segmentation model needs to be trained with a large number of manually annotated scans, which can be accumulated throughout time. When novel WM tracts—i.e., tracts not included in the existing annotated WM tracts—are to be segmented, additional annotations of these novel WM tracts need to be collected. Since tract annotation is time-consuming and costly, it is desirable to make only a few annotations of novel WM tracts for training the segmentation model, and previous work has addressed this problem by transferring the knowledge learned for segmenting existing WM tracts to the segmentation of novel WM tracts. However, accurate segmentation of novel WM tracts can still be challenging in the one-shot setting, where only one scan is annotated for the novel WM tracts. In this work, we explore the problem of one-shot segmentation of novel WM tracts. Since in the one-shot setting the annotated training data is extremely scarce, based on the existing knowledge transfer framework, we propose to further perform extensive data augmentation for the single annotated scan, where synthetic annotated training data is produced. We have designed several different strategies that mask out regions in the single annotated scan for data augmentation. To avoid learning from potentially conflicting information in the synthetic training data produced by different data augmentation strategies, we choose to perform each strategy separately for network training and obtain multiple segmentation models. Then, the segmentation results given by these models are ensembled for the final segmentation of novel WM tracts. Our method was evaluated on public and in-house datasets. The experimental results show that our method improves the accuracy of one-shot segmentation of novel WM tracts.

Keywords: White matter tract segmentation · One-shot learning · Data augmentation

L. Wang et al. (Eds.): MICCAI 2022, LNCS 13431, pp. 133–142, 2022.
https://doi.org/10.1007/978-3-031-16431-6_13

1 Introduction

The segmentation of *white matter* (WM) tracts based on *diffusion magnetic resonance imaging* (dMRI) provides an important tool for the understanding of brain wiring [12,20]. It allows identification of different WM pathways and has benefited various brain studies [13,22]. Since manual delineations of WM tracts can be time-consuming and subjective, automated approaches to WM tract segmentation are developed [3,18,19], and methods based on *convolutional neural networks* (CNNs) have achieved state-of-the-art performance [9,17,23]. The success of CNN-based WM tract segmentation relies on a large number of annotated scans that are accumulated throughout time for network training. However, in a new study, novel WM tracts that are not included in the existing annotated WM tracts may be of interest [1,11,14] and need to be segmented. Repeating the annotation for the novel WM tracts on a large number of scans can be very laborious and prohibitive, and accurate segmentation of novel WM tracts becomes challenging when only a few annotations are made for them.

Previous work has addressed this few-shot segmentation problem with a transfer learning strategy, where the knowledge learned for segmenting existing WM tracts with abundant annotated data is transferred to the segmentation of novel WM tracts [10]. In [10], a CNN-based segmentation model pretrained for segmenting existing WM tracts is used to initialize the target network for segmenting novel WM tracts, so that even with only a few annotations of novel WM tracts the network can learn adequately for the segmentation during fine-tuning. In addition, instead of using classic fine-tuning that discards the pretrained task-specific weights, an improved fine-tuning strategy is developed in [10] for more effective knowledge transfer, where all weights in the pretrained model can be exploited for initializing the target segmentation model. Despite the promising results achieved in [10] for few-shot segmentation of novel WM tracts, when the number of annotated scans for novel WM tracts decreases to one, the segmentation is still challenging. Since fewer annotations are preferred to reduce the annotation time and cost, the development of accurate approaches to one-shot segmentation of novel WM tracts needs further investigation.

In this work, we seek to improve one-shot segmentation of novel WM tracts. We focus on volumetric WM tract segmentation [9,17], where voxels are directly labeled without necessarily performing tractography [2]. Since in the one-shot setting annotated training data is extremely scarce, based on the pretraining and fine-tuning framework developed in [10], we propose to address the one-shot segmentation problem with extensive data augmentation. Existing data augmentation strategies can be categorized into those based on basic image transformation [6,17], generative models [5], image mixing [21,24], and image masking [4]. Basic image transformation is already applied by default in CNN-based WM tract segmentation [9,17], yet it is insufficient for the one-shot segmentation due to the limited diversity of the augmented data. The training of generative models usually requires a large amount of annotated data, or at least a large amount of unannotated data [5], which is not guaranteed in the one-shot setting. Image mixing requires at least two annotated images [24], which is also infeasible in the one-shot setting. Therefore, we develop several strategies based on image masking for data augmentation, where the single annotated image is manipulated by masking out regions in different ways to synthetize additional training data.

The masking is performed randomly either with uniform distributions or according to the spatial location of novel WM tracts, and the annotation of the synthetic image can also be determined in different ways. The augmented data is used to fine-tune the model for segmenting novel WM tracts. To avoid learning from potentially conflicting information in the synthetic data produced by different strategies, we choose to perform each data augmentation strategy separately to train multiple segmentation models, and the outputs of these models are ensembled for the final segmentation. We evaluated the proposed method on two brain dMRI datasets. The results show that our method improves the accuracy of one-shot segmentation of novel WM tracts. The code of our method is available at https://github.com/liuwan0208/One-Shot-Extensive-Data-Augmentation.

2 Methods

2.1 Background: Knowledge Transfer for Segmenting Novel WM Tracts

Suppose we are given a CNN-based model $\mathcal{M}_{\mathrm{e}}$ that segments M existing WM tracts, for which a large number of annotations have been accumulated for training. We are interested in training a CNN-based model $\mathcal{M}_{\mathrm{n}}$ for segmenting N novel WM tracts, for which only one scan is annotated due to annotation cost.

Existing work [10] has attempted to address this problem with a transfer learning strategy based on the pretraining and fine-tuning framework, where $\mathcal{M}_{\mathrm{e}}$ and $\mathcal{M}_{\mathrm{n}}$ share the same network structure for feature extraction, and their last task-specific layers are different. In classic fine-tuning, the network weights of the learned feature extraction layers of $\mathcal{M}_{\mathrm{e}}$ are used to initialize the feature extraction layers of $\mathcal{M}_{\mathrm{n}}$, and the task-specific layer of $\mathcal{M}_{\mathrm{n}}$ is randomly initialized. Then, all weights of $\mathcal{M}_{\mathrm{n}}$ are fine-tuned with the single scan annotated for novel WM tracts. However, the classic fine-tuning strategy discards the information in the task-specific layer of $\mathcal{M}_{\mathrm{e}}$. As different WM tracts cross or overlap, existing and novel WM tracts can be correlated, and the task-specific layer of $\mathcal{M}_{\mathrm{e}}$ for segmenting existing WM tracts may also bear relevant information for segmenting novel WM tracts. Therefore, to exploit all the knowledge learned in $\mathcal{M}_{\mathrm{e}}$, in [10] an improved fine-tuning strategy is developed, which, after derivation, can be conveniently achieved with a warmup stage. Specifically, the feature extraction layers of $\mathcal{M}_{\mathrm{n}}$ are first initialized with those of $\mathcal{M}_{\mathrm{e}}$. Then, in the warmup stage, the feature extraction layers of $\mathcal{M}_{\mathrm{n}}$ are fixed and only the last task-specific layer of $\mathcal{M}_{\mathrm{n}}$ (randomly initialized) is learned with the single annotated image. Finally, all weights of $\mathcal{M}_{\mathrm{n}}$ are jointly fine-tuned with the single annotated image.

2.2 Extensive Data Augmentation for One-Shot Segmentation of Novel WM Tracts

Although the transfer learning approach in [10] has improved the few-shot segmentation of novel WM tracts, when the training data for novel WM tracts is extremely scarce with only one annotated image, the segmentation is still challenging. Therefore, we continue to explore the problem of one-shot segmentation of novel WM tracts. Based on the pretraining and fine-tuning framework developed in [10], we propose to more effectively exploit the information in the single annotated image with extensive data augmentation for network training. Suppose the annotated image is $\mathbf{X}$ and its annotation is $\mathbf{Y}$ (0 for background and 1 for foreground); then we obtain a set of synthetic annotated training images $\widetilde{\mathbf{X}}$ and the synthetic annotations $\widetilde{\mathbf{Y}}$ by transforming $\mathbf{X}$ and $\mathbf{Y}$. We develop several data augmentation strategies for the purpose, which are described below.

Random Cutout. First, motivated by the Cutout data augmentation method [4] that has been successfully applied to image classification problems, we propose to obtain the synthetic image $\widetilde{\mathbf{X}}$ by transforming $\mathbf{X}$ with region masking:

$$\widetilde{\mathbf{X}} = \mathbf{X} \odot (1 - \mathbf{M}), \tag{1}$$

where $\odot$ represents voxelwise multiplication and $\mathbf{M}$ is a binary mask representing the region that is masked out. $\mathbf{M}$ is designed as a 3D box randomly selected with uniform distributions. Mathematically, suppose the ranges of $\mathbf{M}$ in the x-, y-, and z-direction are $(r_x, r_x + w_x)$, $(r_y, r_y + w_y)$, and $(r_z, r_z + w_z)$, respectively; then we follow [21] and select the box as

$$r_x \sim U(0, R_x),\ r_y \sim U(0, R_y),\ r_z \sim U(0, R_z), \tag{2}$$

$$w_x = R_x\sqrt{1-\lambda},\ w_y = R_y\sqrt{1-\lambda},\ w_z = R_z\sqrt{1-\lambda},\ \lambda \sim \mathrm{Beta}(1,1), \tag{3}$$

where $U(\cdot,\cdot)$ represents the uniform distribution, R_x, R_y, and R_z are the image dimensions in the x-, y-, and z-direction, respectively, and λ is sampled from the beta distribution $\mathrm{Beta}(1,1)$ to control the size of the masked region.

The voxelwise annotation $\widetilde{\mathbf{Y}}$ for $\widetilde{\mathbf{X}}$ also needs to be determined. Intuitively, we can obtain $\widetilde{\mathbf{Y}}$ with the same masking operation for $\widetilde{\mathbf{X}}$:

$$\widetilde{\mathbf{Y}} = \mathbf{Y} \odot (1 - \mathbf{M}). \tag{4}$$

The strategy that obtains synthetic training data using Eqs. (1) and (4) with the sampling in Eqs. (2) and (3) is referred to as *Random Cutout One* (RC1). Besides RC1, it is also possible to keep the original annotation $\mathbf{Y}$ for the masked image $\widetilde{\mathbf{X}}$, so that the network learns to restore the segmentation result in the masked region. In this case, the synthetic annotation $\widetilde{\mathbf{Y}}$ is simply determined as

$$\widetilde{\mathbf{Y}} = \mathbf{Y}. \tag{5}$$

The use of Eqs. (1) and (5) for obtaining synthetic training data with the sampling in Eqs. (2) and (3) is referred to as *Random Cutout Two* (RC2).

Tract Cutout. Since we perform data augmentation for segmenting novel WM tracts, in addition to RC1 and RC2, it is possible to obtain $\mathbf{M}$ with a focus on the novel WM tracts. To this end, we design the computation of $\mathbf{M}$ as

$$\mathbf{M} = \left\lceil \frac{1}{N} \sum_{j=1}^{N} a^j \mathbf{Y}^j \right\rceil, \tag{6}$$

where $\mathbf{Y}^j$ denotes the annotation of the j-th novel WM tract in $\mathbf{Y}$, a^j is sampled from the Bernoulli distribution Bernoulli(0.5) to determine whether $\mathbf{Y}^j$ contributes to the computation of $\mathbf{M}$, and $\lceil \cdot \rceil$ represents the ceiling operation. In this way, $\mathbf{M}$ is the union of the regions of a randomly selected subset of the novel WM tracts, and thus the masked region depends on the novel WM tracts.

With the masking strategy in Eq. (6), we can still use Eq. (4) or (5) to determine the synthetic annotation. When Eq. (4) or (5) is used, the data augmentation strategy is named *Tract Cutout One* (TC1) or *Tract Cutout Two* (TC2), respectively. No duplicate synthetic images are allowed in TC1 or TC2.

Network Training with Augmented Data. By repeating the region masking in each data augmentation strategy, a set of synthetic annotated images can be produced. Since the synthetic images can appear unrealistic, they are used only in the warmup stage of the improved fine-tuning framework in [10], where the last layer of $\mathcal{M}_{\mathrm{n}}$ is learned. In the final fine-tuning step that updates all network weights in $\mathcal{M}_{\mathrm{n}}$, only the real annotated training image is used. In addition, to avoid that the network learns from potentially conflicting information in the synthetic data produced by different strategies, we choose to perform each data augmentation strategy separately and obtain four different networks for segmenting novel WM tracts. At test time, the predictions of the four networks are ensembled with majority voting[1] to obtain the final segmentation.

2.3 Implementation Details

We use the state-of-the-art TractSeg architecture [17] for volumetric WM tract segmentation as our backbone segmentation network, which takes fiber orientation maps as input. Like [17], the fiber orientation maps are computed with *constrained spherical deconvolution* (CSD) [15] or *multi-shell multi-tissue CSD* (MSMT-CSD) [7] for single-shell or multi-shell dMRI data, respectively, and three fiber orientations are allowed in the network input [17]. We follow [17] and perform 2D WM tract segmentation for each image view separately, and the results are fused to obtain the final 3D WM tract segmentation.

The proposed data augmentation is performed offline. Since given N novel WM tracts TC1 or TC2 can produce at most $2^N - 1$ different images, we set the number of synthetic scans produced by each data augmentation strategy to $\min(2^N - 1, 100)$. Note that traditional data augmentation, such as elastic

[1] The tract label is set to one when the votes are tied.

deformation, scaling, intensity perturbation, etc., is applied online in TractSeg to training images. Thus, these operations are also applied to the synthetic training data online. The training configurations are set according to TractSeg, where Adamax [8] is used to minimize the binary cross entropy loss with a batch size of 56, an initial learning rate of 0.001, and 300 epochs [17]. The model corresponding to the epoch with the best segmentation accuracy on a validation set is selected.

3 Experiments

3.1 Data Description and Experimental Settings

For evaluation, experiments were performed on the publicly available *Human Connectome Project* (HCP) dataset [16] and an in-house dataset. The dMRI scans in the HCP dataset were acquired with 270 diffusion gradients (three b-values) and a voxel size of 1.25 mm isotropic. In [17] 72 major WM tracts were annotated for the HCP dataset, and the annotations are also publicly available. For the list of the 72 WM tracts, we refer readers to [17]. The dMRI scans in the in-house dataset were acquired with 270 diffusion gradients (three b-values) and a voxel size of 1.7 mm isotropic. In this dataset, only ten of the 72 major WM tracts were annotated due to the annotation cost.

Following [10], we selected the same 60 WM tracts as existing WM tracts, and a segmentation model was pretrained for these tracts with the HCP dMRI scans, where 48 and 15 annotated scans were used as the training set and validation set, respectively. To evaluate the performance of one-shot segmentation of novel WM tracts, we considered a more realistic and challenging scenario where novel WM tracts are to be segmented on dMRI scans that are acquired differently from the dMRI scans annotated for existing WM tracts. Specifically, instead of using the original HCP dMRI scans for segmenting novel WM tracts, we generated clinical quality scans from them. The clinical quality scans were generated by selecting only 34 diffusion gradients at $b = 1000\,\mathrm{s/mm^2}$ and downsampling the selected diffusion weighted images in the spatial domain by a factor of two to the voxel size of 2.5 mm isotropic. The tract annotations were also downsampled accordingly. Since the dMRI scans in the in-house dataset were acquired differently from the original HCP dMRI scans, they were directly used for evaluation together with their original annotations.

The two datasets were used to evaluate the accuracy of one-shot segmentation of novel WM tracts separately based on the model pretrained on the original HCP dataset for segmenting existing WM tracts. We considered three cases of novel WM tracts for each dataset. For the clinical quality scans, the three cases are referred to as CQ1, CQ2, and CQ3, and for the in-house dataset, the three cases are referred to as IH1, IH2, and IH3. The details about these cases are summarized in Table 1. For each dataset, only one scan was selected from it for network training, together with the corresponding annotation of novel WM tracts.[2] For the clinical quality dataset and the in-house dataset, 30 and 16 scans

[2] The single annotated scan was also used as the validation set for model selection.

Table 1. A summary of the cases (CQ1, CQ2, CQ3, IH1, IH2, and IH3) of novel WM tracts considered in the experiments and the tract abbreviations (abbr.). The checkmark (✓) indicates that the tract was included in the case.

WM Tract Name	Abbr	CQ1	CQ2	CQ3	IH1	IH2	IH3
Corticospinal tract left	CST_left	✓	✓	✓	✓	✓	✓
Corticospinal tract right	CST_right	✓	✓	✓	✓	✓	✓
Fronto-pontine tract left	FPT_left			✓			✓
Fronto-pontine tract right	FPT_right			✓			✓
Inferior longitudinal fascicle left	ILF_left			✓			
Inferior longitudinal fascicle right	ILF_right			✓			
Optic radiation left	OR_left	✓	✓	✓	✓	✓	✓
Optic radiation right	OR_right	✓	✓	✓	✓	✓	✓
Parieto-occipital pontine tract left	POPT_left		✓	✓		✓	✓
Parieto-occipital pontine tract right	POPT_right		✓	✓		✓	✓
Uncinate fascicle left	UF_left			✓			✓
Uncinate fascicle right	UF_right			✓			✓

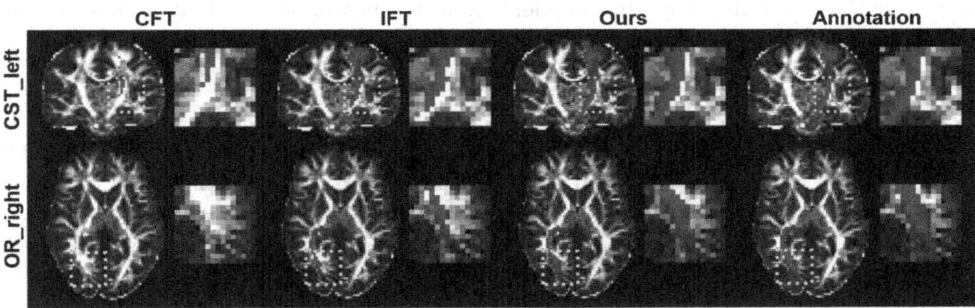

Fig. 1. Examples of the segmentation results (red) for novel WM tracts. The results were obtained in the case of CQ1 and are overlaid on the fractional anisotropy maps. The annotations are also shown for reference. Note the highlighted regions and their zoomed views for comparison. (Color figure online)

were used for testing, respectively, where the annotations of novel WM tracts were available and only used to measure the segmentation accuracy.

3.2 Evaluation of Segmentation Results

We compared the proposed method with two competing methods, which are the classic fine-tuning strategy and the improved fine-tuning strategy [10] described in Sect. 2.1 with the same pretrained model and single annotated training scan used by the proposed method. Both competing methods were integrated with TractSeg [17] like the proposed method. For convenience, the classic fine-tuning strategy and the improved fine-tuning strategy are referred to as CFT and IFT, respectively. Note that as shown in [10], in the one-shot setting directly training a model that segments novel WM tracts from scratch without the pretrained model would lead to segmentation failure. Thus, this strategy was not considered.

Table 2. The average Dice coefficient of each novel WM tract for each method for the cases of CQ1 and IH1. The best results are highlighted in bold. Asterisks indicate that the difference between the proposed method and the competing method is significant using a paired Student's t-test. ($^{***}p < 0.001$)

WM Tract	CQ1					IH1				
	CFT		IFT		Ours	CFT		IFT		Ours
	Dice	p	Dice	p	Dice	Dice	p	Dice	p	Dice
CST_left	0.123	***	0.463	***	**0.644**	0.208	***	0.455	***	**0.569**
CST_right	0.120	***	0.564	***	**0.692**	0.129	***	0.416	***	**0.542**
OR_left	0.000	***	0.281	***	**0.492**	0.118	***	0.504	***	**0.548**
OR_right	0.000	***	0.401	***	**0.533**	0.100	***	0.462	***	**0.518**

Table 3. The mean value of the average Dice coefficients of the novel WM tracts for each case and each method. The best results are highlighted in bold. The results achieved with each individual data augmentation strategy in our method are also listed.

Method	CQ1	CQ2	CQ3	IH1	IH2	IH3
CFT	0.061	0.097	0.092	0.139	0.280	0.192
IFT	0.427	0.351	0.396	0.459	0.518	0.531
Ours	**0.590**	**0.611**	**0.567**	**0.544**	**0.639**	**0.662**
RC1	0.574	0.552	0.519	0.529	0.630	0.637
RC2	0.552	0.555	0.544	0.514	0.625	0.651
TC1	0.524	0.605	0.536	0.509	0.592	0.637
TC2	0.582	0.610	0.552	0.524	0.629	0.654

We first qualitatively evaluated the proposed method. Examples of the segmentation results for novel WM tracts are shown in Fig. 1 for the proposed and competing methods, where the annotations are also displayed for reference. For demonstration, here we show the results obtained in the case of CQ1 for CST_left and OR_right. We can see that the segmentation results of our method better resemble the annotations than those of the competing methods.

Next, we quantitatively evaluated our method. The Dice coefficient was computed for the segmentation result of each novel WM tract on each test scan for each case. For demonstration, we have listed the average Dice coefficient of each novel WM tract in Table 2 for the cases of CQ1 and IH1. For each tract and each case, our method has a higher average Dice coefficient than the competing methods, and the improvement is statistically significant. We have also summarized the mean of the average Dice coefficients of the novel WM tracts for all the cases in Table 3 (upper half table). In all cases our method outperforms the competing methods with higher mean Dice coefficients.

Finally, we confirmed the benefit of each proposed data augmentation strategy, as well as the benefit of ensembling their results. For each case, the mean value of the average Dice coefficients of all novel WM tracts was computed for the segmentation results achieved with RC1, RC2, TC1, or TC2 individually, and the results are also given in Table 3 (lower half table). Compared with the results of IFT that did not use the proposed data augmentation, the integration of IFT with RC1, RC2, TC1, or TC2 led to improved segmentation accuracy, which indicates the individual benefit of each proposed data augmentation strategy. In addition, the Dice coefficients of the proposed method achieved with ensembling are higher than those achieved with a single data augmentation strategy, which confirms the benefit of ensembling. Note that there is not a data augmentation strategy that is better or worse than the others in all cases, which is possibly because of the randomness in RC1 and RC2 and the dependence of TC1 and TC2 on the spatial coverages of the novel WM tracts that vary in different cases.

4 Conclusion

We have proposed an approach to one-shot segmentation of novel WM tracts based on an improved pretraining and fine-tuning framework via extensive data augmentation. The data augmentation is performed with region masking, and several masking strategies are developed. The segmentation results achieved with these strategies are ensembled for the final segmentation. The experimental results on two brain dMRI datasets show that the proposed method improves the accuracy of novel WM tract segmentation in the one-shot setting.

Acknowledgements. This work is supported by Beijing Natural Science Foundation (L192058).

References

1. Banihashemi, L., et al.: Opposing relationships of childhood threat and deprivation with stria terminalis white matter. Hum. Brain Mapp. **42**(8), 2445–2460 (2021)
2. Basser, P.J., Pajevic, S., Pierpaoli, C., Duda, J., Aldroubi, A.: In vivo fiber tractography using DT-MRI data. Magn. Reson. Med. **44**(4), 625–632 (2000)
3. Bazin, P.L., et al.: Direct segmentation of the major white matter tracts in diffusion tensor images. Neuroimage **58**(2), 458–468 (2011)
4. DeVries, T., Taylor, G.W.: Improved regularization of convolutional neural networks with cutout. arXiv preprint arXiv:1708.04552 (2017)
5. Ding, Y., Yu, X., Yang, Y.: Modeling the probabilistic distribution of unlabeled data for one-shot medical image segmentation. In: AAAI Conference on Artificial Intelligence, pp. 1246–1254. AAAI (2021)
6. Isensee, F., Jaeger, P.F., Kohl, S.A., Petersen, J., Maier-Hein, K.H.: nnU-Net: a self-configuring method for deep learning-based biomedical image segmentation. Nat. Methods **18**(2), 203–211 (2021)
7. Jeurissen, B., Tournier, J.D., Dhollander, T., Connelly, A., Sijbers, J.: Multi-tissue constrained spherical deconvolution for improved analysis of multi-shell diffusion MRI data. Neuroimage **103**, 411–426 (2014)
8. Kingma, D.P., Ba, J.: Adam: a method for stochastic optimization. arXiv preprint arXiv:1412.6980 (2014)

9. Liu, W., et al.: Volumetric segmentation of white matter tracts with label embedding. Neuroimage **250**, 118934 (2022)
10. Lu, Q., Ye, C.: Knowledge transfer for few-shot segmentation of novel white matter tracts. In: Feragen, A., Sommer, S., Schnabel, J., Nielsen, M. (eds.) IPMI 2021. LNCS, vol. 12729, pp. 216–227. Springer, Cham (2021). https://doi.org/10.1007/978-3-030-78191-0_17
11. MacNiven, K.H., Leong, J.K., Knutson, B.: Medial forebrain bundle structure is linked to human impulsivity. Sci. Adv. **6**(38), eaba4788 (2020)
12. O'Donnell, L.J., Pasternak, O.: Does diffusion MRI tell us anything about the white matter? An overview of methods and pitfalls. Schizophr. Res. **161**(1), 133–141 (2015)
13. Stephens, R.L., Langworthy, B.W., Short, S.J., Girault, J.B., Styner, M.A., Gilmore, J.H.: White matter development from birth to 6 years of age: a longitudinal study. Cereb. Cortex **30**(12), 6152–6168 (2020)
14. Toescu, S.M., Hales, P.W., Kaden, E., Lacerda, L.M., Aquilina, K., Clark, C.A.: Tractographic and microstructural analysis of the dentato-rubro-thalamo-cortical tracts in children using diffusion MRI. Cereb. Cortex **31**(5), 2595–2609 (2021)
15. Tournier, J.D., Calamante, F., Connelly, A.: Robust determination of the fibre orientation distribution in diffusion MRI: non-negativity constrained super-resolved spherical deconvolution. Neuroimage **35**(4), 1459–1472 (2007)
16. Van Essen, D.C., Smith, S.M., Barch, D.M., Behrens, T.E., Yacoub, E., Ugurbil, K.: Wu-Minn HCP consortium: the WU-Minn human connectome project: an overview. Neuroimage **80**, 62–79 (2013)
17. Wasserthal, J., Neher, P., Maier-Hein, K.H.: TractSeg - fast and accurate white matter tract segmentation. Neuroimage **183**, 239–253 (2018)
18. Wu, Y., Hong, Y., Ahmad, S., Lin, W., Shen, D., Yap, Pew-Thian.: Tract dictionary learning for fast and robust recognition of fiber bundles. In: Martel, A.L., et al. (eds.) MICCAI 2020. LNCS, vol. 12267, pp. 251–259. Springer, Cham (2020). https://doi.org/10.1007/978-3-030-59728-3_25
19. Ye, C., Yang, Z., Ying, S.H., Prince, J.L.: Segmentation of the cerebellar peduncles using a random forest classifier and a multi-object geometric deformable model: application to spinocerebellar ataxia type 6. Neuroinformatics **13**(3), 367–381 (2015)
20. Yeatman, J.D., Dougherty, R.F., Myall, N.J., Wandell, B.A., Feldman, H.M.: Tract profiles of white matter properties: automating fiber-tract quantification. PLoS ONE **7**(11), e49790 (2012)
21. Yun, S., Han, D., Oh, S.J., Chun, S., Choe, J., Yoo, Y.: CutMix: regularization strategy to train strong classifiers with localizable features. In: International Conference on Computer Vision, pp. 6023–6032. IEEE (2019)
22. Zarkali, A., McColgan, P., Leyland, L.A., Lees, A.J., Rees, G., Weil, R.S.: Fiber-specific white matter reductions in Parkinson hallucinations and visual dysfunction. Neurology **94**(14), 1525–1538 (2020)
23. Zhang, F., Karayumak, S.C., Hoffmann, N., Rathi, Y., Golby, A.J., O'Donnell, L.J.: Deep white matter analysis (DeepWMA): fast and consistent tractography segmentation. Med. Image Anal. **65**, 101761 (2020)
24. Zhang, X., et al.: CarveMix: a simple data augmentation method for brain lesion segmentation. In: de Bruijne, M., et al. (eds.) MICCAI 2021. LNCS, vol. 12901, pp. 196–205. Springer, Cham (2021). https://doi.org/10.1007/978-3-030-87193-2_19

Accurate Corresponding Fiber Tract Segmentation via FiberGeoMap Learner

Zhenwei Wang, Yifan Lv, Mengshen He, Enjie Ge, Ning Qiang, and Bao Ge(✉)

School of Physics and Information Technology, Shaanxi Normal University, Xi'an, China
bob_ge@snnu.edu.com

Abstract. Fiber tract segmentation is a prerequisite for the tract-based statistical analysis and plays a crucial role in understanding brain structure and function. The previous researches mainly consist of two steps: defining and computing the similarity features of fibers, and then adopting machine learning algorithm for clustering or classification. Among them, how to define similarity is the basic premise and assumption of the whole method, and determines its potential reliability and application. The similarity features defined by previous studies ranged from geometric to anatomical, and then to functional characteristics, accordingly, the resulting fiber tracts seem more and more meaningful, while their reliability declined. Therefore, here we still adopt geometric feature for fiber tract segmentation, and put forward a novel descriptor (FiberGeoMap) for representing fiber's geometric feature, which can depict effectively the shape and position of fiber, and can be inputted into our revised Transformer encoder network, called as FiberGeoMap Learner, which can well fully leverage the fiber's features. Experimental results showed that the FiberGeoMap combined with FiberGeoMap Learner can effectively express fiber's geometric features, and differentiate the 103 various fiber tracts, furthermore, the common fiber tracts across individuals can be identified by this method, thus avoiding additional image registration in preprocessing. The comparative experiments demonstrated that the proposed method had better performance than the existing methods. The code and more details are openly available at https://github.com/Garand0o0/FiberTractSegmentation.

Keywords: Fiber clustering · Diffusion MRI · Fiber tract segmentation · FiberGeoMap

1 Introduction

The human brain has about 100 billion neurons that connect with each other by more synapses, forming a highly complex brain network. Diffusion MRI can be used to explore the structure of the brain, trace the fiber streamlines, and thus obtain fiber connections of the whole brain. These fiber streamlines provide rich information on the connection,

---

Supplementary Information The online version contains supplementary material available at https://doi.org/10.1007/978-3-031-16431-6_14.

L. Wang et al. (Eds.): MICCAI 2022, LNCS 13431, pp. 143–152, 2022.
https://doi.org/10.1007/978-3-031-16431-6_14

morphology and structure. However, the fiber tracking technique can produce about 10^4 -10^7 fibers streamlines for a whole brain, it is difficult to directly analyze, understand and leverage so many fiber streamlines and their inherent abundant information, therefore, these fibers need to be clustered or segmented into a relatively small number of fiber tracts, the fibers within each tract are similar and each fiber tract should have the relatively independent meaning, namely fiber tract segmentation or fiber clustering. It can facilitate diagnosis and follow-up of brain diseases in tract-based statistical analysis, and has important significance for understanding the structure and function of the brain, such as comparing the statistical differences of fiber tracts between the disease and the control group [1], or analyzing the relationship between the structural characteristics of fiber bundles and the functional connections of corresponding brain regions related to language [2].

A typical fiber tract segmentation framework can be divided into two steps. Firstly, the similarity features between fibers are defined and represented. Then, a variety of clustering and classification algorithms including the current deep learning methods [3, 4, 9] are applied to obtain fiber tracts. Previous studies have focused on three categories of features, which are geometrical [4, 5], anatomical [6, 7] and functional [8, 9] features in chronological order, and seem more and more reasonable and in line with the requirement of fiber clustering, but the uncertainty also increased in sequence, for example, anatomical feature based method depend on anatomical segmentation and registration, while the anatomical atlases are various, registration techniques are also not mature. For functional feature based methods, in addition to the error of registration between diffusion MRI and fMRI, the measures from fMRI data are also not very reliable [10]. So, in this paper we intend to adopt geometric features, which are more reliable, and don't be affected by registration, segmentation and unreliability of fMRI data. Different from the previous methods, we put forward a novel geometrical descriptor (FiberGeoMap) which can effectively depict fiber's geometrical features, including the shape, position and oriental information of a fiber, meanwhile this descriptor can effectively reduce the differences across individuals, and doesn't need a registration step. Accordingly, for FiberGeoMap, we proposed a revised Transformer network, called as FiberGeoMap Learner, which can efficiently explore the FiberGeoMap features, and then we trained the model with the all fibers from 205 HCP subjects [12], the experimental results showed that our method can obtain the accurate and corresponding fiber tract segmentation across individuals.

2 Materials and Methods

2.1 Overview

As shown in Fig. 1, the framework of the whole algorithm includes two parts. First, the global and local FiberGeoMaps of each fiber are calculated respectively, then the two FiberGeoMaps are concatenated as the input of the FiberGeoMap Learner, which is trained by taking the fiber tract atlas provided in [11, 17] as training and test set. The detailed steps to compute the global and local FiberGeoMap are shown in Fig. 2, and Fig. 4 shows the architecture of FiberGeoMap Learner.

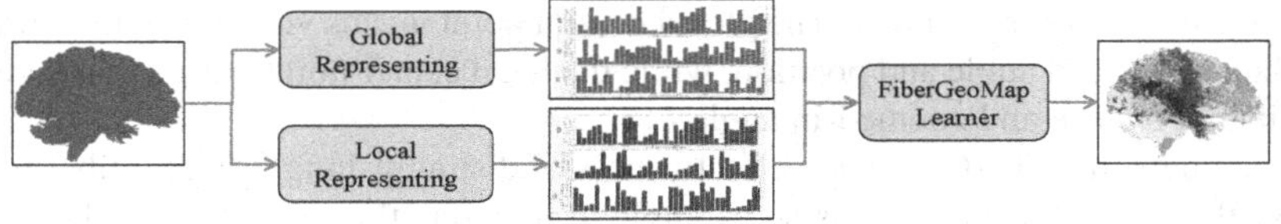

Fig. 1. The overall block diagram of the algorithm.

2.2 Materials and Preprocessing

The experimental data came from the publicly available Human Connectome Project (HCP) dataset [12], and the preprocessing of the data included brain skull removal, motion correction, and eddy current correction. After the pre-processing, the whole-brain streamline tractography was performed. More preprocessing details are referred to [11], we used the same preprocessing script command as [11] for convenient comparison later.

2.3 Fiber Representation by a FiberGeoMap

Although the brain fibers are very intricate and the number of them is huge, there are still some rules to follow. On the one hand, there exist some commonality, that is, the fibers belonging to the same tract are relatively close in position and similar in shape. On the other hand, there still exist some individuality, that is, those corresponding fiber tracts among individuals still have some differences in spatial distribution and shape. In order to distinguish different fiber tracts, and identify the common fiber tracts across subjects, we proposed a novel geometric description method, called FiberGeoMap to represent the spatial distribution and shape of fibers.

The process of computing FiberGeoMap is shown in Fig. 2. Taking the global FiberGeoMap as an example, we computed the geometrical central point according the range of all fibers' coordinate, which were then transformed from Cartesian coordinate (x, y, z) into the standard spherical coordinate (r, θ, φ) (Fig. 3(a)), and normalized to [0, 1], finally, we divided the value range of (r, θ, φ) into 36 equal intervals, respectively, and counted the normalized frequencies that the voxel points on each fiber streamline falling into every equal interval, thus obtained the three statistical histograms, that is, the 3*36 two-dimensional vector, denoted as FiberGeoMap.

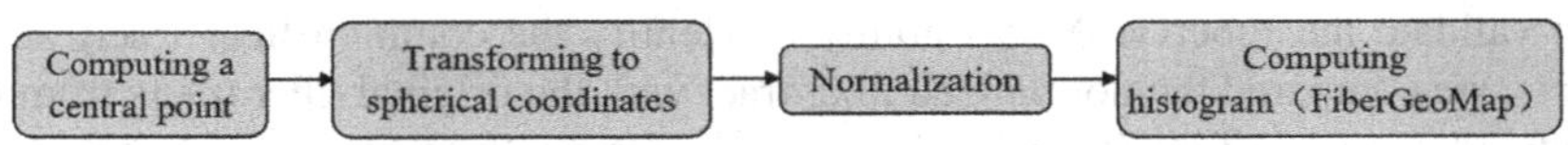

Fig. 2. Schematic diagram to computing FiberGeoMap

Two different center points are used as the origin of spherical coordinate system in this work, they are global and local center point, respectively. The global center point, as the name indicates, is the central point of the whole brain, which is the center of the maximum and minimum coordinates in X, Y and Z directions. The local center point is calculated only according to each fiber's coordinates, that is, each fiber in the brain should have a local central point. The global central point means viewing all fiber from a global unprejudiced angle and position, while the local center point means that viewing each fiber at a close and zoomed-in angle.

To validate the FiberGeoMap's effectiveness of distinguishing different fiber tracts, we take the global center point as an example to illustrate that. We selected the 3 fiber tracts (STR_right, STR_left, and MCP) from one subject to show in Fig. 3(b). Accordingly, the 3 corresponding FiberGeoMaps are shown in Fig. 3(c) with the same colors. We can see that the spatial distributions and shapes of STR_right and STR_left in Fig. 3(b) are approximate and symmetrical about global center, accordingly, their FiberGeoMap histograms in Fig. 3(c) are similar for the component r and θ, and the histograms for component φ are symmetrical about the 10th bin. On the other side, MCP is very different from the above two tracts in the spatial distribution and shape, shown in Fig. 3(b), accordingly, it is easy to see in Fig. 3(c) that the histograms of the three components of MCP are also very different from that of STR_right and STR_left. These results show that FiberGeoMap can effectively represent the fiber shape, orientation and spatial distribution, thus can distinguish different fiber tracts.

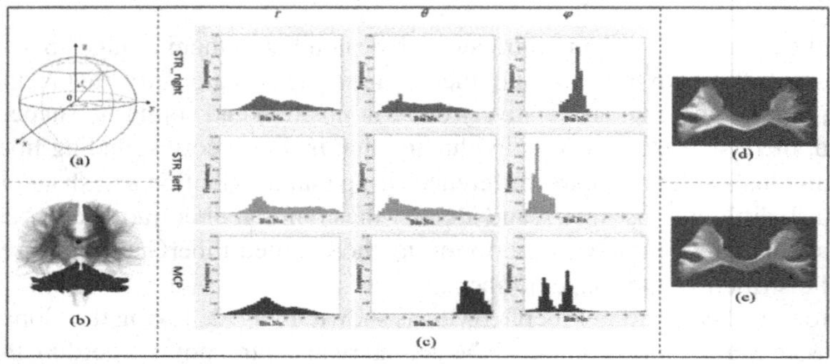

Fig. 3. (a) The Spherical coordinate system. (b) The 3 fiber tracts (STR_right, STR_left and MCP) from one randomly selected subject. The central point is represented by the black ball. The fibers of whole brain are shown gray as background. (c)The 3 corresponding FiberGeoMaps of the 3 fiber tracts. (d)The original spatial distribution of the two CC_3 tracts from two subjects. The two tracts are shown in white and green, respectively. (e) The spatial distribution of the two CC_3 tracts after by transforming to the spherical coordinate and normalization. (Color figure online)

To validate the FiberGeoMap's ability to identify the common fiber tracts across subjects, we compared the normalized fiber tract with the original fiber tracts from two subjects, as shown in Fig. 3(d) (e). The two CC_3 tracts from two subjects are shown in white and green color. It can be seen that the overlap of the original two tracts in Fig. 3(d) are less than that of the normalized two tracts in Fig. 3(e), especially for the part that the orange oval indicates. It illustrates that FiberGeoMap, to a certain extent, can weakened individual differences among different subjects, that is, FiberGeoMap can effectively help to identify the common fiber tracts across subjects.

2.4 Fiber Tract Segmentation via FiberGeoMap Learner

Transformer is a popular machine translation model based on neural network proposed by Ashish Vaswani and et al. [13]. For both the encoder and decoder of Transformer, they mainly rely on the multi-head self-attention module, which can learn the correlation between each time point and does not rely on the output of the decoder. Positional encodings are configured to add position information to the sequence data. A residual connection is also employed in the encoder, so the encoder can be stacked in multiple layers to obtain stronger learning capabilities.

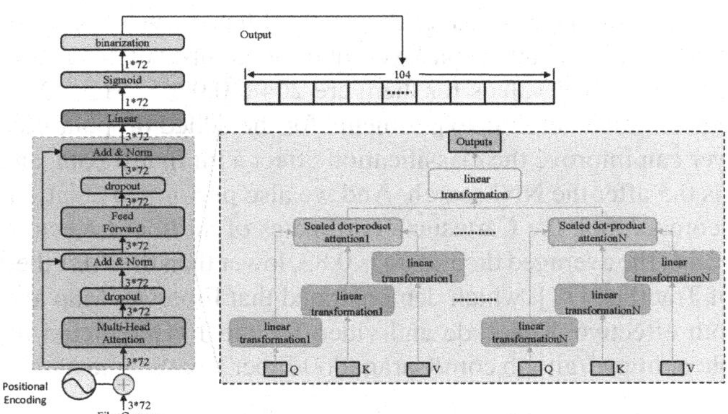

Fig. 4. The framework of FiberGeoMap Learner.

We proposed a revised computational framework entitled FiberGeoMap Learner based on Transformer model and multi-head self-attention, as shown in Fig. 4. We took FiberGeoMap $\in \mathbb{R}^{3\times n}$ as input, where 3 is the number of spherical axes (r, θ, φ), n is the number of bins on each spherical axis, the original n is 36, because we concatenated the global and local FiberGeoMaps, FiberGeoMap data has a size of 3×72. For a fiber, its FiberGeoMap data contains information about the direction and spatial distribution of the fiber. Differently from the typical Transformer network, firstly, we only adopted the encoder part, because we have labels for training data, and don't need a decoder to reconstruct the input. Secondly, because one fiber may have multiple labels, e.g., the fibers in tract CC_1 belong to both CC_1 and CC tract. To achieve multi-label segmentation, Sigmoid and binarization were added in the last layer. Thirdly, several dropout layers were added in the encoder to improve the generalization ability of the model.

3 Results and Discussion

We selected 205 subjects from HCP dataset as training and test set, which included 105 subjects exactly same as those in [11] and 100 subjects exactly same as those in [17]. TractSeg atlas in [11] and WMA atlas in [17] includes 72 major fiber tracts and 73 major fiber tracts, respectively, both were generated by a semi-automatic method with expert's involving. We then combined the two atlases, the combined atlas has 103 fiber tracts and because they have 42 common fiber tract classes. We divided the 205 subjects into training set, validation set, and test set according to the ratio of 3:1:1, and used the whole brain fibers of those subjects to train, validate and test. Actually we divided all fibers into 104 fiber tracts but not 103 tracts because we added one category including all the other fibers not belonging to the 103 major fiber tracts. There are over 100 million fibers for training in our method.

For the combined data, we adopted Neural Network Intelligence (NNI) (https://github.com/microsoft/nni) to automatically search the optimal hyper-parameters for the FiberGeoMap Learner. The hyper-parameters of mainly include: batch size, the learning rate of the Adamax, the number of hidden layers of the feed forward network, the number of attention heads of the self-attention layer, and the number of stacks of Transformer encoding layer. The optimal values for them are 2048, 0.0005, 512, 10, and 8, respectively. We also performed ablation experiments for the added dropout layer, and found the dropout layer can improve the classification effect a lot in test data, and the optimal dropout value is 0.5 after the NNI search. And we also performed ablation experiments for the FiberGeomap, took the Cartesian coordinates of all fiber trajectores as training and predicting data, the averaged dice score is 0.83, lower than 0.93 of FiberGeomap and close to 0.84 of TractSeg [11], which demonstrated that FiberGeoMap and transformer encoder are both effective. The code and video for the 104 predicted fiber tracts are openly available at https://github.com/Garand0o0/FiberTractSegmentation.

3.1 Results for Fiber Tract Segmentation

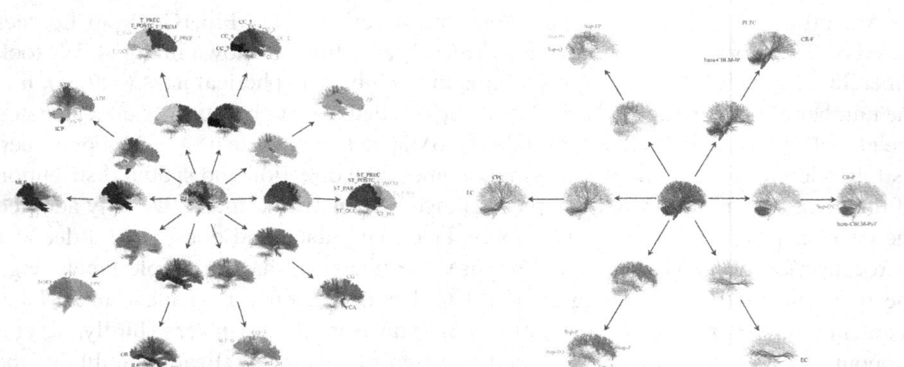

Fig. 5. The results of fiber tract segmentation. The results from our method are shown on the inner ring, and the reference tracts on the outer ring. The whole brain's fibers are shown gray as background. For those fiber tracts such as AF_left and AF_right, only one shown here. (a) The 72 predicted fiber tracts and TractSeg atlas. (b) The 31 predicted fiber tracts WMA atlas.

In order to quantitatively demonstrate the results, we used dice score [14], accuracy, precision and recall as the evaluation criterion, and computed these values between each predicted fiber tract and the corresponding fiber tract atlas from each subject in the test set, and averaged the values among fiber tracts and individuals. The averaged dice score, accuracy, precision and recall on the test set of the combined data (205 subjects) are 0.93, 0.94, 0.93 and 0.93 respectively. For the combined data, we chose randomly two subjects (one is from the 105 subjects and another from the 100 subjects, because the labels are from the two Atlas) in the test set to show the final fiber tract segmentation, as shown in Fig. 5. Each fiber tract is represented by different colors. The results from our method are shown on the inner ring, and the fiber tract atlas on the outer ring. There are 72 predicted and reference fiber tracts in (a) and the remaining 31 predicted and reference fiber tracts different from the 72 tracts are shown in (b). It can be seen that our results are very close to the reference fiber tracts.

3.2 Comparison with Other Methods

We compared the proposed method with TractSeg [11] and WMA [15] methods, their averaged dice scores with the two reference fiber tracts on the common 42 classes are 0.97, 0.85 and 0.54, respectively. Obviously, our method has much better performance than the two methods. For a more detailed comparison, we list the dice scores on 102 fiber bundles, as shown in Fig. 6(a). We found that dice scores on most fiber tracts from our method were very high, and close to 1, while dice scores on some were lower than those of TractSeg. Further, we found that the fiber tracts with high dice scores did not overlap with the other fiber tracts. While those with lower scores had much overlap with others, such as SLF_I_right and AF_right, etc., as shown in Fig. 6(b). The reason lies in that our method classified each fiber, whereas TractSeg classified each voxel. Considering an extreme situation, for a fiber streamline, our method can only classify the all voxels on the fiber streamline as one fiber tract, but TractSeg may classify voxel #1 as tract #1, and voxel #2 as tract #2, and so on, but actually these voxels should belong to the same class, this apparently did not conform to the common sense in neuroscience and may resulted in the higher dice score than ours. Moreover, TractSeg does not generate fiber streamlines, and fiber tracking needs to be carried out later according to the classification information of the voxels. Notably, the WMA method trained its model by using the all 100 subjects same as WMA atlas, so we cannot use the WMA method to do predicting on its own training data. Therefore we can only use the test set selected from the 105 subjects, accordingly, the WMA method only has 42 dice scores corresponding to the common fiber tracts with the TractSeg atlas, so 0.54 is the averaged dice score among the 42 fiber tracts for the WMA method.

3.3 Application on Autism

We applied the proposed method to autism spectrum disorder (ASD) data so as to inspect abnormal fiber tracts and partially validate the reproducibility of the method. The ASD data from the Trinity Centre for Health Sciences of Autism Brain Imaging Data Exchange II (ABIDE II), includes 21 individuals with ASD (10–20 years) and 21 typical controls (TC) (15–20 years). We segmented all fibers of each brain from the dataset into the 103 fiber tracts. Instead of computing FA, MD, etc., we then calculated the proportion of fibers contained in each fiber bundle to the whole brain fibers. In order to check which fiber bundles of ASD have abnormalities, we did a t-test on the ratio of ASD and TC group,

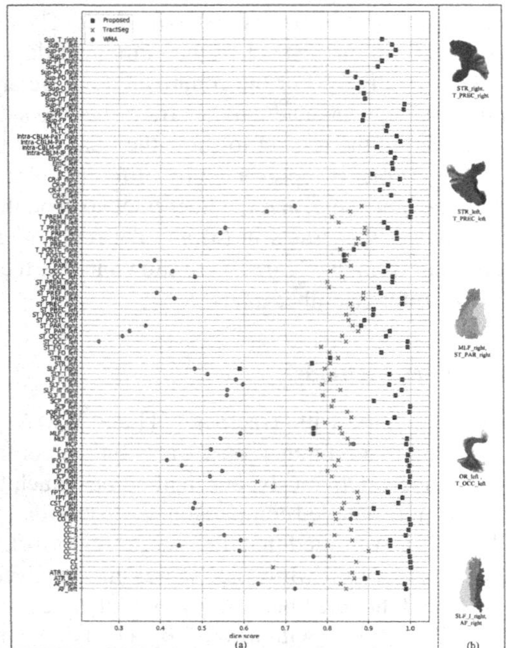

Fig. 6. (a) The dice scores on 103 fiber bundles for our method, TractSeg and WMA. The averaged dice scores for them are 0.93, 0.84 and 0.54, respectively. (b) Some examples that two fiber tracts partially overlap in the reference fiber tracts. The abbreviated name of these tracts are shown below each sub-figure, and the full names can be referred to [11, 17].

and found that CC_6, CR-P_left, EmC_left, OR_right, POPT_left, SLF_III_right, Sup-OT_right and Sup-O_left of ASD group have the statistically significant reduced ratio values than that of TC group. Our fiber tract ratio is very similar to the fiber density (FD) in [16], which reported that FD was statistically significant reduced on the CC and IFO, which is partially consistent with our results but we identified the more discriminating fiber tracts. The 103 predicted fiber tracts and corresponding 103 ratio values of the two groups can be referred to Supplementary Material.

4 Conclusion

In this paper, we proposed an effective fiber tract segmentation method, which can identify the accurate and common fiber tracts across individuals based on a novel representation of fiber's geometrical feature, called as FiberGeoMap, and we also tailored Transformer neural network to meet the input FiberGeoMap. The experimental results showed that the method has higher accuracy than the existing methods, meanwhile, FiberGeoMap can effectively depict the direction and spatial distribution of fibers, and can reduce the differences of corresponding fiber bundles across individuals, so that there is no need for additional image registration between different brains. Due to the large overlap between some fiber tracts in the reference fiber tracts, the segmentation accuracy of our method on these fiber tracts is relatively low. Our future work will focus on improving this, optimizing the parameters and training our model by using more reference tracts.

Acknowledgement. The work was supported by the National Natural Science Foundation of China (NSFC61976131 and NSFC61936007).

References

1. Mandl, R.C., et al.: Altered white matter connectivity in never-medicated patients with schizophrenia. Hum. Brain Mapp. **34**(9), 2353–2365 (2013)
2. Propper, R.E., et al.: A combined fMRI and DTI examination of functional language lateralization and arcuate fasciculus structure: effects of degree versus direction of hand preference. Brain Cogn. **73**(2), 85–92 (2010)
3. Zhang, F., Cetin Karayumak, S., Hoffmann, N., Rathi, Y., Golby, A.J., O'Donnell, L.J.: Deep white matter analysis (DeepWMA): fast and consistent tractography segmentation. Med. Image Anal. **65**, 101761 (2020)
4. Prince, D., Lam, N., Gaetan, et al.: TRAFIC: fiber tract classification using deep learning. In: Proceedings of SPIE–The International Society for Optical Engineering, vol. 10574, p. 1057412(2018)
5. Gerig, G., Gouttard, S., Corouge, I.: Analysis of brain white matter via fiber tract modeling. In: The 26th Annual International Conference of the IEEE Engineering in Medicine and Biology Society, pp. 4421–4424 (2004)
6. Tunç, B., et al.: Multinomial probabilistic fiber representation for connectivity driven clustering. In: Gee, J.C., Joshi, S., Pohl, K.M., Wells, W.M., Zöllei, L. (eds.) IPMI 2013. LNCS, vol. 7917, pp. 730–741. Springer, Heidelberg (2013). https://doi.org/10.1007/978-3-642-38868-2_61
7. Wassermann, D., Nikos, M., Rathi, Y., et al.: The white matter query language: a novel approach for describing human white matter anatomy. Brain Struct. Funct. **221**(9), 1–17 (2016)
8. Ge, B., Guo, L., Zhang, T., Hu, X., Han, J., Liu, T.: Resting state fMRI-guided fiber clustering: methods and applications. Neuroinformatics **11**(1), 119–133 (2013)
9. Wang, H., Qiang, N., Ge, B., Liu, T.: Task fMRI guided fiber clustering via a deep clustering method. In: IEEE 17th International Symposium on Biomedical Imaging (ISBI), Iowa City, IA, USA (2020)
10. Elliott, M., et al.: What is the test-retest reliability of common task-fMRI measures? New empirical evidence and a meta-analysis. Biol. Psychiatry **87**(9), S132–S133 (2020)
11. Wasserthal, J., Neher, P., et al.: TractSeg-fast and accurate while matter tract segmentation. Neuroimage **183**, 239–253 (2018)
12. Essen, D., Smith, S.M., Barch, D.M., Behrens, T., Ugurbil, K.: The WU-Minn Human Connectome Project: an overview. Neuroimage **80**, 62–79 (2013)
13. Vaswani, A., Shazeer, N., Parmar, N., et al.: Attention is all you need. arXiv (2017)
14. Taha, A.A., Hanbury, A.: Metrics for evaluating 3D medical image segmentation: analysis, selection, and tool. BMC Med. Imaging **15**, 29 (2015)
15. O'Donnell, L.J., Suter, Y., Rathi, Y., et al.: Automated white matter fiber tract identification in patients with brain tumors. Neuroimage **13**, 138–153 (2016)

16. Dimond, D., Schuetze, M., Smith, R.E., et al.: Reduced white matter fiber density in autism spectrum disorder. Cereb Cortex. **29**(4), 1778–1788 (2019)
17. Zhang, F., et al.: An anatomically curated fiber clustering white matter atlas for consistent white matter tract parcellation across the lifespan. Neuroimage **179**, 429–447 (2018)

An Adaptive Network with Extragradient for Diffusion MRI-Based Microstructure Estimation

Tianshu Zheng[1], Weihao Zheng[1], Yi Sun[2], Yi Zhang[1], Chuyang Ye[3], and Dan Wu[1](✉)

[1] College of Biomedical Engineering and Instrument Science, Zhejiang University, Hangzhou, Zhejiang, China
danwu.bme@zju.edu.cn
[2] MR Collaboration, Siemens Healthineers Ltd., Shanghai, China
[3] School of Integrated Circuits and Electronics, Beijing Institute of Technology, Beijing, China

Abstract. Diffusion MRI (dMRI) is a powerful tool for probing tissue microstructural properties. However, advanced dMRI models are commonly nonlinear and complex, which requires densely sampled q-space and is prone to estimation errors. This problem can be resolved using deep learning techniques, especially optimization-based networks. In previous optimization-based methods, the number of iterative blocks was selected empirically. Furthermore, previous network structures were based on the *iterative shrinkage-thresholding algorithm* (ISTA), which could result in instability during sparse reconstruction. In this work, we proposed *an* ***adaptive*** *network with* ***extragradient*** *for diffusion MRI-based* ***microstructure estimation*** (AEME) by introducing an additional projection of the extragradient, such that the convergence of the network can be guaranteed. Meanwhile, with the adaptive iterative selection module, the sparse representation process can be modeled flexibly according to specific dMRI models. The network was evaluated on the *neurite orientation dispersion and density imaging* (NODDI) model on a public 3T and a private 7T dataset. AEME showed superior improved accuracy and generalizability compared to other state-of-the-art microstructural estimation algorithms.

Keywords: Extragradient · Adaptive mechanism · Diffusion MRI · q-Space acceleration · Microstructure estimation

1 Introduction

Diffusion MRI (dMRI) is one of the most widely used medical imaging modalities for studying tissue microstructures in a noninvasive manner, based on the

Supplementary Information The online version contains supplementary material available at https://doi.org/10.1007/978-3-031-16431-6_15.

L. Wang et al. (Eds.): MICCAI 2022, LNCS 13431, pp. 153–162, 2022.
https://doi.org/10.1007/978-3-031-16431-6_15

restricted diffusion of the water molecules in the local microstructural environment [10]. With advanced dMRI models, specific microstructural properties such as cell size, permeability, axonal diameter, axonal orientation, and density could be mapped, typically by multi-compartment models [12] such as AxCaliber [1], *neurite orientation dispersion and density imaging* (NODDI) model [21], *soma and neurite density imaging* (SANDI) model [13], and so on.

Because advanced dMRI models are highly non-linear and consist of multiple compartments, they commonly need dense sampling in the q-space to better describe the microstructure properties. Dense sampling in q-space, namely the acquisition of multiple diffusion-weighted images with different gradient orientations and strengths, is time-consuming and prone to motion artifacts. Moreover, the optimization-based methods in the estimation of advanced dMRI models, such as the nonlinear least squares (NLLS) method [5], the Bayesian method [11], and the Markov chain Monte Carlo (MCMC) method [2], are easily prone to estimation errors.

Deep learning techniques have been employed to solve this problem, by learning the mapping between undersampled q-space data and microstructure indices. Golkov et al. [7] firstly employed a multilayer perceptron (MLP) to estimate the diffusion kurtosis and microstructure parameters using a subset of the q-space. Then various kinds of networks are proposed for dMRI model estimation [6,22]. Chen et al. [3] and Sedlar et al. [15] proposed Graph Neural Network (GNN) based structures to estimate the NODDI model parameters. Ye [17] proposed a model-driven deep neural network MEDN to estimate the NODDI model parameters through sparse coding, and further improved it by incorporating the historical information and extended it into various dMRI models [19]. However, the hyper-parameters in previous networks were often selected empirically, which would be an inefficient strategy for different dMRI models. Since different dMRI models had different nonlinearities, the hyper-parameters used in the optimization-based networks should also be different.

To overcome these problems, in this study, we mimicked the process of forming the sparse dictionary and proposed *an* ***adaptive*** *network with* ***extragradient*** *for diffusion MRI-based* ***microstructure estimation*** (AEME). Compared with the previous optimization-based neural networks, our proposed network AEME takes into account the termination criteria of the iterative process in traditional optimization procedures, including the error tolerance and maximum iteration number. By adaptively selecting the number of iterative units, we can avoid using a higher-complexity network structure for dMRI models with lower nonlinearity. Moreover, we introduced an extragradient into the dictionary reconstruction process and formed an extragradient unit (EG-Unit), so that the convergence of our network structure can be mathematically guaranteed [9].

In this study, our network was evaluated on the NODDI model, and the datasets were obtained from the public dataset the *Human Connectome Project* (HCP) collected on 3T [16], and a 7T dataset collected from our center. Both the results indicated that our method yielded microstructural estimates with considerably improved accuracy and generalizability.

2 Method

2.1 Theory

Previously, Ye has exploited the performance of the optimization-based neural network [17–20] in NODDI-based microstructure estimation. In this study, we further exploited the effect of the extragradient method and the adaptive mechanism of the iteration block in the microstructure estimation. A sparse representation can be formulated as follows:

$$\mathbf{Y} = \mathbf{\Gamma X \Upsilon}^{\mathrm{T}} + \mathbf{H} \tag{1}$$

where $\mathbf{Y} = (y_1 \cdots y_V) \in \mathbb{R}^{K \times V}$, y_v ($v \in \{1 \cdots V\}$ with V being the size of the patch) is the diffusion signal vector, $\mathbf{X} \in \mathbb{R}^{N_\Gamma \times N_\Upsilon}$ is the matrix of the sparse dictionary coefficients, and $\mathbf{H}$ represents noise. $\Gamma \in \mathbb{R}^{K \times N_\Gamma}$(with K being the number of the diffusion signal) and $\Upsilon \in \mathbb{R}^{V \times N_\Upsilon}$ are decomposed dictionaries that encode the information in the angular domain and the spatial domain [14]. Then the objective function Eq. (1) can be presented as follows:

$$\min_{\mathbf{X}} \left\| \mathbf{Y} - \Gamma \mathbf{X} \mathbf{\Upsilon}^{\mathrm{T}} \right\|_{\mathrm{F}}^{2} + \beta \|\mathbf{X}\|_1 \tag{2}$$

where β is a constant, and this function can be optimized through the following steps by introducing an extragradient method:

$$\begin{aligned} \mathbf{X}^{n+\frac{1}{2}} &= H_M\left(\mathbf{X}^n - \mathbf{\Gamma}^{\mathrm{T}}\left(\mathbf{\Gamma X}^n \mathbf{\Upsilon}^{\mathrm{T}} - \mathbf{Y}\right)\mathbf{\Upsilon}\right) \\ &= H_M\left(\mathbf{X}^n + \mathbf{\Gamma}^{\mathrm{T}}\mathbf{Y \Upsilon} - \mathbf{\Gamma}^{\mathrm{T}}\mathbf{\Gamma X}^n \mathbf{\Upsilon}^{\mathrm{T}}\mathbf{\Upsilon}\right) \end{aligned} \tag{3}$$

$$\begin{aligned} \mathbf{X}^{n+1} &= H_M\left(\mathbf{X}^n - \mathbf{A_1}\Gamma^{\mathrm{T}}\left(\mathbf{\Gamma X}^{n+\frac{1}{2}}\mathbf{\Upsilon}^{\mathrm{T}} - \mathbf{Y}\right)\mathbf{\Upsilon}\right) \\ &= H_M\left(\mathbf{X}^n + \mathbf{A_1}\mathbf{\Gamma}^{\mathrm{T}}\mathbf{Y\Upsilon} - \mathrm{A_1}\mathbf{\Gamma}^{\mathrm{T}}\mathbf{\Gamma X}^{n+\frac{1}{2}}\mathbf{\Upsilon}^{\mathrm{T}}\mathbf{\Upsilon}\right) \end{aligned} \tag{4}$$

where, $\mathrm{A_1}$ denotes a scalar matrix and H_M denotes a nonlinear operator:

$$H_M\left(\mathbf{X}_{ij}\right) = \begin{cases} 0, & \text{if } \mathbf{X}_{ij} < \lambda \\ \mathbf{X}_{ij}, & \text{if } \mathbf{X}_{ij} \geq \lambda \end{cases} \tag{5}$$

So far, we have obtained the sparse representation of the signal by an extended extragradient method.

2.2 Network Construction

In this part, we will describe how we construct our network based on Eqs. (3) and (4) through the unfolded iterations. Equations (3) and (4) can be rewritten as follows:

$$\mathbf{X}^{n+\frac{1}{2}} = H_M\left(\mathbf{X}^n + \mathbf{W}^{\mathrm{a1}}\mathbf{Y}\mathbf{W}^{\mathrm{s1}} - \mathbf{S}^{\mathrm{a1}}\mathbf{X}^n\mathbf{S}^{\mathrm{s1}}\right) \tag{6}$$

$$\mathbf{X}^{n+1} = H_M\left(\mathbf{X}^n + \mathbf{W}^{\mathrm{a2}}\mathbf{Y}\mathbf{W}^{\mathrm{s2}} - \mathbf{S}^{\mathbf{a2}}\mathbf{X}^{n+\frac{1}{2}}\mathbf{S}^{\mathrm{s2}}\right) \tag{7}$$

where, $W^{a1} \in \mathbb{R}^{N_\Gamma \times K}$,$W^{a2} \in \mathbb{R}^{N_\Gamma \times K}$, $W^{s1} \in \mathbb{R}^{N_\Upsilon \times V}$, $W^{s2} \in \mathbb{R}^{N_\Upsilon \times V}$, $S^{a1} \in \mathbb{R}^{N_\Gamma \times N_\Gamma}$, $S^{a2} \in \mathbb{R}^{N_\Gamma \times N_\Gamma}$, $S^{s1} \in \mathbb{R}^{N_\Upsilon \times N_\Upsilon}$, and $S^{s2} \in \mathbb{R}^{N_\Upsilon \times N_\Upsilon}$ are the matrices that need to be learned through the neural network, where **a** and **s** correspond to the angular domain and the spatial domain, respectively. Then we incorporate the historical information into the iterations:

$$\tilde{\mathbf{C}}^{n+\frac{1}{2}} = \mathbf{W}^{a1}\mathbf{Y}\mathbf{W}^{s1} + \mathbf{X}^{n} - \mathbf{S}^{a1}\mathbf{X}^{n}\mathbf{S}^{s1} \tag{8}$$

$$\mathbf{C}^{n+\frac{1}{2}} = \mathbf{F}^{n+\frac{1}{2}} \circ \mathbf{C}^{n} + \mathbf{G}^{n+\frac{1}{2}} \circ \tilde{\mathbf{C}}^{n+\frac{1}{2}} \tag{9}$$

$$\mathbf{X}^{n+\frac{1}{2}} = H_M\left(\mathbf{C}^{n+\frac{1}{2}}\right) \tag{10}$$

$$\tilde{\mathbf{C}}^{n+1} = \mathbf{W}^{a2}\mathbf{Y}\mathbf{W}^{s2} + \mathbf{X}^{n} - \mathbf{S}^{a2}\mathbf{X}^{n+\frac{1}{2}}\mathbf{S}^{s2} \tag{11}$$

$$\mathbf{C}^{n+1} = \mathbf{F}^{n+1} \circ \mathbf{C}^{n+\frac{1}{2}} + \mathbf{G}^{n+1} \circ \tilde{\mathbf{C}}^{n+1} \tag{12}$$

$$\mathbf{X}^{n+1} = H_M\left(\mathbf{C}^{n+1}\right) \tag{13}$$

where, $\tilde{\mathbf{C}}^{n+\frac{1}{2}}$, $\mathbf{C}^{n+\frac{1}{2}}$, $\tilde{\mathbf{C}}^{n+1}$, and $\mathbf{C}^{n+1}$ are four intermediate terms. The historical information is encoded through $\mathbf{F}^{n+\frac{1}{2}}$,$\mathbf{F}^{n+1}$,$\mathbf{G}^{n+\frac{1}{2}}$, and $\mathbf{G}^{n+1}$and they can be defined below:

$$\mathbf{F}^{n+\frac{1}{2}} = \sigma\left(\mathbf{W}^{a1}_{FX}\mathbf{X}^{n}\mathbf{W}^{s1}_{FX} + \mathbf{W}^{a1}_{FY}\mathbf{Y}\mathbf{W}^{s1}_{FY}\right) \tag{14}$$

$$\mathbf{F}^{n+1} = \sigma\left(\mathbf{W}^{a2}_{FX}\mathbf{X}^{n+\frac{1}{2}}\mathbf{W}^{s2}_{FX} + \mathbf{W}^{a2}_{FY}\mathbf{Y}\mathbf{W}^{s2}_{FY}\right) \tag{15}$$

$$\mathbf{G}^{n+\frac{1}{2}} = \sigma\left(\mathbf{W}^{a1}_{\mathbf{GX}}\mathbf{X}^{n}\mathbf{W}^{s1}_{\mathbf{GX}} + \mathbf{W}^{a1}_{\mathbf{GY}}\mathbf{Y}\mathbf{W}^{s1}_{\mathbf{GY}}\right) \tag{16}$$

$$\mathbf{G}^{n+1} = \sigma\left(\mathbf{W}^{a2}_{\mathbf{GX}}\mathbf{X}^{n+\frac{1}{2}}\mathbf{W}^{s2}_{\mathbf{GX}} + \mathbf{W}^{a2}_{\mathbf{GY}}\mathbf{Y}\mathbf{W}^{s2}_{\mathbf{GY}}\right) \tag{17}$$

where, $\mathbf{W}^{a1}_{FX}$, $\mathbf{W}^{s1}_{FX}$, $\mathbf{W}^{a1}_{GX}$, $\mathbf{W}^{s1}_{GX}$, $\mathbf{W}^{a1}_{FY}$, $\mathbf{W}^{s1}_{FY}$, $\mathbf{W}^{a1}_{GY}$, $\mathbf{W}^{s1}_{GY}$, $\mathbf{W}^{a2}_{FX}$, $\mathbf{W}^{s2}_{FX}$, $\mathbf{W}^{a2}_{GX}$, $\mathbf{W}^{s2}_{GX}$, $\mathbf{W}^{a2}_{FY}$, $\mathbf{W}^{s2}_{FY}$, $\mathbf{W}^{a2}_{GY}$, and $\mathbf{W}^{s2}_{GY}$ are learned matrices through the network. $\sigma(\cdot)$ applies for a sigmoid function. Following Eqs. (5)–(17), the EG-Unit can be constructed in Fig. 1(a).

We further exploit the mechanism of adaptively selecting the number of EG-Units during optimization. The flow chart can be shown in supplementary material Fig. A1. The network will stop optimizing when the validation loss is stable or the number of iterative units reaches its maximum. The overall network can be constructed by repeating the EG-Unit in Fig. 1(a) by N times, with N being the adaptive number of optimal iteration, and then the output will be sent to the mapping networks for mapping the microstructure parameters [20], which consists of three fully connected layers of the feed-forward networks. The overall structure can be shown in Fig. 1(b).

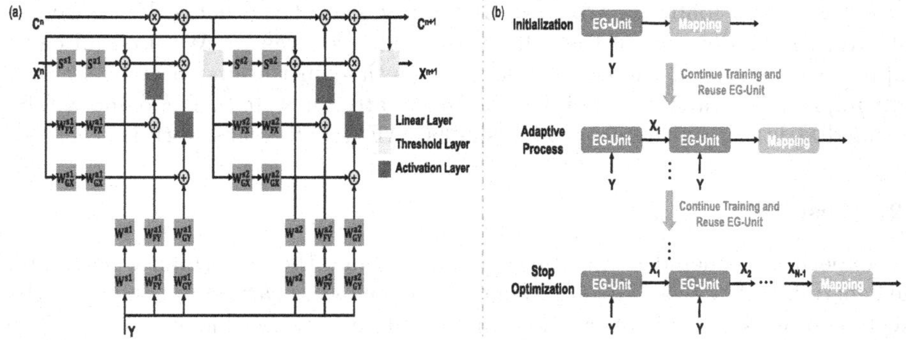

Fig. 1. Model overview. **(a)** An EG-Unit of the network, for sparse representation of dMRI signals with extragradient. **(b)** The overall structure of the AEME consists of an adaptive number of iteration blocks.

3 Experiments

In this section, we evaluated the proposed network to estimate NODDI parameters, including intra-cellular volume fraction (v_{ic}), isotropic volume fraction (v_{iso}), orientation dispersion (OD) [21] and compared the network performance with the existing algorithms developed for NODDI, including the NODDI Matlab Tool Box [7], a sparse coding based method AMICO [4], three learning-based methods q-DL [7], MEDN+ [18], and MESC_Sep_Dict (abbreviated as MESC2) [20]. Meanwhile, MEDN+ and MESC2 are two optimization-based neural networks.

3.1 Data and Training

The multi-shell dMRI from HCP Young Adult data [16] were acquired on a 3T MR scanner with 3 b-values (b = 1000, 2000, 3000 s/mm^2) and 90 diffusion directions in each b-value and 1.25 mm isotropic voxels. We randomly selected 26 adult subjects (age range, 22–35 years) and used 5 of them for training (with 10% of the training samples as validation [7]), and the remaining 21 subjects for testing.

Another dataset including 10 subjects (age range, 24–65 years) was scanned on a Siemens Magnetom 7 T scanner with the following parameters: 1.25 mm isotropic voxels, direction = 33 (for b = 1000 s/mm^2), and 66 (for b = 2000 s/mm^2). Both the non-diffusion-weighted and diffusion-weighted images were scanned with opposite phase-encoding directions for the purpose of EPI distortion correction [8]. Similar to the 3T dataset, We randomly selected 3 of them for training (with 10% of the training samples as validation [7]), and 7 subjects for testing. The gold standard microstructural parameters were computed by the NODDI Matlab Tool Box [7] (abbreviated as Zhang), using the full q-space data for both datasets.

In this work, following [20], the dictionary size of the network was set at 301, and the hidden size of the fully connected layer was 75. We used an early stopping strategy and the cosine warm-up method in the first 5 epochs and a reducing learning rate with an initial learning rate of 1×10^{-4}. The demo will be provided at https://github.com/Tianshu996/AEME after this work is accepted.

3.2 Results

The experiments included the ablation experiments, the performance tests, and the generalizability test. The ablation experiments were carried out to tested the effects of the extragradient method and adaptive iteration block strategy. The performance experiment included the tests on 3T datasets with 30, 18, and 12 diffusion gradient directions per shell, and comparisons with other algorithms. The generalizability experiment was tested on a 7T dataset.

1) **Network performance with or without extragradient** was tested on 30 diffusion directions per shell ($b = 1000{,}2000\,s/mm^2$). In the estimation accuracy test, the number of the iteration block was set to 8 following [18] and [20] then we compared the mean squared error (MSE) of the different algorithms. Table 1 indicated that the v_{ic}, v_{iso}, and OD estimated from AMICO were overall worse than other learning-based methods. The proposed extragradient-based method can achieve the best results compared to other learning algorithms (Table 1). The estimated errors using the proposed extragradient method were statistically lower than all other methods via paired Student's t-test.

Table 1. Evaluation of estimation errors on NODDI parameters using different methods on the downsampled q-space data with 30 diffusion directions. *$p < 0.05$, **$p < 0.01$, ***$p < 0.001$ between the other algorithms and AEME by paired t-test.

	AMICO	q-DL	MEDN+	MESC2	MESC2 + EG
v_{ic}	68 ± 11 (***)	16 ± 2.1 (***)	11 ± 1.5 (***)	9.3 ± 1.3 (***)	8.3 ± 1.2
v_{iso}	31 ± 3.2 (***)	7.0 ± 0.7 (***)	4.5 ± 0.5 (***)	3.8 ± 0.3 (***)	3.5 ± 0.3
OD	187 ± 31 (***)	23 ± 2.1 (***)	18 ± 1.7 (***)	12 ± 1.1 (***)	11 ± 0.9

2) **Network performance with or without the adaptive mechanism of the iteration block.** In this test, we also evaluated on 30 diffusion directions per shell ($b = 1000{,}2000\,s/mm^2$). Figure 2 showed the results of MESC2 with the fixed iteration block, MESC2 with the adaptive mechanism, and extragradient based MESC2 with the adaptive mechanism. It can be found in Fig. 2 with the iteration blocks increased from 0 to 8, validation loss (v_{ic}+v_{iso}+OD) was in the decreasing trend, and the validation loss increased trend after 8 blocks. MESC2 with the adaptive mechanism showed the lowest validation loss was achieved at 6 adaptive iteration blocks, and the validation loss using the adaptive mechanism was overall lower than the fixed approach using the same iteration blocks due to the different ways of concatenating the blocks. Compared with the

strategy of training with fixed-length iteration blocks from scratch, the adaptive mechanism is based on the previous training when each iteration block increases. Therefore, training efficiency can be improved. It was shown that each increment of iteration unit increased the FLOPs by $\sim 2.1 \times 10^{10}$, suggesting that the use of fewer iteration units could improve the computational efficiency. Therefore, the proposed adaptive methods could achieve higher efficiency with fewer iteration blocks and also lower estimation loss. The extragradient based MESC2 with adaptive mechanism showed the lowest variation loss compared with the non-extragradient method in each number of iteration blocks and had the same optimal setup with the adaptive mechanism only structure.

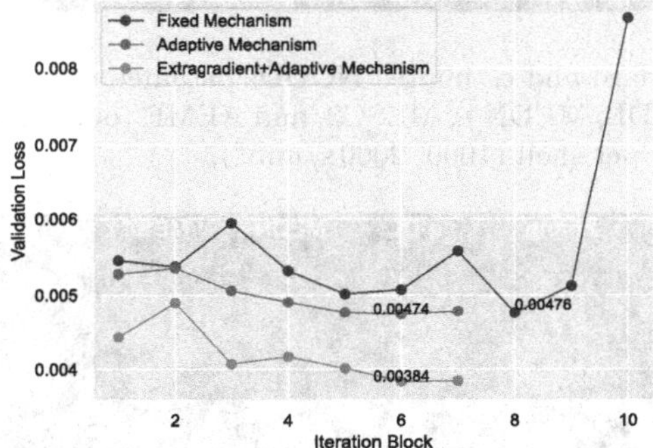

Fig. 2. Validation loss in the fixed mechanism versus the adaptive mechanism versus the extragradient method with the adaptive mechanism.

3) **Comparison between different microstructure estimation algorithms.** The comparison was first tested on 30 diffusion directions per shell (b = 1000,2000 s/mm^2). The estimated microstructural maps can be found in Fig. 3. And error maps were illustrated in Fig. 4, where we observed that AMICO was worse than all the learning-based methods and the AEME method had the least errors. The statistical comparison in Fig. 5 confirmed that AEME resulted in significantly lower estimation errors than all other methods via paired Student's t-test across 21 HCP test subjects.

4) **Performance on fewer diffusion gradients.** In the previous experiments, the number of diffusion directions was set to 30 for each shell (1000, 2000 s/mm^2). In this part, we further reduced the number of directions to 18 and 12 for each shell (1000, 2000 s/mm^2). The results in supplementary material Fig. A2 (a, b) demonstrated the proposed method achieved minimal estimation errors compared to other algorithms for all choices of gradient numbers and the differences were statistically significant via paired Student's t-test.

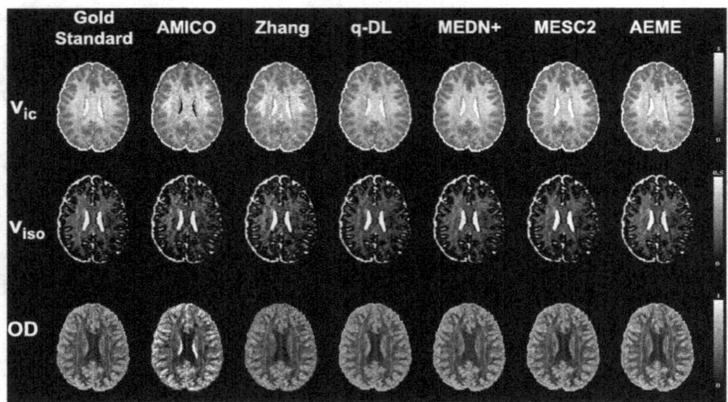

Fig. 3. The gold standard and estimated NODDI parameters v_{ic}, v_{iso}, and OD based on AMICO, Zhang, q-DL, MEDN+, MESC2, and AEME (ours) in a test subject using 30 diffusion directions per shell (1000, 2000 s/mm^2).

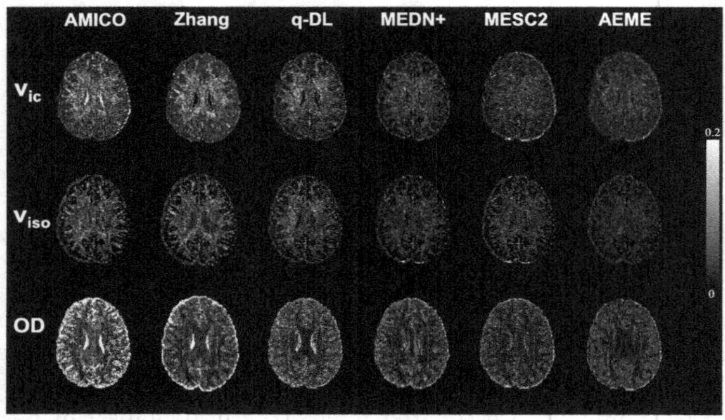

Fig. 4. The estimation errors of NODDI parameters v_{ic}, v_{iso}, and OD based on AMICO, Zhang, q-DL, MEDN+, MESC2, and AEME (ours) compared with the gold standard in a test subject with 30 diffusion directions per shell (1000, 2000 s/mm^2)

5) **Generalizability test on 7T data.** To evaluate the generalizability of the proposed method, we tested our method on the 7T dataset. In these tests, the downsample schemes were the same as in the 3T HCP Young Adult datasets. We tested the on 30, 18, and 12 directions per shell (b = 1000,2000 s/mm^2). The results in supplementary material Table A1 showed that the estimated errors from AEME were significantly lower than all other methods via paired Student's t-test, regardless of the number of diffusion directions used.

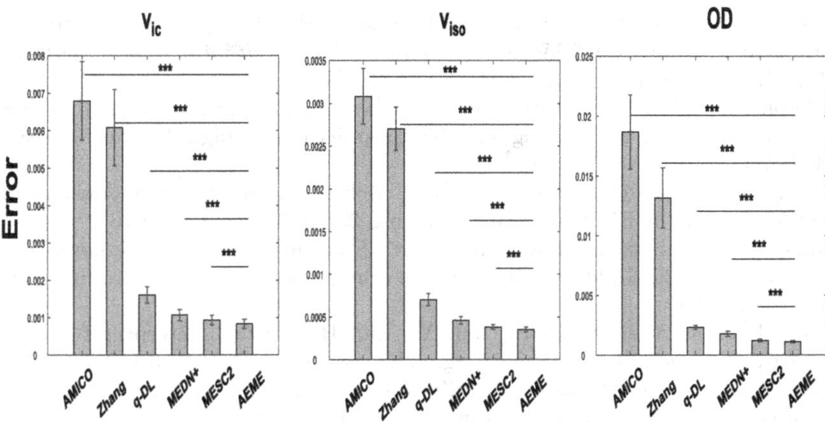

Fig. 5. The means and standard deviations of the estimation errors of NODDI parameters in test subjects (n = 21), using 30 (a), 18 (b), and 12 (c) diffusion directions, respectively. *$p < 0.05$, **$p < 0.01$, ***$p < 0.001$ by paired t-test.

4 Conclusion and Future Work

In this work, we proposed an adaptive network for microstructure properties estimation with an additional projection to estimate the microstructure properties. We tested our network architecture through ablation experiments, public HCP Young Adult dataset, and private 7T dataset tests, demonstrating that our network has the best performance compared to other network architectures and achieves up to 11.25 (270/24) times speedup in q-space sampling.

In the future, we will test our network on additional models to demonstrate the necessity of adaptive adjustment of AEME blocks. Meanwhile, we will test our network for more robustness and generalizability, such as different downsampling strategies and multicenter centers tests.

References

1. Assaf, Y., Blumenfeld-Katzir, T., Yovel, Y., Basser, P.J.: Axcaliber: a method for measuring axon diameter distribution from diffusion MRI. Magn. Reson. Med. Offi. J. Int. Soc. Magn. Reson. Med. **59**(6), 1347–1354 (2008)
2. Behrens, T.E., et al.: Characterization and propagation of uncertainty in diffusion-weighted MR imaging. Magn. Reson. Med. Off. J. Int. Soc. Magn. Reson. Med. **50**(5), 1077–1088 (2003)
3. Chen, G., et al.: Estimating tissue microstructure with undersampled diffusion data via graph convolutional neural networks. In: Martel, A.L., et al. (eds.) MICCAI 2020. LNCS, vol. 12267, pp. 280–290. Springer, Cham (2020). https://doi.org/10.1007/978-3-030-59728-3_28
4. Daducci, A., Canales-Rodríguez, E.J., Zhang, H., Dyrby, T.B., Alexander, D.C., Thiran, J.P.: Accelerated microstructure imaging via convex optimization (AMICO) from diffusion MRI data. Neuroimage **105**, 32–44 (2015)

5. Federau, C., O'Brien, K., Meuli, R., Hagmann, P., Maeder, P.: Measuring brain perfusion with intravoxel incoherent motion (IVIM): initial clinical experience. J. Magn. Reson. Imaging **39**(3), 624–632 (2014)
6. Gibbons, E.K., et al.: Simultaneous NODDI and GFA parameter map generation from subsampled q-space imaging using deep learning. Magn. Reson. Med. **81**(4), 2399–2411 (2019)
7. Golkov, V., et al.: Q-space deep learning: twelve-fold shorter and model-free diffusion MRI scans. IEEE Trans. Med. Imaging **35**(5), 1344–1351 (2016)
8. Holland, D., Kuperman, J.M., Dale, A.M.: Efficient correction of inhomogeneous static magnetic field-induced distortion in echo planar imaging. Neuroimage **50**(1), 175–183 (2010)
9. Kong, L., Sun, W., Shang, F., Liu, Y., Liu, H.: Learned interpretable residual extragradient ISTA for sparse coding. arXiv preprint arXiv:2106.11970 (2021)
10. Mori, S., Zhang, J.: Principles of diffusion tensor imaging and its applications to basic neuroscience research. Neuron **51**(5), 527–539 (2006)
11. Neil, J.J., Bretthorst, G.L.: On the use of Bayesian probability theory for analysis of exponential decay date: an example taken from intravoxel incoherent motion experiments. Magn. Reson. Med. **29**(5), 642–647 (1993)
12. Novikov, D.S., Fieremans, E., Jespersen, S.N., Kiselev, V.G.: Quantifying brain microstructure with diffusion MRI: theory and parameter estimation. NMR Biomed. **32**(4), e3998 (2019)
13. Palombo, M., et al.: Sandi: a compartment-based model for non-invasive apparent soma and neurite imaging by diffusion MRI. Neuroimage **215**, 116835 (2020)
14. Schwab, E., Vidal, R., Charon, N.: Spatial-angular sparse coding for HARDI. In: Ourselin, S., Joskowicz, L., Sabuncu, M.R., Unal, G., Wells, W. (eds.) MICCAI 2016. LNCS, vol. 9902, pp. 475–483. Springer, Cham (2016). https://doi.org/10.1007/978-3-319-46726-9_55
15. Sedlar, S., Alimi, A., Papadopoulo, T., Deriche, R., Deslauriers-Gauthier, S.: A spherical convolutional neural network for white matter structure imaging via dMRI. In: de Bruijne, M., et al. (eds.) MICCAI 2021. LNCS, vol. 12903, pp. 529–539. Springer, Cham (2021). https://doi.org/10.1007/978-3-030-87199-4_50
16. Van Essen, D.C., et al.: The Wu-Minn human connectome project: an overview. Neuroimage **80**, 62–79 (2013)
17. Ye, C.: Estimation of tissue microstructure using a deep network inspired by a sparse reconstruction framework. In: Niethammer, M., et al. (eds.) IPMI 2017. LNCS, vol. 10265, pp. 466–477. Springer, Cham (2017). https://doi.org/10.1007/978-3-319-59050-9_37
18. Ye, C.: Tissue microstructure estimation using a deep network inspired by a dictionary-based framework. Med. Image Anal. **42**, 288–299 (2017)
19. Ye, C., Li, X., Chen, J.: A deep network for tissue microstructure estimation using modified LSTM units. Med. Image Anal. **55**, 49–64 (2019)
20. Ye, C., Li, Y., Zeng, X.: An improved deep network for tissue microstructure estimation with uncertainty quantification. Med. Image Anal. **61**, 101650 (2020)
21. Zhang, H., Schneider, T., Wheeler-Kingshott, C.A., Alexander, D.C.: NODDI: practical in vivo neurite orientation dispersion and density imaging of the human brain. Neuroimage **61**(4), 1000–1016 (2012)
22. Zheng, T., et al.: A model-driven deep learning method based on sparse coding to accelerate IVIM imaging in fetal brain. In: ISMRM 2021: The 29th International Society for Magnetic Resonance in Medicine (2021)

Shape-Based Features of White Matter Fiber-Tracts Associated with Outcome in Major Depression Disorder

Claire Cury[1(✉)], Jean-Marie Batail[2], and Julie Coloigner[1]

[1] Univ Rennes, CNRS, Inria, Inserm, IRISA UMR 6074, Empenn ERL U-1228, 35000 Rennes, France
claire.cury@inria.fr

[2] Academic Psychiatry Department, Centre Hospitalier Guillaume Régnier, 35703 Rennes, France

Abstract. Major depression is a leading cause of disability due to its trend to recurrence and treatment resistance. Currently, there are no biomarkers which could potentially identify patients with risk of treatment resistance. In this original paper, we propose a two-level shape analysis of the white matter bundles based on the Large Diffeomorphic Deformation Metric Mapping framework, to study treatment resistant depression. Fiber bundles are characterised via the deformation of their center line from a centroid shape. We developed two statistical analyses at a global and a local level to identify the most relevant bundles related to treatment resistant depression. Using a prospective longitudinal cohort including 63 patients. We applied this approach at baseline on 50 white matter fiber-tracts, to predict the clinical improvement at 6 months. Our results show a strong association between three bundles and the clinical improvement 6 months after. More precisely, the right-sided thalamo-occipital fascicle and optic radiations are the most robust followed by the splenium. The present study shows the interest in considering white matter shape in the context of depression, contributing to improve our understanding of neurobiological process of treatment resistance depression.

Keywords: Computational anatomy · Depression · Fibre bundles

1 Introduction

Depression is a worldwide leading cause of disability due to its trend to recurrence and treatment resistance [10,15]. The Sequenced Treatment Alternatives to Relieve Depression (STAR*D) trial has indeed demonstrated that 30% of patients remained depressed after 4 trials of antidepressants (ATD) [21]. The

Supplementary Information The online version contains supplementary material available at https://doi.org/10.1007/978-3-031-16431-6_16.

L. Wang et al. (Eds.): MICCAI 2022, LNCS 13431, pp. 163–173, 2022.
https://doi.org/10.1007/978-3-031-16431-6_16

remission rate dropped dramatically after the second ATD trial. The last two decades have been marked by an increasing interest in research on biomarkers that could help to identify at risk patients of Treatment Resistant Depression (TRD) and to guide in the therapeutic strategies [16,28].

Advances in neuroimaging have improved our understanding of the neural circuit involved in depression. The structural connectivity from diffusion-weighted Magnetic Resonance Imaging (dMRI) has provided valuable insights on the white matter abnormalities associated with depression and different depressive phenotypes [4]. Therefore, to date, there is no consensual imaging biomarkers of TRD which could potentially improve our understanding the pathophysiology of this pejorative outcome [8]. Only few studies examined the relation between neuroimaging information and TRD [1]. Thus, cross-sectional works have emphasized that white matter abnormalities affecting networks related to emotion regulation such as cortico-subcortical circuits as well as subgenual anterior cingulate cortex are involved in the TRD pathophysiology [23]. However, these studies have reported only moderate correlations between TRD and white matter microstructure metrics derived from dMRI. While the large majority of dMRI studies in depression focuses on voxel-based analysis of microstructure metrics, we propose a shape-based analysis of white matter fiber-tracts. The shape variability of those bundles (3D curves) may yield new insights into normal and pathologic brain development [11,12].

In this original paper, we propose a two levels shape analysis of the white matter bundles using the Large Diffeomorphic Deformation Metric Mapping framework (LDDMM), now well-known in computational anatomy, that has the great advantage to embed shapes into a Riemannian manifold. In LDDMM Shapes can be characterised via the deformations of their center line to a centroid shape. Our original approach is to analyse those deformations by estimating local and global features, used to model the association between white matter fiber bundles and TRD. The goal is to address the question: is there baseline shape-based biomarkers on white matter tracts that allow to determine the outcome of a depressed patient 6 months after? Using a prospective longitudinal cohort, we applied this approach at baseline on 50 white matter fiber-tracts, to predict the clinical improvement at 6 months.

2 Data and Pre-processings

Population and Imaging Acquisition: Sixty-three depressed patients were recruited from routine care units in the psychiatric university hospital of Rennes. They were enrolled in a prospective longitudinal cohort study, which was approved by an ethic committee and registered in www.clinicaltrial.gov (NCT 02286024). Written informed consents were obtained from all subjects. Depressed patients underwent clinical interview and examination including routine neuropsychological testing at baseline and 6 months after. The Clinical Global Impression - improvement scale (CGI-I) was used at 6-months to measure the treatment response [13]. This measure is a well-established rating tool applicable to all psychiatric disorders and well correlated with other standard scales

[2]. It provides a global rating of illness severity and improvement, taking into accounts the patient's history, social circumstances, symptoms and the impact of the illness on the patient's ability to function: The CGI-I measure is rated from 1 (very much improved) to 7 (very much worse) [2]. A demographic table is given in supplementary material. At baseline, patients were scanned on a 3T Verio Siemens MR scanner with a 32-channel head coil. The 3D T1-weighted image was acquired covering the whole brain (176 sagittal slices) with TR = 1.9 s, TE = 2.26 ms, flip angle = 9°, resolution = 1 mm× 1 mm× 1 mm, FOV = 256 mm ×256 mm). The dMRI data were gathered on 60 slices using an interleaved slice acquisition, no gap, resolution= 2 mm×2 mm×2 mm and in a 256 mm ×256 mm field of view. The acquisition and reconstruction matrices were 128 × 128, using 30 directions and a b-value of 1000 s/mm^2. TR/TE $= 11,000/99$ ms, flip angle was 90° and pixel bandwidth was 1698 Hz.

Diffusion Image Processing: Diffusion images were corrected, using the open source Anima toolbox[1], for eddy current-induced image distortion using a block-matching distortion correction method ensuring an opposite symmetric transformation [14]. Then, the individual B0 image was co-registered with the structural image with 1 mm isotropic resolution, using a linear block-matching algorithm [5,19]. For EPI distortion correction, the B0 image was non-linearly registered to the structural image with a linear block-matching algorithm [5]. A rigid realignment was performed between the 30 dMRIs to compensate for subject motion. Then, denoising step using blockwise non-local means filtering was applied [6]. Skull stripping was also performed using the Anima toolbox. Diffusion images were transformed into the Montreal Neurological Institute (MNI) template space, in two steps: The structural image was non-linearly transformed to the MNI template space, using block-matching algorithms [5,19]; Then, we applied those transformations to the pre-processed diffusion images to transform them into the MNI space.

Center Line Estimation: Single-shell single-tissue constrained spherical deconvolution was used to extract the fibre orientation distributions [24] and then anatomically constrained probabilistic tractography was performed using MRtrix toolbox [25] to estimate the white matter bundles. Then, the automated pipeline TractSeg was used to generate 50 bundle-specific tractograms [29]. This tool is based on convolutional neural networks trained to create tract orientation maps, enabling the creation of accurate bundle-specific tractograms. For each bundle, a center line was determined using QuickBundles [9]. A distance measure called minimum average direct flip was used to calculate the center line of 100 points, for each bundle of each patient.

[1] https://github.com/Inria-Visages/Anima-Public/wiki.

3 Proposed Methodology

3.1 Shape Representation and Framework

To analyse the center lines we used the deformation-based Large Diffeomorphic Deformation Metric Mapping (LDDMM) framework, that generates a time-dependent flow of diffeomorphisms between two shapes [26]. In the LDDMM framework, the shortest diffeomorphism between two objects happens to be a geodesic, embedding them into a Riemannian manifold and allowing the analysis of shapes on the tangent space at a given point of the manifold.

In brief (for more details one can refer to [20, 26]), and to introduce some notations, a deformation is defined by integrating time-varying velocity vector fields $v(t)$ (with time $t \in [0;1]$), belonging to the reproducing kernel Hilbert space V. This is the closed span of vectors fields of the form $K_V(x,\cdot)\alpha$ with K_V a kernel and $\alpha_m(t) \in \mathbb{R}^3$ a time-dependent vector called momentum vector defined at point x_m and time t. The optimal deformation Φ between the shape of bundle i of subject j_1 (i.e. a centre line) and subject j_2 is estimated by minimising the functional $J(\Phi) = \gamma Reg(v(t)) + \|\Phi(B_{i,j_1}) - B_{i,j_2}\|_2^2$, where γ weights the regularisation and the data fidelity terms, and $Reg(v(t))$ is the energy enforcing $v(t)$ to stay in the space V. Here, we selected the sum of squared distance for the fidelity data term, since center lines (one smooth line per subject) can be considered as a set of landmarks. This choice also speed up computation time and avoid extra parameter tuning. Here, the optimal deformation Φ is a geodesic from B_{i,j_1} to B_{i,j_2} and is then fully encoded by the set of initial momentum vectors $\boldsymbol{\alpha}^{j_2}_{B_{i,j_1}}(0) = \{\alpha^{j_2}_m(0)\}_{m\in\{1,\ldots,M\}}$ defined at landmark positions x_m of shape $B_{i,j_1} = \{x_m\}_{m\in\{1,\ldots,M\}}$. The initial momentum vectors allow to analyse deformations on the tangent space at a given position, then we analyse a population of shape through their deformations from a centroid point on the manifold.

3.2 Shape Features Extraction

For a given bundle i and for each subjects j, the features we are interested in this paper are the set of initial momentum vectors $\{\boldsymbol{\alpha}^j_{B_i}(0)\}_j$ going from a centroid point (B_i) on the manifold of the i-th bundle towards subjects j. To do so, we first estimate this centroid shape of the i-th bundle $B_i = \{x_m\}_{m\in\{1,\ldots,M\}}$ as proposed in [7]. A centroid shape is computed by iteratively adding a new subject of the population at each step. This method only requires $N-1$ deformation estimations (i.e. minimisation of $J(\Phi)$). Then, we computed the deformations from the centroid shape of a bundle B_i to each subject's shape j to obtain the set of initial momentum vectors $\{\boldsymbol{\alpha}^j_{B_i}(0)\}_{j\in\{1,\ldots,N\}} = \{\alpha^j_m(0)\}_{m\in\{1,\ldots,M\}}, \forall j \in \{1,\ldots,N\}$, for a total of $2N-1$ shape registrations. For each centroid bundle B_i, to analyse the corresponding $\{\boldsymbol{\alpha}^j_{B_i}(0)\}_j$ for all subjects j, we propose two types of features: local and global.

Local Features describe initial momentum vectors at each landmark positions across all subjects. We define those local features of each initial momentum vectors, located at each point on the centroid shape B_i, as spherical coordinates using the average vector as the pole: amplitude, polar and azimuthal angles. The amplitude of $\{\alpha_m^j(0)\}$ captures the amount of deformation which is the distance to the centroid of the population. To capture vector orientations, we calculated, for each landmark x_m of the bundle i the polar angle between the vector $\alpha_m^j(0)$ and $v_1 = \sum_{j=1}^N \alpha_m^{i,j}(0)/N$, the average of initial momentum vectors of the population, and its azimuthal angle. To build an orthogonal vector v_3 to v_1, where $v_3 = v_1 \times v_2$, we arbitrarily choose a vector $v_2 = v_1 + (1,1,1)$ to preserve the connection between v_1 and v_2 across all x_m along a bundle i. The azimuth is the angle between $\alpha_m^j(0)$ and the vector v_3.

Global Features are defined to capture global shape changes and therefore do not focus on momenta $\alpha_m^j(0)$ independently. Global features are estimated through a Kernel Principal Component Analysis (K-PCA) [22] on $\{\boldsymbol{\alpha}_{B_i}^j(0)\}_{j\in\{1,\ldots,N\}}$ in the tangent space at the centroid shape B_i position. We used the same kernel K_V of the space of deformations. For the global analysis, we used the 5 first eigenvectors, explaining the most important deformations of each bundles.

3.3 Analysis

We defined two linear models, one using the global features (i.e. 5 principal components), and the other one including the local features (i.e. M amplitudes, and $2 \times M$ angular distances for the orientations), to assess the association between those features and the clinical improvement score 6 months after, called CGI-I. As co-factors in both models, we used for each patient: disease duration, age, gender, and the medication load including four types of drugs (anti-depressant, mood stabiliser, anti-psychotic, Benzodiazepine).

To improve and challenge the robustness of the different models, especially for the local features, as they depend on the number and position of landmarks, we used 10 different numbers of landmarks with $M \in \{5, 6, 7, 8, 9, 10, 11, 13, 15, 20\}$. We did not add more landmarks to keep the computation times reasonable, and to avoid having a too high number of co-factors, since for the local model, the number of features depend on the number of landmarks. We also augmented the data, by adding to each subjects 4 noisy versions of each bundles, increasing the size of the dataset from 63 to 315. To do so, a Gaussian noise on each central line was added. The dataset augmentation and the use of different landmarks lead to more than 300000 minimisation of functional J for deformation estimations. Finally, to gain in robustness and minimise false discoveries, each of the $2 \times 10 \times 50$ models are estimated 63 times using a leave-one-out approach. The number 63 corresponds to the original number of patients used in this study. To avoid introducing bias, the leave-one-out is actually a leave-5-out, removing at each step a patient's bundle and its 4 noisy versions. To estimate comparable coefficients, we standardised all variables and co-factors, and to assess the performance of the model, we use the adjusted R^2 that represents the amount of the variation in CGI-I explained by variables and co-factors, adjusted for the number of terms in the model. From the leave-5-out approach, we derived mean and 95% confidence interval for adjusted R^2, for model p-values and for variables coefficients.

4 Results

Figure 1 shows all the observations used and the 50 centroid shapes, one per bundle, used as tangent location to estimate $\{\boldsymbol{\alpha}^j_{B_i}(0)\}_j$, $\forall i \in \{1, ..., 50\}$ and $j \in \{1, ..., 315\}$. For each bundle we have 10 local models and 10 global models. First, to identify bundles of interests, we consider a model to be a good candidate when 1/ the 95% confidence interval of the model's p-value is under the threshold 5.10^{-5} (i.e. 0.05 corrected for multiple comparisons), 2/ the lower part of the 95% confidence interval of adjusted R^2 is higher than 0.3, meaning we consider only models that explain at least 30% of the variance. Then, we consider that a bundle has significant shape features, and is therefore of interest, when 3/ at least one of the approach (local or global) passes condition 1/ and 2/ for 10 out of 10 and the other approach for at least 5 out 10. Figure 2 displays, for each of the 10 models the adjusted R^2 value when passing conditions 1/ and 2/. The mean

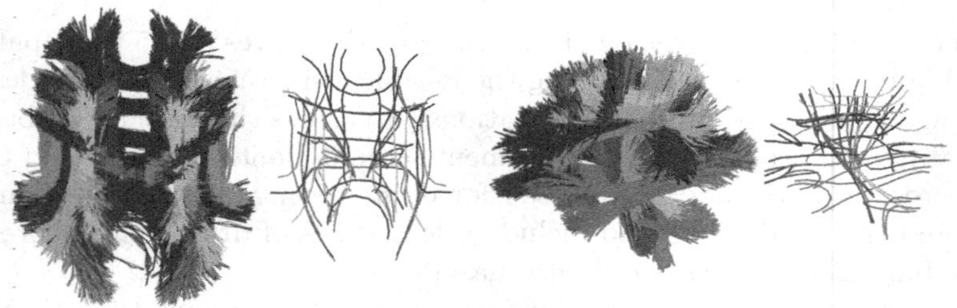

Fig. 1. Axial and sagittal views of all 50 $*$ 315 bundles center lines of the 315 observations (i.e. 63 patients and 4 noisy versions for each bundle) on the left side of each view and the corresponding 50 centroid shapes on the right side of each view.

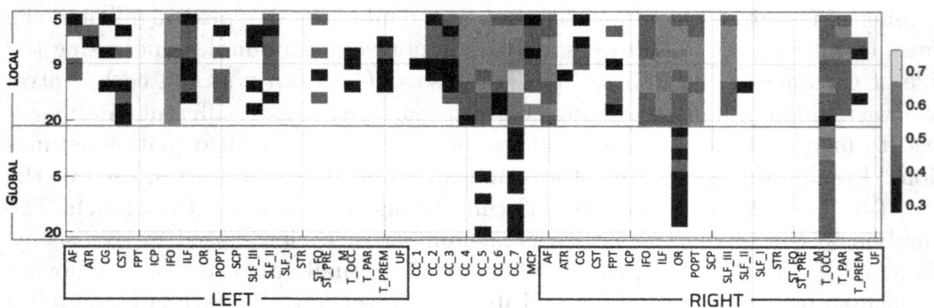

Fig. 2. The colorbar indicates the mean adjusted R^2 for models passing conditions 1/ and 2/. The y-axis indicates the number of landmark used for the 10 local model (on top) and the 10 global model (bottom).

adjusted R^2 values ranges from 0.33 to 0.76. Bundles passing all 3 conditions are the splenium of the corpus callosum (CC7), the right optic radiation (OR) and the right thalamo-occipital fascicule (T-OCC). For all models passing conditions 1/ and 2/ (Fig. 2), except two (local models, $M = 5$ for OR and T-OCC left), shape features contribution significantly to the model. There is no consistent bundles on the left hemisphere. The maximum adjusted R^2 value is obtained by the CC7 with a mean value of 0.76 (confidence interval $[0.74, 0.78]$). More details on statistics can be found in supplementary materials.

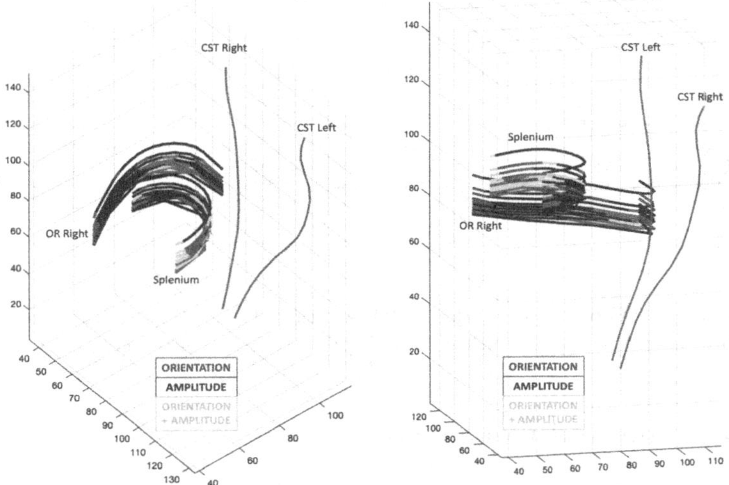

Fig. 3. The 10 different samplings of the right optic radiation (OR) and splenium of the corpus callosum, with significant landmarks indicated in red for orientation, green for amplitude, and yellow when both features are significant. Each centroid shape is shift in z-axis in function of the number of landmarks for display purposes. The cortico-spinal tract (CST) is used as a 3D referential. (Color figure online)

For the local model, Fig. 3 shows, for each sampling, landmarks with significant features involved in the model and the color depends on the type of feature: amplitude, orientation (polar or azimuth) or both. The T-OCC is not displayed in the figure as it is very close to the OR, therefore difficult to display together. The location along the T-OCC and OR centroid lines of significant features are consistent across samplings, furthermore the "corner" of the optic radiation seems to be where shape differences are associated to CGI-I.

For the global model, the right T-OCC is the only one to have 10 significant models. Principal Components (PC) 1, 4 and 5 are relevant to those 10 models, and PC 1 always has the strongest coefficient. From the principal eigenvectors E we computed the k-th principal mode of deformations $\mu^k = \overline{\boldsymbol{\alpha}(0)} + \sum_{j=1}^{N} E_j^k(\boldsymbol{\alpha}^j(0) - \overline{\boldsymbol{\alpha}(0)})$. We then can display the principal mode of variations μ^k by computing the exponential map of μ^k at B_i, for different times $t = -2$ to $t = 2$ along the geodesic. Figure 4 illustrates the 5 first principal modes of deformation of right T-OCC for 7, 10 and 15 landmarks. Each PC captures the same deformations, with a reduced explained variance when increasing the number of landmarks. This result is expected since different samplings do not affect the global shape of a center line, and increasing the sampling adds small information along the curve that reduces the explained variance.

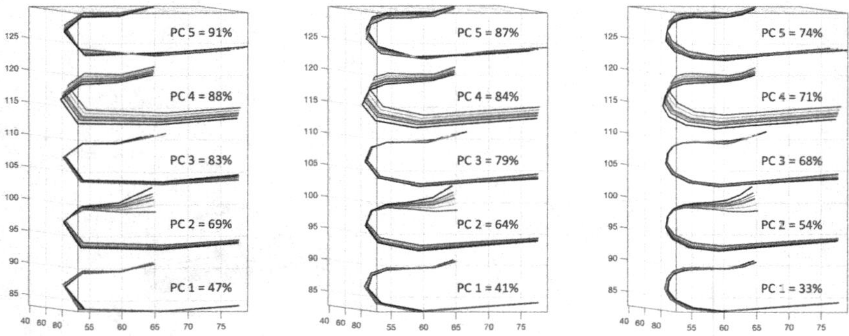

Fig. 4. Right T-OCC principal modes of deformations for 7, 10 and 15 landmarks (from left to right), with the corresponding cumulative explained variance.

5 Discussion and Conclusion

In this paper, we analyse local and global shape features of 50 bundles in order to identify the bundles associated with pejorative outcome at 6 months in patients suffering from depression. First, our results show that right-sided T-OCC and OR are the most significant bundles, followed by CC7, associated with depression outcome at 6 months. These principal results are consistent with prior finding [4, 17] where white matter micro-structures changes were observed in frontal-limbic circuits, such as T-OCC and CC7 (splenium), involved in the pathophysiological mechanisms of depression. Widespread abnormalities in MDD have been reported [27] specifically in CC7 which has been linked with anxiety [4]. Besides, abnormalities of cortico-subcortical projections (such as in OR oand T-OCC) are involved in cognitive and emotional regulation in depression [18, 30]. Another study also highlighted the role of the thalamus in pejorative outcome in depression [1]. Moreover, our models show more coherence across sampling on the right hemisphere (Fig. 2). Specifically, a recent study [3] showed a left-right asymmetry, where resistant patients had lower fractional anisotropy in right afferent fibers of the fronto-limbic circuit, and reduced connectivity between the regions connected to the right basal ganglia (i.e. involving thalamic connections), compared to responders. Thus, together with the literature, our results suggest that the shape of three white matter bundles are associated to resistance severity and to the outcome of depression.

Second, our study focuses on two types of models: local and global shape analysis. Some local models seem to be sensitive to sampling effect, reflecting small variations in shape not captured by the global models. Finally, coherent global models imply coherent local models. To avoid false discoveries - and mainly because we artificially increased our dataset by adding noise to the bundles because of a too high number of covariates, hence reducing p-values - we had a conservative approach by asking the models to be coherent for local and global approaches. Note that for the bundles of interest: (CC7/OR/T-OCC) has for the global model a median p-value of (10e−11/10e−14/10e−18) and (10e−18/10e−16/10e−13) for the local model (see supplementaries for more details). The study of bundles with coherent local models across samplings only (see Fig. 2) is left for another study where we will have to investigate the optimal amount of noise to be added minimising the impact on statistics.

As literature validates our study on the outcome of depressed patients, in future work, we can study the association between those shape-based feature and the neurocognitive performance. Also, as it is suggested in a recent study [11] we will consider the entire fiber bundles segmentation as the shape, instead of there center lines.

Acknowledgment. This work was partly funded by Fondation de France, Rennes Metropole and defi scientifique from CNRS-Inserm. MRI data acquisition was supported by the Neurinfo MRI research facility from Rennes University. Neurinfo is granted by the European Union (FEDER), the French State, the Brittany Council, Rennes Metropole, Inria, Inserm and the Rennes University Hospital. This work has been funded by Institut des Neurosciences Cliniques de Rennes (INCR).

References

1. Batail, J.M., Cologner, J., Soulas, M., Robert, G., Barillot, C., Drapier, D.: Structural abnormalities associated with poor outcome of a major depressive episode: the role of thalamus. Psychiatry Res. Neuroimaging **305** (2020)
2. Busner, J., Targum, S.D.: The clinical global impressions scale: applying a research tool in clinical practice. Psychiatry (Edgmont) **4**(7) (2007)
3. Cho, S.E., et al.: Left-right asymmetric and smaller right habenula volume in major depressive disorder on high-resolution 7-T magnetic resonance imaging. PloS ONE **16**(8) (2021)
4. Cologner, J., et al.: White matter abnormalities in depression: a categorical and phenotypic diffusion MRI study. Neuroimage: Clin. **22** (2019)
5. Commowick, O., Wiest-Daesslé, N., Prima, S.: Block-matching strategies for rigid registration of multimodal medical images. In: 2012 9th IEEE International Symposium on Biomedical Imaging (ISBI). IEEE (2012)
6. Coupe, P., Yger, P., Prima, S., Hellier, P., Kervrann, C., Barillot, C.: An optimized blockwise nonlocal means denoising filter for 3-D magnetic resonance images. IEEE Trans. Med. Imaging **27**(4), 425–441 (2008)
7. Cury, C., Glaunès, J.A., Colliot, O.: Diffeomorphic iterative centroid methods for template estimation on large datasets. In: Nielsen, F. (ed.) Geometric Theory of Information. SCT, pp. 273–299. Springer, Cham (2014). https://doi.org/10.1007/978-3-319-05317-2_10

8. de Diego-Adelino, J., Pires, P., Gomez-Anson, B., Serra-Blasco, M., et al.: Microstructural white-matter abnormalities associated with treatment resistance, severity and duration of illness in major depression. Psychol. Med. (2014)
9. Garyfallidis, E., Brett, M., Correia, M.M., Williams, G.B., Nimmo-Smith, I.: Quickbundles, a method for tractography simplification. Front. Neurosci. **6** (2012)
10. GBD 2019, M.D.C., et al.: Global, regional, and national burden of 12 mental disorders in 204 countries and territories, 1990–2019: a systematic analysis for the global burden of disease study 2019. Lancet Psychiatry (2022)
11. Glozman, T., Bruckert, L., Pestilli, F., Yecies, D.W., Guibas, L.J., Yeom, K.W.: Framework for shape analysis of white matter fiber bundles. NeuroImage (2018)
12. Gori, P., et al.: Parsimonious approximation of streamline trajectories in white matter fiber bundles. IEEE Trans. Med. Imaging **35**(12) (2016)
13. Guy, W.: ECDEU assessment manual for psychopharmacology. US Department of Health, Education, and Welfare, Public Health Service, ... (1976)
14. Hedouin, R., et al.: Block-matching distortion correction of echo-planar images with opposite phase encoding directions. IEEE Trans. Med. Imaging **36**(5), 1106–1115 (2017)
15. Herrman, H., et al.: Time for united action on depression: a lancet-world psychiatric association commission. Lancet (2022)
16. Kennis, M., Gerritsen, L., van Dalen, M., Williams, A., Cuijpers, P., Bockting, C.: Prospective biomarkers of major depressive disorder: a systematic review and meta-analysis. Mol. Psychiatry **25**(2) (2020)
17. Liao, Y., et al.: Is depression a disconnection syndrome? meta-analysis of diffusion tensor imaging studies in patients with MDD. J. Psychiatry Neurosci. **38** (2013)
18. Long, Y., et al.: Altered resting-state dynamic functional brain networks in major depressive disorder: findings from the REST-meta-MDD consortium. NeuroImage: Clin. **26** (2020)
19. Ourselin, S., Roche, A., Prima, S., Ayache, N.: Block matching: a general framework to improve robustness of rigid registration of medical images. In: Delp, S.L., DiGoia, A.M., Jaramaz, B. (eds.) MICCAI 2000. LNCS, vol. 1935, pp. 557–566. Springer, Heidelberg (2000). https://doi.org/10.1007/978-3-540-40899-4_57
20. Pennec, X., Sommer, S., Fletcher, T.: Riemannian Geometric Statistics in Medical Image Analysis. Elsevier, Amsterdam (2020)
21. Rush, A.J., et al.: Acute and longer-term outcomes in depressed outpatients requiring one or several treatment steps: a star* d report. Am. J. Psychiatry **163**(11) (2006)
22. Schölkopf, B., Smola, A., Müller, K.R.: Nonlinear component analysis as a kernel eigenvalue problem. Neural Comput. **10**(5), 1299–1319 (1998)
23. Serafini, G., et al.: The role of white matter abnormalities in treatment-resistant depression: a systematic review. Curr. Pharm. Des. **21**(10) (2015)
24. Tournier, J.D., Calamante, F., Connelly, A.: Robust determination of the fibre orientation distribution in diffusion MRI: non-negativity constrained super-resolved spherical deconvolution. Neuroimage **35**(4) (2007)
25. Tournier, J.D., Calamante, F., Connelly, A., et al.: Improved probabilistic streamlines tractography by 2nd order integration over fibre orientation distributions. In: Proceedings of the International Society for Magnetic Resonance in Medicine, vol. 1670. Wiley, Hoboken (2010)
26. Trouvé, A.: Diffeomorphisms groups and pattern matching in image analysis. Int. J. Comput. Vis. **28**(3), 213–221 (1998)

27. van Velzen, L.S., Kelly, S., Isaev, D., Aleman, A., Aftanas, L.I., et al.: White matter disturbances in major depressive disorder: a coordinated analysis across 20 international cohorts in the ENIGMA MDD working group. Mol. Psychiatry **25**(7), 1511–1525 (2020)
28. Wager, T.D., Woo, C.W.: Imaging biomarkers and biotypes for depression. Nat. Med. **23**(1) (2017)
29. Wasserthal, J., Neher, P., Maier-Hein, K.H.: TractSeg-fast and accurate white matter tract segmentation. NeuroImage **183** (2018)
30. Yan, B., et al.: Quantitative identification of major depression based on resting-state dynamic functional connectivity: a machine learning approach. Front. Neurosci. **14**, 191 (2020)

White Matter Tracts are Point Clouds: Neuropsychological Score Prediction and Critical Region Localization via Geometric Deep Learning

Yuqian Chen[1,2], Fan Zhang[1], Chaoyi Zhang[2], Tengfei Xue[1,2], Leo R. Zekelman[1], Jianzhong He[4], Yang Song[3], Nikos Makris[1], Yogesh Rathi[1], Alexandra J. Golby[1], Weidong Cai[2], and Lauren J. O'Donnell[1](✉)

[1] Harvard Medical School, Boston, MA, USA
odonnell@bwh.harvard.edu
[2] The University of Sydney, Sydney, NSW, Australia
[3] The University of New South Wales, Sydney, NSW, Australia
[4] Zhejiang University of Technology, Zhejiang, China

Abstract. White matter tract microstructure has been shown to influence neuropsychological scores of cognitive performance. However, prediction of these scores from white matter tract data has not been attempted. In this paper, we propose a deep-learning-based framework for neuropsychological score prediction using microstructure measurements estimated from diffusion magnetic resonance imaging (dMRI) tractography, focusing on predicting performance on a receptive vocabulary assessment task based on a critical fiber tract for language, the arcuate fasciculus (AF). We directly utilize information from all points in a fiber tract, without the need to average data along the fiber as is traditionally required by diffusion MRI tractometry methods. Specifically, we represent the AF as a point cloud with microstructure measurements at each point, enabling adoption of point-based neural networks. We improve prediction performance with the proposed Paired-Siamese Loss that utilizes information about differences between continuous neuropsychological scores. Finally, we propose a Critical Region Localization (CRL) algorithm to localize informative anatomical regions containing points with strong contributions to the prediction results. Our method is evaluated on data from 806 subjects from the Human Connectome Project dataset. Results demonstrate superior neuropsychological score prediction performance compared to baseline methods. We discover that critical regions in the AF are strikingly consistent across subjects, with the highest number of strongly contributing points located in frontal cortical regions (i.e., the rostral middle frontal, pars opercularis, and pars triangularis), which are strongly implicated as critical areas for language processes.

We acknowledge funding provided by the following National Institutes of Health (NIH) grants: R01MH125860, R01MH119222, R01MH074794, and P41EB015902.

L. Wang et al. (Eds.): MICCAI 2022, LNCS 13431, pp. 174–184, 2022.
https://doi.org/10.1007/978-3-031-16431-6_17

Keywords: White matter tract · Neuropsychological assessment score · Point cloud · Deep learning · Region localization

1 Introduction

The structural network of the brain's white matter pathways has a strong but incompletely understood relationship to brain function [21,22]. To better understand how the brain's structure relates to function, one recent avenue of research investigates how neuropsychological scores of cognitive performance relate to features from diffusion magnetic resonance imaging (dMRI) data. It was recently shown that structural neuroimaging modalities (such as dMRI) strongly contribute to prediction of cognitive performance measures, where measures related to language and reading were among the most successfully predicted [10]. Other works have shown that measurements of white matter tract microstructure derived from diffusion MRI tractography relate to neuropsychological measures of language performance [29,31]. However, we believe that there are no studies to predict individual cognitive performance based on microstructure measurements of the white matter fiber tracts. Performing accurate predictions of neuropsychological performance could help improve our understanding of the function of the brain's unique white matter architecture. Furthermore, localizing regions along fiber tracts that are important in the prediction of cognitive performance could be used to understand how specific regions of a white matter tract differentially contribute to various cognitive processes.

One important challenge in the computational analysis of white matter fiber tracts is how to represent tracts and their tissue microstructure. Tractography of one anatomical white matter tract, such as the arcuate fasciculus (Fig. 1a), may include thousands of streamlines (or "fibers"), where each streamline is composed of a sequence of points. One tract can therefore contain several hundred thousand points, where each point has one or more corresponding microstructure measurement values. How to effectively represent measurements from points within the tract has been a challenge. One typical approach is to calculate the mean value of measurements from all points within the tract (Fig. 1b). However, this ignores the known spatial variation of measurements along fiber tracts such as the arcuate fasciculus [16,29]. An alternative is to perform "tractometry" or along-tract analysis (Fig. 1c) to investigate the distribution of the microstructure measures along the fiber pathway [16,28]. This method preserves information about the average diffusion measurements at different locations along the tract and thus enables along-tract critical region localization; however, fiber-specific or point-wise information is still obscured.

In contrast to these traditional representations, we propose to represent a complete white matter tract with its set of raw points for microstructure analysis using a point cloud (Fig. 1d). Point cloud is an important type of geometric data structure. In recent years, deep learning has demonstrated superior performance in computer vision tasks [25]. Point-based neural networks have demonstrated successful applications on processing geometric data [2,19]. PointNet [19] is a pioneering effort that enables direct processing of point clouds with deep neural networks. It is a simple, fast and effective framework for point cloud classification and segmentation and has been adapted to various point cloud-related tasks [20, 30,32]. In previous neuroimaging studies, point cloud representations have been applied to tractography-related learning tasks such as tractography segmentation and tractogram filtering [1,27]. However, to our knowledge, no previous studies have investigated the effectiveness of representing a whole white matter tract and its microstructure measurements using a point cloud for learning tasks. By formulating white matter tract analysis in a point-based fashion using point clouds, we can directly utilize tissue microstructure and positional information from all points within a fiber tract and avoid the need for along-tract feature extraction.

In this study, we propose a framework for neuropsychological assessment score prediction using pointwise diffusion microstructure measurements from white matter tracts. The framework is a supervised deep learning pipeline that performs a regression task. This paper has four contributions. First, we represent white matter tracts as point clouds to preserve diffusion measurement information from all fiber points. To our knowledge, this is the first investigation to study microstructure measurements within white matter tracts using point clouds for a learning task. Second, for regression, we propose the Paired-Siamese Loss to utilize information about the differences between continuous labels, which are neuropsychological scores in our study. Third, our framework is able to perform the task of critical region localization. We propose a Critical Region Localization (CRL) algorithm to recognize critical regions within the tract where points that make important contributions to the prediction task are located. Fourth, using data from 806 subjects from the Human Connectome Project, our approach demonstrates superior performance for neuropsychological score prediction via evaluations on a large-scale dataset and effectiveness in predictively localizing critical language regions along the white matter tract.

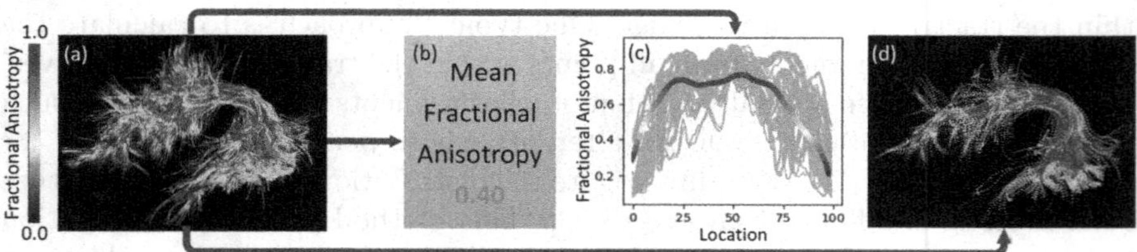

Fig. 1. (a) Three ways to represent a white matter tract and its microstructure measurements. (b) Representation with a single mean value; (c) Along-tract representation; (d) Representation with a point cloud.

2 Methods

2.1 dMRI Dataset, Neuropsychological Assessment, and Tractography

To evaluate performance of our proposed method, this study utilizes neuropsychological assessment and dMRI data from the Human Connectome Project (HCP), a large multimodal dataset composed of healthy young adults (downloaded from http://www.humanconnectomeproject.org/) [24]. We study data from 806 unrelated subjects (380 males and 426 females, aged 28.6 ± 3.7) with 644 (80%) for training and 162 (20%) for testing. Scores from the NIH Toolbox Picture Vocabulary Test (TPVT) are investigated. This computerized neuropsychological assessment measures receptive vocabulary and semantic memory, key components of language functioning [8]. TPVT scores range from 90.69 to 153.09 in this subject cohort. The investigated dMRI data comes from the HCP minimally preprocessed dataset (b = 1000, 2000 and 3000 s/mm^2, TE/TR = 89/5520 ms, resolution = $1.25 \times 1.25 \times 1.25$ mm^3) [24]. For each subject, the b = 3000 shell of 90 gradient directions and all b = 0 scans are extracted for tractography because this single shell can reduce computation time and memory usage while providing the highest angular resolution for tractography [18,19]. Whole brain tractography is generated from each subject's diffusion MRI data using a two-tensor unscented Kalman filter (UKF) method [14,17]. White matter tracts (left AF tract in our study) are identified with a robust machine learning approach that has been shown to consistently identify white matter tracts across the human lifespan and is implemented with the WMA package [33]. For each individual's tract, two measurements are calculated: the tract-specific fractional anisotropy (FA) of the first tensor, which corresponds to the tract being traced, and the number of streamlines (NoS). These measurements of the left AF tract have been found to be significantly related to TPVT scores [31]. Tracts are visualized in 3D Slicer via the SlicerDMRI project [15].

2.2 Point Cloud Construction

The left arcuate fasciculus (AF) tract is selected for the task of predicting TPVT scores because TPVT performance was shown to be significantly associated with left AF microstructure measurements [31]. Each AF tract is represented as a point cloud and N points are randomly sampled from the point cloud as input. Considering the large number of points within the AF tract (on average approximately 300,000 per subject), random sampling serves as a natural but effective data augmentation strategy. Each point has 5 input channels including its spatial coordinates as well as white matter tract measurements (FA and NoS). Therefore, a $N \times 5$ point cloud forms the input of the neural network.

2.3 Network Architecture and Proposed Paired-Siamese Loss

In this work, we propose a novel deep learning framework (Fig. 2) that performs a regression task for TPVT score prediction. The most important aspect of our framework's design is the supervision of neural network backpropagation using information from the difference of continuous labels (TPVT scores). Unlike a classification task where different categories of labels are independent from each other, a regression task has continuous labels as ground truth. We hypothesize that the relationships between inputs with continuous labels can be used to improve the performance of regression.

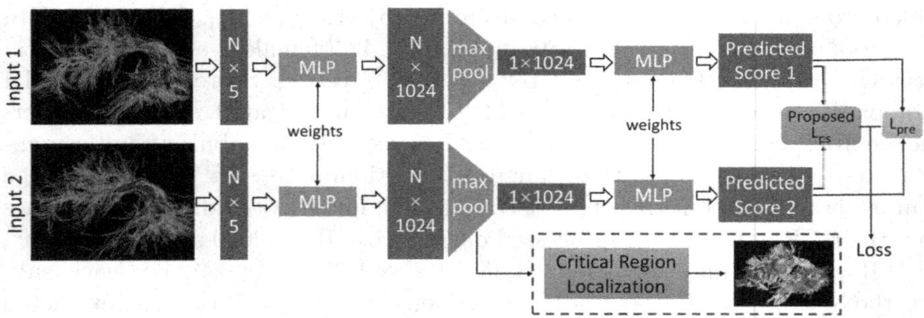

Fig. 2. Overall pipeline. A pair of AF tracts are represented as a pair of point clouds and fed into a point-based Siamese Network. During training, predicted TPVT scores are obtained from the network and a prediction loss (L_{pre}) is calculated as the mean of MSE losses between the predicted scores and ground truth scores. Paired-Siamese loss (L_{ps}) is calculated as the MSE between two differences: the difference between predicted scores and the difference between ground truth scores of input pairs. L_{ps} is added to the total loss with weight w. During inference, critical regions are recognized with our proposed CRL algorithm, as shown in the green dashed box. (Color figure online)

For this goal, we propose to adopt a Siamese Network [3] that contains two subnetworks with shared weights. The subnetwork of our Siamese Network is developed from PointNet [19]. To perform the regression task, the output dimension of the last linear layer is set to 1 to obtain one predicted score. T-Net (the spatial transformation layer) is removed from PointNet to preserve anatomically informative information about the spatial position of tracts [27]. During training, a pair of point cloud sets are used as input to the neural network, and a pair of predicted TPVT scores as well as the difference between them are obtained from the network. For inference, one subnet of the Siamese Network is retained with one point cloud as input.

A new loss function, the Paired-Siamese Loss (L_{ps}), is proposed to guide the prediction of TPVT scores. It takes the relationship between the input pair into consideration when training the model. L_{ps} is defined as the mean squared error (MSE) between two differences, where the first is the difference between predicted TPVT scores of an input pair and the second is the difference between ground truth scores of an input pair, as follows:

$$L_{ps} = \frac{1}{N}\sum_{i}((y_{i1} - y_{i2}) - (\hat{y_{i1}} - \hat{y_{i2}}))^2 \tag{1}$$

where y_{i1} and y_{i2} are labels of the input pair and $\hat{y_{i1}}$ and $\hat{y_{i2}}$ are predicted scores of the input pair. In addition to the proposed difference loss, the prediction losses (L_{pre1} and L_{pre2}) of each input are calculated as the MSE loss between the predicted score and ground truth score. Then the overall prediction loss of the input pair, L_{pre}, is the mean of L_{pre1} and L_{pre2}. Therefore, the total loss of our network is $Loss = L_{pre} + wL_{ps}$, where w is the weight for L_{ps}.

2.4 Critical Region Localization

We propose a Critical Region Localization (CRL) algorithm to identify critical regions along fiber tracts that are important in the prediction of cognitive performance. The algorithm steps are explained in detail as follows. First, to ensure that critical points are identified from the whole point set of the AF tract, we divide the tract into multiple point sets (N points per set) without replacement and sequentially feed these point sets to the network. Second, we propose a Contributing Point Selection (CPS) module to obtain critical points and their weights for each input point set. In CPS, we intuitively take advantage of the max-pooling operation in the network as in [19] to obtain point sets that contribute to the max-pooled features (referred to as contributing points in our algorithm). However, instead of taking contributing points directly as critical points as in [19], our method considers the number of max-pooled features of each contributing point and assigns this number to the point as its weight. The lists of contributing points and their weights are then concatenated across all point sets. Third, the previous two steps are repeated M times (so that every point in the AF tract is fed into the network M times for CPS). The weights of each contributing point are summed across all repetitions. Finally, we define critical regions along the AF tract as the points that have top-ranking weights for each subject (the top 5% of points are chosen in this paper to identify highly important regions). These regions are then interpreted in terms of their anatomical localization by identifying the Freesurfer [5] regions in which they are located.

2.5 Implementation Details

In the training stage, our model is trained for 500 epochs with a learning rate of 0.001. The batchsize of training is 32 and Admax [13] is used for optimization. We use a weight decay with the corresponding coefficient 0.005 to avoid overfitting. All experiments are performed on a NVIDIA RTX 2080 Ti GPU using Pytorch (v1.7.1) [18]. The weight of difference loss w is empirically set to be 0.1 [6,26]. The number of input points N is set to be 2048 considering memory limitation. On average, each epoch (training and validation) takes 4 s with 2 GB GPU memory usage. Number of iteration M in the CRL algorithm is set to 10.

3 Experiments and Results

3.1 Evaluation Metrics

Two evaluation metrics were adopted to quantify performance of our method and enable comparisons among approaches. The first metric is the mean absolute error (MAE), which is calculated as the averaged absolute value of the difference between the predicted TPVT score and the ground truth TPVT score across testing subjects. The second metric is the Pearson correlation coefficient (r), which measures the linear correlation between the predicted and ground truth scores and has been widely applied to evaluating performance of neurocognitive score prediction [10,12,23].

Table 1. Quantitative comparison results.

Methods	Mean+ LR	Mean+ ENR	Mean+ RF	AFQ+ LR	AFQ+ ENR	AFQ+ RF	AFQ+ 1D-CNN	Ours w/o L_{ps}	Ours
Input Features	mean FA, NoS	mean FA, NoS	mean FA, NoS	along-tract FA, NoS	along-tract FA, NoS	along-tract FA, NoS	along-tract FA, NoS	point-wise FA, NoS	point-wise FA, NoS
MAE	6.546 (4.937)	6.548 (4.955)	6.58 (4.812)	7.364 (5.744)	6.928 (5.233)	7.034 (5.447)	6.537 (4.975)	6.639 (5.174)	**6.512 (5.081)**
r	0.062	0.089	0.134	0.135	0.225	0.138	0.114	0.316	**0.361**

3.2 Evaluation Results

Comparison with Baseline Method. We compared our proposed approach with several baseline methods. For all methods, features of FA and NoS were used for prediction. First, the three representations of FA measurements of the AF tract, namely mean FA, along-tract FA and pointwise FA (Fig. 1) were compared. To analyze the mean FA representation, Linear Regression (LR), Elastic-Net Regression (ENR) and Random Forest (RF) were performed to predict TPVT scores. Next, along-tract FA was obtained with the Automated Fiber Quantification (AFQ) algorithm [28] implemented in Dipy v1.3.0 [7]. FA features from 100 locations along the tract were generated from AFQ. FA features and NoS were concatenated into the input feature vector. For along-tract FA, performance of LR, ENR, RF and 1D-CNN models were investigated. We adopted the 1D-CNN model proposed in a recent study that performs the age prediction task with microstructure measurements [11]. The parameters of all methods were fine-turned to obtain the best performance. All results are reported using data from all testing subjects. As shown in Table 1, our proposed method outperforms all baseline methods in terms of both MAE and r, where the r value of our method shows an obvious advantage over the other methods. The reason for the largely improved r but slightly improved MAE is that several baseline methods give predictions narrowly distributed around the mean, producing a reasonable MAE but a low r. The results suggest that point clouds can better represent microstructure measurements of white matter fiber tracts for learning tasks.

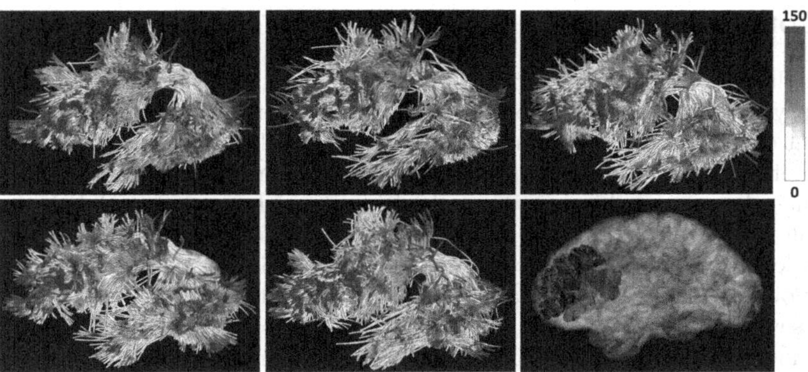

Fig. 3. The weights for contributing points within the AF tract show strong consistency of critical predictive regions across five randomly selected subjects. Lower right image shows the top three cortical areas intersected by critical AF regions (blue: rostralmiddlefrontal; yellow: parsopercularis; red: parstriangularis). (Color figure online)

Ablation Study. We performed an ablation study to evaluate the effectiveness of our proposed Paired-Siamese Loss. Model performances with (Ours) and without the L_{ps} (Ours w/o L_{ps}) were compared. As shown in Table 1, adding the proposed loss obviously improved performance, indicated by a larger r value and a smaller MAE. The large improvement of r is likely because the proposed loss takes the relationship between the input pair into consideration during training.

Interpretation of Critical Regions. Critical points and their corresponding importance values (weights) were obtained for all 162 testing subjects during inference by applying our proposed CRL algorithm. The obtained critical regions within the tract are visualized in Fig. 3. It is apparent that the anatomical locations of the critical regions are strikingly consistent across subjects, demonstrating the potential utility of our proposed method for investigating the relationship between local white matter microstructure and neurocognitive assessment performances. For prediction of TPVT, we find that most critical region points are located near the interface of the white matter and the cortex, where 34.6% are in the white matter and 64.2% in the cortex, on average across all testing subjects. Investigating the critical regions of the left AF that are important for language performance as measured by TPVT, we find that on average 16.5% of the critical region points are located in the rostral middle frontal cortex, 9.1% in the cortex of the pars opercularis, and 7.6% in the cortex of the pars triangularis. This finding is in line with the understanding that frontal cortical regions are critical for semantic language comprehension and representational memory [4,9].

4 Conclusion

To conclude, we propose a novel deep learning based framework for predicting neuropsychological scores. In our framework, we represent white matter tracts as point clouds, to preserve point information about diffusion measurements and enable efficient processing using point-based deep neural networks. We propose to utilize the relationship between paired inputs with continuous labels during training by adding a Paired-Siamese Loss. In addition, we propose a critical region localization (CRL) algorithm to obtain the importance distribution of tract points and identify critical regions within the tract. Our method outperforms several compared algorithms on a large-scale HCP dataset. We find that points located near the interface between white matter and cortex, and especially in frontal cortical regions, are important for prediction of language performance. Potential future work includes improving our framework by addressing some existing limitations such as ignorance of continuous streamline information and lack of density correction of tractograms.

References

1. Astolfi, P., et al.: Tractogram filtering of anatomically non-plausible fibers with geometric deep learning. In: Martel, A.L., et al. (eds.) MICCAI 2020. LNCS, vol. 12267, pp. 291–301. Springer, Cham (2020). https://doi.org/10.1007/978-3-030-59728-3_29
2. Chen, X., Ma, H., Wan, J., Li, B., Xia, T.: Multi-view 3D object detection network for autonomous driving. In: Proceedings of the IEEE Conference on Computer Vision and Pattern Recognition, pp. 1907–1915 (2017)
3. Chopra, S., Hadsell, R., LeCun, Y.: Learning a similarity metric discriminatively, with application to face verification. In: 2005 IEEE Computer Society Conference on Computer Vision and Pattern Recognition (CVPR 2005), vol. 1, pp. 539–546. IEEE (2005)
4. Dronkers, N.F., et al.: The neural architecture of the language comprehension network: converging evidence from lesion and connectivity analyses. Front. Syst. Neurosci. **5**, 1 (2011)
5. Fischl, B.: Freesurfer. Neuroimage **62**(2), 774–781 (2012)
6. Fu, S., et al.: Domain adaptive relational reasoning for 3D multi-organ segmentation. In: Martel, A.L., et al. (eds.) MICCAI 2020. LNCS, vol. 12261, pp. 656–666. Springer, Cham (2020). https://doi.org/10.1007/978-3-030-59710-8_64
7. Garyfallidis, E., et al.: Dipy, a library for the analysis of diffusion MRI data. Front. Neuroinform. **8**, 8 (2014)
8. Gershon, R.C., et al.: Iv. NIH toolbox cognition battery (CB): measuring language (vocabulary comprehension and reading decoding). In: Monographs of the Society for Research in Child Development, vol. 78, no. 4, pp. 49–69 (2013)
9. Goldman-Rakic, P.S.: Circuitry of primate prefrontal cortex and regulation of behavior by representational memory. Compr. Physiol. 373–417 (2011)
10. Gong, W., Beckmann, C.F., Smith, S.M.: Phenotype discovery from population brain imaging. Med. Image Anal. **71**, 102050 (2021)
11. He, H., et al.: Model and predict age and sex in healthy subjects using brain white matter features: a deep learning approach. In: IEEE 19th International Symposium on Biomedical Imaging (ISBI) (2022)

12. Kim, M., et al.: A structural enriched functional network: an application to predict brain cognitive performance. Med. Image Anal. **71**, 102026 (2021)
13. Kingma, D.P., Ba, J.: Adam: a method for stochastic optimization. In: International Conference for Learning Representations (ICLR) (2015)
14. Malcolm, J.G., Shenton, M.E., Rathi, Y.: Filtered multitensor tractography. IEEE Trans. Med. Imaging **29**(9), 1664–1675 (2010)
15. Norton, I., Essayed, W.I., Zhang, F., Pujol, S., Yarmarkovich, A., et al.: SlicerDMRI: open source diffusion MRI software for brain cancer research. Can. Res. **77**(21), e101–e103 (2017)
16. O'Donnell, L.J., Westin, C.F., Golby, A.J.: Tract-based morphometry for white matter group analysis. Neuroimage **45**(3), 832–844 (2009)
17. O'Donnell, L.J., Wells, W.M., Golby, A.J., Westin, C.-F.: Unbiased groupwise registration of white matter tractography. In: Ayache, N., Delingette, H., Golland, P., Mori, K. (eds.) MICCAI 2012. LNCS, vol. 7512, pp. 123–130. Springer, Heidelberg (2012). https://doi.org/10.1007/978-3-642-33454-2_16
18. Paszke, A., et al.: Pytorch: an imperative style, high-performance deep learning library. In: Advances in Neural Information Processing Systems, vol. 32 (2019)
19. Qi, C.R., Su, H., Mo, K., Guibas, L.J.: Pointnet: deep learning on point sets for 3D classification and segmentation. In: Proceedings of the IEEE Conference on Computer Vision and Pattern Recognition, pp. 652–660 (2017)
20. Qi, C.R., Yi, L., Su, H., Guibas, L.J.: Pointnet++: deep hierarchical feature learning on point sets in a metric space. In: Advances in Neural Information Processing Systems, vol. 30 (2017)
21. Sarwar, T., Tian, Y., Yeo, B.T., Ramamohanarao, K., Zalesky, A.: Structure-function coupling in the human connectome: a machine learning approach. Neuroimage **226**, 117609 (2021)
22. Suárez, L.E., Markello, R.D., Betzel, R.F., Misic, B.: Linking structure and function in macroscale brain networks. Trends Cogn. Sci. **24**(4), 302–315 (2020)
23. Tian, Y., Zalesky, A.: Machine learning prediction of cognition from functional connectivity: are feature weights reliable? Neuroimage **245**, 118648 (2021)
24. Van Essen, D.C., et al.: The Wu-Minn human connectome project: an overview. Neuroimage **80**, 62–79 (2013)
25. Voulodimos, A., Doulamis, N., Doulamis, A., Protopapadakis, E.: Deep learning for computer vision: a brief review. Comput. Intell. Neurosci. **2018** (2018)
26. Wang, K., Liu, X., Zhang, K., Chen, T., Wang, G.: Anterior segment eye lesion segmentation with advanced fusion strategies and auxiliary tasks. In: Martel, A.L., et al. (eds.) MICCAI 2020. LNCS, vol. 12265, pp. 656–664. Springer, Cham (2020). https://doi.org/10.1007/978-3-030-59722-1_63
27. Xue, T., et al.: Supwma: consistent and efficient tractography parcellation of superficial white matter with deep learning. In: IEEE 19th International Symposium on Biomedical Imaging (ISBI) (2022)
28. Yeatman, J.D., Dougherty, R.F., Myall, N.J., Wandell, B.A., Feldman, H.M.: Tract profiles of white matter properties: automating fiber-tract quantification. PLoS ONE **7**(11), e49790 (2012)
29. Yeatman, J.D., et al.: Anatomical properties of the arcuate fasciculus predict phonological and reading skills in children. J. Cogn. Neurosci. **23**(11), 3304–3317 (2011)
30. Yu, J., et al.: 3D medical point transformer: introducing convolution to attention networks for medical point cloud analysis. arXiv preprint arXiv:2112.04863 (2021)

31. Zekelman, L.R., et al.: White matter association tracts underlying language and theory of mind: an investigation of 809 brains from the human connectome project. Neuroimage **246**, 118739 (2022)
32. Zhang, C., Yu, J., Song, Y., Cai, W.: Exploiting edge-oriented reasoning for 3D point-based scene graph analysis. In: Proceedings of the IEEE/CVF Conference on Computer Vision and Pattern Recognition, pp. 9705–9715 (2021)
33. Zhang, F., et al.: An anatomically curated fiber clustering white matter atlas for consistent white matter tract parcellation across the lifespan. Neuroimage **179**, 429–447 (2018)

Segmentation of Whole-Brain Tractography: A Deep Learning Algorithm Based on 3D Raw Curve Points

Logiraj Kumaralingam[1], Kokul Thanikasalam[2], Sittampalam Sotheeswaran[1], Jeyasuthan Mahadevan[3], and Nagulan Ratnarajah[4](✉)

[1] Faculty of Science, Eastern University, Chenkaladi, Sri Lanka
[2] Faculty of Science, University of Jaffna, Jaffna, Sri Lanka
[3] Base Hospital, Tellipalai, Sri Lanka
[4] Faculty of Applied Science, University of Vavuniya, Vavuniya, Sri Lanka
rnagulan@univ.jfn.ac.lk
https://github.com/NeuroImageComputingLab/3D_Curve_CNN

Abstract. Segmentation of whole-brain fiber tractography into anatomically meaningful fiber bundles is an important step for visualizing and quantitatively assessing white matter tracts. The fiber streamlines in whole-brain fiber tractography are 3D curves, and they are densely and complexly connected throughout the brain. Due to the huge volume of curves, varied connection complexity, and imaging technology limitations, whole-brain tractography segmentation is still a difficult task. In this study, a novel deep learning architecture has been proposed for segmenting whole-brain tractography into 10 major white matter bundles and the "other fibers" category. The proposed PointNet based CNN architecture takes the whole-brain fiber curves in the form of 3D raw curve points and successfully segments them using a manually created large-scale training dataset. To improve segmentation performance, the approach employs two channel-spatial attention modules. The proposed method was tested on healthy adults across the lifespan with imaging data from the ADNI project database. In terms of experimental evidence, the proposed deep learning architecture demonstrated solid performance that is better than the state-of-the-art.

Keywords: Whole-brain tractography · Segmentation · Major fiber bundles · Deep learning · 3D curve points

1 Introduction

Diffusion MRI based whole-brain fiber tractography streamlines model the white matter neuroanatomy comprehensively [1]. From a single brain, whole-brain tractography generates millions of densely and complexly connected 3D fiber curves. These curves are not meaningful to clinicians or researchers for viewing fiber

L. Wang et al. (Eds.): MICCAI 2022, LNCS 13431, pp. 185–195, 2022.
https://doi.org/10.1007/978-3-031-16431-6_18

bundles or analyzing white matter quantitatively. As a result, whole-brain tractography should be segmented into anatomically relevant fiber bundles. Through tract quantification and visualization, these fiber bundles provide resources for studying white matter anatomy, brain functions, and planning brain surgeries.

White matter bundles are conceptually known pathways, and neuroanatomists and researchers create segmentation algorithms that include anatomical markers and directions for separating the required bundles from whole-brain tractography. Anatomy specialists choose tractography streamlines to extract the specific bundles in the brain interactively by utilizing manually developed ROIs in traditional segmentation methods [1,2]. These manual segmentation methods are difficult to reproduce due to human errors and personal decisions, and they require a great deal of time and substantial anatomical understanding. In recent years, numerous automated approaches for segmenting whole-brain tracts into meaningful fiber bundles have been presented.

Automated methods [3,4] based on the atlases of brain anatomy make use of past knowledge in the form of atlases. These methods show that a large number of bundles can be extracted, but the results are typically non-reproducible and user-biased, necessitate anatomical knowledge, and rely heavily on precise registration. Fiber clustering [5,6] is used to segment the set of fiber tracts based on the characteristics of 3D fiber streamlines in another category of segmentation approach. These algorithms employ classic machine learning approaches and can handle large across-subject anatomical heterogeneity. However, these algorithms frequently necessitate many time-consuming processing procedures. These unsupervised approaches have been criticized for their inconsistency in producing anatomically accurate fiber bundles and an absence of reproducibility. In segmentation, supervised traditional machine learning algorithms [7,8] take into account prior anatomical information, resulting in more reliable and relevant findings. These methods rely on conventional machine learning techniques, which limits their accuracy even though they use supervised machine learning algorithms.

Recent methods segment white matter fiber tracts into voxel-based and fiber-based categories using deep learning-based algorithms. Voxel-based approaches [9–11] partition the fiber tracts into bundles based on each voxel capturing the presence and orientation of a fiber tract. In several cases, local anatomical variances limited the voxel-based segmentation algorithms in subjects who have a lot of anatomical variation. Fiber tractography is performed first, followed by segmentation by classifying fiber streamlines into anatomically meaningful bundles in fiber-based segmentation techniques. Various deep learning algorithms were used for the fiber-based segmentation in the literature, such as a CNN model with a 2D-multi channel feature descriptor [12], a graph neural network [13], and the shape characteristics of fiber bundles and a CNN model [14,15]. Although the deep learning algorithms produce excellent outcomes, they are not appropriate for clinical circumstances since they are computationally expensive. In particular, it is necessary to perform a feature extraction step or some pre-processing on the data because they cannot utilize raw 3D curves, which adds to the complexity of segmentation models. At the time, the community did not employ a standardized approach for the segmentation of whole-brain tracts. We believe that it is essential to harmonize whole-brain tract segmentation and create strategies that can be used in clinical settings if diffusion MRI tractography is to be considered seriously as a clinical tool.

In this paper, a robust whole-brain tractography segmentation approach with a powerful deep learning architecture has been developed, and the approach successfully identifies meaningful anatomical divisions from the curves of whole-brain tractography in real-time with higher accuracy. In particular, the architecture consumes the raw curves directly from the tractography process. There are three primary contributions in this paper. First, we developed a large-scale training dataset of 25 healthy adult subjects' labeled tractography fibers as 10 major fiber bundles and an "other fibers" category. The participants range in age from 45 to 75 years old, both males and females, ensuring segmentation across-the-lifespan. Second, we proposed a novel 3D raw curve points based CNN architecture, and third, we demonstrated successfully unlabeled whole-brain tractography segmentation on a random test dataset.

2 Methodology

The overall methodology of the proposed whole-brain tractography segmentation approach is illustrated in Fig. 1, which includes the creation of training fiber curve points (Sect. 2.1), 3D curve data sampling (Sect. 2.2), a novel PointNet based CNN architecture (Sect. 2.3), and whole-brain tract segmentation for unlabeled test subjects (Sect. 2.4).

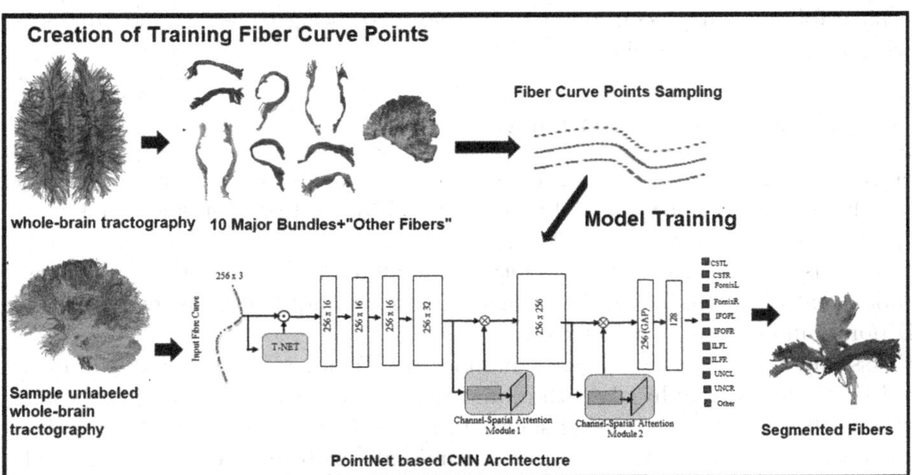

Fig. 1. Method overview. First row: training tractography dataset creation for a sample subject and fiber curve sampling. Second row: proposed PointNet based CNN tract classification model to segment the unlabeled whole-brain tracts of a test subject.

2.1 Training Tractography Data

The Alzheimer's Disease Neuroimaging Initiative (ADNI) database [16] provided the neuroimaging data for this study. The study used diffusion MR images and T1-weighted MR images from 25 healthy subjects (males:females = 14:11, age range: 45–75). The ADNI website (http://adni.loni.usc.edu/) has detailed information regarding the MR acquisition techniques, parameters, and ethical approval. All of the participants' cross-sectional imaging data was subjected to image preparation processes using FSL [17], including the registration based on the MNI152 coordinate system [18].

On diffusion tensor images, whole-brain tractography was performed using a deterministic algorithm (FACT) for each subject. The training 3D fiber curve points were labeled into 11 categories consisting of 10 major fiber bundles (Corticospinal Tract Left, Right (CSTL, CSTR), Inferior Longitudinal Fasciculus Left, Right (ILFL, ILFR), Inferior Fronto Occipital Fasciculus Left, Right (IFOFL, IFOFR), Uncinate fasciculus Left, Right (UNCL, UNCR), Fornix Left, Right (FornixL, FornixR), and "other fibers" category (rest of the fibers after removing 10 major bundles from whole-brain tracts) were manually extracted for all 25 subjects. The segmentation of 10 white matter fiber bundles from the 25 subjects' whole-brain fibers was done using the interactive dissection technique [8]. An experienced neuro-radiologist visually examined the ROI's boundaries. Each extracted fiber bundle was cleaned, aggregated, and trimmed based on the atlas and the primary vector color map of diffusion tensor imaging, which offers additional information about the local orientation and eliminates undesirable paths [8].

This study employed 95,056 curves, with 75% of them used for training and the remainder for testing. Since the number of curves in each class is not equal, we have followed an up-sampling technique in training to set 11,250 curves in each major fiber bundle class.

2.2 Data Sampling of 3D Curves

Let N be the number of three-dimensional (3D) curves in a major fiber bundle class. Each curve C_i $(1 \leqslant i \leqslant N)$ is represented by a set of 3D points. In a curve C_i, a point P_i is known by its 3D coordinates $(x_i, y_i, z_i) \in \mathbb{R}^3$. Since the length of curves is different from one to another, the number of 3D points in each curve is not equal. We have followed a data sampling technique to keep the number of 3D points in each curve at a fixed size.

Assume that a curve C_i has n number of 3D points $P_i(1 \leqslant i \leqslant n)$ generated by the tractography algorithm, and the midpoints of these 3D points were obtained by $P_i^{mid}(1 \leqslant i \leqslant n)$ as shown in Fig. 1. In the next stage of sampling, by choosing every point from set pi and randomly choosing the remaining points (l) from set P_i^{mid} such that $l = m - n$, a fixed number (m) of 3D sample points were identified. Based on this sampling technique, all curves are sampled by a fixed number of 3D points.

2.3 Proposed PointNet Based CNN Architecture

We have proposed a novel CNN architecture to segment the whole-brain fiber tracts into major bundles and "other fibers". The proposed architecture takes fiber curves as 3D points with 256 × 3 dimensions, and outputs segmentation of ten anatomically meaningful bundles and "other fibers". The overview of the proposed architecture is illustrated in Fig. 2.

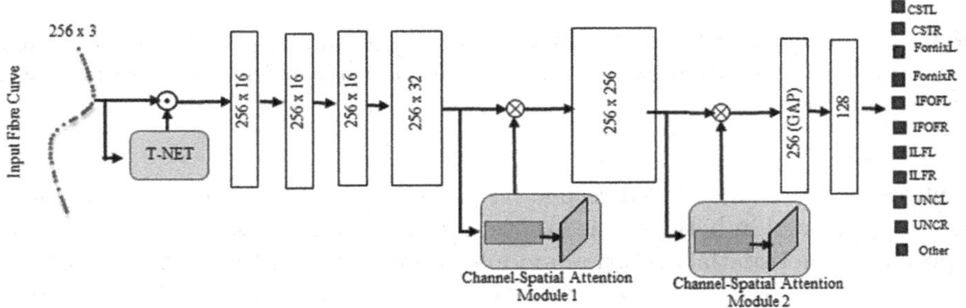

Fig. 2. The outline of the proposed network architecture. Five 1D convolutional layers, two fully connected layers, two channel-spatial attention modules, and input transform network (T-NET) are integral parts of the architecture.

In the proposed architecture, the PointNet [19] model is utilized as the baseline since it is able to segment the unordered data points in an efficient way. An input transformation network module called as T-NET and a multi-layer perception network called as MLP are included in the architecture for learning the canonical representation of 3D data points and segmenting the white matter fiber tracts, respectively. We have modified the T-NET network of the baseline model by including three 1D-CNN layers with a kernel size of 1, one 1D global max pooling layer, and a Dense layer. Similar to the baseline model, the objective of the T-NET is to produce permutation invariant and data dependent transformation parameters from the unordered 3D data points. The proposed T-NET module inputs the 3D coordinates of points on a curve and outputs a global feature vector. The output of T-NET is then multiplied with the input data to create a permutation-invariant global feature vector.

The feature vector is obtained from T-NET by the MLP network of the proposed architecture, which then outputs the segmentation scores of the main fiber bundles. It has five 1D-CNN layers with a kernel size of 1, one 1D global max pooling layer, and two Dense layers. In this MLP network, batch normalization and dropout layers are placed in between CNN and Dense layers to improve the regularization of the network, respectively.

We have used the attention mechanism [20] to boost the segmentation capability of the architecture. To identify which features are important and which parts are informative in the input 3D curve data points, channel and spatial

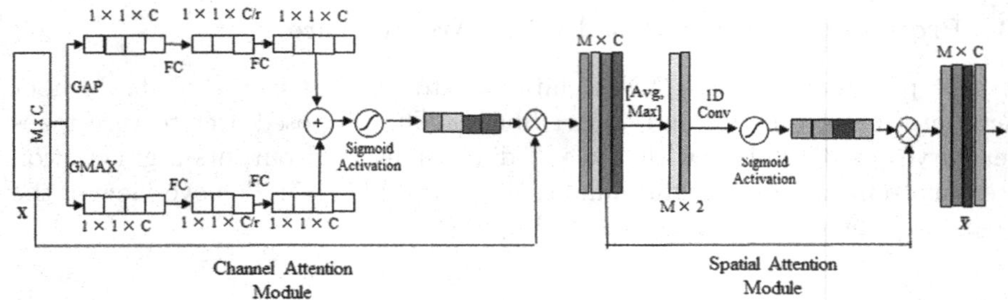

Fig. 3. The overview of the channel-spatial attention module. An $M \times C$ size input feature are fed to the attention module and then the boosted features are produced with the same size.

attention modules are utilized respectively. Two channel-spatial attention modules are positioned in between the last two 1D-CNN layers, as seen in Fig. 1. The proposed channel-spatial attention module is illustrated in Fig. 3. An $M \times C$ size input feature are obtained from a CNN layer and then fed to the Channel attention module. First, global max pooling (GMAX) and global average pooling (GAP) operations are performed on the input features simultaneously. Then these features are forwarded to an MLP network with one hidden layer to learn the channel-wise dependencies. The neurons of the hidden layer are set to C/r, where r is a parameter and it is determined experimentally. Finally, channel attention is computed by adding the two output features using an element-wise addition operation and then a Sigmoid activation function. The output of the channel attention module is used to boost the important features by an element-wise multiplication operation with the input feature.

The proposed spatial attention module inputs the refined features of the channel attention module. As a first step, it computes the channel-wise average and maximum and then concatenates them to produce a 1D spatial attention map with a size of $M \times 2$. Then this feature map is convolved by a 1D-CNN layer with a kernel size of 5. Then the Sigmoid activation function is used to produce the spatial attention between adjacent data points. Finally, element-wise multiplication is used to refine the features with spatial attention values.

2.4 Whole-Brain Unlabeled Tract Segmentation for Test Subjects

The unlabeled whole-brain tracts of test subjects are used to train the trained network to recognize 10 anatomical tracts. To begin, subject-specific imaging data is registered to MNI152 space, where training tractography data is in the same space as fiber streamline coordinates. All 3D fiber curves of a whole-brain tract are classified into one of the 10 anatomical tract categories or the "other fibers" category based on the trained network model. Finally, the qualitative outcomes were visually examined by an anatomical specialist.

3 Results and Discussion

3.1 Implementation Details and Evaluation Criteria

Python and Keras-TensorFlow are used to implement the proposed approach. The Google Colab cloud environment is used to execute the programs. The proposed CNN model was trained for 40 iterations, and the validation data was used to determine the hyper parameters. The Adam optimizer is used in the proposed approach with a learning rate of 0.004. Based on the experimental results, the dropout parameter and r in the first and second channel-spatial modules are set to 0.5, 4, 16, respectively. The source code of this study is available at https://github.com/NeuroImageComputingLab/3D_Curve_CNN.

On the test set of the curve dataset, the proposed approach is assessed, and the model's performance is measured using average accuracy, precision, and recall:

$$Accuracy = \frac{TP + TN}{TP + TN + FP + FN} \quad (1)$$

$$Precision = \frac{TP}{TP + FP} \quad (2)$$

$$Recall = \frac{TP}{TP + FN} \quad (3)$$

where TP, TN, FP, and FN are true positive, true negative, false positive, and false negative, respectively.

3.2 Classification Results

Recently, a variety of methodologies for segmenting white matter fiber tracts into bundles have been developed, and the performance of these methods is normally assessed on their own datasets. The effectiveness of whole-brain tract segmentation is influenced by the amount of fiber bundle classes, MRI capturing techniques and properties, diffusion sampling techniques, and the brain's nature, such as aging, gender, and neurological condition. This makes it impossible to compare our approach's accuracy to that of existing approaches. Furthermore, we could not compare the current dataset with comparable works because the source codes for comparable works were not accessible. However, in order to show the value of the suggested approach, we contrasted our technique's performance with the stated findings of related studies. The comparison results are summarized in detail in Table 1.

Table 1. Performance comparison of our approach with related methods.

Method	Number of classes	Number of subjects	Data format	Average accuracy
Proposed method	11	25	3D Raw curve points	98.8%
[14]	10	04	Image data	97.0%
[13]	12	25	Graph data	95.6%
[21]	08	03	Gradient points	95.0%
[22]	08	03	Curvature points	93.0%
[23]	08	03	Gradient features	92.0%
[10]	12	55	Diffusion MRI	90.0%
[3]	11	28	3D tracts	82.0%
[24]	72	100	2D image data	82.0%
[8]	13	20	3D image data	76.4%
[11]	25	5286	2D image data	74.0%
[9]	3	30	2D image data	84.0%

We visually checked the fibers in the training tractography data that were misclassified by our deep learning architecture. We discovered that the majority of misclassified fibers were in the boundary regions. Due to a lack of ground truth, it is difficult to reliably identify such fibers in tractography. As a result, we believe that the 98.8% average classification accuracy obtained from testing data demonstrates good tractography segmentation performance. Table 2 shows the experimental precision and recall evaluation measures of each category segmentation.

3.3 Whole-Brain Unlabeled Tract Segmentation for Test Subjects

We tested the ability of our proposed approach on ten randomly selected adult individuals from the ADNI project, ranging in age and gender. For experimental evaluation, the trained model was used for segmenting the whole-brain tractography data of these subjects. The whole-brain segmentation results for the five individuals are illustrated in the Fig. 4. According to an anatomist, the visual assessment revealed that the segmented major fiber bundles were visually comparable among the subjects, and we can see that the proposed method created more consistent fiber bundle segmentations across the population.

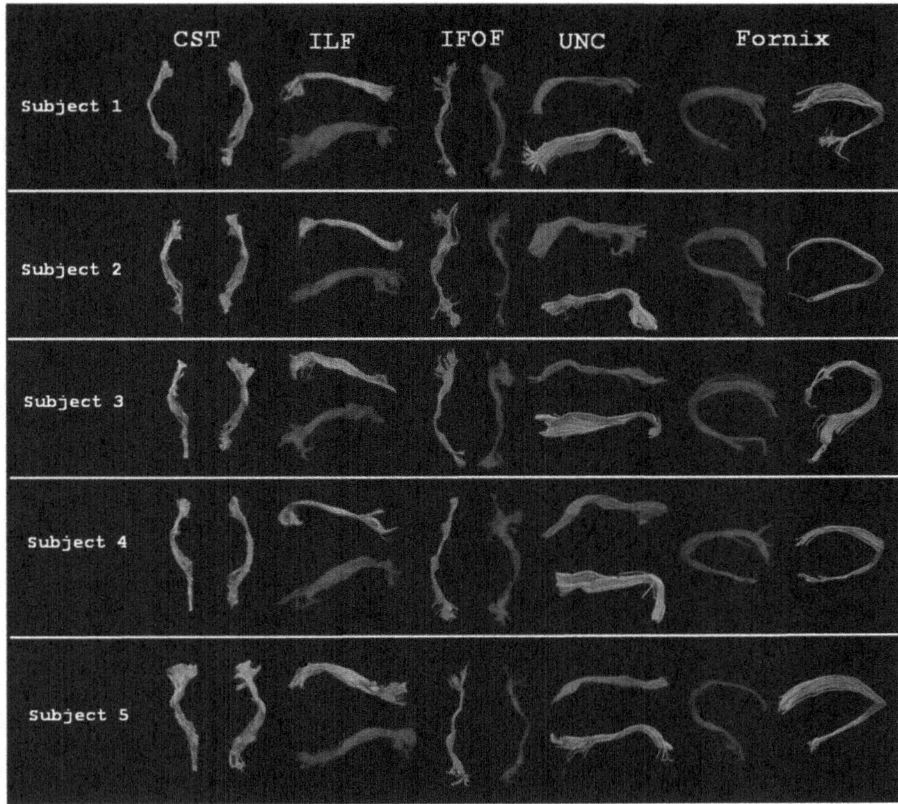

Fig. 4. The whole-brain segmentation results (Major Fiber bundles (Corticospinal Tract (CST), Inferior Longitudinal Fasciculus (ILF), Inferior Fronto Occipital Fasciculus (IFOF), Uncinate fasciculus (UNC), Fornix (Fornix))) for the five randomly selected adult individuals.

Table 2. Precision and recall of 10 Major fiber bundles and "Other Fibers" (OF).

Bundle	CSTL	CSTR	FornixL	FornixR	IFOFL	IFOFR	ILFL	ILFR	UNCL	UNCR	OF
Precision	0.97	0.96	1.00	0.99	1.00	1.00	1.00	1.00	1.00	0.99	1.00
Recall	1.00	1.00	1.00	1.00	1.00	1.00	1.00	1.00	1.00	1.00	0.93

4 Conclusion

We have presented a deep learning approach for segmenting anatomical fiber bundles from whole-brain tractography. We used 25 healthy adult subjects from ADNI's database to manually construct a large-scale labeled tractography fiber curves as a training dataset. A PointNet based CNN deep learning architecture was developed to segment the curves of the whole-brain tracts. The proposed

CNN architecture uses 3D raw curve points to consume the curves and segment the ten major fiber bundles and "other fibers" types. The proposed method is quick and consistent, and whole-brain tract segmentation has been successfully used across diverse adult populations (across genders and ages) according to the experimental data. The proposed approach could segment a curve in two seconds and had an average accuracy of 99%.

References

1. Catani, M., De Schotten, M.T.J.C.: A diffusion tensor imaging tractography atlas for virtual in vivo dissections. Cortex **44**, 1105–1132 (2008)
2. Wakana, S., et al.: Reproducibility of quantitative tractography methods applied to cerebral white matter. Neuroimage **36**, 630–644 (2007)
3. Hua, K., Zhang, J., Wakana, S., Jiang, H., Li, X., Reich, D.S., et al.: Tract probability maps in stereotaxic spaces: analyses of white matter anatomy and tract-specific quantification. Neuroimage **39**, 336–347 (2008)
4. Wassermann, D., et al.: The white matter query language: a novel approach for describing human white matter anatomy. Brain Struct. Funct. **221**(9), 4705–4721 (2016). https://doi.org/10.1007/s00429-015-1179-4
5. Zhang, S., Correia, S., Laidlaw, D.: Identifying white-matter fiber bundles in DTI data using an automated proximity-based fiber clustering method. IEEE Trans. Vis. Comput. Graph. **14**, 1044 (2008)
6. Ratnarajah, N., Simmons, A., Hojjatoleslami, A.: Probabilistic clustering and shape modelling of white matter fibre bundles using regression mixtures. In: Fichtinger, G., Martel, A., Peters, T. (eds.) MICCAI 2011. LNCS, vol. 6892, pp. 25–32. Springer, Heidelberg (2011). https://doi.org/10.1007/978-3-642-23629-7_4
7. Ye, C., Yang, Z., Ying, S.H., Prince, J.L.J.N.: Segmentation of the cerebellar peduncles using a random forest classifier and a multi-object geometric deformable model: application to spinocerebellar ataxia type 6. Neuroinformatics **13**, 367–381 (2015)
8. Ratnarajah, N., Qiu, A.J.N.: Multi-label segmentation of white matter structures: application to neonatal brains. NeuroImage **102**, 913–922 (2014)
9. Wasserthal, J., Neher, P., Maier-Hein, K.H.J.N.: TractSeg-fast and accurate white matter tract segmentation. NeuroImage **183**, 239–253 (2018)
10. Reisert, M., Coenen, V.A., Kaller, C., Egger, K., Skibbe, H.J.: HAMLET: hierarchical harmonic filters for learning tracts from diffusion MRI. arXiv preprint arXiv:01068 (2018)
11. Li, B., et al.: Neuro4Neuro: a neural network approach for neural tract segmentation using large-scale population-based diffusion imaging. NeuroImage **218**, 116993 (2020)
12. Zhang, F., Karayumak, S.C., Hoffmann, N., Rathi, Y., Golby, A.J., O'Donnell, L.J.J.M.I.A.: Deep white matter analysis (DeepWMA): fast and consistent tractography segmentation. Med. Image Anal. **65**, 101761 (2020)
13. Liu, F., et al.: DeepBundle: fiber bundle parcellation with graph convolution neural networks. In: International Workshop on Graph Learning in Medical Imaging, pp. 88–95 (2019)
14. Gupta, V., Thomopoulos, S.I., Corbin, C.K., Rashid, F., Thompson, P.M.: Fibernet 2.0: an automatic neural network based tool for clustering white matter fibers in the brain. In: 2018 IEEE 15th International Symposium on Biomedical Imaging (ISBI 2018), pp. 708–711 (2018)

15. Gupta, V., Thomopoulos, S.I., Rashid, F.M., Thompson, P.M.: FiberNET: an ensemble deep learning framework for clustering white matter fibers. In: International Conference on Medical Image Computing and Computer-Assisted Intervention, pp. 548–555 (2017)
16. "ADNI": Usc.edu. http://adni.loni.usc.edu/. Accessed 27 Jan 2021
17. Smith, S.M., Jenkinson, M., Woolrich, M.W., Beckmann, C.F., Behrens, T.E., JohansenBerg, H., et al.: Advances in functional and structural MR image analysis and implementation as FSL. Neuroimage **23**(Suppl 1), S208-19 (2004)
18. Jenkinson, M., Bannister, P., Brady, M., Smith, S.: Improved optimization for the robust and accurate linear registration and motion correction of brain images. Neuroimage **17**, 825–841 (2002)
19. Qi, C.R., Su, H., Mo, K., Guibas, L.J.,: Pointnet: deep learning on point sets for 3D classification and segmentation. In: Proceedings of the IEEE Conference on Computer Vision and Pattern Recognition, pp. 652–660 (2017)
20. Woo, S., Park, J., Lee, J., Kweon, I.S.: CBAM: convolutional block attention module. In: Proceedings of the European Conference on Computer Vision (ECCV), pp. 3–19 (2018)
21. Gupta, T., Patil, S.M., Tailor, M., Thapar, D., Nigam, A.J.: BrainSegNet: a segmentation network for human brain fiber tractography data into anatomically meaningful clusters. arXiv preprint arXiv:05158 (2017)
22. Patel, V., Parmar, A., Bhavsar, A., Nigam, A.: Automated brain tractography segmentation using curvature points. In: Proceedings of the Tenth Indian Conference on Computer Vision, Graphics and Image Processing, pp. 1–6 (2016)
23. Patil, S.M., Nigam, A., Bhavsar, A., Chattopadhyay, C.J.,: Siamese LSTM based fiber structural similarity network (FS2Net) for rotation invariant brain tractography segmentation. arXiv preprint arXiv:09792 (2017)
24. Lu, Q., Li, Y., Ye, C.J.M.I.A.: Volumetric white matter tract segmentation with nested self-supervised learning using sequential pretext tasks. Med. Image Anal. **72**, 102094 (2021)

TractoFormer: A Novel Fiber-Level Whole Brain Tractography Analysis Framework Using Spectral Embedding and Vision Transformers

Fan Zhang[1](✉), Tengfei Xue[1,2], Weidong Cai[2], Yogesh Rathi[1], Carl-Fredrik Westin[1], and Lauren J. O'Donnell[1]

[1] Harvard Medical School, Boston, USA
fzhang@bwh.harvard.edu
[2] The University of Sydney, Sydney, NSW, Australia

Abstract. Diffusion MRI tractography is an advanced imaging technique for quantitative mapping of the brain's structural connectivity. Whole brain tractography (WBT) data contains over hundreds of thousands of individual fiber streamlines (estimated brain connections), and this data is usually parcellated to create compact representations for data analysis applications such as disease classification. In this paper, we propose a novel parcellation-free WBT analysis framework, *TractoFormer*, that leverages tractography information at the level of individual fiber streamlines and provides a natural mechanism for interpretation of results using the attention mechanism of transformers. TractoFormer includes two main contributions. First, we propose a novel and simple 2D image representation of WBT, *TractoEmbedding*, to encode 3D fiber spatial relationships and any feature of interest that can be computed from individual fibers (such as FA or MD). Second, we design a network based on vision transformers (ViTs) that includes: 1) data augmentation to overcome model overfitting on small datasets, 2) identification of discriminative fibers for interpretation of results, and 3) ensemble learning to leverage fiber information from different brain regions. In a synthetic data experiment, TractoFormer successfully identifies discriminative fibers with simulated group differences. In a disease classification experiment comparing several methods, TractoFormer achieves the highest accuracy in classifying schizophrenia vs control. Discriminative fibers are identified in left hemispheric frontal and parietal superficial white matter regions, which have previously been shown to be affected in schizophrenia patients.

Keywords: Diffusion MRI · Tractography · ViT · Disease classification

1 Introduction

Diffusion MRI (dMRI) tractography is an advanced imaging technique that enables *in vivo* reconstruction of the brain's white matter (WM) connections [1]. Tractography provides an important tool for quantitative mapping of the brain's connectivity

This work is supported by the following NIH grants: R01MH119222, R01MH125860, P41EB015902, R01MH074794.

L. Wang et al. (Eds.): MICCAI 2022, LNCS 13431, pp. 196–206, 2022.
https://doi.org/10.1007/978-3-031-16431-6_19

using measures of connectivity or tissue microstructure [2]. These measures have shown promise as potential biomarkers for disease classification using machine learning [3–5], which can improve our understanding of the brain in health and disease [6].

Defining a good data representation of tractography for machine learning is still an open challenge, especially at the fiber level. Performing whole brain tractography (WBT) on one individual subject can generate hundreds of thousands (or even millions) of fiber streamlines. WBT data is usually parcellated to create compact representations for data analysis applications. While most popular analyses of the brain's structural connectivity rely on coarse-scale WM parcellations [2], recent studies have demonstrated the power of analyzing WBT at much finer scales of parcellation using high-resolution connectomes [7, 8]. While such approaches enable WBT analysis at a very high resolution (e.g., a 32k × 32k connectivity matrix), they are still quite high-dimensional and not able to represent information directly extracted from individual fibers.

Another challenge in machine learning for tractography analysis is the limited sample size (number of subjects) of many dMRI datasets. Developing data augmentation methods to increase sample size is a known challenge in structural connectivity research [9]. Small sample sizes limit the use of recently proposed advanced learning techniques such as Transformers [10] and Vision Transformers (ViTs) [11], which are highly accurate [12] but usually require a large number of samples to avoid overfitting [13].

Finally, an important challenge in deep learning for neuroimaging is to be able to pinpoint location(s) in the brain that are predictive of disease or affected by disease [14]. While interpretability is a well-known challenge in deep learning [15, 16], newer methods such as ViTs have shown advances in interpretability for vision tasks [17, 18].

In this paper, we propose a novel parcellation-free WBT analysis framework, *TractoFormer*, that leverages tractography information at the level of individual fiber streamlines and provides a natural mechanism for interpretation of results using the self-attention scheme of ViTs. TractoFormer includes two main contributions. First, we propose a novel 2D image representation of WBT, referred to as *TractoEmbedding*, based on a spectral embedding of fibers from tractography. Second, we propose a ViT-based network that performs effective and interpretable group classification. In the rest of this paper, we first describe the TractoFormer framework, then we illustrate its performance in two experiments: classification of synthetic data with true group differences, and disease classification between schizophrenia and control.

2 Methods

2.1 Diffusion MRI Datasets and Tractography

We use two dMRI datasets. The first dataset is used to create the embedding space and includes data from 100 subjects (29.1 ± 3.7 years; 54 F, 46 M) from the Human Connectome Project (www.humanconnectome.org) [19], with 18 b = 0 and 90 b = 3000 images, TE/TR = 89/5520 ms, resolution = 1.25 × 1.25 × 1.25 mm^3. The second dataset is used for experimental evaluations and includes data from 103 healthy controls (HCs) (31.1 ± 8.7; 52 F, 51 M) and 47 schizophrenia (SCZ) patients (35.8 ± 8.8; 36 F and 11 M) from the Consortium for Neuropsychiatric Phenomics (CNP) (https://openfmri.org/dataset/ds000030) [20], with 1 b = 0 and 64 b = 1000 images, TE/TR = 93/9000 ms, resolution

$= 2 \times 2 \times 2$ mm^3. WBT is performed using the two-tensor unscented Kalman filter (UKF) method [21, 22] (via SlicerDMRI [23, 24]) to generate about one million fibers per subject. UKF has been successful in neuroscientific applications such as disease classification [25] and population statistical comparison [26], and it allows estimation of fiber-specific microstructural properties (including FA and MD). Fiber tracking parameters are as in [27]. Tractography from the 100 HCP are co-registered, followed by alignment of each CNP WBT using a tractography-based registration [28].

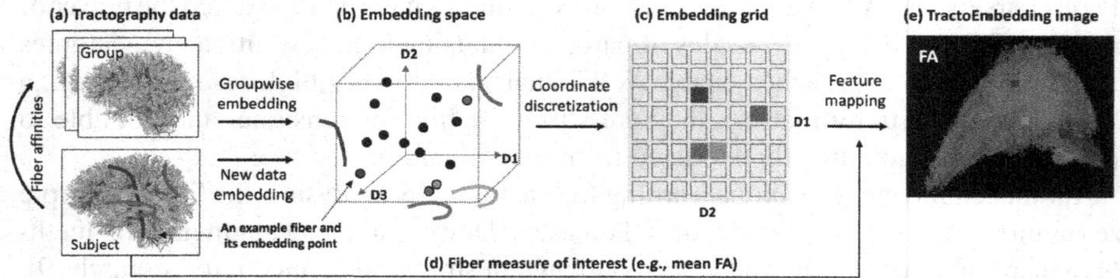

Fig. 1. TractoEmbedding overview. Each input fiber in WBT (a) is represented as a point in a latent embedding space (b), where nearby points correspond to spatially proximate fibers. Then, embedding coordinates of all points (fibers) are discretized onto a 2D grid, where points with similar coordinates are mapped to the same or nearby pixels (c). Next, features of interest from each fiber (e.g., mean fiber FA) are mapped (d) as the intensity of the pixel corresponding to that fiber. This generates a 2D image representation, i.e., a TractoEmbedding image (e).

2.2 TractoEmbedding: A 2D Image Representation of WBT

The TractoEmbedding process includes three major steps (illustrated in Fig. 1). First, we perform spectral embedding to represent each fiber in WBT as a point in a latent space. Spectral embedding is a learning technique that performs dimensionality reduction based on the relative similarity of each pair of points in a dataset, and it has been successfully used for tractography computing tasks such as fiber segmentation [29], fiber clustering [30], and tract atlas creation [27]. To enable a robust and consistent embedding of WBT data from different subjects for population-wise analysis, we first create a groupwise embedding space using a random sample of fibers from co-registered tractography data from 100 subjects (see Sect. 2.1 for data details). This process uses spectral embedding [31] with a pairwise fiber affinity based on mean closest point distance [30, 32]. Next, to embed new WBT data, it is aligned to the 100-subject data [28], followed by computing pairwise fiber affinities to the population tractography sample. Then, each fiber of the new WBT data is spectrally embedded into the embedding space, resulting in an embedding coordinate vector for each fiber. We note that our process of spectral embedding is similar to that used for tractography clustering [30] and we refer the readers to [30] for details.

In the second step, the coordinates of each fiber of the new WBT data are discretized onto a 2D grid for creation of an image. Each dimension of the embedding coordinate vector corresponds to the eigenvectors of the affinity matrix sorted in descending order. A higher order indicates a higher importance of the coordinate to locate the point in the embedding space. A previous work has applied embedding coordinates for effective visualization of tractography data [33]. In our study, we choose the first two dimensions for each point and discretize them onto a 2D embedding grid[1]. A grid size parameter defines the image resolution.

In the third step, we map the measure of interest associated with each fiber to the corresponding pixel on the embedding grid as its intensity value. This generates a 2D image, i.e., the TractoEmbedding image. When multiple fibers that are spatially proximate are mapped to the same voxel, we can compute summary statistics from these fibers, such as max, min, and mean (mean is used in our experiments).

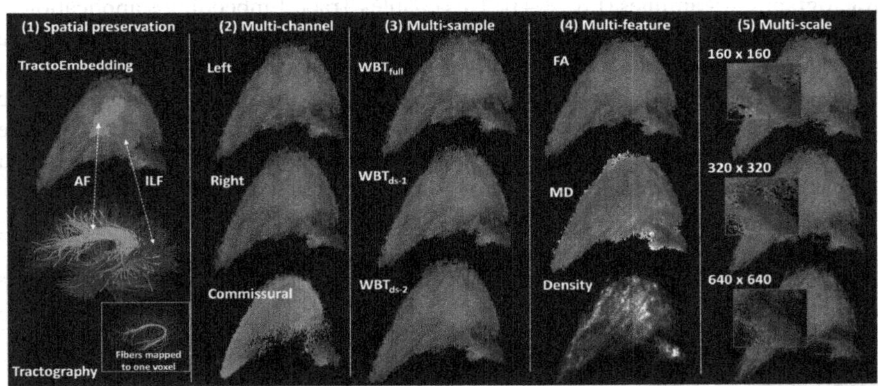

Fig. 2. TractoEmbedding images generated from the left hemisphere data of one randomly selected CNP subject. (1) Spatially proximate fibers from the same anatomical tracts are mapped to nearby pixels using TractoEmbedding. (2) Fibers from the left hemisphere, right hemisphere and commissural regions can be used individually to create a multi-channel image. (3) Multiple TractoEmbedding images are generated using the full WBT and two random samples (80% of the full WBT). (4) Multiple TractoEmbedding images are generated using different features of interest, including the mean FA per fiber, the mean MD per fiber, and the number of fibers mapped to each voxel. (5) Multiple TractoEmbedding images are generated at different resolutions (scales). Inset images give a zoomed visualization of a local image region.

TractoEmbedding has several advantages (as illustrated in Fig. 2). *First*, TractoEmbedding is a 2D image that preserves the relative spatial relationship of every fiber pair in WBT in terms of the pixel neighborship in the 2D image (Fig. 2(1)). In this way, TractoEmbedding enables image-based computer vision techniques such as CNNs and

[1] While embeddings from the first 3 dimensions can be used to generate 3D TractoEmbedding images, our unpublished results show that this decreases group classification performance potentially due to the data sparsity where many voxels on the 3D grid do not have any mapped fibers.

ViT to leverage fiber spatial similarity information. (In the case where multiple fibers are mapped to the same voxel, to quantify the similarity of such fibers, we computed the mean pairwise fiber distance (MPFD) across the fibers. The average of MPFDs across all voxels with multiple fibers is 5.7 mm, which is a low value representing highly similar fibers through the same voxel.) *Second*, TractoEmbedding enables a multi-channel representation where each channel represents fibers from certain brain regions. This allows independent and complementary analysis of WBT anatomical regions, such as the left hemispheric, the right hemispheric and the commissural fibers in our current study. Thus, the TractoEmbedding is a 3-channel 2D image (Fig. 2(2)). *Third*, multiple TractoEmbedding images are generated by performing random downsampling of each subject's input WBT (Fig. 2(3)). This naturally and effectively increases training sample size for data augmentation for learning-based methods, which is particularly important for methods that require a large number of samples. *Fourth*, TractoEmbedding can be generally used to encode any possible features of interest that can be computed at the level of individual tractography streamlines (Fig. 2(4)). This enables TractoEmbedding's application in various tractography-based neuroscientific studies where particular WM properties are of interest. *Fifth*, TractoEmbedding allows a WBT representation at different scales in terms of the resolution of the embedding grid (Fig. 2(5)). With a low resolution, multiple fibers tend to be mapped into the same voxel, enabling WBT analysis at a coarse-scale fiber parcel level; with a high resolution, an individual fiber (or a few fibers) is mapped to any particular voxel, enabling WBT analysis at a fine-scale individual fiber level.

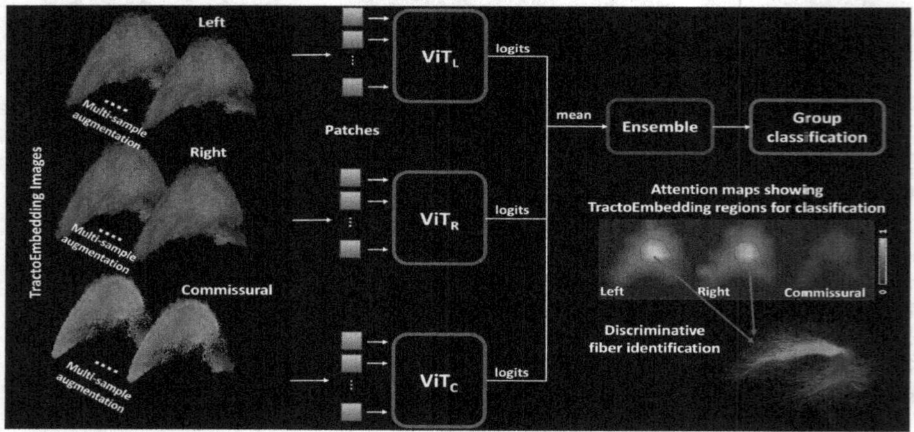

Fig. 3. TractoFormer framework including an ensemble ViT network with input multi-channel TractoEmbedding images using multi-sample data augmentation. Attention maps are computed from ViTs for identification of fibers that are discriminative for classification.

2.3 TractoFormer: A ViT-Based Framework for Group Classification

Figure 3 shows the proposed TractoFormer architecture, which leverages an ensemble of three ViTs to process the three-channel input TractoEmbedding images. Our design aims to address the aforementioned challenges of sample size/overfitting and interpretability. First, we leverage the multi-sample data augmentation (Fig. 2(3)) to reduce the known overfitting issue of ViTs on small sample size datasets [13]. Second, we leverage the self-attention scheme in ViT to identify discriminative fibers that are most useful to differentiate between groups. The interpretation of the ViT attention maps [11] is aided by our proposed multi-channel architecture, which can enable inspection of the independent contributions of different brain regions.

In detail, for each input channel, we use a light-weight ViT architecture (see Sect. 2.4 for details). An ensemble of the predictions is performed by averaging the logit outputs across the ViTs. For data augmentation, for each input subject, we create 100 TractoEmbedding images using randomly downsampled WBT data (80% of the fibers). For interpretation of results, in each ViT we compute the average of the attention weights for each token across all heads per layer, then recursively multiply the averaged weights for the first to the last layer, and finally map the joint token attention scores back to the input image space[2]. This generates an attention score map where the values indicate the importance of the corresponding pixels when classifying the TractoEmbedding image (as shown in Fig. 3). We identify the pixels with higher scores using a threshold T, and then identify the fibers that are mapped to these pixels when performing TractoEmbedding. These fibers are thus the ones that are highly important when classifying the TractoEmbedding image. We refer to the identified fibers as *the discriminative fibers*.

2.4 Implementation Details and Parameter Setting

Our method is implemented using PyTorch (v1.7.1) [34]. For each ViT, we use 3 layers with 8 heads, a hidden size of 128, and a dropout rate of 0.2 (grid search for {3, 4, 5}, {4, 6, 8}, {128, 256}, and {0.2, 0.3}, respectively). Adam [35] is used for optimization with a learning rate 1e−3 and a batch size 64 for a total of 200 epochs. Early stopping is adopted when there is no accuracy improvement in 20 continuous epochs. 5-fold cross-validation is performed for each experiment below and the mean accuracy and F1 scores are reported. T is set to be the mean + 2 stds of the scores in an attention map. The computation is performed using NVIDIA GeForce 1080 Ti. On average, each epoch (training and validation) takes ~30 s with 2 GB GPU memory usage when using data augmentation and 160 × 160 resolution. The code will be made available upon request.

2.5 Experimental Evaluation

Exp 1: Synthetic Data. The goal is to provide a proof-of-concept evaluation to assess if the proposed TractoFormer can 1) successfully classify groups with true WM differences and 2) identify the fibers with group differences in the WBT data for interpretation. To do so, we create a realistic synthetic dataset with true group differences, as follows. From the 103 CNP HC data, we add white Gaussian noise (signal-to-noise ratio at 1 [36, 37]) to the actual measured mean FA value of each fiber in the WBT data. Repeating this process twice generates two synthetic groups of G1 and G2, each with 103 subjects. We then modify the mean FA of the fibers belonging to the corticospinal tract (CST)

[2] Following instructions from: https://github.com/jeonsworld/ViT-pytorch.

(a random tract selected for demonstration) in G2 to have a true group difference. To do so, we decrease the mean FA of each CST fiber in G2 by 20%, a synthetic change suggested to introduce a statistically significant difference in tractography-based group comparison analysis [36]. We apply the TractoFormer to this synthetic data to perform group classification and identify the discriminative fibers.

Exp 2: Disease Classification Between HC and SCZ. The goal is to evaluate the proposed TractoFormer in a real neuroscientific application for brain disease classification. Previous studies have revealed widespread WM changes in SCZ patients using dMRI techniques [38]. In our paper, we apply TractoFormer to investigate the performance of using tractography data to classify between HC and SCZ in the CNP dataset. For interpretation purposes, we compute a group-wise attention map by averaging the attention maps from all subjects that are classified as SCZ, from which the discriminative TractoEmbedding pixels and discriminative fibers are identified. We compare our method with three baseline methods. The first one performs group classification using fiber parcel level features and a 1D CNN network [39], referred to as the *FC-1DCNN* method. Briefly, for each subject, WBT parcellation is performed using a fiber clustering atlas [27], resulting in a total of 1516 parcels per subject. The mean feature of interest (i.e., FA or MD) along each parcel is computed, leading to a 1D feature vector with 1516 values per subject. Then, a 1D CNN is applied to the feature vectors to perform group classification. For parameters, we follow the suggested settings in the author's implementation[3]. The second method performs group classification using track-density images (TDI) [40] and 3D ResNet [41], referred to as the *TDI-3DResNet* method. Briefly, a 3D TDI, where each voxel represents streamline count, is generated per subject and fed into a 3D ResNet for group classification. The third baseline method performs group classification using TractoEmbedding images, but instead of using the proposed ViT, it applies ResNet [41], a classic CNN architecture that has been shown to be highly successful in many applications. We refer to this method as *ResNet*. For the ResNet and TractoFormer methods, we perform classification with and without data augmentation. We also provide interpretability results using Class Activation Maps (CAMs) [42] in ResNet.

3 Results and Discussion

Exp 1: Synthetic Data. TractoFormer achieved, as expected, 100% group classification accuracy because of the added synthetic feature changes to G2. Figure 4 shows the identified discriminative fibers in one example G2 subject based on its subject-specific attention map and the G2-group-wise attention map. The discriminative fibers are generally similar to the CST fibers with synthetic changes.

[3] https://github.com/H2ydrogen/Connectome_based_prediction.

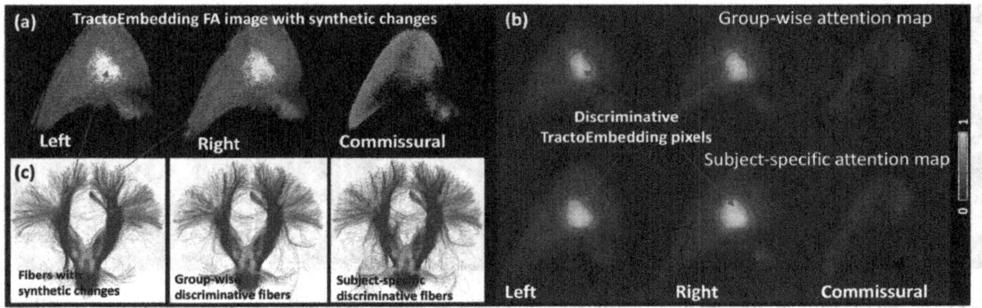

Fig. 4. (a) TractoEmbedding FA images of one example G2 subject (320 × 320). (b) G2-group-wise and subject-specific attention maps (discriminative threshold in red). (c) Identified discriminative fibers, with comparison to the CST fibers with synthetic changes.

Exp 2: Disease Classification Between HC and SCZ. Table 1 shows the classification results of each compared method. In general, the FA measure gives the best result. The FC-1DCNN method generates lower accuracy and F1 scores than the methods that benefit from data augmentation. Regarding the 3 TractoEmbedding-based methods, we can observe that including data augmentation greatly improves the classification performance. The ensemble architecture gives the best overall result (at resolution 160 × 160 with FA feature), with a mean accuracy of 0.849 and a mean F1 of 0.770. Figure 5 gives a visualization of the discriminative fibers from group-wise and subject-specific attention maps. In general, our results suggest that the superficial fibers in the frontal and parietal lobes have high importance when classifying SCZ and HC under study. Multiple studies have suggested these white matter regions are affected in SCZ [43–45]. In *ResNet* (at resolution 160 × 160 with FA feature), CAM identifies the fibers related to the brainstem and cerebellum. The ViT- and ResNet-based methods focus on different brain regions, possibly explaining the accuracy difference of the two methods.

Table 1. Comparison across different methods: the mean accuracy and the mean F1 (the first and second values, respectively, per cell) across the cross-validation are reported.

Method	Resolution	FA		MD		Density	
FC-1DCNN	n.a.	0.808/0.669		0.780/0.636		0.767/0.603	
TDI-3DCNN	n.a.	n.a.		n.a.		0.764/0.589	
Data Augmentation		*no aug*	*with aug*	*no aug*	*with aug*	*no aug*	*with aug*
ResNet	80×80	.719/.491	.819/.751	.758/.607	.753/.659	.630/.506	.744/.634
	160×160	.712/.544	**.829/.753**	.764/.525	.769/.662	.742/.580	.761/.674
	320×320	.653/.524	.778/.685	.761/.572	.703/.604	.728/.533	.694/.524
TractoFormer (stack input)	80×80	.804/.751	.808/.738	.649/.606	.774/.695	.741/.702	.774/.701
	160×160	.716/.506	**.816/.732**	.724/.616	.733/.682	.708/.486	.808/.754
	320×320	.783/.480	.783/.653	.791/.710	.691/.635	.783/.480	.811/.721
TractoFormer (ensemble)	80×80	.788/.623	.824/.758	.716/.600	.766/.724	.791/.533	.808/.743
	160×160	.783/.480	**.849/.770**	.808/.698	.841/.732	.733/.525	.799/.689
	320×320	.783/.480	.758/.645	.816/.619	.845/.742	.741/.489	.801/.726

Fig. 5. Discriminative fibers identified in the disease classification (SCZ vs HC) experiment, corresponding to the best performing results using FA and resolution 160 × 160.

4 Conclusion

We present a novel parcellation-free WBT analysis framework, *TractoFormer*, which leverages tractography information at the level of individual fiber streamlines and provides a natural mechanism for interpretation of results using attention. We propose random sampling of tractography as an effective data augmentation strategy for small sample size WBT datasets. Future work could include an investigation of ensembles of different fiber features in the same network, multi-scale learning to use TractoEmbedding images with different resolutions together, and/or combination with advanced computer vison data augmentation methods. Overall, TractoFormer suggests the potential for deep learning analysis of WBT represented as images.

References

1. Basser, P.J., Pajevic, S., Pierpaoli, C., Duda, J., Aldroubi, A.: In vivo fiber tractography using DT-MRI data. Magn. Reson. Med. **44**, 625–632 (2000)
2. Zhang, F., Daducci, A., He, Y., et al.: Quantitative mapping of the brain's structural connectivity using diffusion MRI tractography: a review. Neuroimage **249**, 118870 (2022)
3. Zhan, L., et al.: Comparison of nine tractography algorithms for detecting abnormal structural brain networks in Alzheimer's disease. Front. Aging Neurosci. (2015). https://doi.org/10.3389/fnagi.2015.00048
4. Deng, Y., et al.: Tractography-based classification in distinguishing patients with first-episode schizophrenia from healthy individuals. Prog. Neuropsychopharmacol. Biol. Psychiatry **88**, 66–73 (2019)
5. Hu, M., X., et al.: Structural and diffusion MRI based schizophrenia classification using 2D pretrained and 3D naive Convolutional Neural Networks. Schizophr. Res. (2021). https://doi.org/10.1016/j.schres.2021.06.011
6. Brown, C.J., Hamarneh, G.: Machine Learning on human connectome data from MRI. arXiv [cs.LG] (2016). http://arxiv.org/abs/1611.08699
7. Mansour, L.S., Tian, Y., Yeo, B.T.T., Cropley, V., Zalesky, A.: High-resolution connectomic fingerprints: mapping neural identity and behavior. Neuroimage **229**, 117695 (2021)
8. Cole, M., Murray, K., et al.: Surface-Based Connectivity Integration: an atlas-free approach to jointly study functional and structural connectivity. Hum. Brain Mapp. **42**, 3481–3499 (2021)
9. Barile, B., Marzullo, A., Stamile, C., Durand-Dubief, F., Sappey-Marinier, D.: Data augmentation using generative adversarial neural networks on brain structural connectivity in multiple sclerosis. Comput. Methods Programs Biomed. **206**, 106113 (2021)

10. Vaswani, A., et al.: Attention is all you need. In: Advances in Neural Information Processing Systems 30 (2017)
11. Dosovitskiy, A., Beyer, L., et al.: An image is worth 16 × 16 words: transformers for image recognition at scale. In: ICLR (2021)
12. Han, K., et al.: A survey on vision transformer. IEEE Trans. Pattern. Anal. Mach. Intell. (2022). https://doi.org/10.1109/TPAMI.2022.3152247
13. Steiner, A., Kolesnikov, A., et al.: How to train your ViT? Data, augmentation, and regularization in vision transformers. TMLR (2022)
14. Hofmann, S.M., Beyer, F., Lapuschkin, S., et al.: Towards the interpretability of deep learning models for human neuroimaging. bioRxiv, p. 2021.06.25.449906 (2021)
15. Zhang, Q.-S., Zhu, S.-C.: Visual interpretability for deep learning: a survey. Front. Inf. Technol. Electron. Eng. **19**(1), 27–39 (2018). https://doi.org/10.1631/FITEE.1700808
16. Lombardi, A., Diacono, D., Amoroso, N., et al.: Explainable deep learning for personalized age prediction with brain morphology. Front. Neurosci. **15**, 674055 (2021)
17. Chefer, H., Gur, S., Wolf, L.: Generic attention-model explainability for interpreting bi-modal and encoder-decoder transformers. In: ICCV, pp. 397–406 (2021)
18. Abnar, S., Zuidema, W.: Quantifying attention flow in transformers. In: ACL 2020. pp. 4190–4197 (2020)
19. Van Essen, D.C., et al.: The WU-Minn Human Connectome Project: an overview. Neuroimage **80**, 62–79 (2013)
20. Poldrack, R.A., et al.: A phenome-wide examination of neural and cognitive function. Sci. Data **3**, 160110 (2016)
21. Malcolm, J.G., Shenton, M.E., Rathi, Y.: Filtered multitensor tractography. IEEE Trans. Med. Imaging **29**, 1664–1675 (2010)
22. Reddy, C.P., Rathi, Y.: Joint multi-fiber NODDI parameter estimation and tractography using the unscented information filter. Front. Neurosci. **10**, 166 (2016)
23. Norton, I., Essayed, W.I., Zhang, F., Pujol, S., et al.: SlicerDMRI: open source diffusion MRI software for brain cancer research. Cancer Res. **77**, e101–e103 (2017)
24. Zhang, F., Noh, T., et al.: SlicerDMRI: diffusion MRI and tractography research software for brain cancer surgery planning and visualization. JCO Clin. Can. Inform. **4**, 299–309 (2020)
25. Zhang, F., Savadjiev, P., Cai, W., et al.: Whole brain white matter connectivity analysis using machine learning: an application to autism. Neuroimage **172**, 826–837 (2018)
26. Hamoda, H.M., et al.: Abnormalities in thalamo-cortical connections in patients with first-episode schizophrenia: a two-tensor tractography study. Brain Imaging Behav. **13**(2), 472–481 (2018). https://doi.org/10.1007/s11682-018-9862-8
27. Zhang, F., Wu, Y., et al.: An anatomically curated fiber clustering white matter atlas for consistent white matter tract parcellation across the lifespan. Neuroimage **179**, 429–447 (2018)
28. O'Donnell, L.J., Wells, W.M., Golby, A.J., Westin, C.-F.: Unbiased groupwise registration of white matter tractography. In: Ayache, N., Delingette, H., Golland, P., Mori, K. (eds.) MICCAI 2012. LNCS, vol. 7512, pp. 123–130. Springer, Heidelberg (2012). https://doi.org/10.1007/978-3-642-33454-2_16
29. Vercruysse, D., Christiaens, D., Maes, F., Sunaert, S., Suetens, P.: Fiber bundle segmentation using spectral embedding and supervised learning. In: CDMRI, pp. 103–114 (2014)
30. O'Donnell, L.J., Westin, C.-F.: Automatic tractography segmentation using a high-dimensional white matter atlas. IEEE Trans. Med. Imaging. **26**, 1562–1575 (2007)
31. Fowlkes, C., Belongie, S., Chung, F., Malik, J.: Spectral grouping using the Nyström method. IEEE Trans. Pattern Anal. Mach. Intell. **26**, 214–225 (2004)
32. Moberts, B., Vilanova, A., van Wijk, J.J.: Evaluation of fiber clustering methods for diffusion tensor imaging. In: IEEE Conference on Visualization, pp. 65–72 (2005)

33. Jianu, R., Demiralp, C., Laidlaw, D.H.: Exploring 3D DTI fiber tracts with linked 2D representations. IEEE Trans. Vis. Comput. Graph. **15**, 1449–1456 (2009)
34. Paszke, A., et al.: PyTorch: an imperative style, high-performance deep learning library. In: Advances in Neural Information Processing Systems 32 (2019)
35. Kingma, D.P., Ba, J.: Adam: a method for stochastic optimization. In: ICLR (2015)
36. Zhang, F., Wu, W., Ning, L., et al.: Suprathreshold fiber cluster statistics: Leveraging white matter geometry to enhance tractography statistical analysis. Neuroimage **171**, 341–354 (2018)
37. Smith, S.M., Nichols, T.E.: Threshold-free cluster enhancement: addressing problems of smoothing, threshold dependence and localisation in cluster inference. Neuroimage **44**, 83–98 (2009)
38. Kelly, S., et al.: Widespread white matter microstructural differences in schizophrenia across 4322 individuals: results from the ENIGMA Schizophrenia DTI Working Group. Mol. Psychiatry **23**, 1261–1269 (2018)
39. He, H., Zhang, F., et al.: Model and predict age and sex in healthy subjects using brain white matter features: a deep learning approach. In: ISBI, pp. 1–5 (2022)
40. Calamante, F., Tournier, D., et al.: Track-density imaging (TDI): super-resolution white matter imaging using whole-brain track-density mapping. Neuroimage **53**(4), 1233–1243 (2010)
41. He, K., Zhang, X., Ren, S., Sun, J.: Deep residual learning for image recognition. In: CVPR, pp. 770–778 (2016)
42. Zhou, B., Khosla, A., Lapedriza, A., Oliva, A., Torralba, A.: Learning deep features for discriminative localization. In: CVPR, pp. 2921–2929 (2016)
43. Nazeri, A., Chakravarty, M., et al.: Alterations of superficial white matter in schizophrenia and relationship to cognitive performance. Neuropsychopharmacology **38**, 1954–1962 (2013)
44. Makris, N., Seidman, L.J., Ahern, T., Kennedy, D.N., et al.: White matter volume abnormalities and associations with symptomatology in schizophrenia. Psychiatry Res. **183**, 21–29 (2010)
45. Ji, E., Guevara, P., et al.: Increased and decreased superficial white matter structural connectivity in schizophrenia and bipolar disorder. Schizophr. Bull. **45**, 1367–1378 (2019)

Multi-site Normative Modeling of Diffusion Tensor Imaging Metrics Using Hierarchical Bayesian Regression

Julio E. Villalón-Reina[1(✉)], Clara A. Moreau[2], Talia M. Nir[1], Neda Jahanshad[1], Simons Variation in Individuals Project Consortium, Anne Maillard[3], David Romascano[3], Bogdan Draganski[4,5], Sarah Lippé[6], Carrie E. Bearden[7], Seyed Mostafa Kia[8,9], Andre F. Marquand[9], Sebastien Jacquemont[10], and Paul M. Thompson[1]

[1] Imaging Genetics Center, Mark and Mary Stevens Neuroimaging and Informatics Institute, Keck School of Medicine, University of Southern California, Marina del Rey, CA, USA
julio.villalon@ini.usc.edu

[2] Institut Pasteur, Université de Paris, Paris, France

[3] Service des Troubles du Spectre de l'Autisme et apparentés, Département de Psychiatrie, Lausanne University Hospital (CHUV), Lausanne, Switzerland

[4] Laboratory for Research in Neuroimaging LREN, Centre for Research in Neuroscience, Department of Clinical Neurosciences, Lausanne University Hospital and University of Lausanne, Lausanne, Switzerland

[5] Neurology Department, Max-Planck-Institute for Human Cognitive and Brain Sciences, Leipzig, Germany

[6] Sainte-Justine Centre de Recherche, Université de Montréal, Montréal, Canada

[7] Departments of Psychiatry and Biobehavioral Sciences and Psychology, University of California, Los Angeles, CA, USA

[8] Department of Psychiatry, University Medical Center Utrecht, 3584 CX Utrecht, The Netherlands

[9] Donders Institute for Brain, Cognition and Behaviour, Radboud University Medical Centre, Nijmegen, The Netherlands

[10] Department of Pediatrics, Université de Montréal, Montréal, Canada

Abstract. Multi-site imaging studies can increase statistical power and improve the reproducibility and generalizability of findings, yet data often need to be harmonized. One alternative to data harmonization in the normative modeling setting is Hierarchical Bayesian Regression (HBR), which overcomes some of the weaknesses of data harmonization. Here, we test the utility of three model types, i.e., linear, polynomial and *b*-spline - within the normative modeling HBR framework - for multi-site normative modeling of diffusion tensor imaging (DTI) metrics of the brain's white matter microstructure, across the lifespan. These models of age dependencies were fitted to cross-sectional data from over 1,300 healthy subjects (age range: 2–80 years), scanned at eight sites in diverse geographic locations. We found that the polynomial and *b*-spline fits were better suited for modeling

Supplementary Information The online version contains supplementary material available at https://doi.org/10.1007/978-3-031-16431-6_20.

L. Wang et al. (Eds.): MICCAI 2022, LNCS 13431, pp. 207–217, 2022.
https://doi.org/10.1007/978-3-031-16431-6_20

relationships of DTI metrics to age, compared to the linear fit. To illustrate the method, we also apply it to detect microstructural brain differences in carriers of rare genetic copy number variants, noting how model complexity can impact findings.

1 Introduction

Clinical neuroimaging studies often need to pool data from multiple sources and scanners, especially if the conditions studied are rare. Consequently, worldwide efforts have been initiated in the past decade to combine existing legacy data from multiple sites to increase statistical power and test how well findings generalize across diverse populations. One such effort is the ENIGMA Consortium [1]. In parallel, large-scale scanning initiatives such as UKBB [2] and HCP [3] have been launched to prospectively scan thousands of subjects at multiple sites. Multi-site analyses have introduced new challenges to the field, particularly cross-site harmonization of the MRI scans, and normative modeling of measures derived from them. These approaches are used to model site effects, including differences in scanning protocols and sample characteristics across sites.

Commonly used harmonization algorithms, such as ComBat [4], cannot fully remove site-associated variance if age and site are confounded, making it difficult to correctly model the age effect when computing a harmonized set of MRI metrics. Here we test a recently proposed method for multi-site normative modeling of derived MRI measures to create a normative model (NM) [5]. Normative modeling is a technique that estimates the normative range and statistical distribution for a metric, (e.g., brain MRI-derived measure) as a function of specific covariates, e.g., age, sex, IQ, etc. Specifically, an NM based on hierarchical Bayesian regression (HBR) can be used to model site-dependent variability in the imaging data [6]. As a Bayesian method, it infers a posterior distribution rather than a single maximum likelihood estimate for the MRI metric, conditioned on the values of the known covariates. As noted by Kia et al. [6], HBR can model the different levels of variation in the data. Structural dependencies among variables are modeled by coupling them via a shared prior distribution. The frequent statistical dependence in multi-site neuroimaging projects that occurs between age and site, caused by different inclusion criteria across cohorts, may be better tackled with HBR.

Here we apply NM with HBR to metrics derived from diffusion tensor imaging (DTI). DTI is sensitive to alterations in brain microstructure and it has proven to be valuable in the diagnosis and characterization of many brain diseases [7, 8]. Nevertheless, the derived measures are highly sensitive to differences in acquisition parameters, such as voxel size and the number of gradient directions. This causes non-trivial differences in the raw scans as well as in their derived metrics [9]. Hence, the harmonization of DTI metrics is essential when performing multi-site studies of psychiatric and neurological diseases.

We set out to establish the normative range for DTI metrics in the brain's white matter (WM) across the lifespan, based on data from 1,377 controls across six datasets, each with a different age range. Although only limited data exists on DTI norms from limited age ranges [10], lifespan normative charts have recently been computed for morphometric measures from T1-weighted MRI using ComBat-GAM [11] and NM [12]. We tackle the problem of fitting the age trajectory for DTI metrics by using HBR with three different fitting strategies, i.e., linear, polynomial and *b*-spline. We aimed to determine which of these models better fits age effects across the lifespan by using standardized evaluation metrics. Finally, we compare the three model types in an anomaly detection experiment on a multi-site sample with a large age range (2–80 years) of individuals with 16p11.2 deletion syndrome (16pDel) - a rare neurogenetic syndrome caused by the deletion of 29 genes on chromosome 16 [13]. These deletions increase the risk for a myriad of neuropsychiatric disorders, including neurodevelopmental delay, autism spectrum disorder and attention deficit hyperactivity disorder [14]. As such, there is keen interest in establishing the profile of brain abnormalities in 16pDel, which in turn requires an accurate reference model for WM microstructure across the lifespan. Additionally, because it is a rare condition (1:3000 live births) [20], multisite data must inevitably be combined to achieve adequate statistical power. Rare genetic variants are an illustrative and extreme case of the ubiquitous problem where clinical sites can only scan a few dozen patients each, making multi-site data pooling crucial and beneficial.

2 Methods

2.1 Diffusion MRI Processing

Image acquisition parameters are shown in Table 1. We combined 16pDel data from six sites (8 scanners in total), resulting in the largest neuroimaging sample assessing people with these CNVs to date (see Table 2). All six sites used Pulsed Gradient Spin Echo protocols. All images were identically preprocessed with DIPY [15] and FSL [16]. DTI-based fractional anisotropy (FA), mean diffusivity (MD), radial diffusivity (RD), and axial diffusivity (AD) were computed with DIPY on the 1000 s/mm^2 shell. All subjects' FA maps were nonlinearly registered to the ENIGMA-FA template with ANTs [17]; these deformations were applied to the other DTI maps. Mean DTI measures were extracted from 21 bilateral regions of interest (ROIs) from the Johns Hopkins University WM atlas (JHU-WM) [18] using the ENIGMA-DTI protocol [19].

JHU-WM ROIs. PCR = Posterior *corona radiata*, CGH = Cingulum of the hippocampus, CGC = Cingulum of the cingulate gyrus, UNC = Uncinate fasciculus, RLIC = Retrolenticular part of internal capsule, SCR = Superior *corona radiata*, ACR = Anterior *corona radiata*, EC = external capsule, PLIC = Posterior limb of internal capsule, GCC = Genu, SS = Sagittal stratum, ALIC = Anterior limb of internal capsule, FXST = Fornix crus/Stria terminalis, BCC = Body of corpus callosum, TAP = Tapetum of the corpus callosum, CST = Corticospinal tract, SLF = Superior longitudinal fasciculus, SFO = Superior fronto-occipital fasciculus, SCC = Splenium, FX = Fornix, PTR = Posterior thalamic radiation.

2.2 Multi-site Normative Modeling

Let $X \in R^{n \times p}$ be a matrix of p clinical covariates and n the number of subjects. Here, we denote the dependent variable as $y \in R$. Typically, NM assumes a Gaussian distribution over y, i.e., $y \sim N(\mu, \sigma^2)$, and it aims to find a parametric or non-parametric form for μ and σ given $\mathbf{X}$. Then, μ and σ are respectively parameterized on $f_\mu(\mathbf{X}, \theta_\mu)$ and $f_\sigma^+(\mathbf{X}, \theta_\sigma)$, where θ_μ and θ_σ are the parameters of f_μ and f_σ^+. f_σ^+ is a non-negative function that estimates the standard deviation of the noise. Consequently, in a multi-site scenario, a separate set of model parameters could be estimated for each site, or batch, i as follows:

$$y_i = f_{\mu_i}(\mathbf{X}, \theta_{\mu_i}) + \epsilon_i \qquad i \in \{1, ..., m\} \tag{1}$$

However, an assumption in HBR is that θ_{μ_i} (and θ_{σ_i}) across different batches come from the same joint prior distribution that functions as a regularizer and prevents over-fitting of small batches. HBR is described as a partial pooling strategy in the multi-site scenario, compared to a complete pooling approach such as ComBat [4]. Similar to the no-pooling scenario, the parameters for f_μ and $f_{\sigma_i}^+$ are estimated separately for each batch. Then, in the NM framework, the deviations from the norm can be quantified as Z-scores for each subject in the ith batch:

$$z_i = \frac{y_i - f_{\mu_i}(\mathbf{X}_i, \theta_{\mu_i})}{f_{\sigma_i}^+(\mathbf{X}_i, \theta_{\sigma_i})} \tag{2}$$

Table 1. Acquisition protocols. **LREN:** Laboratoire de Recherche en Neuroimagerie; **UMon:** Université de Montréal; **CHUV:** Centre Hospitalier Universitaire Vaudois; **SVIP:** Simons Variation in Individuals Project; **UKBB:** United Kingdom BioBank; **HCP:** Human Connectome Project - Young Adult.

Study cohort	Scanner	Field strength	Resolution voxel size (mm)	b-values (s/mm^2)	Gradient directions	TR/TE (ms)	No. of scanners
LREN	Siemens Prisma	3T	96 × 106 2 × 2 × 2	1000	30	71/4100	1
UMon	Siemens Prisma	3T	150 × 154 1.7 × 1.7 × 1.7	1000	30	71/4100	1
CHUV	Siemens Prisma	3T	150 × 154 1.7 × 1.7 × 1.7	1000	30	71/4100	1
SVIP	Siemens Trio Tim	3T	128 × 128 2 × 2 × 2	1000	30	80/10000	2
UKBB	Siemens Skyra	3T	104 × 104 2.01 × 2.01 × 1	1000	53	92/3600	2

(continued)

Table 1. (*continued*)

Study cohort	Scanner	Field strength	Resolution voxel size (mm)	b-values (s/mm^2)	Gradient directions	TR/TE (ms)	No. of scanners
HCP	Siemens Prisma	3T	168 × 144 1.25 × 1.25 × 1.25	1000	111	89.5/5520	1

These z-scores are then adjusted for additive and multiplicative batch effects using the estimated f_{μ_i} and $f_{\sigma_i}^{+}$ for each batch. The harmonization of Z-scores happens at this stage. The predictions of HBR for f_{μ} are not harmonized but the Z-scores are. Consequently, the model does not yield a set of "corrected" data, i.e., with the batch variability removed, as in ComBat. It instead preserves the sources of biological variability that correlate with the batch effects. This approach may be better able to model the heterogeneity of disease that is not captured by individual case-control studies [6]. We used the PCN-toolkit package to fit all HBR models[1].

Data from 80% of the 1,377 control participants were used as the training sample to estimate the norms. The remaining control subjects were used as a test set. Train-test splitting was stratified for the sites. Z-scores were calculated, measuring their deviations from the reference distribution. Z-scores for the 16pDel sample were also computed separately from the same estimated norm to detect abnormalities in the WM DTI metrics. We calculated probabilities of abnormality (p-values) from the Z-scores for the controls and the 16pDel subjects:

$$P_{abn}(z) = \frac{2}{\sqrt{2\pi}} \int_{-\infty}^{|z|} e^{\frac{-t^2}{2}} dt - 1 \tag{3}$$

ROI-wise areas under the ROC curves (AUCs) were calculated to determine the classification accuracy of the computed deviations, using a binary threshold on the Z-scores. To achieve stability, we repeated the same procedure 10 times. Subsequently, we performed permutation tests with 1,000 random samples to derive permutation p-values for each DTI metric per ROI, and applied a false discovery rate (FDR) correction on each DTI metric separately across ROIs to identify those that showed significant group differences. To improve robustness, the ROIs that showed significance in 9 out of the 10 experimental iterations were retained for comparison [12].

Model Type Comparison. We used three evaluation metrics to compare the three model fits across DTI metrics: 1) the Pearson's correlation coefficient (Rho) between observed and estimated DTI metrics; 2) the standardized mean squared error (SMSE); and 3) the mean standardized log-loss (MSLL) [21]. The higher the Rho, and the lower the SMSE, the better the predicted mean. The MSLL (more negative scores being better) also assesses the quality of the estimated variance. After estimating the three models

[1] https://github.com/amarquand/PCNtoolkit.

Table 2. Demographics of the cohorts. LREN, UMon, CHUV, SVIP, UKBB and HCP are independent studies.

Study cohort	Diagnostic group	No. of subjects	Age (years)	Sex
LREN	16p11.2 Del	11	22.87, SD 13.64	M: 7/F: 4
	Healthy controls	29	40.68, SD 11.85	M: 16/F: 13
UMon	16p11.2 Del	2	42.09, SD 19.06	M: 2/F: 0
	Healthy controls	43	35.73, SD 15.07	M: 19/F: 24
CHUV	16p11.2 Del	12	6.17, SD 1.70	M: 9/F: 3
	Healthy controls	33	6.23, SD 1.83	M: 12/F: 21
SVIP	16p11.2 Del	41	14.03, SD 10.03	M: 25/F: 16
	Healthy controls	106	25.92, SD 14.76	M: 58/F: 48
UKBB	16p11.2 Del	4	65.65, SD 3.25	M: 3/F: 1
	Healthy controls	767	62.76, SD 7.32	M: 360/F: 407
HCP	Healthy controls	399	29.07, SD 3.77	M: 182/F: 217
TOTAL	**16p11.2 Del**	**70**	**17.82, SD 16.65**	**M: 46/F: 24**
	Healthy controls	**1377**	**47.50, SD 19.25**	**M: 647 /F: 730**

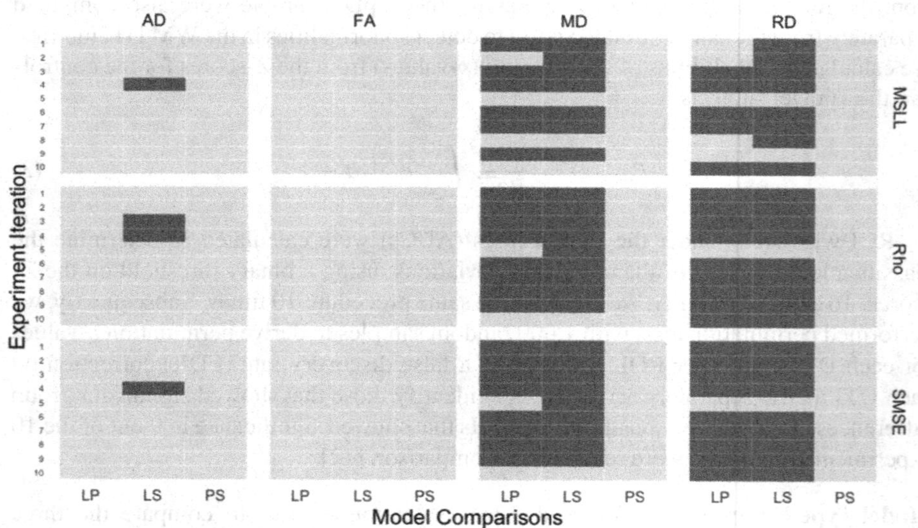

Fig. 1. Model comparisons for the DTI metrics (AD, FA, MD, RD) on three evaluation metrics: Rho, SMSE, MSLL. Red cells indicate significant differences between the model types. Orange cells indicate non-significant differences. LP: linear vs. polynomial; LS: linear vs. *b*-spline; PS: polynomial vs. *b*-spline. (Color figure online)

10 times, we compared these model types by using the percentile bootstrap of the *M*-estimators [22], determining significant differences in three cases: linear vs polynomial (LP), linear vs. *b*-spline (LS) and polynomial vs. *b*-spline (PS).

3 Results

In all experiments, the linear fits had lower Rho, higher SMSE and higher MSLL than the polynomial and the *b*-spline fits. Figure 1 summarizes the results of the model comparisons: LP, LS, PS for all 10 experimental iterations. We found no significant differences between the three model types in any of the iterations for FA. There were no significant differences between the polynomial and *b*-spline models for any of the DTI metrics. Nor were there any significant differences between the linear and the polynomial models (LP) for AD, but there were differences between the linear and the *b*-spline models (LS) for MSLL, Rho and SMSE. MD and RD showed the most significant differences between the linear and the polynomial (LP), and the linear and *b*-spline models (LS) in most of the experiments and for the evaluation metrics. Nevertheless, the LS comparison yielded more significant differences across tests than LP for both MD and RD.

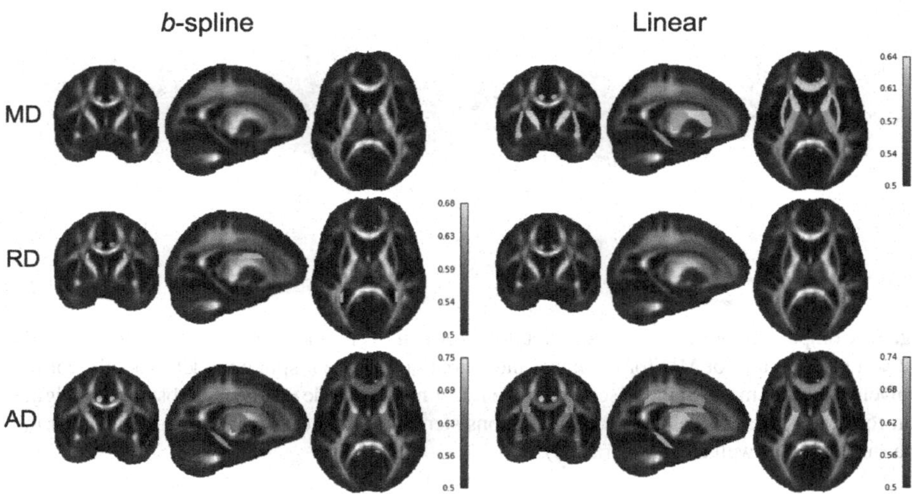

Fig. 2. Comparison of the significant ROIs between the linear and the *b*-spline model. Colored ROIs passed FDR in at least 9 out of 10 experimental iterations. Color bars show the mean AUC for the ROIs across iterations. MD = mean diffusivity, AD = axial diffusivity, RD = radial diffusivity.

For the anomaly detection experiment, the polynomial and *b*-spline models showed significant ROIs for FA, RD and AD; only three ROIs for RD: CGC, SFO and TAP, and seven ROIs for AD: CGC, CGH, GCC, PCR, PLIC, SCR, SFO. Coincidences were found for FA between the three models in the CGH, CGC, PLIC, TAP. Interestingly, the linear model produced the greatest number of significant ROIs for MD and AD (Fig. 2), despite having significantly worse evaluation metrics than the polynomial (MD and RD) and the *b*-spline fits (MD, RD, AD). The most dramatic case was observed for MD, where neither the *b*-spline and polynomial fits yielded significant differences in any of the ROIs between healthy controls and 16pDel, but the linear model detected significant differences in six ROIs (i.e., ALIC, CGC, CGH, EC, GCC, PLIC). To explain these results, we performed an additional analysis of the Z-scores for those six significant MD ROIs in the linear model. We made bee-swarm plots (Fig. 3) for the Z-scores of these 6 ROIs for the linear and *b*-spline fits, for the controls and the 16pDel separately, and calculated the mean and mode for the Z-scores. For the 16pDel group, the Z-scores for linear fit had higher mean and mode (mean = 0.82, mode = 1.05), than those of the *b*-spline fit (mean = 0.37, mode = 0.27). For the controls, the mean and mode of the Z-scores for linear and *b*-spline fits were closer to each other. For the linear model, the tails of the distributions are longer than for the *b*-spline model which yields scores that appear closer to Gaussian, probably due to the better fits. For additional data, please see the Supplements.

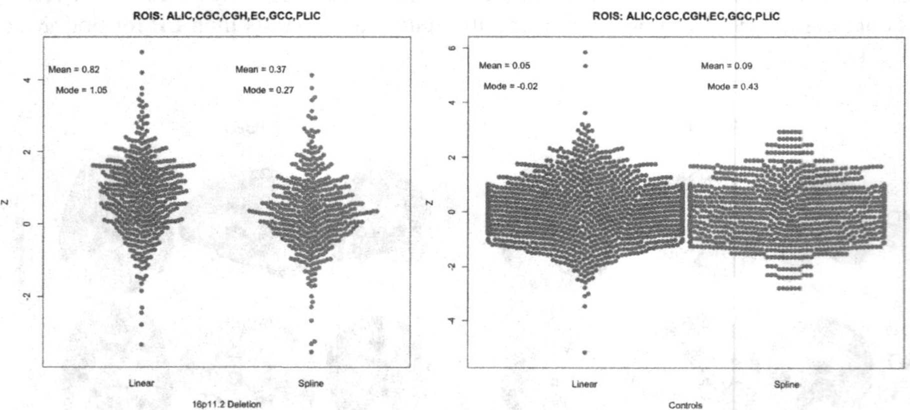

Fig. 3. Comparison of the Z-score distribution between linear and *b*-spline models for the ROIs that were significant for MD for the linear model but not for the *b*-spline model. Notably, for the 16pDel group, the mode of the Z-scores for the linear model (mode = 1.05) is substantially higher than for the *b*-spline model (mode = 0.27), consistent with the notion that it is a poor fit for the data (high bias, as well as high variance).

4 Discussion

Here we used HBR normative modeling to infer the distributional properties of the brain's WM microstructure based on large datasets from healthy subjects, spanning a wide age range. We were able to map the normal range of variation for four DTI metrics and to detect the deviations from this range in 16pDel. Based on the results of the *b*-spline fits, the abnormal deviations in FA and AD may be associated with abnormal axonal density, dispersion and/or dispersion secondary to altered cell migration failures or aberrant pruning, arising from the deletion of genes crucial for these processes.

This is the first study adapting nonlinear HBR theory to study rare neurogenetic conditions, and the first to analyze multi-site brain DTI. It is crucially important to test new open-source medical imaging algorithms on new data modalities (e.g., DTI), and in novel contexts (rare genetic variants), to offer a roadmap to generate rigorous, reproducible findings. Our work aims to ameliorate the reproducibility crisis in the field, as rare variant effects detected in small samples would be unlikely to be robust or reproducible. By adapting the mathematics of normative models beyond structural MRI to DTI, we thoroughly compare alternative models and parameterization choices when merging diverse international data into a single coherent model. Our results reveal how model selection and diverse reference data affect the conclusions regarding abnormalities, in a field (i.e., rare variant genetics) where secure knowledge is lacking.

We were also able to determine which type of model better fits the lifespan trajectory of the DTI metrics in the WM. All our experiments indicated that the polynomial and *b*-spline models offered better fits to the data than the linear fits, with significant differences (in fit, across models) for MD, RD, AD but not for FA. The *b*-spline model was slightly better, on average across all 10 experiments, than the polynomial model. In general, a more highly parameterized model will give a better fit to the test data so long as it is not overfitting the training data, which is less likely in the $n > > p$ regime (where our number of samples exceeds the model complexity). Additionally, several prior publications show the non-linear effect of age on DTI metrics [10, 23]. Even so, this is the first report, to our knowledge, to model age effects on WM DTI metrics using HBR. Although the linear model performed significantly more poorly than the other models, it nonetheless yielded a higher number of significant ROIs for anomaly detection for 16pDel subjects versus healthy controls. A possible explanation for this may be that the *b*-spline models are more complex; thus, given the same amount of data, they may be less stable than linear models. Such a hypothesis will be testable with more data. Hence, in the *b*-spline model, fewer ROIs survived the stringent stability test (showing a significant group difference in at least 9 out of 10 random draws from the data). These findings may serve as a warning to be cautious with any linear fitting strategy for DTI metrics, especially when the age range is large. In summary, we show the feasibility of fitting hierarchical Bayesian models to multi-cohort diffusion data, noting that the model complexity may influence the achievable fit to the data and the brain regions where anomalies can be confidently detected. Further work with an expanded dataset will be valuable to confirm these observations.

Acknowledgements. Canada Research Chair in Neurodevelopmental Disorders, CFREF (Institute for Data Valorization - IVADO), Brain Canada Multi-Investigator Research Initiative (MIRI). Simons Foundation Grant Nos. SFARI219193 and SFARI274424. Swiss National Science Foundation (SNSF) Marie Heim Vögtlin Grant (PMPDP3_171331). Funded in part by NIH grants U54 EB020403, T32 AG058507 and RF1AG057892.

References

1. Thompson, P.M., et al.: The ENIGMA Consortium: large-scale collaborative analyses of neuroimaging and genetic data. Brain Imaging Behav. **8**(2), 153–182 (2014). https://doi.org/10.1007/s11682-013-9269-5
2. Miller, K.L., et al.: Multimodal population brain imaging in the UK Biobank prospective epidemiological study. Nat. Neurosci. **19**, 1523–1536 (2016)
3. Van Essen, D.C., et al.: The WU-Minn human connectome project: an overview. Neuroimage **80**, 62–79 (2013)
4. Fortin, J.-P., et al.: Harmonization of multi-site diffusion tensor imaging data. Neuroimage **161**, 149–170 (2017)
5. Marquand, A.F., Rezek, I., Buitelaar, J., Beckmann, C.F.: Understanding heterogeneity in clinical cohorts using normative models: beyond case-control studies. Biol. Psychiatry **80**, 552–561 (2016)
6. Kia, S.M., et al.: Hierarchical Bayesian regression for multi-site normative modeling of neuroimaging data. In: Martel, A.L., et al. (eds.) MICCAI 2020. LNCS, vol. 12267, pp. 699–709. Springer, Cham (2020). https://doi.org/10.1007/978-3-030-59728-3_68
7. Concha, L.: A macroscopic view of microstructure: using diffusion-weighted images to infer damage, repair, and plasticity of white matter. Neuroscience **276**, 14–28 (2014)
8. Basser, P.J., Mattiello, J., LeBihan, D.: MR diffusion tensor spectroscopy and imaging. Biophys. J. **66**, 259–267 (1994)
9. Landman, B.A., Farrell, J.A.D., Jones, C.K., Smith, S.A., Prince, J.L., Mori, S.: Effects of diffusion weighting schemes on the reproducibility of DTI-derived fractional anisotropy, mean diffusivity, and principal eigenvector measurements at 1.5T. NeuroImage **36**, 1123–1138 (2007)
10. Lawrence, K.E., et al.: Advanced diffusion-weighted MRI methods demonstrate improved sensitivity to white matter aging: percentile charts for over 15,000 UK Biobank participants. Alzheimer's Dement. **17**, e051187 (2021)
11. Pomponio, R., Erus, G., Habes, M., Doshi, J., Srinivasan, D., Mamourian, E., et al.: Harmonization of large MRI datasets for the analysis of brain imaging patterns throughout the lifespan. Neuroimage **208**, 116450 (2020)
12. Rutherford, S., et al.: Charting brain growth and aging at high spatial precision. eLife **11**, e72904 (2022)
13. Jacquemont, S., et al.: Mirror extreme BMI phenotypes associated with gene dosage at the chromosome 16p11.2 locus. Nature **478**, 97–102 (2011)
14. Walsh, K.M., Bracken, M.B.: Copy number variation in the dosage-sensitive 16p11.2 interval accounts for only a small proportion of autism incidence: a systematic review and meta-analysis. Genet. Med. **13**, 377–384 (2011)
15. Garyfallidis, E., et al.: Dipy, a library for the analysis of diffusion MRI data. Front. Neuroinform. **8** (2014). https://pubmed.ncbi.nlm.nih.gov/24600385/
16. Andersson, J.L.R., Sotiropoulos, S.N.: An integrated approach to correction for off-resonance effects and subject movement in diffusion MR imaging. Neuroimage **125**, 1063–1078 (2016)
17. Avants, B.B., Tustison, N.J., Song, G., Cook, P.A., Klein, A., Gee, J.C.: A reproducible evaluation of ANTs similarity metric performance in brain image registration. Neuroimage **54**, 2033–2044 (2011)
18. Mori, S., et al.: Stereotaxic white matter atlas based on diffusion tensor imaging in an ICBM template. NeuroImage **40**, 570–582 (2008)
19. Jahanshad, N., et al.: Multi-site genetic analysis of diffusion images and voxelwise heritability analysis: a pilot project of the ENIGMA–DTI working group. Neuroimage **81**, 455–469 (2013)

20. Gillentine, M.A., Lupo, P.J., Stankiewicz, P., Schaaf, C.P.: An estimation of the prevalence of genomic disorders using chromosomal microarray data. J. Hum. Genet. **63**, 795–801 (2018)
21. Rasmussen, C.E., Williams, C.K.I.: Gaussian Processes for Machine Learning. MIT Press, Cambridge (2006)
22. Wilcox, R.R.: Modern Statistics for the Social and Behavioral Sciences: A Practical Introduction. CRC Press, Boca Raton (2017)
23. Lebel, C., Walker, L., Leemans, A., Phillips, L., Beaulieu, C.: Microstructural maturation of the human brain from childhood to adulthood. Neuroimage **40**, 1044–1055 (2008)

Functional Brain Networks

Contrastive Functional Connectivity Graph Learning for Population-based fMRI Classification

Xuesong Wang[1](✉), Lina Yao[1], Islem Rekik[2], and Yu Zhang[3]

[1] University of New South Wales, Sydney 2052, Australia
{xuesong.wang1,lina.yao}@unsw.edu.au
[2] BASIRA Lab, Istanbul Technical University, 34469 Maslak, Turkey
irekik@itu.edu.tr
[3] Lehigh University, Bethlehem, PA 18015, USA
yuzi20@lehigh.edu

Abstract. Contrastive self-supervised learning has recently benefited fMRI classification with inductive biases. Its weak label reliance prevents overfitting on small medical datasets and tackles the high intraclass variances. Nonetheless, existing contrastive methods generate resemblant pairs only on pixel-level features of 3D medical images, while the functional connectivity that reveals critical cognitive information is under-explored. Additionally, existing methods predict labels on individual contrastive representation without recognizing neighbouring information in the patient group, whereas interpatient contrast can act as a similarity measure suitable for population-based classification. We hereby proposed contrastive functional connectivity graph learning for population-based fMRI classification. Representations on the functional connectivity graphs are "repelled" for heterogeneous patient pairs meanwhile homogeneous pairs "attract" each other. Then a dynamic population graph that strengthens the connections between similar patients is updated for classification. Experiments on a multi-site dataset ADHD200 validate the superiority of the proposed method on various metrics. We initially visualize the population relationships and exploit potential subtypes. Our code is available at https://github.com/xuesongwang/Contrastive-Functional-Connectivity-Graph-Learning.

Keywords: Functional connectivity analysis · Population-based classification · Contrastive learning

1 Introduction

Recently, self-supervised deep models such as contrastive learning have shown promising results in 3D medical image classification [1,8,19,20]. The pillar of

Supplementary Information The online version contains supplementary material available at https://doi.org/10.1007/978-3-031-16431-6_21.

L. Wang et al. (Eds.): MICCAI 2022, LNCS 13431, pp. 221–230, 2022.
https://doi.org/10.1007/978-3-031-16431-6_21

contrastive learning is to augment a similar 3D image from one patient to form a "homogeneous" pair and use images from other patients to construct multiple "heterogeneous" pairs. The "homo-" contrasts are minimized while "heter-" contrasts are maximized. As it incorporates such inductive bias to generate pseudo-labels for augmented samples, contrastive learning enables training on small medical datasets in an unsupervised manner [12]. Such weak reliance on ground truths prevents overfitting in large deep networks [21], and helps tackle high intraclass variances in brain disorder diseases induced by potential disease subtypes [14].

Existing contrastive learning frameworks process 3D fMRI images like typical 2D datasets such as ImageNet [7], where they augment "homo-" pairs with transformations including random cropping, colour distortion, and downsampling [5,10]. Functional Connectivity (FC), which quantifies the connection strengths between anatomical brain regions of interests (ROIs), may be neglected in those transformations and impair the brain disorder classification [15,17], as they are related to cognitive behaviours focusing on perception and vigilance tasks [6]. Such negligence calls for a novel formalization of "homo-" and "heter-" pairs meanwhile preserving the FC information. Additionally, existing frameworks treat contrastive learning as a self-supervised encoder where a single patient embedding is an input to the classification network to obtain a single output [8,11]. They tend to overlook that interpatient contrast itself is a similarity measure, which can intrinsically enable a population-based classification. In the population graph, every patient can represent a node and the similarity between two patients forms a weighted edge [2,16]. Despite the fruitful benefits population-based classification brings to medical datasets, for instance, it alleviates high-intraclass variances by forming sub-clusters of the same class in the graph [22,24], and also stabilizes the prediction metrics [3], population-based classification on contrastive features has not been fully explored.

We aim to classify brain disorders on a population graph using contrastive FC features to tackle the aforementioned challenges. We primitively enable contrastive learning on FC by defining a "homo-" pair with two FC graphs generated from the non-overlapping ROI time series of the same patient, based on which a spectral graph network capable of convolving full ROI connections is optimized. The population-based classifier then adopts the contrastive embedding to dynamically the update population graph structure and utilize neighbouring information of each patient for prediction. Our major contributions are detailed as follows:

- Formalization of the "homogeneous" and "heterogeneous" pairs which primitively enables contrastive learning optimization on FC graphs.
- Contrastive features for population-based classification. We utilize dynamic graph convolution networks to enhance patient clusters within the graph.
- Population graph visualization with the implication of subtypes. We initially visualize the ADHD population graph and reveal two potential subtypes for the health control group.

2 Method

2.1 Problem Formalization

Given a dataset of P patients $\{(\mathcal{G}_i^j, y_i)\}_{i=1}^P$, where $y_i \in \{0, 1\}$ represents the class for the i-th patient, $\mathcal{G}_i^j = (\mathcal{V}_i^j, \mathcal{E}_i^j)$ is the j-th view of the individual FC graph where Pearson correlations for each ROI is calculated for node features $\mathcal{V}_i^j$ and edges $\mathcal{E}_i^j$ are constructed using partial correlations between ROIs. Multiple views are generated by slicing ROI timeseries into non-overlapping windows. As illustrated in Fig. 1, Our objective is to construct a new population graph $\mathcal{G}^{\mathcal{P}}$ and build a model $\boldsymbol{h}$ to predict node labels in $\mathcal{G}^{\mathcal{P}}$: $\boldsymbol{h}(\mathcal{G}^{\mathcal{P}}) \in \mathbb{R}^{P \times class}$. $\mathcal{G}^{\mathcal{P}} = (\mathcal{V}^{\mathcal{P}}, \mathcal{E}^{\mathcal{P}})$ is constructed with $\mathcal{V}^{\mathcal{P}} = \boldsymbol{f}(\mathcal{G}_{i=1,2,\cdots,P}) \in \mathbb{R}^{P \times d}$ that embeds each patient to a d-dimensional vector and $\mathcal{E}^{\mathcal{P}}$ represents the attractiveness between patients. Labels are provided for training nodes in $\mathcal{G}^{\mathcal{P}}$ while the testing labels are masked.

2.2 Overall Framework

The framework overflow is shown in Fig. 1. Multiple non-overlapping views of N of patients are initially sampled to obtain FC graph inputs. The motivation of contrastive learning is that two inputs coming from the same patient should form a "homo-" pair and "attract" each other; otherwise, two from different patients should form a "heter-" pair and "repel". A spectral graph convolutional network $\boldsymbol{f}$ is then trained in a self-supervised manner to acquire contrastive embedding of each patient. A population graph is further defined by a collection of patient contrastive embeddings, based on which a dynamic graph classifier is trained to predict a label for each patient node. It applies a dynamic edge convolution network $\boldsymbol{h}$ to node features. Since K-nearest neighbours establish new edges with adaptive node features, the population graph structure is dynamically modelled to strengthen connections between "attracted" patients.

2.3 Contrastive Functional Connectivity Graph Learning (CGL)

We propose contrastive FC learning to train deep networks on small medical datasets in a self-supervised manner, while preserving FC node-edge relationships. FC graph input of the i-th patient is represented by a functional connectivity graph $\mathcal{G}_i = (\mathcal{V}_i, \mathcal{E}_i)$. Pearson correlations between ROIs are used for node features, where the r-th ROI is represented by $v_i^{(r)} \in \mathbb{R}^{ROIs}$ indicating Person coefficients of all connecting ROIs. Edges are constructed using partial correlations to create sparser connections and simplify the graph structure. The module is detailed as follows:

Spectral Graph Convolution $\boldsymbol{f}$ on FC Graphs. Unlike conventional graph convolutions that pass messages to a single ROI node from its adjacency, spectral graph networks simultaneously convolve full node connections by incorporating the Laplacian matrix $\mathcal{L} = I - D^{-1/2}AD^{-1/2}$, where I is the identity matrix, $A \in \mathcal{R}^{ROIs \times ROIs}$ is the adjacency matrix and $D \in \mathcal{R}^{ROIs \times ROIs}$ refers to the

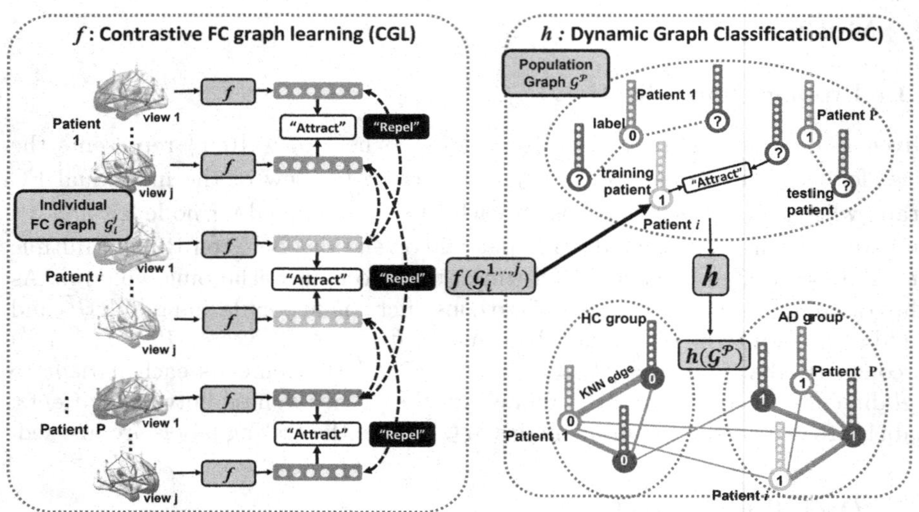

Fig. 1. The framework of our method. Contrastive FC graph learning (CGL) extracts contrastive embeddings from multiview of each patient, based on which a population-graph is constructed for dynamic graph classification (DGC).

diagonal degree matrix. This matches with the domain knowledge about brain disorders that certain isolated FCs cooperate to achieve a cognitive task [6], therefore incorporating all node features will induce a robust representation. A spectral graph convolution block is defined as:

$$\mathcal{V}_i' = \sum_{k=1}^{K} Z^{(k)}(\mathcal{V}_i) \cdot \theta^{(k)} \tag{1}$$

where k is the filter size, $Z^{(k)}(\mathcal{V}_i)$ is recursively computed as: $Z^{(k)}(\mathcal{V}_i) = 2\tilde{\mathcal{L}}Z^{(k-1)}\cdot(\mathcal{V}_i) - Z^{(k-2)}(\mathcal{V}_i)$, with $Z^{(1)}(\mathcal{V}_i) = \mathcal{V}_i, Z^{(2)}(\mathcal{V}_i) = \tilde{\mathcal{L}}\mathcal{V}_i$. $\tilde{\mathcal{L}}$ is the scaled and normalized Laplacian $\tilde{\mathcal{L}} = \frac{2\mathcal{L}}{\lambda_{max}} - I$ with λ_{max} denoting the largest eigenvalue of $\mathcal{L}$. $\theta^{(k)}$ are learnable parameters. In practice, K = 3, hence every spectral convolutional block includes a single node projection $\mathcal{V}_i\theta^{(1)}$, all nodes projection $\tilde{\mathcal{L}}\mathcal{V}_i\theta^{(2)}$ and a combination of both with $Z^{(3)}(\mathcal{V}_i)\theta^{(3)}$. To prevent overfitting, the CGL module $\boldsymbol{f}$ is comprised of 2 spectral graph convolution blocks with top-K graph pooling blocks. Each top-K graph pooling deactivates a subset of nodes to form a smaller graph. The resulting embedding is optimized using contrastive loss and then input to the classification.

Contrastive Optimization. After the spectral graph convolutions $\boldsymbol{f}$, each view of a patient is embedded as: $\boldsymbol{v}_i^j = \boldsymbol{f}(\mathcal{G}_i^j)$. The motivation of contrastive optimization is to enforce attractions among embeddings that are resemblant without referring to the ground truths. For a batch of N patients with 2 views per patient, we can construct an attraction matrix $\mathcal{M} \in 2N \times 2N$ where its element is a similarity measure between a pair of random two views:

$$\mathcal{M}(m,n) = \frac{\boldsymbol{v}_m^\top \boldsymbol{v}_n}{\| \boldsymbol{v}_m \| \| \boldsymbol{v}_n \|} \tag{2}$$

We define a "homogeneous" pair as two embeddings v_m and v_n coming from the same patient. As they are generated from two non-overlapping ROI time series, their original FC graph inputs can be diverse. However, semantically they are representing the same patient hence should have similar embeddings. The contrastive loss function on this "homo-" pair is defined as:

$$\mathscr{L}(m,n) = -\log \frac{\exp\left(\mathcal{M}(m,n)/\tau\right))}{\sum_{i=1}^{2N} \mathbb{1}_{i \neq m} \exp\left(\mathcal{M}(m,i)/\tau\right)} \tag{3}$$

where τ is a temperature factor to control the desired attractiveness strength. $\mathbb{1}(\cdot) = \{0,1\}$ is an indicator function that zeros self-attraction $\mathcal{M}(m,m)$. During training, the numerator is maximized by increasing the "homo-" attraction and the denominator is minimised with "heter-" pairs "repeling" each other. The loss function is not commutative as the denominator will compute "heter-" pairs differently. The overall contrastive loss is the sum of $2N$ "homo-" pairs in the batch: $L_{CGL} = \frac{1}{2N} \sum_{i=1}^{N} (\mathscr{L}(2i-1, 2i) + \mathscr{L}(2i, 2i-1))$. The contrastive loss function is different from typical similarity loss and softmax loss because it simultaneously optimizes "Multiple-Homo-Multiple-Heter" pairs rather than "Single-Homo-Single-Heter" pairs (similarity loss) and "Single-Homo-Multiple-Heter" pairs (softmax loss). This avoids bias towards a single pair, hence stabilizing the representation.

2.4 Population-Based Dynamic Graph Classification (DGC)

Now that we have the embedding $\boldsymbol{v}_i^j = \boldsymbol{f}(\mathcal{G}_i^j)$ that can measure patient similarity, the motivation is to build an attraction graph $\mathcal{G}^P$ for population-based classification. When predicting the label for an unknown patient v_i^P in the graph, the model can infer from its labelled neighbours. In order to construct a graph with an adaptive structure, we propose dynamic edge convolution for classification.

Dynamic Edge Convolution $\boldsymbol{h}$ on the Population Graph. Dynamic edge convolutions form an initial graph with isolated patient nodes, then project node features to a new space and dynamically construct new edges on the new feature space with top-K connection strengths defined as:

$$v_i^{P'} = \sum_{m \in \mathcal{E}_{(i,\cdot)}^P} \phi(v_i^P \parallel v_m^P - v_i^P) \quad , \quad \mathcal{E}^P = KNN(\mathcal{V}^P)_{topK} \tag{4}$$

where $\parallel$ is the concatenation function, $\phi(\cdot)$ is a fully connected network. Initially, the edge-convolution layer ϕ projects the node features v_i^P as well as its connecting edge features$(v_m^P - v_i^P)$ to update node features. New edges are built by selecting top-K similarities. For the next update, connections between closer patients are strengthened while edges on distant patients are abandoned. The final classifier layer is a fully connected network to predict each node's class probability Pr where training nodes are labelled. We adopted focal loss for the classification: $L_{DGC} = -(1 - Pr)^{\gamma} \log(Pr)$. As suggested by [23], the focal loss lowers the threshold of accepting a class, inducing a higher recall rate for disorder groups. As KNN is a non-parametric model, trainable parameters come only from node projection, narrowing the parameter searching space and preventing overfitting on the population graph.

3 Experiments

3.1 Dataset and Experimental Details

A multi-site fMRI dataset ADHD200[1] is used for comparing the proposed method with baseline models. We collect 596 patients data on AAL90 ROIs to construct individual FC graphs. As implied in [23], incorporating personal characteristic data (PCD) such as age helps to stabilize the metrics. We accordingly keep the three sites KU, KKI, and NYU with no missing values on the 7 PCD features: age, gender, handedness, IQ Measure, Verbal IQ, Performance IQ, and Full4 IQ. Data from the rest sites are excluded due to the lack of PCD information or few ADHD samples. Patients are sampled within each site and combined afterwards. The ratio of site training/validating/testing split is 7: 1: 2, and we select patients proportionally to the class-imbalance ratio.

We compare four baseline models and two variants of our methods on four metrics: AUC, accuracy, sensitivity, and specificity. KNN is trained with raw vectorized FC features. MLP utilizes two views of FC vectors for classification [4]. BrainGNN [13] adopts the same FC graph as ours and introduces a novel GNN structure. A population-based method SpectralGCN [16] uses supervised learning to train a spectral graph convolution for similarity measures. For ablation studies, we test dynamic graph classification on a population graph using raw FC features (DGC) and perform contrastive graph learning (CGL) with a KNN classifier to enable unsupervised learning. Regarding implementation details, we run the model with a batch size of 100 for 150 epochs. Adam optimizer is selected, and learning rates for CGL and DGC are 0.001 and 0.005, respectively. PCD features are concatenated to each ROI feature. temperature factor τ in contrastive loss function is 0.1. The filter size for spectral convolution is 3. For dynamic edge convolution, the structure resembles [18] where 20 nearest neighbours are used for edges. All methods with implemented with PyTorch and trained with GPU TITAN RTX.

3.2 Classification Results

Table 1 shows the four metrics of the comparing methods on the three sites. Each method is tested for five runs with the mean and standard values. Our method

[1] http://preprocessed-connectomes-project.org/adhd200/.

significantly outperforms baselines on AUC and achieves the best average accuracy. Due to class imbalance in site KKI with fewer ADHD samples, baseline methods tend to have larger variations on sensitivity and specificity. For example, BrainGNN and CGL achieve higher sensitivity and specificity in the sacrifice of the other metric, whereas our method achieves relatively stabilized and balanced metrics. It is worth mentioning that our variant CGL achieves promising AUC and ACC as a completely self-supervised model without reliance on labels, manifesting the effectiveness of the contrastive training framework. The introduction of focal loss to our framework balances the metrics and increases the recall rate of ADHD, which is medically significant.

Table 1. Classification results on 3 sites (PKU, KKI, NYU) of ADHD dataset. DGC and CGL are two variants of the proposed model. ($*$: $p < 0.05$)

Models	AUC	ACC (%)	SEN (%)	SPEC (%)
KNN [9]	0.6816 ± 0.0305 *	63.43 ± 3.42	56.67 ± 9.94	69.59 ±4.86
MLP [4]	0.6266 ± 0.0190 *	55.43 ± 3.90 *	52.54 ± 27.09	58.08 ± 26.27
BrainGNN [13]	0.6410 ± 0.0659 *	62.00 ± 6.20	73.43 ± 13.26	51.51 ± 23.77
SpectralGCN [16]	0.6321 ± 0.0199 *	59.00 ± 3.05 *	61.19 ± 12.63	56.99 ± 11.18 *
DGC (Var-1)	0.6647 ± 0.0328 *	61.00 ± 3.02	58.21 ± 6.47	63.56 ± 6.80
CGL (Var-2)	0.7101 ± 0.0306	66.71 ± 3.08	53.13 ± 4.59 *	**79.18 ± 3.05**
CGL+DGC (ours)	**0.7210 ± 0.0263**	**67.00 ± 3.79**	61.49 ± 5.93	72.10 ± 9.86

3.3 Discussion

Contrastive FC Graph Learning. To verify the effectiveness of the contrastive FC graph learning, we aim to compare the patient attraction. The distributional similarity of "homo-" and "heter-" pairs is compared in Fig. 2 on raw vectorized FC features and contrastive features. The results on the raw features group show no substantial differences, indicating equal similarity between homo- and heter-pairs. After the contrastive representation, similarities on the homo-pairs are significantly higher with tighter variance, meaning that the loss function managed to enforce homo-attraction. Meanwhile, the attractions on the heter-pairs are averaged around zero with large variance, suggesting that some heter-views can still be similar. Those similar heter-pairs can form the population graph.

Population Graph Visualization. We construct the population graph with edges representing the patient similarities to unveil hidden patient relationships. Only two edges are visualized per patient to construct a cleaner graph. The results of raw FC vectors, self-supervised contrastive features, and the last hidden layer of our method are illustrated in Fig. 3. Two classes are denoted with different colours. Both classes are cluttered with raw FC features, whereas contrastive

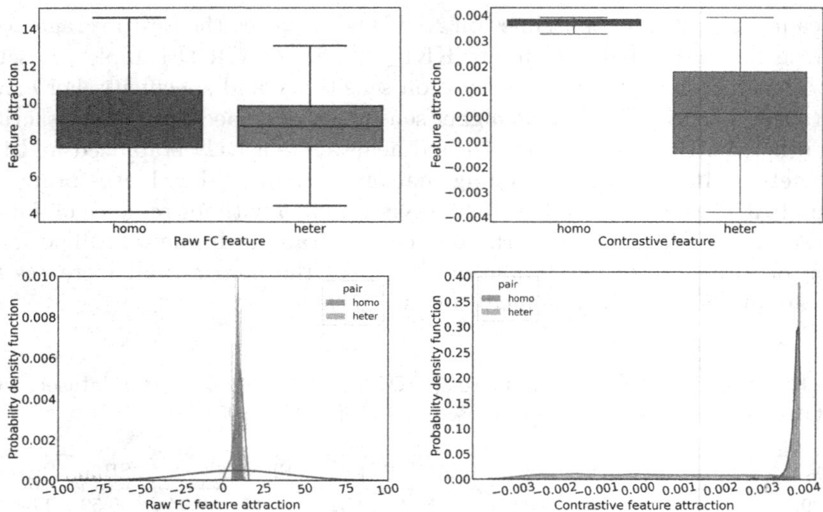

Fig. 2. Boxplot and probability density function of attractions on raw FC and contrastive features.

features reveal limited patterns with self-supervised learning; for instance, the lower-left tends to have more ADHD patients connecting each other while more HC patients are clustered on the right. Our method presents the most distinct patterns with four clear clusters, each formed by multiple edges sharing mutual nodes, i.e., representative patients. It also suggests two subtypes in HC, potentially leading to intraclass variance and deteriorating the metrics. We will leave the exploration and validation of subtypes for future work.

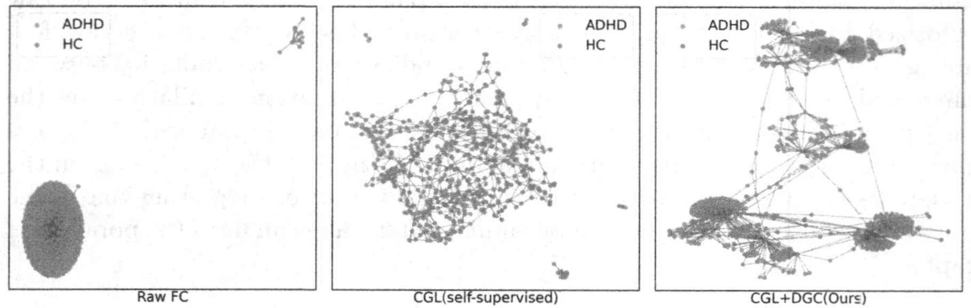

Fig. 3. Population graph visualization on three features.

4 Conclusion

We have addressed two critical questions in fMRI classification: Can contrastive learning work on Functional Connectivity graphs? Can contrastive embeddings benefit population-based classification? We hereby define a "homogeneous" pair in the contrastive loss as two FC graphs generated from non-overlapping ROI time series of the same patient. A "heterogeneous" pair is formalized with two FC graphs from different patients. A spectral graph network capable of full FC connection convolution is optimized on multiple-homo-multiple heter pairs for contrastive learning. The resulting contrastive embeddings on an ADHD200 dataset show that stronger attractions are enforced on homo-pairs, indicating a similarity measure that enables a population graph classification. We then employ a dynamic edge convolution classifier to utilize such similarities in the population graph. Our dynamic graph classification achieves significant improvements on AUC and the highest average accuracy, meanwhile balancing sensitivity and specificity. Our visualization of the resulting population dynamic graph structure originally implies two potential subtypes of the health control group.

References

1. Azizi, S., et al.: Big self-supervised models advance medical image classification. In: Proceedings of the IEEE/CVF International Conference on Computer Vision, pp. 3478–3488 (2021)
2. Bessadok, A., Mahjoub, M.A., Rekik, I.: Graph neural networks in network neuroscience. arXiv preprint arXiv:2106.03535 (2021)
3. Chen, C., Li, K., Wei, W., Zhou, J.T., Zeng, Z.: Hierarchical graph neural networks for few-shot learning. IEEE Trans. Circuits Syst. Video Technol. **32**(1), 240–252 (2021)
4. Chen, M., Li, H., Wang, J., Dillman, J.R., Parikh, N.A., He, L.: A multichannel deep neural network model analyzing multiscale functional brain connectome data for attention deficit hyperactivity disorder detection. Radiol. Artif. Intell. **2**(1), e190012 (2019)
5. Chen, T., Kornblith, S., Norouzi, M., Hinton, G.: A simple framework for contrastive learning of visual representations. In: International Conference on Machine Learning, pp. 1597–1607. PMLR (2020)
6. Cohen, J.R.: The behavioral and cognitive relevance of time-varying, dynamic changes in functional connectivity. Neuroimage **180**, 515–525 (2018)
7. Deng, J., Dong, W., Socher, R., Li, L.J., Li, K., Fei-Fei, L.: ImageNet: a large-scale hierarchical image database. In: 2009 IEEE Conference on Computer Vision and Pattern Recognition, pp. 248–255. IEEE (2009)
8. Dufumier, B., et al.: Contrastive learning with continuous proxy meta-data for 3D MRI classification. In: de Bruijne, M., et al. (eds.) MICCAI 2021. LNCS, vol. 12902, pp. 58–68. Springer, Cham (2021). https://doi.org/10.1007/978-3-030-87196-3_6
9. Eslami, T., Saeed, F.: Similarity based classification of ADHD using singular value decomposition. In: Proceedings of the 15th ACM International Conference on Computing Frontiers, pp. 19–25 (2018)

10. Hu, X., Zeng, D., Xu, X., Shi, Y.: Semi-supervised contrastive learning for label-efficient medical image segmentation. In: de Bruijne, M., et al. (eds.) MICCAI 2021. LNCS, vol. 12902, pp. 481–490. Springer, Cham (2021). https://doi.org/10.1007/978-3-030-87196-3_45
11. Konkle, T., Alvarez, G.A.: Instance-level contrastive learning yields human brain-like representation without category-supervision. BioRxiv, pp. 2020-06 (2020)
12. Li, J., et al.: Multi-task contrastive learning for automatic CT and X-ray diagnosis of COVID-19. Pattern Recogn. **114**, 107848 (2021)
13. Li, X., et al.: BrainGNN: interpretable brain graph neural network for FMRI analysis. Med. Image Anal. **74**, 102233 (2021)
14. Liu, Y., Wang, W., Ren, C.-X., Dai, D.-Q.: MetaCon: meta contrastive learning for microsatellite instability detection. In: de Bruijne, M., et al. (eds.) MICCAI 2021. LNCS, vol. 12908, pp. 267–276. Springer, Cham (2021). https://doi.org/10.1007/978-3-030-87237-3_26
15. Mueller, S., et al.: Individual variability in functional connectivity architecture of the human brain. Neuron **77**(3), 586–595 (2013)
16. Parisot, S., et al.: Spectral graph convolutions for population-based disease prediction. In: Descoteaux, M., Maier-Hein, L., Franz, A., Jannin, P., Collins, D.L., Duchesne, S. (eds.) MICCAI 2017. LNCS, vol. 10435, pp. 177–185. Springer, Cham (2017). https://doi.org/10.1007/978-3-319-66179-7_21
17. Rodriguez, M., et al.: Cognitive profiles and functional connectivity in first-episode schizophrenia spectrum disorders-linking behavioral and neuronal data. Front. Psychol. **10**, 689 (2019)
18. Wang, Y., Sun, Y., Liu, Z., Sarma, S.E., Bronstein, M.M., Solomon, J.M.: Dynamic graph CNN for learning on point clouds. ACM Trans. Graph.(ToG) **38**(5), 1–12 (2019)
19. Xing, X., Hou, Y., Li, H., Yuan, Y., Li, H., Meng, M.Q.-H.: Categorical relation-preserving contrastive knowledge distillation for medical image classification. In: de Bruijne, M., et al. (eds.) MICCAI 2021. LNCS, vol. 12905, pp. 163–173. Springer, Cham (2021). https://doi.org/10.1007/978-3-030-87240-3_16
20. Zeng, D., et al.: Positional contrastive learning for volumetric medical image segmentation. In: de Bruijne, M., et al. (eds.) MICCAI 2021. LNCS, vol. 12902, pp. 221–230. Springer, Cham (2021). https://doi.org/10.1007/978-3-030-87196-3_21
21. Zeng, J., Xie, P.: Contrastive self-supervised learning for graph classification. In: Proceedings of the AAAI Conference on Artificial Intelligence, vol. 35, pp. 10824–10832 (2021)
22. Zhang, Z., Luo, C., Wu, H., Chen, Y., Wang, N., Song, C.: From individual to whole: reducing intra-class variance by feature aggregation. Int. J. Comput. Vis. 1–20 (2022)
23. Zhao, K., Duka, B., Xie, H., Oathes, D.J., Calhoun, V., Zhang, Y.: A dynamic graph convolutional neural network framework reveals new insights into connectome dysfunctions in ADHD. Neuroimage **246**, 118774 (2022)
24. Zhong, H., et al.: Graph contrastive clustering. In: Proceedings of the IEEE/CVF International Conference on Computer Vision, pp. 9224–9233 (2021)

Joint Graph Convolution for Analyzing Brain Structural and Functional Connectome

Yueting Li[1(✉)], Qingyue Wei[1], Ehsan Adeli[1], Kilian M. Pohl[1,2], and Qingyu Zhao[1]

[1] Stanford University, Stanford, CA 94305, USA
lyt1314@stanford.edu
[2] SRI International, Menlo Park, CA 94025, USA

Abstract. The white-matter (micro-)structural architecture of the brain promotes synchrony among neuronal populations, giving rise to richly patterned functional connections. A fundamental problem for systems neuroscience is determining the best way to relate structural and functional networks quantified by diffusion tensor imaging and resting-state functional MRI. As one of the state-of-the-art approaches for network analysis, graph convolutional networks (GCN) have been separately used to analyze functional and structural networks, but have not been applied to explore inter-network relationships. In this work, we propose to couple the two networks of an individual by adding inter-network edges between corresponding brain regions, so that the joint structure-function graph can be directly analyzed by a single GCN. The weights of inter-network edges are learnable, reflecting non-uniform structure-function coupling strength across the brain. We apply our Joint-GCN to predict age and sex of 662 participants from the public dataset of the National Consortium on Alcohol and Neurodevelopment in Adolescence (NCANDA) based on their functional and micro-structural white-matter networks. Our results support that the proposed Joint-GCN outperforms existing multi-modal graph learning approaches for analyzing structural and functional networks.

1 Introduction

The "human connectom" refers to the concept of describing the brain's structural and functional organization as large-scale complex brain networks [19]. An accurate description of such connectome heavily relies on neuroimaging methods. Specifically, Diffusion Tensor Imaging (DTI) characterizes white-matter fiber bundles connecting different gray-matter regions (Structural Connectivity or SC), while resting-state functional Magnetic Resonance Imaging (rs-fMRI) measures spontaneous fluctuations in BOLD signal giving rise to the Functional Connectivity (FC) across brain regions [11]. Central to the connectome research is understanding how the underlying SC supports FC for developing high-order

L. Wang et al. (Eds.): MICCAI 2022, LNCS 13431, pp. 231–240, 2022.
https://doi.org/10.1007/978-3-031-16431-6_22

cognitive abilities [6] and how the structure-function relationship develops over certain ages and differs between sexes [1].

To answer these questions, an emerging approach is using SC and FC of the brain to predict factors of interest, such as age, sex, and diagnosis labels [10,18]. Since brain networks can be treated as graphs, one of the most popular architectures for building such prediction models is Graph Convolution Networks (GCN) [7]. However, the majority of GCN-based studies on brain connectivity focus on a single imaging modality [4,5,9] but fall short in leveraging the relationship between FC and SC. Existing works that perform multi-modal fusion [21] either use two GCNs to extract features from SC and FC graphs separately [23] or directly discard the graph structure from one modality (e.g., regarding FC as features defined on the SC graph) [10]. These simple fusion methods ignore the inter-network dependency that white-matter fiber tracts provide an anatomic foundation for high-level brain function, thereby possibly leading to sub-optimal feature extraction from the SC-FC profile.

To address this issue, we propose coupling the SC and FC graphs by inserting inter-network edges between corresponding brain regions. The resulting joint graph then models the entire structural and functional connectome that can be analyzed by a single GCN. Given prior evidence that the structure-function relationships markedly vary across the neocortex [16], we propose learning the weights (quantifying the coupling strength of SC and FC patterns for each region) of the inter-network edges during the end-to-end training. We tested our proposal, called Joint-GCN, on the public data collected by the National Consortium on Alcohol and Neurodevelopment in Adolescence (NCANDA) [2]. By predicting age and sex of the participants from SC matrices (extracted from DTI) and FC matrices (extracted from rs-fMRI), our joint graph has a higher prediction accuracy than unimodal analysis (based on either SC or FC networks alone) and than existing multi-modal approaches that disjoint the SC and FC graphs. The learned weights of inter-network edges highlight the non-uniform coupling strength between SC and FC over the human cortex.

2 Method

We first review the traditional graph convolution defined on a single network (SC or FC). We then describe the construction of the joint SC-FC network and model the SC-FC coupling as learnable weights of inter-network edges. Figure 1 provides an overview of the proposed Joint-GCN.

Graph Convolution for Individual Graphs. We assume a brain is parcellated into N regions of interest (ROIs), so that a tractography procedure (using a DTI scan) quantifies the strength of the SC between all pairs of ROIs. This SC network is characterized by a graph $\mathcal{G}^S(\mathcal{V}^S, \mathcal{E}^S)$ with $\mathcal{V}^S$ being the node set (N ROIs) and $\mathcal{E}^S$ being the edge set. Edge weights between ROI pairs are encoded by an adjacency matrix $A^S \in \mathbb{R}^{N \times N}$. Let $X^S \in \mathbb{R}^{N \times M}$ be M-dimensional node

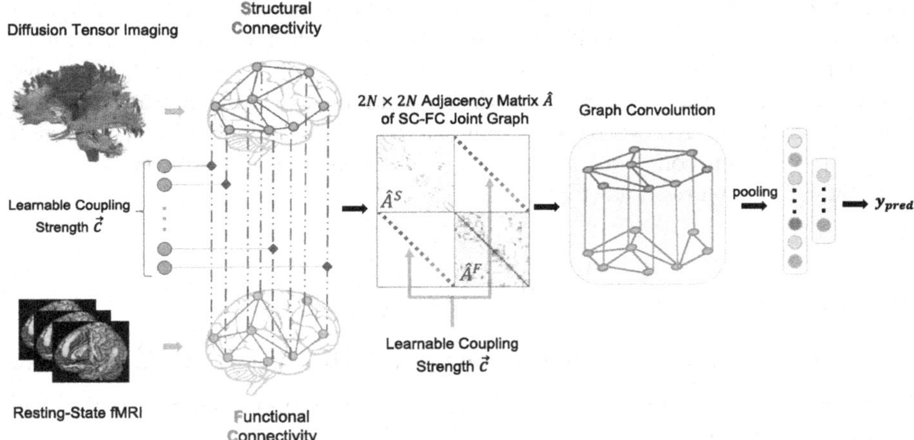

Fig. 1. Proposed joint-GCN method: structural and functional graphs are extracted from DTI and rs-fMRI scans respectively. The two networks are then joined via inter-network edges, whose edge weights are learnable parameters. The adjacency matrix of the joint SC-FC graph is then analyzed by a single graph convolutional network.

features defined for the N ROIs, and $\hat{A}^S = D^{-\frac{1}{2}} A^S D^{-\frac{1}{2}}$ be the normalized adjacency matrix, where D is the degree matrix with $D_{ii} = \sum_j A^S_{ij}$. A typical graph convolution f is

$$f(X^S; A^S) = \sigma(\hat{A}^S X^S W^S), \tag{1}$$

where $\sigma(\cdot)$ is a non-linear activation function and W^S is the convolutional weights to be learned.

On the other hand, we quantify the strength of FC between a pair of ROIs by correlating their BOLD signals recorded in the rs-fMRI scan. As such, the FC network is characterized by another graph $\mathcal{G}^F(\mathcal{V}^F, \mathcal{E}^F)$, and the corresponding graph convolution is defined with respect to the normalized FC adjacency matrix being $\hat{A}^F$. Note, $\mathcal{V}^S$ and $\mathcal{V}^F$ represent the same set of ROIs, whereas $\hat{A}^S$ and $\hat{A}^F$ are generally different, reflecting the divergence between SC and FC patterns. Given the SC and FC graphs of a subject, our goal is to use a GCN model to predict the target label (e.g., age or diagnosis) of the subject.

Joint SC-FC Graph. Instead of using separate GCNs to analyze the two graphs, we merge $\mathcal{G}^S$ and $\mathcal{G}^F$ into a joint graph $\mathcal{G} = \{\mathcal{V}, \mathcal{E}\}$ encoding the entire SC-FC connectome profile. The joint node set $\mathcal{V} = \{\mathcal{V}^S, \mathcal{V}^F\}$ has a size of $2N$, i.e., the duplication of the N ROIs. We then add N inter-network edges $\mathcal{E}^C$ to connect corresponding nodes in $\mathcal{V}^S$ and $\mathcal{V}^F$. The joint edge set is the combination of $\mathcal{E} = \{\mathcal{E}^S, \mathcal{E}^F, \mathcal{E}^C\}$ (Fig. 1). Using this combination, the SC and FC profiles associated with the same brain region are coupled together so that the joint SC-FC graph is analyzed by a single GCN.

Learnable SC-FC Coupling Strength. Recent studies suggest the strength of SC-FC coupling is not uniform across the brain but exhibits variation in parallel to certain cortical hierarchies [1,16]. To support this hypothesis, we set the weights of the N inter-network edges to be learnable parameters. Let $\vec{c} = \{c_1, ..., c_N\}$ be a row vector encoding the coupling strength of $\mathcal{E}^C$ with each $c_i \in [0, 1]$. We then construct the joint $2N \times 2N$ normalized adjacency matrix of the SC-FC joint graph as (Fig. 1)

$$\hat{A} = \begin{bmatrix} \hat{A}^S & \text{diag}(\vec{c}) \\ \text{diag}(\vec{c}) & \hat{A}^F \end{bmatrix}, \tag{2}$$

where diag(·) denotes the diagonalization of a vector. Let $X = \begin{bmatrix} X^S \\ X^F \end{bmatrix}$ denote the vertical concatenation between structural and functional node features. The graph convolution for the joint graph is defined by

$$f(X, A) = \sigma(\hat{A}XW) = \sigma(\begin{bmatrix} \hat{A}^S X^S + \vec{c} \otimes X^F \\ \hat{A}^F X^F + \vec{c} \otimes X^S \end{bmatrix} W). \tag{3}$$

where W defines the convolution weights for the joint graph and $\otimes$ is element-wise multiplication. Equation 3 suggests that through our joint graph convolution, the functional information encoded in X^F is propagated into the structural network through $\mathcal{E}^C$ and the amount of propagation aligns with the SC-FC coupling strength encoded in $\vec{c}$. Vice versa, structural features can also propagate into the functional network through the convolution.

Joint-GCN. To embed the learnable $\vec{c}$ into the end-to-end network, we construct a "dummy" network that contains a single fully connected layer applied to a dummy input scalar. With a sigmoid activation, the N-dimensional output of the dummy network is then viewed as $\vec{c}$ and enters the diagonal entries in the off-diagonal blocks of the joint normalized adjacency matrix. We simply use the sum of the node degree and centrality within the SC and FC networks as node features, which are then fed into a standard GCN with the aforementioned adjacency matrix. The GCN contains one layer of graph convolution with input dimension 80 and output dimension 40 followed by SELU [8] activation, mean pooling, batch-normalization, and two fully connected layers with dimension 1600 and 128 respectively (Fig. 1).

3 Experimental Settings

Using machine learning approaches to predict age and sex of youths based on their neuroimaging data is becoming a popular approach for understanding the dynamic neurodevelopment and the emerging sexual differences during adolescence. We illustrate the potential of Joint-GCN for these prediction tasks based on the brain connectome data provided by the NCANDA study [2].

Data. We used longitudinal rs-fMRI and DTI data from 662 NCANDA participants (328 males and 334 females), who were 12 to 21 years old at their baseline visits and were scanned annually for up to 7 years. Data were acquired on GE and Siemens scanners. Scanner type did not significantly ($p > 0.05$) correlate with age (two-sample t-test) and sex (Fisher's exact test). For each participant, we selected visits where the participant had no-to-low alcohol intake over the past year according to the Cahalan scale [25]. This selection simultaneously resulted in 1976 pairs of rs-fMRI and DTI scans (3 visits per participant on average).

Preprocessing. rs-fMRI data were preprocessed using the publicly available NCANDA pipeline [24], which consists of motion correction, outlier-detection, detrending, physiological noise removal, and both temporal (low pass frequency: 0.1, high pass frequency: 0.01) and spatial smoothing. We then computed the mean BOLD signal within 80 cortical regions by first aligning the mean BOLD image to the subject-specific T1-weighted MRI and then non-rigidly registering the T1 MRI to the SRI24 atlas [17]. Finally, an 80×80 FC connectivity matrix was constructed as the Fisher-transformed Pearson correlation of the BOLD signal between pairwise ROIs. Negative connectivities were set to zero to remove potentially artificial anti-correlation introduced in the rsfMRI preprocessing [20].

The public NCANDA DTI preprocessing pipeline included skull stripping, bad single shots removal, and both echo-planar and Eddy-current distortion correction [25]. The whole-brain boundary of gray-matter white-matter was extracted by the procedure described in [1] and was then parcellated into the same set of 80 ROIs defined in the rs-fMRI preprocessing procedure. Probabilistic tractography was performed by FSL bedpostx and FSL probtrackx, which initiated 10,000 streamlines from each boundary ROI. Finally, each entry in the 80×80 SC connectivity matrix recorded the log of the number of probabilistic streamlines connecting a pair of brain regions.

Model Implementation. We implemented Joint-GCN in Pytorch 1.10 and trained the model using the SGD optimizer for 1000 epochs with a batch size of 64. We used mean squared error as the training loss for age prediction and the binary cross-entropy as the loss for sex prediction after applying a sigmoid activation to the network output. We used a learning rate of 0.01 in the sex prediction task and 0.001 in age prediction. To mitigate differences in the input data associated with the scanner type, a binary variable encoding the scanner type was concatenated with learned features before the last fully connected layer.

Evaluation. We evaluated our proposed Joint-GCN model on the 1976 pairs of DTI and rs-fMRI by running 5-fold cross-validation for 10 times. The five folds were divided on the subject level to ensure data of each subject belonged to the same fold. Although our approach can be extended to a longitudinal setting, we focused on a cross-sectional analysis so the across-visit dependency within an individual was not considered. For sex experiments, we chose the classification accuracy (ACC) and Area under the ROC Curve (AUC) scores as evaluation metrics. For age prediction, we evaluated the Mean Absolute Error (MAE) and the Pearson's Correlation Coefficient (PCC) between predicted and ground truth ages. Average and standard deviation of these metrics over the 10 cross-validation runs were reported. Lastly, we derived the coupling strength of the 80 ROIs for either prediction task by averaging the learned $\vec{\epsilon}$ over the five folds and then averaging the corresponding two regions across left and right hemispheres.

Baselines. To examine the effectiveness of the our Joint-GCN model, we first compared it with single-modality models that only analyzed one type of network (either FC or SC). In this setting, we chose GCN, Multi Layer Perceptron (MLP), and Support Vector Machine (SVM) as the baselines. The MLP was a 2-layer network with 20 hidden units making predictions based on node features. The SVM used either FC or SC adjacency matrices as input features. Next, we compared our model with several existing multi-modal GCN approaches that took both SC and FC networks as inputs. We first implemented a multi-modal version of MLP, which concatenated outputs from the two modality-specific MLPs before entering another fully connected layer of dimension 40. The second approach [3] used a Matrix Auto-Encoder (Matrix-AE) to map FC to SC resulting in a low dimensional manifold embedding that can be used for classification. The third approach [22], UBNfs, used a multi-kernel method to fuse FC and SC to produce a unified brain network and used a standard SVM for classification. Next, we implemented Multi-View GCN (MV-GCN) proposed in [23], which applied two separate GCNs to extract features from SC and FC networks and then used the merged feature for the final prediction. Lastly, we tested SCP-GCN [10] (using their code) that defined adjacency matrix with respect to the SC graph and treated the FC matrix as node features. To align with our Joint-GCN implementation, scanner type was concatenated to the features before the final prediction in all baselines.

4 Results

Predictions by Joint-GCN. Based on 10 runs of 5-fold cross validation, Joint-GCN resulted in an average of 84.9% accuracy for sex classification and an MAE of 1.95 years for age prediction. Figure 2(a)(b) shows the predicted values of the NCANDA participants by our model. We observe that the predicted age non-linearly correlated with ground truth; that is, our model could only successfully stratify the age differences at younger ages. Notably, the MAE for the younger cohort (MAE = 1.86, age < 18 years) was significantly lower ($p < 0.01$, two-sample t-test) than that for the older cohort (MAE = 2.12, age $\geq$ 18 years). This result comports with adolescent neurodevelopment being more pronounced in the younger age but generally slows upon early adulthood [15]. The cubic fitting between predicted age and ground truth in Fig. 2(a) was also inline with the "inverted U-shape" of the developmental trajectory of functional and structural connectome during adolescence [25].

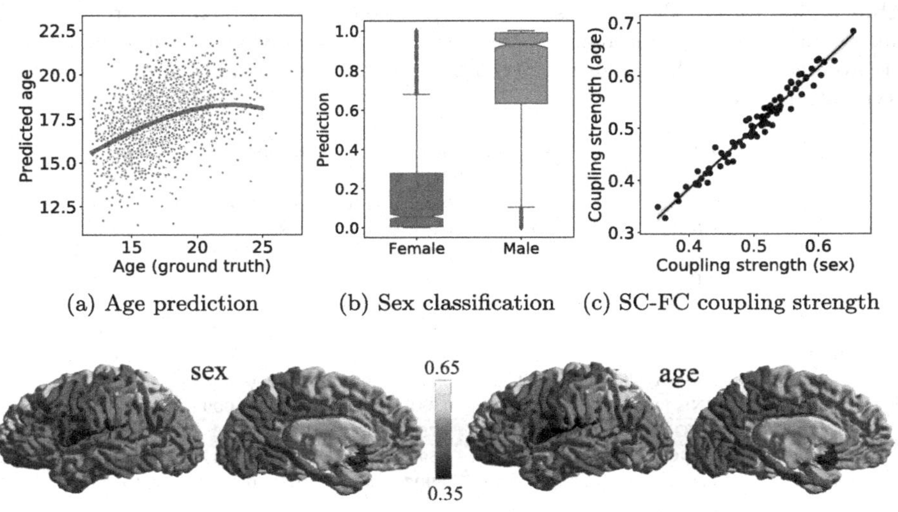

(a) Age prediction (b) Sex classification (c) SC-FC coupling strength

(d) Regional coupling strength displayed on the cortex

Fig. 2. Predicted values of (a) age and (b) sex for the NCANDA participants by Joint-GCN; (c) The ROI-specific weights quantifying the coupling strength between SC-FC learned in the sex prediction task significantly correlated with the weights learned for age prediction ($p < 0.01$, Pearson's r); (d) The learned coupling strength displayed on the brain cortex.

Comparison with Baselines. Table 1 shows the accuracy of sex classification and age prediction for all the comparison methods. GCN-based (including ours) approaches generally resulted in higher accuracy than the others (i.e. MLP and SVM). The higher accuracy of GCN-based methods shows the efficacy of graph convolution in extracting informative features from network data. Compared to the single GCN applied to FC alone and SC alone, multi-view GCN and our Joint-GCN resulted in more accurate sex classification as well as higher PCC in age prediction. This phenomenon indicates SC and FC networks contain complementary information about the brain connectome that is useful for identifying sex and age differences. Notably, if we modeled the coupling by adding inter-network edges between every pair of ROIs (instead of using ROI-specific edges), the prediction accuracy was significantly lowered (sex ACC: 77%, age PCC: 0.34). Lastly, our proposed Joint-GCN resulted in the highest scores in three metrics and the second best in MAE. The improvement in PCC over the second best approach (MV-GCN) was statistically significant ($p < 0.01$, two-sample t-test), and so was the improvement in ACC for sex prediction ($p = 0.015$, McNemar's test).

Learned SC-FC Coupling Strength. Figure 2(d) displays the coupling strength of the 40 bilateral brain regions learned in the age and sex prediction

Table 1. Accuracy of all comparison models trained on single modality or multi-modal brain connectome for predicting sex or age labels of the NCANDA participants. Results were averaged across the 5 folds for 10 times of random dataset splits. Highest accuracy is typeset in bold.

Modality	Method	Sex classification		Age prediction	
		ACC (%) ↑	AUC ↑	MAE ↓	PCC ↑
FC	MLP	60.0 ± 0.1	0.612 ± 0.016	2.03 ± 0.006	0.117 ± 0.018
	GCN [7]	80.9 ± 0.6	0.870 ± 0.007	2.08 ± 0.023	0.316 ± 0.018
	SVM	76.2 ± 0.6	0.838 ± 0.005	2.72 ± 0.042	0.230 ± 0.015
SC	MLP	57.8 ± 2.0	0.598 ± 0.032	2.03 ± 0.005	0.154 ± 0.009
	GCN [7]	78.4 ± 0.6	0.838 ± 0.009	1.98 ± 0.026	0.315 ± 0.015
	SVM	74.1 ± 1.0	0.816 ± 0.010	2.11 ± 0.027	0.316 ± 0.016
Multi-modal	MLP	61.9 ± 1.5	0.647 ± 0.018	2.03 ± 0.007	0.145 ± 0.014
	Matrix-AE [3]	68.8 ± 3.4	0.772 ± 0.031	2.09 ± 0.036	0.246 ± 0.019
	UBNfs [22]	80.4 ± 0.5	0.884 ± 0.003	$\mathbf{1.93 \pm 0.009}$	0.304 ± 0.009
	SCP-GCN [10]	70.8 ± 0.7	0.759 ± 0.011	1.96 ± 0.023	0.303 ± 0.014
	MV-GCN [23]	83.1 ± 0.7	0.893 ± 0.005	2.03 ± 0.034	0.351 ± 0.017
	Joint-GCN(Ours)	$\mathbf{84.9 \pm 0.6}$	$\mathbf{0.907 \pm 0.005}$	1.95 ± 0.006	$\mathbf{0.386 \pm 0.008}$

tasks. We observe that learned coupling strength highly coincided between the two prediction tasks despite the training of the two models being independent. This indicates our learning framework can extract intrinsic SC-FC relationships in the adolescent brain, which contributes to the identification of age and sex differences. Supported by prior studies [1,16], the learned coupling strength was not uniform across the neocortex but varied from 0.35 to 0.65 (0.5 ± 0.05). The top 3 regions with the strongest SC-FC coupling are the superior and the inferior parietal lobe and superior frontal lobe. The lowest coupling strength was measured for the rolandic operculum, olfactory cortex, and frontal inferior operculum. As recent findings have identified macroscopic spatial gradients as the primary organizing principle of brain networks [1,16], we computed the principal gradient of functional connectivity for each ROI [16]. These gradients negatively correlated with the learned coupling strength on a trend level ($p = 0.075$) with unimodal regions generally having stronger and transmodal regions weaker coupling strength. We hypothesize that the trend-level correlation would become more pronounced once we replace our 80-ROI parcellation with a finer one as in [1,16]. Nevertheless, the above result points to the potential of Joint-GCN in learning useful structural-functional properties of the brain.

5 Conclusion and Future Work

In this paper, we proposed a novel framework, called Joint-GCN, for analyzing brain connectome quantified by multi-modal neuroimaging data from DTI and rs-fMRI. Extending prior studies on applying GCN to structural and functional graphs, our work underscored the importance of modeling the coupling between structural and functional connectivity for prediction tasks based on brain connectome. Based on the adolescent neuroimaging dataset provided by

NCANDA, we showed the potential of our framework to accurately characterize the protracted development of structural and functional brain connectivity and the emerging sex differences during youth. One limitation of our study is that we did not focus on modeling the across-visit relationships within the longitudinal data of NCANDA participants (an orthogonal research direction). Rather, we only considered network metrics (degree and centrality) as node features to focus on exploring the model's capability in analyzing network topology. Nevertheless, our model presents a useful data-driven approach to model the complex hierarchical neural systems that have broad relevance for healthy aging and abnormalities associated with neuropsychological diseases.

Acknowledgment. This research was supported in part by NIH U24 AA021697, K99 AA028840, and Stanford HAI GCP Credit. The data were part of the public NCANDA data releases NCANDA_PUBLIC_6Y_REDCAP_V04 [13], NCANDA_PUBLIC_6Y_DIFFUSION_V01 [12], and NCANDA_PUBLIC_6Y_RESTINGSTATE_V01 [14], whose collection and distribution were supported by NIH funding AA021697, AA021695, AA021692, AA021696, AA021681, AA021690, and AA02169.

References

1. Baum, G., et al.: Development of structure-function coupling in human brain networks during youth. Proc. Natl. Acad. Sci. **117**(1), 771–778 (2019)
2. Brown, S., et al.: The national consortium on alcohol and NeuroDevelopment in Adolescence (NCANDA): a multisite study of adolescent development and substance use. J. Stud. Alcohol Drugs **76**(6), 895–908 (2015)
3. D'Souza, N., et al.: A matrix autoencoder framework to align the functional and structural connectivity manifolds as guided by behavioral phenotypes, **12907**, 625–636 (2021)
4. Gadgil, S., Zhao, Q., Pfefferbaum, A., Sullivan, E.V., Adeli, E., Pohl, K.M.: Spatio-temporal graph convolution for resting-state fMRI analysis. In: Martel, A.L., et al. (eds.) MICCAI 2020. LNCS, vol. 12267, pp. 528–538. Springer, Cham (2020). https://doi.org/10.1007/978-3-030-59728-3_52
5. Hanik, M., Demirtaş, M.A., Gharsallaoui, M.A., Rekik, I.: Predicting cognitive scores with graph neural networks through sample selection learning. Brain Imaging Behav. **16**, 1–16 (2021)
6. Jung, J., Cloutman, L., Binney, R., Ralph, M.: The structural connectivity of higher order association cortices reflects human functional brain networks. Cortex **97**, 221–239 (2016)
7. Kipf, T.N., Welling, M.: Semi-supervised classification with graph convolutional networks. arXiv preprint arXiv:1609.02907 (2016)
8. Klambauer, G., Unterthiner, T., Mayr, A., Hochreiter, S.: Self-normalizing neural networks. In: Advances in Neural Information Processing Systems, vol. 30 (2017)
9. Li, X., et al.: BrainGNN: interpretable brain graph neural network for fMRI analysis. Med. Image Anal. **74**, 1–13 (2021)
10. Liu, J., Ma, G., Jiang, F., Lu, C.T., Yu, P., Ragin, A.: Community-preserving graph convolutions for structural and functional joint embedding of brain networks. In: International Conference on Big Data (Big Data), pp. 1163–1168, November 2019

11. Moody, J., Adluru, N., Alexander, A., Field, A.: The connectomes: methods of white matter tractography and contributions of resting state fMRI. Semin. Ultrasound CT and MRI **42**(5), 507–522 (2021)
12. Pohl, K.M., et al.: The 'NCANDA_PUBLIC_6Y_DIFFUSION_V01' data release of the national consortium on alcohol and neurodevelopment in adolescence (NCANDA). Sage Bionetworks Synapse (2022). https://doi.org/10.7303/syn27226988
13. Pohl, K.M., et al.: The 'NCANDA_PUBLIC_6Y_REDCAP_V04' data release of the national consortium on alcohol and neurodevelopment in adolescence (NCANDA). Sage Bionetworks Synapse (2022). https://doi.org/10.7303/syn26951066
14. Pohl, K.M., et al.: The 'NCANDA_PUBLIC_6Y_RESTINGSTATE_V01' data release of the national consortium on alcohol and neurodevelopment in adolescence (NCANDA). Sage Bionetworks Synapse (2022). https://doi.org/10.7303/syn32303917
15. Pujol, J., Vendrell, P., Junqué, C., Martí-Vilalta, J.L., Capdevila, A.: When does human brain development end? Evidence of corpus callosum growth up to adulthood. Ann. Neurol. Off. J. Am. Neurol. Assoc. Child Neurol. Soc. **34**(1), 71–75 (1993)
16. Rodriguez-Vazquez, B., et al.: Gradients of structure-function tethering across neocortex. PNAS **116**(42), 21219–21227 (2019)
17. Rohlfing, T., Zahr, N., Sullivan, E., Pfefferbaum, A.: The SRI24 multichannel atlas of normal adult human brain structure. Hum. Brain Mapp. **31**(5), 798–819 (2009)
18. Song, T.A., et al.: Graph convolutional neural networks for Alzheimer's disease classification. In: IEEE International Symposium on Biomedical Imaging, vol. 2019, pp. 414–417, April 2019
19. Sporns, O., Tononi, G., Kötter, R.: The human connectome: a structural description of the human brain. PLoS Comput. Biol. **1**(4), 245–251 (2005)
20. Weissenbacher, A., Kasess, C., Gerstl, F., Lanzenberger, R., Moser, E., Windischberger, C.: Correlations and anticorrelations in resting-state functional connectivity MRI: a quantitative comparison of preprocessing strategies. Neuroimage **47**(4), 1408–1416 (2009)
21. Yalcin, A., Rekik, I.: A diagnostic unified classification model for classifying multi-sized and multi-modal brain graphs using graph alignment. J. Neurosci. Methods **348**, 1–14 (2021)
22. Yang, J., Zhu, Q., Zhang, R., Huang, J., Zhang, D.: Unified brain network with functional and structural data, **12267**, 114–123 (2020)
23. Zhang, X., He, L., Chen, K., Luo, Y., Zhou, J., Wang, F.: Multi-view graph convolutional network and its applications on neuroimage analysis for Parkinson's disease. In: AMIA Annual Symposium Proceedings, vol. 2018, pp. 1147–1156. American Medical Informatics Association (2018)
24. Zhao, Q., et al.: Longitudinally consistent estimates of intrinsic functional networks. Hum. Brain Mapp. **40**(8), 2511–2528 (2019)
25. Zhao, Q., et al.: Association of heavy drinking with deviant fiber tract development in frontal brain systems in adolescents. JAMA Psychiatry **78**(4), 407–415 (2020)

Decoding Task Sub-type States with Group Deep Bidirectional Recurrent Neural Network

Shijie Zhao, Long Fang, Lin Wu, Yang Yang(✉), and Junwei Han

School of Automation, Northwestern Polytechnical University, Xi'an 710072, China
Tp030ny@gmail.com

Abstract. Decoding brain states under different task conditions from functional magnetic resonance imaging (tfMRI) data has attracted more and more attentions in neuroimaging studies. Although various methods have been developed, existing methods do not fully consider the temporal dependencies between adjacent fMRI data points which limits the model performance. In this paper, we propose a novel group deep bidirectional recurrent neural network (Group-DBRNN) model for decoding task sub-type states from individual fMRI volume data points. Specifically, we employed the bidirectional recurrent neural network layer to characterize the temporal dependency feature from both directions effectively. We further developed a multi-task interaction layer (MTIL) to effectively capture the latent temporal dependencies of brain sub-type states under different tasks. Besides, we modified the training strategy to train the classification model in group data fashion for the individual task. The basic idea is that relational tfMRI data may provide external information for brain decoding. The proposed Group-DBRNN model has been tested on the task fMRI datasets of HCP 900 subject's release, and the average classification accuracy of 24 sub-type brain states is as high as 91.34%. The average seven-task classification accuracy is 95.55% which is significantly higher than other state-of-the-art methods. Extensive experimental results demonstrated the superiority of the proposed Group-DBRNN model in automatically learning the discriminative representation features and effectively distinguishing brain sub-type states across different task fMRI datasets.

Keywords: Brain decoding · Brain states · RNN · Deep learning · Functional magnetic resonance imaging

1 Introduction

Decoding brain states from functional brain imaging data [1–6] has received more and more attention in neuroimaging community. Various strategies have been developed to identify brain states [5–8] using machine learning methods with the objective of understanding how information is represented in the brain [9]. These methods mainly focus on two directions to improve the decoding accuracy. One way is to identify informative functional signatures for distinguishing different brain states [3, 8] and there has been extensive work in this direction. For example, the general linear model (GLM) framework is widely adapted to select informative brain voxels [10]. Besides, a few

L. Wang et al. (Eds.): MICCAI 2022, LNCS 13431, pp. 241–250, 2022.
https://doi.org/10.1007/978-3-031-16431-6_23

studies proposed to select distinctive regions of interest (ROIs) for brain decoding based on expert knowledge [4, 11, 12]. In order to account for the multivariate relationships between fMRI images, multi-voxel pattern analysis (MVPA) [13] is proposed to enhance the feature representation ability. To further improve the general application of selected features, whole-brain based functional signatures have received more and more attention. In [7, 14], whole-brain functional connectivity patterns derived from independent component analysis (ICA) or Graphs are employed for decoding different brain states. Besides, Jang [2] proposed to adopt the deep belief neural network (DBN) to learn a representation of 3D fMRI volume for decoding study. Yong [3] proposed to adopt subject-specific intrinsic functional networks and functional profiles as features.

Another way is to develop and employ better classifiers/models. The most commonly used classifier in conventional classification techniques is the linear classifier [1, 8, 15]. Among all of these classifiers, the support vector machine (SVM) [16] is the most popular one due to its simpleness and efficiency. Despite its popularity, the SVM classifier struggles to perform well on high-dimensional data, and needs further design techniques for feature selection [4]. To improve the prediction accuracy in multi-label classification, the multinomial logistic regression which is also known as Softmax classifier is introduced in bran state decoding [5, 17]. Most recently, convolutional neural network (CNN) based classifier is proposed to decoding task states [4] which suggests that the deep neural network based classifiers may further improve the decoding performance. Although above mentioned classifiers achieved considerable decoding performance, these classifiers do not fully take into account the temporal dependencies between adjacent fMRI volumes which limited the model performance. Besides, existing methods usually training and testing the model in the same type tfMRI dataset which limited the robustness and performance of new data. Thus, novel deep neural network models which can well capture temporal dependencies and advanced training strategies which can improve the model robustness may further boost the brain decoding model performance.

Recently, recurrent neural networks (RNNs) have shown outstanding performance in characterizing temporal dependencies [18, 19, 20] in sequential data which quite match the requirement in brain state decoding. Motivated by these studies, we develop a novel group deep bidirectional recurrent neural network model for decoding task subtype states from individual fMRI volume data points. Specifically, the deep bidirectional recurrent neural network is designed as the backbone network to differentiate brain states with excellent temporal dependency feature modeling ability. Then, we developed the Group-DBRNN model by modifing the training strategy in a group fashion. The basic idea is that relational tfMRI data may provide external information for brain decoding. We further proposed a multi-task interaction layer (MTIL) to encode the correlation and difference between different brain activation states to enhance the model performance. Extensive experimental results demonstrated the superiority of the proposed Group-DBRNN model in automatically learning the discriminative representation features and effectively distinguishing brain sub-type states across task fMRI datasets.

2 Materials and Methods

2.1 Overview

In this paper, we proposed a novel Group-DBRNN model to decode brain sub-type states from individual fMRI volume data. First, by constructing training samples in the form of multi-tasks, the proposed Group-DBRNN introduces multi task information into the classification model, which helps to improve the discriminative ability of the learned features for functional brain states under different sub-type tasks. Then, we developed a bidirectional GRU based network (with two recurrent layers) to capture temporal-spatial-wise contextual information and temporal-task-wise information between multiple brain activity states. The proposed network contains two GRU layers. The first GRU layer captures the temporal-spatial-wise contextual information in each brain activity state. The second layer captures the temporal-task-wise contextual information between brain sub-type states under different tasks. To facilitate the network to perceive temporal task context information from multi-tasks fMRI data, we propose a multi-task interaction layer (MTIL) which embedded between the two-layer GRUs of DBRNN to capture the latent temporal dependencies of multi-tasks fMRI data. The experimental results show that our proposed brain decoding method automatically learns the discriminative representation features by exploring the essential temporal correlations of multi-brain functional states under multiple external stimuli.

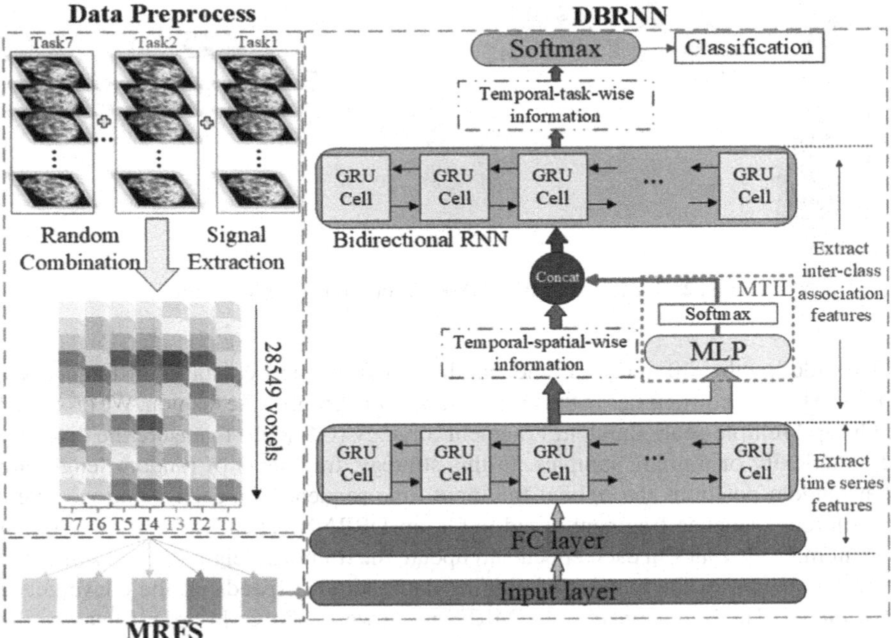

Fig. 1. Flow chart of the proposed Group-DBRNN model. It consists of five layers: a full-connected layer, two GRU layers, a multi-task interaction layer, and a softmax layer.

2.2 Data Preprocessing

We implement our group-wise brain sub-type state decoding task on the HCP 900 subject's release which is one of the most systematic and comprehensive neuroimaging datasets. HCP 900 subject's release dataset contains more than 800 subjects with seven tasks, including Working-Memory, Gambling, Motor, Language, Social, Relational, and Emotion. The seven tasks include 24 different task stimuli. We defined each task stimuli type as a task subtype. All the task-fMRIs in HCP are pre-processed by skull removal, motion correction, slice time correction, spatial smoothing, and global drift removal (high-pass filtering).

To train a brain functional state classification model in the group-wise fashion, we randomly selected 7 fMRI data under the different sub-type tasks to constitute a group-wise training sample. Specifically, we spread out the voxels in each tfMRI volume data to form a two-dimensional matrix with size time points * voxels. The tfMRI data contains 28549 voxels at each moment. The tfMRI data matrices of different tasks are spliced along the time dimension to form the group-wise training samples.

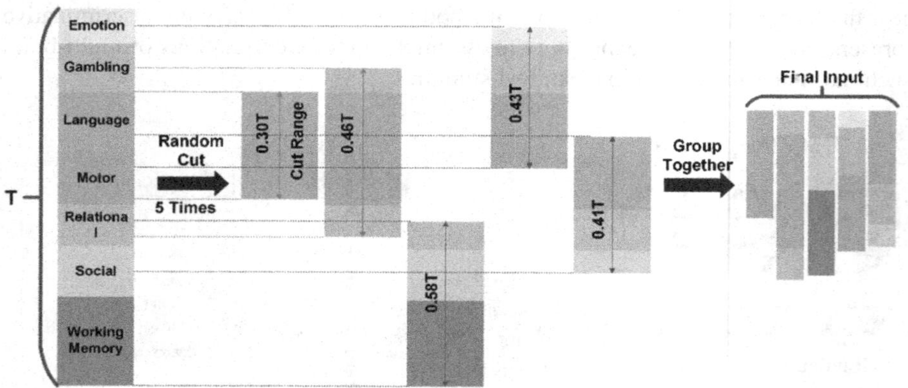

Fig. 2. Basic concepts of GMRFS for group-wise learning

Considering the difference in time length of fMRI sequence data for different tasks will cause learning difficulties in the training model in a group-wise fashion. We proposed the Group Multiple-scale Random Fragment Strategy (GMRFS) to ensure the diversity and complexity of training samples. In this strategy, fragments of random length are cut at random positions throughout the input time sequence. In each iteration, a new and different sequence fragment is fed to Group-DBRNN. Then, we split the random segment multiple times in each epoch and update the training weight of each split data. It not only increases the amount of training data, but also speeds up the convergence of the model. The basic concept of GMRFS is shown in Fig. 2. In order to ensure the randomness and diversity of the training data when training the group-DBRNN model, the same training fragment is randomly cut 5 times. Each training epoch contains training samples consisting of five different randomly cut fragments, and the fragment length varies randomly between 30%-60% of the entire sequence length.

2.3 Group Deep Bidirectional Recurrent Neural Network

The proposed Group-DBRNN consists of five layers: a full-connected layer, two GRU layers, a multi-task interaction layer, and a Softmax layer. The fully connected layer extracts the low-level features of each fMRI volume data sequence. The first GRU layer captures the temporal-spatial-wise contextual information in each brain activity state, and the second GRU layer captures the temporal-task-wise contextual information between brain states. The multi-task interaction layer (MTIL) captures the latent temporal dependencies of brain sub-type states under different tasks. Finally, a Softmax layer outputs the class probability vector.

Bidirectional GRU Layers. We use the bidirectional GRU to extract the contextual information in fMRI volume data from both forward and backward directions. Compared with the feature extraction method of the one-way RNN network, the two-way (Bidirectional) GRU is more capable of perceiving the temporal context information in the fMRI signal, and naturally and effectively integrates the spatial context information. The hidden state of GRU is defined as follows:

$$\boldsymbol{h}^t = \left(1 - z^t\right) \odot \boldsymbol{h}^{t-1} + z^t \odot \boldsymbol{h}^{-t}$$

$$z^t = \sigma(\boldsymbol{U}_z h^{t-1} + \boldsymbol{W}_z x^t + \boldsymbol{b}_z)$$

$$\boldsymbol{r}^t = \sigma(\boldsymbol{U}_r \boldsymbol{h}^{t-1} + \boldsymbol{W}_r x^t + \boldsymbol{b}_r)$$

$$\boldsymbol{h}^{-t} = \tanh(\boldsymbol{U}_h\left(r^t \times \boldsymbol{h}^{t-1}\right) + \boldsymbol{W}_h x^t + \boldsymbol{b}_h)$$

where $\boldsymbol{z}^t$ is update gate activation, $\boldsymbol{r}^t$ is reset gate activation, and $\boldsymbol{h}^{-t}$ is an auxiliary variable. These GRU unit can control the information flow.

The Contextual Information Extracting. For the group-wise brain sub-type states decoding task, the contextual information should consist of the temporal-spatial-wise information and the temporal-task-wise information. The temporal-spatial-wise contextual information describes the temporal and spatial constructs of the input multi-task fMRI volume data. To this end, we use a Fully-connected layer and a bidirectional GRU Layer to extract the temporal-spatial features of the input group tfMRI data (as shown in Fig. 1). The fully-connected layer encodes the spatial information of each input fMRI data and generates the spatial feature sequence. The bidirectional GRU layer takes the spatial feature sequence as input and outputs the temporal-spatial features of each input data by hybrid encoding the temporal and spatial information into one feature vector.

The temporal-task-wise contextual information mainly describes the potential association between different sub-type tasks, which will help to improve the discrimination of the features extracted by our method for different sub-type tasks. To achieve this goal, we proposed a multi-task interaction layer (MTIL) to extract the latent temporal dependencies of brain sub-type states under different tasks and a bidirectional GRU layer to generate the inter-tasks associational features, as shown in Fig. 1. Finally, the inter-tasks associational features will be fed to a Softmax for sub-type tasks decoding.

The Multi-task Interaction layer (MTIL). We developed a simple multilayer perceptron layer to implement a multi-task interaction layer consisting of a multilayer perceptron layer and a softmax layer. The multilayer perceptron layer takes the temporal-spatial feature of multiple brain activity states under different external stimuli tasks as input. Itextracts task-wise semantic features of each brain activity state. The semantic features contain the task information of the corresponding external stimuli. We employ a Softmax to map the task information as a one-hot vector, which will guide the network to learn the differences and commonalities between brain activation states under different subtasks. After that, we concatenate spatiotemporal features and one-hot feature vectors along the channel direction to facilitate the network to perceive potentially relevant information for different tasks.

3 Experimental Results

3.1 Implement

We randomly selected 120 subjects from HCP 900 subject's release for training and 60 subjects for verification. Each subject participated in 7 tasks, and these 7 tasks are spliced in a random order, and the corresponding label file is generated in the form of one-hot encoding. Before entering the model, we performed z-score normalization on all the data to eliminate the influence of the difference in the size and distribution of the data itself on the results. The batch size is set to 20, and batch normalization processing is performed to speed up the training speed and reduce over-fitting. At the same time, in order to further reduce the risk of overfitting and speed up the convergence, as described in Sect. 2.6, we use GMRFS between the input layer and the fully connected layer. The initial learning rate is set to 0.01. After 160 iterations, the learning rate is reduced to 1/4 of the previous round. The training process will be stopped when the learning rate is less than 10–5. We use the two-dimensional cross entropy function to calculate the loss, and use the Adam optimizer to find the optimal solution. The Group-DBRNN model was run on the computing platform with two GPUs (Nvidia GTX 1080 Ti, 12G Graphic Memory) and 64 G memory.

3.2 Decoding Performance on Brain Sub-type Task States

After the training was completed, we randomly selected another 120 subjects to test the performance of the model. And we employ classification accuracy to evaluate the performance of the brain state decoding model. The corresponding relationship between each label ID, sub-type task events, and its classification accuracy is shown in Table 1. From the table, we can see that. The average classification accuracy of 24 sub-type states is 91.34%. Except for the shape accuracy in Emotion, which is slightly lower, the classification accuracy of the other 23 events is around 90%. Among them, the match accuracy in Relational reaches the highest 96.79%. It can be seen that our Group-DBRNN model has very good classification performance and can accurately identify sub-type brain states.

Table 1. The label ID corresponding to each sub-type states and their classification accuracy

Event	Face	Shape	Win	Loss	Math	Story	Left foot	Right foot	Left hand	Right hand	Tongue	Match
ID	E1	E2	G1	G2	L1	L2	M1	M2	M3	M4	M5	R1
Acc	0.8827	0.7211	0.9114	0.9487	0.9224	0.9468	0.9015	0.8914	0.8943	0.9044	0.9191	0.9679
Event	Relational	Social	Random	2back of tools	2back of places	2back of faces	2back of body	0back of tools	0back of places	0back of faces	0back of body	Cue & interval
ID	R2	S1	S2	W1	W2	W3	W4	W5	W6	W7	W8	Cue
Acc	0.9572	0.9581	0.9614	0.9103	0.9321	0.9038	0.9423	0.9205	0.9192	0.8795	0.9038	0.9207

3.3 Comparison of Decoding Performance on Brain Sub-type Task States

In order to prove the advantages of the proposed Group-DBRNN model, several state-of-the-art (SOTA) methods were selected to compare the decoding performance on seven tasks. Specifically, we compare our approach with DNN, Multi-Voxel Pattern Analysis (MVPA), SVM, and softmax methods. DNN [4] is a deep learning network model that includes 5 layers of convolutional neural network layers and 2 layers of fully connected layers. It obtains high-dimensional information by extracting features of pre-processed task-fMRI images and achieves good performance. We reproduced the DNN model and applied it to our 24 classification tasks. Multi-Voxel Pattern Analysis (MVPA) is considered as a supervised classification problem where a classifier attempts to capture the relationships between spatial pattern of fMRI activity and experimental conditions. Figure 3 illustrats the comparison result of different methods. It can be seen that in the classification tasks of all 24 sub-type brain states, the classification accuracy of our Group-DBRNN model is significantly ahead of other methods (Fig. 3).

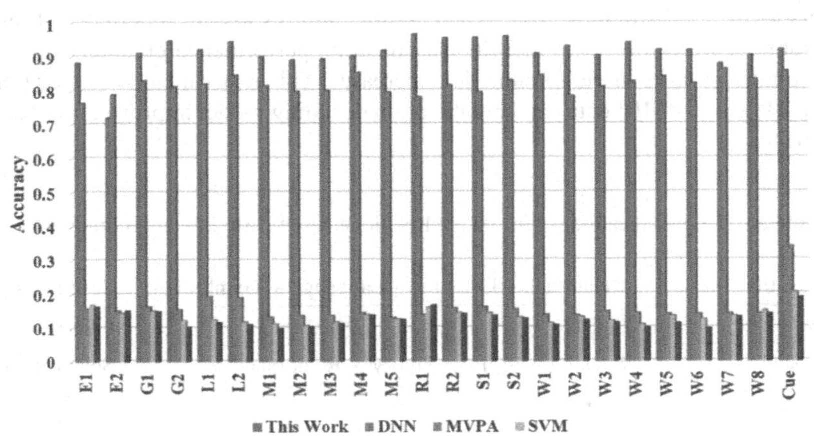

Fig. 3. Decoding sub-type task states performance of different SOTA methods.

3.4 Comparison of Decoding Performance on Seven Brain Task States

To further analyze the decoding performance of our proposed group-wise training fashion, we test our method on the seven decoding tasks as traditional studies. The classification results are reported in Table 2. Except for the Emotion and Language tasks, the classification performance on the other five tasks outperforms all other SOTA methods. The average classification accuracy of our method (95.55%) is the best.

Table 2 also reported the classification results of a baseline (w/o MTIL) that does not use the proposed multi-task interaction layer. The classification accuracy of our method is higher than the baseline method on seven tasks. This experiment demonstrates that our proposed MTIL helps to improve the classification performance of the brain state decoding.

Table 2. Classification results of Group-DBRNN and comparing methods on 7 tasks. The Emt refers to emotion, Gmb refers to Gambling, Lng refers to Language, the Mtr refers to Motor, the Rlt refers to Relational, Scl refers to Social, WM refers to Working-Memory.

Task/Method	Emt	Gmb	Lng	Mtr	Rlt	Scl	WM	Average
DNN [4]	0.929	0.829	0.977	0.956	0.912	0.977	0.905	0.9264
DSRNN [20]	0.9327	0.9472	0.9342	0.9465	0.8706	0.9209	0.9407	0.9275
w/o MTIL	0.8913	0.9468	0.8909	0.9566	0.9907	0.9943	0.9799	0.9501
Ours	0.8926	0.9690	0.8917	0.9702	0.9909	0.9951	0.9892	0.9570

To further analyze the association between the group-task training and the individual task training for the brain states decoding. We evaluate the classification performance of different brain state classification networks, which are trained by combining different numbers of external stimulus tasks. The classification results are reported in Table 3. From Table 3, we can observe that the classification performance gradually improves as the number of tasks in the training data increases. This experiment shows that brain states under group tfMRI data can help the brain decoding model improve classification performance.

Table 3. The classification results on the Relational tasks with different training data.

Only Relational task	Relational + Emotion + Language + Gambling	Seven tasks
0.8913	0.9311	0.9909
Only Social task	Social + working memory + Relational	Seven tasks
0.9156	0.9414	0.9952

4 Conclusion

This paper proposes a novel group deep bidirectional recurrent neural network (Group-DBRNN) explore the essential correlations of multi-brain states under multiple external stimuli. The proposed GMRFS strategy prevents our method from overfitting training data. The proposed multi-task interaction layer helps our approach to explore the temporal-task-wise contextual information of multiple brain activity states. Compared with the state-of-the-art methods, such as DNN, DSRNN, MVPA and SVM, the classification accuracy of Group-DBRNN has achieved the best. The experimental results show that our proposed brain decoding method automatically learns the discriminative representation features by exploring the essential correlations of multi-brain functional states under multiple external stimuli. It effectively distinguishes brain sub-type states on all task fMRI datasets.

Acknowledgements. This work was supported in part by the National Key R&D Program of China under Grant 2020AAA0105701; the National Natural Science Foundation of China under Grant 82060336, 62136004, 62036011, U1801265, 6202781, Guangdong Basic and Applied Basic Research Foundation (2214050008706), Science and Technology Support Project of Guizhou Province under Grant [2021]432, and Guangdong Key Laboratory for Biomedical Measurements and Ultrasound Imaging, Health Science Center, School of Biomedical Engineering, Shenzhen University, Shenzhen 518060, China.

References

1. Haynes, J., Rees, G., et al.: Decoding mental states from brain activity in humans. Nat. Rev. Neurosci. **7**, 523–534 (2006)
2. Jang, H., Plis, S.M., Calhoun, V.D., et al.: Task-specific feature extraction and classification of fMRI volumes using a deep neural network initialized with a deep belief network: evaluation using sensorimotor tasks. Neuroimage **145**, 314–328 (2017)
3. Li, H., Fan, Y.: Interpretable, highly accurate brain decoding of subtly distinct brain states from functional MRI using intrinsic functional networks and long short-term memory recurrent neural networks. Neuroimage **202**, 116059 (2019)
4. Wang, X., et al.: Decoding and mapping task states of the human brain via deep learning. Human Brain Mapping, pp. 1–15 (2019)
5. Zhang, A., Fang, J., Liang, F., Calhoun, V.D., Wang, Y.: Aberrant brain connectivity in schizophrenia detected via a fast gaussian graphical model. IEEE J. Biomed. Health Inform. **23**, 1479–1489 (2019)
6. Cohen, J.D., et al.: Computational approaches to fMRI analysis. Nat. Neurosci. **20**, 304–313 (2017)
7. Shirer, W.R., Ryali, S., Rykhlevskaia, E., Menon, V., Greicius, M.D.: Decoding subject-driven cognitive states with whole-brain connectivity patterns. Cereb. Cortex **22**, 158–165 (2012)
8. Naselaris, T., Kay, K., Nishimoto, S., Gallant, J.L.: Encoding and decoding in fMRI. Neuroimage **56**, 400–410 (2011)
9. Karahanoglu, F.I., De Ville, D.V.: Dynamics of large-scale fMRI networks: deconstruct brain activity to build better models of brain function. Current Opinion in Biomedical Eng. **3**, 28–36 (2017)

10. Mumford, J.A., Turner, B.O., Ashby, F.G., Poldrack, R.A.: Deconvolving BOLD activation in event-related designs for multivoxel pattern classification analyses. Neuroimage **59**, 2636–2643 (2012)
11. Huth, A.G., Lee, T., Nishimoto, S., Bilenko, N.Y., Vu, A.T., Gallant, J.L.: Decoding the semantic content of natural movies from human brain activity. Front. Syst. Neurosci. **10**, 81 (2016)
12. Yousefnezhad, M., Zhang, D.: Decoding visual stimuli in human brain by using anatomical pattern analysis on fMRI images. International Conference on Brain Inspired Cognitive Systems, pp. 47–57 (2016)
13. Norman, K.A., Polyn, S.M., Detre, G.J., Haxby, J.V.: Beyond mind-reading: multi-voxel pattern analysis of fMRI data. Trends in cognitive sciences **10**(9), 424–430 (2006)
14. Richiardi, J., Eryilmaz, H., Schwartz, S., Vuilleumier, P., De Ville, D.V.: Decoding brain states from fMRI connectivity graphs. Neuroimage **56**, 616–626 (2011)
15. Kay, K., Gallant, J.L.: I can see what you see. Nat. Neurosci. **12**, 245 (2009)
16. Cortes, C., Vapnik, V.: Support-vector networks. Mach. Learn. **20**, 273–297 (1995)
17. Xu, W., Li, Q., Liu, x., Zhen, Z., Wu, X.: Comparison of feature selection methods based on discrimination and reliability for fMRI decoding analysis-journal of neuroscience methods. J. Neurosci. Methods **335**, 1–10 (2020)
18. Graves, A., Liwicki, M., Fernandez, S., Bertolami, R., Bunke, H., Schmidhuber, J.: A novel connectionist system for unconstrained handwriting recognition. IEEE Trans. Pattern Anal. Mach. Intell. **31**, 855–868 (2009)
19. Cho, K., et al.: Learning phrase representations using RNN Encoder--decoder for statistical machine translation. arXiv preprint arXiv: 1724–1734 (2014)
20. Wang, H., et al.: Recognizing brain states using deep sparse recurrent neural network. IEEE Trans. Medical Imaging **38**(4), 1058–1068 (2018)

Hierarchical Brain Networks Decomposition via Prior Knowledge Guided Deep Belief Network

Tianji Pang[1], Dajiang Zhu[3], Tianming Liu[2], Junwei Han[1], and Shijie Zhao[1(✉)]

[1] School of Automation, Northwestern Polytechnical University, Xi'an 710072, China
shijiezhao666@gmail.com

[2] Cortical Architecture Imaging and Discovery Lab, Department of Computer Science and Bioimaging Research Center, The University of Georgia, Athens, GA, USA

[3] Department of Computer Science and Engineering, The University of Texas at Arlington, Arlington, TX 76019, USA

Abstract. Modeling and characterizing functional brain networks from task-based functional magnetic resonance imaging (fMRI) data has been a popular topic in neuroimaging community. Recently, deep belief network (DBN) has shown great advantages in modeling the hierarchical and complex task functional brain networks (FBNs). However, due to the unsupervised nature, traditional DBN algorithms may be limited in fully utilizing the prior knowledge from the task design. In addition, the FBNs extracted from different DBN layers do not have correspondences, which makes the hierarchical analysis of FBNs a challenging problem. In this paper, we propose a novel prior knowledge guided DBN (PKG-DBN) to overcome the above limitations when conducting hierarchical task FBNs analysis. Specifically, we enforce part of the time courses learnt from DBN to be task-related (in either positive or negative way) and the rest to be linear combinations of task-related components. By incorporating such constraints in the learning process, our method can simultaneously leverage the advantages of data-driven approaches and the prior knowledge of task design. Our experiment results on HCP task fMRI data showed that the proposed PKG-DBN can not only successfully identify meaningful hierarchical task FBNs with correspondence comparing to traditional DBN models, but also converge significantly faster than traditional DBN models.

Keywords: Hierarchical organization · Supervise · Human connectome project · Deep neural network · Functional brain networks

1 Introduction

Task-based functional magnetic resonance imaging (fMRI) analysis has been widely used to identify and locate functionally distinct brain networks [1–7]. It has been a critical and powerful approach for better understanding of brain function. GLM (general linear model) has traditionally been the dominant method for task related functional brain networks (FBNs) inference from fMRI data [8–12]. To better take full spatial

L. Wang et al. (Eds.): MICCAI 2022, LNCS 13431, pp. 251–260, 2022.
https://doi.org/10.1007/978-3-031-16431-6_24

patterns of brain activities from distributed brain regions into account simultaneously, plenty data-driven based approaches, such as independent component analysis (ICA) [13–16], and sparse representations [17, 18], are developed. Though meaningful results have been achieved, these traditional MVPA methods are limited by their shallow nature. The hierarchical structure of brain networks, which is considered as an intrinsic nature of the brain, cannot be well captured.

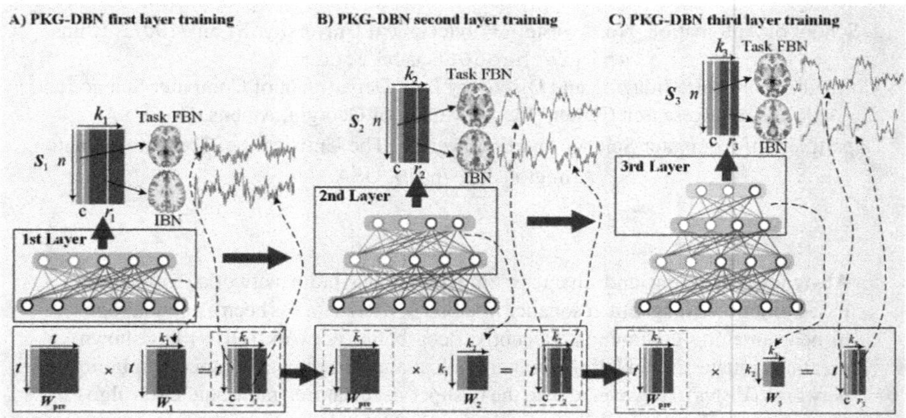

Fig. 1. Illustration of the proposed computational framework of PKG-DBN. FMRI Data of a randomly selected subject is extract into a 2D signal matrix $\boldsymbol{X} \in \mathbb{R}^{t \times n}$, where t is the number of time points in fMRI sequence, and n is the number of voxels of the brain. Each column of $\boldsymbol{X}$, i.e. the time series of each voxel, is taken as a sample for PKG-DBN training. The orange units in PKG-DBN are task-related units, each of whose output represents task functional brain network (task FBN). The green units in PKG-DBN are task mixture units, each of whose output represents intrinsic brain network (IBN). The output of each PKG-DBN layer is taken as the input of the next PKG-DBN layer. $\boldsymbol{W}_1$, $\boldsymbol{W}_2$, and $\boldsymbol{W}_3$ is the learnt weight matrix of the first, second, and the third PKG-DBN layer, respectively. $\boldsymbol{W_{pre}}$ is the identity matrix for the first PKG-DBN layer. The columns of the product of $\boldsymbol{W_{pre}}$ and the learnt weight matrix of each layer represent the time series of corresponding functional brain networks.

Due to the superb ability in capturing mid-level and high-level information from data, deep learning based methods have attracted more and more attention for hierarchical task FBNs inference. For instance, Dong et al. [19] proposed a volumetric sparse DBN method for hierarchical task FBNs inference. Zhang et al. [20] proposed a hybrid spatiotemporal deep learning method which used DBN to extract hierarchical spatial features followed by a deep lasso model for multiple scale temporal information capturing. Zhang et al. [21] investigated the hierarchical organization of brain networks across multiple modality, i.e. fMRI and DTI, by using a multi-model DBN. All of these studies have successfully revealed the hierarchical organization of task FBNs and multiscale responses of brain activities. However, with all of great advantages mentioned above, there are two limitations in current deep learning based data driven methods: 1) The prior knowledge of task paradigm is not used during feature learning. DBN is a greedy algorithm, which is capable of fitting any data distribution theoretically. Without the guidance of prior knowledge, DBN can overfit irrelevant and redundant signals thereby reducing the performance of the model. 2) The FBNs extracted by different DBN layers do not have correspondences, which makes the hierarchical analysis of FBNs a challenging problem.

To address the limitations mentioned above, in this paper, we propose a novel prior knowledge guided DBN (PKG-DBN) algorithm for more efficient feature learning and hierarchical FBNs analysis. Specifically, for task fMRI, we hypothesize brain activities of FBNs should be related to the task paradigm. With this hypothesis and the inspiration from prior study [22], we infuse the prior knowledge of task diagram into the traditional DBN by enforcing part of time courses learnt from DBN to be task related (in either positive or negative) and the rest to be linear combinations of task-related components. The specific way of task diagrams to be combined, i.e., the amplitude and the sign of each task time course, are all learned automatically in a data-driven manner. By doing this, task FBN analysis based on the proposed PKG-DBN algorithm can take advantages from both the data-driven nature of DBN and the prior knowledge of task design. The computational framework of PKG-DBN is illustrated in Fig. 1.

We tested the proposed novel PKG-DBN on the Human Connectome Project (HCP) 900 subject's fMRI dataset. Meaningful functional networks including both model-driven brain networks and intrinsic brain networks (resting state brain networks) have been successfully identified. Experiment results showed the proposed PKG-DBN can converge faster than the traditional DBN and help overcoming the overfitting problem of the traditional DBN. PKG-DBN can successfully infer hierarchical task FBNs. Task related time courses can also be learnt clearly. In addition, the proposed PKG-DBN can learn better intrinsic brain networks compared to the traditional DBN.

2 Methods

2.1 The PKG-DBN

DBN can be viewed as a deep neural network composed by stacking multiple restricted Boltzmann machines (RBMs). To construct PKG-DBN, we first design prior knowledge guided RBM (PKG-RBM). RBM is energy based generative model which approximates a closed-form representation of the essential probability distribution of the input data. Supposing visible variables and hidden variables of an RBM are $\boldsymbol{v}$ and $\boldsymbol{h}$ respectively, and the energy of RBM for continuous inputs can be formulated as follows:

$$E(\boldsymbol{v}, \boldsymbol{h}) = -\boldsymbol{b}^T\boldsymbol{v} - \boldsymbol{a}^T\boldsymbol{h} - \boldsymbol{v}^T\boldsymbol{W}\boldsymbol{h} \tag{1}$$

where $E(\boldsymbol{v}, \boldsymbol{h})$ follows Gaussian-Bernoulli distribution. $\boldsymbol{a}$ and $\boldsymbol{b}$ denotes visible bias and hidden bias. $\boldsymbol{W}$ is the weight between adjacent RBM layers.

Given the model $\boldsymbol{\theta} = (\boldsymbol{a}, \boldsymbol{b}, \boldsymbol{W})$, RBM maximizes the likelihood of the probability of input data as follows:

$$\max_{\boldsymbol{\theta}} lnP(\boldsymbol{v}|\boldsymbol{\theta}) = \max_{\boldsymbol{\theta}} ln\frac{1}{Z}\sum_{\boldsymbol{h}} e^{-E(\boldsymbol{v},\boldsymbol{h})} \tag{2}$$

where Z is the normalizing constant partition function:

$$Z = \sum_{v} \sum_{h} e^{-E(v,h)} \tag{3}$$

By adjusting the model parameter $\boldsymbol{\theta}$, the probability of the input data can be raised.

Algorithm 1: Algorithm for solving the PKG-RBM

Input: Training data $\boldsymbol{X}$, batch size B, epoch number E, learning rate ε, weight decay rate β, the mismatch threshold ϵ_1 and ϵ_2, the Lagrange multiplier β_1 and β_2, the sparsity regularization parameter λ.

Output: $\boldsymbol{\theta}$

Initialize $\boldsymbol{W}, \boldsymbol{a}, \boldsymbol{b}, \boldsymbol{A}_r$ to $\mathrm{N}(0, 0.01)$, initialize $\boldsymbol{A}_t$ and $\boldsymbol{W}_{pre}$ as identity matrix

for $e = 1{:}E$:

  for $b = 1{:}B$:

    Sample batch data $\boldsymbol{X}_{train\_batch}$ from $\boldsymbol{X}$

    Calculate $\boldsymbol{W}$ by $\boldsymbol{W} \leftarrow \boldsymbol{W} + \varepsilon(\langle \boldsymbol{vh}^T\rangle_{data} - \langle \boldsymbol{vh}^T\rangle_{recon})$

    if $\left\|\boldsymbol{W}_{pre}\boldsymbol{W}_t - \boldsymbol{W}_{ct}\boldsymbol{A}_t\right\|_F^2 \le \varepsilon_1$ and $\left\|\boldsymbol{W}_{pre}\boldsymbol{W}_r - \boldsymbol{W}_c\boldsymbol{A}_r\right\|_F^2 + \lambda\|\boldsymbol{A}_r\|_{1,1} \le \varepsilon_2$:

      Update $\boldsymbol{W}$ by $\boldsymbol{W} \leftarrow \boldsymbol{W} + \varepsilon(\langle \boldsymbol{vh}^T\rangle_{data} - \langle \boldsymbol{vh}^T\rangle_{recon})$

    elif $\left\|\boldsymbol{W}_{pre}\boldsymbol{W}_t - \boldsymbol{W}_{ct}\boldsymbol{A}_t\right\|_F^2 > \varepsilon_1$ and $\left\|\boldsymbol{W}_{pre}\boldsymbol{W}_r - \boldsymbol{W}_c\boldsymbol{A}_r\right\|_F^2 + \lambda\|\boldsymbol{A}_r\|_{1,1} \le \varepsilon_2$:

      Update $\boldsymbol{W}$ by $\boldsymbol{W}_t \leftarrow \boldsymbol{W}_t + \varepsilon(\langle \boldsymbol{vh}_t^T\rangle_{data} - \langle \boldsymbol{vh}_t^T\rangle_{recon}) - \boldsymbol{\beta}_1(2\boldsymbol{W}_{pre}^T\boldsymbol{W}_{pre}\boldsymbol{W}_t - 2\boldsymbol{W}_{pre}^T\boldsymbol{W}_{ct}\boldsymbol{A}_t)$,

      $\boldsymbol{W}_r \leftarrow \boldsymbol{W}_r + \varepsilon(\langle \boldsymbol{vh}_r^T\rangle_{data} - \langle \boldsymbol{vh}_r^T\rangle_{recon})$

      where $\boldsymbol{h}_t \in \mathbb{R}^{c\times 1}$ and $\boldsymbol{h}_r \in \mathbb{R}^{(k-c)\times 1}$ is the first c row and the last $(k - c)$ row of $\boldsymbol{h}$, respectively.

      Update $\boldsymbol{A}_t$ by $\boldsymbol{A}_{t_{ii}} \leftarrow \boldsymbol{A}_{t_{ii}} - \boldsymbol{\beta}_1(2\boldsymbol{A}_{t_{ii}}(\boldsymbol{W}_{ct:i})^T\boldsymbol{W}_{ct:i} - 2(\boldsymbol{W}_{ct:i})^T\boldsymbol{W}_{pre}\boldsymbol{W}_{t:i})$

      where $\boldsymbol{A}_{t_{ii}}$ is the i-th diagonal element of $\boldsymbol{A}_t$. $\boldsymbol{W}_{ct:i}$ and $\boldsymbol{W}_{t:i}$ is the i-th column of $\boldsymbol{W}_{ct}$ and $\boldsymbol{W}_t$, respectively.

    elif $\left\|\boldsymbol{W}_{pre}\boldsymbol{W}_t - \boldsymbol{W}_{ct}\boldsymbol{A}_t\right\|_F^2 \le \varepsilon_1$ and $\left\|\boldsymbol{W}_{pre}\boldsymbol{W}_r - \boldsymbol{W}_c\boldsymbol{A}_r\right\|_F^2 + \lambda\|\boldsymbol{A}_r\|_{1,1} > \varepsilon_2$:

      Update $\boldsymbol{W}$ by $\boldsymbol{W}_t \leftarrow \boldsymbol{W}_t + \varepsilon(\langle \boldsymbol{vh}_t^T\rangle_{data} - \langle \boldsymbol{vh}_t^T\rangle_{recon})$,

      $\boldsymbol{W}_r \leftarrow \boldsymbol{W}_r + \varepsilon(\langle \boldsymbol{vh}_r^T\rangle_{data} - \langle \boldsymbol{vh}_r^T\rangle_{recon}) - \boldsymbol{\beta}_2(2\boldsymbol{W}_{pre}^T\boldsymbol{W}_{pre}\boldsymbol{W}_r - 2\boldsymbol{W}_{pre}^T\boldsymbol{W}_c\boldsymbol{A}_r)$

      Update $\boldsymbol{A}_r$ by $\boldsymbol{A}_r \leftarrow \boldsymbol{A}_r - \beta_2(2\boldsymbol{W}_c^T\boldsymbol{W}_c\boldsymbol{A}_r - 2\boldsymbol{W}_c^T\boldsymbol{W}_{pre}\boldsymbol{W}_r + \lambda sign(\boldsymbol{A}_r))$

    else:

      Update $\boldsymbol{W}$ by $\boldsymbol{W}_t \leftarrow \boldsymbol{W}_t + \varepsilon(\langle \boldsymbol{vh}_t^T\rangle_{data} - \langle \boldsymbol{vh}_t^T\rangle_{recon}) - \beta_1(2\boldsymbol{W}_{pre}^T\boldsymbol{W}_{pre}\boldsymbol{W}_t - 2\boldsymbol{W}_{pre}^T\boldsymbol{W}_{ct}\boldsymbol{A}_t)$,

      $\boldsymbol{W}_r \leftarrow \boldsymbol{W}_r + \varepsilon(\langle \boldsymbol{vh}_r^T\rangle_{data} - \langle \boldsymbol{vh}_r^T\rangle_{recon}) - \boldsymbol{\beta}_2(2\boldsymbol{W}_{pre}^T\boldsymbol{W}_{pre}\boldsymbol{W}_r - 2\boldsymbol{W}_{pre}^T\boldsymbol{W}_c\boldsymbol{A}_r)$

      Update $\boldsymbol{A}_t$ by $\boldsymbol{A}_{t_{ii}} \leftarrow \boldsymbol{A}_{t_{ii}} - \boldsymbol{\beta}_1(2\boldsymbol{A}_{t_{ii}}(\boldsymbol{W}_{ct:i})^T - 2(\boldsymbol{W}_{ct:i})^T\boldsymbol{W}_{pre}\boldsymbol{W}_{t:i})$

      Update $\boldsymbol{A}_r$ by $\boldsymbol{A}_r \leftarrow \boldsymbol{A}_r - \beta_2(2\boldsymbol{W}_c^T\boldsymbol{W}_c\boldsymbol{A}_r - 2\boldsymbol{W}_c^T\boldsymbol{W}_{pre}\boldsymbol{W}_r + \lambda sign(\boldsymbol{A}_r))$

    Update $\boldsymbol{a}$ by $\boldsymbol{a} \leftarrow \boldsymbol{a} + \varepsilon(\langle \boldsymbol{v}\rangle_{data} - \langle \boldsymbol{v}\rangle_{recon})$

    Update $\boldsymbol{b}$ by $\boldsymbol{b} \leftarrow \boldsymbol{b} + \varepsilon(\langle \boldsymbol{h}\rangle_{data} - \langle \boldsymbol{h}\rangle_{recon})$

  end

end

In a previous study, Hu et al. [22] have demonstrated that each column of the learnt weight matrix of RBM can be viewed as the time course of an intrinsic brain network.

Inspired by this, the basic idea of PKG-RBM algorithm is to restrict the learnt weight matrix of RBM to be related to some pre-defined brain activities based on prior knowledge. In this paper, we hypothesize brain activities of FBNs should be related to the task paradigm for task fMRI. With this hypothesis, we enforce part of time courses learnt from RBM to be task related (in either positive or negative) and the rest to be linear combinations of task-related components.

In this paper, we formulate the energy function of PKG-RBM as follows:

$$E(\boldsymbol{v}, \boldsymbol{h}) = -\boldsymbol{b}^T \boldsymbol{v} - \boldsymbol{a}^T \boldsymbol{h} - \boldsymbol{v}^T \boldsymbol{W} \boldsymbol{h} \tag{4}$$

where $\boldsymbol{W} = [\boldsymbol{W_t}, \boldsymbol{W_r}] \in \mathbb{R}^{t \times k}$. t is the number of visible unit which equals to the length of task time series. k is the number of hidden unit which equals to the number of FBN to be inferred, $\boldsymbol{W_{ct}} \in \mathbb{R}^{t \times c}$ is the task design matrix whose column is a task timeseries. $\boldsymbol{A_t} \in \mathbb{R}^{c \times c}$ is a diagonal matrix where each diagonal element, which can be learnt during the training procedure, represents the amplitude of the corresponding task timeseries. $\boldsymbol{W_c} \in \mathbb{R}^{t \times (n \times c)}$ is the task related matrix in which each column is either a task time series or the transformation, i.e., integral, differential, and delay, of a task time series. n denotes the number of transformation type. $\boldsymbol{A_r} \in \mathbb{R}^{(n \times c) \times (k-c)}$ is the coefficient matrix which can be learnt during the training procedure. The elements of each column of $\boldsymbol{A_r}$ are the linear combination coefficients of the time series in $\boldsymbol{W_c}$. We add an l_1 regularization to yield a sparse resolution of the column of $\boldsymbol{A_r}$. Notably, instead of forcing the learnt weight matrix of RBM to be task time series and the linear combination of their transformations strictly, we model the mismatch between the learnt weight with task time series and the linear combination of their transformations, i.e., $||\boldsymbol{W_t} - \boldsymbol{W_{ct}} \boldsymbol{A_t}||_F^2$ and $||\boldsymbol{W_r} - \boldsymbol{W_c} \boldsymbol{A_r}||_F^2$. By doing this, the advantages of the data-driven nature of traditional RBM method can be kept. ε_1 and ε_2 are mismatch thresholds which can be adjusted. λ is the regularization parameter which can balance the mismatch residual and the sparsity of $\boldsymbol{A_r}$.

Similar to the traditional DBN, PKG-RBM is also constructed by stacking the PKG-RBM. However, the task design matrix $\boldsymbol{W_{ct}}$ of deeper PKG-RBMs cannot be simply set as task time series as the first PKG-RBM layer. This is because the input of a deeper PKG-RBM is the output of the PKG-RBM layer before instead of the original fMRI data. Columns of the learnt weight matrix of deeper PKG-RBMs cannot be viewed as time courses of intrinsic brain networks. In [23], researchers used $\boldsymbol{W}_1 \times \boldsymbol{W}_2$ as time series of intrinsic brain networks inferred by the second RBM and $\boldsymbol{W}_1 \times \boldsymbol{W}_2 \times \boldsymbol{W}_3$ as time series of intrinsic brain networks inferred by the third RBM, where $\boldsymbol{W}_1$, $\boldsymbol{W}_2$, and $\boldsymbol{W}_3$ are the learnt weight matrix of the first, second, and third RBM, respectively. Inspired by this, for deeper layers, we calculate the mismatch between $\boldsymbol{W_{pre}} \boldsymbol{W_t}$ and task time series or the linear combination of their transformations, where $\boldsymbol{W_{pre}}$ is the identity matrix for the first PKG-RBM layer and $\boldsymbol{W}_1$ for the second PKG-RBM layer, $\boldsymbol{W}_1 \times \boldsymbol{W}_2$ for the third PKG-RBM layer, and so on. Thus, the PKG-RBM can be reformulated as follows:

$$\begin{aligned} & E(\boldsymbol{v}, \boldsymbol{h}) = -\boldsymbol{b}^T \boldsymbol{v} - \boldsymbol{a}^T \boldsymbol{h} - \boldsymbol{v}^T \boldsymbol{W} \boldsymbol{h} \\ & s.t \left|\left|\boldsymbol{W_{pre}} \boldsymbol{W_t} - \boldsymbol{W_{ct}} \boldsymbol{A_t}\right|\right|_F^2 \leq \varepsilon_1, \\ & \left|\left|\boldsymbol{W_{pre}} \boldsymbol{W_r} - \boldsymbol{W_c} \boldsymbol{A_r}\right|\right|_F^2 + \lambda ||\boldsymbol{A_r}||_{1,1} \leq \varepsilon_2 \end{aligned} \tag{5}$$

Thus, given the model $\boldsymbol{\theta} = (\boldsymbol{a}, \boldsymbol{b}, \boldsymbol{W})$, the PKG-RBM maximizes the likelihood of the probability of input data as follows:

$$\begin{aligned} &\max_{\boldsymbol{\theta}} lnP(\boldsymbol{v}|\boldsymbol{\theta}) = \max_{\boldsymbol{\theta}} ln\frac{1}{Z}\sum_{\boldsymbol{h}} e^{-E(\boldsymbol{v},\boldsymbol{h})} \\ &s.t\left|\left|\boldsymbol{W}_{\mathbf{pre}}\boldsymbol{W}_{\mathbf{t}} - \boldsymbol{W}_{\mathbf{ct}}\boldsymbol{A}_{\mathbf{t}}\right|\right|_F^2 \le \varepsilon_1, \\ &\left|\left|\boldsymbol{W}_{\mathbf{pre}}\boldsymbol{W}_{\mathbf{r}} - \boldsymbol{W}_{\mathbf{c}}\boldsymbol{A}_{\mathbf{r}}\right|\right|_F^2 + \lambda||\boldsymbol{A}_{\mathbf{r}}||_{1,1} \le \varepsilon_2 \end{aligned} \quad (6)$$

where Z is the same as Eq. (3). In this paper, we use Lagrange multipliers and Karush-Kuhb-Tucker condition to optimize Eq. (6). The algorithm for solving the proposed PKG-RBM is shown in Algorithm 1.

2.2 Spatial Overlap Rate Between FBNs

To quantitatively compare the results of our proposed methods with other methods, we evaluate the similarity of FBNs derived by different methods by calculating their spatial overlap rate defined in [19]. Specifically, the spatial overlap rate between FBN N^a and N^b derived from method a and method b respectively are calculated as follows:

$$OR\left(N^a, N^b\right) = \frac{\sum_{i=1}^{n}\left|N_i^a \cap N_i^b\right|}{\sum_{i=1}^{n}\left|N_i^a \cup N_i^b\right|} \quad (7)$$

where n is the voxel number of a fMRI volume. This spatial overlap rate calculates the ratio of the area activated by both methods with respect to the area activated by either method.

3 Experimental Results

We applied the proposed PKG-DBN on each individual task fMRI data of HCP 900 for two randomly selected tasks, i.e., gambling and language. We trained PKG-DBN NVIDIA GTX 1080 GPU with learning rate 0.001, weight decay 0.0005, and batch size 200 for 1000 epoch. We set both mismatch threshold ϵ_1 and ϵ_2 to be 0.01. The Lagrange multiplier β_1, β_2, and the sparsity regularization parameter λ is set to be 0.01, 0.01, and 0.05, respectively. We will release the code after this work getting published since the review process is double blind.

We first compared the training loss of PKG-DBN with the traditional DBN by applying the traditional DBN on the same data and with the same experiment setting. The training losses of each RBM layer of PKG-DBN and traditional DBN of each epoch for a randomly selected subject are plotted in Fig. 2. The mismatches between the learnt task FBN time courses by PKG-DBN and task paradigms are also plotted.

As can be seen, for the first RBM layer, the training loss of PKG-DBN decreased faster than that of traditional DBN for the first few epochs. This indicates the prior knowledge from task paradigm can help DBN converge faster at the beginning of training during FBN inference for task fMRI data. However, the traditional DBN can obtain lower data reconstruction loss than PKG-DBN. Notably, the mismatches between the learnt task FBN time courses and task paradigm of PKG-DBN decreased at beginning and then increased during training procedure. This indicates the representation ability of the training model of task FBNs can become better and then become worse along training. The turning point of the mismatch loss always appears at the very beginning of the training procedure and usually before the training loss getting converged of the traditional DBN. Thus, though traditional DBN can obtain lower loss after getting converged, the model can be overfitted. Considering the input data can be different for the deeper layer of PKG-DBN and traditional DBN, the values of training loss are not comparable for these two models. However, it can still be seen, and the PKG-DBN can converge faster than the traditional DBN.

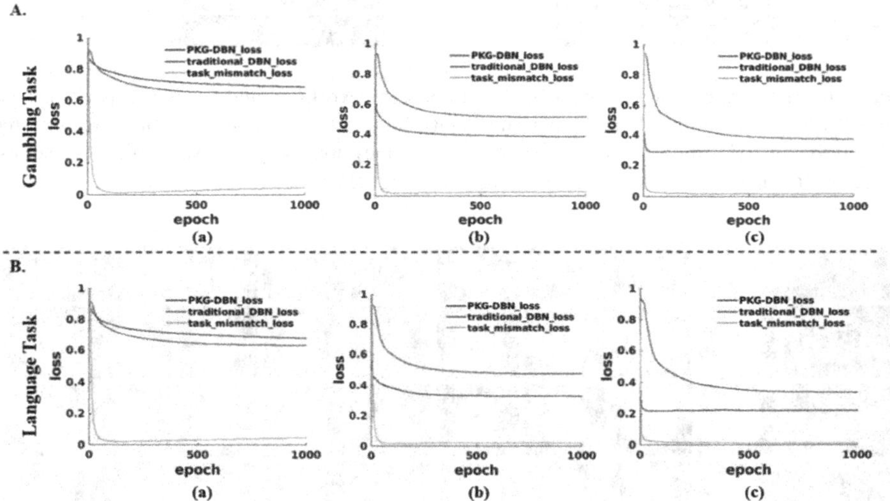

Fig. 2. The training loss of traditional DBN (red line) and PKG-DBN (blue line), and the task paradigm mismatch loss of PKG-DBN (yellow line) for the gambling (A) and language (B) task for a randomly selected subject. (a), (b), and (c) shows the loss plot for the first, second, and the third layer, respectively (Color figure online).

The task FBNs and corresponding learnt time series inferred by the PKG-DBN of a randomly selected subject are visualized (Fig. 3). In this work, the learnt weights of PKG-DBN can have clear and specific correspondence with task paradigms with the guidance of the prior knowledge, i.e. columns of $\boldsymbol{W}_{\mathbf{pre}}\boldsymbol{W}_{\mathbf{t}}$ corresponds to task paradigms specified in $\boldsymbol{W}_{\mathbf{ct}}$. Corresponding task FBNs can be generated by using $\boldsymbol{W}_{\mathbf{pre}}\boldsymbol{W}_{\mathbf{t}}$ and the input data. It can be seen that task FBNs can be consistently extracted from all three layers of PKG-DBN for both tasks. The time series of corresponding task FBNs are clear and their patterns are very similar to task paradigm. This result indicates that PKG-DBN can successfully learn the task paradigm by using our method. The task FBNs can successfully be inferred (Fig. 4).

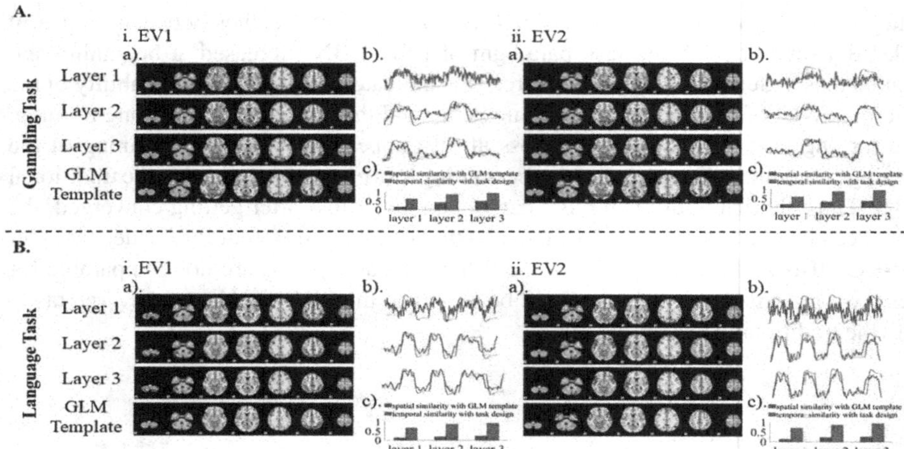

Fig. 3. Identified hierarchical task related functional brain works (a) and their corresponding time series (b) of the gambling (A) and emotion (B) task for a randomly selected subject. (c) Inferred task FBNs' spatial similarity with GLM template (blue) and their corresponding time series' temporal similarity with task paradigms (red) (Color figure online).

Fig. 4. Identified RSNs of the gambling and emotion task for an exemplar subject.

The inferred intrinsic brain network from a randomly selected subject is also shown (Fig. 3). It can also be seen that nine typical resting state networks (RSNs) well defined by previous studies (Smith, Fox et al. 2009, Laird, Fox et al. 2011) can be consistently extracted by the proposed PKG-DBN (Fig. 3) for both tasks, including 'visual network' (RSN 1–3), 'default mode network' (RSN 4), 'sensori-motor network' (RSN 5), 'auditory network' (RSN 6), 'executive control network' (RSN 7), 'frontoparietal network' (RSN 8, 9). These results suggest the effectiveness of the proposed PKG-DBN method.

We also compared the inferred RSNs between the PKG-DBN and the traditional DBN quantitatively. We calculated the mean and the standard deviation of overlap rate between DBN inferred RSNs and their corresponding templates across subjects. This was done for both traditional DBN and the proposed PKG-DBN (Table 1). It can be seen that the RSNs inferred by the PKG-DBN have larger mean and smaller standard deviation of spatial overlap rate with corresponding RSN templates, compared with the traditional DBN models. This indicates PKG-DBN is capable of inferring better intrinsic RSNs compared to the traditional DBN.

Table 1. The mean and standard deviation of overlap rate between DBN inferred RSNs and their corresponding RSN templates across subjects.

GAMBLING									
	RSN1	RSN2	RSN3	RSN4	RSN5	RSN6	RSN7	RSN8	RSN9
Traditional DBN	0.270 ± 0.045	0.365 ± 0.052	0.281 ± 0.048	0.249 ± 0.031	0.196 ± 0.046	0.248 ± 0.069	0.152 ± 0.052	0.294 ± 0.087	0.294 ± 0.062
PKG-DBN	**0.386 ± 0.023**	**0.497 ± 0.031**	**0.389 ± 0.026**	**0.317 ± 0.010**	**0.244 ± 0.032**	**0.308 ± 0.038**	**0.175 ± 0.023**	**0.346 ± 0.048**	**0.388 ± 0.035**
LANGUAGE									
	RSN1	RSN2	RSN3	RSN4	RSN5	RSN6	RSN7	RSN8	RSN9
Traditional DBN	0.310 ± 0.052	0.349 ± 0.048	0.257 ± 0.042	0.290 ± 0.039	0.196 ± 0.041	0.299 ± 0.058	0.186 ± 0.047	0.303 ± 0.075	0.322 ± 0.065
PKG-DBN	**0.379 ± 0.026**	**0.405 ± 0.029**	**0.289 ± 0.015**	**0.302 ± 0.012**	**0.221 ± 0.028**	**0.367 ± 0.042**	**0.212 ± 0.031**	**0.335 ± 0.046**	**0.374 ± 0.027**

4 Conclusion

In this paper, we proposed a novel PKG-DBN method which identify meaningful hierarchical task FBNs with correspondence and converge faster comparing to traditional DBN. By introducing the prior knowledge to the model, our method can simultaneously leverage the advantages of data-driven approaches and the prior knowledge of task design. The proposed PKG-DBN can learn better intrinsic brain networks compared to the traditional DBN. This indicates the proposed PKG-DBN can learn better representation of probability distribution of the input task fMRI data compared to the traditional DBN. Our work contributed a novel approach of training unsupervised deep models with the guidance of prior knowledge for task fMRI analysis.

Acknowledgements. This work was supported in part by the National Key R&D Program of China under Grant 2020AAA0105702; the National Natural Science Foundation of China under Grant 82060336, 62136004, 62036011, U1801265, 6202781, Guangdong Basic and Applied Basic Research Foundation (2214050008706), Science and Technology Support Project of Guizhou Province under Grant [2021]432, and Guangdong Key Laboratory for Biomedical Measurements and Ultrasound Imaging, Health Science Center, School of Biomedical Engineering, Shenzhen University, Shenzhen 518060, China.

References

1. Just, M.A., et al.: Functional and anatomical cortical underconnectivity in autism: evidence from an FMRI study of an executive function task and corpus callosum morphometry. **17**(4), 951–961 (2007)
2. Logothetis, N. K. J. N. (2008). "What we can do and what we cannot do with fMRI." 453(7197): 869–878
3. Friston, K.J.J.S.: Modalities, modes, and models in functional neuroimaging. **326**(5951), 399–403 (2009)
4. Barch, D.M., et al.: Function in the human connectome: task-fMRI and individual differences in behavior. Neuroimage **80**, 169–189 (2013)
5. Lv, J., et al.: Holistic atlases of functional networks and interactions reveal reciprocal organizational architecture of cortical function. **62**(4), 1120–1131 (2014)
6. Cheng, H.-J., et al.: Task-related brain functional network reconfigurations relate to motor recovery in chronic subcortical stroke. **11**(1), 1–12 (2021)
7. Kucyi, A., et al. (2021). Prediction of stimulus-independent and task-unrelated thought from functional brain networks. **12**(1), 1–17
8. Friston, K.J., et al.: Statistical parametric maps in functional imaging: a general linear approach. **2**(4), 189–210 (1994)
9. Friston, K.J., et al.: Event-related fMRI: characterizing differential responses. **7**(1), 30–40 (1998)
10. Beckmann, C.F., et al.: General multilevel linear modeling for group analysis in FMRI. **20**(2), 1052–1063 (2003)
11. Poline, J.-B., Brett, M.J.N.: The general linear model and fMRI: does love last forever? **62**(2), 871–880 (2012)
12. Eklund, A., et al.: A Bayesian heteroscedastic GLM with application to fMRI data with motion spikes. **155**, 354–369 (2017)
13. De Martino, F., et al.: Combining multivariate voxel selection and support vector machines for mapping and classification of fMRI spatial patterns. **43**(1), 44–58 (2008)
14. Song, S., et al.: Comparative study of SVM methods combined with voxel selection for object category classification on fMRI data. **6**(2), e17191 (2011)
15. Calhoun, V.D., et al.: ICA of functional MRI data: an overview. In: Proceedings of the International Workshop on Independent Component Analysis and Blind Signal Separation, Citeseer (2003)
16. Calhoun, V.D., et al.: A review of group ICA for fMRI data and ICA for joint inference of imaging, genetic, and ERP data. **45**(1), S163-S172 (2009)
17. Lee, K., et al.: A data-driven sparse GLM for fMRI analysis using sparse dictionary learning with MDL criterion. **30**(5), 1076–1089 (2010)
18. Lv, J., et al.: Sparse representation of whole-brain fMRI signals for identification of functional networks. **20**(1), 112–134 (2015)
19. Dong, Q., et al.: Modeling Hierarchical Brain Networks via Volumetric Sparse Deep Belief Network (VS-DBN). IEEE Transactions on Biomedical Engineering (2019)
20. Zhang, W., et al.: Hierarchical organization of functional brain networks revealed by hybrid spatiotemporal deep learning. **10**(2), 72–82 (2020)
21. Zhang, S., et al.: Discovering hierarchical common brain networks via multimodal deep belief network. **54**, 238–252 (2019)
22. Hu, X., et al.: Latent source mining in FMRI via restricted Boltzmann machine. Hum. Brain Mapp. **39**(6), 2368–2380 (2018)

Interpretable Signature of Consciousness in Resting-State Functional Network Brain Activity

Antoine Grigis[1(✉)], Chloé Gomez[1,2], Vincent Frouin[1], Lynn Uhrig[2], and Béchir Jarraya[2,3]

[1] Université Paris-Saclay, CEA, NeuroSpin, 91191 Gif-sur-Yvette, France
antoine.grigis@cea.fr

[2] Cognitive Neuroimaging Unit, Institut National de la Santé et de la Recherche Médicale U992, Gif-sur-Yvette, France

[3] Neurosurgery Department, Foch Hospital, University of Versailles, Université Paris-Saclay, Suresnes, France

Abstract. A major challenge in medicine is the rehabilitation of brain-injured patients with poor neurological outcomes who experience chronic impairment of consciousness, termed minimally conscious state or vegetative state. Resting-state functional Magnetic Resonance Imaging (rs-fMRI) is easy-to-acquire and holds the promise of large-range biomarkers. Previous rs-fMRI studies in monkeys and humans have highlighted that different consciousness levels are characterized by the relative prevalence of different functional connectivity patterns - also referred to as brain states - which conform closely to the underlying structural connectivity. Results suggest that changes in consciousness lead to changes in connectivity patterns, not only at the level of the co-activation strength between regions but also at the level of entire networks. In this work, a four-stage framework is proposed to identify interpretable spatial signature of consciousness, by i) defining brain regions of interest (ROIs) from atlases, ii) filtering and extracting the time series associated with these ROIs, iii) recovering disjoint networks and associated connectivities, and iv) performing pairwise non-parametric tests between network activities grouped by acquisition conditions. Our approach yields tailored networks, spatially consistent and symmetric. They will be helpful to study spontaneous recovery from disorders of consciousness known to be accompanied by a functional restoration of several networks.

Keywords: Disorders of consciousness · Interpretability · Explainability · Resting-state functional MRI

Supplementary Information The online version contains supplementary material available at https://doi.org/10.1007/978-3-031-16431-6_25.

L. Wang et al. (Eds.): MICCAI 2022, LNCS 13431, pp. 261–270, 2022.
https://doi.org/10.1007/978-3-031-16431-6_25

1 Introduction

A major challenge in medicine is the rehabilitation of severely brain-injured patients. The challenge culminates when it comes to poor neurological outcome patients who experience chronic impairment of consciousness, termed Minimally Conscious State (MCS) or Vegetative State (VS). Although substantial progress has been made toward a more precise diagnosis of Disorders of Consciousness (DoC), there are no validated therapeutic interventions to ameliorate the consciousness state of these patients. From a theoretical point of view, the Global Neuronal Workspace (GNW) is a model of conscious access, which stipulates that a piece of information becomes conscious when it is available to a widely distributed prefronto-parietal and cingulate cortical networks [4,7–9]. Decoding these levels of consciousness using resting-state functional Magnetic Resonance Imaging (rs-fMRI) is a major challenge in neuroscience. Neuroimaging studies have demonstrated that the GNW is disorganized in patients with disorders of consciousness [11,14,18], and that spontaneous recovery from disorders of consciousness is accompanied by a functional restoration of a broad fronto-parietal network [16] and its cortico-thalamo-cortical connections [15].

A large section of literature has emerged using rs-fMRI to characterize the brain and associated dysfunctions, featuring a wide range of biomarkers [10,20]. rs-fMRI is easy-to-acquire and has gained widespread neuroimaging applications. Instead of a bottom-up investigation starting from molecules and cells, it leverages an information-processing analysis of brain functional imaging. Previous studies in monkeys and humans have highlighted that different consciousness levels are characterized by the relative prevalence of different functional connectivity patterns - also referred to as brain states - which conform fairly closely to the underlying structural connectivity [3,21]. More specifically, a set of seven brain states corresponding to different functional configurations of the brain at a given time are approximated by using sliding windows, and dynamic Functional Connectivity (dFC) (Fig. 1) [1,21]. Going further, each brain state can be characterized by a set of discriminative connections [12]. Results highlight important changes driven by the level of consciousness reflected in the connectivity patterns, not only at the level of the co-activation strength between regions but also at the level of entire networks. However, these interesting findings remain very complex to interpret (see circular saliency maps in Fig. 1). On the other hand, sliding windows synchronization patterns analysis provides an easy-to-interpret physiological proxy of the level of consciousness by investigating the frequency of occurrence of synchronization levels present in the rs-fMRI signal [13]. Reproducing and inspecting the results for the positive and negative peaks (Appendix S1), the only significant difference occurs between the awake state and all anesthetic conditions. These results are in line with the supplementary material presented in [13]. In addition to this low-recognition ability, a brain-wide metric is certainly not adequate to model fine-grain processes which, when reasonably chosen, increase the model interpretability.

In this work, a comprehensive analysis discriminates between different levels of consciousness while providing the essential possibility of a direct interpreta-

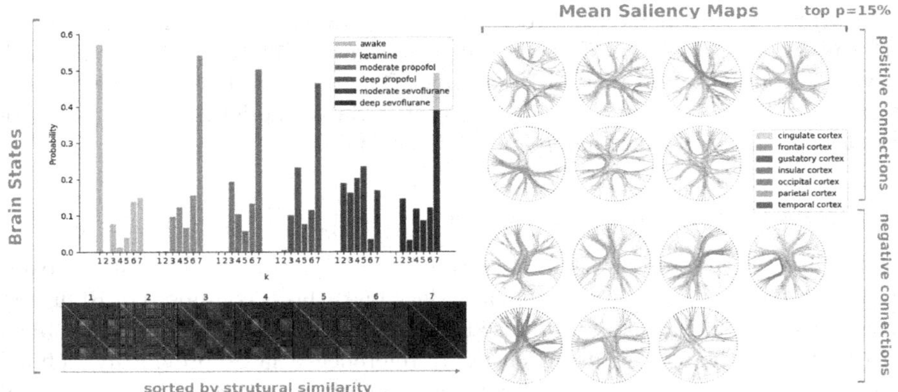

Fig. 1. Illustration of the prevalence of different functional connectivity patterns: a) the brain states derived from dFC matrices ranked according to structural similarity, b) the within-condition probability of occurrence of a state, and c) the within-brain states marker of consciousness depicted by circular saliency maps.

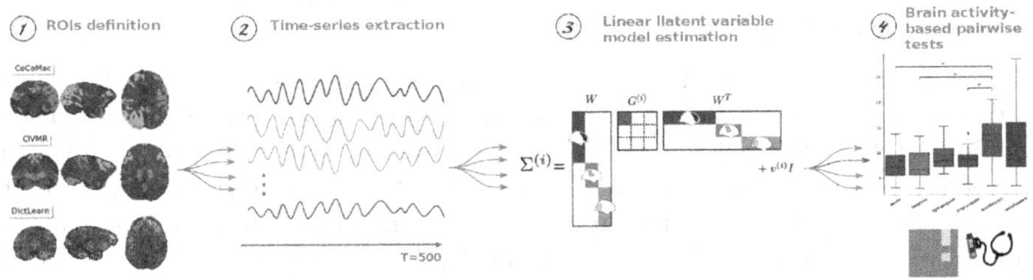

Fig. 2. Illustration of the four steps involved in the proposed rs-fMRI framework applied to characterize different levels of consciousness. Note that boxplot 4 displays the distribution of the functional activities $G^{(i)}$ associated with the network 1 across all acquisition conditions.

tion. In Sect. 2, a four-stage framework is proposed to identify interpretable and robust spatial signatures of consciousness. In Sect. 3, experiments are conducted on a monkey dataset where data have been acquired under different conditions: either in the awake state or loss of consciousness induced by various anesthetics. As a result, we show how network activity can be expressed on an identifiable and disjoint regions basis to facilitate the interpretation and propose a new spatial biomarker of consciousness.

2 Method

Figure 2 depicts our framework for the characterization of different levels of consciousness. It consists of four steps: i) defining brain regions of interest (ROIs) from atlases, ii) filtering and extracting the time series associated with these

ROIs, iii) recovering disjoint networks, and associated connectivities [17], and iv) performing pairwise non-parametric Wilcoxon signed-rank tests between network activities grouped by acquisition conditions. The novelty of this framework lies in the adoption of a constraint linear latent variable model delivering network activity on a basis of identifiable and disjoint ROIs.

2.1 ROIs Definition

The first step of the framework reduces the data dimension by aggregating voxels based on some knowledge encoded in an atlas. The atlas is composed of ROIs and forms a strong a priori of the proposed framework. The atlas selection problem is further investigated by using either i) reference atlases previously defined on other structural or functional datasets sometimes mixed with histological sections and microscopy, or ii) adaptive atlases directly learned from the data [6].

Reference Atlases: we selected two rhesus macaque reference atlases: the CoCoMac atlas, composed of 82 cortical regions (41 cortical regions within each hemisphere) [2], and the CIVMR atlas, composed of a selection of 222 cortical and sub-cortical regions [5].

Atlases Learned from the Data: we use the Online Dictionary Learning in nilearn, shortly named DictLearn in the sequel, as an exploratory method to define ROIs directly from rs-fMRI. A comprehensive analysis of different state-of-the-art strategies ranks the DictLearn method among the best with high robustness and accuracy [6]. The analysis is conducted on a mask defined as the cortical/sub-cortical spatial support of the CIVMR atlas. Because we want to study the connectivity between spatially contiguous ROIs, a connected components analysis is further applied to generate 246 adaptive ROIs. The performance of the atlas estimation is not evaluated in the cross-validation scheme. Thus, the learned atlas is considered - like the CoCoMac and CIVMR atlases - as a priori knowledge of the proposed framework.

2.2 Time-Series Filtering and Extraction

Let $I^{(i)} \in \mathbb{R}^{x \times y \times z \times t}, i \in [1, N]$ be a rs-fMRI time-series volume which is collected across a cohort of N subjects. x, y, and z represent the spatial dimension of each volume, and t is the number of time points. Let p be the fixed number of ROIs defined by the atlas. The feature matrix $X^{(i)} \in \mathbb{R}^{p \times t}$ contains the average time-series computed across all voxels within each ROI. Regarding the high number of ROIs in some atlases, the intersection between the atlas and the rs-fMRI data is performed in the high-resolution template space. In this way, topology issues, from the down-sampling of the atlas, are avoided (i.e. the vanishing of a region due to high contraction in the deformation field). Specific time-series denoising operations are also applied [3,21]. Specifically, voxel time-series are detrended, filtered with a low-pass (0.05-Hz cutoff), a high-pass (0.0025-Hz cutoff), and a zero-phase fast-Fourier notch (0.03 Hz, to remove an artefactual pure frequency

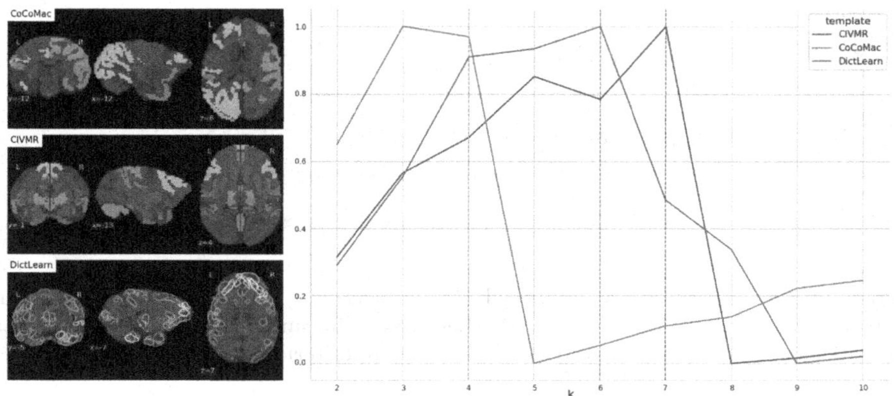

Fig. 3. Illustration of the CoCoMac, CIVMR, and DictLearn atlases, and associated 0–1 normalized log-likelihood on the unseen validation set for $k \in [2, 10]$. The optimal number of networks is represented by a vertical dashed line.

present in all the data) filters, regressed out from motion confounds, and z-score standardized.

2.3 Linear Latent Variable Model Estimation

Linear latent variable model allows the interpretation of the latent variables in terms of functional activity within connectivity networks. In a small number of networks regime, the Modular Hierarchical Analysis (MHA) linear latent variable model has proven to be more efficient than others such as Principal Component Analysis (PCA) or Independent Component Analysis (ICA) [17]. The rs-fMRI features for each subject $X^{(i)}$ is approximated with a stationary multivariate Gaussian distribution, $X^{(i)} \sim \mathcal{N}(0, \Sigma^{(i)})$, where $\Sigma^{(i)} \in \mathbb{R}^{p \times p}$ denotes the functional connectivity over p regions for subject i. Let $Z^{(i)} \in \mathbb{R}^k$ denotes the low-dimensional latent variables taken to follow as well a multivariate Gaussian distribution $Z^{(i)} \sim \mathcal{N}(0, G^{(i)})$. We obtain the following generative model for observed data [17]:

$$z^{(i)} = X^{(i)} | Z^{(i)} \sim \mathcal{N}(W z^{(i)}, v^{(i)} \mathbb{I}) \tag{1}$$

where $G^{(i)} \in \mathbb{R}^{k \times k}$, k denotes the number of disjoint networks, and $v^{(i)} \in \mathbb{R}_+$ denotes the noise. The MHA model captures low-rank covariance structure via the shared across subjects loading matrix W as follows:

$$\Sigma^{(i)} = W G^{(i)} W^T + v^{(i)} \mathbb{I} \tag{2}$$

To compute the model parameters, the optimization maximizes the model log-likelihood $\mathcal{L}$ across all subjects as follows:

$$\hat{W} = \underset{W: W^T W = \mathbb{I}; W \geq 0}{\operatorname{argmax}} \mathcal{L} \tag{3}$$

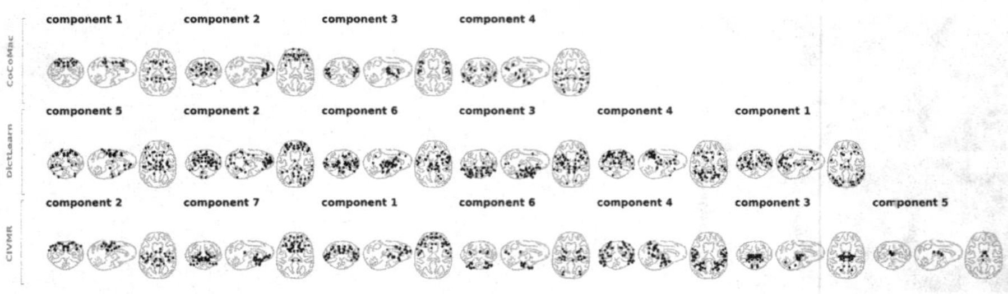

Fig. 4. Inferred networks are composed of sets of unique ROIs represented by their centroids for the CoCoMac ($k = 4$), DictLearn ($k = 6$), and CIVMR ($k = 7$) atlases. Networks are sorted based on the d_{MoC} geometrical criterion.

Over PCA, the MHA model introduces a combination of non-negativity and orthonormal constraints to uncover disjoint networks in W, and associated activities in $G^{(i)}$ for all subjects i. W has a block structure and is uniquely defined and identifiable. W can be interpreted as a shared basis of k non-overlapping networks across all subjects. The only hyper-parameter of the model is the number of networks k that can be determined by maximizing the log-likelihood over unseen data.

To compare models with varying k, a proximity measure between two networks $R \in \mathbb{R}^{m \times 3}$ and $Q \in \mathbb{R}^{n \times 3}$ is essential. Each ROI is represented by its centroid, and m and n denote the number of ROIs in R and Q, respectively. The proximity is defined as the Mean of Closest (MoC) distances between the set of points Q and R:

$$d_{MoC}(Q, R) = \frac{d(Q, R) + d(R, Q)}{2} \tag{4}$$

where $d(Q, R) = mean_{q \in Q} min_{r \in R} ||q - r||$, q and r are points of Q and R, respectively, and $||.||$ is the euclidean norm. To avoid inconsistencies between sets of different sizes, the d_{MoC} proximity is symmetric.

2.4 Statistical Analysis of Functional Network Activity

Each diagonal element of $G^{(i)}$, denoted as functional activities in the sequel, is obtained in an unsupervised manner and represents the extent to which the associated network is expressed in subject i. Samples contain activities grouped by acquisition conditions. Applying the Shapiro-Wilk test reveals that the samples do not meet normal assumptions. Therefore, pairwise non-parametric Wilcoxon signed-rank tests are applied between paired of grouped network activities. The null hypothesis tests if two paired samples come from the same distribution. p-values are adjusted for multiple comparisons using the Benjamini/Yekutieli False Discovery Rate (FDR) correction. Finally, for each network, the pairs of significant tests are encoded in a binary matrix $F \in \mathbb{B}^{c \times c}$, where c is the number of acquisition conditions. F is interpreted as a fingerprint of the underlying consciousness process.

3 Experiments

3.1 Dataset

This study uses the dataset described in [21]. Data were acquired either in the awake state or under different anesthetic conditions (ketamine, propofol, or sevoflurane anesthesia) in five macaque monkeys. Loss of consciousness is thus induced by the injection of anesthetics. Levels of anesthesia were defined by a clinical score (the monkey sedation scale) and continuous electroencephalography monitoring. 156 rs-fMRI runs with corresponding structural MRI were acquired on a Siemens 3-Tesla with a customized single transmit-receiver surface coil. Regarding the limited number of subjects, the index i introduced in Sect. 2.2 represents in the following a run regardless of the acquisition condition or the monitored macaque monkey. The spatial prepossessing is performed with the NSM pipeline described in [19].

3.2 Functional Network Activity

The selection of the optimal number of disjoint networks k in the model is treated as a hyper-parameter tuning. A leave-one-subject-out splitting is performed to generate a train and a validation set. The MHA model is fitted on the train set, and the log-likelihood $\mathcal{L}$ is computed across the unseen validation set. Remember that the input number of regions p is driven by the choice of the atlas. Maximizing $\mathcal{L}$ gives salient $k = 4$, $k = 6$, and $k = 7$ optimal networks for the CoCoMac, DictLearn, and CIVMR atlases, respectively (Fig. 3). For the CoCoMac atlas the $k = 3$, and $k = 4$ configurations are really close, and we choose $k = 4$ as it is a more realistic configuration. In order to visually compare the generated networks the similarity metric d_{MoC} is applied in a bottom-up fashion (see Sect. 2.3). All networks are spatially consistent and symmetric (Fig. 4). Visually, the first networks share common spatial locations while others are tailored to the data.

3.3 Spatial Fingerprints of Consciousness

Keeping the same networks ordering, the most significant functional activity differences between acquisition conditions are highlighted using pairwise statistics (see Fig. 5, and Appendix S2). The fingerprint F summarizes these network-based differences (see Sect. 2.4). In contrast to the method proposed in [13], a wider range of differences is highlighted with a major discrepancy appearing between the awake state and all others anesthetic conditions (the network indicated by a star in Fig. 5, and Appendix S2). Focusing on this network for the CoCoMac atlas (component 1 in Fig. 5), the ROIs underpinning this difference are perfectly symmetric and fits with the macaque GNW theory and its remaining areas CCp, CCa, PCip, FEF, PFCdl, PMCdl, which includes sensory regions M1, S1, V1, A1 [21] (Table 1). While exhibited networks are obtained in an unsupervised - but constrained - manner, they closely support the GNW theory. Contrary to previous works, the networks do not rely on FC or dFC modelization nor some sliding window hyper-parameters.

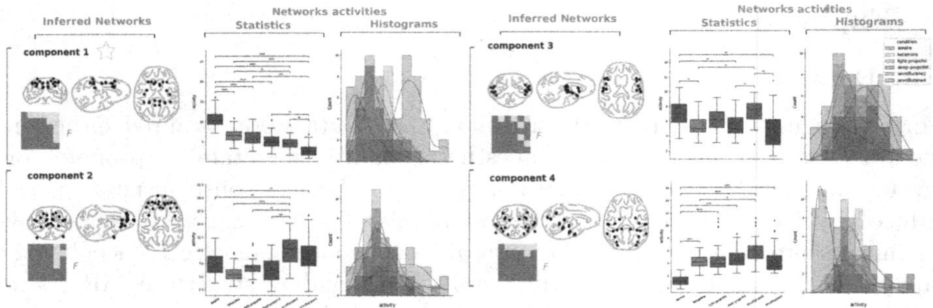

Fig. 5. Inferred networks from the CoCoMac ($k = 4$) atlas, and associated pairwise statistical analysis on network activities. The samples distribution is also displayed, as well as the fingerprint of the underlying consciousness process F.

Table 1. Listing of the network highlighting the difference between the wakefulness and anesthesia-induced loss of consciousness inferred from the CoCoMac atlas (the network indicated by a star in Fig. 5). The detected GNW areas are depicted in blue, and the associated sensory areas in green.

	name	hemi	location
CCp	posterior cingulate cortex	left, right	cingulate cortex
CCa	anterior cingulate cortex	left, right	cingulate cortex
S1	primary somatosensory cortex	left, right	parietal cortex
M1	primary motor cortex	left, right	frontal cortex
PCi	inferior parietal cortex	left, right	parietal cortex
PCm	medial parietal cortex	left, right	parietal cortex
PCip	intraparietal cortex	left, right	parietal cortex
PCs	superior parietal cortex	left, right	parietal cortex
FEF	frontal eye field	left, right	frontal cortex
PMCm	medial premotor cortex	left, right	frontal cortex
PMCdl	dorsolateral premotor cortex	left, right	frontal cortex

4 Discussion and Future Work

In the proposed work, we have presented a four-stage framework to generate a coherent, interpretable, and robust model of consciousness. Despite the small dataset size, discovered networks are tailored, spatially consistent, and symmetric. The optimal number of networks choice depends on the input atlas, but a clear decision appears regarding the log-likelihood. Statistics on networks' activities grouped by levels of consciousness highlight strong differences directly related to interacting ROIs. Contrary to previous works, ROIs linked with the GNW theory are leveraged in an unsupervised manner with no need for a FC, or dFC modelization. Such an interpretable and robust model of functional network brain activity may be helpful to study spontaneous recovery from DOC known to be accompanied by a functional restoration of several networks [15,16]. As a next step, taking advantage of this model, we want to investigate the effect of thalamic DBS as a tool to restore behavioral indices. We are convinced that DBS is a perfect tool for the rehabilitation of brain-injured MCS and VS patients.

References

1. Allen, E.A., Damaraju, E., Plis, S.M., Erhardt, E.B., Eichele, T., Calhoun, V.D.: Tracking whole-brain connectivity dynamics in the resting state. Cerebral Cortex **24**(3), 663–676 (2014). https://doi.org/10.1093/cercor/bhs352
2. Bakker, R., Wachtler, T., Diesmann, M.: Cocomac 2.0 and the future of tract-tracing databases. Front. Neuroinform. **6**, 30 (2012). https://doi.org/10.3389/fninf.2012.00030
3. Barttfeld, P., Uhrig, L., Sitt, J.D., Sigman, M., Jarraya, B., Dehaene, S.: Signature of consciousness in the dynamics of resting-state brain activity. Proc. Natl. Acad. Sci. **112**(3), 887–892 (2015). https://doi.org/10.1073/pnas.1418031112
4. Bernard, J.B., Steven, L.: One, not two, neural correlates of consciousness. Trends Cogn. Sci. **9**(6), 269 (2005). https://doi.org/10.1016/j.tics.2005.04.008
5. Calabrese, E., et al.: A diffusion tensor MRI Atlas of the postmortem rhesus macaque brain. Neuroimage **117**, 408–416 (2015). https://doi.org/10.1016/j.neuroimage.2015.05.072
6. Dadi, K., Abraham, A., Rahim, M., Thirion, B., Varoquaux, G.: Comparing functional connectivity based predictive models across datasets. In: 2016 International Workshop on Pattern Recognition in Neuroimaging (PRNI), pp. 1–4 (2016). https://doi.org/10.1109/PRNI.2016.7552359
7. Dehaene, S., Changeux, J.P.: Experimental and theoretical approaches to conscious processing. Neuron **70**(2), 200–227 (2011). https://doi.org/10.1016/j.neuron.2011.03.018
8. Dehaene, S., Charles, L., King, J.R., Marti, S.: Toward a computational theory of conscious processing. Current Opinion Neurobiol. **25**, 76–84 (2014). https://doi.org/10.1016/j.conb.2013.12.005, theoretical and computational neuroscience
9. Dehaene, S., Kerszberg, M., Changeux, J.P.: A neuronal model of a global workspace in effortful cognitive tasks. Proc. Natl. Acad. Sci. **95**(24), 14529–14534 (1998). https://doi.org/10.1073/pnas.95.24.14529
10. Drysdale, A., et al.: Resting-state connectivity biomarkers define neurophysiological subtypes of depression. Nature Med. **23**, 28–38 (2016). https://doi.org/10.1038/nm.4246
11. Faugeras, F., et al.: Probing consciousness with event-related potentials in the vegetative state. Neurology **77**(3), 264–268 (2011). https://doi.org/10.1212/WNL.0b013e3182217ee8
12. Grigis, A., Tasserie, J., Frouin, V., Jarraya, B., Uhrig, L.: Predicting cortical signatures of consciousness using dynamic functional connectivity graph-convolutional neural networks. bioRxiv (2020). https://doi.org/10.1101/2020.05.11.078535
13. Hahn, G., et al.: Signature of consciousness in brain-wide synchronization patterns of monkey and human fMRI signals. Neuroimage **226**, 117470 (2021). https://doi.org/10.1016/j.neuroimage.2020.117470

14. King, J.R., et al.: Information sharing in the brain indexes consciousness in non-communicative patients. Curr. Biol. **23**(19), 1914–1919 (2013). https://doi.org/10.1016/j.cub.2013.07.075
15. Laureys, S., Faymonville, M., Luxen, A., Lamy, M., Franck, G., Maquet, P.: Restoration of thalamocortical connectivity after recovery from persistent vegetative state. The Lancet **355**(9217), 1790–1791 (2000). https://doi.org/10.1016/S0140-6736(00)02271-6
16. Laureys, S., Lemaitre, C., Maquet, P., Phillips, C., Franck, G.: Cerebral metabolism during vegetative state and after recovery to consciousness. J. Neurolo. Neurosurgery Psychiatry **67**(1), 121–122 (1999). https://doi.org/10.1136/jnnp.67.1.121
17. Monti, R.P., et al.: Interpretable brain age prediction using linear latent variable models of functional connectivity. PLOS ONE **15**(6), 1–25 (2020). https://doi.org/10.1371/journal.pone.0232296
18. Sitt, J.D., et al.: Large scale screening of neural signatures of consciousness in patients in a vegetative or minimally conscious state. Brain J. Neurol. **137**(Pt 8), 2258–2270 (2014). https://doi.org/10.1093/brain/awu141
19. Tasserie, J., Grigis, A., Uhrig, L., Dupont, M., Amadon, A., Jarraya, B.: Pypreclin: an automatic pipeline for macaque functional MRI preprocessing. Neuroimage **207**, 116353 (2020). https://doi.org/10.1016/j.neuroimage.2019.116353
20. Taylor, J.J., Kurt, H.G., Anand, A.: Resting state functional connectivity biomarkers of treatment response in mood disorders: a review. Front. Psychiatry **12** (2021). https://doi.org/10.3389/fpsyt.2021.565136
21. Uhrig, L., et al.: Resting-state dynamics as a cortical signature of anesthesia in monkeys. Anesthesiology **129**(5), 942–958 (2018). https://doi.org/10.1097/ALN.0000000000002336

Nonlinear Conditional Time-Varying Granger Causality of Task fMRI via Deep Stacking Networks and Adaptive Convolutional Kernels

Kai-Cheng Chuang[1,2(✉)], Sreekrishna Ramakrishnapillai[2], Lydia Bazzano[3], and Owen Carmichael[2(✉)]

[1] Medical Physics Graduate Program, Louisiana State University, Baton Rouge, LA, USA
[2] Biomedical Imaging Center, Pennington Biomedical Research Center, Baton Rouge, LA, USA
{Kai.Chuang,Owen.Carmichael}@pbrc.edu
[3] Department of Epidemiology, Tulane School of Public Health and Tropical Medicine, New Orleans, LA, USA

Abstract. Time-varying Granger causality refers to patterns of causal relationships that vary over time between brain functional time series at distinct source and target regions. It provides rich information about the spatiotemporal structure of brain activity that underlies behavior. Current methods for this problem fail to quantify nonlinear relationships in source-target relationships, and require ad hoc setting of relationship time lags. This paper proposes deep stacking networks (DSNs), with adaptive convolutional kernels (ACKs) as component parts, to address these challenges. The DSNs use convolutional neural networks to estimate nonlinear source-target relationships, ACKs allow these relationships to vary over time, and time lags are estimated by analysis of ACKs coefficients. When applied to synthetic data and data simulated by the STANCE fMRI simulator, the method identified ground-truth time-varying causal relationships and time lags more robustly than competing methods. The method also identified more biologically-plausible causal relationships in a real-world task fMRI dataset than a competing method. Our method is promising for modeling complex functional relationships within brain networks.

Keywords: Granger causality · Adaptive convolutional kernels · fMRI

1 Introduction

Effective connectivity refers to the influence that functional activity in one brain system exerts over functional activity in another [1]. It has become an important tool to understand the organization of human neural circuitry underlying perception and cognition in health and disease based on time series data from functional neuroimaging methods such as functional magnetic resonance imaging (fMRI) [2, 3]. Granger causality, the

Supplementary Information The online version contains supplementary material available at https://doi.org/10.1007/978-3-031-16431-6_26.

L. Wang et al. (Eds.): MICCAI 2022, LNCS 13431, pp. 271–281, 2022.
https://doi.org/10.1007/978-3-031-16431-6_26

predominant method for quantifying effective connectivity, assesses the degree to which time series data at a current time point in a target region is predicted by time series data at the current or earlier time points from a different (source) region, after accounting for the influence of other sources and target regions data from previous time points [4–6]. This method has proven useful for clarifying various aspects of brain dynamics [7, 8].

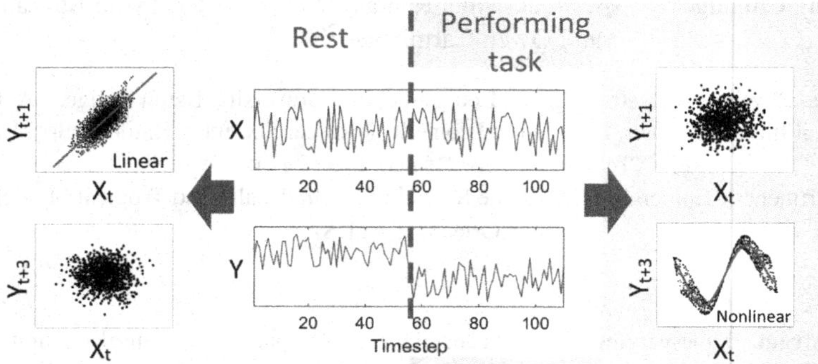

Fig. 1. A hypothetical time-varying causal relationship, where source X_t has a strong linear relationship with target Y_{t+1} during rest (lefthand side), and X_t has a nonlinear relationship with Y_{t+3} during task performance (righthand side). This paper presents a method for automatically discovering such time-varying causal relationships.

Because connectivity relationships between brain regions are believed to change dynamically over the course of task performance [9–11], and even during periods of rest [12], extensions of Granger causality that quantify time-varying causal relationships (Fig. 1) have the potential for high impact. To date, three solutions to this problem have been presented, all based on the vector autoregressive (VAR) model, which allows modeling of linear relationships between source and target [9–11]. Time-varying VAR parameters were estimated using wavelet functions [11], generalized VAR (GVAR) [10], and particle filtering (PF) [9].

This paper seeks to overcome two key limitations of prior time-varying Granger causality methods. First, the prior methods were only able to model linear relationships between source and target signals, thus precluding modeling of nonlinear causal relationships that are expected to arise from complex neural dynamics [7, 13–15]. Second, prior methods for time-varying Granger causality were limited in their ability to handle time lags: the number of timesteps that elapse between causal brain activity at the source and resulting brain activity at the target. One prior method limited the time lag to exactly one time point to reduce computational complexity [9], while the other methods required the user to specify the time lag a priori [10, 11]. This is an important limitation because time lags are not expected to be known a priori. Other methods, including convergent cross mapping [16, 17], dynamic causal model [4], perturbation based methods [18], and neural network based methods [19–21] methods estimated time-varying causality using a sliding temporal window. This sliding window approach has two key limitations. First, the user must specify the size of the sliding window *a priori*, with no clear guidelines for how to do so [9, 22]. Secondly, estimation may be inaccurate when the temporal window is short [16].

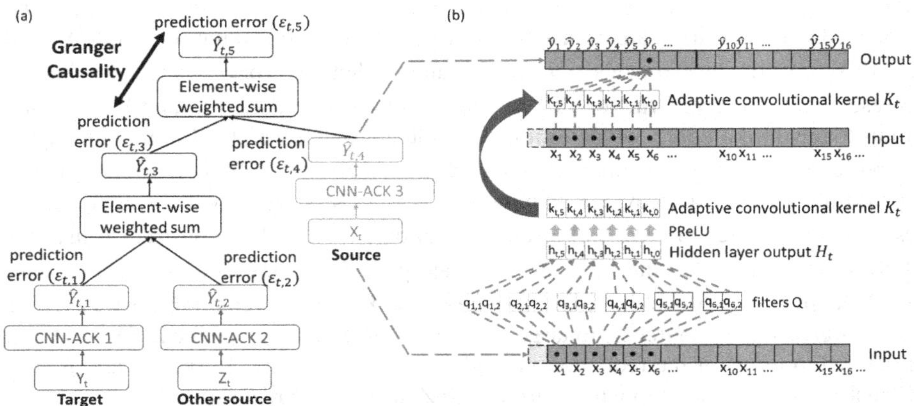

Fig. 2. **(a)** Proposed DSNs with CNN-ACKs to estimate nonlinear time-varying Granger causality between source X_t and target Y_t, conditioned on source Z_t. **(b)** **Top:** Target time series Y_t is modeled in terms of a source time series X_t that is convolved by an ACK K_t, , whose values change over the course of time (t). **Bottom:** K_t is estimated at training time through estimation of six (1 x 2) convolutional filters Q which are applied to the input source time series X_t, followed by application of a PReLU activation function.

Therefore, we propose to use deep stacking networks (DSNs) to overcome these limitations. DSNs allow estimation of nonlinear Granger causality between source (X_t) and target (Y_t), after accounting for the influence of activity in other source regions (Z_t), using convolutional neural network (CNN) modules; stacking multiple such modules allows modeling among multiple sources and targets and each CNN module efficiently capture temporally-localized features between a pair of source and target (Fig. 2 (a)). Within each CNN is an adaptive convolutional kernel (ACK), whose estimated kernel coefficients reveal time-varying causal relationships and time lags at each time point in the time series (Fig. 2 (b)). This approach extends our previous study, which used DSNs to estimate nonlinear Granger causality but was unable to handle time-varying causal relationships [23]. We show that the method identifies time-varying causal relationships, including time-varying time lags, when applied to synthetic datasets and simulated data from a public-domain fMRI simulator. We also show that it provides richer information about causal structures in a real-world task fMRI dataset than traditional non-time-varying causal modeling does.

2 Materials and Methods

2.1 Nonlinear Conditional Time-Varying Granger Causality via DSNs

Given sources X_t and Z_t, and target Y_t, we train a DSNs whose CNNs with ACKs (CNN-ACKs) use X_t, Y_t, and Z_t to reconstruct Y_t. Conditional Granger causality is assessed in terms of how much better X_t reconstructs Y_t, after accounting for how well previous time points of Y_t, along with Z_t, jointly reconstruct Y_t (Fig. 2 (a)). The conditional Granger causality we estimate between X_t and Y_t can be thought of as the "direct causality" between X_t and Y_t after accounting for the "indirect causality" from X_t, to Z_t, to Y_t. First, CNN-ACKs 1 and 2 are trained to transform previous time points of Y_t into Y_t, and Z_t into Y_t, resulting in estimates $\widehat{Y}_{t,1}$ and $\widehat{Y}_{t,2}$ and prediction errors $\varepsilon_{t,1}$ and $\varepsilon_{t,2}$. Then, to represent the best reconstruction of Y_t based on both of Y_t and Z_t, $\widehat{Y}_{t,1}$ and $\widehat{Y}_{t,2}$ provide inputs to the third module, which estimates an element-wise weighted sum of the inputs to predict Y_t, resulting in estimate $\widehat{Y}_{t,3}$ and prediction error $\varepsilon_{t,3}$. To reconstruct Y_t based on X_t, time series X_t is provided as input to CNN-ACK 3, again with Y_t as the target, resulting in predicted time series $\widehat{Y}_{t,4}$ and prediction error $\varepsilon_{t,4}$. To reconstruct Y_t in terms of all of X_t, Y_t, and Z_t jointly, $\widehat{Y}_{t,3}$ and $\widehat{Y}_{t,4}$ are provided as inputs to an element-wise weighted sum to produce the final estimate of Y_t, $\widehat{Y}_{t,5}$, and prediction error $\varepsilon_{t,5}$. The Granger causality of source X_t to target Y_t, conditioned on other source Z_t ($X \rightarrow Y|Z$), is defined in terms of the reduction in modeling error when X_t, Y_t, and Z_t are used to reconstruct Y_t, compared to when only Y_t and Z_t are used to reconstruct Y_t:

$$GC_{indexX \rightarrow Y|Z} = \ln \frac{|\varepsilon_{t,3}|}{|\varepsilon_{t,5}|} \tag{1}$$

If incorporating X_t improves the reconstruction of Y_t after accounting for effects of Y_t and Z_t, $GC_{indexX \rightarrow Y|Z}$ will be a large positive number, providing evidence for conditional Granger causality. Complex causal relationships among several time series can be disentangled by calculating conditional Granger causality with differing assignments of time series to the roles of X_t, Y_t, and Z_t.

Inspired by [24, 25], we used CNN-ACKs in our DSNs architecture to estimate time-varying causal relationships. An ACK is defined by a dynamic filter that changes its weights automatically depending on the data in the input time series. The ACK K_t is generated by convolving filters Q with input time series X_t and using an activation function to transform the result into K_t. The first step is that at each timestep (t), the (1 x 6) hidden layer output $H_t = [h_{t,5}, \ldots, h_{t,0}]$ is calculated as the dot product of six 1 x 2 filter $Q = [q_{1,1}, q_{1,2}; \ldots; q_{6,1}, q_{6,2}]$ with the input time series $X_t = [x_{t-6}, \ldots, x_t]$ (Fig. 2 (b) Bottom).

$$h_{t,5} = \left[q_{1,1}, q_{1,2}\right] \bullet \left[x_{t-6}, x_{t-5}\right]; \ldots; h_{t,0} = \left[q_{6,1}, q_{6,2}\right] \bullet \left[x_{t-1}, x_t\right] \tag{2}$$

Then, the Parametric Rectified Linear Unit (PReLU) activation function is applied to each element of hidden layer output H_t to generate the ACK K_t,

$$K_t = \left[k_{t,5}, \ldots, k_{t,0}\right] = PReLU(H_t) = \left[PReLU_{t,5}\left(h_{t,5}\right), \ldots, PReLU_{t,0}\left(h_{t,0}\right)\right] \tag{3}$$

The coefficients of K_t can be interpreted as evidence of Granger causality at specific time lags at time t of the time series. For example, if $k_{t,5}$ (lag 5 causality) has a large magnitude, it suggests that at time t, there is a causal relationship between the source at time $t - 5$ and the target at time t. The estimate of the target time series, $\widehat{Y}_t$, is the dot product of K_t with the input time series $X_t = [x_{t-5}, \ldots, x_t]$ (Fig. 2 (b) Top).

$$\widehat{Y}_t = K_t^T X_t \tag{4}$$

In each CNN-ACK, the six filters Q (2 weights and 1 bias terms for each filter) and the parameters of PReLU (6 weights for each timestep) are the learnable parameters. We used the TensorFlow and Keras software packages to build our network architecture and optimized it with the Adam optimizer ($\beta 1 = 0.9$, $\beta 2 = 0.999$) with a learning rate of 0.001 to minimize the loss function of mean squared error between the predicted $\widehat{Y}_t$ and the actual Y_t[26, 27].

2.2 Design of Experiments

We applied the proposed method to synthetic time series data, simulated task fMRI data from the public-domain STANCE simulator [28], and a real-world task fMRI dataset. For each synthetic and simulated dataset, 100 X_t, Y_t, Z_t time series triples were generated as described in subsequent sections, and the real-world task fMRI dataset included 100 fMRI scans. For each dataset, ten-fold cross validation was used to repeatedly trains and quantify causal relationships within the testing set of the fold. Conditional Granger causality between source and target, independent of other sources, was considered evident when the mean GC index over the ten folds of cross validation was significantly greater than 0 according to a one-sample t-test (p-value < 0.05). A specific time lag was selected when its time lag causality coefficients over the ten folds were significantly different from 0 via a one-sample t-test. Identified causal relationships were compared to those programmed into the synthetic and simulated data sets, and causal relationships identified in the real-world data were compared to published data about the brain functional underpinnings of the task.

Synthetic Datasets. We designed two synthetic datasets that focused on testing the method's ability to model nonlinear causal relationships whose functional form differed between well-defined epochs of the time series, but whose time lag was constant; and testing the ability to model relationships whose functional form is constant, but whose time lags differ between epochs. In *synthetic dataset 1*: identical time lags in all epochs, each time series had 110 timesteps generated according to the equations in Table 1. In *synthetic dataset 2*: different time lags in different epochs, each time series had 110 timesteps generated according to the equations in Table 2. $N(0,0.1)$ represents Gaussian noise with zero mean and 0.1 standard deviation. The application of the approach to a more complex 5-node network with bi-directional causalities is shown in the Supplemental Material.

Table 1. Synthetic dataset 1: identical time lags in all epochs.

	X-Y relationship	Equation	Timestep (t)
Epoch 1	Linear	$Y_t = 5X_{t-1} + 0.5\text{N}(0, 0.1)$	1–22
Epoch 2	Quadratic	$Y_t = -0.5(X_{t-1})^2 + 0.5\text{N}(0, 0.1)$	23–44
Epoch 3	Exponential	$Y_t = 0.5e^{X_{t-1}} + 0.5\text{N}(0, 0.1)$	45–66
Epoch 4	Cubic	$Y_t = -5(X_{t-1})^3 + 0.5\text{N}(0, 0.1)$	67–88
Epoch 5	None	$Y_t = 0.5\text{N}(0, 0.1)$	89–110

$X_t \sim N(0, 0.1)$; $Z_t \sim N(0, 0.1)$

Table 2. Synthetic dataset 2: different time lags in different epochs.

	X-Y relationship	Equation	Timestep (t)
Epoch 1	Time lag 2	$Y_t = 0.5X_{t-2} + 0.5N(0, 0.1)$	1–55
Epoch 2	Time lag 5	$Y_t = -0.5X_{t-5} + 0.5N(0, 0.1)$	56–110

$X_t \sim N(0, 0.1)$; $Z_t \sim N(0, 0.1)$

Simulated Task fMRI Datasets. For each simulated dataset, triples of 130-timestep time series, each of which contained single-timestep-duration events, were produced, with causal relationships existing between events in one time series, and events in another. Each time series of events was convolved with a canonical hemodynamic response function, followed by addition of simulated system and physiological noise at a magnitude of 1% of the event-related fMRI signal (i.e., a similar noise magnitude to that of real-world task fMRI data [29, 30]). In *simulated dataset 1*: identical time lags in all epochs, each X_t, Y_t, Z_t time series triple was initially generated with 52 randomly placed events. Then, time series Y_t, between time points 1 and 65, was edited so that an event at Y_t was added if there was also an event at X_{t-1}; i.e., X had an excitatory effect on Y during this epoch. Similarly, time series Y_t, between time points 66 and 130, was edited so that an event at Y_t was deleted if there were events at X_{t-1} and Y_t (Table 3); i.e., X had an inhibitory effect on Y. In *simulated dataset 2*: different time lags in different epochs, each X_t, Y_t, Z_t time series triples were generated initially with 52 randomly placed events. Then, as for simulated dataset 1, events in the Y_t were edited to reflect the differing excitatory and inhibitory effects of X_t at differing time lags within each epoch (Table 3).

Table 3. *Simulated dataset 1:* identical time lags in all epochs and *simulated dataset* 2: different time lags in different epochs.

	Simulated dataset 1				*Simulated dataset 2*		
	X-Y relationship	Time lag	Time step (t)		X-Y relationship	Time lag	Time step (t)
Epoch 1	Excitation	1	1–65	Epoch 1	Excitation	1	1–26
Epoch 2	Inhibition	1	66–130	Epoch 2	Inhibition	1	27–52
				Epoch 3	Excitation	3	53–78
				Epoch 4	Inhibition	4	79–104
				Epoch 5	None	-	105–130

Real-World Task fMRI Dataset. We applied the proposed method to task fMRI data collected from the Bogalusa Heart Study [31]. One hundred participants performed a Stroop task during fMRI on a GE Discovery 3T scanner at Pennington Biomedical Research Center. Acquisition of T1-weighted structural MPRAGE and axial 2D gradient echo EPI BOLD fMRI data was described previously [32]. Preprocessing of fMRI included slice timing correction, head motion correction, smoothing, co-registration to the T1-weighted image, and warping of T1-weighted data to a Montreal Neurological Institute (MNI) coordinate. Cardiac and respiratory time series were regressed out of the data using RETROICOR [33]. The regions of interest (ROI) in MNI coordinate previously identified as activated by the Stroop task (fusiform gyrus, occipital gyrus, precuneus, and thalamus) were extracted [34]. The proposed method was applied to all possible assignments of ROIs to the roles of source, target, and other source (i.e., to X_t, Y_t, and Z_t) to explore time-varying Granger causalities. Differences in causal relationships between rest and task conditions were assessed by testing whether the difference in Granger causality coefficients between conditions was significantly different from 0 by permutation testing with 10,000 permutations.

3 Results

Synthetic Datasets. In synthetic dataset 1: identical time lags in all epochs, the proposed method correctly identified the true Granger causality $X \rightarrow Y|Z$ (p-value < 0.0001). The other possible assignments of X_t, Y_t, and Z_t to sources and target had no evidence of Granger causality (minimum p-value $= 0.7437$). In addition, the time dependence of the $X \rightarrow Y|Z$ causal relationship was correctly tracked by the Granger causality coefficients (the coefficients of ACK K_t) that quantified each time lag (Fig. 3 & Fig. 4 (a)). Specifically, Granger causality coefficients corresponding to ground-truth causal relationships (e.g., the lag 1 causality coefficient during epoch 1) were nonzero while other Granger causality coefficients were correctly estimated to be close to the nominal null value. Both GVAR and PF successfully identified the linear causal relationship in the linear epoch as expected; but they failed to identify other causal relationships in other epochs, likely due to the linear assumptions they make. (Fig. 4 (a)). In synthetic dataset 2: different time lags in different epochs, the proposed method correctly identified the true Granger causality $X \rightarrow Y|Z$ (p-value < 0.0001). As above, Granger causality coefficients correctly tracked changes in time lags across epochs (e.g., the low errors in the lag 2 and lag 5 causality coefficients during the epochs they exerted causal effects, Fig. 4 (b)), and other coefficients were correctly estimated to be null. GVAR and PF failed to correctly track time-varying causalities.

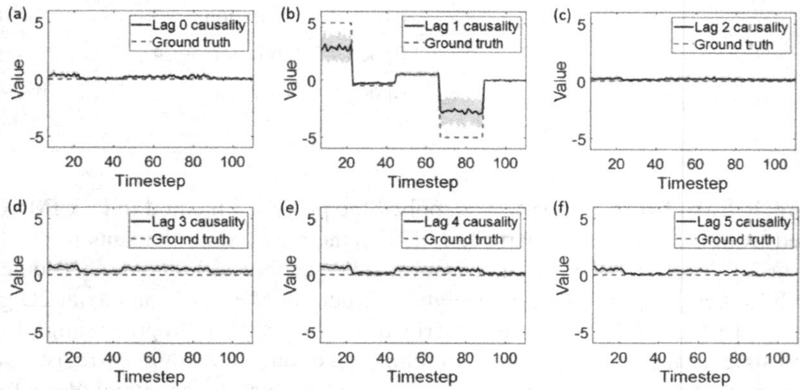

Fig. 3. Time courses of ground-truth Granger causality coefficients (red) and estimates (K_t) from DSN-ACK (mean in black, values within 1 standard deviation of the mean in gray) for differing time lags in synthetic dataset 1.

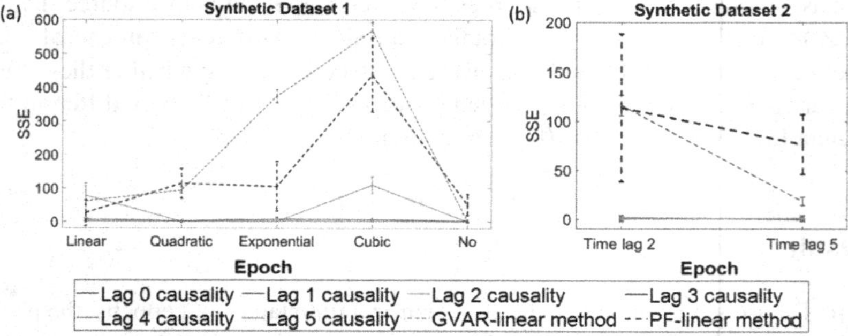

Fig. 4. Sum of squared errors (SSE) differences between ground truth Granger causality coefficients and estimates of time lag 0–5 Granger causality coefficients provided by DSN-ACK, GVAR, and PF for **(a)** synthetic dataset 1 and **(b)** synthetic dataset 2.

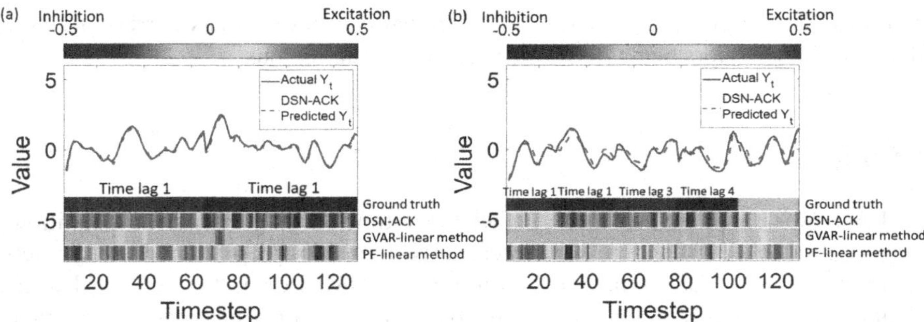

Fig. 5. Time courses of actual time series Y_t and the predicted $\widehat{Y}_t$ by DSN-ACK for simulated dataset 1 **(a)** and 2 **(b)**. Ground truth Granger causality coefficients at time lags relevant to each time series epoch and estimates of those Granger causality coefficients generated by DSN-ACK, GVAR, and PF are shown below the time series via the color bar.

Simulated Task fMRI Datasets. The proposed method correctly identified the true Granger causality $X \rightarrow Y|Z$ in simulated dataset 1 and 2 (p-value = 0.0011 and 0.0015, respectively). Also, the time-varying Granger causality coefficients for $X \rightarrow Y|Z$ were correctly estimated across all epochs by DSN-ACKs (Fig. 5). All the other causality relationships among X_t, Y_t, and Z_t were correctly determined to be null (minimum p-value = 0.1382 and 0.2229 for simulated dataset 1 and 2, respectively). Both GVAR and PF failed to estimate Granger causality coefficients accurately in both simulated datasets.

Real-World Task fMRI Dataset. Lag 0 Granger causality coefficients were estimated to be nonzero among the four Stroop task ROIs during rest and task execution (Supplemental Fig. 2). Reciprocal causal relationships between the occipital gyrus and thalamus were identified independent of the fusiform gyrus, in agreement with previous findings [35, 36]. In addition, a causal relationship from the fusiform gyrus to occipital gyrus, independent of precuneus, has been identified. Reciprocal Granger causalities between the occipital gyrus and precuneus independent of fusiform gyrus have also been identified. No causal relationship has been identified by GVAR and PF. These lag 0 Granger causality coefficients were statistically different between task and rest conditions. This suggests differences in functional causal dynamics corresponding to differences in behaviors being performed by those brain regions.

4 Conclusion

Our DSN-ACK architecture that characterizes time-varying nonlinear conditional Granger causality identifies time-varying causal relationships programmed into synthetic and simulated fMRI data. When applied to real task fMRI data, the method identifies plausible causal brain functional relationships among brain regions that prior methods were unable to identify. The breadth of applicability of the current method includes any data type with multiple time series, both biomedical (EEG, MEG, fNIRS, dynamic PET) and non-biomedical (economic data, geoscience measurements). Future work should extend this approach to account for spatially- and temporally variable hemodynamic response functions that could impact discovery of causal relationships [9, 37].

References

1. Friston, K.: Functional and effective connectivity: a review. Brain connectivity **1**(1), 13–36 (2011)
2. Deshpande, G., et al.: Multivariate Granger causality analysis of fMRI data. Hum. Brain Mapp. **30**(4), 1361–1373 (2009)
3. Seth, A.K., Barrett, A.B., Barnett, L.: Granger causality analysis in neuroscience and neuroimaging. J. Neurosci. **35**(8), 3293–3297 (2015)
4. Friston, K., Moran, R., Seth, A.K.: Analysing connectivity with Granger causality and dynamic causal modelling. Curr. Opin. Neurobiol. **23**(2), 172–178 (2013)
5. Goebel, R., et al.: Investigating directed cortical interactions in time-resolved fMRI data using vector autoregressive modeling and Granger causality mapping. Magn. Reson. Imaging **21**(10), 1251–1261 (2003)
6. Granger, C.W.: Investigating causal relations by econometric models and cross-spectral methods. Econometrica: J. Econometric Society 424–438 (1969)
7. Liao, W., et al.: Kernel Granger causality mapping effective connectivity on fMRI data. IEEE Trans. Med. Imaging **28**(11), 1825–1835 (2009)
8. Zhou, Z., et al.: A conditional Granger causality model approach for group analysis in functional magnetic resonance imaging. Magn. Reson. Imaging **29**(3), 418–433 (2011)
9. Ambrosi, P., et al.: Modeling Brain Connectivity Dynamics in Functional Magnetic Resonance Imaging via Particle Filtering. bioRxiv (2021)
10. Marcinkevičs, R., Vogt, J.E.: Interpretable Models for Granger Causality Using Self-explaining Neural Networks. arXiv preprint arXiv:2101.07600 (2021)
11. Sato, J.R., et al.: A method to produce evolving functional connectivity maps during the course of an fMRI experiment using wavelet-based time-varying Granger causality. Neuroimage **31**(1), 187–196 (2006)
12. Cekic, S., Grandjean, D., Renaud, O.: Time, frequency, and time-varying Granger-causality measures in neuroscience. Stat. Med. **37**(11), 1910–1931 (2018)
13. Marinazzo, D., et al.: Nonlinear connectivity by Granger causality. Neuroimage **58**(2), 330–338 (2011)
14. Príncipe, J.C., Liu, W., Haykin, S.: Kernel adaptive filtering: a comprehensive introduction. John Wiley & Sons (2011)

15. Schoukens, J., Ljung, L.: Nonlinear system identification: a user-oriented road map. IEEE Control Syst. Mag. **39**(6), 28–99 (2019)
16. Ge, X., Lin, A.: Dynamic causality analysis using overlapped sliding windows based on the extended convergent cross-mapping. Nonlinear Dyn. **104**(2), 1753–1765 (2021)
17. Schiecke, K., et al.: Brain–heart interactions considering complex physiological data: processing schemes for time-variant, frequency-dependent, topographical and statistical examination of directed interactions by convergent cross mapping. Physiol. Meas. **40**(11), 114001 (2019)
18. Paus, T.: Inferring causality in brain images: a perturbation approach. Philosophical Trans. Royal Society B: Biological Sci. **360**(1457), 1109–1114 (2005)
19. Antonacci, Y., et al.: Estimation of Granger causality through Artificial Neural Networks: applications to physiological systems and chaotic electronic oscillators. PeerJ Computer Science **7**, e429 (2021)
20. Tank, A., et al.: Neural granger causality. arXiv preprint arXiv:1802.05842 (2018)
21. Wismüller, A., et al.: Large-scale nonlinear Granger causality for inferring directed dependence from short multivariate time-series data. Sci. Rep. **11**(1), 1–11 (2021)
22. Leonardi, N., Van De Ville, D.: On spurious and real fluctuations of dynamic functional connectivity during rest. Neuroimage **104**, 430–436 (2015)
23. Chuang, K.-C., Ramakrishnapillai, S., Bazzano, L., Carmichael, O.T.: Deep stacking networks for conditional nonlinear granger causal modeling of fMRI data. In: Abdulkadir, A., et al. (eds.) MLCN 2021. LNCS, vol. 13001, pp. 113–124. Springer, Cham (2021). https://doi.org/10.1007/978-3-030-87586-2_12
24. Jia, X., et al.: Dynamic filter networks. Adv. Neural. Inf. Process. Syst. **29**, 667–675 (2016)
25. Zamora Esquivel, J., et al. *Adaptive convolutional kernels.* in *Proceedings of the IEEE/CVF International Conference on Computer Vision Workshops.* 2019
26. Abadi, M., et al.: Tensorflow: A system for large-scale machine learning. In: 12th {USENIX} Symposium on Operating Systems Design and Implementation ({OSDI} 16) (2016)
27. Chollet, F.: keras (2015)
28. Koprowski, R.: Image processing. In: Processing of Hyperspectral Medical Images. SCI, vol. 682, pp. 39–82. Springer, Cham (2017). https://doi.org/10.1007/978-3-319-50490-2_4
29. Murphy, K., Bodurka, J., Bandettini, P.A.: How long to scan? the relationship between fMRI temporal signal to noise ratio and necessary scan duration. Neuroimage **34**(2), 565–574 (2007)
30. Triantafyllou, C., et al.: *C*omparison of physiological noise at 1.5 T, 3 T and 7 T and optimization of fMRI acquisition parameters. Neuroimage **26**(1), 243–250 (2005)
31. Berenson, G.S.: Bogalusa Heart Study: a long-term community study of a rural biracial (black/white) population. Am. J. Med. Sci. **322**(5), 267–274 (2001)
32. Carmichael, O., et al.: High-normal adolescent fasting plasma glucose is associated with poorer midlife brain health: bogalusa heart study. J. Clin. Endocrinol. Metab. **104**(10), 4492–4500 (2019)
33. Glover, G.H., Li, T.Q., Ress, D.: Image-based method for retrospective correction of physiological motion effects in fMRI: RETROICOR. Magnetic Resonance in Medicine: An Official J. Int. Society for Magnetic Resonance in Medicine **44**(1), 162–167 (2000)
34. Sheu, L.K., Jennings, J.R., Gianaros, P.J.: Test–retest reliability of an fMRI paradigm for studies of cardiovascular reactivity. Psychophysiology **49**(7), 873–884 (2012)
35. Guido, W.: Development, form, and function of the mouse visual thalamus. J. Neurophysiol. **120**(1), 211–225 (2018)
36. Usrey, W.M., Alitto, H.J.: Visual functions of the thalamus. Annual Review of Vision Sci. **1**, 351–371 (2015)
37. Duggento, A., Guerrisi, M., Toschi, N.: Echo state network models for nonlinear granger causality. Phil. Trans. R. Soc. A **379**(2212), 20200256 (2021)

fMRI Neurofeedback Learning Patterns are Predictive of Personal and Clinical Traits

Rotem Leibovitz[1(✉)], Jhonathan Osin[1], Lior Wolf[1], Guy Gurevitch[2,4], and Talma Hendler[2,3,4,5]

[1] School of Computer Science, Tel Aviv University, Tel Aviv, Israel
rotemcz1@gmail.com

[2] Sagol Brain Institue, Tel-Aviv Sourasky Medical Center, Tel Aviv, Israel

[3] School of Psychological Sciences, Tel Aviv University, Tel Aviv, Israel

[4] Sackler Faculty of Medicine, Tel Aviv University, Tel Aviv, Israel

[5] Sagol School of Neuroscience, Tel Aviv University, Tel Aviv, Israel

Abstract. We obtain a personal signature of a person's learning progress in a self-neuromodulation task, guided by functional MRI (fMRI). The signature is based on predicting the activity of the Amygdala in a second neurofeedback session, given a similar fMRI-derived brain state in the first session. The prediction is made by a deep neural network, which is trained on the entire training cohort of patients. This signal, which is indicative of a person's progress in performing the task of Amygdala modulation, is aggregated across multiple prototypical brain states and then classified by a linear classifier to various personal and clinical indications. The predictive power of the obtained signature is stronger than previous approaches for obtaining a personal signature from fMRI neurofeedback and provides an indication that a person's learning pattern may be used as a diagnostic tool. Our code has been made available, (Our code is available via https://github.com/MICCAI22/fmri_nf.) and data would be shared, subject to ethical approvals.

1 Introduction

An individual's ability to learn to perform a specific task in a specific context is influenced by their transient task demand and context-specific mental capacity [6], as well as their motivation [24], all of which vary considerably between individuals [1]. We hypothesize that the learning pattern is highly indicative of both personal information, such as age and previous experience in performing similar tasks, and personality and clinical traits, such as emotion expressivity, anxiety levels, and even specific psychiatric indications.

L. Wang et al. (Eds.): MICCAI 2022, LNCS 13431, pp. 282–294, 2022.
https://doi.org/10.1007/978-3-031-16431-6_27

Neurofeedback (NF), a closed-loop self-neuromodulation learning procedure, provides a convenient environment for testing our hypothesis. This is because the learning task is well-defined, yet individualized, is presented in a controlled and repeatable manner, and the level of success is measured on a continuous scale of designated neural changes. NF is a reinforcement learning procedure guided by feedback presented depending on self-acquired association between mental and neural states. Mental states that happened to be associated with online modulation in the neural target (e.g. lower or higher activity) are rewarded, and eventually result in a desired modification of the brain signal (i.e. learning success) [20,23]. We hypothesize that the established association between internally-generated mental process and neural signal modulation closely signifies personal brain-mind relation and could therefore serve as an informative marker for personality and/or psychopathology.

fMRI-based neurofeedback enables precise modulation of specific brain regions in real time, leading to sustained neural and behavioral changes. However, the utilization of this method in clinical practice is limited due to its high cost and limited availability, which also hinder further research into the sustained benefit [14,22]. In the NF task we consider, one learns to reduce the activity of the Amygdala, while observing a signal that is directly correlated with it. We consider the activity of the rest of the brain as the context or states in which the learning task takes place, and discretize this space by performing clustering.

A personal signature is constructed by measuring the progress of performing the NF task in each of these clusters. Progress is obtained by comparing the Amygdala activity at a first training session with that observed at the second session for the most similar brain state. Specifically, we consider the difference between the second activity and the one predicted by a neural network that is conditioned on the brain state in the first.

The representation obtained by aggregating these differences across the brain-state clusters is shown to be highly predictive of multiple psychiatric traits and conditions in three datasets: (i) individuals suffering from PTSD, (ii) individuals diagnosed with Fibromyalgia, and (iii) a control dataset of healthy individuals. This predictive power is demonstrated with linear classifiers, in order to demonstrate that the personal information is encoded in an explicit way and to reduce the risk of overfitting by repeating the test with multiple hyperparameters [2].

2 Related Work

fMRI is widely used in the study of psychiatric disorders [5,16]. Recent applications of deep learning methods mostly focus on fully-supervised binary classification of psychopathology-diagnosed versus healthy subjects in resting state [7,27], i.e., when not performing a task. Contributions that perform such diagnosis while performing a task, e.g., [4,10,11,13,19], focus on comparing entire segments that correspond to phases of the task, and have shown improved ability to predict subjects' traits, w.r.t to resting-state fMRI [9]. Our analysis is based on aggregating statistics across individual time points along the acquired fMRI.

The work closest related to ours applies self-supervised learning to the same fMRI NF data in order to diagnose participants suffering from various psychopathologies and healthy controls [17]. There are major differences in the approaches. First, while our method is based on a meaningful signature (it accumulates meaningful statistics) that indicates learning patterns, their work is based on an implicit embedding obtained by training a deep neural network. Second, while we focus on modeling the success in performing the task over a training period, their method is based on the self-supervised task of next frame prediction, which involves both the preparation ("passive") and training ("active") periods (they require more data). Third, while our method compares progress between two active NF sessions, their method is based on mapping a passive session, in which the participant does not try to self-modulate, and the subsequent active NF session. The methods are, therefore, completely different. Finally, in a direct empirical evaluation, our method is shown to outperform [17] by a sizable gap across all datasets and prediction tasks.

3 Data

Real-time blood-oxygen-level-dependent (BOLD) signal was measured from the right Amygdala region during an interactive neurofeedback session performed inside the fMRI. The data fed into the model went through preprocessing steps using the CONN toolbox. The preprocessing steps are detailed in supplementary Fig. 4. In specific parts of the task, subjects were instructed to control the speed of an avatar riding a skateboard using only mental strategies (active phase), while in other parts, subjects passively watched the avatar on the screen (passive phase). During the active phase, local changes in the signal were translated into changing speed, displayed via a speedometer and updated every three seconds.

The neurofeedback datasets used in this experiment were part of larger intervention experiments applying multiple training sessions outside the fMRI by using an EEG statistical model of the right Amygdala. The subjects went through pre/post fMRI scans with the model region as target in order to test for changes in the ability to self-regulate this area [8,12].

fMRI Data. Each subject performed several cycles of the paradigm in a single session, lasting up to one hour, in a similar fashion to common studies in the field [18]. Following each active phase, a bar indicating the average speed during the current cycle was presented for six seconds. Each subject performed $M = 2$ cycles of Passive/Active phases, where each passive phase lasted one minute. Each active phase lasted one minute (for healthy controls and PTSD patients) or two minutes (Fibromyalgia patients). Following previous findings, instructions given to the subjects were not specific to the Amygdala, to allow efficient adoption of individual strategies [15].

The active sessions were comprised of T temporal samples of the BOLD signal, each a 3D box with dimensions $\mathcal{H}[\text{voxels}] \times \mathcal{W}[\text{voxels}] \times \mathcal{D}[\text{voxels}]$. We distinguish BOLD signals of the Amygdala signals from signals of other brain regions, each with a different spatial resolution, $\mathcal{H}_\mathcal{A} \times \mathcal{W}_\mathcal{A} \times \mathcal{D}_\mathcal{A}$, and

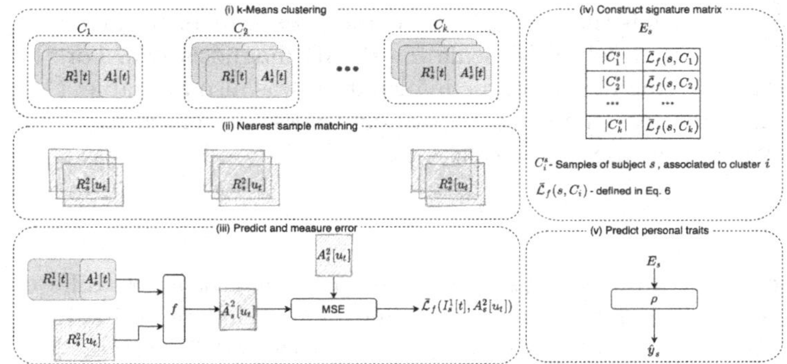

Fig. 1. The training steps: (i) Clustering the frames from the first session, based on the regions outside the Amygdala; (ii) Matching each frame from the previous step to the most similar frame from the second session; (iii) Learning to predict the Amygdala activity in the matched frame of the second session, based on the information from the matching frame in the first session and the brain activity outside the Amygdala in the second session; (iv) constructing a signature based on the prevalence of each cluster and the error of prediction for the frames of each cluster; and (v) a linear classifier on top of the obtained signature.

$\mathcal{H}_{\mathcal{R}} \times \mathcal{W}_{\mathcal{R}} \times \mathcal{D}_{\mathcal{R}}$, respectively. Partitioning was done using a pre-calculated binary Region-of-Interest matrix.

ROI for Rest-of-Brain covers the entire gray matter of the right hemisphere, excluding the right amygdala. This mask was generated with an SPM based segmentation of the MNI brain template. This region was used for providing the feedback during the real-time fMRI experiments. A visualisation of the selected brain regions, is shown in supplementary Fig. 6. The Amygdala region of interest was defined in SPM as a 6mm sphere located at MNI coordinates $[21, -1, -22]$. This region was used for providing the feedback during the real-time fMRI experiments. The matrix contains a positive value for voxels that are part of the Amygdala, and a negative value for all others. Our data is, therefore, comprised of per-subject tuples of tensors: $(\mathbb{R}^{T\times\mathcal{H}_{\mathcal{A}}\times\mathcal{W}_{\mathcal{A}}\times\mathcal{D}_{\mathcal{A}}}, \mathbb{R}^{T\times\mathcal{H}_{\mathcal{R}}\times\mathcal{W}_{\mathcal{R}}\times\mathcal{D}_{\mathcal{R}}})$.

In our setting, $T = 18$, and we use three datasets in our experiments: (i) **PTSD**- 51 subjects, (ii) **Fibromyalgia**- 24 subjects, and (iii) **Healthy Control**- 87 subjects. The Amygdala parameters and rest-of-brain parameters are through datasets: $(\mathcal{H}_{\mathcal{A}}, \mathcal{W}_{\mathcal{A}}, \mathcal{D}_{\mathcal{A}}) = (6, 5, 6)$ and $(\mathcal{H}_{\mathcal{R}}, \mathcal{W}_{\mathcal{R}}, \mathcal{D}_{\mathcal{R}}) = (91, 109, 91)$, respectively. Further information on the data acquisition scheme is provided in supplementary Fig. 4.

Clinical Data. Clinical information about each subject s, denoted as $y_s \in \mathbb{R}^l$ was available in addition to the fMRI sequences, consisting of the following information: (1) **Toronto Alexithymia Scale (TAS-20)**, which is a self-report questionnaire measuring difficulties in expressing and identifying emotions [3], (2) **State-Trait Anxiety Inventory (STAI)**, which is measured using a validated 20-item inventory [21], and (3) **Clinician Administered PTSD Scale**

(CAPS-5) 1, which is the outcome of a clinical assessment by a trained psychologist based on this widely-used scale for PTSD diagnosis [25]. For the healthy controls, the following demographic information was also available: (1) **Age** and (2) **Past experience in neuro-feedback tasks**, presented as a binary label (i.e., experienced/inexperienced subject).

4 Method

Our method aims at obtaining subjects' learning patterns across different "brain states" in order to predict their demographic and psychiatric traits. Identification of said brain states was done by clustering all the fMRI frames. The error in predicting the amygdala in each of the states is used as a unique signature of the subjects' learning patterns.

Our network receives processed fMRI samples as inputs, and uses them to predict subjects' demographic and psychiatric criteria. We consider two types of fMRI signals: Amygdala, and rest-of-brain, and train our networks in five steps, as depicted in Fig. 1: **(i)** identification of prototypical brain states, C, by applying a k-means clustering scheme to the fMRI signals; **(ii)** consider the rest of the brain regions, and identify for each fMRI frame from the first session the most similar frame from the second session; **(iii)** given a subject's complete brain state (Amygdala and greater-brain) at each time-step from the first session, we train a neural network, f, to predict the subject's Amygdala state in the matched closest frame from the second session;**(iv)** create a subject signature by aggregating the prediction error of f in each of the k prototypes; **(v)** train a linear regression network, ρ, to predict subjects' criteria, based on the obtained signature.

Data Structure. For every subject $s \in S$, the dataset contains a series of $M = 2$ active sessions, each with $T = 18$ samples, denoted as $I_s = \{I_s^m[t]\}_{t=1,m=1}^{T,M}$. We treat each sample as a pair: (i) an Amygdala sample, $A_s^m[t]$, which is cropped out of $I_s^m[t]$ using a fixed ROI matrix, as explained in Sect. 3, and (ii) a sample of other brain parts, $R_s^m[t]$.

K-means (Step i). We learn a set of k cluster centroids, $C = \{\mu_1, \ldots, \mu_k\}$, based on rest-of-brain training samples from the first session, minimizing the within-cluster-sum-of-squares: $\underset{C}{\operatorname{argmin}} \sum_{s,t} \min_{\mu \in C} ||R_s^1[t] - \mu||^2$. Note that clustering as well as cluster assignment are carried out independently of the Amygdala samples, and only once for the entire training set. Each cluster represents a different prototypical brain state and the number of clusters is selected such that for most training subjects, no cluster is underutilized. See Sect. 5.

Amygdala State Prediction (Steps ii and iii). For every subject s and time step $t \le T$ in the first session, we identify u_t, the time step during the second session in which the most similar sample was taken: $u_t = \underset{t' \le T}{\operatorname{argmin}} ||R_s^1[t] - R_s^2[t']||$.

The obtained pairs $\{(t, u_t)\}_{t=1}^{T}$ indicate tuples of similar samples $\{(I_s^1[t], I_s^2[u_t])\}$, which we use to train a neural-network, f, aimed to predict $A_s^2[u_t]$ (see implementation details in supplementary Fig. 5): $\hat{A}_s^2[u_t] = f(R_s^1[t], A_s^1[t], R_s^2[u_t])$.

f is trained independently of the centroids, minimizing the MSE loss.

Building a Signature Matrix (step iv) and Predicting Personal Traits (step v). With the group of centroids, C, and the Amygdala-state predictor, f, we construct a signature matrix, $E_s \in \mathbb{R}^{|C|\times 2}$, for every subject $s \in S$. Each row of this signature corresponds to a specific brain state prototype, and the two columns correspond to: (1) the number of samples in a cluster; and (2) the mean prediction error for that cluster which quantifies the distribute of states for every subject and the deviation from the predicted amygdala activation in each state. For every subject s and cluster $C_i \in C$, we define C_i^s as the set of the subject's samples associated with the cluster: We then calculate the average prediction loss of f with respect to each (s, C_i) pair: $\bar{\mathcal{L}}_f(s, C_i) = \frac{1}{|C_i^s|} \cdot \sum_{t\in C_i^s} ||f(R_s^1[t], R_s^2[u_t], A_s^1[t]) - A_s^2[u_t]||^2$, where $|C_i^s|$ indicates the number of visits of subject s in cluster i. The signature matrix E_s has rows of the form $E_s[i] = \left[|C_i^s|, \bar{\mathcal{L}}_f(s, C_i)\right]$.

E_s is then fed into ρ, a linear regression network with an objective to predict y_s. Since E_s is a matrix, we use a flattened version of it, denoted as $e_s \in \mathbb{R}^{2\cdot|C|}$. We then predict $\hat{y}_s = \rho(e_s) = G^\top e_s + b$, where G is a matrix and b is a vector. G, b are learned using the least squared loss over the training set.

Using a low-capacity linear classifier is meant to reduce the effect of overfitting in this step, which is the only one with access to the target prediction labels. The neural network f is trained on a considerably larger dataset, with samples of a much higher dimension (as every fMRI frame from every subject is a sample). Additionally, its task is a self-supervised one, and is therefore less prone to overfitting to the target labels.

The Inference Pipeline. Given an unseen subject r, we assign each fMRI frame, t, in the first session, to a cluster by finding the prototypes in C closest to $R_r^1[t]$. We also match a frame u_t from the second session to each frame t by minimizing $||R_r^1[t] - R_r^2[u_t]||^2$. A signature e_r is then constructed. Finally, the linear predictor ρ is applied to this signature. Implementation details are provided in supplementary Fig. 5.

5 Experiments

Each experiment was repeated five times on random splits. We report the mean and SD. The data is partitioned using a cross-validation scheme between train, validation and test sets, each composed of different subjects, with a 60-20-20 split. The same partition holds for all training stages in each dataset. For each dataset (i.e., Healthy, PTSD, Fibromyalgia), we trained a separate network.

Setting k. In order to assure that all k chosen brain-state-prototypes are visited by all subjects, we evaluate the ratio $\bar{n}_i = \frac{\sum_{s\in S}|C_i^s|}{T\cdot|S|}$ of brain states assigned to each of the clusters, when applying k-means on the training data. We chose the largest k for which the variance between all $\left\{\bar{n}_i\right\}_{i=1}^{k}$ is relatively small. In order to show that the clusters we got hold meaningful information, we performed two experiments: (a) train an LSTM based network, which receives as input the temporal signal of transitions between clusters, which proved to have predictive power w.r.t subjects' traits (results are shown in Table 1); and (b) train a similar LSTM network to predict the next cluster brain state, given past visited clusters. The network accurately predicted the cluster for 41% of the frames, compared to guessing the mean cluster which yields accuracy of 21%. To better understand the prototypical brain states that are obtained through the clustering process, we have performed an anatomical visualization of the cluster centroids, which represent the prototypical brain states found during the NF task for the Healthy subgroup. See supplementary Fig. 7.

Amygdala State Prediction. We trained the network f until its validation loss converged for each of the three datasets - healthy, PTSD and Fibromyalgia. The MSE error obtained is presented in supplementary Table 2. The network is quite successful in performing its prediction task, compared to the simple baseline of predicting the network's input, $A_s^1[t]$.

Predicting a Subject's Psychiatric and Demographic Criteria. We test whether our learned representation, trained only with fMRI images, has the ability to predict a series of psychiatric and demographic criteria not directly related to the neurofeedback task. We used our method to predict (i) STAI and (ii) TAS-20 for PTSD, Fibromyalgia and control subjects, (iii) CAPS-5 for PTSD subjects. Demographic information, (iv) age, and (v) past neuroFeedback experience were predicted for the control subjects.

Our linear regression scheme, applied to the learned signature vectors, is compared to the following baselines, which all receive the fMRI sequence as input, denoted as x: (1) **Mean prediction** simply predicts the mean value of the training set, (2) **Conditional LSTM-** The Amygdala sections of the passive and active temporal signals are fed to a neural network, which learns a personal representation for each subject. This learned representation is later used to predict the subject's criteria [17]. In contrast to our method, the learned representation of the conditional-LSTM also employs the passive "watch" data, which our method ignores. (3) **CNN-** A convolutional network with architecture identical to our f network, except for two modifications: (a) the input signal to this network is the entire second sample, $I_s^2[u_t]$ (instead of $R_s^2[u_t]$). This way, the network has access to the same signals our proposed method has; and (b) the decoder is replaced with a fully connected layer, which predicts the label. (4) **clinical prediction-** an SVM regression with the RBF kernel performed on every trait, according to the other traits. (5) **Raw difference-** A similar signature matrix, with the $\bar{\mathcal{L}}_f$ value replaced by the average norm of differences between amygdala signals for every pair $\{(t, u_t)\}$. Rows of the resulting matrix are: $\tilde{E}_s[i] = \left[|C_i^s|, \frac{1}{|C_i^s|} \cdot \sum_{t \in C_i^s} ||A_s^1[t] - A_s^2[u_t]||^2\right]$; and (6, 7) **partial** E_s**-** A network trained using only one of $|C_i^s|$ or $\bar{\mathcal{L}}_f$; (8) **ClusterLSTM-** an LSTM based network, which receives as input the sequence of cluster memberships per frame, and predicts the subject's traits according to it. To show that the neurofeedback learning that occurs across sessions is what is important, rather than the

Table 1. Traits prediction results (MSE)

	Healthy			Fibromyalgia		PTSD		
	Age↓	TAS↓	STAI↓	TAS↓	STAI↓	TAS↓	STAI↓	CAPS-5↓
Mean	13.7 ± 1	121.7 ± 6	79.3 ± 6	98.6 ± 7	78.5 ± 5	153.0 ± 11	159.5 ± 13	119.3 ± 10
[17]	10.1 ± 1	81.6 ± 7	67.4 ± 6	44.0 ± 5	73.1 ± 3	99.2 ± 5	132.7 ± 8	85.4 ± 8
CNN	17.2 ± 1	110.3 ± 12	81.6 ± 7	65.2 ± 10	90.1 ± 9	105.0 ± 11	166.3 ± 33	98.4 ± 16
SVM	13.0 ± 2	100.0 ± 12	78.2 ± 8	74.3 ± 7	77.9 ± 11	148.3 ± 12	150.0 ± 11	117.4 ± 14
Ours	**9.3** ± 1	**79.0** ± 3	**55.2** ± 7	**35.3** ± 8	**62.3** ± 7	**89.1** ± 11	**97.0** ± 10	**84.3** ± 10
Ablation $\hat{E}_s$	13.2 ± 1	98.4 ± 15	73.2 ± 9	77.8 ± 7	77.4 ± 11	102.2 ± 20	194.8 ± 6	136.0 ± 14
Ablation $\mathcal{L}_f$	10.3 ± 1	91.6 ± 4	70.0 ± 9	84.4 ± 13	81.9 ± 14	119.6 ± 11	141.0 ± 9	119.2 ± 19
Ablation $\|C_i\|$	11.0 ± 2	84.2 ± 9	68.1 ± 9	113.6 ± 32	109.4 ± 20	137.5 ± 18	137.8 ± 20	151.5 ± 19
ClusterLSTM	12.1 ± 1	96.8 ± 6	68.8 ± 8	75.7 ± 9	90.0 ± 6	118.1 ± 12	155.0 ± 14	108.0 ± 11
No Feedback	13.7 ± 2	125.0 ± 9	75.0 ± 8	90.0 ± 8	77.7 ± 7	165.3 ± 16	167.6 ± 18	139.0 ± 11
Alt. clustering	12.3 ± 1	114 ± 5	72.5 ± 9	70.2 ± 8	80.3 ± 8	98.3 ± 12	174.0 ± 20	103.6 ± 12
Alt. ROI	13.5 ± 1	125.5 ± 7	79.0 ± 10	96.3 ± 9	84.5 ± 8	135.6 ± 16	135.8 ± 19	128.7 ± 5

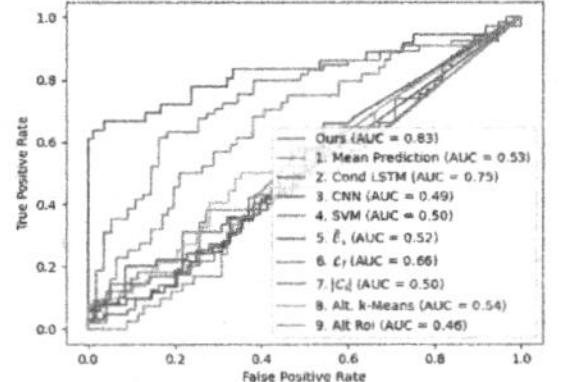

Fig. 2. Experience ROC

expected state of amygdala based on"Rest-of-Brain" state, we implemented (9) **No Feedback-** which predicts the Amygdala state given "Rest-of-Brain" state for the sample, without pairing samples from different sessions. (10) **Alternative clustering-** In step (i), the clustering objective is changed, such that it depends on the subjects' complete brain state: $\underset{C}{\operatorname{argmin}} \sum_{s,t} \min_{\mu \in C} ||I_s^1[t] - \mu||^2$. Lastly, to demonstrate the importance of using the Amygdala itself, we run baseline (11) **Alternative ROI-** a framework identical to ours, but with neural area of focus shifted from the Amygdala to the primary motor cortex, an area of dimensions $\mathcal{H} = \mathcal{W} = \mathcal{D} = 8$, which is presumed not to take part in the performance of the NF task.

The full results are shown in Table 1 for performing regression on age, TAS, STAI and CAPS-5. Our results are statistically significant, and the p-values of the corrected re-sampled t-test between our method and the baseline methods is always lower than 0.01. As can be seen, the baseline of [17] greatly outperforms the mean prediction and both the CNN classifier and the one based on the clinical data. However, it is evident that across all three datasets our method outperforms this method, as well as all ablations, by a very significant margin in predicting the correct values for both demographic and psychiatric traits.

Figure 2 presents classification results for past-experience information, which is only available for the healthy control subjects. Here, too, our method outperforms the baselines and ablation methods. Specifically, it obtains an AUC of 0.83 ± 0.03, while the method presented by [17] obtains an AUC of 0.75 ± 0.03.

The ablation experiments provide insights regarding the importance of the various components. First, modeling based on an irrelevant brain region, instead of the Amygdala, leads to results that are sometimes worse than a mean prediction. Similarly, predicting using raw differences in the Amygdala activity (without performing prediction), is not effective. It is also important to remove the Amygdala from the clustering procedure, keeping this region and outside regions separate. The variant based on the prediction error alone seems to be more informative than that based only on cluster frequency. However, only together do they outperform the strong baseline of [17].

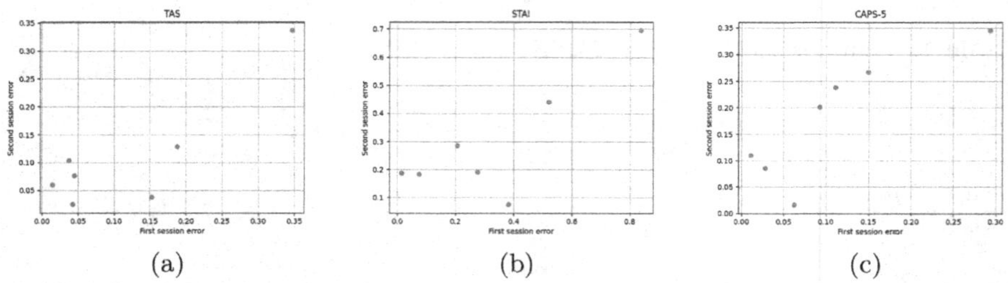

(a) (b) (c)

Fig. 3. The normalized error in the second session as a function of that of the first session. Each point reflects a single patient of the PTSD cohort. (a) TAS, (b) STAI, (c) CAPS-5. The normalized error is computed as the absolute error divided by the ground truth value.

Longitudinal Data. In order to show the robustness of our method over time, we performed inference of our network (which was trained using data from the first session of each subject) on fMRI scans that were obtained a few months later. As can be seen from Fig. 3, while there are not many subjects that have both sessions, there seems to be a correlation between the errors in the first session and the errors in the second one. More data is requires in order to further study the validity of the model beyond the first collection point.

6 Conclusion

NF data offers unique access to individual learning patterns. By aggregating the deviation between actual and predicted learning success across clusters of brain activities, we obtain a signature that is highly predictive of the history of a person, as well as of their clinical test scores and psychiatric diagnosis.

The presented method provides a sizable improvement in performance over previous work. Perhaps even more importantly, the obtained signature is based on explicit measurements that link brain states to the difference between actual and expected learning success, while previous work was based on an implicit embedding that is a by-product of training a network to predict a loosely related task of predicting transient signal dynamics. By accurately predicting the psychiatric tests score, our method provides a step toward an objective way to diagnose various clinical conditions.

Acknowledgments. This project has received funding from the European Research Council (ERC) under the European Union's Horizon 2020 research and innovation programme (grant ERC CoG 725974), and the ISRAEL SCIENCE FOUNDATION (grant No. 2923/20) within the Israel Precision Medicine Partnership program.

Appendix

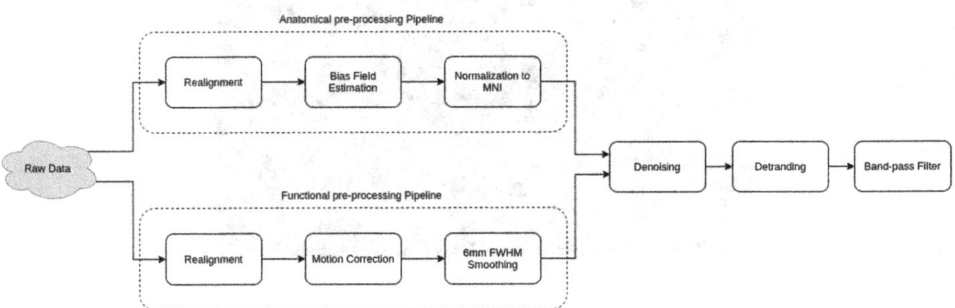

Fig. 4. Structural and functional scans were obtained with a 3.0T Siemens PRISMA MRI system. The CONN MATLAB toolbox [26] was used for functional volumes realignment, motion correction, normalization to MNI space and spatial smoothing with an isotropic 6-mm FWHM Gaussian kernel. Subsequently de-noising and de-trending regression algorithms were applied, followed by bandpass filtering in the range of 0.008–0.09 Hz. The frequencies in the bandpass filter reflect the goal of modeling the individual throughout the session, while removing the effects of the fast paced events that occur during the neurofeedback session. This filtering follows previous work [17] for a fair comparison.

Table 2. The MSE error of network f in comparison to the simple baselines of predicting the activations in the network's input $A_s^1[t]$ and predicting the mean activation of each Amygdala's voxels

Method	Healthy	Fibromyalgia	PTSD
Mean Prediction	$0.1151 \pm 3 \cdot 10^{-3}$	$0.0780 \pm 1 \cdot 10^{-3}$	$0.0770 \pm 1 \cdot 10^{-3}$
Predicting $A_s^1[t]$	$0.1087 \pm 2 \cdot 10^{-3}$	$0.0833 \pm 4 \cdot 10^{-3}$	$0.0781 \pm 3 \cdot 10^{-3}$
$f(I_s^1[t], R_s^2[u_t])$	$\mathbf{0.0735} \pm 2 \cdot 10^{-3}$	$\mathbf{0.0561} \pm 2 \cdot 10^{-3}$	$\mathbf{0.0550} \pm 1 \cdot 10^{-3}$

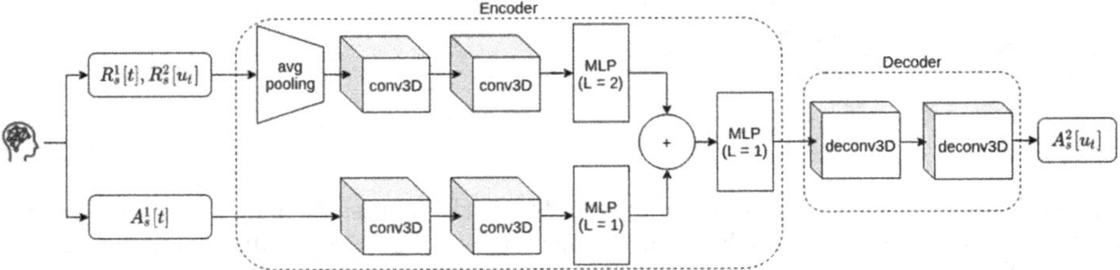

Fig. 5. Architecture of f, our Amygdala prediction Neural Network. We trained the network f, until convergence of its validation loss for each of the three datasets. The hyper-parameters of the network f were selected according to a grid search using the cross validation scores on the validation set. For training, we used an Adam optimizer, with initial learning rate of 0.001, and a batch sizes of 16.

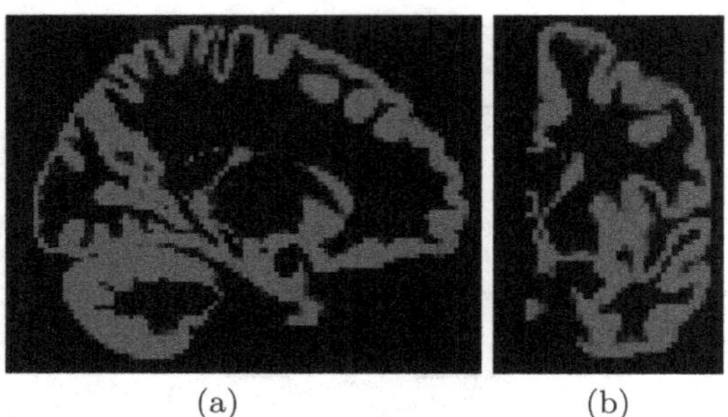

(a) (b)

Fig. 6. (a) Mid-sagittal slice of Rest of brain mask; and (b) a coronal slice of it.

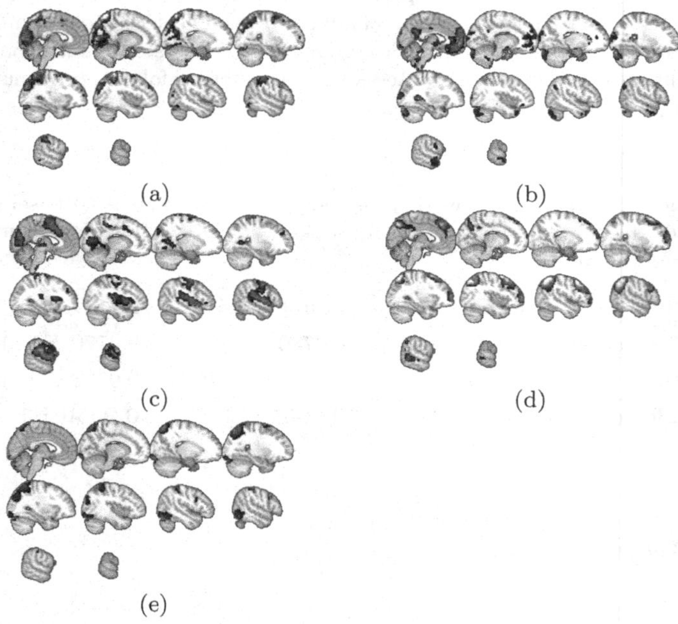

(a) (b) (c) (d) (e)

Fig. 7. Visualization of the cluster centroids which represent the prototypical brain states found during the NF task for the Healthy subgroup. The resulting maps show the main activated nodes in each state. Clusters are distinct in their spatial arrangement, which supports the relevance of using clustering for this purpose. In this figure, the five prototypical clusters obtained on the healthy individuals dataset are presented. (a) Sensorimotor Network and Visual Cortex. (b) Main nodes of the Default Mode network (c) Main nodes of the Salience Network. (d) Dorsal attention network. (e) High visual areas (Ventral and Dorsal Stream).

References

1. Ackerman, P.L.: Determinants of individual differences during skill acquisition: cognitive abilities and information processing. J. Exp. Psychol. Gen. **117**(3), 288 (1988)
2. Asano, Y., Rupprecht, C., Vedaldi, A.: A critical analysis of self-supervision, or what we can learn from a single image. In: International Conference on Learning Representations (2019)
3. Bagby, R.M., Parker, J.D., Taylor, G.J.: The twenty-item Toronto alexithymia scale-i. item selection and cross-validation of the factor structure. J. Psychosomatic Res. **38**(1), 23–32 (1994)
4. Bleich-Cohen, M., Jamshy, S., Sharon, H., Weizman, R., Intrator, N., Poyurovsky, M., Hendler, T.: Machine learning fmri classifier delineates subgroups of schizophrenia patients. Schizophr. Res. **160**(1–3), 196–200 (2014)
5. Calhoun, V.D., Miller, R., Pearlson, G., Adalı, T.: The chronnectome: time-varying connectivity networks as the next frontier in FMRI data discovery. Neuron **84**(2), 262–274 (2014)
6. Crump, M.J., Vaquero, J.M., Milliken, B.: Context-specific learning and control: the roles of awareness, task relevance, and relative salience. Conscious. Cogn. **17**(1), 22–36 (2008)
7. Dvornek, N.C., Ventola, P., Pelphrey, K.A., Duncan, J.S.: Identifying Autism from resting-State fMRI using long short-term memory networks. In: Wang, Q., Shi, Y., Suk, H.-I., Suzuki, K. (eds.) MLMI 2017. LNCS, vol. 10541, pp. 362–370. Springer, Cham (2017). https://doi.org/10.1007/978-3-319-67389-9_42
8. Fruchtman-Steinbok, T., et al.: Amygdala electrical-finger-print (amygefp) neurofeedback guided by individually-tailored trauma script for post-traumatic stress disorder: Proof-of-concept. NeuroImage: Clinical **32**, 102859 (2021)
9. Gal, S., Tik, N., Bernstein-Eliav, M., Tavor, I.: Predicting individual traits from unperformed tasks. NeuroImage, 118920 (2022)
10. Hendler, T., et al.: Social affective context reveals altered network dynamics in schizophrenia patients. Transl. Psychiatry **8**(1), 1–12 (2018)
11. Jacob, Y., Shany, O., Goldin, P., Gross, J., Hendler, T.: Reappraisal of interpersonal criticism in social anxiety disorder: a brain network hierarchy perspective. Cereb. Cortex **29**(7), 3154–3167 (2019)
12. Keynan, J.N., et al.: Electrical fingerprint of the amygdala guides neurofeedback training for stress resilience. Nat. Hum. Behav. **3**(1), 63–73 (2019)
13. Lerner, Y., et al.: Abnormal neural hierarchy in processing of verbal information in patients with schizophrenia. NeuroImage: Clinical **17**, 1047–1060 (2018)
14. Lubianiker, N., et al.: Process-based framework for precise neuromodulation. Nat. Hum. Behav. **3**(5), 436–445 (2019)
15. Marxen, M., et al.: Amygdala regulation following fMRI-neurofeedback without instructed strategies. Front. Hum. Neurosci. **10**, 183 (2016)
16. Oksuz, I., et al.: Magnetic resonance fingerprinting using recurrent neural networks. In: 2019 IEEE 16th International Symposium on Biomedical Imaging (ISBI 2019), pp. 1537–1540. IEEE (2019)
17. Osin, J., et al.: Learning personal representations from fMRI by predicting neurofeedback performance. In: Martel, A.L., et al. (eds.) MICCAI 2020. LNCS, vol. 12267, pp. 469–478. Springer, Cham (2020). https://doi.org/10.1007/978-3-030-59728-3_46

18. Paret, C., et al.: Current progress in real-time functional magnetic resonance-based neurofeedback: methodological challenges and achievements. Neuroimage **202**, 116107 (2019)
19. Raz, G., Shpigelman, L., Jacob, Y., Gonen, T., Benjamini, Y., Hendler, T.: Psychophysiological whole-brain network clustering based on connectivity dynamics analysis in naturalistic conditions. Hum. Brain Mapp. **37**(12), 4654–4672 (2016)
20. Sitaram, R., et al.: Closed-loop brain training: the science of neurofeedback. Nat. Rev. Neurosci. **18**(2), 86–100 (2017)
21. Spielberger, C.D., Gorsuch, R.L.: State-trait anxiety inventory for adults: Manual and sample: Manual, instrument and scoring guide. Consulting Psychologists Press (1983)
22. Sulzer, J., et al.: Real-time fMRI neurofeedback: progress and challenges. Neuroimage **76**, 386–399 (2013)
23. Taschereau-Dumouchel, V., Cushing, C., Lau, H.: Real-time functional MRI in the treatment of mental health disorders. Ann. Rev. Clin. Psychol. **18** (2022)
24. Utman, C.H.: Performance effects of motivational state: a meta-analysis. Pers. Soc. Psychol. Rev. **1**(2), 170–182 (1997)
25. Weathers, F., Blake, D., Schnurr, P., Kaloupek, D., Marx, B., Keane, T.: The clinician-administered ptsd scale for dsm-5 (caps-5). interview available from the national center for ptsd (2013)
26. Whitfield-Gabrieli, S., Nieto-Castanon, A.: Conn: a functional connectivity toolbox for correlated and anticorrelated brain networks. Brain Connect. **2**(3), 125–141 (2012)
27. Yan, W., et al.: Discriminating schizophrenia using recurrent neural network applied on time courses of multi-site fMRI data. EBioMedicine **47**, 543–552 (2019)

Multi-head Attention-Based Masked Sequence Model for Mapping Functional Brain Networks

Mengshen He[1], Xiangyu Hou[1], Zhenwei Wang[1], Zili Kang[1], Xin Zhang[2], Ning Qiang[1], and Bao Ge[1](✉)

[1] School of Physics and Information Technology, Shaanxi Normal University, Xi'an, China
bob_ge@snnu.edu.com

[2] Northwestern Polytechnical University, Xi'an, China

Abstract. It has been of great interest in the neuroimaging community to discover brain functional networks (FBNs) based on task functional magnetic resonance imaging (tfMRI). A variety of methods have been used to model tfMRI sequences so far, such as recurrent neural network (RNN) and Autoencoder. However, these models are not designed to incorporate the characteristics of tfMRI sequences, and the same signal values at different time points in a fMRI time series may rep-resent different states and meanings. Inspired by cloze learning methods and the human ability to judge polysemous words based on context, we proposed a self-supervised a Multi-head Attention-based Masked Sequence Model (MAMSM), as BERT model uses (Masked Language Modeling) MLM and multi-head attention to learn the different meanings of the same word in different sentences. MAMSM masks and encodes tfMRI time series, uses multi-head attention to calculate different meanings corresponding to the same signal value in fMRI sequence, and obtains context information through MSM pre-training. Furthermore this work redefined a new loss function to extract FBNs according to the task de-sign information of tfMRI data. The model has been applied to the Human Connectome Project (HCP) task fMRI dataset and achieves state-of-the-art performance in brain temporal dynamics, the Pearson correlation coefficient between learning features and task design curves was more than 0.95, and the model can extract more meaningful network besides the known task related brain networks.

Keywords: Masked sequence modeling · Multi-head attention · Functional networks

1 Introduction

Identifying functional brain networks (FBNs) is a prerequisite step towards understanding functional organizational architecture of human brain, and is very important to the research of cognition and diseases. FBNs are usually detected by various methods, such as general linear model (GLM) [2, 3], independent component analysis (ICA) [7] and sparse dictionary learning (SDL) [4–6], etc. Although these methods are widely used, their abilities to represent functional networks are all limited by their shallow attributes. In recent years, with the rapid development of deep learning, functional brain networks

L. Wang et al. (Eds.): MICCAI 2022, LNCS 13431, pp. 295–304, 2022.
https://doi.org/10.1007/978-3-031-16431-6_28

were also drawn by using convolutional neural network (CNN) [8–10], recurrent neural network (RNN) [11–14], auto-encoder (AE) and their deformations, such as Convolutional Autoencoder (CAE) [9] and Recur-rent Autoencoder (RAE) [11], Spatiotemporal Attention Autoencoder (STAAE) [19], etc. However, the current deep learning models to detect functional networks still have some limitations. For examples, the CNN-related models are difficult to capture the long-distance features; the RNN-related models cannot be computed in parallel, and both of them have long-range dependency problems [21]. Besides, the training style of these models are not efficiently designed to fully explore the characteristics of the fMRI data. Specifically, the same signal values at different time points in a fMRI time series may be in various states such as rising, falling, and so on, hence they may represent different states/tasks or meanings.

In recent researches, the Transformer model based on the multi-head attention mechanism has shown excellent ability in analyzing and predicting sequence models such as text [16]. The core of Transformer is that uses the multi-head self-attention mechanism to calculate the implicit representation between the input and output. Models such as Bert [17] demonstrate that the multi-head attention mechanism has obtained high performance in representing multiple semantic space information of the same word in different sentences. On the other hand, the success of Masked Language Modeling (MLM) [17] and Masked Image Modeling (MIM) [18] training approaches in Natural Language Processing (NLP) and Computer Vision (CV) fields demonstrates the powerful advantage of mask-based training methods in extracting contextual information. The characteristics of fMRI sequence data and NLP data are similar, and the feature extraction of these two kinds of data is more dependent on the understanding of the sequence context, and there is a phenomenon of polysemy. Therefore, masked model and multi-head attention are more suitable for processing fMRI sequence.

Considering the above two aspects, we propose a Multi-head Attention-based Masked Sequence Model (MAMSM) for mapping FBNs, in which we use MSM to process fMRI time series like sentences in NLP. Meanwhile, we use multi-head attention to estimate the specific state of the voxel signal at different time points. In addition, we design a new loss function that is more suitable for tfMRI data for more accurate training.

To verify the validity of the model, it was applied to the Motor tfMRI dataset of the Human Connectome Project (HCP). The experimental results show that the proposed framework reaches the state-of-the-art in terms of brain temporal dynamics, and found the existence of some of the resting state networks (RSNs) while extracting the task FBNs.

2 Materials and Methods

The proposed computational framework is summarized in Fig. 1. For input, the preprocessed tfMRI data of all subjects are registered to a standard-MNI space and mapped to a 2D spatiotemporal matrix [22] and then masked. Then, the data is fed into the MAMSM model for pre-training, which is based on the multi-transformer encoder layer in combination with the mask training method. Finally, all features obtained from pre-training are inputted into a feature selection layer composed of a pair of encoder and decoder, using the new loss function proposed for downstream training, then the features outputted by the encoder layer are visualized as interpretable FBNs.

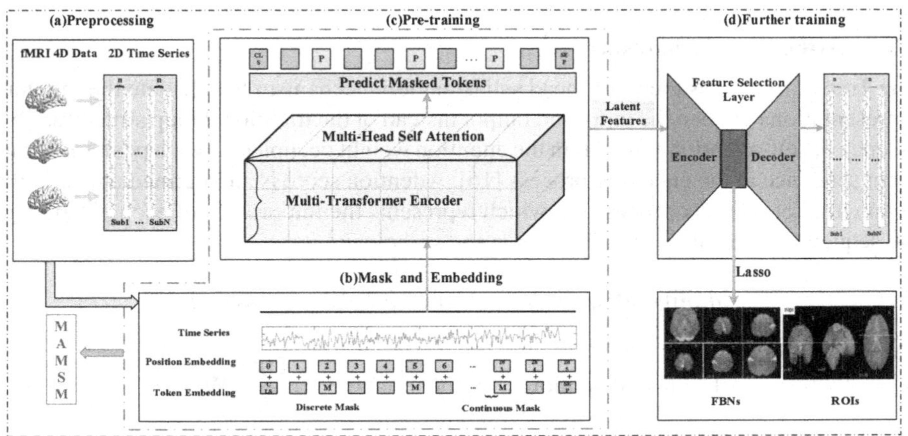

Fig. 1. Illustration of MAMSM computational framework. (a)Preprocessing: the preprocessed tfMRI data of all subjects are mapped to a 2D spatiotemporal matrix. (b) Each time series on the 2D spatiotemporal matrix is masked and embedded. (c) Pre-training: the input and output are embedded data and the predicted tokens of the masked position, respectively. (b) and (c) are the main parts of the MAMSM model. (d)Further training: the high-level features are extracted by the encoder which are used to construct FBNs in this module.

2.1 Dataset and Preprocessing

We selected the motor tfMRI data of 22 young healthy subjects from the HCP Q1 dataset. The participants are presented with visual cues that ask them to tap their left or right fingers, squeeze their left or right toes, or move their tongue. Each block of a movement type lasts 12 s (10 movements), and is preceded by a 3 s cue. There are 13 blocks, with 2 of tongue movements, 4 of hand movements (2 right and 2 left), 4 of foot movements (2 right and 2 left) and three 15 s fixation blocks per run. The pre-processing of tfMRI data included brain skull removal, motion correction, spatial smoothing, temporal pre-whitening, slice time correction, and global drift removal [15]. These steps are implemented by the FSL FEAT toolkit.

2.2 Methods

2.2.1 The Proposed MAMSM Model Structure

The overall framework of the MAMSM model is shown in Fig. 1. There are two major parts. The first part includes the mask and embedding processing for the input time series, where the mask method combines continuous mask and discrete mask, shown in Fig. 1 (b). The second part is to input the processed data into a multi-transformer encoding layer composed of the multi-head self-attention mechanism, shown in Fig 1(c). The model is pre-trained by predicting the token at the input mask position, and extracts latent features after training as the input of the feature selection layer in in Fig. 1(d).

2.2.2 Attention Mechanism

The Transformer uses the multi-head self-attention mechanism to calculate the implicit representation between the input and output instead of the traditional loop structure. The essence of self-attention is to obtain the attention weight occupied by each position of the input sequence in the encoding process [16]. Attention score is an intermediate variable obtained after softmax processing, which represents the importance of each position in the sequence. The calculation formula is shown below:

$$\mathrm{MultiHead}(\mathrm{Q}, \mathrm{K}, \mathrm{V}) = \mathrm{Concat}(\mathrm{head}_1, \ldots, \mathrm{head}_\mathrm{N}) \tag{1}$$

$$\mathrm{head}_\mathrm{i} = \mathrm{Attention}(\mathrm{Q}, \mathrm{K}, \mathrm{V}) = \mathrm{softmax}(\frac{QK^T}{\sqrt{d_k}})\mathrm{V} \tag{2}$$

2.2.3 Masked Sequence Modeling

The MSM model randomly masks some percentage of the tokens on the input, and the objective is to predict those masked tokens. For each sequence input, we mask approximately 10% tokens at random locations, and replace original signal value with the [mask], and then use multi-head attention to predict the token value corresponding to the masked position in the context. During the model training, the error between the tokens at the predicted mask's position and the actual tokens is simplified into a multi-classification problem, and the cross-entropy loss is used to calculate the error. In order to improve the learning ability of the model, different from other mask training methods, this paper uses a combination of random mask and continuous mask methods. The illustration is shown in Fig. 1(b).

2.2.4 Feature Selection and Lasso Regression

To compress the pretrained latent features into more representative high-level features, we set up a feature selection layer. The computational framework of the feature selection layer is shown in Fig. 1(d). The latent feature matrix obtained by pre-training is used as the input of the encoder, and the original 2D matrix is used as a comparison of the output of the decoder to train the feature selection layer. In addition, this paper defines a new loss function for feature selection.

The previous models such as auto-encoders often adopt the reconstruction mean squared error (MSE) while ignoring the potential feature distribution and the relationship with the task design curves, both of which are indispensable to characterize fMRI time series. Therefore, this paper uses a combination of cosine similarity error calculation and mean square error to design a loss function Loss_{task}, which is more conducive to tfMRI data. The calculation formula can be described as follows:

$$\text{Loss}_{mse} = \text{MSE} = \frac{1}{n}\sum_{i=1}^{n}(y_i - \hat{y}_i)^2 \tag{3}$$

$$\text{Loss}_{cos} = l_i = 1 - \cos(a_i, b_i) \tag{4}$$

$$\text{Loss}_{task} = \text{Loss}_{mse} + k * \text{Loss}_{cos} \tag{5}$$

In Eq. 3, Loss_{mse} is the MSE reconstruction error between the decoder output and the original data, where y_i corresponds to the original real input, $\hat{y}_i$ corresponds to the reconstructed data output by the decoder, and n is the number of samples. In Eq. 4, Loss_{cos} is the cosine similarity error between the first six features of the encoder output and the six task design curves, where a_i and b_i respectively correspond to the sequence that needs to calculate the cosine similarity. In Eq. 5, the final training error Loss_{task} is the summation of Loss_{mse} and Loss_{cos} weighted by k, considering the magnitudes of the two errors are different in the training process and cannot be directly added.

After the training of the feature selection layer, the output X of the encoder and the original 2D data **Z** were put together into LASSO regression. Then, the coefficient matrix **W** can be obtained according to the following formula, where λ is the penalty coefficient of the L1 regular term:

$$\mathbf{W} = \min\|\mathbf{Z} - \mathbf{XW}\|_2^2 + \lambda\|\mathbf{W}\|_1 \tag{6}$$

Finally, we mapped the sparse coefficient matrix W to the 3D brain space, thus obtained FBNs [22].

3 Results

3.1 Comparison of Learned Features with Task Design Curves

We compare the attention scores after pre-training with the task design curves here, as shown in Fig. 2. We can see that the learned attention scores are similar to the six task design curves in Fig. 2(a). In order to better show their similarities, we performed a sliding average on the attention scores, the results are shown in Fig. 2(b), it is obvious that the sliding averaged results are more similar to the task design curves. We then compare the features after the feature selection layer and task design curves, shown in Fig. 2(c), we can see that the learned features are advanced further. The quantitative comparison for their Pearson's correlation is listed in Table 1, which demonstrates that the learned feature is very close to the original task design curves and very meaningful.

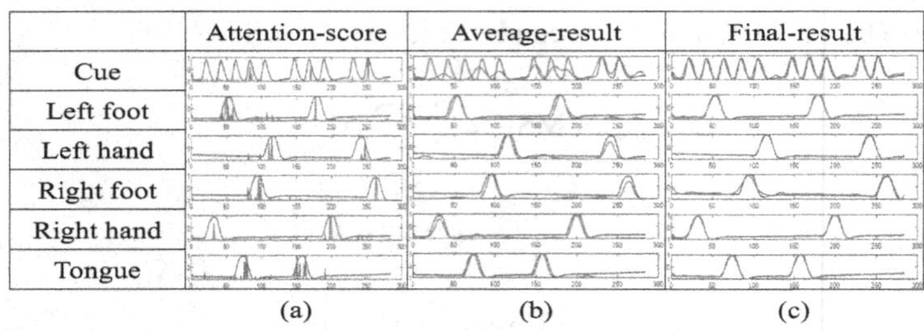

Fig. 2. (a) The comparison between the attention score and the task design curves. (b) The comparison between the attention score after sliding and the task design curves. (c) The comparison between the final result after the feature selection layer and the task design curves.

Table 1. Pearson correlation coefficient between the learned features and task design curves.

	Cue	Left foot	Left hand	Right foot	Right hand	Tongue
Attention-score	0.2633	0.4401	0.3176	0.3540	0.3165	0.4379
Average-result	0.5045	0.9496	0.9432	0.9519	0.9646	0.9691
Final-result	0.9677	0.9972	0.9975	0.9715	0.9958	0.9967

3.2 Comparison of Learn FBNs

3.2.1 Comparison of Task FBNs

In order to verify the advancement of the model proposed in this paper, we use the SDL algorithm and the deep learning algorithm STAAE to process the same dataset, respectively, then use the FBNs obtained by the GLM as the benchmark template for comparative analysis. Due to the limitation of the STAAE calculation framework, the intermediate temporal features cannot be obtained. Therefore, the comparison of temporal features is mainly performed on SDL and MAMSM, shown in Fig. 3. As a quantitative comparison, the temporal signatures produced by MAMSM and SDL are compared with the task design curves using the Pearson correlation coefficient, shown in Table 2. It can be found that the correlation between the features generated by MAMSM and the task design curves (the average value is 0.987) is greater than the correlation between the features generated by SDL and the task design curves (the average value is 0.686). By comparing Table1 and Table2, it can be seen that the correlation between the Average-result generated by MAMSM's unsupervised pre-training and the task design curve (Average is 0.880) is greater than the correlation between the features generated by SDL and the task design curve (Average is 0.686) at a significance level of 0.001.

After mapping the features obtained by these methods back to the brain space to display FBNs, the comparison result is shown in Fig. 4. In order to compare quantitatively the functional network obtained by these three methods, we use the spatial overlap rate to measure the similarity of the two FBNs [5]. The spatial overlap rates of every method with the GLM result are shown in Table 3, it can be seen that the resulting MAMSM FBNs have a higher spatial overlap rate than SDL and STAAE.

Table 2. The Pearson correlation coefficients between the temporal characteristics generated by SDL and MAMSM with the task design curves.

	Cue	Left foot	Left hand	Right foot	Right hand	Tongue	Average
SDL	0.765	0.622	0.633	0.670	0.700	0.725	0.686
MAMSM	0.967	0.997	0.997	0.971	0.995	0.996	0.987

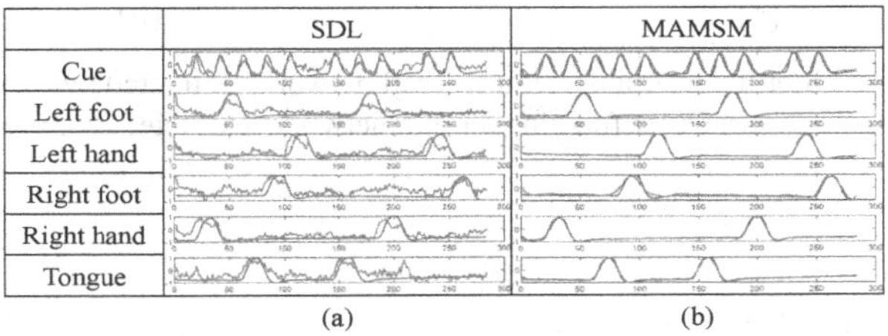

Fig. 3. (a) The features extracted by SDL compared to the task design curves; (b) The features extracted by MAMSM compared to the task design curves.

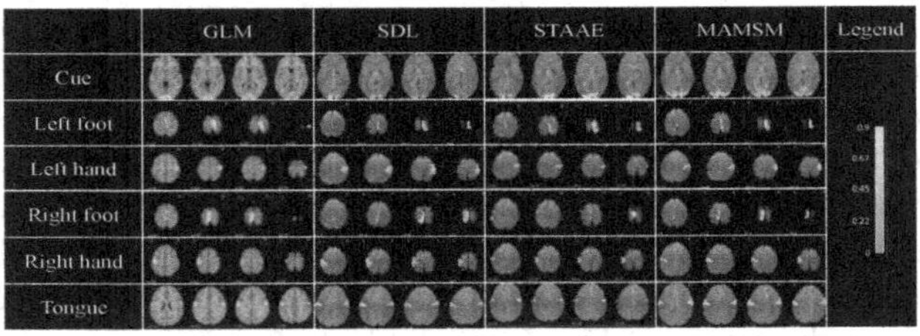

Fig. 4. The Motor task functional brain network obtained by the three different methods (SDL, STAAE, MAMSM), compared with the GLM template.

Table 3. Spatial overlap rate of FBNs of the three methods (SDL, STAAE, and MAMSM) with the GLM template.

	Cue	Left foot	Left hand	Right foot	Right hand	Tongue
SDL	0.265	0.294	0.279	0.324	0.312	0.346
STAAE	0.466	0.305	0.260	0.354	0.281	0.382
MAMSM	0.592	0.421	0.405	0.436	0.385	0.542

3.2.2 Comparison of Other FBNs

We found not only the existence of the expected task related FBNs, but also the existence of other brain networks, which are shown in Fig. 5. After comparing with the previous RSNs by ICA method [7], it is found that some FBNs derived from MAMSM consistent with the RSNs, as shown in Fig. 5(a). Therefore, the MAMSM model can not only extract the task related functional networks, but also extract some resting-state functional networks in the meantime.

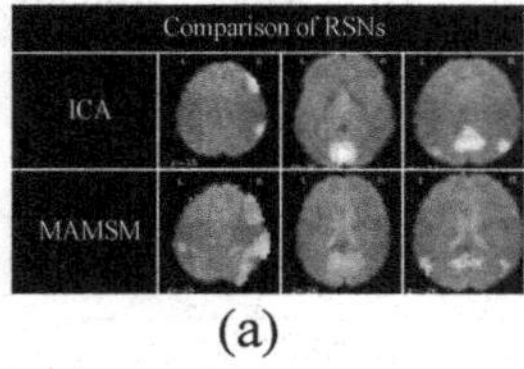

(a)

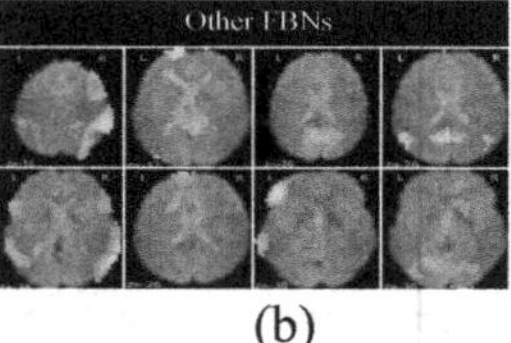

(b)

Fig. 5. (a) Comparison of the resting state functional networks extracted by MAMSM and those extracted by ICA method. (b) Other functional brain networks extracted by the MAMSM model.

3.3 Reconstruction Loss

The MAMSM model has three transformer encoder layers and each layer contains six-head attention. We trained MAMSM model on a computer platform with a NVIDIA 1080ti GPU. The pre-train process converged within 12 h with a learning rate of 0.0002 and 300 epochs. The feature selection model was trained with a learning rate of 0.0001 for 2000 epochs, and the convergence result was achieved within 15 min.

The loss variation of the feature selection model using different loss functions is shown in Fig. 6. In order to quantitatively compare the results, the final error obtained by using different loss functions under the same parameter setting is given in Table 4. Cos-error is the cosine similarity error between the extracted features and the task design curves, and Mse-error is the mean squared error of the output of the decoder and the original data. From the results in Fig. 6 and Table 4, we can see the obtained feature matrix has relatively small cosine similarity error and reconstruction mean square error by training the model using Loss_{task}.

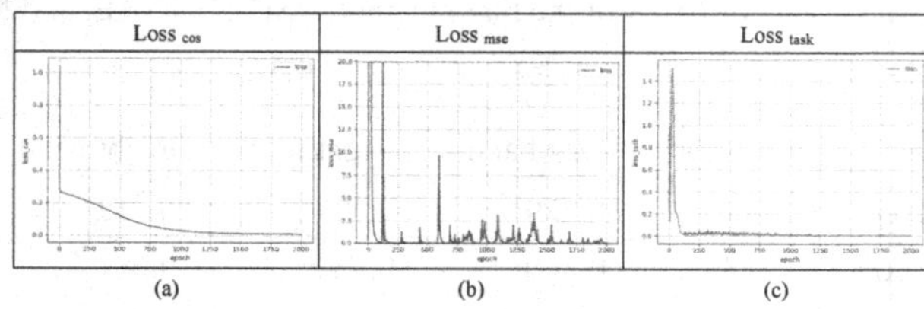

Fig. 6. (a) The loss variation of the feature selection model using Loss_{cos}. (b) The loss variation of the feature selection model using Loss_{mse}. (c) The loss variation of the feature selection model using Loss_{task}.

Table 4. Comparison of the final error obtained by using different loss functions

	$Loss_{cos}$	$Loss_{mse}$	$Loss_{task}$
Cos-error	0.003	0.899	0.003
Mse-error	9021.529	0.032	0.038

4 Conclusions

To best of our knowledge, this paper first adopted the masked model method to reconstruct the functional brain network of tfMRI, and to design a new loss function that is more suitable for this field. With a group-wise experiments on tfMRI data, the MAMSM model showed its capability to learn the temporal dynamics features of fMRI. A comparison study with SDL and STAAE shows that the features learned by MAMSM are meaningful and have reached the current advanced level. Moreover, some of the RSNs were found while extracting the task FBNs, suggesting that the MAMSM model is very helpful for understanding how the brain works.

However, there is still some space for improvement in our method, we only explored the temporal characteristics of fMRI signals but not leveraged the space information fully. In our future work, we plan to add the attention on space information and train a more efficient model to represent fMRI signals and reconstruct functional brain networks.

Acknowledgement. The work was supported by the National Natural Science Foundation of China (NSFC61976131 and NSFC61936007).

References

1. Cabral, J., Kringelbach, M.L., Deco, G.: Exploring the network dynamics underlying brain activity during rest. Prog. Neurobiol. **114**, 102–131 (2014)
2. Kanwisher, N.: Functional specificity in the human brain: a window into the functional architecture of the mind. Proc. Natl. Acad. Sci. **107**(25), 11163–11170 (2010)
3. Beckmann, C.F., et al.: General multilevel linear modeling for group analysis in FMRI. Neuroimage **20**(2), 1052–1063 (2003)
4. Jiang, X., et al.: Sparse representation of HCP grayordinate data reveals novel functional architecture of cerebral cortex. Hum. Brain Mapp. **36**(12), 5301–5319 (2015)
5. Lv, J., et al.: Holistic atlases of functional networks and interactions reveal reciprocal organizational architecture of cortical function. IEEE Trans. Biomed. Eng. **62**(4), 1120–1131 (2015)
6. Li, X., et al.: Multple-demand system identification and characterization via sparse representations of fMRI data. In: 2016 IEEE 13th International Symposium on Biomedical Imaging(ISBI). IEEE (2016)
7. Smith, S.M., et al.: Correspondence of the brain's functional architecture during activation and rest. Proc. Natl. Acad. Sci. **106**(31), 13040–13045 (2009)
8. Huang, H., et al.: Modeling task fMRI data via mixture of deep expert networks. In: 2018 IEEE 15th International Symposium on Biomedical Imaging (ISBI 2018). IEEE (2018)

9. Huang, H., et al.: Modeling task fMRI data via deep convolutional autoencoder. IEEE Trans. Med. Imaging **37**(7), 1551–1561 (2018)
10. Zhao, Y., et al.: Automatic recognition of fMRI-derived functional networks using 3-D convolutional neural networks. IEEE Trans. Biomed. Eng. **65**(9), 1975–1984 (2018)
11. Li, Q., et al.: Simultaneous spatial-temporal decomposition of connectome-scale brain networks by deep sparse recurrent auto-encoders. In: Chung, A.C.S., Gee, J.C., Yushkevich, P.A., Bao, S. (eds.) IPMI 2019. LNCS, vol. 11492, pp. 579–591. Springer, Cham (2019). https://doi.org/10.1007/978-3-030-20351-1_45
12. Sak, H., et al.: Long short-term memory recurrent neural network architectures for large scale acoustic modeling. In: Fifteenth Annual Conference of the International Speech Communication Association (2014)
13. Wang, H., et al.: Recognizing brain states using deep sparse recurrent neural network. IEEE Trans. Med. Imaging **38**, 1058–1068 (2018)
14. Barch, D.M., et al.: Function in the human connectome: task-fMRI and individual differences in behavior. Neuroimage **80**, 169–189 (2013)
15. Glasser, M.F., et al.: The minimal preprocessing pipelines for the human Connectome project. Neuroimage **80**, 105–124 (2013)
16. Vaswani, A., et al.: Attention is all you need. In: Advances in Neural Information Processing Systems (2017)
17. Devlin, J., et al.: Bert: pre-training of deep bidirectional transformers for language understanding (2018)
18. Xie, Z., et al.: Simmim: A simple framework for masked image modeling. arXiv preprint arXiv:2111.09886 (2021)
19. Dong, Q., et al.: Spatiotemporal Attention Autoencoder (STAAE) for ADHD Classification. In: International Conference on Medical Image Computing and Computer-Assisted Intervention. Springer, Cham, (2020) https://doi.org/10.1007/978-3-030-59728-3_50
20. He, K., et al.: Masked autoencoders are scalable vision learners. arXiv preprint arXiv:2111.06377 (2021)
21. Tang, G., et al.: Why self-attention? a targeted evaluation of neural machine translation architectures. In: Proceedings of the 2018 Conference on Empirical Methods in Natural Language Processing, Brussels, Belgium, pp. 4263–4272 (2018)
22. Abraham, A., et al.: Machine learning for neuroimaging with scikit-learn. Front. Neuroinform. **8**, 14 (2014)

Dual-HINet: Dual Hierarchical Integration Network of Multigraphs for Connectional Brain Template Learning

Fatih Said Duran, Abdurrahman Beyaz, and Islem Rekik(✉)

BASIRA Lab, Faculty of Computer and Informatics, Istanbul Technical University, Istanbul, Turkey
irekik@itu.edu.tr
http://basira-lab.com

Abstract. A connectional brain template (CBT) is a normalized representation of a population of brain multigraphs, where two anatomical regions of interests (ROIs) are connected by multiple edges. Each edge captures a particular type of interaction between pairs of ROIs (e.g., structural/functional). Learning a well-centered and representative CBT of a particular brain multigraph population (e.g., healthy or atypical) is a means of modeling complex and varying ROI interactions in a holistic manner. Existing methods generate CBTs by locally integrating heterogeneous multi-edge attributes (e.g., weights and features). However, such methods are agnostic to brain network modularity as they ignore the hierarchical structure of neural interactions. Furthermore, they only perform node-level integration at the individual level without learning the multigraph representation at the group level in a layer-wise manner. To address these limitations, we propose Dual Hierarchical Integration Network (Dual-HINet) for connectional brain template estimation, which simultaneously learns the node-level and cluster-level integration processes using a dual graph neural network architecture. We also propose a novel loss objective to jointly learn the clustering assignment across different edge types and the centered CBT representation of the population multigraphs. Our Dual-HINet significantly outperforms state-of-the-art methods for learning CBTs on a large-scale multigraph connectomic datasets. Our source code can be found at https://github.com/basiralab/Dual-HINet.

Keywords: Brain multigraph population · Connectional brain templates · Graph neural networks · Hierarchical multigraph embedding

1 Introduction

Network Neuroscience [1] is focused on developing tools that model and analyze complex and multi-type relationships between pairs of anatomical brain

L. Wang et al. (Eds.): MICCAI 2022, LNCS 13431, pp. 305–314, 2022.
https://doi.org/10.1007/978-3-031-16431-6_29

regions of interest (ROIs). Recently, the brain multigraph has gained momentum in *jointly* representing different connectivity facets of the brain derived from unimodal as well as multimodal non-invasive magnetic resonance imaging (MRI) [2]. A brain multigraph is endowed with the capacity to model different types of brain connectivity (e.g., functional/morphological) since two ROIs can be connected by multiple edges. Several projects such as the Connectome Related to Human Disease (CRHD) [3] and the Human Connectome Project (HCP) [4] probed the generation of large-scale brain multigraph datasets with the aim to investigate the connectivity patterns in both healthy and atypical populations [5]. To chart out the connectome landscape with all its typical and atypical variability across individuals, the connectional brain template (CBT) was proposed in [6] as a way to transfer the knowledge from a population of brain multigraphs into a single, typically compact network (i.e., single graph), thereby performing a type of data fingerprinting as well as compression. In fact, leveraging the brain template enables not only the integration of complementary information of a given connectomic population but also the generation of new connectomes for synthesizing brain graphs when minimal resources exist [2]. A population-driven CBT can be used to guide brain graph classification as well as evolution prediction [2]. Furthermore, the estimation of a population CBT provides an excellent tool for extracting the integral connectional fingerprint of each population holding its most specific traits, which is an essential step for group comparison studies (e.g., investigating gender differences [7]).

Specifically, [6] presents a landmark work in this field, which first selects the most representative edge connectivity features in a population of brain multigraphs, then leverages similarity network fusion [8] for the final CBT generation. Still, such approach remains dichotomized by patching together different methods, which might accumulate errors throughout the learning pipeline. Besides, it resorts to vectorizing the brain multigraph, encoded in a connectivity tensor, into a single vector which violates the non-Euclidean nature of the brain multigraph and overlooks its topology. To address the issue of topology preservation, [9] proposed a supervised multi-topology network cross-diffusion (SM-Net Fusion) method for learning a population CBT from a weighted combination of multi-topological matrices, that encapsulate the various topologies in a heterogeneous graph population. Nevertheless, the proposed solution cannot support a population of multigraphs as it was primarily designed for graphs.

To address both issues, [10] recently proposed Deep Graph Normalizer (DGN), a state-of-the-art method that compresses a population of brain multigraphs into a CBT using an end-to-end trained graph neural network (GNNs). GNNs are advantageous for learning graph representations from brain networks [2,11] since they can grasp the heterogeneous topology and learn edge-wise and graph-wise complex relationships in an end-to-end fashion. Specifically, DGN uses edge conditioned convolutions [12] to learn node embeddings of ROIs *locally*. The CBTs are then generated from the learned embeddings, capturing well the non-linear relationships within a brain multigraph population. However, DGN integrates multigraph edge attributes locally *at the node-level*, failing to capture the high-order relationship between ROIs in the multigraph population fusion task. As such, DGN is agnostic to both brain modularity and hierarchy in connectivity, which produces a CBT that does not necessarily model the relationship between brain modules (i.e., cluster of nodes) in a hierarchical manner [13,14]. In fact, the hierarchical nestedness of neural interactions [14] is ignored. Besides, connectivities between brain ROIs share cluster-level connectional information [15]. DGN loses the ability to gain knowledge from possible connections among clusters of ROIs due to the limited node-level integration.

To the best of our knowledge, there is no CBT method that captures brain connectivity modularity and hierarchy *at the group level*. Here, we propose **Dual Hierarchical Integration Network (Dual-HINet)**, a novel graph convolutional neural network architecture for integrating a brain multigraph population into a CBT, that considers the duality between node-level and hierarchical cluster-level properties of the brain network. Given a population of brain multigraphs, we learn a representative CBT from each sample using dual blocks jointly trained in an end-to-end fashion: a first block that acts locally at the node-level (Fig. 1-B) and a second dual block that acts globally capturing modular and hierarchical interactions (Fig. 1-C-D). *First*, given a training brain multigraph, we define a first block that integrates multi-edge connectivity weights using an edge-conditioned graph convolutional neural network (GCN), thereby producing node embeddings capturing local node-level ROI interactions. *Second*, we unprecedentedly design a dual GCN block (parallelized to the first one) that learns how to group the brain *multigraph* nodes based on their edge features (i.e., connectivity weights). Specifically, we unprecedentedly propose a multigraph clustering network which groups the nodes through hierarchical layers by examining their multi-edge interactions. In each layer, we learn a coarser representation of the brain multigraph where 'highly interactive' ROIs fuse together (Fig. 1-C), and their embeddings are aggregated from the previous layer through another GCN (Fig. 1-D). *Third*, we concatenate the *node-level* and *cluster-level* embeddings, learned from the dual blocks, into a representative vector from which the *subject-specific* CBT is derived (Fig. 1-E). To this aim, we propose a novel loss objective to jointly learn the clustering assignment across different edge types and the centered CBT representation of the population multigraphs. The representative CBT of the input multigraph population and the multigraph clustering assignments are learned simultaneously. *Finally*, we generate the population CBT by taking the median of all training subject-specific CBTs.

2 Proposed Method

In the following, we explain our dual architecture of the proposed Dual-HINet for estimating CBTs. Figure 1 displays the main components of our framework: **(A)** Our GNN model inputs a population of brain multigraphs. To learn graph representations that capture node-level and cluster-level properties of a multigraph, we use edge-conditioned GCNs 1 to design our blocks (Fig. 1-**B**–**D**). **(C)** Parallel to the node-level block **(B)**, for each input brain multigraph, we propose a hierarchical clustering that groups similar multigraph nodes together based on their

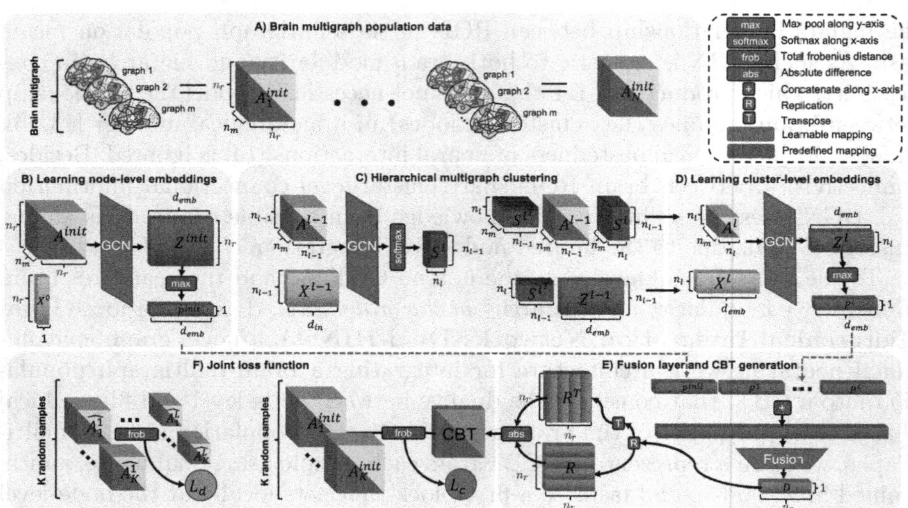

Fig. 1. Proposed Dual-HINet architecture for integrating a population of multigraphs into a centered and representative CBT. **(A) Brain multigraph population data.** Each multigraph sample is represented as a stack of adjacency matrices forming a tensor $\mathcal{A}_s^l \in \mathbb{R}^{n_l \times n_l \times n_m}$. **(B) Learning node-level embeddings.** Multi-edges between brain ROIs are integrated through edge conditioned [12] graph convolution layers.**(C) Hierarchical multigraph clustering.** Parallel to B), a multigraph clustering block groups similar nodes based on their multi-edge attributes. The $\mathbf{S}^l$ soft assignment matrix pools the multigraph node embeddings and multi-edges. **(D) Learning cluster-level embeddings.** We feed the multigraph hierarchical ROI clusters and their heterogeneous connections into GCN layers. Deep cluster node embeddings are then learned and pooled to produce the coarsened multigraph node embeddings. **(E) Fusion layer and CBT generation.** Both learned embeddings from the dual blocks (**B & D**) are concatenated into a feature vector that represents learned attributes for each ROI. We then fuse the feature vectors across layers and transform the resulting vector into a subject-specific CBT matrix for the given brain multigraph sample. **(F) Joint loss function.** To this aim, we jointly optimize a CBT centredness loss (i.e., achieving the minimal distance to all training samples) and the multigraph clustering loss.

multi-edge attributes in a layer-wise manner. As the network gets deeper, the multigraph coarsens from layer to layer. **(D)** We also generate the new embeddings for the coarsened multigraph by learning the cluster-level node features. **(E)** Next, we design a fusion layer for the subject-specific CBT generation by jointly optimizing the CBT centeredness and clustering loss objectives **(F)**. We detail each of these steps below.

Graph Convolutional Networks. Dual-HINet uses edge-conditioned graph convolutional networks with message-passing architecture [16] for integrating brain multigraphs. Specifically, the following graph convolution operation in Eq. 1 is used. In a particular layer k, embeddings $\mathbf{z}_i^k \in \mathbb{R}^{d_k}$ of node i are computed by convolving the learned embeddings $\mathbf{z}_i^{k-1} \in \mathbb{R}^{d_{k-1}}$ from the previous layer with a learnable weight Θ^k and aggregating them using edge-based message passing from local neighboring nodes $\mathcal{N}(i)$. The messages are filtered by $h_\Theta^k(\mathbf{e}_{ij})$, where h_Θ^k is a learnable filter that maps edge features (e.g., connectivity weights) onto a projection matrix: $h_\Theta^k : \mathbb{R}^{n_m} \mapsto \mathbb{R}^{d_k \times d_{k-1}}$.

$$\mathbf{z}_i^k = \Theta^k \cdot \mathbf{z}_i^{k-1} + \frac{1}{|\mathcal{N}(i)|}\left[\sum_{j \in \mathcal{N}(i)} h_\Theta^k(\mathbf{e}_{ij}) \cdot \mathbf{z}_j^{k-1}\right] \tag{1}$$

A) Brain Multigraph Population Data. Given a population of brain multigraphs, each multigraph sample is represented as an initial tensor $\mathcal{A}_s^{init} \in \mathbb{R}^{n_r \times n_r \times n_m}$. Each multigraph has n_r nodes denoting ROIs, which are connected by n_m types of edges (e.g., functional, structural connectivities). More specifically, each edge from node i to node j is a vector $\mathbf{e}_{ij} \in \mathbb{R}^{n_m}$, where n_m connectivity weights are defined as features. For each multigraph, we set the node attributes to ones $[1, \ldots, 1]$ since brain multigraph data do not naturally hold node features.

B) Learning Node-Level Embeddings. Here, we introduce the first block of Dual-HINet (Fig. 1-B), where the most representative attributes of a given brain multigraph are learned and pooled *at the node level.* Specifically, we apply a GCN to the edges of the input subject-specific multigraph. Each GCN layer leverages the local topology of the multigraph ROIs, aggregates the new node embeddings from the previous layer using Eq. 1, and then passes them through a ReLU non-linearity layer. The learned embeddings from each layer are then concatenated into a $\mathbf{Z}^{init} \in \mathbb{R}^{n_r \times d_{emb}}$ matrix. Next, the embedding matrix is pooled node-wise as a vector $\mathbf{P}^{init} \in \mathbb{R}^{1 \times d_{emb}}$. We use max-pooling to identify the most discriminative attributes.

C) Hierarchical Multigraph Clustering. Parallel to block (Fig. 1-B) which operates at the node-level, we further propose a hierarchical multigraph clustering method to model *cluster-level interactions* between nodes for the CBT estimation, thereby capturing its modular aspect [13,14]. While drawing inspiration from ECConv [17] and Diffpool [18], we note that the hierarchical pooling firstly proposed in [18] does not operate on multigraphs and was originally designed for classification where a node denotes a sample to label. Conversely, here we propose a novel multigraph clustering network which learns the optimal clustering to produce the most centered CBT (i.e., achieving the lowest distance to all training multigraph samples) (Fig. 1-C). Specifically, for a given subject, we input its brain multigraph tensor along with the learned node embeddings from the previous layer to a GCN followed by ReLu layers. The GCN here learns embeddings that quantify the assignment scores of each multigraph node to a predefined number of clusters (the node gets assigned to the cluster with the highest score). The learned embeddings pass through a row-wise softmax layer outputting the soft-pooling assignment matrix $\mathbf{S}^l \in \mathbb{R}^{n_{l-1} \times n_l}$. Each element in a row gives the probability of node i belonging to n_l clusters. We replicate the matrix n_m times over z-axis to create a tensor $\mathcal{S}^l \in \mathbb{R}^{n_l \times n_{l-1} \times n_m}$ to pool the multigraph edges into the new coarsened multigraph and clustered nodes are fused into a single node. The multigraph tensor in layer l is derived using the following tensor mapping:

$$\mathcal{A}^l = \mathcal{S}^{l^T} \mathcal{A}^{l-1} \mathcal{S}^l \in \mathbb{R}^{n_l \times n_l \times n_m} \quad (2)$$

The pooled multigraph node embeddings in layer l are updated using the following operation:

$$\mathbf{X}^l = \mathbf{S}^{l^T} \mathbf{Z}^{l-1} \in \mathbb{R}^{n_l \times d_{emb}} \quad (3)$$

Each edge of $\mathcal{A}^l$ represents the connectivity between clusters with n_m features. The new coarsened tensor and node embeddings represent the integrated modular brain multigraph. The hierarchical networks pooled in each layer give us a better holistic representation of the connections between brain ROIs.

D) Learning Cluster-Level Embeddings. With the new clustered multigraph, we move on to the second part of the Dual-HINet, where the modular attributes of ROIs are learned. The multi-edges of the cluster multigraph and the cluster attributes are updated by passing them through a GCN and pooling layers as in part B). Next, we concatenate the output cluster-based embeddings into a matrix $\mathbf{Z}^l \in \mathbb{D}^{n_l \times d_{emb}}$. The pooled node attributes are aggregated into $\mathbf{P}^l \in \mathbb{D}^{1 \times d_{emb}}$.

E) Fusion Layer and CBT Generation. To capture both the brain local and modular interactions in a hierarchical layer-wise manner, we concatenate the learned node-level embeddings $\mathbf{P}^{init}$ and the cluster-level embeddings of layers $l \in \{1, 2, \ldots, L\}$ and pass them through a fusion layer which fuses the hierarchical representations into a single vector $\mathbf{D} \in \mathbb{R}^{1 \times n_r}$ for each ROI, independently. The fusion layer is a linear transformation with learnable weights and biases.

Next, we set out to derive the subject-biased CBT from the final multigraph ROI-based embeddings. To this aim, we replicate $\mathbf{D}$ n_r times creating a matrix $\mathbf{R} \in \mathbb{R}^{n_r \times n_r}$, which is then transposed into a second matrix. We then compute the element-wise absolute difference between $\mathbf{R}$ and its transpose $\mathbf{R}^T$ to produce the final subject-biased CBT matrix $\mathbf{C}_s$. Note that we used the absolute difference since our brain multigraph data was constructed using this metric. Other operations (e.g., correlation) can be used dependently on the data genesis.

F) Joint Loss Function. Here we detail our novel loss function used to jointly learn a well-centered and modular CBT of a given multigraph population. To compute the centeredness loss $\mathcal{L}_c$ [10], we start by randomly sampling K number of subjects from the training population and computing their Frobenius distance to the subject-specific CBT $\mathbf{C}_s$. Random sampling alleviates both computation and memory loads. Next, we sum the distances to the random samples then apply an edge type normalization. The average loss across the selected number of random samples is then taken as the final subject centeredness loss: $\mathcal{L}_c = \sum_{m=1}^{n_m} \sum_{i \in K} \lambda_m \left\| \mathbf{C}_s - \mathbf{A}_i^m \right\|_F ; \lambda_m = \frac{max\{\mu_m\}_{m=1}^{n_m}}{\mu_m}$, where μ_m denotes an edge-type normalizer to prevent the CBT bias towards a particular connectivity type.

On the other hand, we compute the clustering loss $\mathcal{L}_d$ by applying the learned *layer-wise* subject-specific assignment matrices to the K random multigraph samples and compute the average dissimilarity between them. This way we measure the generalizability of the predicted clusters for other subjects. Given a pair of training multigraphs $\mathcal{A}_i^{init}$ and $\mathcal{A}_j^{init}$ of subjects $i, j \in K$, we cluster them using $\mathbf{S}_s^l$ in layers $l \in \{1, 2, ..., L\}$, consecutively. n_d is the total number of distances. The average dissimilarity d_{ij}^l between clustered multigraphs $\hat{\mathcal{A}}^l$ in the lth layer is computed as follows: $\mathcal{L}_d = \sum_{l=1}^{L}(\frac{1}{n_d}\sum_{i,j\in K} d_{ij}^l); d_{ij}^l = \left\|\hat{\mathcal{A}}_i^l - \hat{\mathcal{A}}_j^l\right\|_F$. Last, the final joint loss to optimize is defined as: $\mathcal{L}_J = \mathcal{L}_c + \mathcal{L}_d$. The loss gradient is backpropagated throughout all Dual-HINet blocks.

3 Results and Discussion

Connectomic Datasets. We evaluated our Dual-HINet on 4 large-scale connectomic datasets from the Autism Brain Imaging Data Exchange ABIDE I public dataset [19], including 310 subjects (155 normal control (NC) and 155 subjects with autism spectral disorder (ASD). Each subject of this dataset is represented by 6 cortical morphological brain networks extracted from the maximum principal curvature, mean cortical thickness, mean sulcal depth, average curvature cortical surface area and minimum principle area. For each hemisphere, the cortical surface is reconstructed from T1-weighted MRI using FreeSurfer pipeline [20] and parcellated into 35 ROIs using Desikan-Killiany atlas [21] and its corresponding brain network is derived by computing the pairwise absolute difference in cortical measurements between pairs of ROIs.

Evaluation and Parameter Setting. We trained 4 different variants of Dual-HINet on AD and NC left and right hemisphere datasets: (1) one ablated version where we remove the node-level based block (Fig. 1–B) and (2) the dual version, each trained using $L = \{1, 2\}$ clustering layers, independently. Our Dual-HINet variants were evaluated against the state-of-the-art method DGN [10]. We used 3-fold cross-validation where we derive the CBT from the training folds and evaluate its centeredness and topological soundness against testing multigraph samples. All models were trained for 150 epochs using NVIDIA Tesla V100 GPU. Our graph convolutional networks consisted of 3 edge-conditioned convolutional layers, separated by ReLU activation functions. Each layer's output node embeddings were concatenated altogether. The hidden dimension of the hierarchical embedding layers are set to 25 and 5 dimensions, respectively. The final number of output nodes pooled is 3. We set the learning rate to 0.0005 using Adam optimizer for all models. We set $K = 10$ for the centeredness loss.

CBT Representativeness. To evaluate the representativeness of the learned CBT generated from the training set, we used the average Frobenius distance to all testing multigraph tensors. Figure 2 displays A) the average Frobenius distance computed for the Dual-HINet variants and DGN [10] and B) the mean absolute error in node strength between the learned CBT and training multigraphs. All variants of our proposed model significantly outperformed DGN

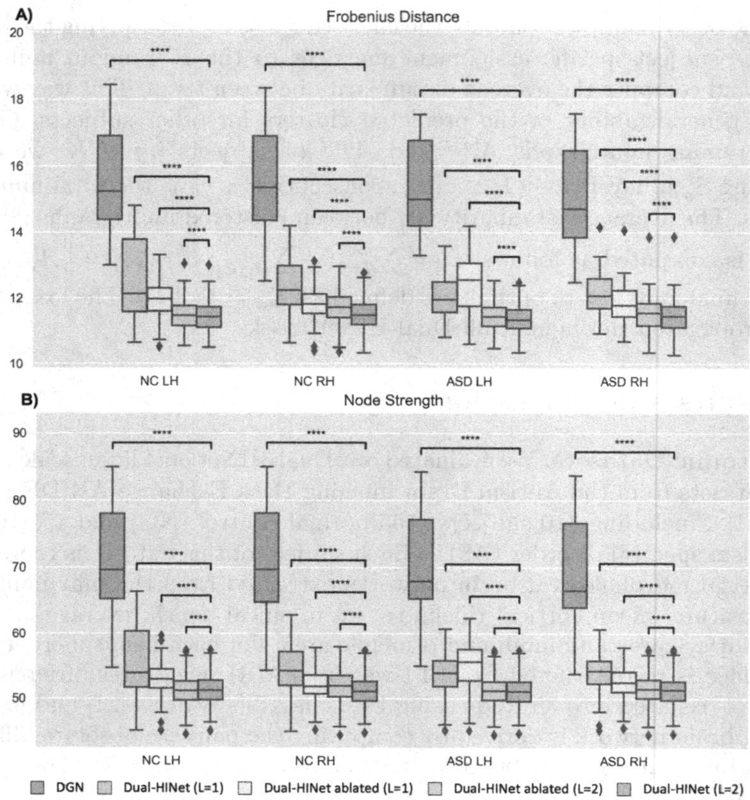

Fig. 2. CBT evaluation by DGN and Dual-HINet variants based on its A) centeredness measure quantifying its representativeness and B) topological soundness using node strength. NC: normal control. ASD: autism spectrum disorder. RH: left hemisphere. LH: right hemisphere.

($p < 0.0001$ using two tailed t-test) in terms of representativeness and topological soundness. So far, we used max-pooling for learning multigraph representations from node embeddings (Fig. 1-B-D). In our future work, we will investigate more advanced pooling strategies [22]. Inspired by [23], we will also investigate the relationships between the training multigraphs in the CBT estimation task.

4 Conclusion

In this paper, we introduced Dual-HINet for learning connectional brain templates from a given population of brain multigraphs. Our model simultaneously learns local node-level and modular cluster-level multigraph representations using a dual graph neural network architecture. We showed that our Dual-HINet significantly outperformed the state-of-the-art and that both dual blocks are capable of synergistically estimating a centered, well-representative and topologically sound CBT. In our future work, we will generalize our framework to handle directed brain connectivity networks [24] as well as predicting the evolution of CBTs [25].

References

1. Bassett, D.S., Sporns, O.: Network neuroscience. Nat. Neurosci. 20, 353–364
2. Bessadok, A., Mahjoub, M.A., Rekik, I.: Graph neural networks in network neuroscience. arXiv preprint arXiv:2106.03535 (2021)
3. Van Essen, D., Glasser, M.: The human connectome project: Progress and prospects. Cerebrum: the Dana Forum on Brain Science 2016 (2016)
4. Essen, D., et al.: The human connectome project: a data acquisition perspective. Neuroimage **62**, 2222–31 (2012)
5. van den Heuvel, M.P., Sporns, O.: A cross-disorder connectome landscape of brain dysconnectivity. Nat. Rev. Neurosci. **20**, 435–446 (2019)
6. Dhifallah, S., Rekik, I., Initiative, A.D.N., et al.: Estimation of connectional brain templates using selective multi-view network normalization. Med. Image Anal. **59**, 101567 (2020)
7. Nebli, A., Rekik, I.: Gender differences in cortical morphological networks. Brain Imaging and Behavior **14**(5), 1831–1839 (2019). https://doi.org/10.1007/s11682-019-00123-6
8. Wang, B., et al.: Similarity network fusion for aggregating data types on a genomic scale. Nat. Methods **11**, 333–337 (2014)
9. Mhiri, I., Mahjoub, M.A., Rekik, I.: Supervised multi-topology network cross-diffusion for population-driven brain network atlas estimation. International Conference on Medical Image Computing and Computer-Assisted Intervention, pp. 166–176 (2020)
10. Gurbuz, M.B., Rekik, I.: Deep graph normalizer: A geometric deep learning approach for estimating connectional brain templates. In: Medical Image Computing and Computer Assisted Intervention, pp. 155–165 (2020)
11. Zhou, J., et al.: Graph neural networks: a review of methods and applications. arXiv preprint arXiv:1812.08434 (2018)
12. Simonovsky, M., Komodakis, N.: Dynamic edge-conditioned filters in convolutional neural networks on graphs. CoRR abs/1704.02901 (2017)
13. Meunier, D., Lambiotte, R., Fornito, A., Ersche, K., Bullmore, E.T.: Hierarchical modularity in human brain functional networks. Front. Neuroinform. **3**, 37 (2009)
14. Zhou, C., Zemanová, L., Zamora, G., Hilgetag, C.C., Kurths, J.: Hierarchical organization unveiled by functional connectivity in complex brain networks. Phys. Rev. Lett. **97**, 238103 (2006)
15. Zhong, H., et al.: Graph contrastive clustering. In: Proceedings of the IEEE/CVF International Conference on Computer Vision, pp. 9224–9233 (2021)
16. Hinrichs, C., Singh, V., Xu, G., Johnson, S.C., Initiative, A.D.N., et al.: Predictive markers for ad in a multi-modality framework: an analysis of mci progression in the adni population. Neuroimage **55**, 574–589 (2011)
17. Simonovsky, M., Komodakis, N.: Dynamic edge-conditioned filters in convolutional neural networks on graphs. In: Proceedings of the IEEE Conference on Computer Vision and Pattern Recognition, pp. 3693–3702 (2017)

18. Ying, Z., You, J., Morris, C., Ren, X., Hamilton, W., Leskovec, J.: Hierarchical graph representation learning with differentiable pooling. Advances in neural information processing systems 31 (2018)
19. Di Martino, A., Yan, C.G., Li, Q., Denio, E., Castellanos, F.X., Alaerts, K., Anderson, J.S., Assaf, M., Bookheimer, S.Y., Dapretto, M., et al.: The autism brain imaging data exchange: towards a large-scale evaluation of the intrinsic brain architecture in autism. Mol. Psychiatry **19**, 659–667 (2014)
20. Fischl, B.: Freesurfer. Neuroimage **62**, 774–781 (2012)
21. Fischl, B., et al.: Automatically parcellating the human cerebral cortex. Cereb. Cortex **14**, 11–22 (2004)
22. Xu, K., Hu, W., Leskovec, J., Jegelka, S.: How powerful are graph neural networks? arXiv preprint arXiv:1810.00826 (2018)
23. Ma, G., He, L., Cao, B., Zhang, J., Yu, P.S., Ragin, A.B.: Multi-graph clustering based on interior-node topology with applications to brain networks. In: Joint European Conference on Machine Learning and Knowledge Discovery in Databases, pp. 476–492 (2016)
24. Shovon, M., Islam, H., Nandagopal, N., Vijayalakshmi, R., Du, J.T., Cocks, B.: Directed connectivity analysis of functional brain networks during cognitive activity using transfer entropy. Neural Process. Lett. **45**, 807–824 (2017)
25. Demirbilek, O., Rekik, I.: Recurrent multigraph integrator network for predicting the evolution of population-driven brain connectivity templates, pp. 584–594 (2021)

RefineNet: An Automated Framework to Generate Task and Subject-Specific Brain Parcellations for Resting-State fMRI Analysis

Naresh Nandakumar[1(✉)], Komal Manzoor[2], Shruti Agarwal[2], Haris I. Sair[2], and Archana Venkataraman[1]

[1] Department of Electrical and Computer Engineering, Johns Hopkins University, Baltimore, USA
nnandak1@jhu.edu

[2] Department of Neuroradiology, Johns Hopkins School of Medicine, Baltimore, USA

Abstract. Parcellations used in resting-state fMRI (rs-fMRI) analyses are derived from group-level information, and thus ignore both subject-level functional differences and the downstream task. In this paper, we introduce RefineNet, a Bayesian-inspired deep network architecture that adjusts region boundaries based on individual functional connectivity profiles. RefineNet uses an iterative voxel reassignment procedure that considers neighborhood information while balancing temporal coherence of the refined parcellation. We validate RefineNet on rs-fMRI data from three different datasets, each one geared towards a different predictive task: (1) cognitive fluid intelligence prediction using the HCP dataset (regression), (2) autism versus control diagnosis using the ABIDE II dataset (classification), and (3) language localization using an rs-fMRI brain tumor dataset (segmentation). We demonstrate that RefineNet improves the performance of existing deep networks from the literature on each of these tasks. We also show that RefineNet produces anatomically meaningful subject-level parcellations with higher temporal coherence.

Keywords: Rs-fMRI · Parcellation refinement · Task optimization

1 Introduction

Resting-state fMRI (rs-fMRI) captures intrinsic neural synchrony, which provides insight into the functional organization of the brain [1]. Due to voxel-level variability, rs-fMRI data is often analyzed at the region level based on a predefined brain parcellation [2,3]. Most parcellation schemes are based on group-level averages [4,5]; however, it is well known that functional landmarks vary from person to person [6,7], particularly for clinical populations [8].

L. Wang et al. (Eds.): MICCAI 2022, LNCS 13431, pp. 315–325, 2022.
https://doi.org/10.1007/978-3-031-16431-6_30

Over the past decade, several methods have been proposed to obtain subject-specific functional boundaries from rs-fMRI data. The popular technique is independent component analysis (ICA), which estimates a set of spatially independent maps based on a linear decomposition of the voxel-wise rs-fMRI time series [9–12]. While ICA provides valuable subject-level information, the user cannot control the number or spatial compactness of the estimated components. For this reason, it may not be possible to match region boundaries across subjects to draw group-level inferences. The more recent work of [13] uses a spatio-temporal deep residual network and ICA template priors to obtain subject-specific parcellations. While this method goes beyond traditional ICA to encode time-varying dynamics into network extraction, it has only been validated on a cohort of three subjects and might not generalize. The authors of [6] take a different approach by casting the problem as one of *refining* an existing parcellation, rather than creating a new parcellation from scratch. The proposed method iteratively reassigns the voxel memberships based on the maximum Pearson's correlation between the voxel time series and the mean time series of the current regions. The method in [8] builds upon this work by using a Markov random field prior to encourage spatial continuity during the iterative reassignment procedure. However, this method was only evaluated on a coarse initial parcellation (17 regions). Finally, the work of [7] uses a group sparsity prior as well as Markov Random Fields to generate subject-specific parcellations using an iterative graph partitioning approach. However, the authors do not show that the method improves performance on downstream tasks such as regression or classification.

In this paper we introduce RefineNet, the first deep learning approach for subject-specific and task-aware parcellation refinement using rs-fMRI data. RefineNet encodes both spatial and temporal information via a weight matrix that learns relationships between neighboring voxels and a coherence module that compares the voxel- and region-level time series. Importantly, RefineNet is designed as an all-purpose module that can be attached to existing neural networks to optimize task performance. We validate RefineNet on rs-fMRI data from three different datasets, each one designed to perform a different task: (1) cognitive fluid intelligence prediction (regression) on HCP [14], (2) autism spectrum disorder (ASD) versus neurotypical control (NC) classification on ABIDE [15], and (3) language localization using an rs-fMRI dataset of brain tumor patients. In each case, we attach RefineNet to an existing deep network from the literature designed for the given task. Overall, RefineNet improves the temporal cohesion of the learned region boundaries and the downstream task performance.

2 Methods

Figure 1 illustrates our RefineNet strategy. The inputs to RefineNet are the 4D rs-fMRI data $\mathbf{Z}$ and the original brain parcellation $\mathbf{X}^{(0)}$. We formulate a pseudo-prior, pseudo-likelihood and MAP style inference model to obtain the refined parcellation $\mathbf{X}^{(e)}$. Following this procedure, RefineNet can be attached to an existing deep network to fine-tune $\mathbf{X}^{(e)}$ for downstream task performance.

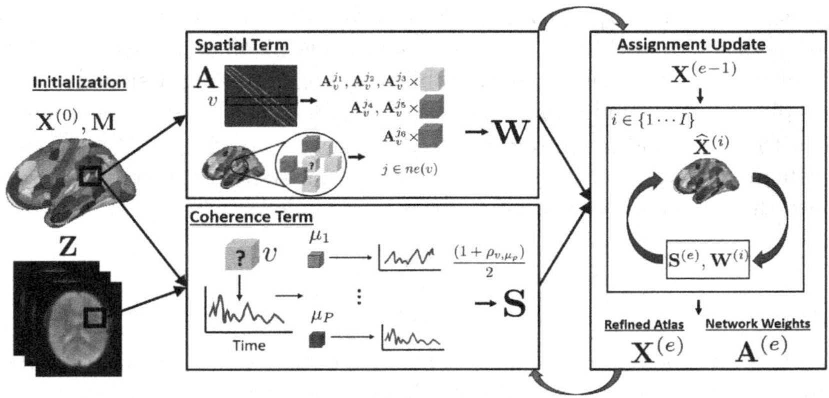

Fig. 1. Inputs: rs-fMRI $\mathbf{Z}$, existing parcellation $\mathbf{X}^{(0)}$ and neighbor mask $\mathbf{M}$. **Top:** We show a six neighbor model for clarity. Our network parameter $\mathbf{A}$ learns voxel neighbor weights while the product $(\mathbf{AX})$ multiplies these weights with the labels of the neighboring voxels. **Bottom:** The coherence term $\mathbf{S}$ uses the pearson correlation coefficient with each mean time series μ_p. **Right:** We obtain intermediate labels $\widehat{\mathbf{X}}^{(i)}$ I times before taking the mode and producing the next epoch's parcellation $\mathbf{X}^{(e)}$, which is used during backpropagation to obtain $\mathbf{A}^{(e)}$.

2.1 Spatial and Temporal Coherence Terms

Let V be the number of voxels in the rs-fMRI scan, and P be the number of regions in the original parcellation. We define $\mathbf{X} \in \mathbb{R}^{V \times P}$ to RefineNet as a one-hot encoded label matrix, where $\mathbf{X}_{v,p} = 1$ when voxel v is assigned to region p and $\mathbf{X}_{v,p} = 0$ otherwise. The core assumption of RefineNet is that voxels in close spatial proximity to each other are likely to belong to the same region [6,8]. We encode this information via the intermediate activation $\mathbf{W} \in \mathbb{R}^{V \times P}$

$$\mathbf{W} = \text{ReLU}(\mathbf{AX}), \tag{1}$$

where the matrix $\mathbf{A} \in \mathbb{R}^{V \times V}$ enforces the local structure of the data. Formally, we obtain $\mathbf{A}$ as the Hadamard product of a sparse binary adjacency matrix $\mathbf{M} \in \mathbb{R}^{V \times V}$ that is nonzero only when the voxels are spatial neighbors and a learnable weight matrix $\hat{\mathbf{A}} \in \mathbb{R}^{V \times V}$ to encode spatially varying dependencies. Figure 1 shows the nonzero weights in $\mathbf{A}_v$ being multiplied by the current labels of the neighbors of voxel v, where $ne(v)$ denotes neighbors of voxel v.

At a high level, Eq. (1) acts as a proxy for the prior probability that voxel v belongs to region p based on the contribution of its neighbors currently assigned to region p, as governed by the spatially varying weights in $\mathbf{A}$. Thus, our pseudo-prior term is designed to identify which neighbors are more important for voxel reassignment, which is important for boundary areas. Note that $\mathbf{A}$ is sparse by construction, which reduces both memory and computational overhead.

It is generally accepted that highly correlated voxels are more likely to be involved in similar functional processes, and if near each other, should be grouped into the same region [6,7]. Let $\mathbf{Z} \in \mathbb{R}^{V \times T}$ denote the voxel-wise time series, where T is the duration of rs-fMRI scan. Thus, our pseudo-likelihood matrix $\mathbf{S} \in \mathbb{R}^{V \times P}$ that captures the un-normalized probability of voxel v being assigned to region p is simply a shifted and scaled version of the Pearson's correlation coefficient between the voxel and mean region-wise time series, i.e., $\mathbf{S}_{v,p} = \frac{\rho_{\mathbf{z}_v,\mu_p}+1}{2}$.

Mathematically, given the voxel-to-region membership captured in $\mathbf{X}$, we can compute the region-wise mean time series μ_p as follows:

$$\mu_p = \frac{\sum_v^V \mathbf{Z} \cdot \mathbf{X}_{v,p}}{\sum_{v=1}^V \mathbf{X}_{v,p}}. \tag{2}$$

The correlation coefficient $\rho_{\mathbf{z}_v,\mu_p}$ can also be obtained via matrix operations, allowing us to integrate the pseudo-likelihood term directly into a deep network.

2.2 RefineNet Training and Optimization

We adopt an iterative max product approach to derive our assignment updates. For convenience, let the index e denote the main epochs and the index i denote the refinement iterate. For each epoch e, we initialize the intermediate variable $\widehat{\mathbf{X}}^{(1)}$ with the assignment matrix $\mathbf{X}^{(e-1)}$ from the previous iterate and compute the pseudo-likelihood matrix $\mathbf{S}^{(e)}$ via the mean time series defined in Eq. (2). We then iteratively update $\widehat{\mathbf{X}}^{(i)}$ based on neighborhood information as follows:

$$\widehat{\mathbf{X}}_{v,p}^{(i+1)} = \begin{cases} 1 & \text{argmax}_p \left\{ \mathbf{S}_{v,:}^{(e)} \odot \mathbf{W}_{v,:}^{(i)} \right\} \\ 0 & \text{else}, \end{cases} \tag{3}$$

where $\mathbf{W}^{(i)} = \text{ReLU}(\mathbf{A}\widehat{\mathbf{X}}^{(i)})$ as defined in Eq. (1), and the operator $\odot$ is the Hadamard product. The term $\mathbf{S}^{(e)}$ remains constant throughout this iterative process from $i = \{1 \cdots I\}$ to act as the previous stationary point. The refined parcellation $\mathbf{X}^{(e)}$ for epoch e is given by the majority vote over the intermediate region assignments $\{\widehat{\mathbf{X}}^{(i)}\}_{i=1}^I$. We employ this iterative approach over the pseudo-prior term to leverage the space of intermediate label distributions for a robust re-assignment. We fix $I = 20$ in this work, as we empirically observed that this was large enough to provide robust reassignment.

We optimize the weights $\mathbf{A}$ in RefineNet via stochastic gradient descent to maximize the average temporal coherence with the newly assigned regions. Let V_p be the set of voxels assigned to region p. Our loss for backpropagation is

$$\mathcal{L}_{RN} = -\frac{1}{P}\sum_{v=1}^{V} \frac{1}{|V_p|} \sum_{v \in V_p} \frac{(1 + \rho_{\mathbf{z}_v,\mu_p})}{2} \tag{4}$$

For clarity, our full training procedure is described in Algorithm 1.

Algorithm 1. RefineNet Training Procedure

procedure RefineNet($\mathbf{X}, \mathbf{Z}, \mathbf{M}, E, I = 20$)
  $\mathbf{X}^{(0)} \leftarrow \mathbf{X}$
  $\mathbf{A}^{(0)} \leftarrow \hat{\mathbf{A}}, \mathbf{M}$ ▷ Random initialization of weights in nonzero entries
  $\{\mu_1^{(0)} \cdots \mu_P^{(0)}\}, \mathbf{S}^{(0)} \leftarrow \mathbf{Z}, \mathbf{X}^{(0)}$ ▷ Eq.(2)
  for $e = 1 : E$ **do**
    $\widehat{\mathbf{X}}^{(1)} \leftarrow \mathbf{X}^{(e-1)}$
    for $i = 1 : I$ **do**
      $\mathbf{W}^{(i)} \leftarrow \mathbf{A}^{(e-1)}, \widehat{\mathbf{X}}^{(i)}$ ▷ Eq.(1)
      $\widehat{\mathbf{X}}^{(i+1)} \leftarrow \mathbf{S}^{(e-1)}, \mathbf{W}^{(i)}$ ▷ Eq.(3)
    $\mathbf{X}^{(e)} \leftarrow mode(\{\widehat{\mathbf{X}}\}_{i=1}^{I})$
    $\{\mu_1^{(e)} \cdots \mu_P^{(e)}\}, \mathbf{S}^{(e)} \leftarrow \mathbf{Z}, \mathbf{X}^{(e)}$ ▷ Eq.(2)
    $\mathcal{L}_{RN} \leftarrow \mathbf{X}^{(e)}, \{\mu_1^{(e)} \cdots \mu_P^{(e)}\}$ ▷ Eq.(4)
    $\mathbf{A}^{(e)} \leftarrow \mathcal{L}_{RN}$, SGD ▷ Backpropagation and gradient update
  return $\mathbf{X}^{(E)}$

2.3 Creating Task-Aware Parcellations with RefineNet

Once pretrained using Eq. (4), RefineNet can be attached to existing deep neural networks and re-optimized for performance on the downstream task. Our strategy is to pre-train RefineNet for 50 epochs using a learning rate of 0.001 before jointly training RefineNet with the network of interest. Here, we alternate between training just the network of interest for task performance and training both RefineNet and the network of interest in an end-to-end fashion. Empirically, we observed this strategy provides a good balance of task-optimization and preserving functional cohesion. Our second-stage loss function is a weighted sum of the downstream task and the RefineNet loss in Eq. (4):

$$\mathcal{L}_{total} = \mathcal{L}_{net} + \lambda \mathcal{L}_{RN}, \tag{5}$$

where the hyperparameter λ can be chosen via a grid search or cross validation.

3 Experimental Results

We validate RefineNet on three different rs-fMRI datasets and prediction tasks. In each case, we select an existing deep network architecture from the literature to be combined with RefineNet. These networks take as input a $P \times P$ rs-fMRI correlation matrix. Figure 2 illustrates the combined network architectures for each prediction task. We implement each network in Pytorch and use the hyperparameters and training strategy specified in the respective paper.

Our task-aware optimization (Sect. 2.3) alternates between by training the network of interest for e_a epochs while keeping RefineNet (and the input correlation matrices) fixed. We then jointly train both networks for e_a epochs while refining the parcellation, and thus, the correlation inputs between epochs.

3.1 Description of Networks and Data

M-GCN for Regression using HCP: We use the M-GCN model (rs-fMRI only) from [16] to predict the cognitive fluid intelligence score (CFIS). The dataset contains 300 healthy subjects from the publicly available Human Connectome Project (HCP) S1200 release [14]. Standard rs-fMRI preprocessing was done according to [17], which handles motion, physiological artifacts, and registration to the MNI template. For simplicity, the CFIS values are scaled between $(0-10)$ based on the training data of each fold. We report the mean absolute error (MAE) and correlation coefficient between the predicted and true scores.

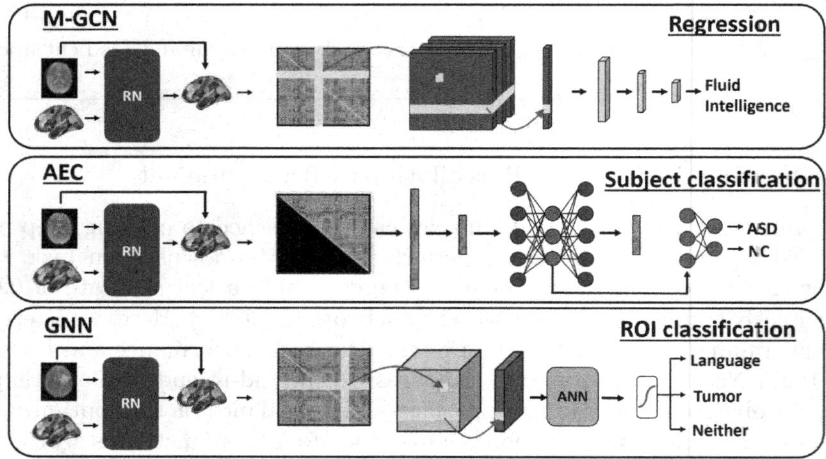

Fig. 2. Top: The M-GCN uses a graph convolution network applied to the connectivity matrix to predict fluid intelligence in HCP subjects. **Middle:** The AEC couples an autoencoder and a single layer perceptron to classify ASD vs. NC on ABIDE data. **Bottom:** The GNN uses graph convolutions to segment the language areas of eloquent cortex on a tumor dataset.

AEC for Classification using ABIDE: We use the autoencoder/classifier (AEC) framework from [18] to predict subject diagnosis. The dataset contains 233 subjects (131 ASD, 102 NC) from the Autism Brain Imaging Data Exchange (ABIDE) II dataset [15]. The data was acquired across six different sites and preprocessed using the Configurable Pipeline for Analysis of Connectomes (CPAC) toolbox [19]. As per [18], the AEC network performs ASD vs. NC (neurotypical control) classification using the upper triangle portion of the rs-fMRI correlation matrix. We report the accuracy and area under the curve (AUC).

GNN for Language Localization in Tumor Patients: We use the GNN proposed by [20] to localize language areas of the brain in a lesional cohort. The dataset contains rs-fMRI and task fMRI data from 60 brain tumor patients. The

Table 1. Results across all experiments considered. Metric 1 represents MAE for regression and AUC for classification and localization while metric 2 represents correlation for regression and overall accuracy for classification and localization.

Task	Model	Atlas	Metric 1	Metric 2	P-value
CFIS Prediction	Original	BNA246	2.20 ± 0.13	0.24 ± 0.029	
		CC200	2.24 ± 0.14	0.27 ± 0.045	
		AAL90	2.22 ± 0.16	0.23 ± 0.048	
	RefineNet Only	BNA246	2.22 ± 0.13	0.19 ± 0.026	0.64
		CC200	2.22 ± 0.18	0.22 ± 0.036	0.293
		AAL90	2.15 ± 0.14	0.25 ± 0.032	0.121
	Combined	BNA246	$\mathbf{1.73 \pm 0.14}$	$\mathbf{0.3 \pm 0.039}$	**0.016****
		CC200	$\mathbf{1.84 \pm 0.12}$	$\mathbf{0.34 \pm 0.046}$	**0.045****
		AAL90	$\mathbf{1.91 \pm 0.11}$	$\mathbf{0.36 \pm 0.04}$	**0.078***
ASD vs. NC	Original	BNA246	0.65 ± 0.017	65.5 ± 1.57	
		CC200	0.66 ± 0.024	64.9 ± 2.12	
		AAL90	0.66 ± 0.029	64.5 ± 2.49	
	RefineNet Only	BNA246	0.63 ± 0.021	63.8 ± 1.78	0.74
		CC200	0.69 ± 0.016	66.6 ± 1.80	0.22
		AAL90	0.70 ± 0.021	67.5 ± 1.94	**0.08***
	Combined	BNA246	$\mathbf{0.69 \pm 0.013}$	$\mathbf{67.8 \pm 1.60}$	**0.062***
		CC200	$\mathbf{0.72 \pm 0.029}$	$\mathbf{69.8 \pm 1.76}$	**0.022****
		AAL90	$\mathbf{0.74 \pm 0.023}$	$\mathbf{71.8 \pm 1.84}$	**0.006****
Localization	Original	BNA246	0.74 ± 0.022	84.6 ± 0.09	
		CC200	0.75 ± 0.021	85.9 ± 0.92	
		AAL90	0.67 ± 0.023	82.32 ± 1.21	
	RefineNet Only	BNA246	0.75 ± 0.023	84.95 ± 0.91	0.261
		CC200	0.75 ± 0.018	84.6 ± 0.71	0.531
		AAL90	0.65 ± 0.021	81.8 ± 1.34	0.834
	Combined	BNA246	$\mathbf{0.77 \pm 0.021}$	$\mathbf{85.9 \pm 0.91}$	**0.065***
		CC200	$\mathbf{0.78 \pm 0.017}$	$\mathbf{86.9 \pm 1.01}$	**0.047****
		AAL90	0.68 ± 0.019	82.63 ± 1.09	0.312

data was acquired on a 3T Siemens Trio Tim system (EPI; TR = 2000 ms, TE = 30 ms, res = $1\,\text{mm}^3$ isotropic). The rs-fMRI was scrubbed using ArtRepair, followed by CompCorr for nuisance regression [21], bandpass filtering, and spatial smoothing. The task fMRI was used to derive "ground-truth" language labels for training and evaluation [22]. The tumor boundaries were obtained via expert segmentation. The GNN outputs a label (language, tumor, or neither) for each region. We report the overall accuracy and AUC for detecting the language class.

3.2 Quantitative Task Performance

We compare three model configurations: (1) no refinement (original), (2) using just RefineNet to maximize temporal coherence (RefineNet only), and integrating RefineNet into an auxiliary network, as described in Sect. 2.3 (combined). We also apply three parcellations to each task: the Brainnetome atlas (BNA246) [23],

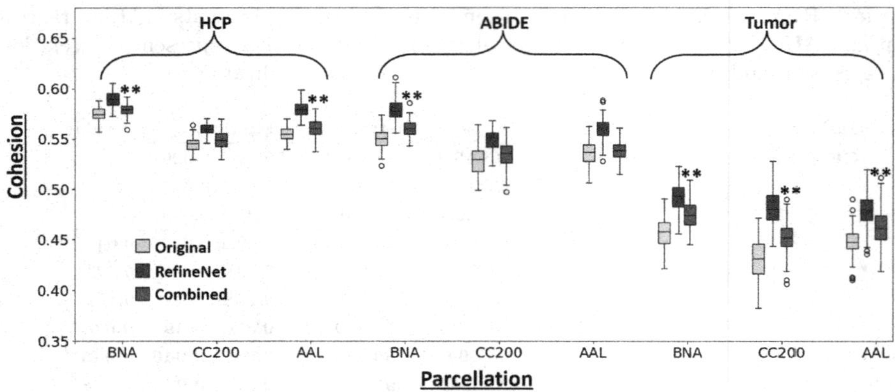

Fig. 3. Boxplots for region cohesion across the nine experiments. Yellow refers to the original model, blue refers to RefineNet only and green refers to combined. (**) denotes a significant increase from the original to combined parcellation.

the Craddocks 200 atlas (CC200) [24], and the Automated Anatomical Labelling (AAL90) atlas [25]. To prevent data leakage, we tune the hyperparameters λ in Eq. (5) and alternating training epoch for regression on 100 additional HCP subjects, yielding $\lambda = 0.2$ and $e_a = 5$.

Table 1 reports the quantitative performance for each model/atlas configuration. Metrics 1/2 refer to MAE/correlation for the regression task and AUC/accurary for the classification and localization tasks, respectively. We employ a ten repeated 10-fold cross validation (CV) evaluation strategy to quantify performance variability. We report mean$\pm$ standard deviation for each metric along with the FDR corrected p-value to indicate statistically improved performance in Metric 1 over the original model using the same parcellation [26]. As seen, the combined model provides statistically significant performance gains in eight out of nine experiments. In contrast, using RefineNet alone to strengthen functional coherence does not necessarily improve performance. Thus, our task-aware optimization procedure is crucial when considering downstream applications. Finally, we note that the AAL90 parcellation is likely too coarse for the language localization task, as reflected in the drastically lower performance metrics.

3.3 Parcellation Cohesion

Figure 3 illustrates the average temporal cohesion of regions in the final parcellation, as computed on the testing data in each repeated CV fold. Once again, let μ_p denote the mean time series in each region p. We define the cohesion C as

$$C = \frac{1}{P} \sum_{v=1}^{V} \frac{1}{|V_p|} \sum_{v \in V_p} \rho_{\mathbf{z}_v, \mu_p}. \tag{6}$$

Unsurprisingly, the parcellations recovered from just using RefineNet (with no downstream task awareness) achieve the highest cohesion. However, as shown in Table 1, these parcellations are not always suited to the prediction task. In contrast, the combined model produces more cohesive parcellations than the original atlas with statistically significant improvement denoted by (**). Taken together, attaching RefineNet to an existing model achieves a good balance between functionally-cohesive grouping and task performance.

4 Conclusion

We present RefineNet, a flexible neural network module capable of obtaining meaningful subject-specific and task-aware parcellations. Our Bayesian-inspired approach considers both spatial contiguity and temporal coherence in reassignment. We show significant performance gains across three different datasets and prediction tasks when RefineNet is appended to existing networks from the literature. Finally, we show that even the task-driven refinement procedure produces more functionally cohesive parcellations than the origial atlas. Our work is a first of its kind, as other parcellation refinement methods are not able to be jointly trained with existing deep networks for task-awareness. In summary, our results show that RefineNet can be a promising tool for rs-fMRI analysis.

Acknowledgements. This work was supported by the National Science Foundation CAREER award 1845430 (PI: Venkataraman) and the Research & Education Foundation Carestream Health RSNA Research Scholar Grant RSCH1420.

References

1. B. Biswal, F. Zerrin Yetkin, V. M. Haughton, and J. S. Hyde, "Functional connectivity in the motor cortex of resting human brain using echo-planar mri," Magnetic resonance in medicine, vol. 34, no. 4, pp. 537–541, 1995
2. van Oort, E.S., et al.: Functional parcellation using time courses of instantaneous connectivity. Neuroimage **170**, 31–40 (2018)
3. Khosla, M., Jamison, K., Kuceyeski, A., Sabuncu, M.R.: Ensemble learning with 3d convolutional neural networks for functional connectome-based prediction. Neuroimage **199**, 651–662 (2019)
4. Fischl, B., et al.: Whole brain segmentation: automated labeling of neuroanatomical structures in the human brain. Neuron **33**(3), 341–355 (2002)
5. Glasser, M.F., et al.: A multi-modal parcellation of human cerebral cortex. Nature **536**(7615), 171–178 (2016)
6. Wang, D., et al.: Parcellating cortical functional networks in individuals. Nat. Neurosci. **18**(12), 1853 (2015)
7. Chong, M., et al.: Individual parcellation of resting FMRI with a group functional connectivity prior. Neuroimage **156**, 87–100 (2017)
8. Nandakumar, N., et al.: Defining patient specific functional parcellations in Lesional Cohorts via Markov random fields. In: Wu, G., Rekik, I., Schirmer, M.D., Chung, A.W., Munsell, B. (eds.) CNI 2018. LNCS, vol. 11083, pp. 88–98. Springer, Cham (2018). https://doi.org/10.1007/978-3-030-00755-3_10

9. Esposito, F., et al.: Independent component model of the default-mode brain function: combining individual-level and population-level analyses in resting-state fmri. Magn. Reson. Imaging **26**(7), 905–913 (2008)
10. Tessitore, A., et al.: Default-mode network connectivity in cognitively unimpaired patients with parkinson disease. Neurology **79**(23), 2226–2232 (2012)
11. Calhoun, V.D., Adali, T.: Multisubject independent component analysis of FMRI: a decade of intrinsic networks, default mode, and neurodiagnostic discovery. IEEE Rev. Biomed. Eng. **5**, 60–73 (2012)
12. Sair, H.I., et al.: Presurgical brain mapping of the language network in patients with brain tumors using resting-state FMRI: comparison with task f MRI. Hum. Brain Mapp. **37**(3), 913–923 (2016)
13. Kazemivash, B., Calhoun, V.D.: A novel 5d brain parcellation approach based on spatio-temporal encoding of resting FMRI data from deep residual learning. J. Neurosci. Methods, 109478 (2022)
14. Van Essen, D.C., et al.: The wu-minn human connectome project: an overview. Neuroimage **80**, 62–79 (2013)
15. Di Martino, A., et al.: Enhancing studies of the connectome in autism using the autism brain imaging data exchange ii. Sci. Data **4**(1), 1–15 (2017)
16. Dsouza, N.S., Nebel, M.B., Crocetti, D., Robinson, J., Mostofsky, S., Venkataraman, A.: M-GCN: a multimodal graph convolutional network to integrate functional and structural connectomics data to predict multidimensional phenotypic characterizations. In: Medical Imaging with Deep Learning, pp. 119–130, PMLR (2021)
17. Smith, S.M., et al.: Resting-state FMRI in the human connectome project. Neuroimage **80**, 144–168 (2013)
18. Zhang, J., Feng, F., Han, T., Gong, X., Duan, F.: Detection of autism spectrum disorder using FMRI functional connectivity with feature selection and deep learning. Cognitive Computation, pp. 1–12 (2022)
19. Craddock, C., et al.: Towards automated analysis of connectomes: the configurable pipeline for the analysis of connectomes (C-PAC). Front. Neuroinform. **42**, 10–3389 (2013)
20. Nandakumar, N., et al.: A novel graph neural network to localize eloquent cortex in brain tumor patients from resting-state fMRI connectivity. In: Schirmer, M.D., Venkataraman, A., Rekik, I., Kim, M., Chung, A.W. (eds.) CNI 2019. LNCS, vol. 11848, pp. 10–20. Springer, Cham (2019). https://doi.org/10.1007/978-3-030-32391-2_2
21. Behzadi, Y., Restom, K., Liau, J., Liu, T.T.: A component based noise correction method (compcor) for bold and perfusion based FMRI. Neuroimage **37**(1), 90–101 (2007)
22. Penny, W.D., Friston, K.J., Ashburner, J.T., Kiebel, S.J., Nichols, T.E.: Statistical parametric mapping: the analysis of functional brain images. Elsevier (2011)
23. Fan, L., et al.: The human brainnetome atlas: a new brain atlas based on connectional architecture. Cereb. Cortex **26**(8), 3508–3526 (2016)
24. Craddock, R.C., James, G.A., Holtzheimer, P.E., III., Hu, X.P., Mayberg, H.S.: A whole brain FMRI atlas generated via spatially constrained spectral clustering. Hum. Brain Mapp. **33**(8), 1914–1928 (2012)

25. Tzourio-Mazoyer, N., et al.: Automated anatomical labeling of activations in SPM using a macroscopic anatomical parcellation of the MNI MRI single-subject brain. Neuroimage **15**(1), 273–289 (2002)
26. Bouckaert, R.R., Frank, E.: Evaluating the replicability of significance tests for comparing learning algorithms. In: Dai, H., Srikant, R., Zhang, C. (eds.) PAKDD 2004. LNCS (LNAI), vol. 3056, pp. 3–12. Springer, Heidelberg (2004). https://doi.org/10.1007/978-3-540-24775-3_3

Modelling Cycles in Brain Networks with the Hodge Laplacian

Sixtus Dakurah(✉), D. Vijay Anand, Zijian Chen, and Moo K. Chung

University of Wisconsin-Madison, Madison, USA
{sdakurah,zijian.chen,mkchung}@wisc.edu

Abstract. Cycles or loops in a network embed higher-order interactions beyond pairwise relations. The cycles are essential for the parallel processing of information and enable feedback loops. Despite the fundamental importance of cycles in understanding the higher-order connectivity, identifying and extracting them are computationally prohibitive. This paper proposes a novel persistent homology-based framework for extracting and modelling cycles in brain networks using the Hodge Laplacian. The method is applied in discriminating the functional brain networks of males and females. The code for modeling cycles through the Hodge Laplacian is provided in https://github.com/laplcebeltrami/hodge.

Keywords: Topological data analysis · Persistent homology · Hodge Laplacian · Cycles

1 Introduction

The human brain network is a complex system that exhibits collective behaviors at multiple spatial and temporal scales [25]. The mechanisms responsible for these behaviors are often attributed to the higher-order interactions that occur across multiple scales. Understanding the higher-order interactions of the brain regions is crucial to modelling the dynamically evolving structural and functional organization of the brain networks.

The brain networks are often analyzed using graph theory methods that provide quantitative measures ranging from local scales at the node level to global scales at the large community level [4]. Despite the success of graph theory approaches, they can only account for pairwise (dyadic) interactions [2]. The representation of complex networks using high-dimensional objects such as triangles to capture triadic interactions has recently gained traction [2,11,19,22]. The mathematical construct used for this purpose is the simplicial complex, which contains basic building blocks referred to as simplices: nodes (0-simplices), edges (1-simplices), triangles (2-simplices) and tetrahedrons (3-simplices). These simplices systematically encode higher-order interactions [12]. The dynamics of

This study is funded by NIH R01 EB022856, EB02875, NSF MDS-2010778.

L. Wang et al. (Eds.): MICCAI 2022, LNCS 13431, pp. 326–335, 2022.
https://doi.org/10.1007/978-3-031-16431-6_31

these interactions across multiple scales are quantified by hierarchically generating a nested sequence of simplicial complexes called the filtration in persistent homology [10]. In a simplicial complex representation of the brain network, some regions are densely connected while others remain sparse leading to the formation of cycles [21]. A cycle in a brain network is the most fundamental higher-order interaction, which allows for the information flow in a closed path and enables feedback [17,21]. However, it is not trivial to extract or explicitly model them [5,9,10].

This paper aims to model the cyclic relationship in brain networks using the Hodge Laplacian. Further, we develop a new topological inference procedure to characterize cycles across subjects and determine the most discriminating cycles between groups. The explicit cycle modelling framework we introduced allows us to localize the connections contributing to this difference, a novelty not attainable by existing persistent homology methods [7,10,16,20]. The method is applied in determining the most discriminating cycles between the male and female brain networks obtained from the resting-state functional magnetic resonance images (fMRI).

2 Methods

2.1 Homology of a Simplicial Complex

A k-simplex $\sigma_k = (v_0, \cdots, v_k)$ is a k-dimensional convex hull (polytope) of nodes $v_0, \cdots, v_k$. A simplicial complex K is a set of simplices such that for any $\tau_i, \tau_j \in K$, $\tau_i \cap \tau_j$ is a face of both simplices; and a face of any simplex $\tau_i \in K$ is also a simplex in K [10]. A 0-skeleton is a simplicial complex consisting of only nodes. A 1-skeleton is a simplicial complex consisting of nodes and edges. Graphs are 1-skeletons. A k-chain is a finite sum $\sum a_i \tau_i$, where the a_i are either 0 or 1. The set of k-chains forms a group and a sequence of these groups is called a chain complex. To relate different chain groups, we use the boundary maps [24]. For two successive chain groups $\mathcal{K}_k$ and $\mathcal{K}_{k-1}$, the boundary operator $\partial_k : \mathcal{K}_k \longrightarrow \mathcal{K}_{k-1}$ for each k-simplex σ_k is given by

$$\partial_k(\sigma_k) = \sum_{i=0}^{k} (-1)^i (v_0, \cdots, \widehat{v_i}, \cdots, v_k),$$

where $(v_0, \cdots, \widehat{v_i}, \cdots, v_k)$ gives the k-1 faces of σ_k obtained by deleting node $\widehat{v_i}$. The matrix representation $\mathbb{B}_k = (\mathbb{B}_k^{ij})$ of the boundary operator is given by

$$\mathbb{B}_k^{ij} = \begin{cases} 1, & \text{if } \sigma_{k-1}^i \subset \sigma_k^j \text{ and } \sigma_{k-1}^i \sim \sigma_k^j \\ -1, & \text{if } \sigma_{k-1}^i \subset \sigma_k^j \text{ and } \sigma_{k-1}^i \nsim \sigma_k^j \\ 0, & \text{if } \sigma_{k-1}^i \not\subset \sigma_k^j \end{cases}, \tag{1}$$

where $\sim$ and $\nsim$ denote similar and dissimilar orientations respectively.

The kernel of the boundary operator is denoted as $\mathcal{Z}_k = ker(\partial_k)$ and its image denoted as $\mathcal{B}_k = img(\partial_{k+1})$. $\mathcal{Z}_k$ and $\mathcal{B}_k$ are the subspaces of $\mathcal{K}_k$. The elements of $\mathcal{Z}_k$ and $\mathcal{B}_k$ are known as k-cycles and k-boundaries respectively [14]. Note that $\mathcal{B}_k \subseteq \mathcal{Z}_k$. The set quotient $\mathcal{H}_k = \mathcal{Z}_k/\mathcal{B}_k$ is termed as the k-th homology group [6,14,24]. The k-th Betti number $\beta_k = rank(\mathcal{H}_k)$ counts the number of algebraically independent k-cycles. The first homology group is $\mathcal{H}_1 = ker(\partial_1)$ since $img(\partial_2) = \varnothing$.

2.2 Spectral Representation of 1-cycles

Hodge Laplacian. The Hodge Laplacian $\mathcal{L}_k$ is a higher dimensional generalization of the graph Laplacian for k-simplices [16]. The k-th Hodge Laplacian $\mathcal{L}_k$ is defined as

$$\mathcal{L}_k = \mathbb{B}_{k+1}\mathbb{B}_{k+1}^\top + \mathbb{B}_k^\top \mathbb{B}_k. \tag{2}$$

The k-th homology group $\mathcal{H}_k$ is the kernel of Hodge Laplacian, i.e., $\mathcal{H}_k = ker\mathcal{L}_k$ [16]. The kernel space of $\mathcal{L}_k$ is spanned by the eigenvectors corresponding to the zero eigenvalues of $\mathcal{L}_k$. The multiplicity of the zero eigenvalues is β_k. The eigen decomposition of $\mathcal{L}_k$ is given by

$$\mathcal{L}_k \mathbf{U}_k = \mathbf{U}_k \boldsymbol{\Lambda}_k, \tag{3}$$

where $\boldsymbol{\Lambda}_k$ is the diagonal matrix consisting of eigenvalues of $\mathcal{L}_k$ with corresponding eigenvectors in the columns of $\mathbf{U}_k$. Brain networks are usually represented as connectivity matrices, from which the 1-skeleton can be obtained by thresholding. For 1-skeletons, the boundary matrix $\mathbb{B}_2 = 0$ and the Hodge Laplacian is reduced to $\mathcal{L}_1 = \mathbb{B}_1^\top \mathbb{B}_1$ [16].

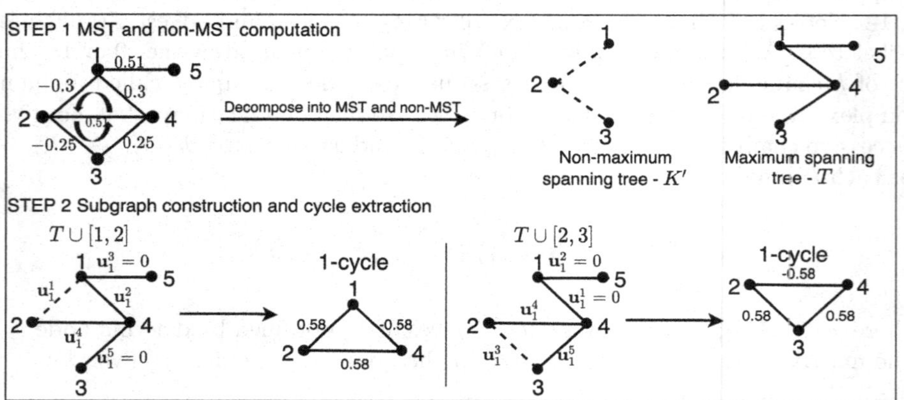

Fig. 1. Construction of 1-cycle basis. STEP 1: A graph is decomposed into MST (T) and non-MST (K'). STEP 2: The subgraphs are formed by adding an edge in K' to T (dotted lines). $ker\mathcal{L}_1$ for each subgraph is computed to extract a 1-cycle.

Basis Representation of Cycles. We partition the edges of a graph into the maximum spanning tree (MST) (T) and non-MST parts (K') (Fig. 1) [23]. If the m-th edge σ_1^m in K' is added to T, a subgraph

$$\mathcal{X}_m = \{T \cup \sigma_1^m : \sigma_1^m \in K'\}$$

with exactly one 1-cycle is formed. The Hodge Laplacian on $\mathcal{X}_m$ will yield the eigen decomposition identifying the 1-cycle. The entries of the corresponding eigenvector will have non-zero values only for those edges that constitute the cycle and the rest of entries are zero. The m-th 1-cycle is given by

$$\mathcal{C}_m = \sum_{j=1}^{|\mathcal{K}_1|} c_m^j, \quad \text{where} \quad c_m^j = \begin{cases} u_m^j \sigma_1^j, & \text{if} \quad \sigma_1^j \in \mathcal{X}_m \\ 0, & \text{otherwise} \end{cases}. \tag{4}$$

Here, u_m^j is the j-th entry of the m-th eigenvector (a column of $\mathbf{U}_1$) corresponding to zero eigenvalue. We can show that $\mathcal{C}_m$ forms a basis [1].

Theorem 1. *1-cycles* $\mathcal{C}_1, \cdots, \mathcal{C}_{|K'|}$ *spans* $ker\mathcal{L}_1$ *and forms a basis over the collection of all possible 1-cycles.*

Proof. Let E_m be the edge set of the cycle $\mathcal{C}_m$. Since E_m and E_n differ at least by two edges, they are algebraically independent. Hence, all the cycles $\mathcal{C}_1, \cdots, \mathcal{C}_{|K'|}$ are linearly independent from each other. Since there should be exactly $\beta_1 = |K'|$ number of independent cycles in the 1-st Homology group $H_1 = ker\mathcal{L}_1$, $\{\mathcal{C}_1, \cdots, \mathcal{C}_{|K'|}\}$ spans $ker\mathcal{L}_1$.

Example 1. We illustrate how to compute the 1-cycle basis. Consider the subgraph $T \cup [1,2]$ in Fig. 1. The boundary matrix $\mathbb{B}_1$ and the corresponding Hodge Laplacian $\mathcal{L}_1$ is computed as

$$\mathbb{B}_1 = \begin{array}{c} \\ [1] \\ [2] \\ [3] \\ [4] \\ [5] \end{array} \begin{array}{c} \begin{array}{ccccc} [1,2] & [1,4] & [1,5] & [2,4] & [3,4] \end{array} \\ \begin{pmatrix} 1 & 1 & 1 & 0 & 0 \\ -1 & 0 & 0 & 1 & 0 \\ 0 & 0 & 0 & 0 & 1 \\ 0 & -1 & 0 & -1 & -1 \\ 0 & 0 & -1 & 0 & 0 \end{pmatrix} \end{array} \quad \mathcal{L}_1 = \mathbb{B}_1^T \mathbb{B}_1 = \begin{pmatrix} 2 & 1 & 1 & -1 & 0 \\ 1 & 2 & 1 & 1 & 1 \\ 1 & 1 & 2 & 0 & 0 \\ -1 & 1 & 0 & 2 & 1 \\ 0 & 1 & 0 & 1 & 2 \end{pmatrix}.$$

The eigen decomposition $\mathcal{L}_1 \mathbf{U}_1 = \mathbf{U}_1 \Lambda_1$ is given by

$$\mathbf{U}_1 = \begin{array}{c} \\ [1,2] \\ [1,4] \\ [1,5] \\ [2,4] \\ [3,4] \end{array} \begin{array}{c} \begin{array}{ccccc} \mathbf{u}_1 & \mathbf{u}_2 & \mathbf{u}_3 & \mathbf{u}_4 & \mathbf{u}_5 \end{array} \\ \begin{pmatrix} \mathbf{0.58} & -0.25 & 0.37 & -0.60 & 0.33 \\ \mathbf{-0.58} & -0.49 & 0.00 & 0.00 & 0.65 \\ \mathbf{0.00} & 0.57 & -0.60 & -0.37 & 0.42 \\ \mathbf{0.58} & -0.25 & -0.37 & 0.60 & 0.22 \\ \mathbf{0.00} & 0.57 & 0.60 & 0.37 & 0.43 \end{pmatrix} \end{array} \quad \Lambda_1 = \begin{pmatrix} \mathbf{0.00} & 0 & 0 & 0 & 0 \\ 0 & 0.70 & 0 & 0 & 0 \\ 0 & 0 & 1.38 & 0 & 0 \\ 0 & 0 & 0 & 3.62 & 0 \\ 0 & 0 & 0 & 0 & 4.30 \end{pmatrix}.$$

The 1-cycle is then represented as

$$\mathcal{C}_1 = 0.58\sigma_1^1 - 0.58\sigma_1^2 + 0.00\sigma_1^3 + 0.00\sigma_1^4 + 0.58\sigma_1^5 + 0.00\sigma_1^6,$$

where $\sigma_1^1 = [1,2]$, $\sigma_1^2 = [1,4]$, $\sigma_1^3 = [1,5]$, $\sigma_1^4 = [2,3]$, $\sigma_1^5 = [2,4]$, and $\sigma_1^6 = [3,4]$. Similarly, the cycle in the second subgraph $T \cup [2,3]$ is represented as

$$\mathcal{C}_2 = 0.00\sigma_1^1 + 0.00\sigma_1^2 + 0.00\sigma_1^3 + 0.58\sigma_1^4 - 0.58\sigma_1^5 + 0.58\sigma_1^6.$$

Any cycle in the graph $\mathcal{X}$ can be represented as a linear combination of basis $\mathcal{C}_1$ and $\mathcal{C}_2$. For large graphs, the coefficients of basis expansion can be vectorized and efficiently stored as a sparse matrix.

2.3 Common 1-Cycles Across Networks

Using the 1-cycle basis, we identify common 1-cycles across different networks. We extend the idea of extracting 1-cycle basis for a single network (individual level) to collection of networks (group level). Let $\bar{\mathcal{S}} = (\bar{\mathcal{S}}_{i,j})$ be the average of all the individual connectivity matrices. $\bar{\mathcal{S}}$ is used to construct a graph $\bar{\mathcal{X}}$ where we assume any two nodes (i,j) in $\bar{\mathcal{X}}$ is incident by an edge if $\bar{\mathcal{S}}_{i,j} > 0$, i.e., positive correlations [3,18,30]. The cycles extracted from $\bar{\mathcal{X}}$ will represent the common cycle basis. To reflect the individual network variability, we model the vectorized individual network connectivity $\mathcal{M}$ as a linear combination of the common 1-cycle basis. This gives the minimization problem

$$\arg\min_{\boldsymbol{\alpha}} \|\mathcal{M} - \boldsymbol{\psi}\boldsymbol{\alpha}\|_2^2, \tag{5}$$

where the basis matrix $\boldsymbol{\psi} = \left[\mathcal{C}_1, \cdots, \mathcal{C}_{|K'|}\right]$, and $\mathcal{C}_m$'s are the 1-cycle basis from $\bar{\mathcal{X}}$. The coefficients vector $\boldsymbol{\alpha} = [\alpha_1, \cdots, \alpha_{|K'|}]^\top$ is estimated in the least squares fashion as $\widehat{\boldsymbol{\alpha}} = (\boldsymbol{\psi}^\top\boldsymbol{\psi})^{-1}\boldsymbol{\psi}^\top\mathcal{M}$.

Let $\bar{\alpha}_m^1$ and $\bar{\alpha}_m^2$ be the mean coefficients corresponding to the m-th 1-cycle basis of networks in group $\mathcal{N}_1$ and $\mathcal{N}_2$ respectively. We propose the following statistic for testing the group difference [1]:

$$\mathcal{T}(\mathcal{N}_1, \mathcal{N}_2) = \max_{1 \le m \le |K'|} |\bar{\alpha}_m^1 - \bar{\alpha}_m^2|. \tag{6}$$

The statistical significance is determined using the permutation test [8].

3 Validation

The proposed methodology is validated on network simulations with different number of cycles. The simulation is done using deltoid, limaçon trisectrix and trifolium [29]. Groups 1 and 2 have three 1-cycles each and are topologically equivalent. Groups 3 and 4 have five 1-cycles each and are topologically equivalent. 50 points were sampled along these curves and perturbed with noise $N(0, 0.025^2)$. 10 networks were generated in each group. All the networks consist of identical number of nodes. For networks with different number of nodes, data augmentation can be done [23]. The simulation code is provided in https://github.com/laplcebeltrami/hodge (Fig. 2).

We validated whether the test statistic (6) can discriminate networks with different topology. 10,000 permutations were used to compute the p-values. Since there are no established methods for modeling cycles, we validated our method against geometric distances $\mathcal{L}_1$, $\mathcal{L}_2$, $\mathcal{L}_\infty$ and the Gromov-Hausdorff (GH) distance [8]. The simulations were performed 50 times and the results are given in Table 1, where the average p-values are reported. Also, we reported the false positive rates computed as the fraction of 50 simulations with p-values below 0.05 and the false negative rates computed as the fraction of 50 simulations with p-values above 0.05 [23].

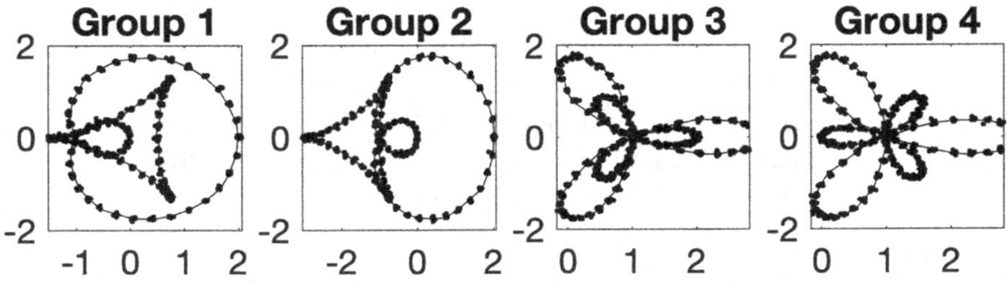

Fig. 2. Random networks used in Validation. Groups 1 and 2 have three 1-cycles and topologically equivalent. Groups 3 and 4 have five 1-cycles and topologically equivalent.

We tested if the proposed method can detect *topological equivalence* by comparing Groups 1 vs. 1, 2 vs. 2, 1 vs. 2, 3 vs. 3, 4 vs. 4 and 3 vs. 4 (first 6 rows). The test procedures should not detect signals and higher p-values and smaller false positive rates are preferred. The proposed method $\mathcal{T}$ performed well compared to the other distances. We also tested if the proposed method can detect *topological difference* by comparing Groups 1 vs. 3, 1 vs. 4, 2 vs. 3 and 2 vs. 4. The test procedures should detect signals and smaller p-values and smaller false negative rates are preferred. $\mathcal{T}$ consistently outperformed the other distances.

This study shows that existing methods will likely produce significant topological false negatives while reasonably good at not producing large false positives. However, the proposed method will not produce large amount of topological false positives and negatives at the same time.

4 Application

Dataset. We used a subset of the resting-state fMRI (rs-fMRI) data in the Human Connectome Project [27,28]. rs-fMRI are collected at 2 mm isotropic

Table 1. The performance results showing average p-values, false positive rates (first 6 rows) and false negative rates (last 4 rows). Group 1 and 2 have three 1-cycles and topologically equivalent. Group 3 and 4 have five 1-cycles and topologically equivalent. Smaller false positive and false negative rates are preferred.

Groups	$\mathcal{L}_1$	$\mathcal{L}_2$	$\mathcal{L}_\infty$	GH	$\mathcal{T}$
1 vs. 1	0.26 ± 0.15 (0.04)	0.27 ± 0.14 (0.04)	0.24 ± 0.13 (0.00)	0.45 ± 0.31 (0.12)	0.60 ± 0.26 (0.00)
2 vs. 2	0.29 ± 0.14 (0.04)	0.31 ± 0.15 (0.00)	0.26 ± 0.14 (0.12)	0.56 ± 0.25 (0.04)	0.53 ± 0.29 (0.04)
1 vs. 2	0.27 ± 0.15 (0.08)	0.25 ± 0.15 (0.08)	0.20 ± 0.12 (0.12)	0.43 ± 0.24 (0.04)	0.38 ± 0.28 (0.16)
3 vs. 3	0.28 ± 0.16 (0.08)	0.27 ± 0.15 (0.04)	0.23 ± 0.15 (0.04)	0.45 ± 0.25 (0.00)	0.52 ± 0.31 (0.08)
3 vs. 4	0.23 ± 0.14 (0.18)	0.23 ± 0.14 (0.12)	0.26 ± 0.13 (0.18)	0.52 ± 0.27 (0.04)	0.52 ± 0.30 (0.03)
4 vs. 4	0.32 ± 0.15 (0.04)	0.30 ± 0.15 (0.04)	0.24 ± 0.14 (0.16)	0.49 ± 0.28 (0.10)	0.41 ± 0.28 (0.04)
1 vs. 3	0.26 ± 0.14 (0.92)	0.25 ± 0.14 (0.88)	0.27 ± 0.13 (0.88)	0.00 ± 0.00 (0.00)	0.00 ± 0.00 (0.00)
1 vs. 4	0.29 ± 0.14 (0.72)	0.28 ± 0.16 (0.80)	0.27 ± 0.16 (0.72)	0.00 ± 0.00 (0.00)	0.00 ± 0.00 (0.00)
2 vs. 3	0.25 ± 0.17 (0.76)	0.24 ± 0.17 (0.76)	0.25 ± 0.15 (0.88)	0.00 ± 0.00 (0.00)	0.00 ± 0.00 (0.00)
2 vs. 4	0.27 ± 0.15 (0.88)	0.23 ± 0.15 (0.80)	0.22 ± 0.14 (0.96)	0.00 ± 0.00 (0.00)	0.00 ± 0.00 (0.00)

voxels and 1200 time points [27]. Data that went through the standard minimal preprocessing pipelines [13] was used. Volumes with framewise displacement larger than 0.5mm and their neighbors were scrubbed [27,28]. Twelve subjects having excessive head movement were excluded from the dataset. Subsequently, the Automated Anatomical Labeling (AAL) template is used to parcellate and average rs-fMRI spatially into 116 non-overlapping anatomical regions [26]. The details on image processing are given in [15]. The final data is comprised of the fMRI of 400 subjects of which 168 are males and 232 are females.

Sexual Dimorphism in 1-Cycles. We constructed the network template by averaging 400 subjects connectivity matrices. The 1-cycle basis was then extracted from this network template. The average network template contains $p = 116$ nodes, hence we expect $q = p(p-1)/2 = 6555$ linearly independent 1-cycles [23]. Each individual connectivity matrix is then expanded and the test carried out using the estimated expansion coefficients resulting in the observed

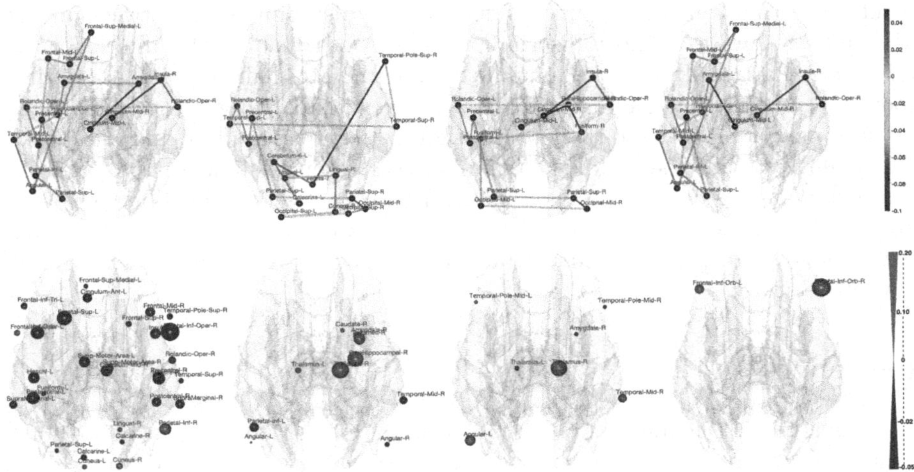

Fig. 3. Top: The four most discriminating cycles corresponding to maximum test statistic values. The edge colors correspond to the edge weight differences (female - male). Bottom: The four most discriminating connected components. The size of nodes correspond to the weighted node degree differences (female - male).

test statistic of 0.41. 0.1 million permutations were used to obtain the p-value of 0.03. We compared the discriminating power of our method against geometric distances $\mathcal{L}_1$, $\mathcal{L}_2$, $\mathcal{L}_\infty$ and GH distance, which respectively give the p-values of 0.014, 0.004, 0.013, 0.54. All the baseline methods performed well except for the GH distance. Nonetheless, only the statistic $\mathcal{T}$ can localize the connections contributing to the difference. Figure 3-top shows the four most discriminating cycles. The edge color represents the average correlation difference between the groups.

Sexual Dimorphism in 0-Cycles. Our method for 1-cycles is also applicable to 0-cycles (connected components) by replacing Hodge Laplacian $\mathcal{L}_1$ with graph Laplacian $\mathcal{L}_0$. The basis will be defined along nodes. The j-th basis vector will have value 1 at node j and 0 in other places. Including the test statistic, the same pipeline can be used. Since no data is defined on nodes, we used the weighted degree (sum of correlations of all the edges connecting at each node). The observed test statistic is 27.41, which gives the p-value of 0.006. Figure 3-bottom shows the four most discriminating 0-cycles. The node size corresponds to the difference (female - male) in the weighted degree between the groups.

5 Conclusion

We proposed a novel Hodge Laplacian framework for explicitly identifying and modelling cycles in brain networks. We were able to show that cycles, both 0- and

1-cycles are topologically significant features in discriminating between males and females. The four most discriminating 1-cycles include the inferior parietal lobule (Parietal-Inf-L), the rolandic operculum (Rolandic-Oper-L, Rolandic-Oper-R), and the amygdala (Amygdala-R, Amygdala-L) (Fig. 3-top). The symmetric connections between the left and right rolandic operculum, superior parietal lobule as well as the middle cingulate are consistently showing up in at least 2 most dominating cycles. The most discriminating 0-cycles (Fig. 3-bottom) also contain most of these brain regions. The method can be extended to higher order connectivity such as 2-cycles using higher order Hodge Laplacians. This is left as a future study.

References

1. Anand, D.V., Dakurah, S., Wang, B., Chung, M.K.: Hodge-Laplacian of brain networks and its application to modeling cycles. arXiv preprint arXiv:2110.14599 (2021)
2. Battiston, F., et al.: Networks beyond pairwise interactions: structure and dynamics. Phys. Rep. **874**, 1–92 (2020)
3. Buckner, R.L.: Cortical hubs revealed by intrinsic functional connectivity: mapping, assessment of stability, and relation to alzheimer's disease. J. Neurosci. **29**, 1860–1873 (2009)
4. Bullmore, E., Sporns, O.: Complex brain networks: graph theoretical analysis of structural and functional systems. Nat. Rev. Neurosci. **10**, 186–198 (2009)
5. Busaryev, O., Cabello, S., Chen, C., Dey, T.K., Wang, Y.: Annotating simplices with a homology basis and its applications. In: Fomin, F.V., Kaski, P. (eds.) SWAT 2012. LNCS, vol. 7357, pp. 189–200. Springer, Heidelberg (2012). https://doi.org/10.1007/978-3-642-31155-0_17
6. Chen, C., Freedman, D.: Measuring and computing natural generators for homology groups. Comput. Geom. **43**, 169–181 (2010)
7. Chung, M.K., Huang, S.G., Songdechakraiwut, T., Carroll, I.C., Goldsmith, H.H.: Statistical analysis of dynamic functional brain networks in twins. arXiv preprint arXiv:1911.02731 (2019)
8. Chung, M.K., Lee, H., DiChristofano, A., Ombao, H., Solo, V.: Exact topological inference of the resting-state brain networks in twins. Netw. Neurosci. **3**, 674–694 (2019)
9. Dey, T.K., Fan, F., Wang, Y.: Computing topological persistence for simplicial maps. In: Proceedings of the Thirtieth Annual Symposium on Computational Geometry, pp. 345–354 (2014)
10. Edelsbrunner, H., Harer, J., et al.: Persistent homology-a survey. Contemp. Math. **453**, 257–282 (2008)
11. Farazi, M., Zhan, L., Lepore, N., Thompson, P.M., Wang, Y.: A univariate persistent brain network feature based on the aggregated cost of cycles from the nested filtration networks. In: 2020 IEEE 17th International Symposium on Biomedical Imaging (ISBI), pp. 1–5. IEEE (2020)
12. Giusti, C., Ghrist, R., Bassett, D.S.: Two's company, three (or more) is a simplex. J. Comput. Neurosci. **41**, 1–14 (2016)
13. Glasser, M.F., et al.: The minimal preprocessing pipelines for the human connectome project. Neuroimage **80**, 105–124 (2013)

14. Hatcher, A., Press, C.U., of Mathematics, C.U.D.: Algebraic Topology. Algebraic Topology. Cambridge University Press, Cambridge (2002)
15. Huang, S.G., Samdin, S.B., Ting, C.M., Ombao, H., Chung, M.K.: Statistical model for dynamically-changing correlation matrices with application to brain connectivity. J. Neurosci. Methods **331**, 108480 (2020)
16. Lee, H., Chung, M.K., Kang, H., Lee, D.S.: Hole detection in metabolic connectivity of Alzheimer's disease using k-Laplacian. In: Golland, P., Hata, N., Barillot, C., Hornegger, J., Howe, R. (eds.) MICCAI 2014. LNCS, vol. 8675, pp. 297–304. Springer, Cham (2014). https://doi.org/10.1007/978-3-319-10443-0_38
17. Lind, P.G., Gonzalez, M.C., Herrmann, H.J.: Cycles and clustering in bipartite networks. Phys. Rev. E **72**, 056127 (2005)
18. Meunier, D., Lambiotte, R., Fornito, A., Ersche, K., Bullmore, E.T.: Hierarchical modularity in human brain functional networks. Front. Neuroinf. **3**, 37 (2009)
19. Petri, G., et al.: Homological scaffolds of brain functional networks. J. Roy. Soc. Interface **11**, 20140873 (2014)
20. Reani, Y., Bobrowski, O.: Cycle registration in persistent homology with applications in topological bootstrap. arXiv preprint arXiv:2101.00698 (2021)
21. Sizemore, A.E., Giusti, C., Kahn, A., Vettel, J.M., Betzel, R.F., Bassett, D.S.: Cliques and cavities in the human connectome. J. Comput. Neurosci. **44**(1), 115–145 (2017). https://doi.org/10.1007/s10827-017-0672-6
22. Sizemore, A.E., Phillips Cremins, J.E., Ghrist, R., Bassett, D.S.: The importance of the whole: topological data analysis for the network neuroscientist. Netw. Neurosci. **3**, 656–673 (2019)
23. Songdechakraiwut, T., Shen, L., Chung, M.: Topological learning and its application to multimodal brain network integration. In: de Bruijne, M., et al. (eds.) MICCAI 2021. LNCS, vol. 12902, pp. 166–176. Springer, Cham (2021). https://doi.org/10.1007/978-3-030-87196-3_16
24. Topaz, C.M., Ziegelmeier, L., Halverson, T.: Topological data analysis of biological aggregation models. PloS one **10**, e0126383 (2015)
25. Torres, L., Blevins, A.S., Bassett, D., Eliassi-Rad, T.: The why, how, and when of representations for complex systems. SIAM Rev. **63**, 435–485 (2021)
26. Tzourio-Mazoyer, N., et al.: Automated anatomical labeling of activations in SPM using a macroscopic anatomical parcellation of the MNI MRI single-subject brain. Neuroimage **15**, 273–289 (2002)
27. Van Essen, D.C., et al.: The WU-MINN human connectome project: an overview. Neuroimage **80**, 62–79 (2013)
28. Van Essen, D.C., et al.: The human connectome project: a data acquisition perspective. Neuroimage **62**, 2222–2231 (2012)
29. Yates, R.C.: Curves and their properties (1974)
30. Zhan, L., et al.: The significance of negative correlations in brain connectivity. J. Comp. Neurol. **525**, 3251–3265 (2017)

Predicting Spatio-Temporal Human Brain Response Using fMRI

Chongyue Zhao[1], Liang Zhan[1], Paul M. Thompson[2], and Heng Huang[1](✉)

[1] Department of Electrical and Computer Engineering, University of Pittsburgh, Pittsburgh, PA, USA
{liang.zhan,heng.huang}@pitt.edu

[2] Imaging Genetics Center, University of Southern California, Los Angeles, CA, USA
pthomp@usc.edu

Abstract. The transformation and transmission of brain stimuli reflect the dynamical brain activity in space and time. Compared with functional magnetic resonance imaging (fMRI), magneto- or electroencephalography (M/EEG) fast couples to the neural activity through generated magnetic fields. However, the MEG signal is inhomogeneous throughout the whole brain, which is affected by the signal-to-noise ratio, the sensors' location and distance. Current non-invasive neuroimaging modalities such as fMRI and M/EEG excel high resolution in space or time but not in both. To solve the main limitations of current technique for brain activity recording, we propose a novel recurrent memory optimization approach to predict the internal behavioral states in space and time. The proposed method uses Optimal Polynomial Projections to capture the long temporal history with robust online compression. The training process takes the pairs of fMRI and MEG data as inputs and predicts the recurrent brain states through the Siamese network. In the testing process, the framework only uses fMRI data to generate the corresponding neural response in space and time. The experimental results with Human connectome project (HCP) show that the predicted signal could reflect the neural activity with high spatial resolution as fMRI and high temporal resolution as MEG signal. The experimental results demonstrate for the first time that the proposed method is able to predict the brain response in both milliseconds and millimeters using only fMRI signal.

Keywords: Brain dynamics · Recurrent neural network · fMRI

1 Introduction

In computational neuroscience, brain state often refers to wakefulness, sleep, and anesthesia. However, the precise and dynamical brain complexity is still

Supplementary Information The online version contains supplementary material available at https://doi.org/10.1007/978-3-031-16431-6_32.

L. Wang et al. (Eds.): MICCAI 2022, LNCS 13431, pp. 336–345, 2022.
https://doi.org/10.1007/978-3-031-16431-6_32

missing. Dynamical neural representation is thought to arise from neural firing, which do not fire in isolation [11]. The development of modern brain measuring techniques makes it possible to infer the large-scale structural and functional connectivity and characterize the anatomical and functional patterns in human cortex. Electrophysiological methods provide a direct and non-invasive way to record brain with milliseconds temporal resolution which is not affected by the problems commonly caused by intermediate processes. fMRI uses brain activity-related blood-oxygen-level-dependent (BOLD) to understand the neural representations in millimeter level, however it is too sluggish in the time domain. None of the current non-invasive brain recording techniques could measure the high resolution dynamics in space and time. This is a major challenge in neuroscience which draws attention of researchers to develop simultaneous simulation models of brain dynamics.

Since each brain region interacts with other regions, it is hard to measure the intraregional brain dynamics with single technique such as fMRI or MEG. With the increase of the strength of brain region interactions, the measured brain dynamical signals decrease. Recent study on complex biological system provides the powerful mathematical functions to model the coupling structural and functional brain activity using multimodal brain measurements [19]. Previous study use principle component analysis (PCA) [3,12,13] or linear models [9,15] to quantify the inherent neural representations in either high dimensional or low dimensional. The high dimensionality is used for information encoding. The low dimensionality is used to encode information in complex cognitive or motor tasks [8]. Modern methods such as representational similarity analysis (RSA) [7] and multivoxel pattern analysis (MVPA) [10] is used to quantify the brain activity across task conditions. Further developments utilize the states defined as the transmission between different task activity to function higher cognitive dynamics [2,3,14,16,18].

The question remains: how to model the change and transmission between neurons with high spatial and temporal resolution? In this paper, we propose a novel method that use fMRI to predict the corresponding neural activity with high resolution in space and time. Our main contributions are summarized as follows:

1) The current non-invasive brain measuring technique could not identify the neural activity with high resolution in both spatial and temporal domain. MEG/EEG could resolve the brain activity in milliseconds but the spatial resolution is low. In contrast, fMRI could represent the neural dynamics in millimeters, but it is too slow for the rapid transmission between cortical sites. To link the mapping between brain regions and time points, we propose a general framework to project the time points onto polynomial basis. In the training step, the proposed method takes the ROI parcellated brain signals (fMRI and MEG) to learn the dynamical representation. In the testing step, the proposed method is the first to use original 4D fMRI signal to predict the internal brain response in space and time.

2) Most of the existing recurrent neural network (RNN) suffers from the vanishing gradient problem when dealing with long-term time series. It is challenging to represent the brain response at milliseconds with current RNN model. The proposed method could discretize and project the time points onto the polynomial basis, which is able to deal with any time length data using the memory compression.
3) The two experimental results demonstrate that the proposed method is able to resolve the neural activity in space and time. Its feasibility is validated when compared with fMRI in structural domain using brain networks of independent component analysis (ICA). The predicted results also show the similar pattern to MEG signal in temporal domain.

The proposed method is related to the fMRI and MEG data fusion research. The difference between the proposed method and existing data fusion method is summarized as follows. 1) The proposed method does not use the MEG data in the testing step. It is able to use 4D fMRI data to predict the brain internal states in space and time. The MEG data is used in the training step as a reference for the proposed method to learn the temporal pattern of brain sates. The research problem could be considered as the "super-resolution" in the temporal domain of fMRI data and keeps the structural details. However, as far as we know, most of the previous methods require both fMRI and MEG data to predict the internal states. It is hard to obtain simultaneous fMRI and MEG data at the same time. 2) Secondly, the proposed method could predict the high resolution temporal signals for each voxel of fMRI data. In the testing step, the proposed method does not require the preprocessing steps like ROI average or beamforming which could introduce noise in the preprocessed data. As far as we know, the proposed method is the first to predict the spatio-temporal brain internal states for each voxel using only fMRI data.

2 Spatio-Temporal Dynamical Modeling of Brain Response

Our approach benefits from the framework of Legendre Memory Unit (LMU) and online function approximation to learn the memory representation [6,17]. The dynamical brain network with N neurons could be modeled as

$$\begin{aligned} \dot{z}(t) &= f(z, l, t) + e_1(t), \\ y(t) &= O(z(t)) + e_2(t), \end{aligned} \tag{1}$$

where $z(t) = [z_1(t), z_2(t), ..., z_N(t)]^T$ represents the internal states of N neuron nodes at time t. $f(\cdot)$ denotes the nonlinear dynamical function of each node. And $l(t) = [l_1(t), l_2(t), ..., l_S(t)]^T$ represents the external stimuli for S neurons. $y(t) = [y_1(t), y_2(t), ..., y_N(t)]^T \in R^N$ is the observed time series of brain measurements (i.e. MEG or fMRI signals). $e_1(t)$ and $e_2(t)$ are time series noise of the internal states and measurements. $O(\cdot)$ is the output function which controls the output of the brain measurements at time t. The nonlinear brain function $f(\cdot)$ is modeled using recurrent neural network (RNN).

The whole framework of the proposed method is shown in Fig. 1. In the training stage, the proposed method consists of two parts, the RNN model with Polynomial Projection and the Siamese network to score the agreement between the network prediction and the original time series of brain activity. Both Region-of-interest (ROI) extracted fMRI and MEG signals are used to train the network. However, in the testing stage, only fMRI signal is included to predict the internal states in space and time. The experiments demonstrate that the proposed method could deal with both the ROI extract fMRI signal and the original 4D fMRI image. The predicted internal behavioral states show high resolution in spatial and temporal domain.

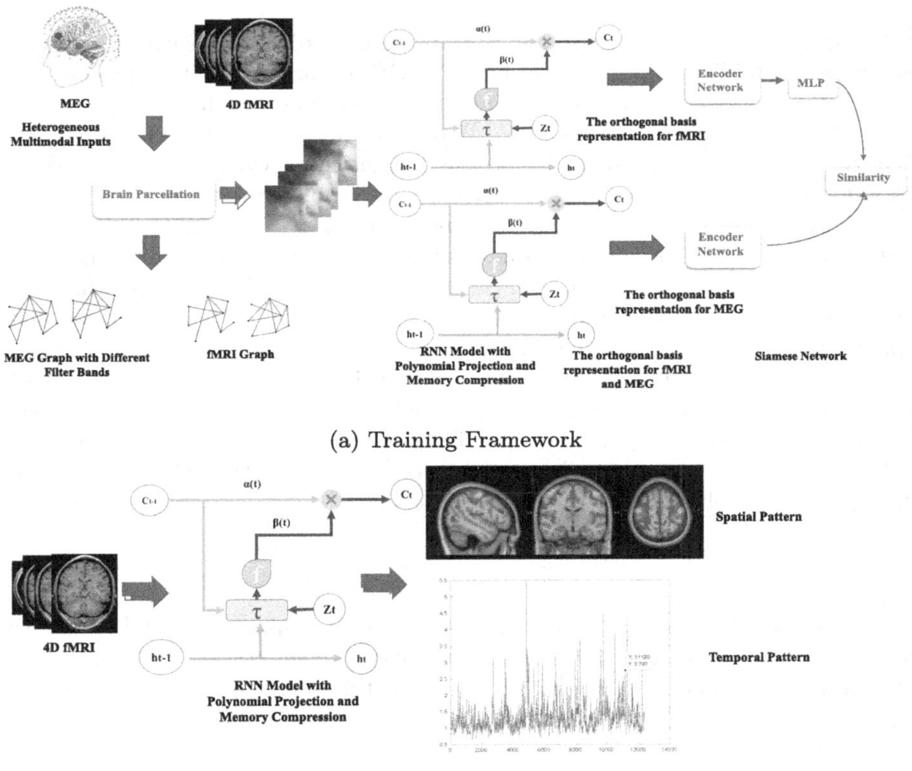

Fig. 1. Illustration of the proposed spatio-temporal brain dynamical network (a) Training framework. Both ROI extracted fMRI and MEG signals are used in the training process. (b) Testing framework for generating the internal states which uses 4D fMRI as input.

We propose a Polynomial Projection Operators together with Recurrent memory to solve the dynamical internal state $z(t)$. Given the input $z(t)$ or $f(z, l, t)$, the proposed method aims to solve the future prediction $z(t+1)$ based on the cumulative history $f_{\leq t}$. However, as temporal resolution for MEG signal is intractably high, there exists vanishing gradient problem when the model evolves over all time states. To solve the vanishing gradient problem, we project the input state onto the subspace and maintain the compressed historical representation. So there are two problems to be solved: how to quantify the approximation, and the way to learn the subspace.

Function Approximation. We introduce the probability measure $\mu \in [0, \infty)$ to define the space of square function $< f, p >_{\mu} = \int_0^{\infty} f(x)p(x)d\mu(x)$. $p(\cdot)$ is defined on a M subspace $\mathcal{P}$. The function $p_t \in \mathcal{P}$ is utilized to minimize the approximation of $f_{\leq t}$ with $\| f_{\leq t} - p_t \|_{L_2(\mu_t)}$, where μ_t is the measure ranges in $(-\infty, t)$.

Polynomial Basis Expansion for Subspace Learning. To learn the suitable subspace, we define the polynomial basis $\mathcal{P}$ with parameter M to represent the projected history, which means the historical dynamics could be represented using M coefficients with the basis $\mathcal{P}$. M represents the size of the compression. However, it is challenging to maintain the parameter μ_t when $t \rightarrow \infty$. We show more details of the suitable polynomial basis in Supplementary material.

The first step is to choose the appropriate basis $p_m \in \mathcal{P}$ in the projection operator. The projection operator takes the historical memory of t and minimize the approximation of $f_{\leq t}$ using $\| f_{\leq t} - p_t \|_{L_2(\mu_t)}$. According to approximation theory, we use the orthogonal polynomials of μ_t as the orthogonal basis and represent the coefficients $c_m(t) := \langle f_{\leq t}, p_m \rangle_{\mu_t}$.

The second key step is to differentiate the projection using the inner product $< f, p >_{\mu} = \int_0^{\infty} f(x)p(x)d\mu(x)$, which will lead to the similar result that $\frac{d}{dt}c_m(t)$ is expressed using $(c_k(t))_{k \in M}$ and $f(t)$. Thus, $c(t) \in R^M$ satisfies the ODE form $\frac{d}{dt}c(t) = \alpha(t)c(t) + \beta(t)f(t)$, where $\alpha(t) \in R^{M \times M}$ and $\beta(t) \in R^{M \times 1}$. For each time t, We could change $c(t)$ to the ODE form,

$$\frac{d}{dt}c(t) = -\frac{1}{t}\alpha(t)c(t) + \frac{1}{t}\beta(t)f(t), \tag{2}$$

The whole derivation of $c(t)$ is shown in Supplementary material. Next we want to solve the ODE problem and obtain the coefficient $c(t)$ by discretizing the continuous function $c(t)$ and $f(t)$, which yields the recurrence $c_{k+1} = -\alpha_k c_k + \beta_k f_k$ using $(f_k)_{k \in M}$. The orthogonal basis is defined as follows,

$$c_{k+1} = (1 - \frac{\alpha}{k})c_k + \frac{1}{k}\beta f_k, \tag{3}$$

2.1 The Siamese Network for Behavioral Prediction in Space and Time

To score the agreement between the time series prediction and original brain measurements, we introduce the Siamese network [1] which is the weight sharing network for comparing two views. It should be noted that the Siamese network is only used in the training process to help learn the structural and functional patterns of fMRI and MEG. We give detailed description of training and testing procedure in Supplementary material. The Siamese network is used to measure the similarity between fMRI and MEG signals. The Siamese network consists of two views, the predicted MEG signal using fMRI and the corresponding ground truth MEG signal. The two input views are preprocessed using the encoder network $\acute{O}(\cdot)$. The encoder network shares the same weights between two views. The encoder network $\acute{O}(\cdot)$ of the predicted MEG-like signal is followed by an MLP head $\Omega(\cdot)$ to match the output of the other view.

Encoder Network with Continuous Convolutions. To predict the multi-modal brain measurements using the orthogonal basis in Eq. 3, We could use the continuous convolutions to represent the output signals. Following the previous work of bilinear method, we converts the state α and β in Eq. 3 into an approximation $\widetilde{\alpha} = (I - \Delta/2 \cdot \alpha)^{-1}(I + \Delta/2 \cdot \alpha)$ and $\widetilde{\beta} = (I - \Delta/2 \cdot \alpha)^{-1}\Delta\beta$. Based on Eq. 3, the continuous convolutions is defined as,

$$\begin{aligned} y_k &= \widetilde{O\alpha}^k \widetilde{\beta} f_0 + \widetilde{O\alpha}^{k-1} \widetilde{\beta} f_1 + \cdots + \widetilde{O\alpha}\widetilde{\beta} f_{k-1} + \widetilde{O\beta} f_k \\ y &= \acute{O} * f \quad \text{and} \quad \acute{O} \in R^K = (\widetilde{O\beta}, \widetilde{O\alpha\beta}, \cdots, \widetilde{O\alpha}^k \beta), \end{aligned} \tag{4}$$

Similarity Measurement. Given two kinds of inputs y_1 and y_2 at time t with $y_1(t) = O(z_1(t))$. $y_1(t)$ represents the predicted brain measurement of the output model $O(\cdot)$ in Eq. 1. $y_2(t)$ is the original time series of brain activity, such as fMRI or MEG signals. The similarity of two view is defined as

$$D(\rho_1(t), \eta_2(t)) = -\frac{\rho_1(t)}{\| \rho_1(t) \|_2} * \frac{\eta_2(t)}{\| \eta_2(t) \|_2}, \tag{5}$$

where $\rho_1(t) = \Omega(\acute{O}(y_1(t)))$ and $\eta_2(t) = \acute{O}(y_2(t))$. $\| \cdot \|_2$ is l_2-norm.

In the experiment, we adopt the stop-gradient operation. The stop-gradient treat the second input $\eta_2(t)$ as constant. The symmetrized loss is denoted by

$$\mathcal{L}_s = \frac{1}{2} D(\rho_1(t), stopgrad(\eta_2(t))) + \frac{1}{2} D(\rho_2(t), stopgrad(\eta_1(t))), \tag{6}$$

3 Experimental Results

We tested the fidelity of the proposed method based on the resting state fMRI and MEG from HCP dataset. The resting-state fMRI was pre-processed following the minimal preprocessing pipeline [5]. Then the pre-processed data was registered into a standard cortical surface using MSMAll [5]. The artefacts were removed using ICA-FIX. The cortical surface was parcellated into N=360 major ROI [4]. In addition, the averaged time course of each ROI was normalized using z-score. The resting-state MEG was pre-processed using ICA to remove out artefacts related to head and eye movement. Sensor-space data were down-sampled

300 Hz using anti-aliasing filter. Next the MEG data were source-reconstructed with a scalar beamformer and registered into the standard space of the Montreal Neuroimaging Institute (MNI). MEG signals were then filtered into 1-30Hz and beamformed onto 6 mm grid. We used theta (4–8 Hz), alpha (8–13 Hz) and beta (13–30 Hz) bands to filter the source-space data. The parcellation atlas and z-score normalization method of MEG were similar to the resting-state fMRI.

3.1 Spatio-Temporal Patterns of the Predicted Results Using 4D fMRI Image

We next applied the proposed method to acquire the dynamics of the behavioral representation for each voxel in fMRI signal. We used the whole 4D fMRI image to generate the corresponding behavioral states in the spatio-temproal domain. We show the spatial map of independent temporal signal using original MEG and the predicted results in Fig. 2. The spatial map was generated using the ICA. With the 25 generated ICA temporal components of MEG and predicted results, we paired the 8 resting state brain networks(RSNs) spatial map with that derived using resting state fMRI. The DMN pattern is shown in Fig. 2a. The nodes is highlighted in the medial frontal cortex and inferior parietal lobules. The patterns of left lateralized frontoparietal and sensorimotor network are shown in Fig. 2b and Fig. 2c. From Fig. 2, we could see that the spatial pattern of predicted results could match that of MEG in all the network. However, the spatial resolution of predicted result is much higher than MEG signal.

The results have shown that the temporal ICA components originate from the brain regions correlated with the RSNs spatial maps. We show more results in the Supplementary material. The proposed method could generate the high resolution behavioral states that are the perfect matches to the spatial pattern of the observed fMRI. Thus, the proposed method could provide fundamental role of brain dynamics related to behavioral measurements.

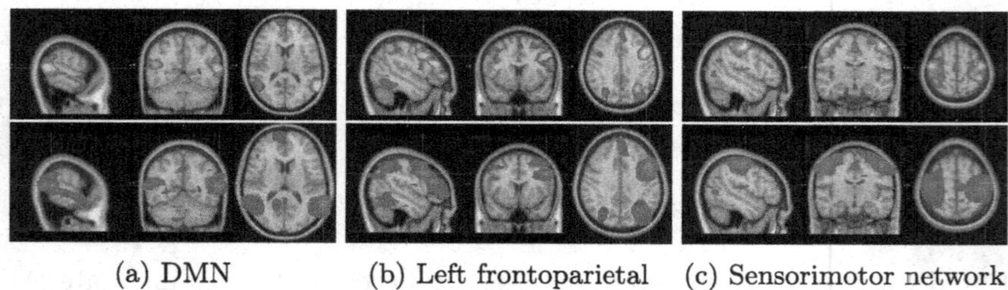

(a) DMN (b) Left frontoparietal (c) Sensorimotor network

Fig. 2. Brain networks acquired with ICA shown in the order predicted results (top) and MEG (bottom). (a) DMN, (b) left frontoparietal network, (c) sensorimotor network.

3.2 Temporal Pattern of Predicted Results

We finally evaluated temporal pattern of the predicted internal states compared with original 4D fMRI and MEG image. Figure 3 shows the averaged neural time series of the predicted results, fMRI and MEG signal in visual network. From Fig. 3, we could see that the proposed method inherits the high temporal resolution of MEG signal in a dynamical system. The proposed method predicts the hidden observation of the dynamical neural transmission locally and globally. In Table 1, we use the mean squared error (MSE) to measure the similarity between the ROI averaged predicted results and ROI averaged MEG signals. In addition, we also compare the proposed method with three baselines LSTM, GRU-D and proposed method without Siamese netwrok. Polynomial Projection Operators could be combined with The introduction of the spatio-temporal constraints into the dynamical system is consistent with the fundamental role in biological system. The proposed method provides chance to understand how behavioral representations evolve over time.

Table 1. Temporal pattern prediction results with different baseline models

Methods	LSTM	GRU-D	Without Siamese network	Proposed method
MSE	0.3199	0.5805	0.1134	0.0577

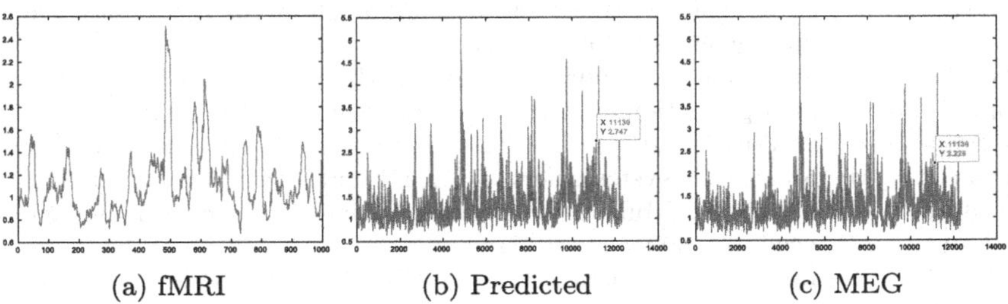

(a) fMRI (b) Predicted (c) MEG

Fig. 3. Averaged neural time series in visual network (a) fMRI, (b) predicted result, (c) MEG.

4 Conclusions

To understand the complex brain dynamics, we need to record the activity in space and time, which could not be solved using current noninvasive techniques of brain measurements. We propose a novel computational model that could combine the information from several techniques and predict the internal brain activity with both high spatial and high temporal resolution. The proposed framework address the vanishing gradient problem by abstracting the long term temporal relationship with functional approximation. In the training step, we use both fMRI and MEG data to learn the brain dynamical representation with Siamese network. While in the testing step, for the first time, the proposed method solves the problem of predicting the internal behavioral states with high resolution in spatial and temporal domain using only fMRI data. The potential of the proposed method to represent the spatio-temporal dynamics has been demonstrated using two experiments with HCP data.

Acknowledgement. This work was partially supported by NIH R01AG071243, R01MH125928, R01AG049371, U01AG068057, and NSF IIS 2045848, 1845666, 1852606, 1838627, 1837956, 1956002, IIA 2040588.

References

1. Chen, X., He, K.: Exploring simple siamese representation learning. arXiv preprint arXiv:2011.10566 (2020)
2. Cornblath, E.J., et al.: Temporal sequences of brain activity at rest are constrained by white matter structure and modulated by cognitive demands. Commun. Biol. **3**(1), 1–12 (2020)
3. Gallego, J.A., Perich, M.G., Naufel, S.N., Ethier, C., Solla, S.A., Miller, L.E.: Cortical population activity within a preserved neural manifold underlies multiple motor behaviors. Nat. Commun. **9**(1), 1–13 (2018)
4. Glasser, M.F., et al.: A multi-modal parcellation of human cerebral cortex. Nature **536**(7615), 171–178 (2016)
5. Glasser, M.F., et al.: The minimal preprocessing pipelines for the human connectome project. Neuroimage **80**, 105–124 (2013)
6. Gu, A., Dao, T., Ermon, S., Rudra, A., Ré, C.: Hippo: Recurrent memory with optimal polynomial projections. arXiv preprint arXiv:2008.07669 (2020)
7. Kriegeskorte, N., Mur, M., Bandettini, P.A.: Representational similarity analysis-connecting the branches of systems neuroscience. Front. Syst. Neurosci. **2**, 4 (2008)
8. McIntosh, A.R., Mišić, B.: Multivariate statistical analyses for neuroimaging data. Annu. Rev. Psychol. **64**, 499–525 (2013)
9. Musall, S., Kaufman, M.T., Juavinett, A.L., Gluf, S., Churchland, A.K.: Single-trial neural dynamics are dominated by richly varied movements. Nat. Neurosci. **22**(10), 1677–1686 (2019)
10. Norman, K.A., Polyn, S.M., Detre, G.J., Haxby, J.V.: Beyond mind-reading: multi-voxel pattern analysis of FMRI data. Trends Cogn. Sci. **10**(9), 424–430 (2006)
11. Saxena, S., Cunningham, J.P.: Towards the neural population doctrine. Curr. Opin. Neurobiol. **55**, 103–111 (2019)
12. Shine, J.M., et al.: Human cognition involves the dynamic integration of neural activity and neuromodulatory systems. Nat. Neurosci. **22**(2), 289–296 (2019)
13. Stringer, C., Pachitariu, M., Steinmetz, N., Carandini, M., Harris, K.D.: High-dimensional geometry of population responses in visual cortex. Nature **571**(7765), 361–365 (2019)
14. Taghia, J., et al.: Uncovering hidden brain state dynamics that regulate performance and decision-making during cognition. Nat. Commun. **9**(1), 1–19 (2018)

15. Tang, E., Mattar, M.G., Giusti, C., Lydon-Staley, D.M., Thompson-Schill, S.L., Bassett, D.S.: Effective learning is accompanied by high-dimensional and efficient representations of neural activity. Nat. Neurosci. **22**(6), 1000–1009 (2019)
16. Tavares, R.M., et al.: A map for social navigation in the human brain. Neuron **87**(1), 231–243 (2015)
17. Voelker, A.R., Kajić, I., Eliasmith, C.: Legendre memory units: Continuous-time representation in recurrent neural networks. In: Proceedings of the 33st International Conference on Neural Information Processing Systems (2019)
18. Zhao, C., Gao, X., Emery, W.J., Wang, Y., Li, J.: An integrated spatio-spectral-temporal sparse representation method for fusing remote-sensing images with different resolutions. IEEE Trans. Geosci. Remote Sens. **56**(6), 3358–3370 (2018)
19. Zhao, C., Li, H., Jiao, Z., Du, T., Fan, Y.: A 3D convolutional encapsulated long short-term memory (3DConv-LSTM) model for denoising fMRI data. In: Martel, A.L., Abolmaesumi, P., Stoyanov, D., Mateus, D., Zuluaga, M.A., Zhou, S.K., Racoceanu, D., Joskowicz, L. (eds.) MICCAI 2020. LNCS, vol. 12267, pp. 479–488. Springer, Cham (2020). https://doi.org/10.1007/978-3-030-59728-3_47

Revealing Continuous Brain Dynamical Organization with Multimodal Graph Transformer

Chongyue Zhao[1], Liang Zhan[1], Paul M. Thompson[2], and Heng Huang[1(✉)]

[1] Department of Electrical and Computer Engineering, University of Pittsburgh, Pittsburgh, PA, USA
{liang.zhan,heng.huang}@pitt.edu

[2] Imaging Genetics Center, University of Southern California, Los Angeles, CA, USA
pthomp@usc.edu

Abstract. Brain large-scale dynamics is constrained by the heterogeneity of intrinsic anatomical substrate. Little is known how the spatio-temporal dynamics adapt for the heterogeneous structural connectivity (SC). Modern neuroimaging modalities make it possible to study the intrinsic brain activity at the scale of seconds to minutes. Diffusion magnetic resonance imaging (dMRI) and functional MRI reveals the large-scale SC across different brain regions. Electrophysiological methods (i.e. MEG/EEG) provide direct measures of neural activity and exhibits complex neurobiological temporal dynamics which could not be solved by fMRI. However, most of existing multimodal analytical methods collapse the brain measurements either in space or time domain and fail to capture the spatio-temporal circuit dynamics. In this paper, we propose a novel spatio-temporal graph Transformer model to integrate the structural and functional connectivity in both spatial and temporal domain. The proposed method learns the heterogeneous node and graph representation via contrastive learning and multi-head attention based graph Transformer using multimodal brain data (i.e. fMRI, MRI, MEG and behavior performance). The proposed contrastive graph Transformer representation model incorporates the heterogeneity map constrained by T1-to-T2-weighted (T1w/T2w) to improve the model fit to structure-function interactions. The experimental results with multimodal resting state brain measurements demonstrate the proposed method could highlight the local properties of large-scale brain spatio-temporal dynamics and capture the dependence strength between functional connectivity and behaviors. In summary, the proposed method enables the complex brain dynamics explanation for different modal variants.

Keywords: Multimodal graph transformer · Graph contrastive representation · Neural graph differential equations

---

Supplementary Information The online version contains supplementary material available at https://doi.org/10.1007/978-3-031-16431-6_33.

L. Wang et al. (Eds.): MICCAI 2022, LNCS 13431, pp. 346–355, 2022.
https://doi.org/10.1007/978-3-031-16431-6_33

1 Introduction

Understanding how our brain dynamically adapts for mind and behaviors helps to extract fine-grained information for typical and atypical brain functioning. But how the microcircuit heterogeneity shapes the structure-function interactions remains an open question in systems neuroscience. Magnetic resonance imaging (MRI) makes it possible to infer the large-scale structural and functional connectivity and characterize the anatomical and functional patterns in human cortex. Electrophysiological methods reveal the dynamical circuit mechanisms at the structural and functional level with higher temporal resolution. Different neuroimaging modalities such as fMRI, dMRI and MEG enable us to estimate both static and functional connectivity during resting state and task experimental paradigms.

Existing studies of large-scale brain dynamics relate the structural and functional connectivity with dynamical circuit mechanisms. The biophysically based dynamical models explore the time-variant function connectivity with excitatory and inhibitory interactions which is interconnected through structural connections [6,20,22]. Microcircuit specialization could be summarized using graph model [3], showing insights into inter-individual brain architecture, development and dysfunction in disease or disorder states. Recently, the graph harmonic analysis based on Laplacian embedding [4] and spectral clustering [19] is introduced to inform the cortical architectural variation. Basically, the previous methods define the nodes of the graph with the harmonic components to quantify the density of anatomical fibers. However, the inter-areal heterogeneity of human cortex has not been widely studied. The next challenging is to decompose the spatio-temporal brain dynamics with multimodal data [21]. Rahim et al. [15] improve the Alzheimer's disease classification performance with fMRI and PET modalities. The stacking method of multimodal neuroimaging data is explored in age prediction task [10]. Representational similarity analysis (RSA) [9] based methods use the common similarity space to associate multivariate modalities. Subsequent research uses Gaussian process to allow complex linking functions [2]. In order to associate the higher temporal resolution of Electrophysiological measurements at millisecond with the higher spatial resolution of MRI and fMRI at millimeter, we introduce the contrastive learning with the Graph Transformer model to learn the heterogeneous graph representation.

Contrastive methods measure the distribution loss with the discriminative structure and achieve the state-of-the-art performance in graph classification. Contrastive multiview coding (CMC) [18], augmented multi-scale DIM (AMDIM) [1] and SimCLR [5] take advantages of multiview mutual information maximization with data augmentation to learn better representations. The graph-level representation is further explored with the extension of the mutual information principle in [17]. Recently, the Graph Transformer based method [16] is proposed to explore the nodes relationship in node embedding learning. However, it is challenging to accurately represent the entire given graph.

Literature on previous multimodal based methods could not be directly applied to link structural connectivity, functional connectivity and behaviors for the following reasons. 1) Most of the existing methods use simple and direct correlational approaches to relate SC and FC. However, the linearity assumption violates the brain spatio-temporal dynamics in many cases. 2) The scanner availability and patient demands may cause the incomplete data problem which may affect the model's performance. 3) As far as we know, little efforts has been made to study how the heterogeneity across human cortex affects the dynamical coupling strength of brain function with structure.

To address these issues, we develop a novel Graph Transformer based framework for associating the heterogeneity of local circuit properties and revealing the dependency of functional connectivity on anatomical structure. The proposed method consists of three parts, the Dynamical Neural Graph Encoder, the graph Transformer pooling with multi-head attention, and the contrastive representation learning model. The proposed method has the following advantages:

- The proposed method provides insights into the brain spatio-temporal dynamical organization related to mind and behavior performance. Existing graph pooling methods may yield similar graph representation for two different graphs. To obtain accurate graph representation, the novel Graph Transformer use multi-head attention to acquire the global graph structure given multimodal inputs. Moreover, we use the contrastive learning model to associate structural and functional details of dMRI, fMRI and MEG.
- The proposed method makes it possible to incorporate the areal heterogeneity map with functional signals using multimodal data from the human connectome project (HCP).
- The proposed method is evaluated with the meta-analysis to explore the behavioral relevance of different brain regions and characterize the brain dynamical organization into low level functions region (i.e. sensory) and the complex function regions (i.e. memory).

2 Methods

To explore the coupling strength of structural and functional connectivity, the heterogeneous Graph Transformer with contrastive learning is trained based on the multimodal brain measurements (i.e. MRI, fMRI and MEG). We use a graph $\mathcal{G}_i = (\mathcal{V}, \mathcal{E}_i)$ to represent the heterogeneous graph representation, with the node type $\mathcal{V} = [v_{t,i}|t = 1, ..., T; i = 1, ..., N]$ with N brain ROIs and T time points. The connection of different brain ROIs in spatial and temporal domain is denoted by the edge mapping $\mathcal{E}$. Given two types of multimodal graph representation $\mathcal{G}_\mathcal{A}$ and $\mathcal{G}_\mathcal{B}$ with different time points $T_\mathcal{A}$ and $T_\mathcal{B}$ and their multivariate value $X_\mathcal{A} = [x^\mathcal{A}_{t_1}, x^\mathcal{A}_{t_2}, ..., x^\mathcal{A}_{t_\mathcal{A}}]$ and $X_\mathcal{B} = [x^\mathcal{B}_{t_1}, x^\mathcal{B}_{t_2}, ..., x^\mathcal{B}_{t_\mathcal{B}}]$. The dynamical neural graph encoder is used to represent the spatio-temporal dynamics within each modality. $Y_\mathcal{A} = [y^\mathcal{A}_{t_1}, y^\mathcal{A}_{t_2}, ..., y^\mathcal{A}_{T_P}]$ and $Y_\mathcal{B} = [y^\mathcal{B}_{t_1}, y^\mathcal{B}_{t_2}, ..., y^\mathcal{B}_{T_P}]$. The adjacency matrices for each view are represented as $A_\mathcal{A}$ and $A_\mathcal{B}$. We use $H^\mathcal{A}$ and $H^\mathcal{B}$ to represent the learned node representation within each modality. Then we use the Graph Transformer pooling layer together with the multi-head attention model to stack the entire node features. The overall framework is illustrated in Fig. 1.

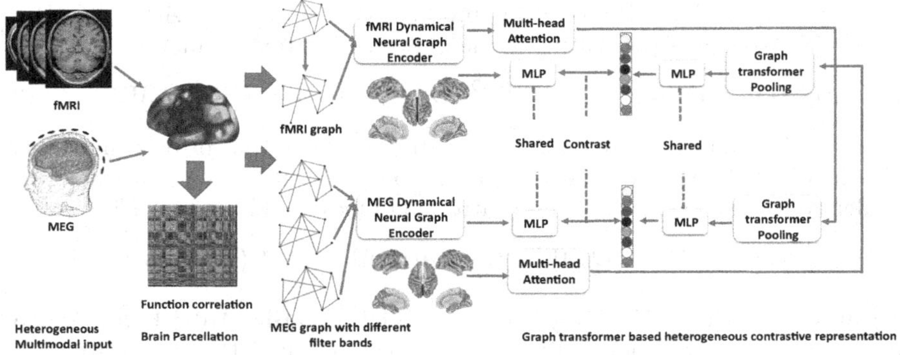

Fig. 1. Schematic illustration of the graph transformer representation learning

2.1 Dynamical Neural Graph Encoder for Single Modal Data

The dynamical brain network with N neurons could be modeled as

$$\dot{z}(t) = f(z, l, t), \tag{1}$$

where $z(t) = [z_1(t), z_2(t), ..., z_N(t)]^T$ represents the internal states of N neuron nodes at time t. $f(\cdot)$ denotes the nonlinear dynamical function of each node. And $l(t) = [l_1(t), l_2(t), ..., l_S(t)]^T$ represents the external stimuli for S neurons.

To represent the dynamics of each single modality, we define a continuous neural-graph differential equation as follows,

$$\dot{Z}(t) = f_{G_{t_k}}(t, Z_t, \theta_t);\ Z_t^+ = \mathcal{L}^j_{G_{t_k}}(Z_t, X_t);\ Y_t = \mathcal{L}^y_{G_{t_k}}(Z_t), \tag{2}$$

where f_G, $\mathcal{L}^j_G$ and $\mathcal{L}^y_G$ are graph encoder networks. Z_t^+ is introduced to represent the value after discrete operation.

2.2 Multimodal Graph Transformer Module

To explore the relationship among different modalities, we introduce the multimodal graph transformer layer. The previous pooling layer ignores the importance of nodes, we design a novel Graph Transformer pooling layer to keep the permutation invariance and injectiveness. The Graph Transformer module consists of a multi-head attention layer and a Graph Transformer pooling layer.

Graph Multi-head Attention. Within each view, the inputs for the multi-head attention consists the terminal state of dynamical neural graph encoder. The inputs are transformed to query $Q \in R^{n_q \times d_k}$, key $K \in R^{n \times d_k}$ and value $V \in R^{n \times d_v}$, where n_q is the number of query vectors and n represents the number of input nodes. d_k and d_v denotes the dimensionlity of corresponding key vector and value vector. The attention dot production is defined as $Att(Q, K, V) = w(QK^T)V$. We define the output of the multi-head attention module (MH) as

$$MH(Q, K, V) = [O_1, ..., O_h]W^O;\ O_i = Att(QW_i^Q, KW_i^K, VW_i^V), \quad (3)$$

where W_i^Q,W_i^K and W_i^V are parameter matrices. The output project matrices is defined as W^O. Using the heterogeneous graph representation learned by graph encoder (GE), the graph multi-head attention block could be denoted by

$$GMH(Q, K, V) = [O_1, ..., O_h]W^O;\ O_i = Att(QW_i^Q, GE_i^K(H, A), GE_i^V(H, A)), \quad (4)$$

Graph Transformer Pooling Together with Graph Multi-head Attention. Inspired by traditional Transformer based method [12], we introduce a novel Graph Transformer pooling layer to learn the global representation of the entire graph, which is defined as follows:

$$GMPool_k(H, A) = LN(Z + rFF(Z));\ Z = LN(S + GMH(S, H, A)), \quad (5)$$

where rFF is any row-wise feed forward layer. S is the seed matrix which could be directly optimized. LN is a layer normalization. In addition, we introduce the self-attention layer to explore the relationship between different nodes.

$$SelfAtt(H) = LN(Z + rFF(Z));\ Z = LN(H + MH(H, H, H)), \quad (6)$$

Together with graph encoder module, the overall framework is defined with the coarsened adjacency matrix A'

$$Pooling(H, A) = GMPool_1(selfAtt(GMPool_k(H, A)), A'). \quad (7)$$

2.3 Contrastive Graph Representation Learning

Finally, we apply the shared projection head $f_\phi(.) \in R^{d_h}$ to the aggregated heterogeneous representation of each view. In the experiment, we use an MLP with two hidden layers as the projection head. The projected representations is defined as, $\vec{h}_g^{\mathcal{A}}$ and $\vec{h}_g^{\mathcal{B}}$. For each view, the node representation are concatenated as follows,

$$\vec{h}_g = \sigma(\overset{L}{\underset{l=1}{\|}} [\sum_{i=1}^{n} \vec{h}_i^l]W), \quad (8)$$

The graph and node representations of the overall Graph Transformer module are defined as $\vec{h} = \vec{h_g^{\mathcal{A}}} + \vec{h_g^{\mathcal{B}}}$ and $\hat{H} = H^{\mathcal{A}} + H^{\mathcal{B}}$. In the training stage, the cross modal mutual information between the node representation and graph representation is defined as,

$$\max_{\theta,\omega,\phi,\psi} \frac{1}{|\mathcal{G}|} \sum_{\mathcal{G}} [\frac{1}{|g|} \sum_{i=1}^{|g|} [MI(\vec{h}_i^{\mathcal{A}}, \vec{h}_g^{\mathcal{B}}) + MI(\vec{h}_i^{\mathcal{B}}, \vec{h}_g^{\mathcal{A}})]], \quad (9)$$

where $\theta, \omega, \phi, \psi$ represent the parameters of heterogeneous graph convolution and projection head. $|\mathcal{G}|$ is the total numbers of graph. $|g|$ is the number of nodes. MI is denoted as the dot production $MI(\vec{h}_i^{\mathcal{A}}, \vec{h}_g^{\mathcal{B}}) = \vec{h}_i^{\mathcal{A}} \cdot (\vec{h}_g^{\mathcal{B}})^T$.

3 Experimental Results

We evaluated the proposed method with resting state fMRI of 1200 subjects from Human Connectome Project (HCP). The resting state fMRI was preprocessed using the HCP minimal preprocessing pipeline [8]. The artefacts of the BOLD signal were further removed using ICA-FIX. The cortical surface was parcellated into N = 360 major ROIs using MMP1.0 parcellation [7]. We excluded 5 subjects with less than 1200 time points for resting-state fMRI data. Additionally, about 95 subjects have resting-state and/or task MEG (tMEG) data. We used 80% of the whole dataset for training and evaluation. The remaining dataset is used for testing. The corresponding resting-state MEG data were acquired in three 6 min runs. The preprocessing of MEG data followed the pipeline provided by HCP data [11]. The source reconstruction was performed using the FieldTrip toolbox. Then sensor data was bandpass filtered into 1.3 55 Hz and projected into source space by synthetic aperture magnetometry. After source reconstruction on the 8k-grid, the time courses were parcellated using MSMAll atlas [8]. The parcellated time courses were z-score normalized for both fMRI and MEG data. In addition, we used the ratio between T1 to T2 weighted maps from HCP dataset as the heterogeneity map. The parcellated diffusion MRI (dMRI) was analysed to generate the structural connectivity (SC) and compute the adjacency matrix $\tilde{A}$ for graph encoder module.

3.1 Heterogeneity Improves the Model Fit to FC

In the first experiment, we tested the similarity between empirical FC and heterogeneous FC patterns acquired by the proposed method compared with the homogeneous model. The empirical group averaged FC, particle-averaged homogeneous FC and heterogeneous FC are shown in Fig. 2. We used a simple non-neural model to introduce self-coupling heterogeneity strength $w_i = w_{min} + w_{scale} s_i$ based on the heterogeneity map s_i, where w_{min} and w_{scale} are heterogeneity parameters.

Synaptic Dynamical Equations. We introduced the biophysically-based computational model to simulate the functional dynamics $\dot{y}_i(t)$ for each node i with the heterogeneity map s_i.

$$\dot{y}_i(t) = -y_i(t) + \sum_j C_{ij} y_j(t) + n_{\nu_i}(t), \tag{10}$$

where $n_{\nu_i}(t)$ is the independent Gaussian white noise. C represents the coupling matrix. $y_i(t)$ is the learned representation using the proposed method. We incorporated the SC matrix S^C and global coupling parameter G^C with $\dot{y}_i(t)$,

$$\dot{y}_i(t) = -\sum_j [(1 - w_i)\delta_{ij} - G^C S^C_{ij}] y_j(t) + n_{\nu_i}(t), \tag{11}$$

We used the squared Pearson correlation coefficient to evaluate the similarity between empirical FC and model fit FC for a single hemisphere. Figure 2

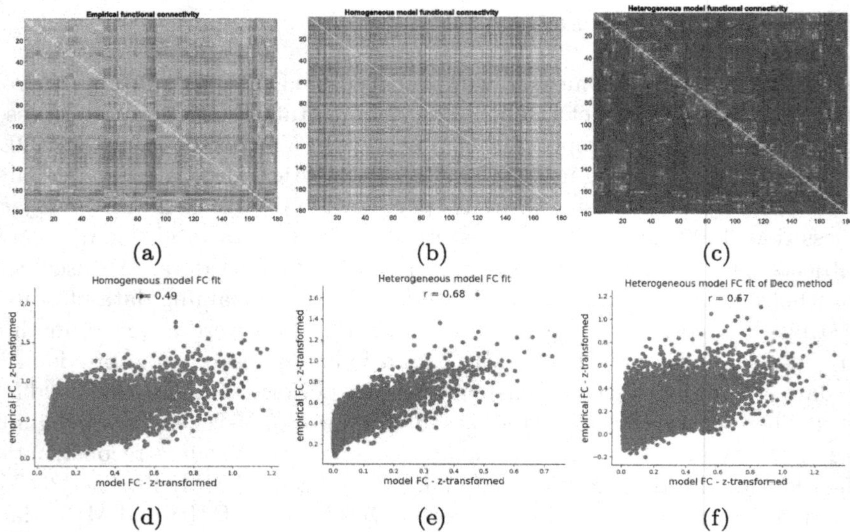

Fig. 2. Heterogeneity map improves the model fit to functional connectivity (FC). (a) Empirical FC. (b) Homogeneous FC. (c) Heterogeneous FC. And (d-f) correlations between empirical FC and model FC, for the Homogeneous (d) Heterogeneous Graph Transformer (e) and Deco's model (f).

shows that the similarity of the proposed graph Transformer model is larger with $r = 0.68$ than the homogeneous model ($r = 0.49$)($p < 10^{-4}$, dependent correlation test). The proposed model also yields higher FC similarity than Deco's model [6] ($r = 0.57$). The experimental result linking multiple modalities demonstrates the hypothesis that the T1w/T2w map shapes the microcircuit properties and spatio-temporal brain dynamics. The introduction of T1w/T2w heterogeneity with the proposed method could capture the dominant neural axis for microcircuit specialization is shown in Fig. 3. In summary, the proposed method with the prior of areal heterogeneity could inform the dynamical relationships among structure, function, and physiology.

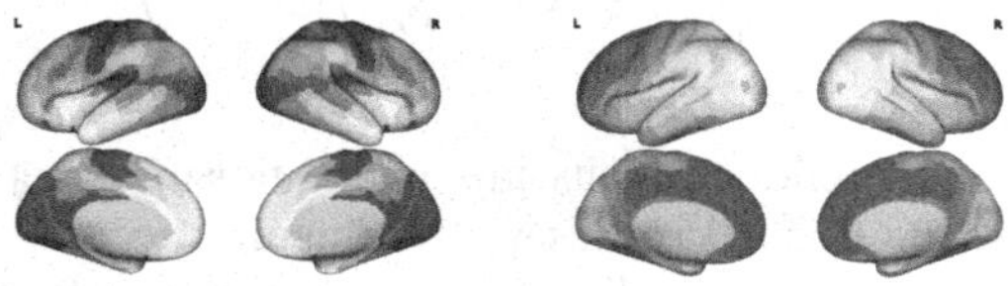

Fig. 3. T1w/T2w heterogeneity map (left) and the example surrogate heterogeneity map (right)

3.2 Functional Connectivity and Behaviors

In the second experiment, we used a NeuroSynth meta-analysis [13] to assess the topic terms with the structural-functional coupling index in Fig. 4. The experimental results demonstrate the existence of behavior related global gradient spanning from lower to higher level cognitive functions. The evidence of global gradient reveals that higher coupling strength in sensory-motor areas which requires fast reacting (i.e. "visual perception", "multisensory processing", "motor/eye movement"). However, the coupling strength in high level cognitive regions (i.e. "autobiographical memory", "emotion" and "reward-based decision making") is low. Similar organization phenomenon could be found in the previous research [13,14].

Fig. 4. Behaviorally relevant gradient shows brain organization

4 Conclusions

In the study, we propose a novel Graph Transformer based method for decoding the brain spatio-temporal dynamics. Different from most of the existing graph convolution method, the Graph Transformer model with multi-head attention guarantees the learning of global graph structure of multimodal data (i.e. dMRI, fMRI and MEG). The contrastive learning model makes it possible to associate multimodal graph representation and reveal how the heterogeneity map shapes the human cortical dynamics. The experimental results demonstrate the importance of regional heterogeneity and the corresponding intrinsic structure-function relationship within brain dynamical organization. Moreover, the proposed method provides insights into brain inter- and intra-regional coupling structure and the relationship between dynamical FC and human behaviors.

Acknowledgement. This work was partially supported by NIH R01AG07 1243, R01MH125928, R01AG049371, U01AG068057, and NSF IIS 2045848, 1845666, 1852606, 1838627, 1837956, 1956002, IIA 2040588.

References

1. Bachman, P., Hjelm, R.D., Buchwalter, W.: Learning representations by maximizing mutual information across views. arXiv preprint arXiv:1906.00910 (2019)
2. Bahg, G., Evans, D.G., Galdo, M., Turner, B.M.: Gaussian process linking functions for mind, brain, and behavior. Proc. Natl. Acad. Sci. **117**(47), 29398–29406 (2020)
3. Bassett, D.S., Sporns, O.: Network neuroscience. Nat. Neurosci. **20**(3), 353–364 (2017)
4. Belkin, M., Niyogi, P.: Laplacian eigenmaps for dimensionality reduction and data representation. Neural Comput. **15**(6), 1373–1396 (2003)
5. Chen, T., Kornblith, S., Norouzi, M., Hinton, G.: A simple framework for contrastive learning of visual representations. In: International Conference on Machine Learning, pp. 1597–1607. PMLR (2020)
6. Deco, G., Ponce-Alvarez, A., Hagmann, P., Romani, G.L., Mantini, D., Corbetta, M.: How local excitation-inhibition ratio impacts the whole brain dynamics. J. Neurosci. **34**(23), 7886–7898 (2014)
7. Glasser, M.F., et al.: A multi-modal parcellation of human cerebral cortex. Nature **536**(7615), 171–178 (2016)
8. Glasser, M.F., et al.: The minimal preprocessing pipelines for the human connectome project. Neuroimage **80**, 105–124 (2013)
9. Guggenmos, M., Sterzer, P., Cichy, R.M.: Multivariate pattern analysis for meg: a comparison of dissimilarity measures. Neuroimage **173**, 434–447 (2018)
10. Jas, M., Engemann, D.A., Bekhti, Y., Raimondo, F., Gramfort, A.: Autoreject: Automated artifact rejection for MEG and EEG data. Neuroimage **159**, 417–429 (2017)
11. Larson-Prior, L.J., et al.: Adding dynamics to the human connectome project with meg. Neuroimage **80**, 190–201 (2013)
12. Lee, J., Lee, Y., Kim, J., Kosiorek, A., Choi, S., Teh, Y.W.: Set transformer: a framework for attention-based permutation-invariant neural networks. In: International Conference on Machine Learning, pp. 3744–3753. PMLR (2019)
13. Margulies, D.S., et al.: Situating the default-mode network along a principal gradient of macroscale cortical organization. Proc. Natl. Acad. Sci. **113**(44), 12574–12579 (2016)
14. Preti, M.G., Van De Ville, D.: Decoupling of brain function from structure reveals regional behavioral specialization in humans. Nat. Commun. **10**(1), 1–7 (2019)
15. Rahim, M., et al.: Integrating multimodal priors in predictive models for the functional characterization of Alzheimer's disease. In: Navab, N., Hornegger, J., Wells, W.M., Frangi, A.F. (eds.) MICCAI 2015. LNCS, vol. 9349, pp. 207–214. Springer, Cham (2015). https://doi.org/10.1007/978-3-319-24553-9_26
16. Rong, Y., et al.: Grover: self-supervised message passing transformer on large-scale molecular data. arXiv preprint arXiv:2007.02835 (2020)
17. Sun, F.Y., Hoffmann, J., Verma, V., Tang, J.: InfoGraph: unsupervised and semi-supervised graph-level representation learning via mutual information maximization. arXiv preprint arXiv:1908.01000 (2019)
18. Tian, Y., Krishnan, D., Isola, P.: Contrastive multiview coding. arXiv preprint arXiv:1906.05849 (2019)
19. Von Luxburg, U.: A tutorial on spectral clustering. Stat. Comput. **17**(4), 395–416 (2007)

20. Yang, G.J., et al.: Altered global brain signal in schizophrenia. Proc. Natl. Acad. Sci. **111**(20), 7438–7443 (2014)
21. Zhao, C., Gao, X., Emery, W.J., Wang, Y., Li, J.: An integrated spatio-spectral-temporal sparse representation method for fusing remote-sensing images with different resolutions. IEEE Trans. Geosci. Remote Sens. **56**(6), 3358–3370 (2018)
22. Zhao, C., Li, H., Jiao, Z., Du, T., Fan, Y.: A 3D convolutional encapsulated long short-term memory (3DConv-LSTM) model for denoising fMRI data. In: Martel, A.L., et al. (eds.) MICCAI 2020. LNCS, vol. 12267, pp. 479–488. Springer, Cham (2020). https://doi.org/10.1007/978-3-030-59728-3_47

Explainable Contrastive Multiview Graph Representation of Brain, Mind, and Behavior

Chongyue Zhao[1], Liang Zhan[1], Paul M. Thompson[2], and Heng Huang[1(✉)]

[1] Department of Electrical and Computer Engineering, University of Pittsburgh, Pittsburgh, PA, USA
liang.zhan@pitt.edu, heng.huang@pitt.edu

[2] Imaging Genetics Center, University of Southern California, Los Angeles, CA, USA
pthomp@usc.edu

Abstract. Understanding the intrinsic patterns of human brain is important to make inferences about the mind and brain-behavior association. Electrophysiological methods (i.e. MEG/EEG) provide direct measures of neural activity without the effect of vascular confounds. The blood oxygenated level-dependent (BOLD) signal of functional MRI (fMRI) reveals the spatial and temporal brain activity across different brain regions. However, it is unclear how to associate the high temporal resolution Electrophysiological measures with high spatial resolution fMRI signals. Here, we present a novel interpretable model for coupling the structure and function activity of brain based on heterogeneous contrastive graph representation. The proposed method is able to link manifest variables of the brain (i.e. MEG, MRI, fMRI and behavior performance) and quantify the intrinsic coupling strength of different modal signals. The proposed method learns the heterogeneous node and graph representations by contrasting the structural and temporal views through the mind to multimodal brain data. The first experiment with 1200 subjects from Human connectome Project (HCP) shows that the proposed method outperforms the existing approaches in predicting individual gender and enabling the location of the importance of brain regions with sex difference. The second experiment associates the structure and temporal views between the low-level sensory regions and high-level cognitive ones. The experimental results demonstrate that the dependence of structural and temporal views varied spatially through different modal variants. The proposed method enables the heterogeneous biomarkers explanation for different brain measurements.

Keywords: Brain dynamics · Spatio-temporal graphs · Explanations on graphs

---

Supplementary Information The online version contains supplementary material available at https://doi.org/10.1007/978-3-031-16431-6_34.

L. Wang et al. (Eds.): MICCAI 2022, LNCS 13431, pp. 356–365, 2022.
https://doi.org/10.1007/978-3-031-16431-6_34

1 Introduction

The brain activity remains latent construct that could not be directly measured with present technologies [18]. Non-invasive electrophysiology such as Magneto- and electo- phencephalography (M/EEG) shows insights into many healthy and diseased brain activity at millisecond but lacks spatial resolution. Additional modalities with higher spatial resolution at millimeter such as Magnetic resonance imaging (MRI), functional MRI(fMRI) and positron emission tomography (PET) have paved the way to human connectomics in clinical practice. However, this kind of technique is sluggish to reveal neuronal activity. The challenge of fusing non-invasive brain measurements is that different technique provides either high spatial or temporal resolution but not both [3,11,21]. The development of fMRI, MEG and MRI made it possible to obtain the system level of structural connectomics and functional connectomics. Many methods have been proposed to associate the connectomics from different modalities. Simple and direct correlational approaches have been commonly used to link SC and FC [9,10,22]. With the prior of SC, dynamic casual model could explain the functional signals in terms of excitatory and inhibitory interactions [5,17]. Graph models allow the extraction of system level connectivity properties associated with brain changes in the life cycle, such as attention and control networks related to late adolescence and aging process, the strength and organization of function connectivity related to neurological diseases and intrinsic brain activity of behavior performance during resting and task state [1,2,4]. Recently, graph harmonic analysis with Laplacian embedding and spectral clustering have been utilized for revealing brain organization [16]. Basically, the graph harmonic model use harmonic components to summarize the spatial patterns with the nodes of the graph. With the structurally informed components, the relationship among structural connectivity, functional connectivity and behavior performance could be decomposed.

Literature on previous graph based methods that utilizes graph theoretical metrics to summarize the function connectivity ignores the high-order interactions between ROIs [12,14,19]. The existing methods are not very suitable for the integration of structural connectivity, functional connectivity and behavior performance for the following reasons:

Lack of Individual and Group-Level Explanation: existing methods especially for fMRI analysis assume that the nodes in the same brain graphs are translation invariant. Ignoring the correspondence of nodes of different brain ROIs limits the explanation in individual and group level.

Incomplete and Missing Data in Clinical Data Collection: due to the scanner availability and patient demands, it is impossible to do multimodal assessment for all patients. Incomplete or missing data hinders the potential of multimodal usage. There are very few databases that provides public access to MEG, MRI and fMRI of the same subjects.

Violation of the Brain Dynamics Information: the existing joint model uses the linearity assumption among latent variables from different modal measurements. The effective usage includes subject specific integration (structural connectivity), modal specific association (i.e. fMRI and MEG). However, the brain is highly dynamic and the linearity assumption is not applicable in many cases.

In the paper, we build a novel heterogeneous contrast subgraphs representation learning based method to exploit the coupling of structural and functional connectivity from different brain modality. The proposed method has the following advantages: 1) The proposed heterogeneous graph representation learning method utilizes the contrastive learning to explore the coupling information of different modality. The proposed method with the semantic attention model enables the complex and dynamics link between structural and functional connectivity within each modal measures. The proposed method is capable of modeling heterogeneous spatio-temporal dynamics and learn the contrast graph structures simultaneously.

2) The proposed method uses a causal explanation model to improve the individual and group-level interpretability. The explanation approach helps to locate the significant brain region with sex difference and neurodevelopment. The experimental results for gender classification with 1200 subjects from HCP have shown the performance of the proposed method with incomplete multimodal data (fMRI and MEG).

3) The proposed method utilizes graph convolution theory to link the brain structure and function. The experimental results with meta-analysis reveal the strength of structural-functional coupling patterns among functional connectivity, structural connectivity and behavior performance.

2 Problem Formulation

To associate heterogeneous multimodal brain measurements, a heterogeneous graph representation learning with semantic attention is introduced based on fMRI and filtered MEG data. Next we introduce dynamical neural graph encoder framework to associate the spatial and temporal patterns from structural and functional connectivity of multimodal brain measurements. Then, we give the details of the multi-view contrastive graph representation learning. The contrastive graph learning method makes sure the maximization of mutual information of the node representation from one view and the graph representation from another view. Finally, we discuss the interpretable causal explanations for the proposed method on graph. The overall framework is illustrated in Fig. 1.

Heterogeneous Graph Representation Learning. We use a graph $\mathcal{G}_i = (\mathcal{V}, \mathcal{E}_i)$ to represent the heterogeneous graph representation, with the node type $\mathcal{V} = [v_{t,i} | t = 1, ..., T; i = 1, ..., N]$ with N brain ROIs and T time points. The edge mapping $\mathcal{E}$ represents the connection of different brain ROIs in spatial and temporal domain. The aim of contrastive graph representation is to explore the spatial and temporal pattern of the fMRI and MEG data. Given two multimodal graph $\mathcal{G}_\mathcal{A}$ and $\mathcal{G}_\mathcal{B}$ with different time points $T_\mathcal{A}$ and $T_\mathcal{B}$ and their correspondence multivariate value $X_\mathcal{A} = [x^\mathcal{A}_{t_1}, x^\mathcal{A}_{t_2}, ..., x^\mathcal{A}_{t_\mathcal{A}}]$ and $X_\mathcal{B} = [x^\mathcal{B}_{t_1}, x^\mathcal{B}_{t_2}, ..., x^\mathcal{B}_{t_\mathcal{B}}]$. To integrate manifest variables of the brain with different spatial and temporal resolution, a dynamical neural graph encoder is proposed to explore the spatio-temporal dynamics. The data augmentation mechanism introduces a $\Phi_{i}{}_{i=1}^{P}$ of P

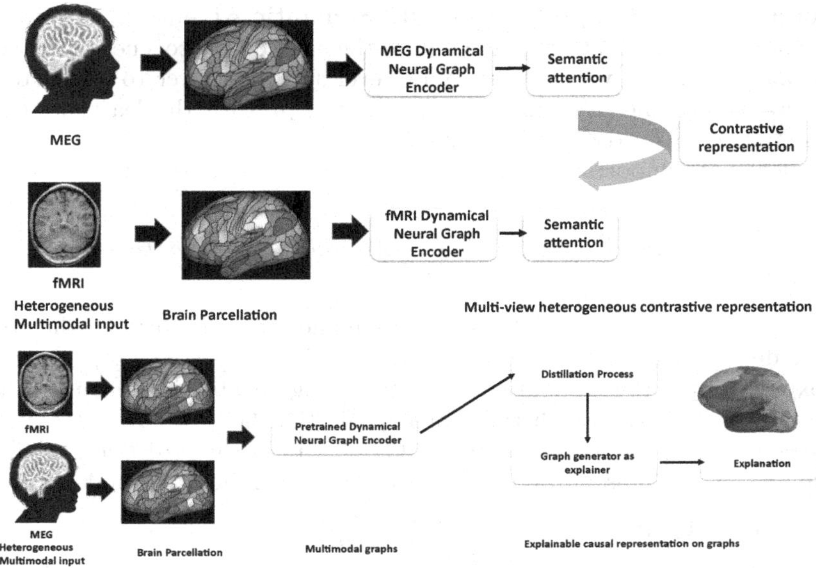

Fig. 1. Schematic illustration of the explainable contrastive graph representation with heterogeneous brain measurements. Top: training process, bottom: explanation process.

set of multiview heterogeneous graph for each modality, where P represents the total view of the brain measurement after data augmentation (i.e. the number of filter bands for MEG). We could define its corresponding adjacency matrices as $[A_{\Phi_i}]_{i=1}^{P}$. The node representation within each view is defined as $Z^{\mathcal{A}}_{\Phi_i}$ and $Z^{\mathcal{B}}_{\Phi_i}$. Then we use the heterogeneous latent attention module to aggregate the graph representation for each modal measurement $H^{\mathcal{A}} = f_{\psi}(L_{att}(Z^{\mathcal{A}}_{\Phi_1}, Z^{\mathcal{A}}_{\Phi_2}, ..., Z^{\mathcal{A}}_{\Phi_P}))$.

2.1 Dynamical Neural Graph Encoder

We define the brain dynamical state with N neurons as $\dot{z}(t) = f(z, l, t)$, where $z(t) = [z_1(t), z_2(t), ..., z_N(t)]^T$ represents the internal states of N neuron nodes at time t. $f(\cdot)$ denotes the nonlinear dynamical function of each node. And $l(t) = [l_1(t), l_2(t), ..., l_S(t)]^T$ represents the external stimuli for S neurons.

Within each single modality, we define a continuous neural-graph differential equation as follows,

$$\dot{Z}(t) = f_{G_{t_k}}(t, Z_t, \theta_t) \quad and \quad Z_t^+ = \mathcal{L}^j_{G_{t_k}}(Z_t, X_t) \tag{1}$$

where f_G, $\mathcal{L}^j_G$ are graph encoder networks. Z_t^+ is introduced to represent the value after discrete operation. Z_t^+ could represent the state 'jump' for brain measurements such as task fMRI.

Heterogeneous Graph Output with Semantic Attention. Within each brain modal measurement, we could obtain the a set of heterogeneous representation $[Z_{\Phi_i}]_{i=1}^{P}$. Then, we use a semantic level attention layer to associate the cross view heterogeneous graph representation L_{att} with the learned weights $\beta_{\Phi_1}, \beta_{\Phi_2}, ..., \beta_{\Phi_p} = Latt(Z_{\Phi_1}, Z_{\Phi_2}, ..., Z_{\Phi_p})$.

We define the importance of each view graph representation as,

$$e_{\Phi_i} = \frac{1}{N}\sum_{n=1}^{N} tanh(\vec{q}^T \cdot [W_{sem} \cdot \vec{Z}_{n,\Phi_i} + \vec{b}]) \quad and \quad \beta_{\Phi_i} = softmax(e_{\Phi_i}), \tag{2}$$

where W_{sem} denotes the linear transformation and $\vec{q}$ is learnable. The heterogeneous node representation could be denoted by $H = f_{\psi}(\sum_{i=1}^{P} \beta_{\Phi_i} Z_{\Phi_i})$.

Next, we apply the readout function to the aggregated heterogeneous representation of each view with the shared projection head $f_{\phi}(.) \in R^{d_h}$. In the experiment, we use an MLP with two hidden layers as the projection head. We get the projected representations $\vec{h}_g^{\mathcal{A}}$ and $\vec{h}_g^{\mathcal{B}}$. For each view, the node representation are concatenated as $\vec{h}_g = \sigma(\overset{L}{\underset{l=1}{\|}} [\sum_{i=1}^{n} \vec{h}_i^l]W)$.

2.2 Mutual Information Based Training Process

In the training process, we maximize the mutual information between the node representation of one modal and the graph representation of another modal, i.e. the node representation of fMRI and the graph representaion of MEG and vice versa. The objective is defined with contrastive learning as follows,

$$\max_{\theta,\phi,\psi} \frac{1}{|G|}\sum_{g\in G}\frac{1}{|g|}\sum_{i=1}^{|g|}[MI(\vec{h}_i^{\mathcal{A}}, \vec{h}_g^{\mathcal{B}}) + MI(\vec{h}_g^{\mathcal{A}}, \vec{h}_i^{\mathcal{B}})], \tag{3}$$

where θ, ϕ, ψ represent the parameters of heterogeneous dynamical neural graph encoder and projection head. $|G|$ denotes the total numbers of graph. $|g|$ is the number of nodes. MI is denoted as the dot production $MI(\vec{h}_i^{\mathcal{A}}, \vec{h}_g^{\mathcal{B}}) = \vec{h}_i^{\mathcal{A}} \cdot (\vec{h}_g^{\mathcal{B}})^T$.

2.3 Explainable Causal Representation on Graphs

To highlight the importance of the brain ROIs, we introduce the explainable causal representation to encourage the reasonable node selection process. We train an explanation model to explain the multmodal graph representation approach based on granger causality. The explanation process are divided into two steps, the distillation process and explainer training process.

In distillation process, we use a subgraph G_s to represent the main cause of the target prediction y. The explainable causal representation does not require the re-training of the dynamical graph encoder which could lower the computation complexity. We use $\delta_{G\backslash e_j}$ to represent the prediction error exclude the edge e_j. The model error is defined as $\Delta_{\delta,e_j} = \delta_{G\backslash e_j} - \delta_G$.

With the ground truth label y, we define the model error as the loss difference $\mathcal{L}(y, \hat{y}_G)$.

$$\delta_G = \mathcal{L}(y, \hat{y}_G) \quad and \quad \delta_{G \setminus e_j} = \mathcal{L}(y, \hat{y}_{G \setminus e_j}), \tag{4}$$

Given the causal contributions of each edge, we could sort the top-K most relevant edges for model explanation. After the model distillation process, we will train a new explainer model based on graph convolutional layer.

$$Z = GCN(A, Z) \quad and \quad \hat{A} = \sigma(ZZ^T), \tag{5}$$

where each value in A represent the contribution of specific edge to prediction. The explanation model generate $\hat{A}$ as an explanation mask. We show more details of the explainable causal model in Supplementary material.

3 Experiments

In the experiments, we use the public available s1200 dataset from Human Connectome Project (HCP), which contains 1096 young adults. Additionally, about 95 subjects have resting-state and/or task MEG (tMEG) data. The resting-state fMRI is pre-processed following the minimal preprocessing pipeline [8]. Then the pre-processed data is registered into a standard cortical surface using MSMAll [8]. The cortical surface was parcellated into N = 22 major ROIs [7]. In addition, the averaged time course of each ROI is normalized using z-score. The resting-state MEG has been pre-processed using ICA to remove out artefacts related to head and eye movement. Sensor-space data were down-sampled 300 Hz using anti-aliasing filter. Next the MEG data were source-reconstructed with a scalar beamformer and registered into the standard space of the Montreal Neuroimaging Institute (MNI). Data were then filtered into 1–30 Hz and beamformed onto 6 mm grid. The parcellation atlas and z-score normalization method of MEG are similar to resting-state fMRI.

3.1 Sex Classification

We first test the performance of the proposed method with sex classification task using HCP data. We adopt 5-fold cross validation on the 1091 subjects. We compare the proposed method with several state-of-the-art methods such as Long-Short-Term Memory (LSTM) [13], graph convolution LSTM (GC-LSTM) [20] and spatio-temporal graph convolution network (ST-GCN) [6]. The hidden state of LSTM was set to 256. A simple Multi-Layer Perception (MLP) with 2 hidden layers and ReLU activation is also included as the baseline method. For the proposed method, we report two kinds of sex classification accuracy. The first single model uses only the fMRI to explore the dynamics within the brain ROIs. The multimodal based method integrates both fMRI and MEG to exploit the spatio-temporal dynamics and achieves better sex classification performance.

The accuracy of sex classification is shown in Table 1. Comparing with the baseline method, the proposed method could learn the dynamic contrast graph representation between fMRI and MEG. The proposed method could take advantage of the high spatial resolution of fMRI and high temporal resolution of MEG to achieve the highest sex classification performance of 85.2%. The importance of the brain regions that contributes to the sex classification is show in Fig. 2. The causal explanation module provides us a new way to find individual-level and group-level biomarkers for sex difference.

Table 1. Sex classification accuracy with different baseline models

Method	Accuracy
LSTM	0.808(0.033)
GC-LSTM	0.811(0.075)
MLP	0.770(0.051)
ST-GCN	0.839(0.044)
Proposed method with only fMRI	0.827(0.061)
Proposed method with multimodal	0.852(0.046)

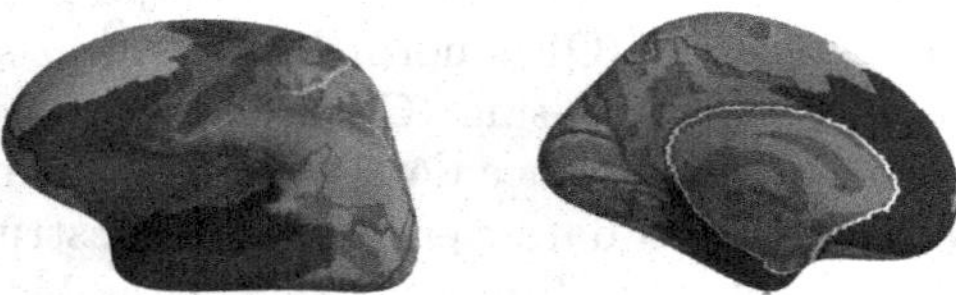

Fig. 2. Top K important brain regions for sex classification

3.2 Brain Activity Decomposed with Functional and Structural Connectivity

In the second experiment, we use the structural-decoupling index which reveals the function and structure relationship to measure the energies of high pass decoupled activity versus low pass coupled activity per brain ROIs. The average structural-decoupling index for surrogate (with or without SC) and function signals is shown in Fig. 3. Without the SC prior knowledge, the surrogate shows significant decoupling patterns. While the knowledge of SC increases the coupling pattern in functional signals. Compared with the functional time courses, the high-level cognition network detaches from the SC. We also use the NeuroSynth meta-analysis on the same topic in [15] to assess the structural-decoupling index. As shown in Fig. 4, the structural-decoupling index associates the behaviorally relevant gradient based on FC data. We could find a macroscale gradient of regions related to low to high level cognition with the learned graph representation. Due to the fact that the functional connectivity comes from MEG data in the second experiment, we compare the gradient learned from the graph representation with the original MEG data. For example, the terms related to acting and perceiving such as "visual perception", "multisensory processing", "reading" and "motor/eye movement" are grouped into the top end. The terms related to complex cognition such as "autobiographical memory", "emotion" and "reward-based decision making" are characterized into the other end. Similar organization phenomenon could be found in the previous research [15,16]. However, the gradient learned by original MEG data lacks the pattern of system organization.

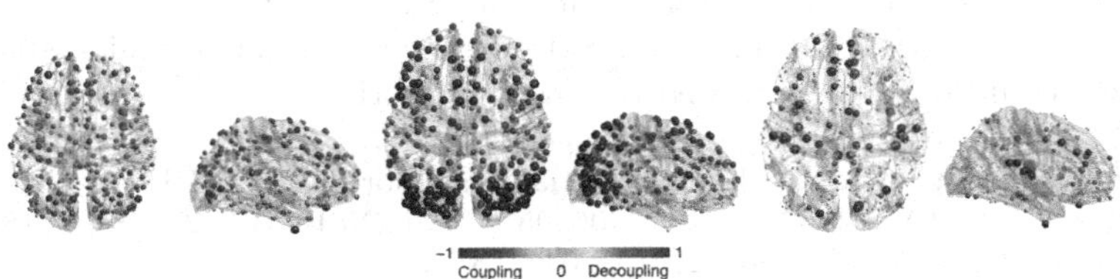

Fig. 3. Structural decoupling index shows brain activity between function and structure. Left: Surrogate brain activity without structural connectome. Middle: surrogate brain activity with structural connectome. Right: brain activity with decoupling difference to the surrogate

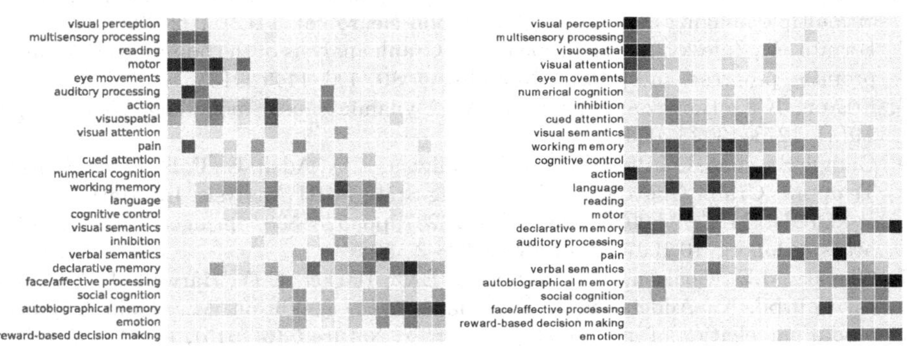

Fig. 4. Behaviorally relevant gradient shows brain organization with Structural decoupling index. Left: the learned graph representation. Right: original MEG data

4 Conclusions

Brain activity is shaped by the anatomical structure. In the paper, we propose an explainable contrastive graph representation learning based model to associate heterogeneous brain measurements such as MRI, functional MRI, MEG and behavior performance. The framework allows key advantages to concentrate brain, mind and behavior in cognitive neuroscience. The proposed method outperforms the state-of-the-art methods in gender classification using fMRI and MEG data. Moreover, the framework could localize the important brain region with sex difference through a causal explanation model. The second experiment with meta analysis demonstrates that the structure-function coupling pattern with the learned contrast graph representation. Future work that links the function connectivity with other modal data (i.e. gene expression and microstructure properties) could be easily adapted to our framework.

Acknowledgement. This work was partially supported by NIH R01AG071243, R01MH125928, R01AG049371, U01AG068057, and NSF IIS 2045848, 1845666, 1852606, 1838627, 1837956, 1956002, IIA 2040588.

References

1. Bassett, D.S., Sporns, O.: Network neuroscience. Nat. Neurosci. **20**(3), 353–364 (2017)
2. Bullmore, E., Sporns, O.: Complex brain networks: graph theoretical analysis of structural and functional systems. Nat. Rev. Neurosci. **10**(3), 186–198 (2009)
3. Dale, A.M., Halgren, E.: Spatiotemporal mapping of brain activity by integration of multiple imaging modalities. Curr. Opin. Neurobiol. **11**(2), 202–208 (2001)
4. Fornito, A., Zalesky, A., Breakspear, M.: Graph analysis of the human connectome: promise, progress, and pitfalls. Neuroimage **80**, 426–444 (2013)
5. Friston, K.J., Harrison, L., Penny, W.: Dynamic causal modelling. Neuroimage **19**(4), 1273–1302 (2003)
6. Gadgil, S., Zhao, Q., Pfefferbaum, A., Sullivan, E.V., Adeli, E., Pohl, K.M.: Spatio-Temporal Graph Convolution for Resting-State fMRI Analysis. In: Martel, A.L., et al. (eds.) MICCAI 2020. LNCS, vol. 12267, pp. 528–538. Springer, Cham (2020). https://doi.org/10.1007/978-3-030-59728-3_52
7. Glasser, M.F., Coalson, T.S., Robinson, E.C., Hacker, C.D., Harwell, J., Yacoub, E., Ugurbil, K., Andersson, J., Beckmann, C.F., Jenkinson, M., et al.: A multi-modal parcellation of human cerebral cortex. Nature **536**(7615), 171–178 (2016)
8. Glasser, M.F., et al.: The minimal preprocessing pipelines for the human connectome project. Neuroimage **80**, 105–124 (2013)
9. Hagmann, P., Cammoun, L., Gigandet, X., Meuli, R., Honey, C.J., Wedeen, V.J., Sporns, O.: Mapping the structural core of human cerebral cortex. PLoS Biol. **6**(7), e159 (2008)
10. Honey, C.J., et al.: Predicting human resting-state functional connectivity from structural connectivity. Proc. Natl. Acad. Sci. **106**(6), 2035–2040 (2009)
11. Jorge, J., Van der Zwaag, W., Figueiredo, P.: EEG-fMRI integration for the study of human brain function. Neuroimage **102**, 24–34 (2014)

12. Kazi, A., et al.: InceptionGCN: receptive field aware graph convolutional network for disease prediction. In: Chung, A.C.S., Gee, J.C., Yushkevich, P.A., Bao, S. (eds.) IPMI 2019. LNCS, vol. 11492, pp. 73–85. Springer, Cham (2019). https://doi.org/10.1007/978-3-030-20351-1_6
13. Li, H., Fan, Y.: Brain decoding from functional MRI using long short-term memory recurrent neural networks. In: Frangi, A.F., Schnabel, J.A., Davatzikos, C., Alberola-López, C., Fichtinger, G. (eds.) MICCAI 2018. LNCS, vol. 11072, pp. 320–328. Springer, Cham (2018). https://doi.org/10.1007/978-3-030-00931-1_37
14. Li, X., Dvornek, N.C., Zhou, Y., Zhuang, J., Ventola, P., Duncan, J.S.: Graph neural network for interpreting task-fMRI biomarkers. In: Shen, D., et al. (eds.) MICCAI 2019. LNCS, vol. 11768, pp. 485–493. Springer, Cham (2019). https://doi.org/10.1007/978-3-030-32254-0_54
15. Margulies, D.S., et al.: Situating the default-mode network along a principal gradient of macroscale cortical organization. Proc. Natl. Acad. Sci. **113**(44), 12574–12579 (2016)
16. Preti, M.G., Van De Ville, D.: Decoupling of brain function from structure reveals regional behavioral specialization in humans. Nat. Commun. **10**(1), 1–7 (2019)
17. Stephan, K.E., Tittgemeyer, M., Knösche, T.R., Moran, R.J., Friston, K.J.: Tractography-based priors for dynamic causal models. Neuroimage **47**(4), 1628–1638 (2009)
18. Turner, B.M., Palestro, J.J., Miletić, S., Forstmann, B.U.: Advances in techniques for imposing reciprocity in brain-behavior relations. Neurosci. Biobehav. Rev. **102**, 327–336 (2019)
19. Yan, Y., Zhu, J., Duda, M., Solarz, E., Sripada, C., Koutra, D.: GroupINN: grouping-based interpretable neural network for classification of limited, noisy brain data. In: Proceedings of the 25th ACM SIGKDD International Conference on Knowledge Discovery & Data Mining, pp. 772–782 (2019)
20. Yu, B., Yin, H., Zhu, Z.: Spatio-temporal graph convolutional networks: a deep learning framework for traffic forecasting. arXiv preprint arXiv:1709.04875 (2017)
21. Zhao, C., Gao, X., Emery, W.J., Wang, Y., Li, J.: An integrated spatio-spectral-temporal sparse representation method for fusing remote-sensing images with different resolutions. IEEE Trans. Geosci. Remote Sens. **56**(6), 3358–3370 (2018)
22. Zhao, C., Li, H., Jiao, Z., Du, T., Fan, Y.: A 3D convolutional encapsulated long short-term memory (3DConv-LSTM) model for denoising fMRI data. In: Martel, A.L., et al. (eds.) MICCAI 2020. LNCS, vol. 12267, pp. 479–488. Springer, Cham (2020). https://doi.org/10.1007/978-3-030-59728-3_47

Embedding Human Brain Function via Transformer

Lin Zhao[1(✉)], Zihao Wu[1], Haixing Dai[1], Zhengliang Liu[1], Tuo Zhang[2], Dajiang Zhu[3], and Tianming Liu[1]

[1] Department of Computer Science, The University of Georgia, Athens, GA, USA
lin.zhao@uga.edu
[2] School of Automation, Northwestern Polytechnical University, Xi'an, China
[3] Department of Computer Science and Engineering, The University of Texas at Arlington, Arlington, TX, USA

Abstract. BOLD fMRI has been an established tool for studying the human brain's functional organization. Considering the high dimensionality of fMRI data, various computational techniques have been developed to perform the dimension reduction such as independent component analysis (ICA) or sparse dictionary learning (SDL). These methods decompose the fMRI as compact functional brain networks, and then build the correspondence of those brain networks across individuals by viewing the brain networks as one-hot vectors and performing their matching. However, these one-hot vectors do not encode the regularity and variability of different brains, and thus cannot effectively represent the functional brain activities in different brains and at different time points. To bridge the gaps, in this paper, we propose a novel unsupervised embedding framework based on Transformer to encode the brain function in a compact, stereotyped and comparable latent space where the brain activities are represented as dense embedding vectors. The framework is evaluated on the publicly available Human Connectome Project (HCP) task based fMRI dataset. The experiment on brain state prediction downstream task indicates the effectiveness and generalizability of the learned embeddings. We also explore the interpretability of the embedding vectors and achieve promising result. In general, our approach provides novel insights on representing regularity and variability of human brain function in a general, comparable, and stereotyped latent space.

Keywords: Brain function · Embedding · Transformer

L. Zhao and Z. Wu–Co-first authors.

Supplementary Information The online version contains supplementary material available at https://doi.org/10.1007/978-3-031-16431-6_35.

L. Wang et al. (Eds.): MICCAI 2022, LNCS 13431, pp. 366–375, 2022.
https://doi.org/10.1007/978-3-031-16431-6_35

1 Introduction

FMRI has been an established neuroimaging technique for studying the human brain's functional organization [10]. However, a major challenge in fMRI-based neuroscience studies is that the number of voxels in 4D spatiotemporal fMRI data is greatly larger than the number of subject brains [12], which is also known as "curse-of-dimensionality" problem [3]. To mitigate negative effects brought by this imbalance, various computational tools have been developed to select the task-relevant features and discard the redundant ones as well as the noises [1,4,11]. For example, principal component analysis (PCA) transforms the correlated voxels into several uncorrelated principal components, which is used as representation of the spatiotemporal fMRI data [1]. Independent component analysis (ICA) based methods assume that the fMRI signals are a "mixture" of spatially or temporally independent patterns (e.g., paradigm-related responses) that could be decomposed from brain fMRI signals [4]. In this way, the analysis can be performed on those much more compactly represented independent patterns rather than the raw voxels in 4D space. In addition to PCA and ICA based matrix decomposition methods, sparse dictionary learning (SDL) was also employed to decompose the fMRI into a over-complete dictionary (temporal activities) and a sparse representation matrix (spatial patterns) [11].

Despite the wide adoption and application of the above-mentioned matrix decomposition techniques, the resulted temporal and/or spatial patterns obtained by those methods are not intrinsically comparable across different individual brains. That is, even with the same hyper-parameters in those matrix decomposition methods like ICA/SDL for different brains, there is no correspondence among the temporal and/or spatial brain network patterns from different subjects. Moreover, even with image registration or pattern matching methods, the huge variability of brain function across individual brains cannot ensure that corresponding brain networks can be identified and matched in different brains. From our perspective, a fundamental difficulty in traditional matrix decomposition methods for fMRI data modeling is that these methods attempt to decompose and represent the brain's functional organization as brain networks and then try to match the correspondences of those networks across individuals and populations. In this process, different brain networks are viewed as one-hot vectors and the mapping or matching are performed on those one-hot vectors. Actually, the one-hot vector representation and matching of brain networks in those matrix decomposition methods do not encode the regularity and variability of different brains, and as a consequence, these one-hot vectors of brain networks do not offer a general, comparable, and stereotyped space for brain function. To address this critical problem, an intuitive way is to encode brain function in a compact, stereotyped and comparable latent space where the brain activities measured by fMRI data in different brains and at different time points can be meaningfully embedded and compactly represented.

As an effective methodology for high-dimensional data embedding, deep learning has been widely employed in fMRI data modeling and achieved superior results over those traditional matrix decomposition methods [6,8,13,17].

However, as far as we know, prior deep learning models of fMRI data were not specifically designed towards the effective embedding of human brain function for the purpose of compact representation of regularity and variability in different brains. Instead, prior methods were designed for specific tasks, such as fMRI time series classification [9], hierarchical brain network decomposition [5], brain state differentiation [15], among others. Therefore, existing deep learning models of fMRI data still do not offer a general, comparable, and stereotyped space for representing human brain function. Importantly, the compact and comparable embeddings can also be easily integrated into other deep learning frameworks, paving the road for multi-modal representation learning such as connecting the text stimuli in semantic space of Natural Language Processing (NLP) and the brain's responses to those stimuli in brain function embedding space.

To bridge the above gaps, in this paper, we formulate the effective and general representation of human brain function as an embedding problem. The key idea is that each 3D volume of fMRI data can be embedded as a dense vector which profiles the functional brain activities at the corresponding time point. The regularity and variability of brain function at different time points and across individual brains can be effectively measured by the distance in the embedding space. That is, our embedding space offers a general, comparable, and stereotyped space for brain function. Specifically, to achieve such effective embedding space of human brain function, we designed a novel Temporal-Correlated Autoencoder (TCAE) based on the Transformer model [14] and self-attention mechanism. The major theoretical and practical advantage of Transformer is that its multi-head self-attention can explicitly capture the correlation between time points, especially for those far away from each other and jointly attending to information from different representation subspaces, which is naturally suitable for our objective of holistic embedding of human brain function. We evaluate the proposed brain function embedding framework on the publicly available Human Connectome Project (HCP) task based fMRI dataset and it achieves state-of-the-art performance on brain state prediction downstream task. The interpretability of the learning embedding is also studied by exploring its relationship with the block-design paradigm. Overall, our method provides a novel and generic framework for representing the human brain's functional architecture.

2 Methods

2.1 Overview

The proposed framework is shown in Fig. 1. We first illustrate the TCAE embedding framework used in this study in Sect. 2.2. Then, the learned embeddings are evaluated on a downstream task of brain state prediction which is introduced in Sect. 2.3.

2.2 Temporal-Correlated Autoencoder

In this section, we introduce the TCAE embedding framework based on the Transformer model and multi-head self-attention [14]. As illustrated in Fig. 1,

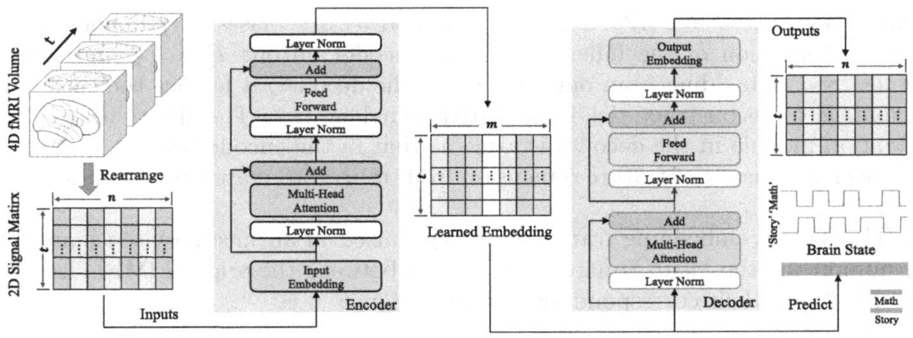

Fig. 1. Illustration of the proposed TCAE embedding framework. The 4D spatiotemporal fMRI data are firstly rearranged into 2D signal matrix and then input into the encoder. The output of the encoder is recognized as the learned embedding, which is used for the brain state prediction downstream task and reconstructing the input signal matrix.

TCAE has an encoder-decoder architecture. For both encoder and decoder, they consist of one embedding layer and one multi-head self-attention module. Specifically, in the encoder, the rearranged 2D fMRI signal matrix $\boldsymbol{S} \in \mathbb{R}^{t \times n}$, where t is the number of time points and n is the number of voxels, is firstly embedded into a new feature matrix $\boldsymbol{S}_f \in \mathbb{R}^{t \times m}$ through a fully-connected (FC) layer, where m is the reduced feature dimension ($m \ll n$). Then, for attention head i, the self-attention map that captures the temporal correlations of different time points is computed as:

$$Attn\_Map_i = \boldsymbol{S}_f\, \boldsymbol{W}_i^Q (\boldsymbol{S}_f\, \boldsymbol{W}_i^K)^T \tag{1}$$

where $\boldsymbol{W}_i^Q \in \mathbb{R}^{m \times k}$ and $\boldsymbol{W}_i^K \in \mathbb{R}^{m \times k}$ are projection matrices and k is the feature dimension of the self-attention operation. With the self-attention maps, the output of attention head i can be computed as:

$$\boldsymbol{S}_{attn_i} = softmax(\frac{Attn\_Map_i}{\sqrt{k}})\boldsymbol{S}_f\, \boldsymbol{W}_i^V \tag{2}$$

where $\boldsymbol{W}_i^V \in \mathbb{R}^{m \times v}$ and v is the feature dimension for the output of attention heads. We then concatenate $\boldsymbol{S}_{attn_i}$ along the feature dimension and transform the concatenated matrix into a new feature matrix $\boldsymbol{S}_{multi} \in \mathbb{R}^{t \times m}$ as:

$$\boldsymbol{S}_{multi} = Concat(head_1, \ldots, head_h)\, \boldsymbol{W}^O \tag{3}$$

where $Concat(\cdot)$ represents the concatenation operation. h is the number of heads and $\boldsymbol{W}^O \in \mathbb{R}^{hv \times m}$ is the projection matrix. $\boldsymbol{S}_{multi}$ is further fed into the feed forward layer to obtain the encoder output $\boldsymbol{E} \in \mathbb{R}^{t \times m}$:

$$\boldsymbol{E} = ReLU(\boldsymbol{S}_{multi}\, \boldsymbol{W}_1 + b_1)\, \boldsymbol{W}_2 + b_2 \tag{4}$$

where $\boldsymbol{W}_1 \in \mathbb{R}^{m \times d_{ff}}$, $\boldsymbol{W}_2 \in \mathbb{R}^{d_{ff} \times m}$, b_1 and b_2 are biases. d_{ff} denotes the feature dimension of the inner layer. The encoder output $\boldsymbol{E}$ is recognized as the learned embedding from our model. For the decoder, it fetches the encoder output $\boldsymbol{E}$ to reconstruct the input fMRI signal matrix. The multi-head self-attention module in the decoder is same as that in the encoder except that the FC layer increases the feature dimension from m to n to match the input signal matrix.

The TCAE embedding framework is optimized in an unsupervised manner by minimizing the Mean Square Error (MSE) between the original fMRI signals $\boldsymbol{S} \in \mathbb{R}^{t \times n}$ and their corresponding reconstruction $\boldsymbol{S}' \in \mathbb{R}^{t \times n}$.

2.3 Prediction of Brain State

In this subsection, we introduce a brain state prediction task to evaluate the learned embedding. Specifically, each time point can be classified into a specific brain state according to the task that the subject participated in, e.g., math calculation or listening to a story. The prediction of brain state is performed by a two-stage manner. In the first stage, we pre-train a TCAE model to learn the embedding as described in Sect. 2.2. In the second stage, we fix the pre-trained model and obtain the embedding for each time point. The embedding is then input into a classifier consisting of two fully-connected (FC) layers with $tanh/softmax$ as activation function, respectively. The classifier is optimized by minimizing the cross-entropy between predictions and labels. In this way, the learned embedding is not task-specific and the effectiveness and generalizability can be fairly evaluated.

3 Experiments

3.1 Dataset and Pre-Processing

We adopt the publicly available HCP task fMRI (tfMRI) dataset of S1200 release (https://www.humanconnectome.org/) [2]. In this paper, among 7 different tasks in HCP tfMRI dataset, Emotion and Language tasks are used as testbeds for our framework due to the space limit. The acquisition parameters of HCP tfMRI data are as follows: 90×104 matrix, 72 slices, TR = 0.72 s, TE = 33.1 ms, 220 mm FOV, flip angle = 52°, BW = 2290 Hz/Px, in-plane FOV = 208×180 mm, 2.0 mm isotropic voxels. The preprocessing pipelines of tfMRI data are implemented by FSL FEAT [16], including skull removal, motion correction, slice time correction, spatial smoothing, global drift removal (high-pass filtering) and registration to the standard MNI 152 4 mm space for reducing the computational overhead. Besides, time series from the voxels of 4D tfMRI data are rearranged into a 2D array with zero mean and standard deviation one. For a total of more than 1000 subjects in HCP S1200 release, we randomly select 600 subjects as training set, 200 subjects as validation set, and another 200 subjects as testing set for both two tasks, respectively. All the experimental results in our study are reported based on the testing set.

3.2 Implementation Details

In our experiments, we uniformly set the embedding dimension as 64, i.e., the embedding has 64 digits. For TCAE model, the m,k,v are set to 64 and d_{ff} is set to 128. For the predictor, the size of two FC layers are 64/32, respectively. The framework is implemented with PyTorch (https://pytorch.org/) deep learning library. We use the Adam optimizer [7] with $\beta_1 = 0.9$ and $\beta_2 = 0.999$. The batch size is 16 and the model is trained for 100 epochs with a learning rate 0.01 for both tasks on a single GTX 1080Ti GPU. It is noted that all experiments were performed with testing set based on model with the lowest loss on validation dataset.

3.3 Brain State Prediction Results

In this subsection, we evaluate the learned embedding on a brain state prediction downstream task. Here, we introduce several baseline methods for comparison: Autoencoder (AE), deep sparse recurrent autoencoder (DSRAE) [8], deep recurrent variational autoencoder (DRVAE) [13], spatiotemporal attention autoencoder (STAAE) [6]. AE denotes an autoencoder model consisting of one FC layer with $tanh$ activation function in both encoder and decoder, which can be considered as a baseline without the multi-head self-attention module of our model. It is noted that DSRAE was designed for decomposing the spatial and temporal patterns as SDL; DRVAE model aimed at the augmentation of fMRI data; STAAE was proposed for the classification of Attention Deficit Hyperactivity Disorder (ADHD). We implement their network architectures and take the encoder's output as embedding. The details about the configuration of those baselines can be found in supplemental materials. In Table 1, we report the brain state classification accuracy as well as the number of parameters (Params) and the number of multiply-accumulate operations (MACs) of those baselines and the proposed method with an embedding size of 64. The results for other embedding size and the hyper-parameter sensitivity analysis can be found in supplemental materials.

Table 1. The prediction accuracy, number of parameters and MACs of baselines and the proposed framework on brain state prediction tasks. The accuracy is averaged among all subjects in testing dataset on both Emotion and Language task. Red and blue denotes the best and the second-best results, respectively.

Methods	Accuracy		Params(M)	MACs(G)
	Emotion	Language		
AE	0.6989	0.8481	3.68	10.29
STAAE [6]	0.6080	0.8165	14.70	41.32
DRVAE [13]	0.6364	0.8418	15.13	42.54
DSRAE [8]	0.6932	0.8006	15.12	42.52
TCAE	0.7557	0.8829	3.75	10.47

It is observed that all baselines have an accuracy over 0.6. The baseline AE outperforms all the other baselines in terms of prediction accuracy with much less parameters and MACs. This is probably because other baselines such as DRVAE are designed for a specific task which requires a specially designed architecture with more parameters. On the other hand, an architecture with more parameters may not be generalizable for our embedding task and thus degenerate the performance. Our proposed TCAE embedding framework introduces an additional self-attention module with slight increases in model parameters and computational operations compared with the baseline AE. But the performance gain is significant with the highest accuracy (two-sample one-tailed un-pair-wise t-test ($p < 0.025$, corrected)). This observation indicates that our model is compact and can learn a more generalizable embedding compared with other baselines.

3.4 Interpretation of the Learned Embedding

To explore the interpretation of the learned embedding, similar to [8], we assume that each digit of embedding corresponds to a temporal brain activity pattern, which can be obtained by extracting the digit value over time. With the extracted temporal pattern of each digit, we compute the Pearson Correlation Coefficient (PCC) between each pattern and the Hemodynamic Response Function (HRF) responses of task stimulus. From all digits, we select the one with the highest PCC value as the task-relevant digit which is an indicator of the embedding' relevance to task stimulus. Here, we randomly select four subjects as examples to show the temporal pattern of task-relevant digit from the TCAE model as well as the corresponding HRF responses in Fig. 2.

Table 2. Mean (± standard deviation) PCC (Pearson Correlation Coefficient) between the temporal pattern of task-relevant digit and the HRF response. E1: 'Faces' stimulus; E2: 'Shapes' stimulus; L1: 'Math' stimulus; L2: 'Story' stimulus. Red and blue denotes the highest and the second-highest PCC, respectively.

Methods	Emotion		Language	
	E1	E2	L1	L2
AE	0.45 ± 0.09	0.51 ± 0.11	0.76 ± 0.10	0.61 ± 0.10
STAAE [6]	0.48 ± 0.11	0.40 ± 0.11	0.73 ± 0.09	0.72 ± 0.09
DRVAE [13]	0.46 ± 0.11	0.40 ± 0.10	0.59 ± 0.08	0.68 ± 0.10
DSRAE [8]	0.48 ± 0.11	0.45 ± 0.10	0.68 ± 0.08	0.71 ± 0.08
TCAE	0.55 ± 0.09	0.53 ± 0.09	0.66 ± 0.09	0.71 ± 0.09

Generally, the temporal pattern of task-relevant digit matches the corresponding HRF response well with PCC value larger than 0.45, indicating that

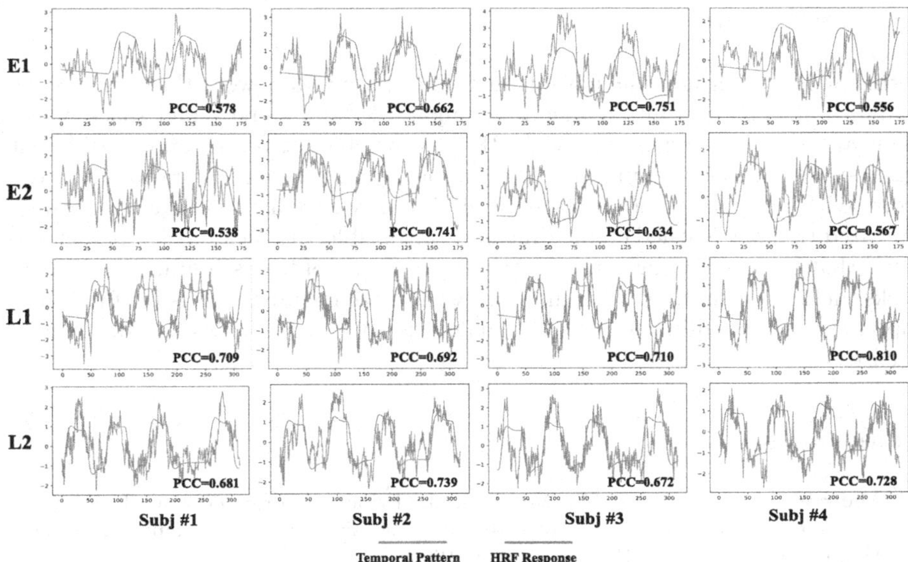

Fig. 2. The temporal pattern of task-relevant digit from TCAE model's embedding compared with HRF responses from 4 randomly selected subjects for each task stimulus, respectively. Abbreviations: E1: 'Faces' stimulus; E2: 'Shapes' stimulus; L1: 'Math' stimulus; L2: 'Story' stimulus.

the digits of the learned embedding are to some extent correlated to task stimulus. To quantitatively measure such correlation, we average the PCC of all subjects and compare it with those from baseline models in Table 2. It is observed that in Emotion task, the averaged PCC of TCAE's task-relevant digit is larger than that of all compared baselines. However, in the Language task, AE and STAAE have the highest PCC for two task designs, respectively. A possible reason is that in Emotion task, the response of task stimulus is more complex and hard to be decoded and decomposed from the raw fMRI data. Our embeddings from TCAE model can better characterize such responses, which is in alignment with the highest brain state prediction accuracy in Table 1. While in the Language task, the responses are quite straightforward and can be easily captured by other deep learning models. It is consistent with overall higher brain state prediction accuracy in Table 1 than Emotion task. The TCAE embedding model may focus on more intrinsic responses and patterns which are still task-relevant but discriminative, resulting in a higher accuracy in brain state prediction but relatively lower PCC than baselines. Overall, the embedding from TCAE framework provides rich and meaningful information which is relevant to the brain's response to task stimulus.

4 Conclusion

In this paper, we proposed a novel transformer-based framework that embeds the human brain function into a general, comparable, and stereotyped space where the brain activities measured by fMRI data in different brains and at different time points can be meaningfully and compactly represented. We evaluated the proposed framework in brain state prediction downstream task, and the results indicated that the learned embedding is generalizable and meaningful. It was also found that the embedding is relevant to the response of task stimulus. Our future works include evaluating the framework with more cognitive tasks in tfMRI and applying the embedding for disease diagnosis such as ADHD and Alzheimer's disease.

References

1. Andersen, A.H., Gash, D.M., Avison, M.J.: Principal component analysis of the dynamic response measured by fMRI: a generalized linear systems framework. Magn. Reson. Imaging **17**(6), 795–815 (1999)
2. Barch, D.M., et al.: Function in the human connectome: task-fMRI and individual differences in behavior. Neuroimage **80**, 169–189 (2013)
3. Bellman, R.E.: Adaptive Control Processes. Princeton University Press, Princeton (2015)
4. Calhoun, V.D., Adali, T.: Unmixing fMRI with independent component analysis. IEEE Eng. Med. Biol. Mag. **25**(2), 79–90 (2006)
5. Dong, Q., et al.: Modeling hierarchical brain networks via volumetric sparse deep belief network. IEEE Trans. Biomed. Eng. **67**(6), 1739–1748 (2019)
6. Dong, Q., Qiang, N., Lv, J., Li, X., Liu, T., Li, Q.: Spatiotemporal attention autoencoder (STAAE) for ADHD classification. In: Martel, A.L., et al. (eds.) MICCAI 2020. LNCS, vol. 12267, pp. 508–517. Springer, Cham (2020). https://doi.org/10.1007/978-3-030-59728-3_50
7. Kingma, D.P., Ba, J.: Adam: a method for stochastic optimization. arXiv preprint arXiv:1412.6980 (2014)
8. Li, Q., Dong, Q., Ge, F., Qiang, N., Wu, X., Liu, T.: Simultaneous spatial-temporal decomposition for connectome-scale brain networks by deep sparse recurrent autoencoder. Brain Imaging Behav. **15**(5), 2646–2660 (2021). https://doi.org/10.1007/s11682-021-00469-w
9. Liu, H., et al.: The cerebral cortex is bisectionally segregated into two fundamentally different functional units of gyri and sulci. Cereb. Cortex **29**(10), 4238–4252 (2019)
10. Logothetis, N.K.: What we can do and what we cannot do with fMRI. Nature **453**(7197), 869–878 (2008)
11. Lv, J., et al.: Holistic atlases of functional networks and interactions reveal reciprocal organizational architecture of cortical function. IEEE Trans. Biomed. Eng. **62**(4), 1120–1131 (2014)
12. Mwangi, B., Tian, T.S., Soares, J.C.: A review of feature reduction techniques in neuroimaging. Neuroinformatics **12**(2), 229–244 (2014)
13. Qiang, N., et al.: Modeling and augmenting of fMRI data using deep recurrent variational auto-encoder. J. Neural Eng. **18**(4), 0460b6 (2021)

14. Vaswani, A., et al.: Attention is all you need. Adv. Neural Inf. Process. Syst. **30** (2017)
15. Wang, H., et al.: Recognizing brain states using deep sparse recurrent neural network. IEEE Trans. Med. Imaging **38**(4), 1058–1068 (2018)
16. Woolrich, M.W., Ripley, B.D., Brady, M., Smith, S.M.: Temporal autocorrelation in univariate linear modeling of fMRI data. Neuroimage **14**(6), 1370–1386 (2001)
17. Zhao, L., Dai, H., Jiang, X., Zhang, T., Zhu, D., Liu, T.: Exploring the functional difference of Gyri/Sulci via hierarchical interpretable autoencoder. In: de Bruijne, M., et al. (eds.) MICCAI 2021. LNCS, vol. 12907, pp. 701–709. Springer, Cham (2021). https://doi.org/10.1007/978-3-030-87234-2_66

How Much to Aggregate: Learning Adaptive Node-Wise Scales on Graphs for Brain Networks

Injun Choi[1], Guorong Wu[2], and Won Hwa Kim[1,3](✉)

[1] Pohang University of Science and Technology, Pohang, South Korea
{surung9898,wonhwa}@postech.ac.kr
[2] University of North Carolina at Chapel Hill, Chapel Hill, USA
grwu@med.unc.edu
[3] University of Texas at Arlington, Arlington, USA

Abstract. Brain connectomes are heavily studied to characterize early symptoms of various neurodegenerative diseases such as Alzheimer's Disease (AD). As the connectomes over different brain regions are naturally represented as a graph, variants of Graph Neural Networks (GNNs) have been developed to identify topological patterns for disease early diagnosis. However, existing GNNs heavily rely on the fixed local structure given by an initial graph as they aggregate information from a direct neighborhood of each node. Such an approach overlooks useful information from further nodes, and multiple layers for node aggregations have to be stacked across the entire graph which leads to an over-smoothing issue. In this regard, we propose a flexible model that learns adaptive scales of neighborhood for individual nodes of a graph to incorporate broader information from appropriate range. Leveraging an adaptive diffusion kernel, the proposed model identifies desirable scales for each node for feature aggregation, which leads to better prediction of diagnostic labels of brain networks. Empirical results show that our method outperforms well-structured baselines on Alzheimer's Disease Neuroimaging Initiative (ADNI) study for classifying various stages towards AD based on the brain connectome and relevant node-wise features from neuroimages.

1 Introduction

Rich bodies of works show that amyloid deposition and neurofibrillary tangles damage neural connections in the brain, suggesting the analysis of brain connectomes in neuroimaging studies to characterize early symptoms of brain disorders such as Autism [1], Parkinson's Disease (PD) [24] and Alzheimer's Disease (AD) [6,20]. The connectome connects different anatomical regions of interest (ROI) in the brain and comprises a brain network for individual subjects. Such a brain

Supplementary Information The online version contains supplementary material available at https://doi.org/10.1007/978-3-031-16431-6_36.

L. Wang et al. (Eds.): MICCAI 2022, LNCS 13431, pp. 376–385, 2022.
https://doi.org/10.1007/978-3-031-16431-6_36

network is mathematically represented as a graph that is defined by a set of nodes and edges, whose nodes are given by the ROIs and the connectomes define edges as a measure of strength among the nodes derived from structural and functional neuroimages, e.g., tractography on Diffusion Tensor Images (DTI).

Such a graph representation of a brain network, together with image-derived measurements at each ROI, naturally justifies the utilization of graph deep learning approaches such as graph neural network (GNN) for disease characterization. GNN [9] and its variants [4,13] incorporate the structure of graphs via message passing among connected nodes in the graph. To obtain more robust features, they aggregate direct neighborhood information and refer to indirectly connected nodes by stacking multiple aggregation layers, which lead to promising results in node classification [7], link prediction [26] under the homophily condition [15], i.e., adjacent nodes have high similarity as in adjacent pixels in natural images.

However, there are still several issues for previous GNNs to be adopted for brain network analysis. First, they often use a single graph merely as the domain of measurements at each node. Also, the graph is often represented as a binary matrix, which does not incorporate exact relationships in the neighborhood of its node. Of course there have been recent efforts to alleviate these problems such as Graph Attention Network (GAT) [22], Graph Convolutional using Heat Kernel (GraphHeat) [25] and Graph Diffusion Convolution (GDC) [14]. However, they cannot incorporate heterogeneous characteristics of brains, where both ROI measures and brain networks are different across subjects. A bigger problem is that these methods are either too local or global: aggregation of information occurs only within the direct neighbors of each node and adding layers to incorporate indirect neighbors triggers the local aggregation across all the nodes.

The issues above naturally lead to an idea of learning adaptive range for individual nodes. As each brain ROI has different biological and topological properties, it is feasible to learn different local receptive fields that provide an understanding of subnetwork structure. For this, we propose a novel flexible framework that learns suitable scales of each node's neighborhood by leveraging a diffusion kernel and a specialized model architecture to update the scale as a parameter in the model. Learning individual scales for each node lets our model find the right spatial range for information propagation for each ROI.

Key Contributions: Our work leads to 1) learning adaptive local neighborhood to aggregate information for better prediction of graph labels, 2) deriving a parametric formulation to perform gradient-based learning on local receptive field of nodes using a diffusion kernel, and 3) validating the developed framework in comparisons to the recent graph neural network models. Experiments on structural brain networks from Diffusion Tensor Imaging (DTI) and ROI measures from functional imaging from Alzheimer's Disease Neuroimaging Initiative (ADNI) study show that the developed framework demonstrates superior graph-level classification results identifying the independence of each ROI.

2 Related Work

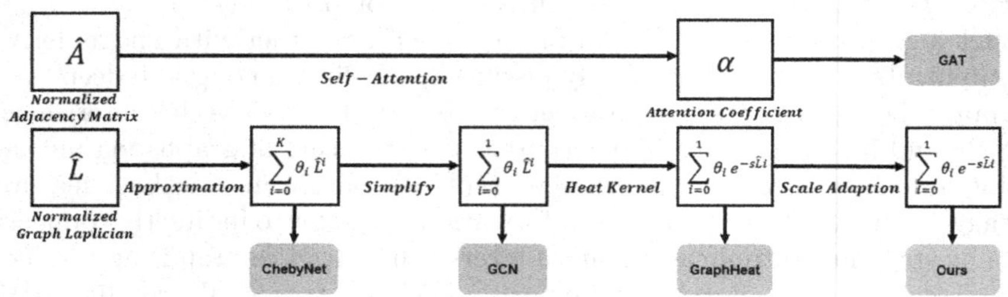

Fig. 1. An overview of node aggregation frameworks. Different from the previous methods, our model can flexibly learn the scale s for each node. Note that s in the GraphHeat is a constant hyperparameter.

To generalize the Convolutional Neural Networks (CNNs) to signals defined on graphs, various spectral methods such as Graph Convolutional Network and ChebyNet were proposed in [2,4,11,13], allowing the use of shared filters. In these models, the importance of each node is given dichotomously, limiting the selection of proper nodes in the neighborhood. To address this issue, the Graph Convolutional Networks using Heat Kernel (GraphHeat) [25] uses heat diffusion to quantify relationships among nodes, Graph Attention Network (GAT) [22] generates the importance score of different nodes by using attention mechanism, and Graph Diffusion Convolution (GDC) [14] uses a transition matrix utilizing personalized PageRank node proximity measure. A streamline of these methods is introduced in Fig. 1, and are discussed in detail in the following section.

3 Preliminaries

Graph Convolution. An undirected graph $G = \{V, E\}$, where V is a node set with $|V| = N$ and E is a set of edges, has an symmetric adjacency matrix $A_{N \times N}$ encoding node connectivity. Graph Laplacian $L = D - A$, where D is a diagonal degree matrix of G with $D_{i,i} = \sum_j A_{i,j}$, and a normalized graph Laplacian is defined as $\hat{L} = I_N - D^{-1/2}AD^{-1/2}$, where I_N is an identity matrix. Since $\hat{L}$ is real and symmetric, $\hat{L}$ has a complete set of orthonormal basis $U = [u_1|u_2|...|u_N]$ and corresponding real and non-negative eigenvalues $0 = \lambda_1 \leq \lambda_2 \leq ... \leq \lambda_N$.

With U, the Graph Fourier transform of signal x is defined as $\hat{x} = U^T x$ and its inverse transform is $x = U\hat{x}$, where $\hat{x}$ is the signal in Graph Fourier space [19]. By the convolution theorem [16], a graph convolution operation is defined as:

$$g * x = U((U^T g) \circ (U^T x)), \tag{1}$$

where g is a filter and $\circ$ is hadamard product.

Graph Convolutional Network. In a spectral graph convolutional network [2], spectral convolution between a filter g and a signal x on a graph was defined as:

$$g * x = U g_\theta U^\mathcal{T} x = (\theta_1 u_1 u_1^\mathcal{T} + \theta_2 u_2 u_2^\mathcal{T} + ... + \theta_N u_N u_N^\mathcal{T})x \tag{2}$$

where $U^\mathcal{T} g$ in Eq. (1) is replaced by a kernel $g_\theta = diag(\{\theta_i\}_{i=1}^N)$. As using Eq. (2) can be computationally expensive, polynomial approximation using Chebyshev expansions was proposed [4]. ChebyNet provides a polynomial filter $g = \sum_{k=0}^{K-1} \theta_k \Lambda^k$ where the parameter $\theta \in \mathbb{R}^K$ with $K \ll N$ is vector of polynomial coefficients. With a polynomial filter g, graph convolution is performed as:

$$g * x \simeq (\theta_0 I + \theta_1 \hat{L} + ... + \theta_{K-1} \hat{L}^{K-1})x. \tag{3}$$

Later, GCN [13] was proposed with only the first-order approximation as:

$$g * x \simeq \theta(I - \hat{L})x, \tag{4}$$

which is a simplified version of ChebyNet.

Heat Kernel on Graphs. In [3], heat kernel between nodes p and q of a graph G is defined in the spectral graph domain as:

$$h_s(p, q) = \sum_{i=1}^{N} e^{-s\lambda_i} u_i(p) u_i(q) \tag{5}$$

where λ_i and u_i are the i-th eigenvalue and eigenvector of the graph Laplacian, and s controls the time/scale of diffusion. Later in [25], the GraphHeat generates the connectivity measure using heat-kernel, and the similarity via the heat diffusion replaces binary adjacency matrix for GNN to capture more precise relationships. Since GraphHeat only retains the first two terms in Eq. (3) for efficiency, the convolution with heat kernel is approximated as:

$$h_s * x \simeq (\theta_0 I + \theta_1 e^{-s\hat{L}}), \tag{6}$$

where h_s acts as a low-pass filter.

4 Method: Learning Node-Wise Adaptive Scales

In this section, we propose a flexible model to learn the range of adaptive neighborhoods for each graph node to capture the optimal local context to improve a downstream prediction task. Figure 2 explains the fundamental idea where nodes n_1 and n_2 aggregate information from their neighborhoods. The left panel shows an example where n_2 (blue) is misclassified as red when the equal receptive field of s_1 is used both n_1 and n_2, whereas n_1 is misclassfied as blue if s_2 is used. Therefore, s_1 and s_2 need to be applied adaptively for n_1 and n_2 so that the right local neighborhoods are selected for individual nodes for data aggregation. Manual selection of these scales can be extremely exhaustive, thus it requires a specialized model to "learn" the scales in a data-driven way.

In the following, we consider a graph classification problem. The objective is to learn to predict a graph-wise label y for input $G = \{A, X\}$, where A is an

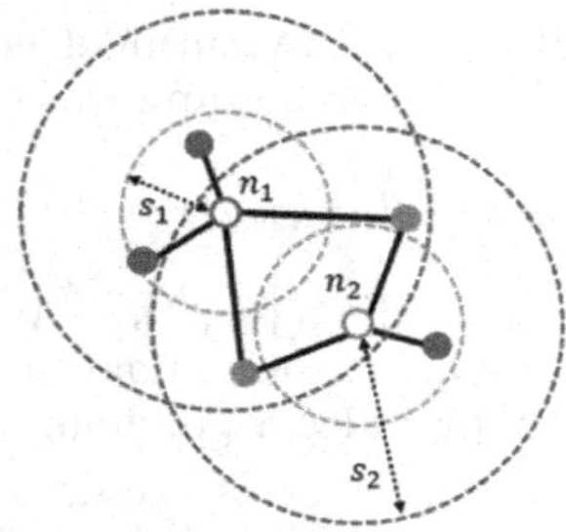

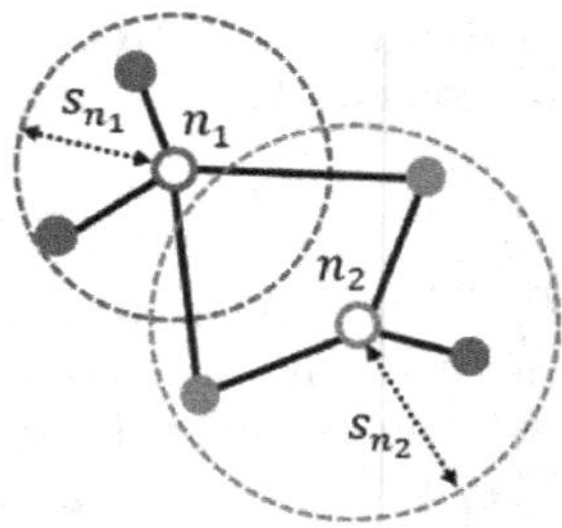

Fig. 2. An example of adaptive neighborhood for nodes n_1 and n_2. Node color (red and blue) denotes node class. Left: Applying the same scale (s_1 or s_2) leads to false aggregation. Right: adaptively applying s_{n_1} and s_{n_2} leads to a proper aggregation. (Color figure online)

adjacency matrix and X is node-wise feature. We first introduce a model that learns a single s globally for a graph, then design a model that learns individual s_i for each node of the graph expecting that adaptively aggregating X through the graph structure will significantly improve the prediction performance.

Adaptive Convolution via Scale. While [25] defines the scale s as a hyper-parameter for an entire graph, we propose a model that trains on the s. The objective function is defined by cross-entropy between the true value Y_{tc} at the c-th class for the t-th sample and the model prediction $\hat{Y}_{tc}$, and a regularizer on s to prevent it from becoming negative:

$$\mathcal{L}(s) = -\frac{1}{T}\sum_{t=1}^{T}\sum_{c\in C} Y_{tc}\log \hat{Y}_{tc}(s) + \lambda|s|[s<0], \tag{7}$$

where λ is a hyperparameter, T is a sample size, and $|s|[s<0]$ takes an absolute value of s when the scale becomes less than 0. Update of the scale is performed as $s \leftarrow s - \alpha_s \frac{\partial \mathcal{L}}{\partial s}$ with a learning rate α_s for s via gradient-based methods along with other learnable parameters W. Derivation of $\frac{\partial \mathcal{L}}{\partial s}$ is shown below.

Forward Propagation. Our model consists of multiple graph convolution layers that adaptively aggregate information for each node and an output layer that predicts a class label for an input graph. From Eq. (6), each of graph convolution layer is defined with a non-linear activation function σ_k as:

$$H_k = \sigma_k(e^{-s\hat{L}} H_{k-1} W_k), \tag{8}$$

where H_k is an output from the k-th convolution layer with $H_0 = X$, and W_k is a matrix of learnable parameters. To obtain a prediction $\hat{Y}_{tc}$, given K graph convolution layers, the output H_K is vectorized and applied with a readout function $\psi(\cdot)$ (e.g., Multi-Layer Perceptron) to obtain integrated values for predicting each class in C. Finally, $Softmax$ is used to get class-wise pseudo-probability as

$$\hat{Y}_{tc} = \frac{\psi(H_K)_{tc}}{\sum_{c'\in C}\psi(H_K)_{tc'}}, \tag{9}$$

which is fed into Eq. (7) for training.

Training on the Scale. From Eq. (5) and Eq. (7)-(9), we can calculate the derivative of $\mathcal{L}$ in terms of scale parameter s. For simplicity, let us consider a single sample case with $T = 1$ and an arbitrary kernel $\mathcal{K}(s)$. In Eq. (7), let the error term be $\mathcal{L}_{ce}$ and the regularization be $\mathcal{L}_{scale}$. First, the derivative of $\mathcal{L}_{scale}$ with respect to s is:

$$\frac{\partial \mathcal{L}_{scale}}{\partial s} = -\lambda[s < 0]. \tag{10}$$

Now, the derivative of $\mathcal{L}_{ce}$ w.r.t. s can be derived using chain rule as:

$$\frac{\partial \mathcal{L}_{ce}}{\partial s} = \frac{\partial \mathcal{L}_{ce}}{\partial \psi(H_K)} \frac{\partial \psi(H_K)}{\partial H_K} \frac{\partial H_K}{\partial s}, \tag{11}$$

where

$$\frac{\partial \mathcal{L}_{ce}}{\partial \psi(H_K)} = \hat{Y} - Y, \frac{\partial H_k}{\partial s} = \sigma'_k(\mathcal{K}(s)H_{k-1}W_k)^T W_k^T (H_{k-1}^T \mathcal{K}'(s) + \frac{\partial H_{k-1}}{\partial s}\mathcal{K}(s)).$$

As $\psi'(H_K)$ depends on the choice of $\psi(H_K)$, the final gradient on the loss is:

$$\frac{\partial \mathcal{L}}{\partial s} = \sum_{i=1}^{N}\sum_{j=1}^{N}(\frac{\partial \mathcal{L}_{ce}}{\partial H_k}\sigma'_k(\mathcal{K}(s)H_{k-1}W_k)^T W_k^T (H_{k-1}^T \mathcal{K}'(s) + \frac{\partial H_{k-1}}{\partial s}\mathcal{K}(s)))_{ij} - \lambda[s < 0] \tag{12}$$

where $i, j \in \{1, ..., N\}$ denoting the i-th and the j-th node, and $\frac{\partial \psi(H_K)}{\partial H_K}$ is embedded in $\frac{\partial \mathcal{L}_{ce}}{\partial H_K}$ for all $k \in \{1, ..., K\}$. Note that s is univariate and covers the entire graph. The full derivation of Eq. (12) is shown in the supplementary.

Localization to Each Node. Figure 2 (right) shows that node n_1 and n_2 having adaptive neighborhood size s_{n_1} and s_{n_2} can capture more precise information. Therefore, we propose a diffusion model to train on the local receptive fields (i.e., scale) for each node. We directly update each scale by removing the marginalization over the nodes (i.e., $\sum_{i=1}^{N}$) in the cross-entropy term of Eq. (12) as:

$$\frac{\partial \mathcal{L}_{ce}}{\partial s_i} = \frac{1}{T}\sum_{t=1}^{T}\sum_{j=1}^{N}(\frac{\partial \mathcal{L}_{ce}}{\partial H_k}\sigma'_k(\mathcal{K}(s_i)H_{k-1}W_k)^T W_k^T (H_{k-1}^T \mathcal{K}'(s_i) + \frac{\partial H_{k-1}}{\partial s_i}\mathcal{K}(s_i)))_{tij} \tag{13}$$

where $i \in \{1, ..., N\}$. The s_i is given for each node n_i and can be trained with gradient-based methods.

5 Experiment

Dataset. Total of 401 subjects with diffusion-weighted imaging (DWI), Amyloid-PET and FDG-PET were taken from pre-selected ADNI cohort. Each brain was partitioned into 148 cortical surface regions using Destrieux atlas [5],

Table 1. Demographics of the ADNI dataset.

Category	CN	SMC	EMCI	LMCI	AD
# of Subjects	89	53	132	55	72
Gender (M/F)	37/52	19/34	84/48	32/23	42/30
Age (Mean ± std)	72.6 ± 4.8	73.4 ± 4.8	70.3 ± 7.1	72.3 ± 6.2	75.8 ± 7.2

and tractography on DWI was applied to calculate the white matter fibers connecting the brain regions to construct 148×148 structural network. On the same parcellation, region-wise imaging features such as SUVR (standard uptake value ratio) of β-amyloid protein from Amyloid-PET, SUVR of metabolism level from FDG-PET, cortical thickness from MRI, and nodal degree were defined. Cerebellum was used as the reference for SUVR normalization. The diagnostic labels for each subject were defined as cognitive normal (CN), significant memory concern (SMC), early mild cognitive impairment (EMCI), late mild cognitive impairment (LMCI) and AD. The demographics of the ADNI dataset is given in Table 1.

Table 2. Performance Comparisons of Various GNN Models on ADNI Data.

Feature	Methods	Accuracy	Precision	Specificity	Sensitivity (Recall)
Degree	SVM	0.532 ± 0.162	0.509 ± 0.109	0.886 ± 0.037	0.616 ± 0.220
	GCN	0.466 ± 0.079	0.422 ± 0.085	0.867 ± 0.019	0.461 ± 0.107
	GAT	0.633 ± 0.093	0.610 ± 0.085	0.908 ± 0.025	0.681 ± 0.101
	GraphHeat	0.641 ± 0.077	0.624 ± 0.071	0.908 ± 0.021	0.672 ± 0.083
	GDC	0.670 ± 0.090	0.660 ± 0.088	0.917 ± 0.024	0.684 ± 0.103
	Ours (global t)	**0.703 ± 0.068**	**0.671 ± 0.079**	**0.925 ± 0.017**	**0.744 ± 0.072**
	Ours (local t)	0.653 ± 0.076	0.620 ± 0.068	0.913 ± 0.022	0.690 ± 0.101
Cortical Thickness	SVM	0.721 ± 0.051	0.669 ± 0.060	0.933 ± 0.016	0.862 ± 0.050
	GCN	0.494 ± 0.076	0.464 ± 0.073	0.867 ± 0.024	0.518 ± 0.098
	GAT	0.865 ± 0.111	0.865 ± 0.098	0.966 ± 0.028	0.874 ± 0.110
	GraphHeat	0.828 ± 0.056	0.843 ± 0.050	0.956 ± 0.014	0.853 ± 0.055
	GDC	0.860 ± 0.063	0.871 ± 0.061	0.965 ± 0.016	0.878 ± 0.055
	Ours (global t)	0.841 ± 0.106	0.848 ± 0.112	0.960 ± 0.024	0.865 ± 0.090
	Ours (local t)	**0.875± 0.043**	**0.873 ± 0.046**	**0.968 ± 0.011**	**0.896 ± 0.032**
β-Amyloid	SVM	0.843 ± 0.093	0.819 ± 0.089	0.961 ± 0.025	0.882 ± 0.084
	GCN	0.526 ± 0.069	0.499 ± 0.074	0.880 ± 0.019	0.535 ± 0.106
	GAT	0.873 ± 0.057	0.876 ± 0.055	0.968 ± 0.015	0.889 ± 0.053
	GraphHeat	0.881 ± 0.081	0.878 ± 0.095	0.970 ± 0.020	0.877 ± 0.114
	GDC	0.893 ± 0.108	0.875 ± 0.151	0.974 ± 0.025	**0.915 ± 0.077**
	Ours (global t)	0.911 ± 0.072	0.912 ± 0.085	0.977 ± 0.017	0.911 ± 0.085
	Ours (local t)	**0.916 ± 0.078**	**0.912 ± 0.093**	**0.979 ± 0.019**	0.914 ± 0.099
FDG	SVM	0.853 ± 0.044	0.829 ± 0.053	0.964 ± 0.011	0.919 ± 0.029
	GCN	0.511 ± 0.066	0.474 ± 0.088	0.876 ± 0.017	0.535 ± 0.108
	GAT	0.678 ± 0.089	0.673 ± 0.102	0.919 ± 0.024	0.685 ± 0.105
	GraphHeat	0.885 ± 0.065	0.893 ± 0.067	0.971 ± 0.016	0.902 ± 0.065
	GDC	0.923 ± 0.089	0.923 ± 0.104	0.980 ± 0.024	0.949 ± 0.052
	Ours (global t)	0.928 ± 0.067	0.931 ± 0.078	0.982 ± 0.017	0.945 ± 0.050
	Ours (local t)	**0.960 ± 0.028**	**0.963 ± 0.031**	**0.990 ± 0.007**	**0.965 ± 0.028**
All Imaging Features	SVM	0.935 ± 0.042	0.917 ± 0.048	0.985 ± 0.010	0.953 ± 0.037
	GCN	0.556 ± 0.074	0.537 ± 0.065	0.888 ± 0.018	0.562 ± 0.126
	GAT	0.726 ± 0.073	0.710 ± 0.070	0.932 ± 0.019	0.746 ± 0.077
	GraphHeat	0.923 ± 0.047	0.923 ± 0.050	0.980 ± 0.012	0.931 ± 0.043
	GDC	0.930 ± 0.066	0.930 ± 0.073	0.983 ± 0.016	0.945 ± 0.052
	Ours (global t)	0.933 ± 0.056	0.933 ± 0.057	0.983 ± 0.015	0.945 ± 0.044
	Ours (local t)	**0.953 ± 0.032**	**0.955 ± 0.035**	**0.988 ± 0.008**	**0.957 ± 0.029**

Setup. We trained a two-layer graph convolution model with 16 hidden units, and used a rectified linear unit (ReLU) for the activation function. ψ was a two-level linear layers, and dropout with predefined rate for each model. Weights were initialized with Xavier initialization [8] and trained with Adam optimizer [12] at the learning rate of 0.01. The α_s was chosen as either 0.1 or 0.01. We used the kernel $\mathcal{K}(\hat{L}, s)$ as Heat kernel defined in Eq. (5). Heat kernel's initial scale was $s = 2$ for both global and local models, and heat kernel values were thresholded at $< 1e-5$. Regularization λ was 1. 10-fold cross-validation (CV) was used and accuracy/precision/specificity/sensitivity in average were computed for evaluation. One-vs-rest scheme was used to compute the evaluation metrics.

Support Vector Machine (SVM) as well as recent GNN models such as GCN [13], GAT [22], GraphHeat [25], and GDC [14] were adopted as the baseline models. Each baseline was set up and trained at our best effort to obtain feasible outcomes for fair comparisons. More details are given in the supplementary.

Quantitative Results. All results are reported in Table 2 at a glance. It shows that our models (training both global and local s) empirically outperform in all experiments except the recall of classification with β-amyloid. The highest accuracy of 96% in classifying the 5 classes was achieved with the metabolism (FDG) on the structural network, which is known to be an effective biomarker for characterizing early AD. The standard deviation on all evaluation measures stayed low across the 10-folds addressing our models' stability. Especially comparing the results from GraphHeat and our models proves that training on the scale definitely improves the results where the scale (hyper-)parameter for the GraphHeat was used as the initialization for our models. As the sample-size was small, SVM worked efficiently, and the GNN models other than GCN showed good performance; this may be because these GNN models extract weighted adjacency matrix but binary adjacency matrix (thresholded from the brain network) was used for the GCN. Adopting all features was mostly better but underperformed FDG measure with training with local scale. This may be because cortical thickness is not a suitable biomarker to discriminate very early stages of AD.

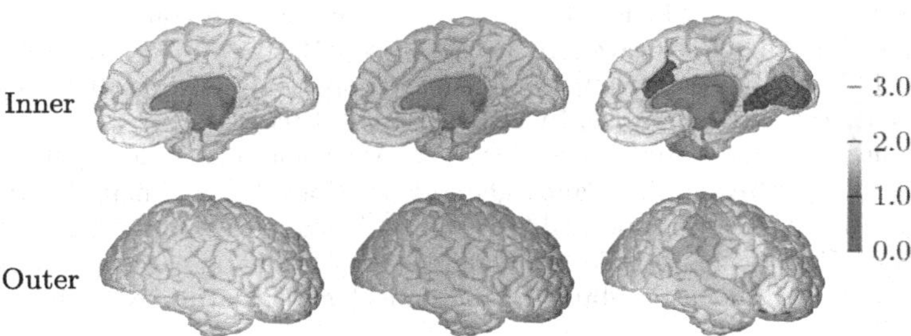

Fig. 3. A visualization of learned scales on the right hemisphere of a brain and localized ROIs. Left: initial scale ($s = 2$), Middle: globally trained scale ($s = 1.59$), Right: locally trained scales, Bottom: Region of interests having lower scale than 1 in the local model.

Qualitative Results. In Fig. 3, we visualize the initial scale ($s = 2$ for GraphHeat), globally trained scale ($s = 1.586$), and adaptively trained scale for each ROI. The ROIs with small scales denote that it is independent: they do not require node aggregation from large neighborhoods to improve the AD classification. Table in Fig. 3 shows those ROIs with the trained $s < 1$. Interestingly, middle-anterior cingulate gyrus and sulcus from both left and right hemispheres showed the smallest scales (0.171 and 0.207) demonstrating the highest locality. It is responsible for cognitive and executive functions reported in [10,21] to be AD-specific. Other ROIs with low scales such as temporal lingual gyrus, inferior temporal and post central regions are also consistently found in preclinical AD [17,18,23], which demonstrate that these were the key ROIs in discriminating even early stages of Alzheimer's disease in our model.

6 Conclusion

In this paper, we proposed a novel model that flexibly learns individual node-wise scales in a brain network to adaptively aggregate information from neighborhoods. The developed model lets one identify which ROIs in the brain behave locally (i.e., independently) on the brain network structure to predict global diagnostic labels. We have derived a rigorous formulation of the scale such that it can be trained via gradient-based method, and validated that our model can accurately classify AD-specific labels of brain networks and detect key ROIs corroborated by other AD literature.

Acknowledgements. This research was supported by NSF IIS CRII 1948510, NIH R03 AG070701 and partially supported by IITP-2019-0-01906 (AI Graduate Program at POSTECH), IITP-2022-2020-0-01461 (ITRC) and IITP-2022-0-00290 funded by Ministry of Science and ICT (MSIT).

References

1. Anderson, J., Nielsen, J., Froehlich, A., et al.: Functional connectivity magnetic resonance imaging classification of autism. Brain **134**(12), 3742–3754 (2011)
2. Bruna, J., Zaremba, W., Szlam, A., Lecun, Y.: Spectral networks and deep locally connected networks on graphs. In: ICLR, Banff, Canada (2014)
3. Chung, F.: Spectral graph theory. American Mathematical Society (1997)
4. Defferrard, M., Bresson, X., Vandergheynst, P.: Convolutional neural networks on graphs with fast localized spectral filtering. In: NIPS, vol. 29, pp. 3844–3852 (2016)
5. Destrieux, C., Fischl, B., Dale, A., Halgren, E.: Automatic parcellation of human cortical gyri and sulci using standard anatomical nomenclature. Neuroimage **53**(1), 1–15 (2010)
6. DeTure, M., Dickson, D.: The neuropathological diagnosis of Alzheimer's disease. Mol. Neurodegeneration **14**(32), 1–18 (2019)
7. Gao, H., Ji, S.: Graph U-Nets. In: ICML, pp. 2083–2092. Macao, China (2019)
8. Glorot, X., Bengio., Y.: Understanding the difficulty of training deep feedforward neural networks. In: AISTATS, vol. 9, p. 249–256 (2010)

9. Gori, M., Monfardini, G., Scarselli, F.: A new model for learning in graph domains. In: IJCNN, pp. 729–734 (2005)
10. Guo, Z., Zhang, J., Liu, X., Hou, H., et al.: Neurometabolic characteristics in the anterior cingulate gyrus of Alzheimer's disease patients with depression: a 1h magnetic resonance spectroscopy study. BMC Psychiatr. **15**(1), 1–7 (2015)
11. Henaff, M., Bruna, J., LeCun, Y.: Deep convolutional networks on graph-structured data. arXiv preprint arXiv:1506.05163 (2015)
12. Kingma, D., Ba, J.: Adam: a method for stochastic optimization. In: ICLR, San Diego, U.S. (2015)
13. Kipf, T., Welling, M.: Semi-supervised classification with graph convolutional networks. In: ICLR, Toulon, France (2017)
14. Klicpera, J., Weißenberger, S., Günnemann, S.: Diffusion improves graph learning. In: NIPS (2019)
15. McPherson, M., Smith-Lovin, L., Cook, J.: Birds of a feather: homophily in social networks. Ann. Rev. Sociol. **27**(1), 415–444 (2001)
16. Oppenheim, A., Schafer, R., Buck, J.: Signal and Systems, 2nd edn. Prentice Hall, Upper Saddle River, N.J. (1999)
17. Scheff, S.W., Price, D.A., Schmitt, F.A., Scheff, M.A., Mufson, E.J.: Synaptic loss in the inferior temporal gyrus in mild cognitive impairment and Alzheimer's disease. J. Alzheimers Dis. **24**(3), 547–557 (2011)
18. Scott, M.R., Hampton, O.L., Buckley, R.F., et al.: Inferior temporal tau is associated with accelerated prospective cortical thinning in clinically normal older adults. Neuroimage **220**, 116991 (2020)
19. Shuman, D., Ricaud, B., Vandergheynst, P.: Vertex-frequency analysis on graphs. ACHA **40**(2), 260–291 (2016)
20. Stam, C.J., de Haan, W., Daffertshofer, A., et al.: Graph theoretical analysis of magnetoencephalographic functional connectivity in Alzheimer's disease. Brain **132**(1), 213–224 (2008)
21. Tekin, S., Mega, M.S., Masterman, D.M., Chow, T., et al.: Orbitofrontal and anterior cingulate cortex neurofibrillary tangle burden is associated with agitation in Alzheimer disease. Ann. Neurol. **49**(3), 355–361 (2001)
22. Velickovic, P., Cucurull, G., Casanova, A., Romero, A., Lio, P., Bengio, Y.: Graph attention networks. In: ICLR, Vancouver, Canada (2018)
23. Venneri, A., McGeown, W.J., Hietanen, H.M., Guerrini, C., Ellis, A.W., Shanks, M.F.: The anatomical bases of semantic retrieval deficits in early Alzheimer's disease. Neuropsychologia **46**(2), 497–510 (2008)
24. Wu, T., Wang, L., Chen, Y., Zhao, C., Li, K., Chan, P.: Changes of functional connectivity of the motor network in the resting state in Parkinson's disease. Neurosci. Lett. **460**(1), 6–10 (2009)
25. Xu, B., Shen, H., Cao, Q., Cen, K., Cheng, X.: Graph convolutional networks using heat Kernel for semi-supervised learning. In: IJCAI, Macao, China (2019)
26. Zhang, M., Chen, Y.: Link prediction based on graph neural networks. In: NIPS (2018)

Combining Multiple Atlases to Estimate Data-Driven Mappings Between Functional Connectomes Using Optimal Transport

Javid Dadashkarimi[1(✉)], Amin Karbasi[2], and Dustin Scheinost[3]

[1] Department of Computer Science, Yale University, New Haven, USA
javid.dadashkarimi@yale.edu

[2] Department of Electrical Engineering, Yale University, New Haven, USA
amin.karbasi@yale.edu

[3] Department of Radiology and Biomedical Imaging, Yale School of Medicine, New Haven, USA
dustin.scheinost@yale.edu

Abstract. Connectomics is a popular approach for understanding the brain with neuroimaging data. Yet, a connectome generated from one atlas is different in size, topology, and scale compared to a connectome generated from another atlas. These differences hinder interpreting, generalizing, and combining connectomes and downstream results from different atlases. Recently, it was proposed that a mapping between atlases can be estimated such that connectomes from one atlas (*i.e.*, source atlas) can be reconstructed into a connectome from a different atlas (*i.e.*, target atlas) without re-processing the data. This approach used optimal transport to estimate the mapping between one source atlas and one target atlas. Yet, restricting the optimal transport problem to only a single source atlases ignores additional information when multiple source atlases are available, which is likely. Here, we propose a novel optimal transport based solution to combine information from multiple source atlases to better estimate connectomes for the target atlas. Reconstructed connectomes based on multiple source atlases are more similar to their "gold-standard" counterparts and better at predicting IQ than reconstructed connectomes based on a single source mapping. Importantly, these results hold for a wide-range of different atlases. Overall, our approach promises to increase the generalization of connectome-based results across different atlases.

Keywords: Optimal transport · Functional connectome · fMRI

Supplementary Information The online version contains supplementary material available at https://doi.org/10.1007/978-3-031-16431-6_37.

L. Wang et al. (Eds.): MICCAI 2022, LNCS 13431, pp. 386–395, 2022.
https://doi.org/10.1007/978-3-031-16431-6_37

1 Introduction

A connectome—a matrix describing the connectivity between any pair of brain regions—is a popular approach used to model the brain as a graph-like structure. They are created by parcellating the brain into distinct regions using an atlas (*i.e.*, the nodes of a graph) and estimating the connections between these regions (*i.e.*, the edges of a graph). As different atlases divide the brain into a different number of regions of varying size and topology, connectomes created from different atlases are not directly comparable. In other words, connectome-based results generated from one atlas cannot be directly compared to connectome-based results generated from a different atlas. This fact hinders not only replication and generalization efforts, but also simply comparing the results from two independent studies that use different atlases. For example, large-scale projects—like the Human Connectome Project (HCP), the Adolescent Brain Cognitive Development (ABCD) study [5], and the UK Biobank [25]—share fully processed connectomes to increase the wider-use of the data, while reducing redundant processing efforts [15]. Yet, several atlases, but no gold standards, exist [2]. As such, released connectomes for each project are based on different atlases, which prevents these datasets being combined without reprocessing data from thousands of participants. Being able to map between these connectomes—without need for raw data—would facilitate existing connectomes to be easily reused in a wide-range of analyses while eliminating wasted and duplicate processing efforts.

To facilitate this mapping, it was shown that an existing connectome could be transformed into a connectome from a different atlas without needing the raw functional magnetic imaging (fMRI) data [8]. This method used optimal transport, or the mathematics of converting a probability distribution from one set to another, to find a spatial mapping between a pair of atlases. This mapping could then be applied the timeseries fMRI data parcellated with the first atlas (source atlas), then creating connectome based on the second atlas (target atlas). While these previous mappings were based on only a single source atlas, most large-scale projects, release data processed data from 2–4 atlases. As such, richer information than that provided by a single source atlas is available and ignored in the current approach. We propose to combine information from multiple source atlases to jointly estimate mappings to the target atlas. Using 6 different atlases, we show that our approach results in significant improvements in the quality of reconstructed connectomes and their performance in downstream analyses.

2 Background

Optimal Transport: The optimal transport problem solves how to transport resources from one location α to another β while minimizing the cost C to do so [12,14,18,26]. Using a probabilistic approach in which the amount of mass located at x_i potentially dispatches to several points in target [17], admissible solutions are defined by a coupling matrix $\mathcal{T} \in \mathbb{R}_+^{n\times m}$ indicating the amount of mass being transferred from location x_i to y_j by $\mathcal{T}_{i,j}$:

$$U(a,b) = \{\mathcal{T} \in \mathbb{R}_+^{n\times m} : \mathcal{T}\mathbb{1}_m = a, \mathcal{T}^T\mathbb{1}_n = b\}, \tag{1}$$

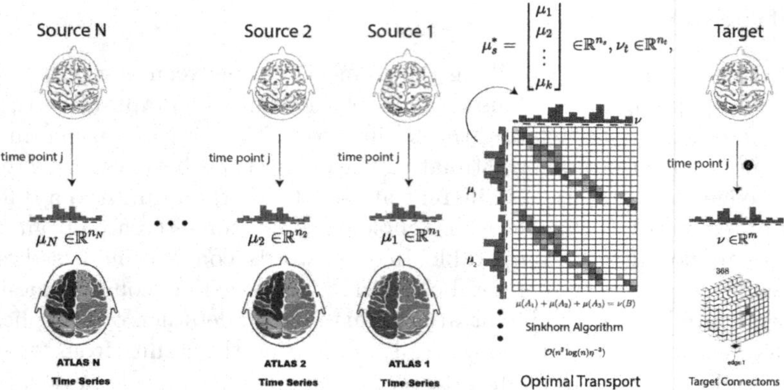

Fig. 1. Our all-way optimal transport algorithm combines timeseries data from multiple source atlases to jointly estimate a single mapping from these atlases to the target atlas. Once this mapping is found, it can be applied to independent data to reconstruct connectomes based on the target data based only timeseries data from the source atlases.

for vectors of all 1 shown with $\mathbb{1}$. An optimum solution is obtained by solving the following problem for a given "ground metric" matrix $C \in \mathbb{R}^{n \times m}$ [21]:

$$L_c(a, b) = \min_{\mathcal{T} \in U(a,b)} < C, \mathcal{T} >= \sum_{i,j} C_{i,j} \mathcal{T}_{i,j}. \tag{2}$$

which is a linear problem and is not guarantee to have a unique solution [19], but always there exists an optimal solution (see proof in [3,4]). Unlike, the KL divergence, optimal transport is one of the few methods that provides a well-defined distance metric when the support of the distributions is different.

Single-Source Optimal Transport: The single-source optimal transport algorithm from Dadashkarimi et al. [8], first, transforming timeseries data from one atlas (labeled the source atlas) into timeseries from an unavailable atlas (labeled the target atlas). Next, the corresponding functional connectomes can be estimated using standard approaches (e.g., full or partial correlation). Formally, it is assumed that we have training timeseries data consisting of T timepoints from the same individuals but from two different atlases (atlas $\mathscr{P}_n$ with n regions and atlas $\mathscr{P}_m$ with m regions). Additionally, let $\mu_t \in \mathbb{R}^n$ and $\nu_t \in \mathbb{R}^m$ to be the vectorized brain activity at single timepoint t based on atlases $\mathscr{P}_n$ and $\mathscr{P}_m$, respectively. For a fixed cost matrix $C \in \mathbb{R}^{n \times m}$, which measures the pairwise distance between regions in $\mathscr{P}_m$ and $\mathscr{P}_n$, this approach aims to find a mapping $\mathcal{T} \in \mathbb{R}^{n \times m}$ that minimizes transportation cost between μ_t and ν_t:

$$L_c(\mu_t, \nu_t) = \min_{\mathcal{T}} C^T \mathcal{T} \textbf{ s.t, } A\underline{\mathcal{T}} = \begin{bmatrix} \mu_t \\ \nu_t \end{bmatrix}, \tag{3}$$

in which $\underline{\mathcal{T}} \in \mathbb{R}^{nm}$ is vectorized version of $\mathcal{T}$ such that the $i+n(j-1)$'s element of $\mathcal{T}$ is equal to $\mathcal{T}_{ij}$ and A is defined as:

$$A = \begin{matrix} m \\ n \end{matrix} \begin{pmatrix} \overset{\mathbf{1}}{\begin{pmatrix} 1 & 0 & \dots & 0 \\ 0 & 1 & \dots & 0 \\ \vdots & \vdots & \ddots & \vdots \\ 0 & 0 & \dots & 1 \end{pmatrix}} & \overset{\mathbf{2}}{\begin{pmatrix} 1 & 0 & \dots & 0 \\ 0 & 1 & \dots & 0 \\ \vdots & \vdots & \ddots & \vdots \\ 0 & 0 & \dots & 1 \end{pmatrix}} & \dots & \overset{\mathbf{n}}{\begin{pmatrix} 1 & 0 & \dots & 0 \\ 0 & 1 & \dots & 0 \\ \vdots & \vdots & \ddots & \vdots \\ 0 & 0 & \dots & 1 \end{pmatrix}} \\ (1\ 1 \dots 1) & (1\ 1 \dots 1) & \dots & (1\ 1 \dots 1) \\ \vdots & & & \\ (1\ 1 \dots 1) & \dots & & \dots (1\ 1 \dots 1) \end{pmatrix}. \tag{4}$$

$\mathcal{T}$ represents the optimal way of transforming the brain activity data from n regions into m regions. Thus, by applying $\mathcal{T}$ to every timepoint from the timeseries data of the source atlas, the timeseries data of the target atlas and corresponding connectomes can be estimated. The cost matrix C was based on the similarity of pairs of timeseries from the different atlases:

$$C = 1 - \begin{pmatrix} \rho(U_{1,.}, N_{1,.}) & \dots & \rho(U_{1,.}, N_{n,.}) \\ \vdots & \ddots & \vdots \\ \rho(U_{m,.}, N_{1,.}) & \dots & \rho(U_{m,.}, N_{n,.}) \end{pmatrix} \in \mathbb{R}^{m \times n} \tag{5}$$

where U_x and N_x are timeseries from $\mathscr{P}_m$ and $\mathscr{P}_n$ and $\rho(U_x, N_y)$ is correlation between them.

3 All-Way Optimal Transport

A key drawback of the single-source optimal transport is that it relies on a single pair of source and target atlases (i.e., one source atlas and one target atlas), which ignores additional information when multiple source atlases exist. To overcome this weakness, we designed a new approach, called all-way optimal transport, that uses a varying number of source atlases to better reconstruct the target atlas. All-way optimal transport combines information from multiple source atlases by using a larger cost matrix generated from stacking the set of region centers in each source atlas (see Fig. 1). In general, assume we have paired time-series, from the same person, but from k different source atlases with a total of n_s regions (where $n_s = n_1 + n_2 + .. + n_k$ from source atlas $\mathscr{P}_{n_1}$ with n_1 regions, $\mathscr{P}_{n_2}$ with n_2 regions, .., $\mathscr{P}_{n_k}$ with n_k regions) and a target atlas $\mathscr{P}_m$ with m regions, lets define $\mu_t \in \mathbb{R}^{n_s}$ and $\nu_t \in \mathbb{R}^m$ to be the distribution of brain activity at single time point t based on atlases $\mathscr{P}_s$ and $\mathscr{P}_m$:

$$\mu_s^* = \begin{bmatrix} \mu_1 \\ \mu_2 \\ \vdots \\ \mu_k \end{bmatrix} \in \mathbb{R}^{n_s}, \nu_t \in \mathbb{R}^{n_t}, C^* = \begin{pmatrix} C_{1,1} & \dots & C_{1,m} \\ \vdots & \ddots & \vdots \\ C_{n_s,1} & \dots & C_{n,m} \end{pmatrix} \in \mathbb{R}^{n_s \times m}, \tag{6}$$

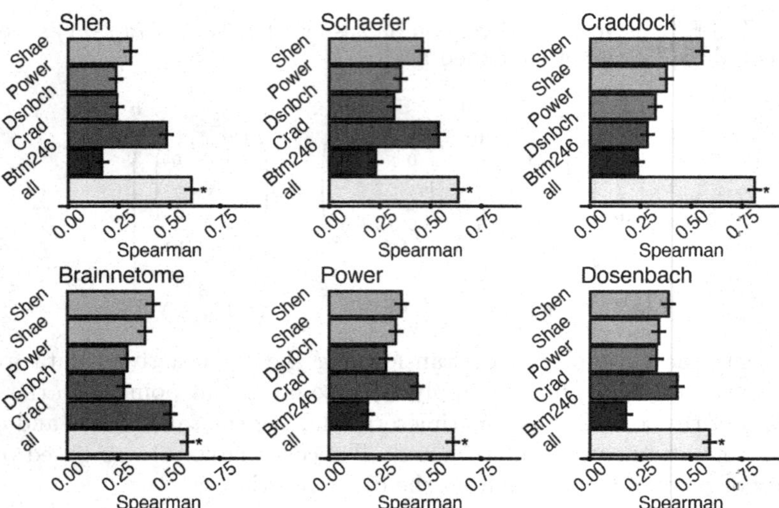

Fig. 2. Using multiple source atlases improves the similarity of reconstructed connectomes. The correlation between the reconstructed connectomes and connectomes generated directly with the target atlases are shown for each pair of source and target atlas as well reconstructed connectomes using all source atlases. For each target atlas, using all source atlases produces higher quality reconstructed connectomes. Error bars are generated from 100 iterations of randomly splitting the data into training and testing data. In all cases, all-way optimal transport resulted in significantly more similar connectomes (indicated by $*$).

and $C_{i,j}$ is based the similarity of pairs of timeseries from nodes i and j from different atlases. Next, we want to minimize distance between μ_s^* and ν_t as:

$$L_c(\mu_t^*, \nu_t^*) = \min_{\mathcal{T}} C^T \mathcal{T} \textbf{ s.t, } A\underline{\mathcal{T}} = \begin{bmatrix} \mu_t^* \\ \nu_t^* \end{bmatrix}, \tag{7}$$

4 Implementation

Solving the large linear program in Eq. 7 is computationally hard [9]. As such for both all-way and single source optimal transport, we used the entropy regularization, which gives an approximation solution with complexity of $\mathcal{O}(n^2 \log(n)\eta^{-3})$ for $\epsilon = \frac{4\log(n)}{\eta}$ [19], and instead solve the following:

$$L_c(\mu_t^*, \nu_t^*) = \min_{\mathcal{T}} C^T \mathcal{T} - \epsilon H(\mathcal{T}) \textbf{ s.t, } A\underline{\mathcal{T}} = \begin{bmatrix} \mu_t^* \\ \nu_t^* \end{bmatrix}. \tag{8}$$

Specifically, we use the Sinkhorn algorithm—an iterative solution for Eq. 8 [1]—to find $\mathcal{T}$ as implemented in the Python Optimal Transport toolbox [11].

5 Results

Datasets: To evaluate our approach, we used data from the Human Connectome Project (HCP) [27], starting with the minimally preprocessed data [13]. First, data with a maximum frame-to-frame displacement of 0.15 mm or greater were excluded, resulting in a sample of 515 resting-state scans. Analyses were restricted only to the LR phase encoding, which consisted of 1200 individual time points. Further preprocessing steps were performed using BioImage Suite [16]. These included regressing 24 motion parameters, regressing the mean white matter, CSF, and grey matter time series, removing the linear trend, and low-pass filtering. After processing, Shen (268 nodes) [24], Schaefer (400 nodes) [22], Craddock (200 nodes) [7], Brainnetome (246 nodes) [10], Power (264 nodes) [20], and Dosenbach (160 nodes) [6] atlases were applied to the preprocessed to create mean timeseries for each node. Connectomes were generated by calculating the Pearson's correlation between each pair of these mean timeseries and then taking the fisher transform of these correlations.

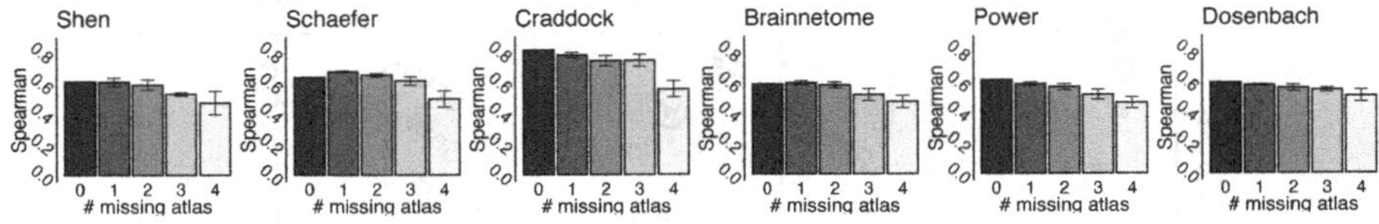

Fig. 3. Bar plots exhibit correlation of estimated connectomes and original connectomes based on $n-k$ samplings of available atlases (*i.e.*, n indicates the number of all available atlases to be transported) for each target atlas for all-way optimal transport. Strong correlations can be observed with less than the maximum number of source atlases.

Similarity Between Reconstructed and Original Connectomes: To validate our approach, we assessed the similarity of connectomes reconstructed using the proposed optimal transport algorithms and the original connectomes generated directly from the raw data. First, We partitioned our sample into 80% for optimal 'parameter estimation'. These optimal parameters were then applied on 20% remaining data for measuring the efficacy of the method. Therefore, we estimated $\mathcal{T}$ using all 1200 time points and 412 participants for each source-target atlas pairs (for single-source optimal transport) as well as using all available source atlases to a single target atlas (for all-way optimal transport).

Next, in the left out partition, we applied the estimated $\mathcal{T}$ to reconstruct the target atlases. Finally, the reconstructed connectomes were compared to the "gold-standard" connectomes (*i.e.*, connectomes generated directly from an atlas) using correlation. Results from all-way were compared to results from the single-source optimal transport algorithm.

As shown in Fig. 2, we observed strong correlation between the reconstructed connectomes and their original counterparts when using the all-way optimal transport algorithm. In every case, these algorithms produce significantly more similar connectomes than the previous single-source optimal transport algorithm (all $\rho's > 0.50$; $p < 0.01$). For most atlases, explained variance is more than tripled using multiple source atlases compare to using a single source atlas.

Effect of Number of Source Atlases: We investigated the impact of using a smaller number of source atlases by only including k random source atlases when creating connectome for the target atlas. This process was repeated with 100 iterations over a range of k =2–6. As shown in Fig. 3, while similarity between reconstructed and original connectomes increases as the number of source atlases increases, strong correlations (e.g., $\rho > 0.6$) can be observed with as little as two or three source atlases, suggesting that a small number of atlases may be sufficient for most applications. Overall, improvements in similarity level off after combining a few atlases, suggesting that adding a greater number of atlases than tested here will have diminished returns.

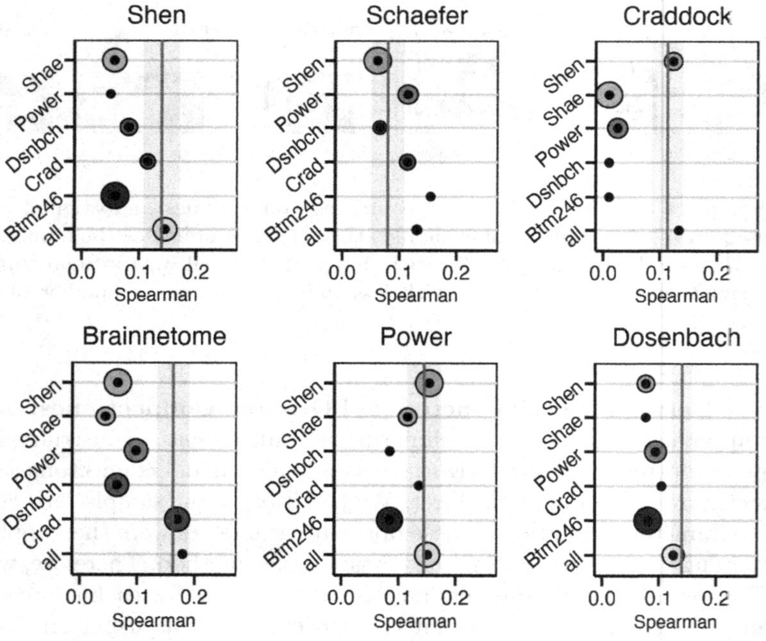

Fig. 4. The reconstructed connectomes using multiple source atlases retain significantly more individual differences than using a single source atlas and predicted IQ as well or better than the original connnectome (red line). Size of circle represents the variance of prediction of 100 iteration of 10-fold cross-validation. (Color figure online)

IQ Prediction: To further evaluated the reconstructed connectomes, we show that reconstructed connectomes can be used to predict fluid intelligence using connectome-based predictive modeling (CPM) [23]. We partitioned the HCP dataset into three groupings: g_1, which consisted of 25% of the participants; g_2, which consisted of 50% of the participants; and, g_3, which consisted of the final 25% of the participants. In g_1, $\mathcal{T}$'s for each algorithm were estimated as above. We then applied $\mathcal{T}$ on g_2 and g_3 to reconstruct connectomes. Finally, for each set of connectomes, we trained a CPM model of fluid intelligence using g_2 and tested this model in g_3. Spearman correlation between observed and predicted values was used to evaluate prediction performance. This procedure was repeated with 100 random splitting of the data into the three groups. In all cases, connectomes reconstructed using all of the source atlases performed as well in prediction as the original connectomes (Fig. 4).

Parameter Sensitivity: We investigated the sensitivity of all-way optimal transport to the free parameters: frame size, training size, and entropy regularization (see Eq. 8). We observe stable correlations with original connectomes using different frame sizes, emphasizing that our cost matrix captures the geometry between the different atlases well. Also, all-way optimal transport is trainable with limited amount of data (see Fig. 5). Finally, increasing entropy regularization ϵ overly penalizes the mapping and degrades the quality of connectomes.

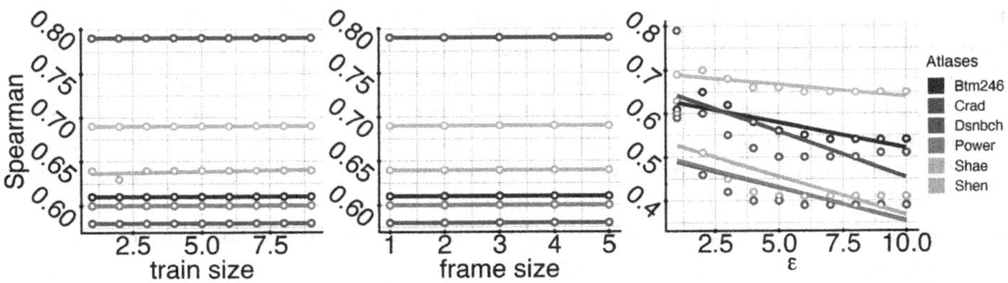

Fig. 5. Parameter sensitivity of frame size (×50), training data (×100), and entropy regularization ϵ for different target atlases using all-way optimal transport.

6 Discussion and Conclusions

Here, we significantly improve upon previous efforts to enable fMRI data, previously processed with one atlas, to be mapped to a connectome generated from a different atlas , without the need for further prepossessing. To accomplish this, we proposed and validate two algorithms that combine information from multiple source atlases to better estimate connectomes for the target atlas. All-ways optimal transport directly estimates a single mapping between multiple source atlases and the target atlas. In contrast, stacking optimal transport combines previously estimated mappings between a single source and target atlas, allowing these previously estimated mappings to be reused. Reconstructed connectomes from both algorithms are more similar to their "gold-standard" counterparts and better at predicting IQ than reconstructed connectomes based on a single source mapping. Importantly, these results hold for a wide-range of different atlases. Future work includes generalizing our framework to other functional timeseries data—*e.g.*, electroencephalography (EEG) and functional near infrared spectroscopy (fNIRS). Overall, our approach is a promising avenue to increase the generalization of connectome-based results across different atlases.

Acknowledgements. Data were provided by the Human Connectome Project, WU-Minn Consortium (Principal Investigators: David Van Essen and Kamil Ugurbil; U54 MH091657) and funded by the 16 NIH Institutes and Centers that support the NIH Blueprint for Neuroscience Research; and by the McDonnell Center for Systems Neuroscience at Washington University. Amin Karbasi is partially supported by NSF (IIS-1845032), ONR (N00014-19-1-2406), and Tata.

Compliance with Ethical Standards. This research study was conducted retrospectively using human subject data made available in open access by the Human Connectome Project. Approval was granted by local IRB. Yale Human Research Protection Program (HIC #2000023326) on May 3, 2018.

References

1. Altschuler, J., Weed, J., Rigollet, P.: Near-linear time approximation algorithms for optimal transport via sinkhorn iteration. arXiv preprint arXiv:1705.09634 (2017)
2. Arslan, S., Ktena, S.I., Makropoulos, A., Robinson, E.C., Rueckert, D., Parisot, S.: Human brain mapping: a systematic comparison of parcellation methods for the human cerebral cortex. NeuroImage **170**, 5–30 (2018). https://doi.org/10.1016/j.neuroimage.2017.04.014, https://www.sciencedirect.com/science/article/pii/S1053811917303026, segmenting the Brain
3. Bertsimas, D., Tsitsiklis, J.: Introduction to linear optimization, Athena scientific (1997). http://athenasc.com/linoptbook.html
4. Birkhoff, G.: Tres observaciones sobre el algebra lineal. Univ. Nac. Tucuman, Ser. A **5**, 147–154 (1946)
5. Casey, B., et al.: The adolescent brain cognitive development (ABCD) study: imaging acquisition across 21 sites. Dev. Cogn. Neurosci. **32**, 43–54 (2018)
6. Cohen, A.L., et al.: Defining functional areas in individual human brains using resting functional connectivity MRI. Neuroimage **41**(1), 45–57 (2008)
7. Craddock, R.C., James, G.A., Holtzheimer, P.E., III., Hu, X.P., Mayberg, H.S.: A whole brain FMRI atlas generated via spatially constrained spectral clustering. Hum. Brain Mapping **33**(8), 1914–1928 (2012)
8. Dadashkarimi, J., Karbasi, A., Scheinost, D.: Data-driven mapping between functional connectomes using optimal transport. In: de Bruijne, M., et al. (eds.) MICCAI 2021. LNCS, vol. 12903, pp. 293–302. Springer, Cham (2021). https://doi.org/10.1007/978-3-030-87199-4_28
9. Dantzig, G.B.: Reminiscences about the origins of linear programming. In: Mathematical Programming The State of the Art, pp. 78–86. Springer, Heidelberg (1983). https://doi.org/10.1007/978-3-642-68874-4_4

10. Fan, L., et al.: The human brainnetome atlas: a new brain atlas based on connectional architecture. Cereb. Cortex **26**(8), 3508–3526 (2016)
11. Flamary, R., Courty, N.: Pot python optimal transport library (2017). https://pythonot.github.io/
12. Gangbo, W., McCann, R.J.: The geometry of optimal transportation. Acta Math. **177**(2), 113–161 (1996)
13. Glasser, M.F., et al.: The minimal preprocessing pipelines for the human connectome project. Neuroimage **80**, 105–124 (2013)
14. Hitchcock, F.L.: The distribution of a product from several sources to numerous localities. J. Math. Phys. **20**(1–4), 224–230 (1941)
15. Horien, C., et al.: A hitchhiker's guide to working with large, open-source neuroimaging datasets. Nat. Hum. Behav. 1–9 (2020)
16. Joshi, A., et al.: Unified framework for development, deployment and robust testing of neuroimaging algorithms. Neuroinformatics **9**(1), 69–84 (2011)
17. Kantorovich, L.: On the transfer of masses In: Doklady Akademii Nauk, vol. 37, pp. 227–229 (1942). (in russian)
18. Koopmans, T.C.: Optimum utilization of the transportation system. Econometrica J. Econ. Soc. **17**, 136–146 (1949)
19. Peyré, G., Cuturi, M., et al.: Computational optimal transport: With applications to data science. Found. Trends® Mach. Learn. **11**(5–6), 355–607 (2019)
20. Power, J.D., et al.: Functional network organization of the human brain. Neuron **72**(4), 665–678 (2011)
21. Rubner, Y., Tomasi, C., Guibas, L.J.: The earth mover's distance as a metric for image retrieval. Int. J. Comput. Vis. **40**(2), 99–121 (2000)
22. Schaefer, A., et al.: Local-global parcellation of the human cerebral cortex from intrinsic functional connectivity MRI. Cereb. Cortex **28**(9), 3095–3114 (2018)
23. Shen, X., et al.: Using connectome-based predictive modeling to predict individual behavior from brain connectivity. Nat. Protoc. **12**(3), 506 (2017)
24. Shen, X., Tokoglu, F., Papademetris, X., Constable, R.T.: Groupwise whole-brain parcellation from resting-state FMRI data for network node identification. Neuroimage **82**, 403–415 (2013)
25. Sudlow, C., et al.: Uk biobank: an open access resource for identifying the causes of a wide range of complex diseases of middle and old age. PLoS Med. **12**(3), e1001779 (2015)
26. Tolstoi, A.: Methods of finding the minimal total kilometrage in cargo transportation planning in space. Trans. Press Natl. Commissariat Transp. **1**, 23–55 (1930)
27. Van Essen, D.C., et al.: The WU-Minn human connectome project: an overview. Neuroimage **80**, 62–79 (2013)

The Semi-constrained Network-Based Statistic (scNBS): Integrating Local and Global Information for Brain Network Inference

Wei Dai, Stephanie Noble, and Dustin Scheinost(✉)

Yale University, New Haven, CT 06511, USA
dustin.scheinost@yale.edu

Abstract. Functional connectomics has become a popular topic over the last two decades. Researchers often conduct inference at the level of groups of edges, or "components", with various versions of the Network-Based Statistic (NBS) to tackle the problem of multiple comparisons and to improve statistical power. Existing NBS methods pool information at one of two scales: within the local neighborhood as estimated from the data or within predefined large-scale brain networks. As such, these methods do not yet account for both local and network-level interactions that may have clinical significance. In this paper, we introduce the "Semi-constrained Network-Based Statistic" or scNBS, a novel method that uses a data-driven selection procedure to pool individual edges bounded by predefined large-scale networks. We also provide a comprehensive statistical pipeline for inference at a large-scale network-level. Through benchmarking studies using both synthetic and empirical data, we demonstrate the increased power and validity of scNBS as compared to traditional approaches. We also demonstrate that scNBS results are consistent for repeated measurements, meaning it is robust. Finally, we highlight the importance of methods designed to achieve a balance between focal and broad-scale levels of inference, thus enabling researchers to more accurately capture the spatial extent of effects that emerge across the functional connectome.

Keywords: Network-based statistics · Large-scale network inference

1 Introduction

Functional connectomes are connectivity matrices based on the pairwise temporal correlation of brain activity between brains regions. They are powerful approaches for analyzing functional neuroimaging data and with the right statistical strategy, can lead to meaningful inferences. These inferences are typically made via "mass-univariate regression", a method in which independent statstical tests are performed for each element, or edge, within the connectome [5].

L. Wang et al. (Eds.): MICCAI 2022, LNCS 13431, pp. 396–405, 2022.
https://doi.org/10.1007/978-3-031-16431-6_38

The main drawback of this approach is that correction for multiple comparisons with such a massive number of tests (>35,000 edges in the connectome) can greatly reduce power. To address this issue, the Network-Based Statistic (NBS) [11] was developed to increase power while still providing control over the family-wise error rate (FWER), by pooling data for inference within connected clusters (consisting of neighboring edges that display statistical effects above a predetermined threshold). NBS relies on the assumption that the effects are localized to a small neighborhood rather than spread across these networks. Yet, effects are more similar across large-scale networks [6]. Increasingly, neuroscience is acknowledging the importance of network-level effects as opposed to small local effects [2]. An extension to NBS, the Constrained Network-Based Statistic (cNBS), explicitly changes the level of inference to widespread networks by averaging all test statistics calculated within a predefined brain network [6]. Respecting this network structure of the brain has been shown to increase power.

While these approaches offer better power than those that evaluate tests at each edge independently, critical gaps remain. First, while edges in known large-scale networks tend to show similar effects, this is not guaranteed for a given experiment. Some effects may even be in opposite directions. Indeed, cNBS may average both positive and negative effects together within a predefined network. Second, while aggregating edge-level effects to a predefined network increases power [6,11], the increased power comes at the price of localizing effects to only the network as a whole, rather than to a subset of edges within the network.

To address these gaps, we develop an innovative network-level inference method that: (1) performs network-level statistical inference by pooling weak edge signals within predefined large-scale networks, (2) allows for positive and negative effects within a network, (3) balances localization between the cluster- and network-level, and (4) improves power while still controlling for error rate.

2 Methods

Semi-constrained Network-Based Statistic (scNBS). The goal of Semi-constrained Network-Based Statistic (scNBS) is to identify edges in the connectome $\boldsymbol{X} \in \mathbb{R}^{n \times p \times p}$ associated with a phenotypic measure $y \in \mathbb{R}^n$, where n is the sample size and p is the number of nodes. scNBS consists of the following steps: network partition, marginal ranking, cut-off selection, and network-level inference (see Fig. 1 for reference of the implementation schematic).

(a) Network partition: Assume that there are predefined brain networks that partition the p nodes into G networks with p_g nodes corresponding with J_g edges in the g-th network ($1 \leq g \leq G$).

(b) Marginal ranking: Consider the following generalization of the linear model: $Y_i = F(\boldsymbol{X}_i^g, Z_i)$, where F is a function not limited to be linear, and Y_i indicates some phenotypic measure, X_i^g includes all functional connectivity edges in network g, Z_i denotes all confounding covariates to be adjusted and $1 \leq i \leq n$ and n is the total number of subjects. The edges are ranked based on some measure of effect size between Y and $\boldsymbol{X}$ estimated from the above model

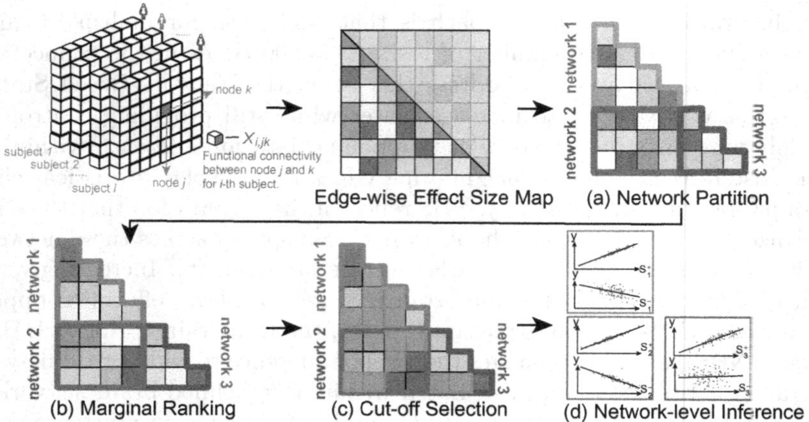

Fig. 1. Overview of Semi-constrained NBS approach. **(a)** The edge-wise map is partitioned into three predefined networks. **(b)** For each network, positive and negative effects are ordered respectively. **(c)** Within each network, top positive and negative edges are selected separately based on the best cut-offs. For examp;e, the cut-offs for network 1 are 4 (positive) and 1 (negative), *i.e.*, $t_1^+ = 4, t_1^- = 1$). **(d)** Within each network, a regression is performed between network-level variant (x-axis) and phenotype (y-axis). The dataset is divided into two sub-datasets. **(a)**–**(c)** steps is performed on the first sub-dataset while the last inference step **(d)** is applied to the second sub-dataset. Each step **(a)**–**(d)** is described in the text with greater detail.

and the estimated effects are placed in both decreasing and increasing orders so that edges with positive or negative effects will be accounted for. For example, if F is chosen as linear model and effects are measured with a t-statistic, then the ranking is based on the t-statistics for each $X^g_{i,jk}, 1 \le j,k \le J_g$. After obtaining the ordering of edges within the set g, let d_{lg} and e_{lg} be the indices of the edges with the l-th largest and smallest effect size (*e.g.*, t-statistic). Next, we define two aggregate measures v as $v^+_{ig}(d_{lg}) = \frac{\sum_{j,k=1}^{d_{lg}} X^g_{i,jk}}{l}$ and $v^-_{ig}(d_{lg}) = \frac{\sum_{i,k=1}^{e_{lg}} X^g_{i,jk}}{l}$. In other words, for each subject, the transformation returns the average values of the first j edges within the ordered edges list of set g.

(c) Cut-off selection: To obtain the network-level variant, all possible cut-off values $c \in \{1, \ldots, J_g\}$ are inspected. For each cut-off value c, a marginal regression is carried out to investigate the effects between variable $V_g^+(c)$ with observations $\{v^+_{1g}(c), \ldots, v^+_{ng}(c)\}$ and y, and between variable $V_g^-(c)$ with observations $\{v^-_{1g}(c), \ldots, v^-_{ng}(c)\}$ and y. The final threshold is the one that gives the largest marginal effect among all possible cut-offs: $t_g^+ = \text{argmax}_c \text{Cor}(V_g^+(c), y)$, $t_g^- = \text{argmax}_c \text{Cor}(V_g^-(c), y)$. This leads to the network-level variant constructed with top edges in the ordered edges list and the total number of edges used to form the network-level variant will be the same as the final threshold. Finally, with the best cut-off for set g, t_g^+ and t_g^-, the final network-level variants will be $s^+_{ig} = v^+_{ig}(t^+_g)$ and $s^-_{ig} = v^-_{ig}(t^-_g)$ for positive or negative effects, respectively.

(d) Network-level inference: After the two network-level variants S_g^+ and S_g^- with observations $\{s_{1g}^+, \ldots, s_{ng}^+\}$ and $\{s_{1g}^-, \ldots, s_{ng}^-\}$, where $1 \leq g \leq G$, are constructed, marginal linear or logistic regression will be performed for statistical inference depending on whether y is continuous or binary. The FWER or FDR will be controlled at network-level with multiple comparison approaches.

Semi-constrained NBS Implementation: This approach begins with network partitioning. We used 10 networks previously defined on the Shen 268 atlas [7]: medial frontal (MF), frontoparietal (FP), default mode network (DMN), motor (Mot), visual I (VI), visual II (VII), visual association (VAs), limbic (LIC), basal ganglia (BG), and cerebellum (CBL). Edges of a 268 × 268 connectome matrix were taken from these networks and assigned to $G = 10 \cdot 11/2 = 55$ subnetworks (10 within and 45 between networks).

As the function F in scNBS is general, we considered simple linear regression of F since it is analogous to regression produces in NBS and cNBS. Note that, the appropriate other functions (like Poisson regression) could be chosen. We highlight that we use the 1st part of data to perform the threshold selection, reserving the 2nd for inference for cross validation and avoid circularity. All the code has been made publicly available at https://github.com/daiw3/scNBS.git.

3 Synthetic Data Evaluations

Synthetic Data Generation for Benchmarking: We used resting-state fMRI data from the Human Connectome Project [10], processed with the same methods in [3]. Connectomes were created using the Shen 268 atlas [7] with 514 (240 males) participants, denoted as matrix $\boldsymbol{X}$. We created two different scenarios for synthetic effects $\boldsymbol{B} = [\beta_{jk}] \in \mathbb{R}^{p \times p}$, localized edge-level effects (scenario 1) and wider-spread network-level effects (scenario 2; Fig. 2b). To produce realistic synthetic phenotypic data, synthetic effects $\boldsymbol{B}$ were added to the real functional connectome $\boldsymbol{X}$ as illustrated in Fig. 2, where $\epsilon_i \sim N(0, 0.1), 1 \leq i \leq n$. Specifically, β_{jk} was generated from $N(1, 0.1)$, $N(-1, 0.1)$ or 0 depending on whether the effect between j-th and k-th edge is positive, negative or zero.

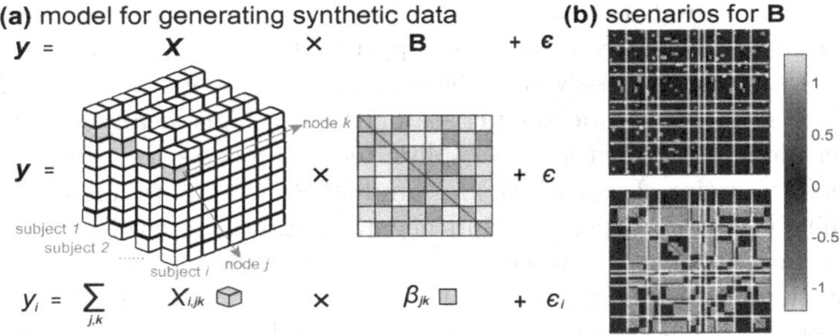

Fig. 2. Synthetic Data. (a) Model for generating synthetic data. (b) True synthetic effects $\boldsymbol{B}$ added at the edge-level (scenario 1) and at the network-level (scenario 2).

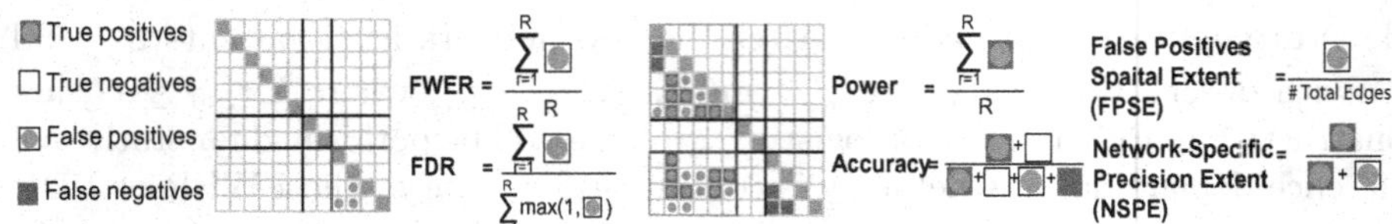

Fig. 3. Summary of measures used to evaluate scNBS. Definitions of power, accuracy, FWER, FDR, false positives spatial extent, and network-specific precision extent, where R is the total number of repetition.

Evaluations: We compared the performance of scNBS against five competing approaches o evaluate specificity and power at the edge-, cluster- or network-level, depending on methods. The competing approaches are the following: ***(i) edge***: a parametric procedure for FWER correction (*i.e.*, Bonferroni procedure [1]); ***(ii) edge-FDR***: a parametric procedure for false discovery rate (FDR) correction (Storey procedure [8]); ***(iii) NBS***: the Network-Based Statistic (NBS) [11] method for cluster-level inference with permutation-based FWER correction; ***(iv) cNBS***: the Constrained NBS (cNBS) [6] method proposed for network-level inference with permutation-based estimation of nulls followed by parametric FWER correction (Bonferroni procedure); ***(v) cNBS-FDR***: the cNBS method with FDR correction; ***(vi) scNBS***: the semi-constrained NBS (scNBS) method we proposed for network-level inference with Bonferroni correction; and ***(vii) scNBS-FDR***: the scNBS with FDR correction. For each 500 repetition, true positives were defined as effects detected during each repetition which are found in the same direction as synthetic effects $\boldsymbol{B}$. False positives were defined as detected effects that are in the opposite direction as the synthetic effects $\boldsymbol{B}$ or not found in the synthetic effects $\boldsymbol{B}$. Seven measures were calculated to explore the balance between true and false positives: power, accuracy, false positives spatial extent (FPSE), network-specific precision extent (NSPE), FDR, and FWER (see Fig. 3). Code for competing methods was modified from [6] and [11]. scNBS and all evaluations were implemented in MATLAB (R2020b)

Consistency Analysis: To assess the extent to which scNBS gives consistent results across repeated measurements, which reflects its reliability, we applied scNBS together with other competing approaches to compared sex-difference (*i.e.*. males vs. females) observed in the first of two resting-state scans (REST1) to with those from the second resting scan (REST2). Specifically, we ran scNBS for 100 iterations with different data splits, recording the selected networks and edges at each iteration. For both REST1 and REST2, we obtained two network-level frequency matrices (positive and negative) and edge-level matrices (positive and negative) for each method, showing the frequency of a network or an edge being significant over iterations. To compare the two results, we calculated the Dice coefficients, Pearson's correlation (r), and Spearman's correlation (ρ) between the two resulting frequency matrices from REST1 and REST2.

4 Results

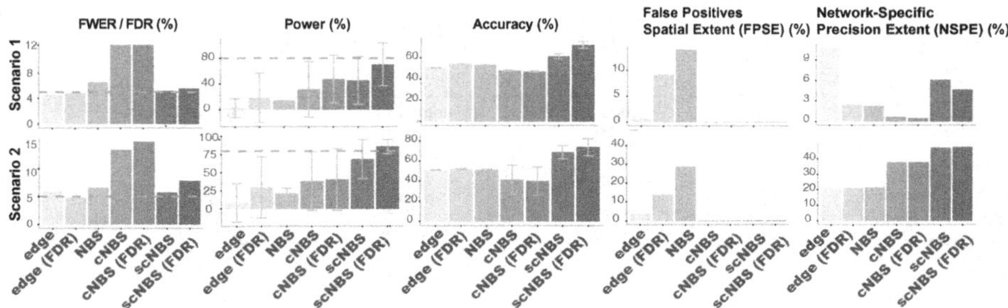

Fig. 4. Comparison of corrections methods for simulated edge-level (scenario 1) and network-level (scenario 2) effects. First column is results of FWER/FDR. The second to fifth column are results of power, accuracy, false positives spatial extent and network-specific precision extent. Each bar represents a correction procedure in the order (from left to right): *(i)* edge, *(ii)* edge-FDR, *(iii)* NBS, *(iv)* cNBS, *(v)* cNBS-FDR, *(vi)* scNBS, and *(vii)* scNBS-FDR. For Power, Accuracy, and NSPE larger values are better; whereas, for FWER, FDR, and FPSE smaller values are better. Dash lines in the FWER, FDR, and Power represent commonly acceptable values of 5% for FWER and FDR and 80% for Power. Error bars are based on 500 repeats of the simulations.

Specificity: FWER and FDR were well controlled for scNBS and scNBS-FDR. In fact, except for cNBS and cNBS-FDR, all approaches similarly controlled FWER and FDR at desired 5% level (1st column in Fig. 4). All network-level inference approaches (scNBS, scNBS-FDR, cNBS, and cNBS-FDR) had less than 0.5% of the edges to be a false positives, while edge-FDR and NBS had significantly more false positives (4th column in Fig. 4). With the exception of the edge approach in scenario 1, scNBS and scNBS-FDR exhibited the largest precision of effects and identified up to 47% of the edges within each network to overlap with the true effects (5th column in Fig. 4). Overall, these results indicated that scNBS and scNBS-FDR controlled for false positives while detecting spatially localized effects at significantly better rates than the competing approaches.

Power: Only scNBS-FDR achieved desirable power, defined commonly as 80%, which was nearly double the power of the best competing method (2nd column in Fig. 4). The competing methods had average power ranging from 2% (edge) to 48% (cNBS-FDR). scNBS and scNBS-FDR achieved highest accuracy compared to other methods (3rd column in Fig. 4). We also visualized the detected edges with over 80% power for all procedures under two scenarios (Fig. 5) and compared these results with the ground-truth using the Jaccard index, $\frac{A \cap B}{A \cup B}$, [4]. A larger Jaccard index indicated higher similarity between matrices. With the

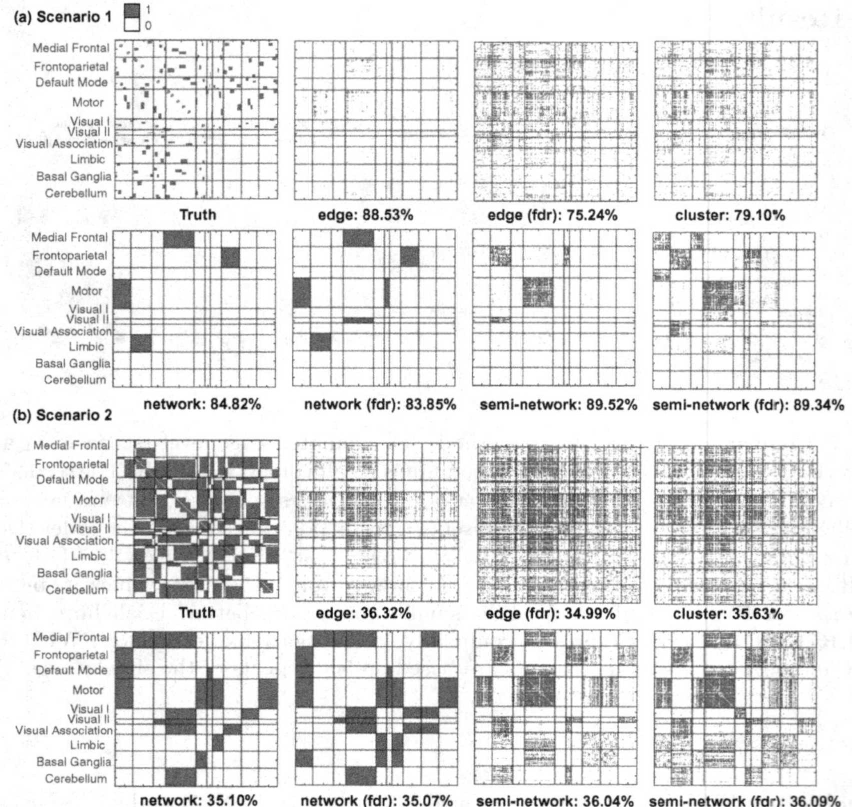

Fig. 5. Detected Edges with over 80% power across methods for (a) scenario 1 and (b) scenario 2 (*e.g.*, for scNBS, one indicates that this edge has been reported to be positive over 80% of iterations). The Jaccard index between the ground-truth and the estimated map is shown under each map.

exception of the edge approach in scenerio 1, scNBS and scNBS-FDR achieved the highest Jaccard index, compared to the competing approaches. Overall, in terms of detecting effects, scNBS demonstrates significant advantages.

Consistency: For all similarity measures, scNBS produced significantly similar results across repeated measures (Fig. 6). It also performed significantly better compared to the other inference methods. The consistent performance of scNBS points to its robustness and stability, important properties of a measure.

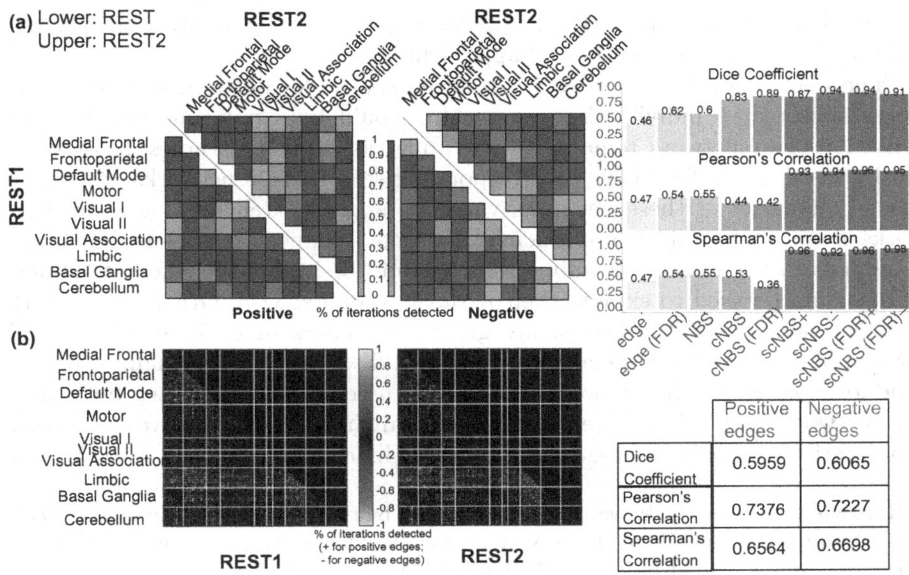

	Positive edges	Negative edges
Dice Coefficient	0.5959	0.6065
Pearson's Correlation	0.7376	0.7227
Spearman's Correlation	0.6564	0.6698

Fig. 6. Consistency of detected effects with scNBS. (a) Detected frequency at network-level, defined as the number of iterations that detected an effect over total iterations (*e.g.*, for Motor-Motor network, positive effect was detected for 98 times out of 100 iterations while negative effect was only observed for 46 times out of 100 iterations). Frequency of REST1 and REST2 were displayed in lower and upper triangle matrix respectively and effects directions were indicated by colors. The Dice coefficients, Pearson's correlation (r), and Spearman's correlation (ρ) between the two resulting frequency matrices from REST1 and REST2 were displayed in the bar plot, where scNBS+/scNBS (FDR)+ and scNBS-/scNBS (FDR)- denoted for the positive and negative network-level results under scNBS or scNBS (FDR) procedures. (b) Detected frequency at edge-level, defined as the number of iterations that detected an effect over total iterations. Frequency of REST1 and REST2 were displayed in two separate matrix respectively and effects directions were indicated by colors within each matrix.

5 Conclusion

The present work introduces an approach for performing connectome-based inference that balances both local and global properties, called scNBS. To assess its strengths and weaknesses, we evaluated its sensitivity, specificity, and consistency. As the only approach to reach 80% power, scNBS demonstrated greater sensitivity in detecting true effects by aggregating weak signals and decreasing the noise occurring at the edge-level. More importantly, it balances edge- and network-level inference, resulting in the largest network-specific precision extent (NSPE) of all tested procedures. In empirical data, scNBS showed good reproducibility across repeated analyses. These results from both simulations and empirical applications highlight the advantages of scNBS relative to existing edge- and network-level inferential approaches.

The main limitation of scNBS is reduced spatial localization as only a network, rather than individual edges within, would be significant. Besides, the definition of subnetwork remains an open question in fMRI studies. Most standard atlases in the field have a corresponding set of 10–20 networks [9], while exact boundaries of these networks vary, the spatial extent of these core networks tends to be fairly reproducible. For most connectomes generated from standard atlases, reasonable subnetworks should be available. Different subnetworks could be used to explore their performance of scNBS. Other future work includes using scNBS to investigate group differences in both functional and structural connectome associated with clinical disorders, where smaller samples (and lower corresponding power) are typically used. By incorporating both local and network level effects, we hope scNBS can help improve power to better characterize connectome-based differences in clinical contexts.

Acknowledgments. Data were provided in part by the Human Connectome Project, WU-Minn Consortium (Principal Investigators: David Van Essen and Kamil Ugurbil; U54 MH091657) funded by the 16 NIH Institutes and Centers that support the NIH Blueprint for Neuroscience Research; and by the McDonnell Center for Systems Neuroscience at Washington University. The authors report no conflicts of interest. The work was funded by NIMH P50MH115716.

Compliance with Ethical Standards. This research study was conducted retrospectively using human subject data made available in open access by the Human Connectome Project. Approval was granted by local IRB. Yale Human Research Protection Program (HIC #2000023326) on May 3, 2018.

References

1. Bland, J.M., Altman, D.G.: Multiple significance tests: the Bonferroni method. BMJ **310**(6973), 170 (1995)
2. Cremers, H.R., Wager, T.D., Yarkoni, T.: The relation between statistical power and inference in fMRI. PLoS ONE **12**(11), e0184923 (2017)
3. Finn, E.S., et al.: Functional connectome fingerprinting: identifying individuals using patterns of brain connectivity. Nat. Neurosci. **18**(11), 1664–1671 (2015)
4. Jaccard, P.: The distribution of the flora in the alpine zone. 1. New Phytol. **11**(2), 37–50 (1912)
5. Meskaldji, D.E., Fischi-Gomez, E., Griffa, A., Hagmann, P., Morgenthaler, S., Thiran, J.P.: Comparing connectomes across subjects and populations at different scales. NeuroImage **80**, 416–425 (2013). https://doi.org/10.1016/j.neuroimage.2013.04.084, https://www.sciencedirect.com/science/article/pii/S105381191300431X, mapping the Connectome
6. Noble, S., Scheinost, D.: The constrained network-based statistic: a new level of inference for neuroimaging. In: Martel, A.L., et al. (eds.) MICCAI 2020. LNCS, vol. 12267, pp. 458–468. Springer, Cham (2020). https://doi.org/10.1007/978-3-030-59728-3_45

7. Shen, X., Tokoglu, F., Papademetris, X., Constable, R.T.: Groupwise whole-brain parcellation from resting-state fMRI data for network node identification. Neuroimage **82**, 403–415 (2013)
8. Storey, J.D.: A direct approach to false discovery rates. J. Royal Stat. Soc. Ser. B (Stat. Methodol.) **64**(3), 479–498 (2002)
9. Uddin, L.Q., Yeo, B., Spreng, R.N.: Towards a universal taxonomy of macro-scale functional human brain networks. Brain Topogr. **32**(6), 926–942 (2019)
10. Van Essen, D.C., et al.: The WU-Minn human connectome project: an overview. Neuroimage **80**, 62–79 (2013)
11. Zalesky, A., Fornito, A., Bullmore, E.T.: Network-based statistic: identifying differences in brain networks. Neuroimage **53**(4), 1197–1207 (2010)

Unified Embeddings of Structural and Functional Connectome via a Function-Constrained Structural Graph Variational Auto-Encoder

Carlo Amodeo[1(✉)], Igor Fortel[1], Olusola Ajilore[2], Liang Zhan[4], Alex Leow[1,2], and Theja Tulabandhula[3]

[1] Department of Biomedical Engineering, University of Illinois Chicago, Chicago, IL, USA
camode2@uic.edu

[2] Department of Psychiatry, University of Illinois Chicago, Chicago, IL, USA

[3] Department of Information and Decision Sciences, University of Illinois Chicago, Chicago, IL, USA

[4] Department of Electrical and Computer Engineering, University of Pittsburgh, Pittsburgh, PA, USA

Abstract. Graph theoretical analyses have become standard tools in modeling functional and anatomical connectivity in the brain. With the advent of connectomics, the primary graphs or networks of interest are structural connectome (derived from DTI tractography) and functional connectome (derived from resting-state fMRI). However, most published connectome studies have focused on either structural or functional connectome, yet complementary information between them, when available in the same dataset, can be jointly leveraged to improve our understanding of the brain. To this end, we propose a function-constrained structural graph variational autoencoder (FCS-GVAE) capable of incorporating information from both functional and structural connectome in an unsupervised fashion. This leads to a joint low-dimensional embedding that establishes a unified spatial coordinate system for comparing across different subjects. We evaluate our approach using the publicly available OASIS-3 Alzheimer's disease (AD) dataset and show that a variational formulation is necessary to optimally encode functional brain dynamics. Further, the proposed joint embedding approach can more accurately distinguish different patient sub-populations than approaches that do not use complementary connectome information.

Keywords: Neuroimaging · Brain networks · Deep learning

1 Introduction

Advances in magnetic resonance imaging (MRI) technology have made very large amounts of multi-modal brain imaging data available, providing us with unparalleled opportunities to investigate the structure and function of the human brain.

L. Wang et al. (Eds.): MICCAI 2022, LNCS 13431, pp. 406–415, 2022.
https://doi.org/10.1007/978-3-031-16431-6_39

Functional magnetic resonance imaging (fMRI), for example, can be used to study the functional activation patterns of the brain based on cerebral blood flow and the Blood Oxygen Level Dependent (BOLD) response [4], whereas diffusion tensor imaging (DTI) can be used to examine the *wiring diagram* of the white matter fiber pathways, i.e., the structural connectivity of the brain [1].

Because of their utility in understanding human brain structure, in examining neurological illnesses, and in developing therapeutic/diagnostic applications, brain networks (also called *connectomes*) have attracted a lot of attention recently. The principal networks of interest are structural brain networks (derived from DTI) and functional brain networks (derived from fMRI). Graph-based geometric machine learning approaches have shown promise in processing these connectomics datasets due to their ability to leverage the inherent geometry of such data. However the majority of these existing studies in brain network analysis tend to concentrate on either structural or functional connectomes [13,15]. Our hypothesis is that both the anatomical characteristics captured by structural connectivity and the physiological dynamics properties that form the basis of functional connectivity can lead to a much improved understanding of the brain's integrated organization, and thus it would be advantageous if both structural and functional networks could be analyzed simultaneously.

To this end, in this study we propose employing a graph variational autoencoder (GVAE) [8] based system to learn low-dimensional embeddings of a collection of brain networks, which jointly considers both the structural and functional information. Specifically, we employ graph convolutions to learn structural and functional joint embeddings, where the graph structure is defined by the structural connectivity and node properties are determined by functional connectivity. The goal here is to capture structural and functional network-level relationships between subjects in a low-dimensional continuous vector space so that inferences about their individual differences, as well as about the underlying brain dynamics, can be made. Experimental results in Sect. 3 show how the embedding space obtained through this preliminary line of research enables comparison of higher-order brain relationships between subjects. We also validate the usefulness of the resulting embedding space via a classification task that seek to predict whether the subject is affected by AD or not, and show how it outperforms single modality baselines.

2 Proposed Framework

2.1 Problem Statement

We are given a set of brain network instances $D = G_1, ..., G_M$, where each instance G_i corresponds to a different subject and is composed of a structural $G_i^{(s)}$ and a functional $G_i^{(f)}$ network. Our goal is to jointly embed the information contained in the two networks in order to obtain a common coordinate space that enables interpretable comparisons between subjects. We will test the quality of the embeddings in this common space via a downstream task that involves classifying AD subjects and healthy subjects.

Note that most existing approaches focus on one of the network types (structural or functional) to learn embeddings. Instead, our objective is to design an unsupervised learning approach that can extract the complimentary information that exists across these two modalities to improve the quality of embeddings, and thus the ability to improve downstream tasks.

2.2 Our Modeling Framework

We propose an augmented function-constrained structural graph variational autoencoder based system, or FCS-GVAE, that involves a GVAE and an Autoencoder (AE), as shown in Fig. 1. We employ graph convolutions to learn the structural and functional joint embeddings, $\mathbf{Z_1} \in \mathbb{R}^{N \times D_1}$, where D_1 represents the dimensionality of each node embedding. The sampling at the GVAE bottleneck is similar to the one of a traditional VAE, with straightforward Gaussian sampling (and the only assumption taken is the one of gaussianity of the latent variables). Given the joint node level embedding matrix, a graph-level embedding $\mathbf{Z_2} \in \mathbb{R}^{D_2}$ is then obtained through an AE, where D_2 is the chosen dimensionality of the graph-level embedding. These resultant embeddings, one per subject, are then visualized via t-distributed stochastic neighbor (*t-sne*) [16].

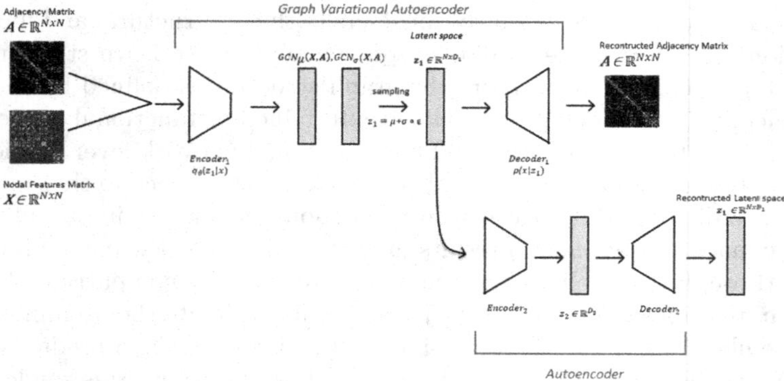

Fig. 1. An overview of the proposed FCS-GVAE model: The adjacency matrix A used as input is represented by the structural connectivity network. An n-dimensional feature vector $\mathbf{x_i}$ is assigned on each node, which is the vector of the corresponding node in the functional connectivity network. The resulting encoded data, $\mathbf{Z_1}$ is then further compressed using an autoencoder, whose latent vector, $\mathbf{Z_2}$, is the one used to compare and contrast different subjects.

Once the parameters of the GVAE and AE are learned, each subject's data is transformed into the structural functional joint embedding as follows. The structural connectivity, SC, is the adjacency input matrix, $A \in \mathbb{R}^{N \times N}$, and the functional connectivity, FC, is the matrix of nodal features, $X \in \mathbb{R}^{N \times C}$, where $C = N$. In Sect. 3, we compare the value of designing our embedding architecture this way against a baseline that removes the FC nodal feature data, as suggested in [12]. The quality of these embeddings is assessed using a downstream classification task involving the detection of AD in subjects. Next, we briefly discuss the notion of graph convolutions and GVAE and justify the need for GVAE vs a regular graph auto-encoder as well as the need for another downstream AE that transforms node embeddings to a graph embedding.

2.3 Graph Convolution

A graph convolutional layer [9] works as follows. Let $G = \{\mathbb{V}, A\}$ be a weighted, undirected graph, where $\mathbb{V}$ is a set of N nodes, and $A \in \mathbb{R}^{N \times N}$ is an adjacent matrix, specifying the inter-nodal connections. The normalized Laplacian is defined as $L = I_N - D^{-1/2}AD^{-1/2}$, with I_N representing the N-dimensional identity matrix and the diagonal degree matrix D having entries $D_{i,i} = \sum_j A_{i,j}$. L can be further decomposed into the form $U \Lambda U^T$, where U is the matrix of eigenvectors of L and Λ is the diagonal matrix of its eigenvalues.

Let $\mathbf{x}$ be the attribute vector corresponding to a node.

Given a filter $\mathbf{h} \in \mathbb{R}^N$, the graph convolutional operation between $\mathbf{x}$ and a filter $\mathbf{h}$ is defined as:

$$\mathbf{x} \circledast \mathbf{h} = U((U^T\mathbf{h}) \odot (U^T\mathbf{x})) = U\hat{H}U^T\mathbf{x}, \tag{1}$$

where $\hat{H} = \text{diag}(\theta)$ replaces $U^T\mathbf{h}$ with $\boldsymbol{\theta} \in \mathbb{R}^N$ parameterizing it in the Fourier domain. Since the evaluation of Eq. 1 requires explicit computation of the eigenvector matrix, it can be prohibitively costly for very large networks; consequently, $\hat{H}$ has been approximated through a truncated expansion in terms of the Chebyshev polynomials [2,5]. Kipf et al. [9] subsequently provided a second order approximation, such that $\mathbf{x} \circledast \mathbf{h} \approx \theta(I_n + D^{-1/2}AD^{-1/2})\mathbf{x}$. Generalizing the $\mathbf{x}$ vector to the matrix $X \in \mathbb{R}^{N \times C}$ with C input channels, the following graph convolutional layer filtering is introduced:

$$Z = \tilde{A}X\tilde{\Theta} \tag{2}$$

With $\tilde{A} = I_n + D^{-1/2}AD^{-1/2}$, $\tilde{\Theta} \in \mathbb{R}^{C \times F}$ a matrix of the F filter parameters to be learned, and $Z \in \mathbb{R}^{N \times F}$ the convolved signal matrix. Stacking numerous convolutional layers (see Eq. 2 for instance), each of them followed by the application of a non-linearity [17], graph convolutional networks are defined as:

$$GCN^{l+1}(\tilde{A}, X) = \sigma(\tilde{A}\, GCN^l(\tilde{A}, X)^l\, \tilde{\Theta}_{l+1}), \tag{3}$$

where $\sigma(\cdot)$ represents the activation function and $GCN^0(\tilde{A}, X) := X$. Important to our setting, note that Eqs. 2 and 3 show how the output of the graph convolution process contains not only the network's topological information (represented by A), but also the nodal properties (represented by X). Next we discuss the GVAE component of our setup.

2.4 Graph Variational Autoencoder

Variational Autoencoder (VAE) [7] is a variant of deep generative models used for learning latent representations from an unlabeled dataset by simultaneously training an encoder and a decoder to maximize the evidence lower bound (ELBO). The encoder maps the data into a low-dimensional latent representation z. The z is then sampled from the approximate posterior distribution $q(z|X)$, typically chosen to be an independent Gaussian distribution $\mathcal{N}(\mu, \mathrm{diag}(\sigma^2))$, where μ and σ are output by the encoder. The decoder reconstructs the original data by deriving the likelihood of data X based on the variable z, $p(X|z)$.

GVAE extends this idea to graph-structured data, and in learning the latent representation of the data on which it is trained, it incorporates both the input network's topological characteristics, as defined by the adjacency matrix, and the network's node features [8]. In a GVAE, the encoder is parameterized by a series of graph convolutional network layers. Technical details about the approximate posterior distribution, as well as the justification for using a non-trainable innerproduct decoder can be found in [8] and [7].

2.5 Justifying the Choice of GVAE and AE Components

As shown in Fig. 1, our model involves a cascade of two auto-encoders, one of which is generative. We need the graph auto-encoder to be generative because we are not only interested in compression but also the generalizability of the encoders and decoders. For instance, we can reuse the decoder to generate/sample new structural connectomes, which is not possible with a vanilla auto-encoder. Further, since the GVAE produces high-granularity node-level embeddings, we use a straight-forward AE to compress that information to a low enough dimension for downstream visualization/supervised learning tasks. While doing so, we are able to obtain a unified graph level embedding (for a given structural functional connectome pair) that can be much better than naive approaches (such as averaging node-embeddings). Note that this second component, the AE, may be replaced, for example by a PCA. However, unlike PCA, a single layer AE with no-nonlinearity does not impose orthogonality, hence reducing the reconstruction error. Nonetheless, in our case we make use of multiple linear layers followed by appropriate non-linearities to achieve maximal compression (which is beyond what PCA can achieve). As can be seen in the following section, these choices indeed show the value of complementary information present in both networks and how they help define better embeddings (as evaluated using the AD classification task).

3 Experimental Evaluation and Results

We discuss the dataset, the hyperparameter choices, and how resulting embeddings improve on existing methods, using both structural and functional network information simultaneously. To evaluate our node and graph embeddings, we use an auxiliary link prediction problem and the AD classification problem respectively. Both these tasks quantitatively show the value of our joint embedding approach compared to natural baselines.

3.1 Dataset

MRI imaging used in this study comes from the OASIS-3 dataset for Normal Aging and Alzheimer's disease [10] and was collected in a 16-channel head coil of a Siemens TIM Trio 3T scanner. OASIS-3 includes the clinical, neuropsychological, neuroimaging, and biomarker data of 1098 participants (age: 42–95 years; www.oasis-brains.org). We analyzed the data from 865 participants with combined structural and functional MRI sessions (N = 1326). The dataset includes 738 Females (112 with AD), and 588 Males (163 with AD). AD has been defined here as having a clinical dementia rating greater than 0.

The brain regions considered for this study, which cover the entire brain, consist of 132 regions: 91 cortical Region of Interests (ROIs) are obtained from the FSL Harvard-Oxford Atlas maximum likelihood cortical atlas, 15 subcortical ROIs are obtained from the FSL Harvard-Oxford Atlas maximum likelihood subcortical atlas [3], and the remaining 26 are cerebellar ROIs from the AAL atlas [14].

The brain structure connectivity graph is generated from combining brain grey matter parcellation extracted from T1-weighted MRI and the white matter fiber tracking obtained from DTI acquisition. The graph is undirected and each node v_i in V denotes a specific brain region of interest (ROI). Element A_{ij} in A, the adjacency matrix, denotes the weight of the connection between the two nodes v_i and v_j. Note that we performed minimal processing on the structural graph and have generated functional graphs via Pearson coefficients of BOLD signals in a particular time window (without thresholding). Thus, the choice of time windows is the only preprocessing step.

3.2 Hyperparameter Choices

We chose $L = 4$ graph convolutional layers to model the encoder of the GVAE component. The number of layers for GVAE is determined by computing the diameter of the adjacency matrix, defined as $\max_{v_i,v_j}(\text{dist}(v_i, v_j)), \forall v_i, v_j \in V$, where $\text{dist}(v_i, v_j)$ represents the shortest path to reach v_j from v_i. The maximum graph diameter for our dataset turned out to be 4, implying that every node reaches/influences every other node's representation after 4 graph convolution layers. Note that this choice is motivated by the need to avoid the *vanishing gradient* phenomena [11].

The dimension D_1 of the GVAE latent space is selected by taking in consideration the Average Precision (AP) and the Area Under the Curve (AUC) of an auxiliary link prediction problem that is solved while maximizing ELBO. Based on Fig. 2b, we set the D_1 parameter to be equal to 6. Note that this result also suggests that the non-linear dynamics of the brain (as defined using structural and functional graphs) can essentially be captured by a 6 dimensional node embedding space. In addition to the methodological justification for choosing GVAE compare to a vanilla graph auto-encoder given in Sect. 2.5, we also show empirically that GVAE performs much better when compared to a vanilla/non-variational graph auto-encoder (GAE) on the link prediction task (see Fig. 2a). The first graph convolutional layer in the encoder of the GVAE component is characterized by 48 filters, the second one by 24, the third by 12 and both $GCN_{\mu}^{4}(A, X)$ and $GCN_{\sigma}^{4}(A, X)$ have 6 filters each. To maximize evidence lower bound (ELBO), we use a gradient descent variant known as Adaptive Moment Estimation (ADAM) [6], setting a learning rate equal to 0.001.

We further compress the latent matrices corresponding to node-level embeddings of a subject, viz., $\mathbf{Z_1} \in \mathbb{R}^{132\times 6}$, into a smaller subject level graph embedding vector, $z_2 \in \mathbb{R}^{D_2}$ through the vanilla AE component (see Fig. 1). This second latent space is obtained through an encoding structure composed of 5 linear layers, with each of them using the ReLU activation function. To learn the parameters of the AE, we employ a mean squared error (MSE) loss function, which we minimize using the same ADAM optimizer mentioned earlier. The learning rate is set to 0.001 and the weight decay is set to 0.00005. The dimensionality D_2 is set to the value 6 based on the convergence of the MSE loss.

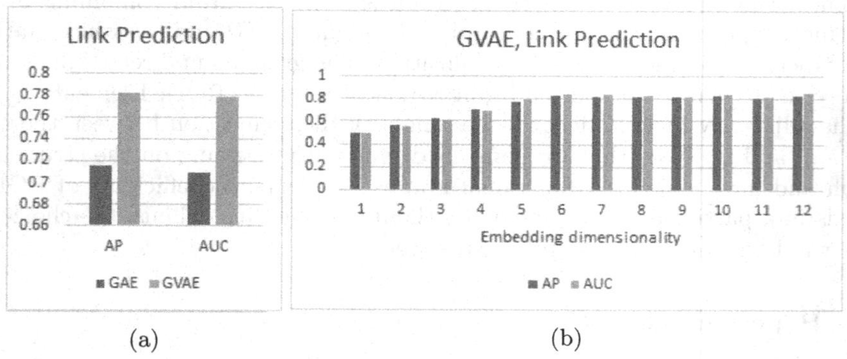

(a) (b)

Fig. 2. Performances of the two different models on the link prediction task: a) FCS-GVAE performs better than the GAE for both the considered metrics. b) Performances obtained by varying the dimensionality of the embedding layer. It can be seen that a plateau is reached when the dimensionality is ≥ 6.

3.3 Results Demonstrating the Quality of Joint Embeddings

A key point of our work is to identify how including the functional connectome (as nodal features) can help further characterize the underlying biology, relative to a baseline model that does not use it (instead using the identity matrix as the nodal feature, a common practice in the graph neural networks research community [12]). Indeed, Fig. 3 qualitatively shows how including the FC (left panel) leads to better clustering in the latent space with respect to diagnostic labels (e.g., the right cluster comprises primarily healthy subjects), likely by forcing the latent embedding to account for the temporal dynamics of FC. By contrast, right panel shows the embedding with structural information alone.

We also evaluate the performance of our learned embeddings for AD detection. Here, the latent embedding vector representations across all brain regions of an individual are used as features to predict their diagnostic label (AD versus not AD) using 2 classic classification models: multi-layer perceptron and random forest. With the hyperparameters set to default values in sklearn, we ran classification by randomly splitting the data into training/test sets (80%/20%), and reported performance over the test set with cross-validation. The results (Table 1) highlight the value of using functional connectome as the nodal features, as evidenced by the performance improvement over the baseline model (identity matrix as the nodal feature) in terms of F1-score, precision, and recall.

Table 1. Classification performances under two different embedding approaches. Using the FC as nodal features (right column) leads to an overall improvement in the classifier performance when compared to simply using the identity matrix as nodal features (left column).

Model	Identity matrix			Functional connectome		
	Precision	Recall	F1-score	Precision	Recall	F1-score
Multi layer perceptron	0.308 ± 0.018	0.045 ± 0.012	0.109 ± 0.031	0.587 ± 0.014	0.762 ± 0.045	0.663 ± 0.023
Random forest	0.304 ± 0.013	0.189 ± 0.042	0.233 ± 0.032	0.573 ± 0.009	0.706 ± 0.002	0.63 ± 0.014

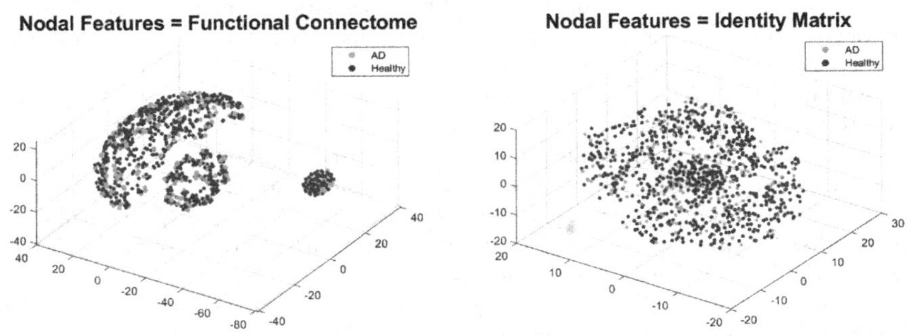

Fig. 3. T-sne projections of the learned embeddings. When using the joint embeddings computed by FCS-GVAE, different sub-populations are captured.

4 Conclusion

We introduced a variational graph auto-encoder framework to unify DTI structural and resting-state functional brain networks, allowing for the definition of a common-coordinate embedding space, i.e., a joint structure-function representation of the brain. When trained on a large AD dataset, this joint embedding framework was able to uncover biologically meaningful sub-populations in an unsupervised manner. Further, we quantitatively demonstrated improvement in classification tasks with our variational formulation (versus a more traditional non-variational graph auto-encoder), suggesting that a variational framework is necessary in optimally capturing functional brain dynamics. In the future, we will aim to further use our approach to uncover the biological underpinnings of different AD sub-populations.

Acknowledgement. This study is partially supported by the NIH (R01AG071243 and R01MH125928) and NSF (IIS 2045848).

References

1. Assaf, Y., Pasternak, O.: Diffusion tensor imaging (DTI)-based white matter mapping in brain research: a review. J. Mol. Neurosci. **34**(1), 51–61 (2008)
2. Defferrard, M., Bresson, X., Vandergheynst, P.: Convolutional neural networks on graphs with fast localized spectral filtering. In: Advances in Neural Information Processing Systems, (Nips), pp. 3844–3852 (2016)
3. Desikan, R.S., et al.: An automated labeling system for subdividing the human cerebral cortex on MRI scans into gyral based regions of interest. Neuroimage **31**(3), 968–980 (2006)
4. Fox, M.D., Raichle, M.E.: Spontaneous fluctuations in brain activity observed with functional magnetic resonance imaging. Nat. Rev. Neurosci. **8**(9), 700–711 (2007)
5. Hammond, D.K., Vandergheynst, P., Gribonval, R.: Wavelets on graphs via spectral graph theory. Appl. Comput. Harmon. Anal. **30**(2), 129–150 (2011)
6. Kingma, D.P., Ba, J.L.: Adam: a method for stochastic optimization. In: 3rd International Conference on Learning Representations, ICLR 2015 - Conference Track Proceedings, pp. 1–15 (2015)
7. Kingma, D.P., Welling, M.: Auto-encoding variational Bayes. arXiv preprint arXiv:1312.6114 (2013)
8. Kipf, T.N., Welling, M.: Variational Graph Auto-Encoders 1 A latent variable model for graph-structured data. In: NIPS workshop, no. 2, pp. 1–3 (2016)
9. Kipf, T.N., Welling, M.: Semi-supervised classification with graph convolutional networks. In: 5th International Conference on Learning Representations, ICLR 2017 - Conference Track Proceedings, pp. 1–14 (2017)
10. LaMontagne, P.J., et al.: Oasis-3: longitudinal neuroimaging, clinical, and cognitive dataset for normal aging and alzheimer disease. MedRxiv (2019)
11. Li, Q., Han, Z., Wu, X.-M.: Deeper insights into graph convolutional networks for semi-supervised learning. In: Thirty-Second AAAI Conference on Artificial Intelligence (2018)
12. Li, Y., Shafipour, R., Mateos, G., Zhang, Z.: Mapping brain structural connectivities to functional networks via graph encoder-decoder with interpretable latent embeddings. In: 2019 IEEE Global Conference on Signal and Information Processing (GlobalSIP), pp. 1–5. IEEE (2019)
13. Sporns, O.: Structure and function of complex brain networks. Dialogues Clin. Neurosci. **15**(3), 247 (2013)

14. Tzourio-Mazoyer, N., et al.: Automated anatomical labeling of activations in SPM using a macroscopic anatomical parcellation of the MNI MRI single-subject brain. Neuroimage **15**(1), 273–289 (2002)
15. Van Den Heuvel, M.P., Hulshoff Pol, H.E.: Exploring the brain network: a review on resting-state fMRI functional connectivity. Eur. Neuropsychopharmacol. **20**(8), 519–534 (2010)
16. Van der Maaten, L., Hinton, G.: Visualizing data using t-SNE. J. Mach. Learn. Res. **9**(11), 2579–2605 (2008)
17. Wu, Z., Pan, S., Chen, F., Long, G., Zhang, C., Yu Philip, S.: A comprehensive survey on graph neural networks. IEEE Trans. Neural Networks Learn. Syst. **32**(1), 4–24 (2020)

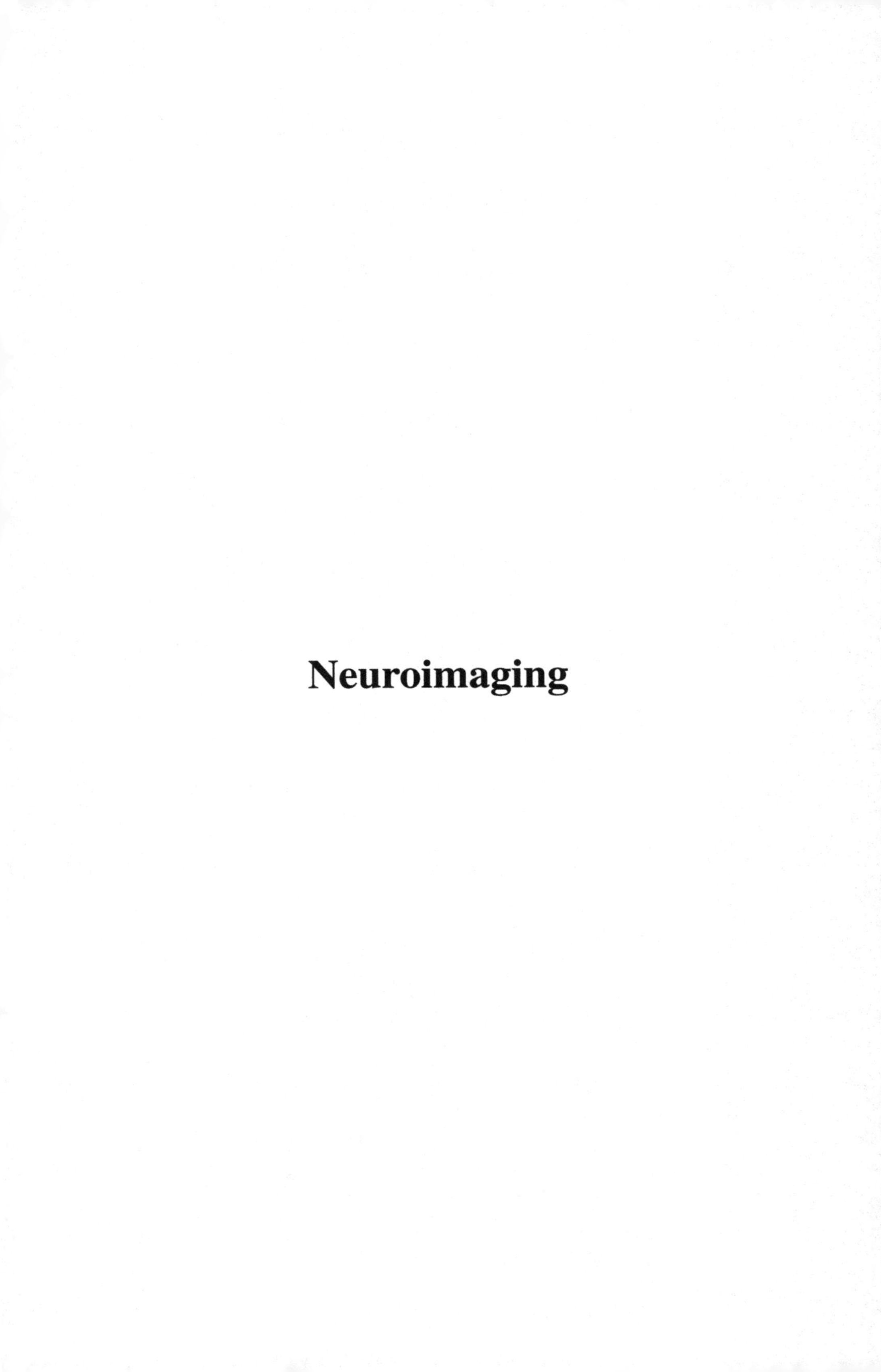

Neuroimaging

Characterization of Brain Activity Patterns Across States of Consciousness Based on Variational Auto-Encoders

Chloé Gomez[1,2(✉)], Antoine Grigis[1], Lynn Uhrig[2], and Béchir Jarraya[2,3]

[1] Université Paris-Saclay, CEA, NeuroSpin, 91191 Gif-sur-Yvette, France
chloe.gomez@cea.fr

[2] Cognitive Neuroimaging Unit, Institut National de la Santé et de la Recherche Médicale U992, Gif-sur-Yvette, France

[3] University of Versailles, Université Paris-Saclay, Neurosurgery Department, Foch Hospital, Suresnes, France

Abstract. Decoding the levels of consciousness from cortical activity recording is a major challenge in neuroscience. The spontaneous fluctuations of brain activity through different patterns across time are monitored using resting-state functional MRI. The different dynamic functional configurations of the brain during resting-state are also called "brain states". The specific structure of each pattern, lifetime, and frequency have already been studied but the overall organization remains unclear. Recent studies showed that low-dimensional models are adequate to capture the correlation structure of neural activity during rest. One remaining question addressed here is the characterization of the latent feature space. We trained a dense Variational Auto-Encoder (dVAE) to find a low two-dimensional representation that maps dynamic functional connectivity to probability distributions. A two-stage approach for latent feature space characterization is proposed to facilitate the results' interpretation. In this approach, we first dissect the topography of the brain states and then perform a receptive field analysis to track the effect of each connection. The proposed framework instill interpretability and explainability of the latent space, unveiling biological insights on the states of consciousness. It is applied to a non-human primate dataset acquired under different experimental conditions (awake state, anesthesia induced loss of consciousness).

Keywords: Dynamic functional connectivity · States of consciousness · Non-human primates · Variational auto-encoder · Generative embeddings

---

Supplementary Information The online version contains supplementary material available at https://doi.org/10.1007/978-3-031-16431-6_40.

L. Wang et al. (Eds.): MICCAI 2022, LNCS 13431, pp. 419–429, 2022.
https://doi.org/10.1007/978-3-031-16431-6_40

1 Introduction

"The stream of our consciousness, [...] like a bird's life, seems to be made of an alternation of flights and perchings", said the philosopher William James [13]. A fundamental observation that still questions many scientists. In the manner of the seasons that transform our landscapes, the spontaneous fluctuations of the brain reveal very different brain configurations. Neuroimaging, such as resting-state functional MRI (rs-fMRI), enables brain activity monitoring and has been widely used to uncover specific states of consciousness both in human and animal models [1,8,10,12]. Handling time in a dynamic Functional Connectivity (dFC) analysis shows that wakefulness and loss of consciousness exemplify strikingly different patterns of brain states repertoire - distinct, repeatable patterns of brain activity [3,5,21,25].

Few studies explore the organization of the dFC patterns, limiting themselves to a lifetime or frequency of occurrence characterization [1,3]. Due to the high-dimensional nature of the dFC, brain states representation and interpretation is still an open question. Recently, several studies proposed to project fMRI data onto a low-dimensional space (two or three dimensions) which remains a good approximation of neuronal signals [9,18,20]. Based on this low-dimensional representation, the topography of brain states repertoire is revealed, independently of any a priori knowledge (such as a clustering technique).

In particular, generative models are promising to investigate the structure of the data. Indeed, the generative modeling power lies in the probabilistic representation of the data and can be opposed to discriminative modeling. Where discriminative models learn a mapping between the data and classes, generative models directly learn probability distributions without external labels. Dimensionality reduction techniques demonstrate that dFC data reflect the interplay of a small number of latent processes [1,19]. In other words, the high-dimensional dFC features come from a lower-dimensional manifold embedded within the high-dimensional input space and possibly corrupted by noise. In the literature [6], latent linear models estimate these underlying processes using a linear interaction model. Unfortunately, when the mapping is non-linear or equivalently, the manifold is curved, such models may fail. Dense Variational Auto-Encoders (dVAE) bypass such limitations and propose a deep learning generative framework for non-linear manifold discovery and low-dimensional structure analysis [14,20,27].

In Sect. 2.2, we use a dVAE model to discover a low-dimensional representation of dFC matrices derived from the dataset acquired in different arousal conditions described in Sect. 2.1. Then, we present a two-fold interpretation framework that dissects the topography of the brain states. In Sect. 2.3, a connection-wise receptive field analysis of the encoder models brain transitions. In Sect. 2.4, dense descriptors and a query engine coherently and stably describe the latent space, leveraging the generative part of the decoder. This querying tool is referred to as the VAE framework for Visualizing and Interpreting the ENcoded Trajectories (VAE-VIENT) between states of consciousness.

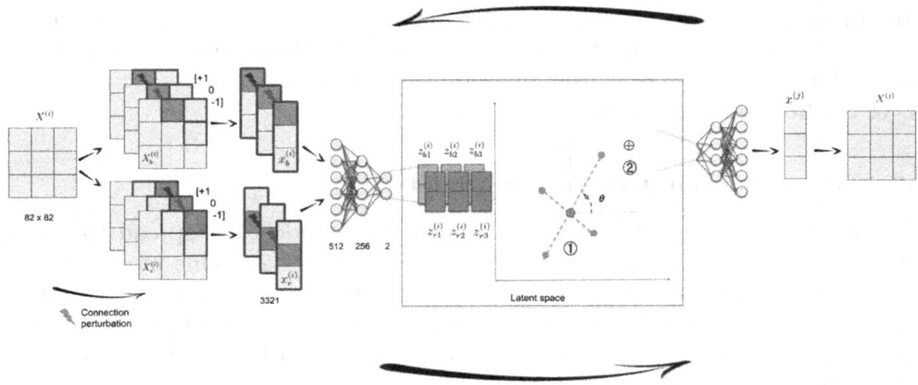

Fig. 1. Vizualization of the VAE-VIENT framework. 1) From a dFC matrix $X^{(i)}$, or equivalently its upper terms $x^{(i)}$, two simulations are performed on connections b (blue) and r (red) using a $[-1, 0, 1]$ traverse ($p = 3$). The corresponding $z_b^{(i)}$ and $z_r^{(i)}$ latent representations follows lines represented by the inclination θ with the x-axis. 2) From a custom latent space location, a dFC matrix $\tilde{X}^{(i)}$ is decoded. (Color figure online)

2 Methods

2.1 Dataset and Preprocessing

Dataset. Data was acquired in five rhesus macaques (Macaca mulatta) in different arousal states: awake, propofol anesthesia (deep and light), ketamine anesthesia, and sevoflurane anesthesia (deep and light) [3,25]. Levels of anesthesia were defined by a clinical score (obtained through the monkey sedation scale) and continuous electroencephalography monitoring. A small number of animals was investigated as it is advised in nonhuman primate studies. All procedures were conducted in accordance with the European Convention for the Protection of Vertebrate Animals used for Experimental and Other Scientific Purposes (Directive 2010/63/EU) and the National Institutes of Health's Guide for the Care and Use of Laboratory Animals. Monkeys were scanned on a 3-Tesla scanner with a single transmit-receiver surface coil customized to monkey. 156 resting-state runs were acquired with a repetition time of 2.400 ms and 500 brain volumes per run. A contrast agent (the MION) was injected into monkey's vein before scanning.

Preprocessing. A spatial preprocessing was performed with a specific monkey pipeline and voxel time series were filtered with low-pass (0.05-Hz cutoff) and high-pass (0.0025-Hz cutoff) filters and a zero-phase fast-Fourier notch filter (0.03 Hz) to remove an artifactual frequency present in all data [3,25]. Voxel time series were averaged for each one of the 82 regions of the CoCoMac atlas [2] and cut in sliding time-windows of 35 TR, with a sliding step of one [3]. Then, a pairwise correlation between averaged time-series was computed for each window [1].

This methodology leads to more than 72,000 functional connectivity matrices (see Appendix S1). We split the dataset in two to perform leave-one-subject-out cross validation: all the runs from one subject were used as the test set and were never used to train the model.

2.2 Low-dimension Generative Model

The emergence of deep learning-based generative models has spread to many different disciplines including medical domain [14,17,22,23]. One model among the prominent families that stand out is the Variational Autoencoder (VAE) [15]. It trains an encoder to convert data as a distribution over the latent space and a decoder. Presenting an efficient architecture on a limited-size dataset, it is therefore chosen in this work.

We optimize the ELBO objective coupling mean-squared error loss and the β-weighted Kullback-Leibler divergence. The mean-squared error reconstruction term makes the encoding/decoding scheme as effective as possible. Latent space regularity is enforced during the training using regularization to avoid overfitting and ensure continuity (two nearby points in the latent space give similar contents once decoded) and completeness (a code sampled from the latent space should give relevant content once decoded). These properties form the core of the generative process. In practice, the regularization term enforces the encoding distributions to be close to a standard normal distribution using the Kulback-Leibler divergence.

To learn disentangled representations and to boost interpretability, a regularization parameter β is further introduced [4,11]. A high value of β puts more emphasis on the statistical independence than on the reconstruction.

In this work, we considered a dVAE with a two-dimensional latent space, mimicking the architecture proposed in [20]. The encoder part uses two fully connected layers (of 512 and 256 units respectively) with ReLU activation functions, and the decoder part is implemented with the same structure. The dimension of the latent space corresponds to common neurobiological assumptions made when studying disorders of consciousness [7,16]. Moreover, in our experimental settings, the proposed model shows similar reconstruction performance to a model with 32 latent variables.

2.3 Connection-wise Receptive Field (RF) Analysis

The characterization of the latent space is essential to build an interpretable model. The idea is to help visualizing and interpreting encoded trajectories between states of consciousness, e.g. from anesthesia-induced loss of consciousness to wakefulness. In particular, a key analysis captures the Receptive Field (RF) of the latent space at the connection level. More than highlighting the model sensitivity, this analysis leverages the regions to perturb for inflecting trajectories and perhaps restoring wakefulness.

Let's consider a dataset $D = \{x^{(1)}, ..., x^{(n)}\}$ with n data samples, where each sample is a vector of d dimensions $x^{(i)} = [x_1^{(i)}, ..., x_d^{(i)}]$. The RF analysis focuses

Metrics	State 1	State 2	State 3	State 4	State 5	State 6	State 7	All (relative to support)
Recall	0.57 +/- 0.11	0.09 +/- 0.12	0.37 +/- 0.14	0.54 +/- 0.04	0.85 +/- 0.02	0.64 +/-0.08	0.85 +/- 0.02	**0,70 +/- 0.03**
SSIM	0.44 +/- 0.05	0.44 +/- 0.01	0.51 +/- 0.05	0.65 +/- 0.02	0.68 +/- 0.01	0.85 +/- 0.01	0.79 +/- 0.01	**0.62 +/- 0.008**
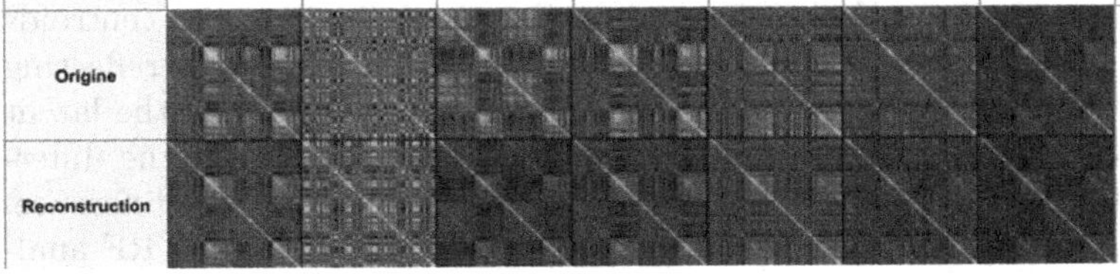 Origine								
Reconstruction								

Fig. 2. Evaluation and illustration of the reconstruction performance of the trained dVAE model using a 5-fold cross-validation.

on the trained dVAE encoder by simulating a perturbation on each connection $x_j^{(i)}$ in the input space and monitoring the impact on the latent variables [17] (Fig. 1). The simulation traverses a single connection j, $j \in [1, d]$, in a [-1, 1] range while keeping the remaining connections fixed, leading to p latent encoded vectors $z_j^{(i)} = [z_{1jk}^{(i)}, z_{2jk}^{(i)}]$, $k \in [1, p]$. Simulations from all connections return a cloud of points describing the RF. The covariance structure of these latent variables presents an ellipse shape. Based on the Pearson correlation coefficient, a confidence ellipse E ($n = 3$ standard deviation) with sorted eigenvalues $[\lambda_1, \lambda_2]$ models this structure[1]. The ellipse reduces to scalar descriptors such as the Fractional Anisotropy (FA), the Mean Diffusivity (MD), or the first eigenvector color-coded FA:

$$MD = \frac{\lambda_1 + \lambda_2}{2}, FA = \sqrt{\frac{3}{2} \frac{(\lambda_1 - MD)^2 + (\lambda_2 - MD)^2}{\lambda_1^2 + \lambda_2^2}} \quad (1)$$

One specificity of our experiments lies in the trajectories described by each $z_j^{(i)}$ set in the latent space. Trajectories are almost linear and can be parametrized by the inclination $\theta_j^{(i)}$ with the x-axis computed as the mean angle across the p samples (Fig. 1). Discretizing the set of $\theta^{(i)}$, resulting angular spreading cones describe which connections/regions relate to a trajectory in particular angular locations. This angular map associates sub-networks with preferential directions, producing effective trajectories.

2.4 Latent Space Querying: Automatic Trajectory Description Based on Dense Descriptors (VAE-VIENT)

In Sect. 2.3, descriptors are sparse over the latent space. The main problem is how to provide dense characteristics for a comprehensive representation of the latent

[1] https://carstenschelp.github.io/2018/09/14/Plot_Confidence_Ellipse_001.html.

space structure. This analysis relies on VAE for Visualizing and Interpreting the ENcoded Trajectories (VAE-VIENT) between states of consciousness. Leveraging the generative capabilities of the dVAE model, it is possible to decode the whole latent space. Let's consider a discrete grid $G \in R^2$ with $g \times g$ samples and the associated decoded dFC $\tilde{x_{ij}}$. The Pearson correlation enables the association between each $\tilde{x_{ij}}$ and the closest brain state, defined as the centroids from a k-means clustering on the dFC matrices [3,25]. Dense labeling, reflecting the brain functional reconfigurations, forms the first characteristic of the latent space structure (Fig. 3G). An associated confidence map carries out the difference between the two largest associations (Fig. 3H). The higher the difference, the more confident the model is. Inputting the decoded dFC into the RF analysis, dense angular spreading and velocity fields form innovative characteristics of the latent space structure (Fig. 3D, Fig. 4). Recall that the angular spreading characteristic depends strongly on the input dataset. A query system describes the latent space dynamic. A query takes as input a location or a trajectory in the latent space and returns a set of descriptors reflecting the regions to perturb for inflecting trajectories and perhaps restoring wakefulness.

3 Experiments

3.1 Evaluations

For model selection and stability evaluation, we perform 5-fold cross-validation on the training set. Stratification further enforces class distribution in each training split. The ground truth labels and pseudo-labels are the conditions of arousal (awake and the different anesthetics) and the brain states ranked in ascending order of similarity to structural connectivity, respectively. As suggested in [3,25], seven clusters represent the different configurations of the brain. This choice comes from a trade-off between biological assumptions and computational evaluations. We train the dVAE model on 3000 epochs and retain the weights at epoch 2500 to avoid over-fitting.

Brain States Classification Accuracy. From the trained dVAE, we compute the embeddings associated with the test set. Taking advantage of the latent space dense descriptors presented in Sect. 2.4, we infer from these locations brain states labels using nearest neighbors interpolation. The brain states classification accuracy compares these labels with the pseudo-labels of the k-means (Fig. 2). The average accuracy over the folds is high (0.7 ± 0.03) compared to the chance level (0.14). The class accuracy varies with the worst case for brain state 2 having only 13 samples support. The models with the highest 74% accuracy correspond to folds 2 and 3.

Brain States Reconstruction Quality. From the trained dVAE, we compute the decoded dFC matrices $\tilde{x_{ij}}$ associated with the test set. Rather than using the MSE as in the training loss, we prefer the Structural SIMilarity (SSIM [26])

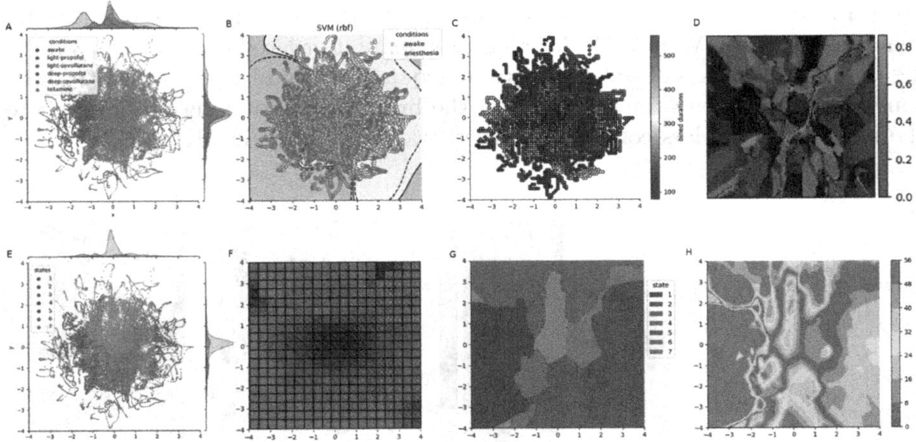

Fig. 3. Sparse latent space topography colored by A) acquisition conditions, C) lifetime duration and E) brain states. B) SVM to discriminate awake and anesthesia conditions. Dense latent space colored by D) fractional anisotropy and G) brain states. F) Reconstructed connectivity matrices. H) Confidence map associated with the brain states matching process.

between the averaged decoded brain states and the true brain states as generated with the k-means (Fig. 2). The average SSIM value is strong (0.62 ± 0.008) with high disparities between the different classes (Fig. 2). The models with the highest 0.63 SSIM again correspond to folds 2 and 3. Therefore, for the rest of the experiments, we select the trained weights associated with fold 3.

Effect of the β Regularization Parameter. We perform a grid search to determine the better choice for $\beta \in [0.5, 20]$ with the following custom steps [0.5, 1, 4, 7, 10, 20]. As presented in Sect. 3.1, monitoring the reconstruction quality with the structural similarity (SSIM) gives relatively stable performances except when β is over 10 (10% drop in the metric). The classification accuracy promotes the use of $\beta = 7$ or $\beta = 10$ (13% drop in the metric otherwise). For the rest of the paper, we use $\beta = 10$, a good trade-off imposing spatial coherence in latent space without degrading too much the reconstruction quality.

3.2 Latent Space Topography

Arousal Conditions and Brain States. We investigate the sparse composition of the latent space using the ground truth labels. We associate the arousal conditions (Fig. 3A) or the brain states labels (Fig. 3E) to the embeddings. The learned latent space is well stratified. A non-linear SVM trained on the latent representations separates the awake condition from the anesthetized conditions (Fig. 3B). The brain states are isolated while no constraint is introduced during training. State 7 corresponds to the underlying brain structure and occupies a central location in the representation space. The dense latent space descriptors, generated with the VAE-VIENT framework, exhibit ordered brain states, reflecting the different levels of wakefulness and acquisition conditions (Fig. 3F-G). The confidence map shows the reliability of the brain state matching process, where the boundaries are less confident than central locations (Fig. 3H).

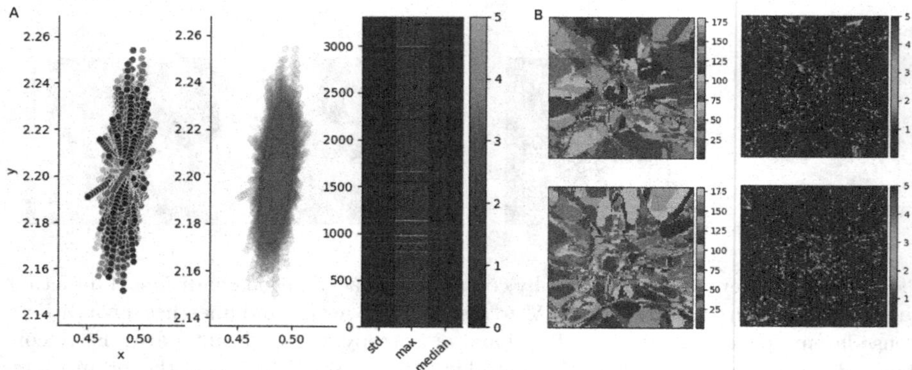

Fig. 4. Angular analysis. A) The proposed connection-wise receptive field analysis is applied to one dFC matrix $X^{(i)}$ (left). The covariance structure of the resulting latent variables presents an ellipse shape (middle). The resulting trajectories follow lines parametrized by their mean inclinations in degrees. The standard, max, and median inclination errors are plotted for each connection (right). B) We focus on two specific connections (top and bottom). We display inclinations across the whole latent space (left), and associated standard error (right). The line hypothesis holds for most latent locations, and besides, the inclination map appears smooth and regular.

Lifetime, Trajectories and Transitions. While the dVAE does not model the acquisition time course, the encoded latent variables exhibit a coherent temporal structure highlighting a successful modelization (see Appendix S2). Meta-transitions between brain configurations, as temporal transitions within a run, reveals stable periods with few jumps. The proposed dVAE generates similar trends to t-SNE without selecting dimension reduction parameters [24]. The arrangement of brain states according to lifetimes (continuous duration in one state) (Fig. 3C) and to similarity to the underlying structural connectivity (Fig. 3E) suggests a quantitative organization of the latent space. The VAE-VIENT framework represents the latent space as angular spreading (Fig. 4) and velocity fields (Fig. 3D). A request in the query engine makes possible to obtain for a given location, the MD, the FA as well as for each angle step, the connections associated with this direction, forming a strong definition of the latent space.

4 Discussion and Perspectives

In this paper, we demonstrate that the dVAE model can i) reconstruct the average pattern of each brain state with high accuracy, ii) generate a latent feature space stratified into a base of brain states, according to the experimental conditions, lifetime, and underlying structure, and iii) reconstruct new brain states coherently and stably, despite the limited dataset size, leveraging the generative part of the model. The VAE-VIENT framework provides a complete definition of the latent space in terms of wakefulness status and dynamic brain trajectories. Interestingly, the proposed dense descriptors associated with the query engine decode the latent space trajectories as a set of evolving connections/regions. The latent space is also separable into awake/anesthetized groups. This suggests restoring consciousness, by specifying a trajectory in the latent space defined by a set of evolving networks. In future work, the proposed model will be tested on a Deep Brain Stimulation (DBS) dataset where DBS is used to restore consciousness. We hypothesize that the latent space structure will be key to dissect the exact mechanisms of DBS and might help to build a general predictive model of the global brain effects of DBS.

References

1. Allen, E.A., Damaraju, E., Plis, S.M., Erhardt, E.B., Eichele, T., Calhoun, V.D.: Tracking whole-brain connectivity dynamics in the resting state. Cerebral Cortex (New York, N.Y.: 1991) **24**(3), 663–676 (2014). https://doi.org/10.1093/cercor/bhs352
2. Bakker, R., Wachtler, T., Diesmann, M.: CoCoMac 2.0 and the future of tract-tracing databases. Front. Neuroinform. **0** (2012). https://doi.org/10.3389/fninf.2012.00030
3. Barttfeld, P., Uhrig, L., Sitt, J.D., Sigman, M., Jarraya, B., Dehaene, S.: Signature of consciousness in the dynamics of resting-state brain activity. Proc. Natl. Acad. Sci. U.S.A. **112**(3), 887–892 (2015). https://doi.org/10.1073/pnas.1418031112
4. Burgess, C.P., et al.: Understanding disentangling in β-VAE. arXiv:1804.03599 [cs, stat] (2018). http://arxiv.org/abs/1804.03599
5. Cao, B., et al.: Abnormal dynamic properties of functional connectivity in disorders of consciousness. NeuroImage: Clin. **24**, 102071 (2019). https://doi.org/10.1016/j.nicl.2019.102071
6. Cunningham, J.P., Yu, B.M.: Dimensionality reduction for large-scale neural recordings. Nat. Neurosci. **17**(11), 1500–1509 (2014). https://doi.org/10.1038/nn.3776
7. Demertzi, A., Laureys, S., Boly, M.: Coma, persistent vegetative states, and diminished consciousness. In: Banks, W.P. (ed.) Encyclopedia of Consciousness, pp. 147–156. Academic Press, Oxford (2009). https://doi.org/10.1016/B978-012373873-8.00017-7
8. Demertzi, A., et al.: Human consciousness is supported by dynamic complex patterns of brain signal coordination. Sci. Adv. **5**(2), eaat7603 (2019). https://doi.org/10.1126/sciadv.aat7603

9. Gao, S., Mishne, G., Scheinost, D.: Nonlinear manifold learning in functional magnetic resonance imaging uncovers a low-dimensional space of brain dynamics. Hum. Brain Mapp. **42**(14), 4510–4524 (2021). https://doi.org/10.1002/hbm.25561
10. Gutierrez-Barragan, D., et al.: Unique spatiotemporal fMRI dynamics in the awake mouse brain. Curr. Biol. **32**(3), 631-644.e6 (2022). https://doi.org/10.1016/j.cub.2021.12.015
11. Higgins, I., et al.: beta-VAE: learning basic visual concepts with a constrained variational framework. In: ICLR (2017). https://openreview.net/forum?id=Sy2fzU9gl
12. Huang, Z., Zhang, J., Wu, J., Mashour, G.A., Hudetz, A.G.: Temporal circuit of macroscale dynamic brain activity supports human consciousness. Sci. Adv. **6**(11), eaaz0087 (2020). https://doi.org/10.1126/sciadv.aaz0087
13. James, W.: The principles of psychology, Vol I. The principles of psychology, Vol I., Henry Holt and Co, New York, NY, US (1890). https://doi.org/10.1037/10538-000
14. Kim, J.H., Zhang, Y., Han, K., Wen, Z., Choi, M., Liu, Z.: Representation learning of resting state fMRI with variational autoencoder. Neuroimage **241**, 118423 (2021). https://doi.org/10.1016/j.neuroimage.2021.118423
15. Kingma, D.P., Welling, M.: Auto-encoding variational Bayes. arXiv:1312.6114 [cs, stat] (2014). http://arxiv.org/abs/1312.6114, arXiv: 1312.6114
16. Laureys, S.: The neural correlate of (un)awareness: lessons from the vegetative state. Trends Cogn. Sci. (2005). https://doi.org/10.1016/j.tics.2005.10.010
17. Liu, R., et al.: A generative modeling approach for interpreting population-level variability in brain structure. bioRxiv p. 2020.06.04.134635 (2020). https://doi.org/10.1101/2020.06.04.134635
18. Misra, J., Surampudi, S.G., Venkatesh, M., Limbachia, C., Jaja, J., Pessoa, L.: Learning brain dynamics for decoding and predicting individual differences. PLoS Comput. Biol. **17**(9), e1008943 (2021). https://doi.org/10.1371/journal.pcbi.1008943
19. Monti, R.P., Lorenz, R., Hellyer, P., Leech, R., Anagnostopoulos, C., Montana, G.: Decoding time-varying functional connectivity networks via linear graph embedding methods. Front. Comput. Neurosci. **11** (2017). https://www.frontiersin.org/article/10.3389/fncom.2017.00014
20. Perl, Y.S., et al.: Generative embeddings of brain collective dynamics using variational autoencoders. Phys. Rev. Lett. **125**(23), 238101 (2020). https://doi.org/10.1103/PhysRevLett.125.238101
21. Preti, M.G., Bolton, T.A., Van De Ville, D.: The dynamic functional connectome: State-of-the-art and perspectives. Neuroimage **160**, 41–54 (2017). https://doi.org/10.1016/j.neuroimage.2016.12.061
22. Qiang, N., Dong, Q., Sun, Y., Ge, B., Liu, T.: Deep variational autoencoder for modeling functional brain networks and ADHD identification. In: 2020 IEEE 17th International Symposium on Biomedical Imaging (ISBI), pp. 554–557 (2020). https://doi.org/10.1109/ISBI45749.2020.9098480, iSSN: 1945-8452
23. Seninge, L., Anastopoulos, I., Ding, H., Stuart, J.: VEGA is an interpretable generative model for inferring biological network activity in single-cell transcriptomics. Nat. Commun. **12**(1), 5684 (2021). https://doi.org/10.1038/s41467-021-26017-0
24. Tseng, J., Poppenk, J.: Brain meta-state transitions demarcate thoughts across task contexts exposing the mental noise of trait neuroticism. Nat. Commun. **11**(1), 3480 (2020). https://doi.org/10.1038/s41467-020-17255-9
25. Uhrig, L., et al.: Resting-state dynamics as a cortical signature of anesthesia in monkeys. Anesthesiology **129**(5), 942–958 (2018). https://doi.org/10.1097/ALN.0000000000002336

26. Wang, Z., Bovik, A., Sheikh, H., Simoncelli, E.: Image quality assessment: from error visibility to structural similarity. IEEE Trans. Image Process. **13**(4), 600–612 (2004). https://doi.org/10.1109/TIP.2003.819861, conference Name: IEEE Transactions on Image Processing
27. Zhao, Q., Honnorat, N., Adeli, E., Pfefferbaum, A., Sullivan, E.V., Pohl, K.M.: Variational autoencoder with truncated mixture of gaussians for functional connectivity analysis. In: Chung, A.C.S., Gee, J.C., Yushkevich, P.A., Bao, S. (eds.) IPMI 2019. LNCS, vol. 11492, pp. 867–879. Springer, Cham (2019). https://doi.org/10.1007/978-3-030-20351-1_68

Conditional VAEs for Confound Removal and Normative Modelling of Neurodegenerative Diseases

Ana Lawry Aguila(✉), James Chapman, Mohammed Janahi, and Andre Altmann

University College London, London WC1E 6BT, UK
ana.aguila.18@ucl.ac.uk

Abstract. Understanding pathological mechanisms for heterogeneous brain disorders is a difficult challenge. Normative modelling provides a statistical description of the 'normal' range that can be used at subject level to detect deviations, which relate to disease presence, disease severity or disease subtype. Here we trained a conditional Variational Autoencoder (cVAE) on structural MRI data from healthy controls to create a normative model conditioned on confounding variables such as age. The cVAE allows us to use deep learning to identify complex relationships that are independent of these confounds which might otherwise inflate pathological effects. We propose a latent deviation metric and use it to quantify deviations in individual subjects with neurological disorders and, in an independent Alzheimer's disease dataset, subjects with varying degrees of pathological ageing. Our model is able to identify these disease cohorts as deviations from the normal brain in such a way that reflect disease severity.

Keywords: Unsupervised learning · VAE · Normative modelling · Confound adjustment

1 Introduction

Normative modelling is becoming a popular method to study brain disorders by quantifying how brain imaging-based measures of individuals deviate from a healthy population [16,17]. The parameters of a normative model are learned such that they characterise healthy brains from a control population and provide a statistical measure of normality. Thus, applying the normative model to a disease cohort allows for quantification of the deviation of subjects within this cohort from the norm.

Regression models such as hierarchical linear models, support vector regression, and gaussian process regression (GPR) have traditionally been used as

Supplementary Information The online version contains supplementary material available at https://doi.org/10.1007/978-3-031-16431-6_41.

L. Wang et al. (Eds.): MICCAI 2022, LNCS 13431, pp. 430–440, 2022.
https://doi.org/10.1007/978-3-031-16431-6_41

normative models [8,16,20,30]. GPRs are particularly popular for learning non-linear relationships. However, it is necessary to train one GPR for each individual brain region which does not incorporate the interactions between brain regions. Furthermore, GPRs are notoriously computationally costly.

Given advances in deep learning technology and the growing availability of large scale datasets, there have been a number of deep learning-based normative models proposed in recent years [15,22,23] that learn complex structures in the data to best capture patterns in healthy brains. These methods use autoencoders; consisting of an encoder network, which encodes the input data to a low dimension latent space, and a decoder network, which reconstructs the input data from the latent representation. These works use a metric derived from the error between original brain volumes and their reconstructions to measure deviations of disease cohorts from the norm. However, typically only a subset of brain regions are affected by a disorder and therefore the need to average across all regions reduces the sensitivity of such reconstruction-based metrics.

As far as we are aware, there has been no derivation of a subject-level quantitative measure on the latent space of autoencoder models. Making use of the parameters of a probabilistic autoencoder, we derive a robust latent deviation metric which incorporates variability in healthy subjects as well as subject level uncertainty. By forming our metric as a z-score, it is easy to interpret and admits standard statistical analysis, similarly to previous work [16,17,30].

A challenge with measuring deviation in the latent representation is that latent vectors can be driven by confounding covariates such as age and intracranial volume (ICV). In neuroimaging, the effects of confounds are commonly removed prior to analysis using confound regression [13,24]. However, this has been shown to introduce bias by removing sources of unknown variation [10]. Recently, there have been some deep learning methods proposed to derive representations invariant to confounding factors [29,31]. Dincer et al. [6] introduced the AD-AE model to remove confounds from the latent vector of an autoencoder with an adversarial loss. However, adversarial training can be difficult to optimise [4] and it is unclear whether the AD-AE model can be extended to multiple confounds. Pinaya et al. [22] used a conditional AAE to disentangle the covariate labels from the latent vectors. Although, they do not provide verification of the removal of covariates from the latent space of the AAE model. Similarly, conditional Normalizing Flows (NFs) [28] could be used for confound adjustment. However, normalizing flows have been shown to struggle with out-of-distribution data [12] which could be problematic given our aim is to perform outlier detection. Here, we explore whether we can control for confounding covariates in the latent vectors using a conditional VAE (cVAE) [26] as a normative model and derive a latent deviation metric. We extend the work of Pinaya et al. [22] and Kumar et al. [15] to introduce a model capable of modelling subtle deviations in disease cohorts whilst controlling for covariates directly in the latent space.

We make the following contributions; (1) A normative cVAE model that encodes patterns of brain variation in a healthy population and separates disease cohorts from controls. (2) We derive a subject-level latent deviation metric to

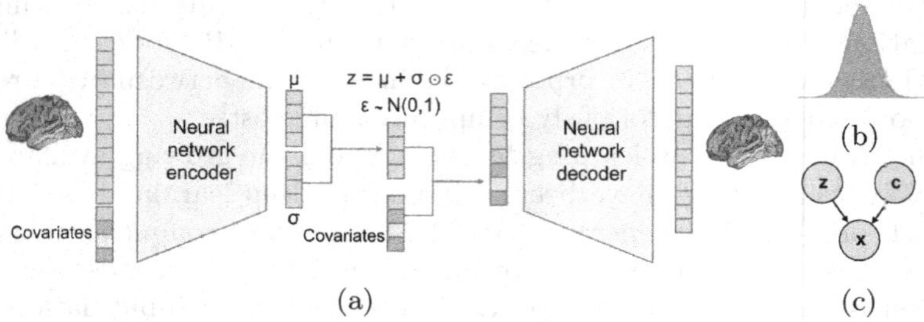

Fig. 1. (a) Schematic of the cVAE model. We used an encoder network with a linear layer of 40 nodes, a non-linear ReLU function, followed by a mean and log variance layer. A latent space size of 10 was used. The decoder consisted of two linear layers separated by a ReLU function. (b) Latent vector positions of the training data approximated by a gaussian distribution. (c) Graphical model denoting dependencies between data x, latent space z and condition vectors c.

measure deviations in the latent vectors which takes into account uncertainty in these deviations and identifies subtle deviations between disease cohorts. (3) By conditioning our model on confounds directly, we generate a latent embedding free of confounding effects thus eliminating the need to remove the effects of confounds prior to analysis.

2 Methods

Variational Autoencoder (VAE). Autoencoders are popular unsupervised machine learning models which learn low dimensional representations of complex data. The autoencoder framework consists of two mappings: the encoder which embeds information from the input space X into a latent space Z, and a decoder which transforms point estimates from the latent space back into in the input space. In the case of VAEs [11], these mappings are probabilistic and the encoder and decoder model the approximate posterior distribution $q_\phi(\mathbf{z} \mid \mathbf{x})$ with parameters ϕ and conditional likelihood distribution $p_\theta(\mathbf{x} \mid \mathbf{z})$ with parameters θ, respectively. Here, ϕ and θ are learnt by neural networks, f_e and f_d. A fixed prior distribution $p(\mathbf{z})$ over Z is assumed [25]. VAEs are trained to maximize the evidence lower bound (ELBO) [11] on the marginal data log-likelihood $p_\theta(\mathbf{x})$:

$$\mathbf{E}_{q_\phi(\mathbf{z}|\mathbf{x})}\left[\log p_\theta(\mathbf{x} \mid \mathbf{z})\right] - D_{KL}\left(q_\phi(\mathbf{z} \mid \mathbf{x}) \| p(\mathbf{z})\right) \tag{1}$$

where the first term corresponds to the reconstruction loss between the input data and the reconstruction and the second the KL divergence between the latent distribution $q(\mathbf{z} \mid \mathbf{x})$ and the prior distribution $p(\mathbf{z})$. However, calculating the gradient of the ELBO with respect to ϕ is problematic. For a gaussian encoder of the form $q_\phi(\mathbf{z} \mid \mathbf{x}) = \mathcal{N}(\mathbf{z} \mid f_e^\mu(\mathbf{x}), f_e^\sigma(\mathbf{x}))$, where the encoder f_e outputs both the mean μ and variance σ of $\mathbf{z}$ [1], we can use the reparameterization trick [11] to sample from $\mathbf{z}$ and overcome this problem.

One of the goals of VAEs is to learn a low dimensional latent space which encodes different and useful information from the input data. Whilst it has been noted that disentanglement of factors of variation in the latent space is difficult to achieve in the VAE framework [14,18], some independence of information across latent vectors is attained implicitly [25].

In this work, we use a VAE with gaussian encoder, decoder, and prior distributions trained on a healthy cohort as a normative model. The VAE encodes patterns in healthy brains by learning the parameters of the neural network layers which minimise the reconstruction loss between input data and reconstructions and the dissimilarity between the latent space and prior. The gaussian prior together with the KL divergence term, encourage the sampled $\mathbf{z}$ to be close to gaussian and thus suitable for normative modelling.

Conditional Variational Autoencoder (cVAE). Without accounting for them beforehand, the effect of confounding covariates present in the imaging data will likely be encoded in the latent space of the VAE. To avoid this problem, we must control for these covariates within our modelling framework. A conditional variational autoencoder (cVAE) [26] is a variant of the VAE which considers an additional variable $\mathbf{c}$ on which the models distributions are conditioned. This model provides a method of controlling the structure of the latent space and condition variable $\mathbf{c}$ independently. The cVAE is trained to maximise the updated ELBO:

$$\mathbb{E}_{q_\phi(\mathbf{z}|\mathbf{x},\mathbf{c})}\left[\log p_\theta(\mathbf{x} \mid \mathbf{z},\mathbf{c})\right] - D_{KL}\left(q_\phi(\mathbf{z} \mid \mathbf{x},\mathbf{c}) \| p(\mathbf{z})\right). \tag{2}$$

Deviation Metrics. For subject j, Pinaya et al. [22] assume variation from the healthy population is encoded in $\widehat{x}_{ij}$, the decoder reconstruction, and use the difference of input data x_{ij} and $\widehat{x}_{ij}$ to derive a metric in the feature space, D_{MSE}:

$$D_{\text{MSE}j} = \frac{1}{I}\sum_{i=1}^{I}\left(x_{ij} - \widehat{x}_{ij}\right)^2, \tag{3}$$

where I is the number of brain regions, $(i = 1, ..., I)$ and x_{ij} is the input value for subject j for the i-th brain region. However, by taking an average across brain regions, D_{MSE} might dilute deviation measures by incorporating MSE values from regions which do not show deviations between cohorts or reconstruction noise from an imperfect model fit. To improve performance, Pinaya et al. [22] measure deviations between bootstrap averages. However, in practice we believe it is more useful to measure deviations at the individual level.

We hypothesise that, given that the latent vectors act as an information bottleneck, a deviation metric based on the latent vectors could better separate healthy controls from disease cohorts compared to average reconstruction error metrics. Furthermore, by modelling the parameters of the encoding distribution we can easily incorporate a measure of uncertainty in the latent position of test samples. We introduce the following metric:

$$D_{\mathrm{L}j} = \frac{1}{K} \sum_{k=1}^{K} \frac{|\mu_{kj} - \bar{\mu}_k|}{\sqrt{\sigma_k^2 + \sigma_{kj}^2}} \tag{4}$$

where K is the number of latent vectors, $(k = 1, ..., K)$, $\bar{\mu}_k$ is the expected latent value from the normative model, the mean of the latent positions of the healthy controls, for latent vector k and σ_k is the corresponding variance. Variables μ_{kj} and σ_{kj} are the position along the latent vector k and encoder variance respectively of subject j. We derive this metric by approximating the point estimates of the training samples as a gaussian, treating test samples as points drawn from this distribution, and using test subject variance as a proxy for uncertainty in the latent position. In a similar approach to other normative models [7,32], we expect disease subjects to sit in the tails of the distribution.

By using a deviation metric on the latent space, we can quantify the deviation of disease subjects for each latent vector separately and thus select only vectors which encode brain patterns exhibiting deviations in disease cohorts. Hence we also introduce the following deviation metric:

$$D_{\mathrm{LS}kj} = \frac{\mu_{kj} - \bar{\mu}_k}{\sqrt{\sigma_k^2 + \sigma_{kj}^2}} \tag{5}$$

which quantifies the magnitude and direction of deviation in a latent vector k for subject j.

3 Experiments

Data Processing. We used 22,528 subjects from the UK Biobank [27](application number 65299) to train the cVAE model. Subjects were selected such that they had no neurological, psychiatric disorders or head trauma. We retained a separate set of healthy controls (HC; N=240) and subjects with multiple sclerosis (MS; N = 129), and with mania or bipolar disorder (BP; N = 96) to test the model.

For testing with an external dataset, we extracted 328 subjects from the Alzheimer's Disease Neuroimaging Initiative (ADNI)[1] [21] dataset with early mild cognitive impairment (EMCL; N = 113), late mild cognitive impairment (LMCL; N = 86), Alzheimer's disease (AD; N = 45) as well as healthy controls (HC; N = 84). T1-weighted MRIs were processes using FreeSurfer [5] and grey-matter volumes for 66 cortical (Desikan-Killiany atlas) and 16 subcortical brain

[1] Data used in preparation of this article were obtained from the Alzheimer's Disease Neuroimaging Initiative (ADNI) database (adni.loni.usc.edu). As such, the investigators within the ADNI contributed to the design and implementation of ADNI and/or provided data but did not participate in analysis or writing of this report. A complete listing of ADNI investigators can be found at: http://adni.loni.usc.edu/wp-content/uploads/how_to_apply/ADNI_Acknowledgement_List.pdf.

Table 1. Associations between latent vectors and covariates for VAE and cVAE models. We trained linear regression models with respect to the un-binned covariates. Entries highlighted in bold are significantly associated to the latent vector considering a Bonferroni adjusted p-value threshold (p = 0.005).

Dataset	Covariate	latent 1	latent 2	latent 3	latent 4	latent 5	latent 6	latent 7	latent 8	latent 9	latent 10
	VAE										
UK Biobank	age	0.343	0.584	**8.84E-16**	0.260	0.663	0.917	7.35E-3	**7.49E-05**	0.871	0.593
UK Biobank	ICV	0.250	0.578	**5.13E-08**	0.364	0.309	0.266	0.159	**3.00E-110**	**1.64E-3**	0.250
ADNI	age	0.244	0.255	**1.46E-09**	0.589	0.821	0.783	0.323	0.048	0.033	0.093
ADNI	ICV	0.101	0.122	**6.82E-12**	0.037	**2.62E-3**	0.513	0.193	**3.49E-60**	0.736	0.030
	cVAE										
UK Biobank	age	0.220	0.965	0.600	0.784	0.203	0.097	0.440	0.499	0.346	0.525
UK Biobank	ICV	0.946	0.431	0.478	0.640	0.193	0.563	0.070	0.071	0.943	0.674
ADNI	age	0.802	0.209	0.476	0.129	0.978	0.414	0.242	**2.59E-3**	0.032	0.648
ADNI	ICV	0.020	**1.27E-4**	0.022	**2.93E-3**	0.682	0.377	0.288	0.118	0.812	0.040

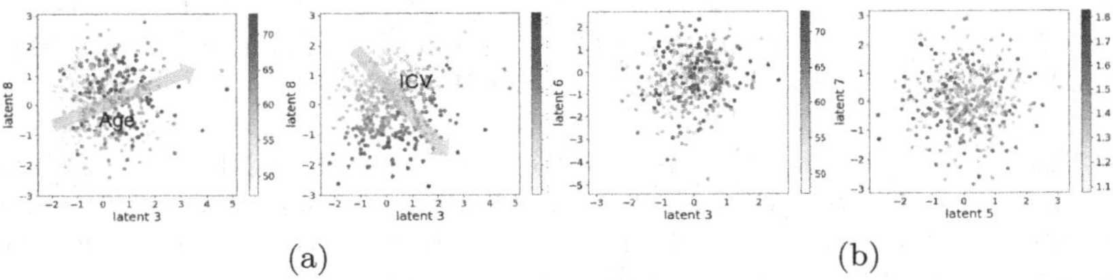

Fig. 2. Plots of two latent vectors with greatest association with covariates for the (a) VAE and (b) cVAE models coloured by age and ICV for the UK Biobank dataset. We can clearly see the cVAE latent vectors do not exhibit separation by the covariates.

regions were extracted. Subjects were restricted to ages 47–73 and ICV 0.95–2.31 and each brain region was standardised by removing the mean and dividing by the standard deviation of the UK Biobank brain regions (Fig. 2 and Table 1).

Parameter Screening. We screened different latent space dimensions (5, 10, 15 and 20) and encoding/decoding layer sizes (20, 40, 60, 80). We monitored the separation of CN vs AD. Our chosen parameters (listed in Fig. 1) gave average performance. Whilst there is room for improvement, changing parameters could lead to worse performance on other disease cohorts.

Conditioning on Confounding Variables. To illustrate the ability of the cVAE to separate the effect of covariates from the latent vectors, we trained VAE and cVAE models on the healthy control cohort from the UK Biobank. Age and ICV were both divided into 10 quantile-based bins and fed into the cVAE model as the conditional variables as one-hot-encoded vectors. We applied the trained models to two separate test sets: case/control subjects from the UK Biobank and subjects from the ADNI dataset. We found that for the association between the

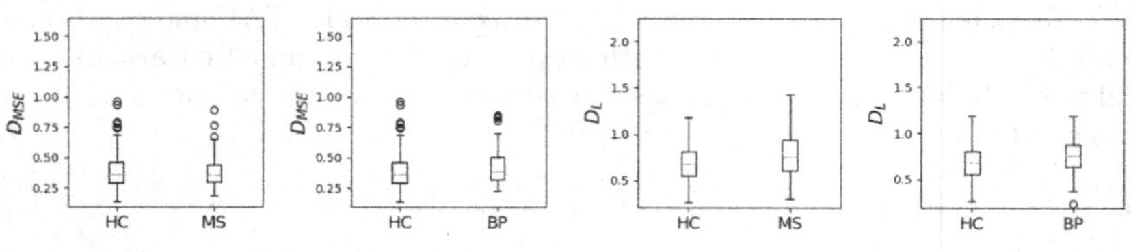

(a) D_{MSE} UK Biobank. HC vs MS; p=0.820, HC vs BP; p=0.0233

(b) D_{L} UK Biobank. HC vs MS; p=2.48E-05, HC vs BP; p=2.19E-3

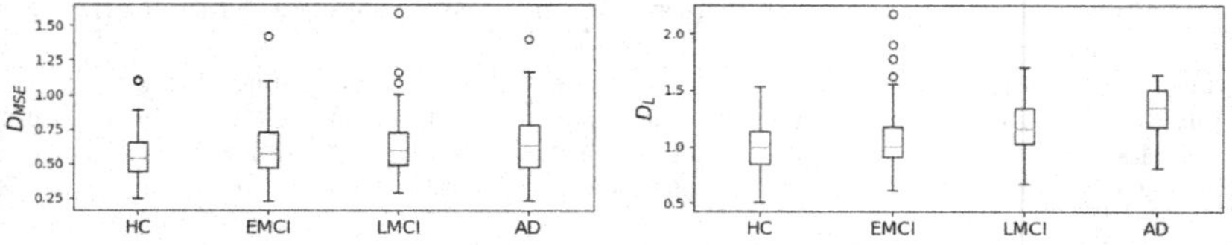

(c) D_{MSE} ADNI. HC vs EMCI; p=0.0721, HC vs LMCI; p=0.0182, HC vs AD; p=0.0176

(d) D_{L} ADNI. HC vs EMCI; p=0.098, HC vs LMCI; p=3.46E-06, HC vs AD; p=2.67E-08

Fig. 3. Box plots of the deviations for the UK Biobank and ADNI test sets. p-values for the associations between HC and the other diagnostic groups were calculated from logistic regression models. In general D_{L} proves a substantially better deviation metric than D_{MSE} evidenced by the more significant p-values.

covariates and latent vectors was greatly reduced for the cVAE model compared to VAE (Fig. 2 and Table 1).

Deviation Metric Results. Figure 3 shows the MSE deviation metric [22], D_{MSE}, and the latent deviation, D_{L}, for disease and control cohorts from the UK Biobank and ADNI test sets. In general, D_{L} showed substantially better separation between controls and disease cohorts than D_{MSE}. Both metrics reflected the increasing disease severity in the ADNI cohorts, again D_{L} outperforming D_{MSE}.

Comparison with GPR. For comparison purposes, we trained GPRs on the raw data using the PCNToolkit. We used extreme value statistics to derive a single deviation measure per subject using the approach from Marquand et al. 2016 [17]. Due to time and computational restrictions, we limited the UK Biobank training data to 1000 subjects. We generated p-values using the same approach as for Fig. 3: UK Biobank: HC vs MS p = 3.19E−3, HC vs BP p = 3.61E−3. ADNI: HC vs EMCI p = 0.0245, HC vs LMCI p = 4.22E−5, HC vs AD p = 1.67E−7. P-values are broadly not as significant as those calculated from D_{L} and training incurred significantly higher computational cost.

Mapping from Latent Deviations to Brain Deviations. Whilst one can use all latent vectors of the cVAE to reconstruct the input data and quantify brain region deviations similar to Pinaya et al. [22], D_{LS} allows for the exploration of healthy brain patterns encoded in only latent vectors which show deviation between controls and disease cohorts. This can provide an interpretation of how deviations between controls and disease subjects in the latent space give rise to deviations in the data space. For the ADNI dataset, we used D_{LS} to identify two latent vectors (5 and 6) which showed statistically significant deviation for both HC vs LMCI and HC vs AD. We passed these latent vectors through the cVAE decoder setting the remaining latent vectors and covariates to zero such that generated values reflected only the information encoded in the select vectors.

We used a Welch's t-test per brain region to test the significance of the separation between cohorts. We found that 43 brain regions showed statistically significant separation (using a Bonferroni corrected threshold $p = 0.00061$) for HC vs LMCI and 63 for HC vs AD. Figure 4 shows the average reconstructions for each cohort relative to average reconstructions for the HC cohort. With increasing disease severity, we see shrinkage relative to the average HC reconstruction most noticeably in the amygdala and hippocampus. These brain regions are amongst consistent findings in neuroimaging studies in both MCI and AD [2,3,19]. It is interesting that, despite encoding patterns learnt from healthy brains, the latent deviations reflect some pathological effects in the data space.

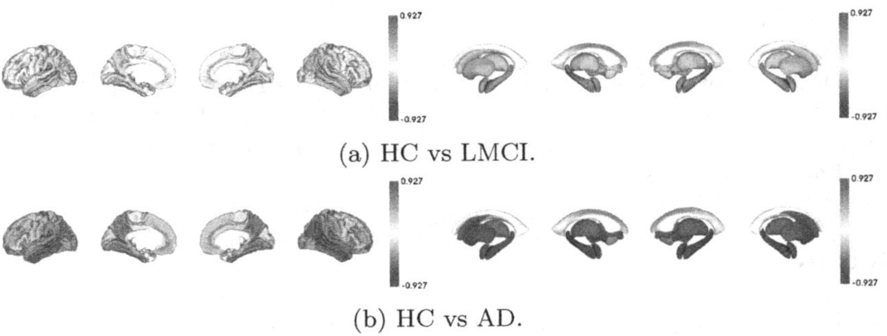

(a) HC vs LMCI.

(b) HC vs AD.

Fig. 4. Average brain region reconstructions relative to the HC cohort using a select two latent vectors. The colour scale depicts the relative reconstructed volumes with red indicating larger values and blue smaller values. (Color figure online)

4 Discussion

In this work we evaluated the performance of a normative cVAE of detecting differences between disease cohorts and healthy controls and identified brain regions associated with deviations. Similarly to previous works [15,22], our model was able to generate deviation values which reflected disease severity. Our deviation metric was better able to identify differences than a metric based on MSE highlighting the potential of this metric to detect subtle pathological effects.

We also explored the ability of the cVAE to disentangle the effects of covariates from the latent vectors. We found that there was no significance association between the latent vectors and covariates for the UK Biobank test set and substantially reduced association compared to the VAE model for the external ADNI dataset. We emphasise that most confound removal methods are not applied to previously unseen datasets [29,31] and thus the ability of the model to generalise well to an external cohort and significantly reduce the effect of covariates highlights the potential of the cVAE model for covariate adjustment. Further work could include fine-tuning a pre-trained model on the ADNI controls cohort in a transfer learning approach so that the model is adjusted to dataset differences such as processing pipelines, acquisition parameters or MRI scanner. Alternatively, further studies could adjust for differences using data harmonisation tools such as ComBat [9] or Hierarchical Bayesian Regression [10].

Deep normative models show substantial potential to the study of heterogeneous diseases. As the model is trained only using healthy controls, large cohorts of healthy subjects can be used for training alleviating the problem of effectively training complex deep learning models from small disease cohorts. In summary, here we have built on recent works [15,22,23] and derived a novel deviation metric based on the latent space of the cVAE which showed substantially increased sensitivity to detect deviations from normality in three disorders. Code and trained models are publicly available at https://github.com/alawryaguila/normativecVAE.

Acknowledgements. This work is supported by the EPSRC-funded UCL Centre for Doctoral Training in Intelligent, Integrated Imaging in Healthcare (i4health) and the Department of Health's NIHR-funded Biomedical Research Centre at University College London Hospitals.

References

1. Alemi, A.A., Fischer, I., Dillon, J.V., Murphy, K.: Deep variational information bottleneck. CoRR abs/1612.00410 (2016). http://arxiv.org/abs/1612.00410
2. Apostolova, L., et al.: Hippocampal atrophy and ventricular enlargement in normal aging, mild cognitive impairment (mci), and Alzheimer disease. Alzheimer Dis. Assoc. Disord. **26**, 17–27 (2012)
3. Baron, J.C., et al.: In vivo mapping of gray matter loss with voxel-based morphometry in mild Alzheimer's disease. NeuroImage **14**, 298–309 (2001)
4. Bojanowski, P., Joulin, A., Lopez-Paz, D., Szlam, A.: Optimizing the latent space of generative networks (2017)
5. Desikan, R.S., et al.: An automated labeling system for subdividing the human cerebral cortex on MRI scans into gyral based regions of interest. Neuroimage **31**(3), 968–980 (2006)

6. Dincer, A.B., Janizek, J.D., Lee, S.I.: Adversarial deconfounding autoencoder for learning robust gene expression embeddings. Bioinformatics **36**, 573–582 (2020)
7. Elad, D., et al.: Improving the predictive potential of diffusion MRI in schizophrenia using normative models-towards subject-level classification. Hum. Brain Mapp. **42**, 4658–4670 (2021)
8. Erus, G., et al.: Imaging Patterns of Brain Development and their Relationship to Cognition. Cerebral Cortex **25**(6), 1676–1684 (2014)
9. Johnson, W.E., Li, C., Rabinovic, A.: Adjusting batch effects in microarray expression data using empirical Bayes methods. Biostatistics **8**(1), 118–127 (2006)
10. Kia, S.M., et al.: Hierarchical Bayesian regression for multi-site normative modeling of neuroimaging data. In: Martel, A.L., et al. (eds.) MICCAI 2020. LNCS, vol. 12267, pp. 699–709. Springer, Cham (2020). https://doi.org/10.1007/978-3-030-59728-3_68
11. Kingma, D., Welling, M.: Auto-encoding variational Bayes (12 2014)
12. Kirichenko, P., Izmailov, P., Wilson, A.G.: Why normalizing flows fail to detect out-of-distribution data (2020). https://doi.org/10.48550/ARXIV.2006.08545. https://arxiv.org/abs/2006.08545
13. Kostro, D., et al.: Correction of inter-scanner and within-subject variance in structural MRI based automated diagnosing. NeuroImage **98**, 405–415 (2014)
14. Kumar, A., Sattigeri, P., Balakrishnan, A.: Variational inference of disentangled latent concepts from unlabeled observations. CoRR abs/1711.00848 (2017). http://arxiv.org/abs/1711.00848
15. Kumar, S.: Normvae: normative modeling on neuroimaging data using variational autoencoders. arXiv e-prints arXiv:2110.04903 (2021)
16. Marquand, A., Kia, S.M., Zabihi, M., Wolfers, T., Buitelaar, J., Beckmann, C.: Conceptualizing mental disorders as deviations from normative functioning. Mol. Psychiatry **24**, 1415–1424 (2019)
17. Marquand, A.F., Rezek, I., Buitelaar, J., Beckmann, C.F.: Understanding heterogeneity in clinical cohorts using normative models: beyond case-control studies. Biol. Psychiat. **80**(7), 552–561 (2016)
18. Mathieu, E., Rainforth, T., Siddharth, N., Teh, Y.W.: Disentangling disentanglement in variational autoencoders (2019)
19. Miller, M.I., et al.: Amygdala atrophy in symptomatic Alzheimer's disease based on diffeomorphometry: the biocard cohort. Neurobiol. Aging **36**, S3–S10 (2015)
20. Ordaz, S.J., Foran, W., Velanova, K., Luna, B.: Longitudinal growth curves of brain function underlying inhibitory control through adolescence. J. Neurosci. **33**(46), 18109–18124 (2013)
21. Petersen, R., et al.: Alzheimer's disease neuroimaging initiative (ADNI): clinical characterization. Neurology **74**(3), 201–209 (2010). https://doi.org/10.1212/wnl.0b013e3181cb3e25,https://europepmc.org/articles/PMC2809036
22. Pinaya, W., et al.: Using normative modelling to detect disease progression in mild cognitive impairment and Alzheimer's disease in a cross-sectional multi-cohort study. Sci. Rep. **11**, 1–13 (2021)
23. Pinaya, W., Mechelli, A., Sato, J.: Using deep autoencoders to identify abnormal brain structural patterns in neuropsychiatric disorders: a large-scale multi-sample study. Hum. Brain Map. **40**, 944–954 (2018)
24. Rao, A., Monteiro, J.M., Mourao-Miranda, J.: Predictive modelling using neuroimaging data in the presence of confounds. Neuroimage **150**, 23–49 (2017)
25. Rolinek, M., Zietlow, D., Martius, G.: Variational autoencoders pursue PCA directions (by accident) (2019)

26. Sohn, K., Lee, H., Yan, X.: Learning structured output representation using deep conditional generative models. In: Advances in Neural Information Processing Systems, vol. 28. Curran Associates, Inc. (2015)
27. Sudlow, C., et al.: UK biobank: an open access resource for identifying the causes of a wide range of complex diseases of middle and old age. PLoS Med. 12, e1001779 (2015)
28. Trippe, B.L., Turner, R.E.: Conditional density estimation with Bayesian normalising flows (2018). https://doi.org/10.48550/ARXIV.1802.04908, https://arxiv.org/abs/1802.04908
29. Wang, H., Wu, Z., Xing, E.P.: Removing confounding factors associated weights in deep neural networks improves the prediction accuracy for healthcare applications. Pacific Symposium on Biocomputing. Pacific Symposium on Biocomputing **24**, 54–65 (2019)
30. Wolfers, T., Beckmann, C.F., Hoogman, M., Buitelaar, J.K., Franke, B., Marquand, A.F.: Individual differences v. the average patient: mapping the heterogeneity in ADHD using normative models. Psychol. Med. **50**(2), 314–323 (2019)
31. Zhao, Q., Adeli, E., Pohl, K.: Training confounder-free deep learning models for medical applications. Nat. Commun. **11**, 1–9 (2020)
32. Ziegler, G., Ridgway, G., Dahnke, R., Gaser, C.: Individualized gaussian process-based prediction and detection of local and global gray matter abnormalities in elderly subjects. NeuroImage **97**, 1–9 (2014)

Semi-supervised Learning with Data Harmonisation for Biomarker Discovery from Resting State fMRI

Yi Hao Chan, Wei Chee Yew, and Jagath C. Rajapakse(✉)

School of Computer Science and Engineering, Nanyang Technological University, Singapore, Singapore
{yihao001,yeww0006,asjagath}@ntu.edu.sg

Abstract. Computational models often overfit on neuroimaging datasets (which are high-dimensional and consist of small sample sizes), resulting in poor inferences such as ungeneralisable biomarkers. One solution is to pool datasets (of similar disorders) from other sites to augment the small dataset, but such efforts have to handle variations introduced by site effects and inconsistent labelling. To overcome these issues, we propose an encoder-decoder-classifier architecture that combines semi-supervised learning with harmonisation of data across sites. The architecture is trained end-to-end via a novel multi-objective loss function. Using the architecture on multi-site fMRI datasets such as ADHD-200 and ABIDE, we obtained significant improvement on classification performance and showed how site-invariant biomarkers were disambiguated from site-specific ones. Our findings demonstrate the importance of accounting for both site effects and labelling inconsistencies when combining datasets from multiple sites to overcome the paucity of data. With the proliferation of neuroimaging research conducted on retrospectively aggregated datasets, our architecture offers a solution to handle site differences and labelling inconsistencies in such datasets. Code is available at https://github.com/SCSE-Biomedical-Computing-Group/SHRED.

Keywords: Auto-encoders · Biomarker discovery · Data harmonisation · Decoding · fMRI · Functional connectome · Semi-supervised learning

1 Introduction

Neuroimaging modalities such as functional magnetic resonance imaging (fMRI) have been extensively studied due to the potential of discovering early disease

Supplementary Information The online version contains supplementary material available at https://doi.org/10.1007/978-3-031-16431-6_42.

L. Wang et al. (Eds.): MICCAI 2022, LNCS 13431, pp. 441–451, 2022.
https://doi.org/10.1007/978-3-031-16431-6_42

biomarkers [18]. However, their analysis pose challenges due to high dimensionality and small dataset sizes [22]. It is common to face a situation where an institution has a small dataset (∼50) and wishes to generate biomarkers from it. While models can be fitted on small datasets to achieve high accuracies, their performance do not generalise across other datasets [26]. Biomarkers obtained from such models are likely to be site-specific ('overfitted') and may not be applicable across populations. While site-specific biomarkers could be useful to identify features important to a particular sub-population, it is desirable to know which of these are site-specific and which are site-invariant (i.e. generalisable features that are in the sub-population and also in other populations).

One way to increase the size of small datasets is to augment it with other datasets on the same disease. In recent years, numerous data consortiums such as the Autism Brain Imaging Data Exchange (ABIDE) initiative and ADHD-200 have been set up to collate datasets from multiple sites and release them as a single open-source data repository [7]. This has enabled many researchers to further our understanding about these complex neurological disorders [8]. Such datasets can be used to augment small datasets too. However, pooling datasets together introduces site differences (scanner variability, different inclusion/exclusion criteria) and label inconsistencies (such as varied diagnostic criteria). Conventional approaches [10,17] train a classifier on the whole dataset without checking for these issues. We hypothesise that doing so might be limiting model performance.

A recent study [11] showed that ComBat [12], which has been popularly used on bioinformatics data to remove batch effects, had been successful in removing site differences between connectivity matrices, leading to improved performance on the ABIDE dataset. Other ways to remove site differences include domain adaption techniques like adversarial alignment [13] or learning common low-rank representations [28]. However, previous works have not considered label inconsistency issues. Being a retrospective aggregation of datasets, assessment protocols used are not consistent across sites [9]. One way to mitigate this is to employ semi-supervised learning (SSL) when incorporating data from other sites. In SSL, latent patterns are learnt from unlabeled data to aid model performance by combining unsupervised and supervised losses [25]. Little research has been done on using SSL for fMRI datasets, with one notable study using resting state fMRI as unlabeled data to improve prediction tasks on task-fMRI data [3].

In this paper, we proposed an architecture, SHRED (SSL with data HaRmonisation via Encoder-Decoder-classifier), which performs SSL and data harmonisation simultaneously. Via a joint loss function, SHRED combines the optimisation of a generalised linear model for data harmonisation with the training of an encoder-decoder-classifier architecture - based on variational auto-encoder (VAE) - that performs SSL. On the ABIDE I and ADHD-200 dataset, SHRED outperformed existing approaches of using auto-encoders for SSL [2,15] or solely using ComBat for data harmonisation [11]. We found that a two-stage version of SHRED was best at improving model performance on small datasets (< 50), while SHRED performed better when more data (~ 200) is available. Biomarkers generated from our models are site-invariant and provide insights into how limitations of site-specific biomarkers generated from supervised learning approaches can be overcome. In sum, the main contributions of this study are as follows:

1. We proposed SHRED, a deep encoder-decoder-classifier architecture that removes site differences and performs SSL to combine neuroimaging data from multiple sites. SHRED outperformed existing approaches, showing the importance of addressing site differences and label inconsistencies.
2. Biomarkers generated by SHRED can help researchers disambiguate between site-specific and site-invariant biomarkers, potentially providing deeper insights into how disorders affect various sub-populations differently.

2 Methods

2.1 Datasets and Preprocessing

ABIDE I [9] contains fMRI scans of subjects with Autism Spectrum Disorder (ASD) and age-matched typical controls (TC), collected from 20 sites. Resting-state fMRI (rs-fMRI) scans from 436 TC subjects and 387 ASD subjects were used in this study. Data from ADHD-200 [14] was used to further validate our approach. It contains rs-fMRI scans of children and adolescents (7–12 years old) from 7 different sites. rs-fMRI scans from 488 typically developing individuals (TC) and 279 subjects diagnosed with Attention Deficit Hyperactivity Disorder (ADHD) were used[1]. Preprocessed rs-fMRI data [4] were downloaded from the Preprocessed Connectomes Project. Data from the C-PAC pipeline was used for ABIDE while data from the Athena pipeline was used for ADHD-200.

In the experiments carried out below, the upper triangular of functional connectivity (FC) matrices was used as the input to the models. To construct FC matrices from rs-fMRI scans, 264 sufficiently extensive and diverse regions of interest (ROI) were selected using the atlas from Power et al. [19], covering the entire cerebral cortex. Then, the mean activation time series for each ROI is extracted by averaging over voxels within a sphere of radius 2.5 mm around the ROI for each timestamp. Pearson correlation between the mean activation time series of each ROI pair were computed to derive the values in the FC matrix.

2.2 Proposed Architecture: SHRED

Let $X \in \mathbb{R}^{N \times V}$ represent a dataset with N subjects and V FC features from I sites. Subjects $S = S_l \cup S_u$ are made up of a set S_l of labelled and a set S_u of unlabelled subjects. For a subject $j \in S$, let $x_j = (x_{jv})_{v \in V}$ denote a FC vector.

[1] Sites with too few data (site CMU in ABIDE) or extreme class imbalance (sites KKI, SBL, SDSU in ABIDE; KKI, PITT, WUSTL in ADHD-200) were excluded.

Harmonising at the Encoder Input. We propose a *harmonisation module* that performs data harmonisation via a generalised linear model. Unlike more complicated alternatives like ComBat, our approach of removing site differences allows biomarkers to be easily derived via computing saliency scores [10] since the implementation is based on linear layers. For a subject j scanned at site i, let x_{ijv} be the value of functional connectivity feature v. We model x_{ijv} as

$$x_{ijv} = \alpha_v + M_{jv}\beta_v + \gamma_{iv} + \delta_{iv}\varepsilon_{ijv} \tag{1}$$

where α_v denotes an approximation to the overall mean of feature v across subjects in S, M_{jv} is a design matrix for the covariates of interest (e.g. age, gender), and β_v is a vector of regression coefficients corresponding to M_{jv}. The residual terms ε_{ijv} arise from a normal distribution with zero mean and variance σ_v^2. The terms γ_{iv} and δ_{iv} represent the additive and multiplicative site effects of site i for feature v, respectively. α_v is implemented as weights initialised as 0 and site effect is one-hot encoded and projected by a set of weights to γ_{iv} and δ_{iv}. The output of the harmonisation module is $x_{res} = \alpha_v + M_{jv}\beta_v + \varepsilon_{ijv}$ (Fig. 1).

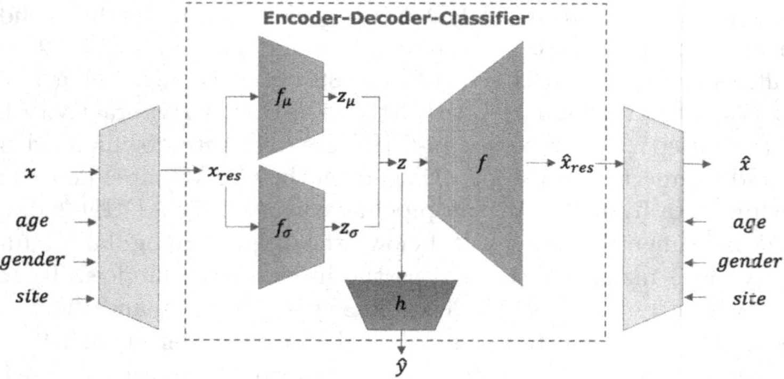

Fig. 1. Illustration of SHRED. The left grey trapezoid represents the harmonisation module, while the one on the right is its reversed form (Eq. 2). (Color figure online)

Encoder-Decoder-Classifier Architecture. The encoder-decoder-classifier (EDC) architecture is designed by combining a VAE with a vanilla deep neural network (DNN) for the classification of functional connectomes. A VAE consists of two encoders f_μ and f_σ which are used to encode the input data x_{res} by learning a normal distribution with mean z_μ and standard deviation z_σ from it. A lower dimensional hidden representation z is then produced by sampling from that distribution. A decoder f reconstructs the input data x_{res} from the hidden representation z, producing $\hat{x}_{res} = f(z)$. The DNN takes in the hidden representation z as its input to perform disease classification tasks. Letting h be the DNN mapping, the classification output is represented as $\hat{y} = \mathrm{softmax}(h(z))$.

Harmonising at the Decoder Output. Finally, $\hat{x}_{res}$ is used to generate $\hat{x}$ (reconstruction of the original input), by reversing the harmonisation module:

$$\hat{x}_{ijv} = \alpha_v + M_{jv}\beta_v + \gamma_{iv} + \delta_{iv}\hat{\varepsilon}_{ijv} \tag{2}$$

where $\hat{\varepsilon}_{ijv} = \hat{x}_{res} - \alpha_v - M_{jv}\beta_v$ is the reconstructed residual derived from $\hat{x}_{res}$.

2.3 End-to-End Training

Unlike previous work [15], we propose to train the whole EDC model in a SSL framework using a single multi-objective loss function which also optimises the weights to remove site differences. The loss function consists of 2 components: the classification loss L_C that compares the output of the classifier $\hat{y}$ with the ground truth disease labels y, and the encoder loss that compares the reconstructed input $\hat{x}$ and the real input x. The classification loss is evaluated only on labelled data whereas the encoder loss is evaluated on both labelled and unlabelled data.

The cross-entropy loss L_C is used as the classification cost:

$$L_C = -\frac{1}{|S_l|}\sum_{j\in S_l}\sum_{c=1}^{C} y_{jc}\log(\hat{y}_{jc}) \tag{3}$$

The encoder loss is made up of 3 components: L_L, L_D and L_R. The likelihood loss L_L for VAE is introduced to maximise the log likelihood of input x being sampled from Gaussian distribution with standard deviation σ_v.

$$L_L = \frac{1}{2N}\sum_{j=1}^{N}\sum_{v=1}^{V}\log(2\pi\sigma_v^2) + \frac{(x_{jv} - \hat{x}_{jv})^2}{2\sigma_v^2} \tag{4}$$

VAE minimises the Kullback-Leibler (KL) divergence L_D between the learned distribution and a Gaussian prior $p(z) \sim N(0, 1)$. This constrains the learned distribution of the hidden representation to have a mean close to 0 and a standard deviation close to 1 to prevent the model from cheating by learning distributions far away from each other for different subjects. We used L_D in the form of Evidence Lower Bound [16].

$$L_D = \frac{1}{2N}\sum_{j=1}^{N}\sum_{v=1}^{V} z_{\sigma_{jv}}^2 + z_{\mu_{jv}}^2 - 2\log(z_{\sigma_{jv}}) - 1 \tag{5}$$

The reconstruction loss L_R is added so as to force ε_{jv} to be minimised and for $\alpha_v + M_{jv}\beta_v$ to be as close as possible to $\bar{x}_{jv}$, the mean feature vector.

$$L_R = \frac{1}{N}\sum_{j=1}^{N}\sum_{v=1}^{V}\varepsilon_{jv}^2 + (\alpha_v + M_{jv}\beta_v - \bar{x}_{jv})^2 \tag{6}$$

The loss function L for training SHRED is defined below, where γ_1, γ_2 and γ_3 are hyperparameters to be decided prior to training. L is minimised via the Adam optimiser during model training.

$$L = L_C + \gamma_1 L_L + \gamma_2 L_D + \gamma_3 L_R \tag{7}$$

2.4 SHRED-II

As a deep encoder-decoder architecture, SHRED is expected to perform better when more data is available. Thus, we propose a two-step variant of SHRED to further improve model performance on very small datasets (<50): site removal is first performed by ComBat [12] (based on empirical Bayes, it is ideal for small datasets), then the harmonised data is passed to our proposed EDC architecture. In this setup, γ_3 in Eq. 7 will be set to 0.

3 Results

3.1 Model Training and Performance

Model performance is evaluated quantitatively and qualitatively (biomarkers). 5-fold cross-validation repeated over 10 seeds was done for all experiments. A fixed set of hyperparameters was used across all experiments for the same model architecture. This is arrived at by performing model tuning on a subset of ABIDE (KKI, CMU, SBL and SDSU) not used in the main experiments. Parameters tuned include number of hidden layers $\{0, 1, 2\}$ and size of hidden layer $\{16, 32, 64, 128\}$ in the VAE, dropout probability $\{0.1, 0.2, 0.3\}$, learning rate $\{5 \times 10^{-4}, 10^{-3}, 2 \times 10^{-3}, 4 \times 10^{-3}\}$ and γ_1, γ_2, γ_3 $\{10^{-5}, 10^{-4}, ..., 10^{0}\}$. Tuning was done with Optuna [1] using Bayesian optimisation for 50 iterations. Gradient descent was done using Adam optimiser with a learning rate of 2×10^{-3} and a weight decay of 10^{-3}. Dropout of 0.2 was used during training across all layers except the output layers of the encoder, decoder and classifier. For the EDC, both encoders f_μ and f_σ have 1 hidden layer with 32 neurons and an output dimension of 32 each. The decoder f has 1 hidden layer with 32 neurons. The classifier h consisted of one fully-connected layer with 2 output nodes. γ_1, γ_2 and γ_3 were set to 10^{-4}, 10^{-3} and 1 respectively.

For the experiments in Table 1, three different training settings were used. (i) Supervised Learning (SL): Models were trained using FC matrices and their disease labels on one specific site only. No unlabeled data was used. (ii) SSL: Models were trained using labeled FC matrices on one specific site and unlabeled FC matrices from the other remaining sites. (iii) SSL + Harmonisation, similar to SSL but with data harmonisation. For the comparison with existing works under SL (ASD-SAENet [2] and GAE-FCNN [15]), the original hyperparameters provided by the papers were used, except that the atlas from [19] was used.

From Table 1, it is evident that SHRED and SHRED-II outperformed existing methods. The use of SSL (without data harmonisation) led to consistent improvements in model performance over existing works [2, 15] and the baseline of DNN (SL). Performing data harmonisation simultaneously with SSL (SHRED and SHRED-II) led to even greater improvements. Notably, the performance of SHRED in Table 1 was often poor compared to SHRED-II whenever the site had very few data (see Figure S1 for subject counts), but for larger sites such as NYU in ADHD, SHRED outperformed SHRED-II (two-sample t-test p-value of 2.1×10^{-5}). On the other hand, the largest improvements from SHRED-II often occur when the size of the site's data is small (<50).

Table 1. Comparison of model performances for individual sites.

ABIDE	SL			SSL	SSL + Harmonisation	
	ASD-SAENet	GAE-FCNN	DNN	EDC-VAE	SHRED	SHRED-II
CALT	77.6 ± 4.5	65.2 ± 7.4	86.0 ± 2.9	88.6 ± 1.9	89.8 ± 2.6	**93.9 ± 1.3**
LEU1	76.5 ± 3.1	79.0 ± 2.5	80.3 ± 2.9	80.4 ± 3.9	**84.7 ± 3.7**	84.5 ± 3.0
LEU2	73.4 ± 2.8	73.0 ± 3.5	81.4 ± 2.4	86.7 ± 3.6	83.29 ± 2.7	**95.7 ± 2.3**
MAXM	74.1 ± 3.5	70.1 ± 4.2	82.8 ± 3.0	84.3 ± 3.7	**87.6 ± 2.8**	86.1 ± 3.4
NYU	70.9 ± 2.2	71.6 ± 2.0	78.3 ± 1.3	79.1 ± 1.0	77.27 ± 1.5	**79.3 ± 1.2**
OHSU	74.3 ± 3.6	77.4 ± 6.5	90.5 ± 4.8	93.0 ± 3.1	**96.4 ± 3.6**	95.4 ± 3.8
OLIN	76.8 ± 6.2	85.2 ± 3.8	88.8 ± 3.2	90.8 ± 4.6	89.2 ± 4.2	**96.0 ± 2.7**
PITT	77.9 ± 4.3	71.8 ± 4.6	84.8 ± 2.2	87.7 ± 3.0	88.4 ± 3.3	**88.5 ± 1.9**
STAN	74.5 ± 4.5	73.4 ± 3.2	84.0 ± 4.9	86.5 ± 4.2	88.4 ± 4.1	**91.0 ± 3.8**
TRIN	74.4 ± 5.5	62.8 ± 2.5	80.6 ± 2.7	83.2 ± 1.8	85.0 ± 2.5	**88.7 ± 2.5**
UCLA1	74.4 ± 2.6	72.6 ± 3.5	78.7 ± 3.6	79.8 ± 3.1	79.3 ± 4.0	**81.3 ± 3.7**
UCLA2	67.5 ± 3.5	76.5 ± 4.7	84.0 ± 7.4	88.5 ± 6.3	86.0 ± 5.2	**90.5 ± 3.7**
UM1	71.3 ± 3.1	70.7 ± 2.9	78.2 ± 3.0	81.1 ± 3.0	80.0 ± 2.8	**82.9 ± 2.7**
UM2	67.2 ± 3.0	68.5 ± 2.7	85.1 ± 3.8	87.7 ± 3.4	87.6 ± 4.3	**94.1 ± 2.7**
USM	81.3 ± 3.1	73.7 ± 2.7	89.2 ± 1.7	90.6 ± 1.6	88.2 ± 2.0	88.1 ± 2.0
YALE	71.4 ± 3.5	65.7 ± 3.4	80.5 ± 3.4	84.8 ± 2.7	83.3 ± 3.5	**88.6 ± 3.0**
ADHD						
NI	80.0 ± 3.3	60.2 ± 2.3	79.1 ± 3.0	80.2 ± 4.9	83.1 ± 4.8	**83.1 ± 4.1**
NYU	63.9 ± 0.9	63.1 ± 1.9	63.6 ± 2.1	64.4 ± 2.1	**68.7 ± 1.6**	64.4 ± 2.2
OHSU	68.4 ± 1.8	67.3 ± 2.3	70.4 ± 3.1	71.4 ± 3.0	**75.0 ± 2.9**	72.6 ± 2.8
PKU	71.4 ± 1.5	67.8 ± 1.0	74.4 ± 1.6	75.3 ± 1.8	76.4 ± 1.0	**76.5 ± 1.5**

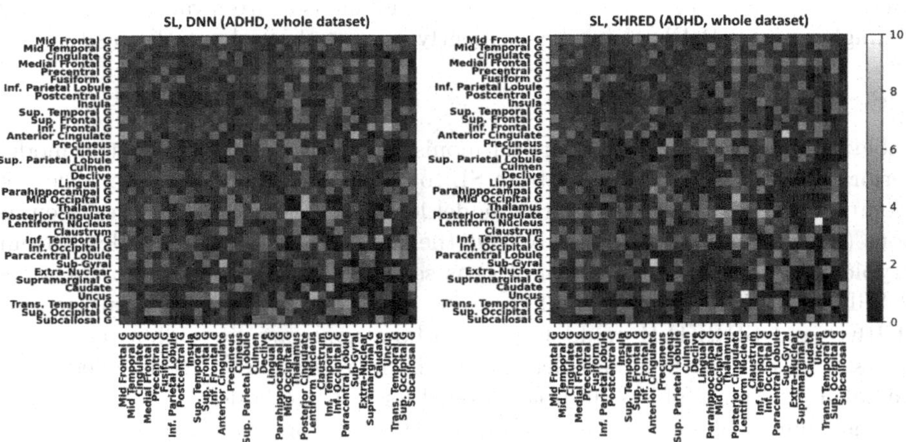

Fig. 2. Comparison of heatmaps showing saliency scores generated from decoding models trained on ADHD data. Regions with higher scores are more important.

3.2 Site-specific vs Site-invariant Biomarkers

Biomarkers were generated based on the approach described in [10], with a difference being that integrated gradients [24] was used to compute the saliency scores instead of DeepLIFT [23]. Briefly, applying such attribution methods on trained deep learning models would generate saliency scores for each feature, which is an indicator of its contribution to the classification task. This is done in 2 ways: models trained on a site and models trained on the whole dataset. Models trained via SL on a single site would produce site-specific biomarkers. The nature of biomarkers from models trained on the whole dataset is uncertain and two hypotheses are examined in this analysis: (1) DNN produces site-specific biomarkers biased towards bigger sites, (2) SHRED produces site-invariant biomarkers.

Figure 2 shows the biomarkers for models trained on the whole ADHD dataset. DNN produced 3 key biomarkers (in bright yellow): connections between uncus-inferior frontal gyrus, uncus-lentiform nucleus, posterior cingulate-thalamus. SHRED, which removes site differences, also highlighted uncus-lentiform nucleus as the most important feature, indicating that it is likely a site-invariant biomarker. The individual site heatmaps in Figure S3 (which shows site-specific biomarkers) confirms this as it remains salient throughout all 4 sites, supporting Hypothesis 2. For the other two DNN biomarkers, Figure S3 shows that uncus-inferior frontal gyrus is specific to OHSU, while posterior cingulate-thalamus is specific to sites NYU and PKU (more than half of the ADHD dataset), supporting Hypothesis 1. Site-specific analysis of ABIDE biomarkers is not shown due to space constrains but Figure S2 shows that 3 biomarkers (transverse temporal gyrus-paracentral lobule, subcallosal gyrus-subgyral, paracentral lobule-postcentral gyrus) were initially identified from the DNN heatmap, out of which transverse temporal gyrus-paracentral lobule is a site-invariant biomarker highlighted by SHRED-II (and consistently present throughout all 16 sites).

3.3 Discussion

Table 1 shows that combining data harmonisation with SSL leads to better model performance. The improvement from SL to SSL is likely due to the inclusion of the unlabelled datasets to help the model learn better intermediate representations (which would otherwise only have the small labelled dataset to learn from - typically less than 50 data samples, as seen in Fig S1), showing the benefits of SSL when dealing with small datasets. The large jump in performance for SHRED and SHRED-II could be due to how removing site differences in an SSL setting caused the unlabelled data to be more similar to the labelled data, validating the manifold assumption needed for SSL methods.

Also, our results outperformed [11], which used ComBat harmonisation with supervised learning in a two-step process. This shows the value of using SSL[2]. While SSL does not directly correct the inconsistency - rather, it removes the problem by not using labels - such an approach has already achieved good results on individual sites. Better approaches could be investigated, but would likely lead to incremental improvements from the high baseline set by SHRED.

Biomarkers generated are consistent with existing literature on ADHD (inferior frontal gyrus [29], lentiform nucleus [6], posterior cingulate [27]) and ASD

[2] In ABIDE I [9], 13 sites used clinical judgement along with the gold standard, while others used gold standards or clinical judgement only. There could be differences in clinical judgement, warranting the need to deal with label inconsistency.

(transverse temporal gyrus/Heschl's gyrus [10], paracentral lobule [5], subcallosal gyrus [20], postcentral gyrus [21]). These biomarkers were obtained by taking the average saliency scores over 50 iterations (10 seeds, 5 folds), thus enhancing their stability. Overall, SHRED provides a way to derive site-invariant biomarkers without having to train separate models on each site.

4 Conclusion

In summary, we proposed a data harmonisation and SSL framework to improve disease classification and biomarker discovery by using data from multiple sites. Two architectures were presented: SHRED and SHRED-II. SHRED-II trains the VAE with a joint loss function, in contrast to existing two-step approaches [15]. SHRED allows end-to-end training in a single stage rather than separating the data harmonisation step from the encoder-decoder training into two separate stages. SHRED-II was shown to be best at improving the model performances on very small datasets (< 50), while SHRED outperforms it when more data (~ 200) is available. Through these experiments, we demonstrated that taking site differences and label inconsistencies into consideration improves model performance. Future work on such datasets pooled together from multiple sites by data consortiums should integrate such techniques into their models. Biomarkers generated from SHRED and SHRED-II were shown to give higher emphasis on a subset of previously discovered site-specific features and these could represent site-invariant biomarkers. One could investigate clinical significance and validity of the existence of site-specific and site-invariant biomarkers in follow-up work.

References

1. Akiba, T., Sano, S., Yanase, T., Ohta, T., Koyama, M.: Optuna: a next-generation hyperparameter optimization framework. In: Proceedings of the 25th ACM SIGKDD International Conference on Knowledge Discovery & Data Mining, pp. 2623–2631 (2019)
2. Almuqhim, F., Saeed, F.: ASD-SAENet: a sparse autoencoder, and deep-neural network model for detecting autism spectrum disorder (ASD) using fMRI data. Front. Comput. Neurosci. **15**, 27 (2021)
3. Bartels, A., Blaschko, M., Shelton, J.: Augmenting feature-driven fMRI analyses: Semi-supervised learning and resting state activity. In: Advances in Neural Information Processing Systems, vol. 22 (2009)
4. Bellec, P., Chu, C., Chouinard-Decorte, F., Benhajali, Y., Margulies, D.S., Craddock, R.C.: The neuro bureau ADHD-200 preprocessed repository. Neuroimage **144**, 275–286 (2017)
5. Chen, C.P., et al.: Diagnostic classification of intrinsic functional connectivity highlights somatosensory, default mode, and visual regions in autism. NeuroImage Clin. **8**, 238–245 (2015)
6. Chen, Y., Li, G., Ide, J.S., Luo, X., Li, C.S.R.: Sex differences in attention deficit hyperactivity symptom severity and functional connectivity of the dorsal striatum in young adults. Neuroimage Rep. **1**(2), 100025 (2021)

7. Craddock, C., et al.: The neuro bureau preprocessing initiative: open sharing of preprocessed neuroimaging data and derivatives. Front. Neuroinform. **7**, 27 (2013)
8. Di Martino, A., et al.: Enhancing studies of the connectome in autism using the autism brain imaging data exchange II. Sci. Data **4**(1), 1–15 (2017)
9. Di Martino, A., et al.: The autism brain imaging data exchange: towards a large-scale evaluation of the intrinsic brain architecture in autism. Mol. Psychiatry **19**(6), 659–667 (2014)
10. Gupta, S., Chan, Y.H., Rajapakse, J.C., Initiative, A.D.N., et al.: Obtaining leaner deep neural networks for decoding brain functional connectome in a single shot. Neurocomputing **453**, 326–336 (2021)
11. Ingalhalikar, M., Shinde, S., Karmarkar, A., Rajan, A., Rangaprakash, D., Deshpande, G.: Functional connectivity-based prediction of autism on site harmonized abide dataset. IEEE Trans. Biomed. Eng. **68**(12), 3628–3637 (2021)
12. Johnson, W.E., Li, C., Rabinovic, A.: Adjusting batch effects in microarray expression data using empirical Bayes methods. Biostatistics **8**(1), 118–127 (2007)
13. Li, X., Gu, Y., Dvornek, N., Staib, L.H., Ventola, P., Duncan, J.S.: Multi-site fMRI analysis using privacy-preserving federated learning and domain adaptation: ABIDE results. Med. Image Anal. **65**, 101765 (2020)
14. Milham, M.P., Fair, D., Mennes, M., Mostofsky, S.H., et al.: The ADHD-200 consortium: a model to advance the translational potential of neuroimaging in clinical neuroscience. Front. Syst. Neurosci. **6**, 62 (2012)
15. Noman, F., et al.: Graph autoencoders for embedding learning in brain networks and major depressive disorder identification. arXiv preprint arXiv:2107.12838 (2021)
16. Odaibo, S.: Tutorial: Deriving the standard variational autoencoder (VAE) loss function. arXiv preprint arXiv:1907.08956 (2019)
17. Parisot, S., et al.: Spectral graph convolutions for population-based disease prediction. In: Descoteaux, M., Maier-Hein, L., Franz, A., Jannin, P., Collins, D.L., Duchesne, S. (eds.) MICCAI 2017. LNCS, vol. 10435, pp. 177–185. Springer, Cham (2017). https://doi.org/10.1007/978-3-319-66179-7_21
18. Parkes, L., Satterthwaite, T.D., Bassett, D.S.: Towards precise resting-state fMRI biomarkers in psychiatry: synthesizing developments in transdiagnostic research, dimensional models of psychopathology, and normative neurodevelopment. Curr. Opin. Neurobiol. **65**, 120–128 (2020)
19. Power, J.D., et al.: Functional network organization of the human brain. Neuron **72**(4), 665–678 (2011)
20. Rakić, M., Cabezas, M., Kushibar, K., Oliver, A., Llado, X.: Improving the detection of autism spectrum disorder by combining structural and functional MRI information. NeuroImage Clin. **25**, 102181 (2020)
21. Ramos, T.C., Balardin, J.B., Sato, J.R., Fujita, A.: Abnormal cortico-cerebellar functional connectivity in autism spectrum disorder. Front. Syst. Neurosci. **12**, 74 (2019)
22. Schnack, H.G., Kahn, R.S.: Detecting neuroimaging biomarkers for psychiatric disorders: sample size matters. Front. Psych. **7**, 50 (2016)
23. Shrikumar, A., Greenside, P., Kundaje, A.: Learning important features through propagating activation differences. In: International Conference on Machine Learning, pp. 3145–3153. PMLR (2017)
24. Sundararajan, M., Taly, A., Yan, Q.: Axiomatic attribution for deep networks. In: International Conference on Machine Learning, pp. 3319–3328. PMLR (2017)
25. van Engelen, J.E., Hoos, H.H.: A survey on semi-supervised learning. Mach. Learn. **109**(2), 373–440 (2019). https://doi.org/10.1007/s10994-019-05855-6

26. Varoquaux, G.: Cross-validation failure: small sample sizes lead to large error bars. Neuroimage **180**, 68–77 (2018)
27. Vogt, B.A.: Cingulate impairments in ADHD: comorbidities, connections, and treatment. Handb. Clin. Neurol. **166**, 297–314 (2019)
28. Wang, M., Zhang, D., Huang, J., Yap, P.T., Shen, D., Liu, M.: Identifying autism spectrum disorder with multi-site fMRI via low-rank domain adaptation. IEEE Trans. Med. Imaging **39**(3), 644–655 (2019)
29. Yu-Feng, Z., et al.: Altered baseline brain activity in children with ADHD revealed by resting-state functional MRI. Brain Develop. **29**(2), 83–91 (2007)

Cerebral Microbleeds Detection Using a 3D Feature Fused Region Proposal Network with Hard Sample Prototype Learning

Jun-Ho Kim, Mohammed A. Al-masni, Seul Lee, Haejoon Lee, and Dong-Hyun Kim(✉)

Department of Electrical and Electronic Engineering, Yonsei University, Seoul, Republic of Korea
donghyunkim@yonsei.ac.kr

Abstract. Cerebral Microbleeds (CMBs) are chronic deposits of small blood products in the brain tissues, which have explicit relation to cerebrovascular diseases, including cognitive decline, intracerebral hemorrhage, and cerebral infarction. However, manual detection of the CMBs is a time-consuming and error-prone process because of the sparse and tiny properties of CMBs. Also, the detection of CMBs is commonly affected by the existence of many CMB mimics that cause a high false-positive rate (FPR), such as calcification, iron depositions, and pial vessels. This paper proposes an efficient single-stage deep learning framework for the automatic detection of CMBs. The structure consists of a 3D U-Net employed as a backbone and Region Proposal Network (RPN). To significantly reduce the FPs, we developed a new scheme, containing Feature Fusion Module (FFM) that greatly detects small candidates utilizing contextual information and Hard Sample Prototype Learning (HSPL) that mines CMB mimics and generates additional loss term called concentration loss using Convolutional Prototype Learning (CPL). The proposed network utilizes Susceptibility-Weighted Imaging (SWI) and phase images as 3D input to efficiently capture 3D information. The proposed model was trained and tested using data containing 114 subjects with 365 CMBs. The performance of vanilla RPN shows a sensitivity of 93.33% and an average number of false positives per subject (FP_{avg}) of 14.73. In contrast, the proposed Feature Fused RPN that utilizes the HSPL outperforms the vanilla RPN and achieves a sensitivity of 94.66% and FP_{avg} of 0.86.

Keywords: Cerebral Microbleeds · Deep learning · Region Proposal Network · Prototype learning · Cerebral small vessel disease

1 Introduction

Cerebral Microbleeds (CMBs) are chronic deposits of small blood products in the brain tissues and are generated due to the damage of the vessel walls. CMBs usually occur close to the arteries and capillaries [1, 2]. Microbleeds are commonly detected in individuals of advancing age and patients with cerebrovascular disease [3]. Recently, CMBs have been

L. Wang et al. (Eds.): MICCAI 2022, LNCS 13431, pp. 452–460, 2022.
https://doi.org/10.1007/978-3-031-16431-6_43

reported to be related to cognitive decline, intracerebral hemorrhage, cerebral infarction, recurrence of transient ischemic attack [4, 5].

Magnetic Resonance Imaging (MRI) is the most widely utilized modality for CMBs detection. Vessel Bleeding with a small diameter of 200 um can be screened utilizing the Susceptibility-Weighted Images (SWI) generated from Gradient-Recalled Echo (GRE) MRI pulse sequences [6]. However, there are challenging factors in CMBs detection. CMBs, which are round or elliptical lesions with a size of 2–10 mm, are very sparse and small compared to the whole brain tissue [7, 8]. Another challenging factor is the presence of many CMB mimics that appear with hypointensities similar to CMBs in SWI images. For these reasons, manual detection is time-consuming, laborious, fault-prone, and the inspection results are subjective among neuroradiologists. These problems can be alleviated using an automated detector as an auxiliary tool, which assists in increasing the time-efficiency of microbleeds detection and agreement between raters [9].

Recent works on CMBs detection mainly employed Convolutional Neural Network (CNN), which extracts the effective features that differentiate between CMBs and CMB mimics effectively [10–14]. Among them, most CMBs detection approaches developed two-stage frameworks to automatically detect CMBs from MR images. The first stage is usually used for screening (i.e., potential candidate detection), while the second stage is responsible for distinguishing true CMBs and CMB mimics (i.e., false positives (FPs) reduction). For example, Dou et al. utilized a 3D fully convolutional network (3D-FCN) for the first stage and 3D-CNN for the second stage, and the framework achieved an overall sensitivity of 91.45%, false positives per subject (FP_{avg}) of 2.74 [11]. Liu et al. exploited a 3D-FRST as the first stage and a 3D residual network (3D-ResNet) as the second stage, and the cascaded network obtained an overall sensitivity of 95.24% and FP_{avg} of 1.6 [12]. Similarly, Chen et al. adopted 2D-FRST for the first stage and 3D-ResNet for the second stage, and reported an overall sensitivity of 81.9% and FP_{avg} of 11.58 [13]. Al-masni et al. utilized YOLO for the first stage and 3D-CNN for the second stage, and the cascaded framework achieved an overall sensitivity of 88.3% and FP_{avg} of 1.42 [14]. However, since the two-stage model is not trained in an end-to-end manner, there is a disadvantage that the second stage depends on how the first stage performs; especially the lost false-negative cases.

In this paper, we propose a single-stage 3D deep CNN for automatic CMBs detection. The proposed work utilizes both the SWI and phase images as 3D input to efficiently capture 3D information. The followings are the main three contributions of this work. First, we eliminated the second stage, which is used for FP reduction by integrating a 3D U-Net [15] as backbone and Region Proposal Network (RPN) of Faster R-CNN in a single end-to-end framework [16]. Second, we added Feature Fusion Module (FFM) to the model so that it can efficiently utilize contextual information. Third, we reduced the number of false positives further by applying the newly proposed Hard Sample Prototype Learning (HSPL) inspired by Convolutional Prototype Learning (CPL) [17].

2 Method

2.1 Data Acquisition and Preprocessing

We collected MR images of patients with CMBs from Gachon University Gil Medical Center. A total of 114 subjects including 365 CMBs were scanned. In 114 subjects, 23 subjects including 75 CMBs were selected for the testing, and the remaining 91 subjects including 290 CMBs composed the training dataset. The size range of CMBs is almost under 5 mm and its size limit reaches 10 mm. The subjects consisted of 59 patients with cognitively normal, 7 patients with mild cognitive impairment, and 48 patients with dementia (e.g., Alzheimer's dementia, frontotemporal dementia, and traumatic brain injury). All subjects were scanned with the following imaging parameters: echo time (TE): 20 ms; repetition time (TR): 27 ms; flip angle (FE): 15°; bandwidth/pixel (BW/pixel): 120 Hz/pixel; resolution: 0.50 × 0.50 mm^2; slice thickness: 2 mm; matrix size: 512 × 448 × 72. Expert neuroradiologists labeled this data, providing the exact coordinates of CMBs' location of each 3D MRI subject.

We trained the model using both the SWI and phase images. Before training the model, the data were preprocessed. Firstly, we applied the brain extraction tool (BET) for brain skull stripping. In addition, since there is a variance in pixel intensity for each subject, all MR images of each subject were normalized using min-max normalization, which brings all voxel intensities in the range of 0 to 1. Moreover, we applied slice interpolation to increase the number of slices in the z-direction from 72 to 224. The reason for this interpolation is to satisfy the input size and increase the resolution. The training and testing data are finally prepared by cropping the whole MR image into 128 × 128 × 128 voxels. During training, random cropped data is utilized. In the case of testing, we utilized a sliding window to infer all regions of each subject.

2.2 Network Architecture and Loss Function

The proposed single-stage CMBs detection model is trained in an end-to-end manner by incorporating the U-Net and Region Proposal Network (RPN) of Faster R-CNN. As shown in Fig. 1, our network does not expand the reduced feature map by original input size because the bounding box regression module finds the fine-tuned center of bounding boxes unlike the vanilla U-Net that expands the reduced feature map by original input size. This method reduced the computational cost. In addition, we added a FFM that involves the contextual information into the final feature map [18, 19]. This module makes the network more robust in distinguishing between CMBs and CMB mimics. The number of final output channels is:

$$N_{out} = (N_{ld} + N_{cd} + N_{cls}) \cdot N_{anchor} \tag{1}$$

where $\mathrm{N_{ld}}$, $\mathrm{N_{cd}}$, $\mathrm{N_{cls}}$, and $\mathrm{N_{anchor}}$ denote the number of lengths for each dimension, dimension of coordinate, classes, and anchor boxes. In this work, the size of bounding box is fixed to 20 × 20 × 20. Therefore, the number of lengths for each dimension is 0, and the number of anchor boxes is 1. Since the input data is three-dimensional, the number of coordinate dimension is 3, and there are two classes, CMB and non-CMB.

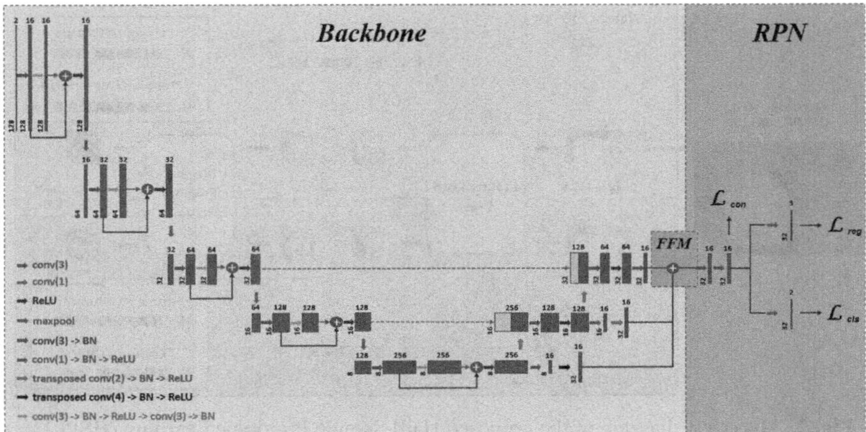

Fig. 1. The architecture of the proposed network. conv(n): n × n × n convolutional layer, BN: batch normalization, transposed conv(n): n × n × n transposed convolutional layer, maxpool(n): n × n × n max pooling layer. The L_{cls}, L_{reg}, and L_{con} are losses for classification, bounding box offset, and concentration learning, respectively.

The final loss function of our network is computed as follows:

$$L_{\text{final}} = \lambda_1 L_{\text{cls}} + \lambda_2 L_{\text{reg}} + \lambda_3 L_{\text{con}} \tag{2}$$

where L_{cls} is the classification loss implemented by focal loss [20], and L_{reg} is regression loss following the bounding box regression of YOLO-v2 [21]. The L_{con} is concentration loss for concentrating on the CMB mimics. The L_{con} is explained in detail in Sect. 2.3. The λ_1, λ_2, and λ_3 denote hyperparameters that weight the L_{reg}, L_{con}, and L_{cls}, respectively. We empirically set λ_1, λ_2, and λ_3 as 1, 0.001, 0.01.

2.3 Hard Sample Prototype Learning (HSPL)

The reason why the second stage of the two-stage models can reduce the false positives is that the second stage is trained on a dataset consisting only of CMBs and CMB mimics, excluding the background and easy samples. However, since the approach of the two-stage frameworks is not end-to-end learning, there is a disadvantage of that the second stage should be trained after finishing the training of the first stage.

To enable our model to concentrate on CMB mimics without the need for employing a second stage, we developed a Hard Sample Prototype Learning (HSPL) that mines CMB mimics and generates concentration loss during training. Firstly, as illustrated in Fig. 2, due to the sparse and tiny properties of CMBs, the HSPL crops the data based on the rule that the number of crops containing CMBs equals the crops that do not contain the CMBs. After the cropped data passes the backbone and RPN, the HSPL finds coordinates of CMB and CMB mimic using label and probability map $P \in \mathbb{R}^{d \times w \times h}$, where d, w, and h are depth, width, and height, respectively. In the case of data containing CMB, the coordinates of the CMB can be inferred directly from the ground-truth label. On

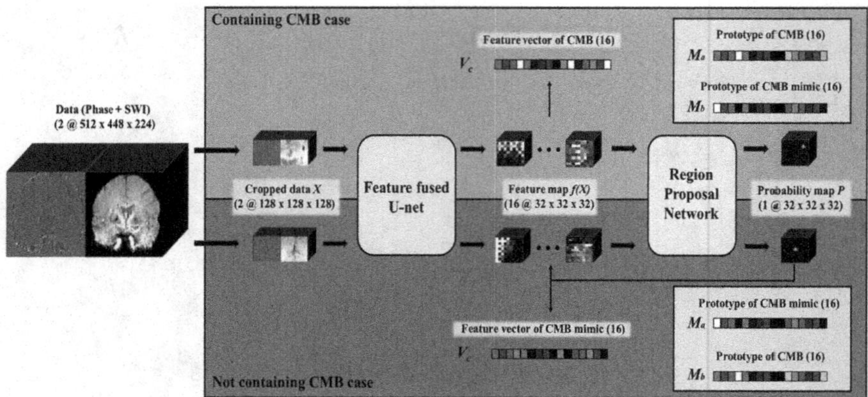

Fig. 2. Overview diagram of the proposed Hard Sample Prototype Learning (HSPL).

the other hand, in the case of data not containing CMB, it is assumed that there is a CMB mimic in the cropped data, and the point with the highest confidence score in its probability map is considered as the point where the CMB mimic is located. The final coordinate of CMB or CMB mimic is defined by:

$$c = \begin{cases} argmax_{i \in S}(P_i), & if\ X\ not\ contain\ CMB \\ i_l, & if\ X\ contain\ CMB \end{cases} \tag{3}$$

where X represents the cropped data. i and i_l denote the coordinates of the highest probability point and the coordinates of CMB, respectively. Utilizing these coordinates, the HSPL creates a feature vector $Vc \in \mathbb{R}^{N_{ch}}$ by collecting the values located at coordinates c in each channel of the feature map $f(X) \in \mathbb{R}^{N_{ch} \times \mathrm{d} \times \mathrm{w} \times \mathrm{h}}$, where N_{ch} indicates the number of feature map's channels. When the V_c is extracted from the data containing CMB, the prototype of CMB becomes M_a, and the prototype of CMB mimic becomes M_b. Conversely, when the V_c is extracted from the data not containing CMB, the prototype of CMB mimic becomes M_a, and the prototype of CMB becomes M_b. The M_a and M_b are trainable parameters, where they are automatically learned from the feature vectors V_c. The concentration loss is formulated as follows:

$$L_{\text{con}} = \frac{\|V_c - M_a\|_2^2 - \|V_c - M_b\|_2^2}{\|V_c - M_a\|_2^2 + \|V_c - M_b\|_2^2} + n \tag{4}$$

where n is the margin and set to 1 so that the concentration loss ranges from 0 to 2. The loss makes the distance between V_c and M_a closer and the distance between V_c and M_b farther in the feature space.

3 Results and Discussion

3.1 FFM Analysis

We qualitatively validated the effect of FFM on CMB detection in Fig. 3. To evaluate the feature map at each level visually, the channels of feature maps at each level are set

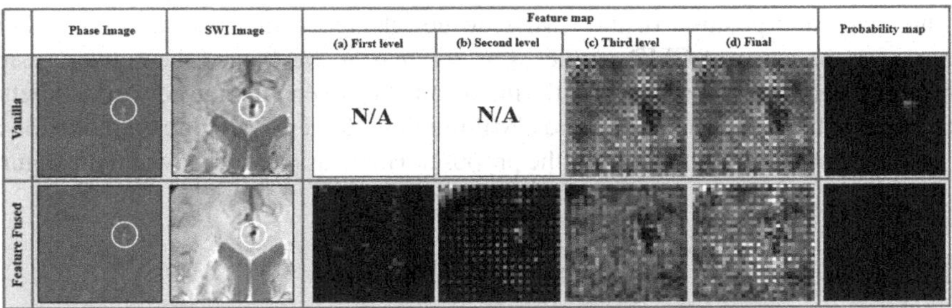

Fig. 3. The feature maps and their generated probability map of the vanilla RPN and Feature Fused RPN. The lesion in red circle is CMB mimic. (a), (b), and (c) show the feature maps from first to third levels. (d) shows the final feature map. (Colour figure online)

to 1. Since the vanilla RPN and the Feature Fused RPN were trained separately, it is difficult to compare the feature maps of the two models ideally. However, we trained both models in the same condition, and the contrast difference in the final feature map affects the probability map in both models. For vanilla RPN, the feature map of the third level becomes the final feature map. As it is shown in the probability map of vanilla RPN, the regions where the CMB mimic is located show a high probability score. In the case of the Feature Fused RPN, the final feature map is computed as an aggregation of the feature maps in three levels. As shown in Fig. 3 (a) and (b) of the Feature Fused RPN, they generate regions where the CMB mimics might exist utilizing contextual information. They are added with Fig. 3 (c) of the Feature Fused RPN, which incorrectly predicted the CMB mimic regions as the CMB regions, to produce a corrected Fig. 3 (d) of the Feature Fused RPN. Therefore, the probability scores of the Feature Fused RPN in the regions of CMB mimic get reduced compared to the vanilla RPN case.

3.2 Quantitative Comparison

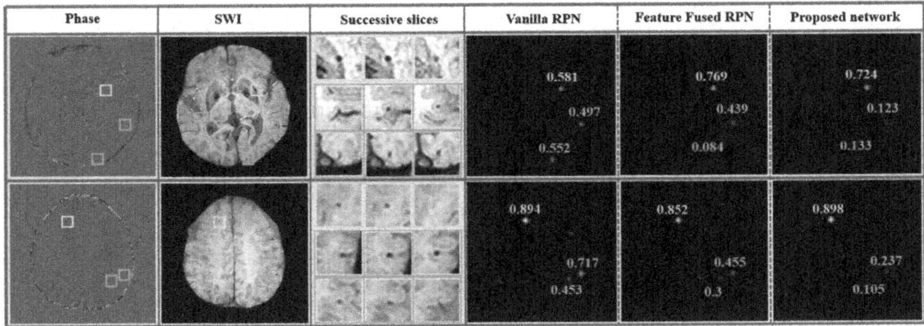

Fig. 4. The lesions in green boxes are CMBs and lesions in red boxes are CMB mimics. The values written over the probability maps indicate the probability scores.

For the quantitative evaluation, we compared the probability maps of the vanilla RPN, Feature Fused RPN, and the proposed network. As shown in Fig. 4, we observed that the shape and position of the CMBs do not change within the successive slices compared to the cases of CMB mimics. For this reason, we trained the network using 3D information. The results show that the inclusion of FFM enables the network for better differentiation between true CMBs and CMB mimics. However, significant reduction of FPs was accomplished through using the proposed HSPL as clearly shown in this figure.

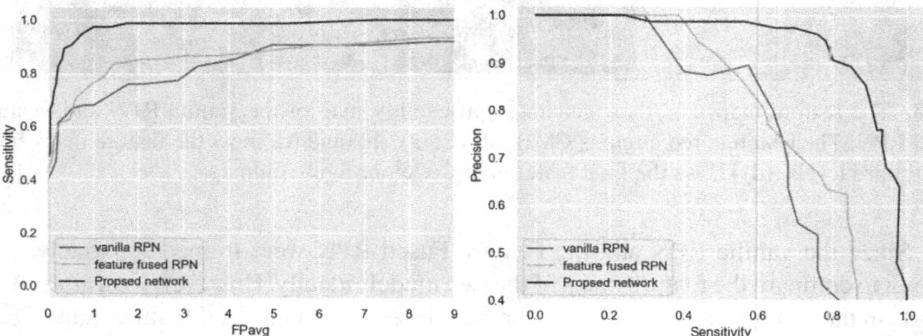

Fig. 5. The left plot shows the sensitivity vs. FP_{avg}, while the right plot presents the PR curve for vanilla RPN, feature fused RPN, and proposed network.

Quantitatively, we evaluated the performance of different models using the following measures; sensitivity, precision, false positives per subject (FP_{avg}), and area under the curve of precision-recall (AUC-PR). Figure 5 presents the sensitivity vs. FP_{avg} and PR plots for the vanilla RPN, Feature Fused RPN, and proposed network. Obviously, the proposed network significantly detected the CMBs with a true positive rate of greater than 0.95 at less than 1 FP_{avg}. The PR curve in Fig. 5 shows that the proposed network achieved the highest AUC-PR of 0.94 compared to the vanilla RPN and Feature Fused RPN of 0.75 and 0.79, respectively. These results prove the ability of the proposed HSPL in reducing the FPs significantly.

Table 1. Performance of three models at the optimal hyperparameters

Method	Sensitivity (%)	Precision (%)	FP_{avg}
Vanilla RPN	93.33	17.11	14.73
Feature fused RPN	93.33	25.64	8.82
Proposed network	94.66	78.02	0.86

As shown in Table 1, the FFM and HSPL modules provide a major improvement rates of 60.91% on Precision and 95.38% on FP_{avg}, demonstrating its ability to capture contextual information and its efficiency to concentrate hard samples for CMB mimics reduction.

3.3 Limitation and Future Work

In fact, the HSPL randomly crops the non-CMB samples from the entire MR brain image to be trained with the CMBs; however, the point with the highest confidence score in the cropped data is considered where the CMB mimic is located. For this reason (i.e., randomly cropped non-CMBs), there is a limitation that the results are varied for each training, which could lead to low reproducibility. To solve this to some extent, enough epochs is set so that the cropped data cover all regions of the entire brain image considering the ratio between the size of the cropped data and the size of the entire brain image.

Further, when using CMBs as a biomarker, not only the number of CMBs is important, but also including the anatomical location of CMBs is a significant information. It may be interesting in the future work to develop a framework that enables to inform not only the existence of CMBs, but also tell where the corresponding CMBs are located anatomically.

4 Conclusion

In this paper, we present a single-stage framework for CMBs detection using Feature Fused Module (FFM) and Hard Sample Prototype Learning (HSPL). The FFM reduced false positives by injecting contextual information into the final feature map. The HSPL allows the model to concentrate on the CMB mimics, which are hard samples, so even if the second stage for false positives reduction is eliminated, the performance gets improved compared to the two-stage models.

Acknowledgements. This research was supported by the Brain Research Program through the National Research Foundation of Korea (NRF) funded by the Ministry of Science, ICT & Future Planning (2018M3C7A1056884) and (NRF -2019R1A2C1090635), Korea Healthcare Technology R&D Project through the Korea Health Industry Development Institute (KHIDI) (HI14C1135), Korea Medical Device Development Fund grant funded by the Korea government (the Ministry of Science and ICT, the Ministry of Trade, Industry and Energy, the Ministry of Health & Welfare, Republic of Korea, the Ministry of Food and Drug Safety) (Project Number: 202011D23).

References

1. Roob, G., Schmidt, R., Kapeller, P., Lechner, A., Hartung, H.-P., Fazekas, F.: MRI evidence of past cerebral microbleeds in a healthy elderly population. Neurology **52**, 991 (1999)
2. Tajudin, A.S., Sulaiman, S.N., Isa, I.S., Karim, N.K.A.: Cerebral microbleeds (CMB) from MRI brain images. In: Proceedings of the 2016 6th IEEE International Conference on Control System, Computing and Engineering (ICCSCE), pp. 534–539. IEEE, (2016)
3. Werring, D., Coward, L., Losseff, N., Jäger, H., Brown, M.: Cerebral microbleeds are common in ischemic stroke but rare in TIA. Neurology **65**, 1914–1918 (2005)
4. Akoudad, S., et al.: Association of cerebral microbleeds with cognitive decline and dementia. JAMA Neurol. **73**, 934–943 (2016)

5. Park, M.Y., Park, H.J., Shin, D.S.: Distribution analysis of cerebral microbleeds in Alzheimer's disease and cerebral infarction with susceptibility weighted MR imaging. J. Korean Neurol. Assoc. **35**, 72–79 (2017)
6. Tanaka, A., Ueno, Y., Nakayama, Y., Takano, K., Takebayashi, S.: Small chronic hemorrhages and ischemic lesions in association with spontaneous intracerebral hematomas. Stroke **30**, 1637–1642 (1999)
7. Greenberg, S.M., et al.: Cerebral microbleeds: A guide to detection and interpretation. Lancet Neurol. **8**, 165–174 (2009)
8. Wardlaw, J.M., et al.: Neuroimaging standards for research into small vessel disease and its contribution to ageing and neurodegeneration. Lancet Neurol. **12**, 822–838 (2013)
9. Gregoire, S., et al.: The microbleed anatomical rating scale (MARS): Reliability of a tool to map brain microbleeds. Neurology **73**, 1759–1766 (2009)
10. Al-Masni, M.A., Kim, W.-R., Kim, E.Y., Noh, Y., Kim, D.-H.: A two cascaded network integrating regional-based YOLO and 3D-CNN for cerebral microbleeds detection. In: Proceedings of the 2020 42nd Annual International Conference of the IEEE Engineering in Medicine & Biology Society (EMBC), pp. 1055–1058. IEEE, (2020)
11. Dou, Q., et al.: Automatic detection of cerebral microbleeds from MR images via 3D convolutional neural networks. IEEE Trans. Med. Imaging **35**, 1182–1195 (2016)
12. Liu, S., et al.: Cerebral microbleed detection using susceptibility weighted imaging and deep learning. Neuroimage **198**, 271–282 (2019)
13. Chen, Y., Villanueva-Meyer, J.E., Morrison, M.A., Lupo, J.M.: Toward automatic detection of radiation-induced cerebral microbleeds using a 3D deep residual network. J. Digit. Imaging **32**, 766–772 (2019)
14. Al-Masni, M.A., Kim, W.-R., Kim, E.Y., Noh, Y., Kim, D.-H.: Automated detection of cerebral microbleeds in MR images: A two-stage deep learning approach. NeuroImage Clin. **28**, 102464 (2020)
15. Ronneberger, O., Fischer, P., Brox, T.: U-net: Convolutional networks for biomedical image segmentation. In: Navab, N., Hornegger, J., Wells, W.M., Frangi, A.F. (eds.) MICCAI 2015. LNCS, vol. 9351, pp. 234–241. Springer, Cham (2015). https://doi.org/10.1007/978-3-319-24574-4_28
16. Ren, S., He, K., Girshick, R., Sun, J.: Faster r-cnn: Towards real-time object detection with region proposal networks. Adv. Neural. Inf. Process. Syst. **28**, 91–99 (2015)
17. Yang, H.-M., Zhang, X.-Y., Yin, F., Liu, C.-L.: Robust classification with convolutional prototype learning. In: Proceedings of the IEEE Conference on Computer Vision and Pattern Recognition, pp. 3474–3482 (2018)
18. Cao, G., Xie, X., Yang, W., Liao, Q., Shi, G., Wu, J.: Feature-fused SSD: Fast detection for small objects. In: Proceedings of the Ninth International Conference on Graphic and Image Processing (ICGIP 2017), pp. 106151E. International Society for Optics and Photonics (2017)
19. Cao, C., et al.: An improved faster R-CNN for small object detection. IEEE Access **7**, 106838–106846 (2019)
20. Lin, T.-Y., Goyal, P., Girshick, R., He, K., Dollár, P.: Focal loss for dense object detection. In: Proceedings of the IEEE International Conference on Computer Vision, pp. 2980–2988 (2017)
21. Redmon, J., Farhadi, A.: YOLO9000: Better, faster, stronger. In: Proceedings of the IEEE Conference on Computer Vision and Pattern Recognition, pp. 7263–7271 (2016)

Brain-Aware Replacements for Supervised Contrastive Learning in Detection of Alzheimer's Disease

Mehmet Saygın Seyfioğlu(✉), Zixuan Liu, Pranav Kamath, Sadjyot Gangolli, Sheng Wang, Thomas Grabowski, and Linda Shapiro

University of Washington, Seattle, WA 98195, USA
{msaygin,zucksliu,pranavpk,sadjyotg,swang,tgrabow,shapiro}@uw.edu

Abstract. We propose a novel framework for Alzheimer's disease (AD) detection using brain MRIs. The framework starts with a data augmentation method called Brain-Aware Replacements (BAR), which leverages a standard brain parcellation to replace medically-relevant 3D brain regions in an anchor MRI from a randomly picked MRI to create synthetic samples. Ground truth "hard" labels are also linearly mixed depending on the replacement ratio in order to create "soft" labels. BAR produces a great variety of realistic-looking synthetic MRIs with higher local variability compared to other mix-based methods, such as CutMix. On top of BAR, we propose using a soft-label-capable supervised contrastive loss, aiming to learn the relative similarity of representations that reflect how mixed are the synthetic MRIs using our soft labels. This way, we do not fully exhaust the entropic capacity of our hard labels, since we only use them to create soft labels and synthetic MRIs through BAR. We show that a model pre-trained using our framework can be further fine-tuned with a cross-entropy loss using the hard labels that were used to create the synthetic samples. We validated the performance of our framework in a binary AD detection task against both from-scratch supervised training and state-of-the-art self-supervised training plus fine-tuning approaches. Then we evaluated BAR's individual performance compared to another mix-based method CutMix by integrating it within our framework. We show that our framework yields superior results in both precision and recall for the AD detection task.

Keywords: Alzheimer's disease · Magnetic resonance imaging · Brain aware · Contrastive learning

1 Introduction

Alzheimer's Disease (AD) is an irreversible neurodegenerative condition, which is characterized by atrophy of brain tissue, with distinctive microscopic changes.

Supplementary Information The online version contains supplementary material available at https://doi.org/10.1007/978-3-031-16431-6_44.

L. Wang et al. (Eds.): MICCAI 2022, LNCS 13431, pp. 461–470, 2022.
https://doi.org/10.1007/978-3-031-16431-6_44

However, AD-related atrophy is hard to detect, because healthy aging also causes some atrophy. Therefore, for a population-level impact, an abundantly available medical modality, MRI, can be used to detect the disease. Lately, deep-learning-based approaches have become common [1,2], mostly using the ADNI dataset. However, much of the early work is hard to reproduce due to data-leakage problems [3]. Thus further research is needed on the topic.

Contrastive learning has been recently shown to be a powerful technique to learn semantics-preserving visual representations in a self-supervised manner [4,5]. Based on SimCLR [5], the idea is to create two differently augmented copies (positives) of the anchor image, while considering the rest of the samples within the batch as negatives. Augmentations are a set of parametric transformations, such as random crops, rotations, etc. that aim to preserve semantics of the data while altering them. These positives are then mapped closer in the latent space, while the negatives become further away. This approach is shown to be very effective in natural images [5], as well as in some medical tasks [6,7]. However, the self-supervised contrastive approach has its drawbacks in AD detection, which is a binary classification problem. The assumption that the anchor and the rest of the batch are equally semantically different is incorrect, because it is highly likely that a batch could contain a false negative sample, thus making the training harder.

One way to fix this problem is to use supervised-contrastive learning, which leverages hard labels [8]. However, this approach has its limitations as using hard labels during pre-training exhausts the entropic capacity of labels, thus leading to sub-optimal fine-tuning performance. Soft labels could be employed during supervised contrastive training, which can be exploited to learn the relative similarity of pairs. CutMix [9] is a technique known to be very effective in creating soft labels by non-linearly combining images to create synthetic images and labels. A slightly modified version of CutMix has recently been applied in a brain lesion segmentation task [10], where instead of using an arbitrary patch for replacement, lesion-based ROIs are utilized according to the lesion location and geometry. We argue that since AD-related atrophy is widely distributed across different parts of the brain, replacing a big patch, as in [9], or focusing on a single ROI, as in [10] is not suitable for AD detection, where global understanding of the entire input MRI is essential. Furthermore, for pixel-wise aligned inputs such as ours, replacing a big patch usually creates an easier task for the model, but for pre-training the whole idea is to create difficult tasks so the model will learn more powerful representations.

We propose an augmentation technique for brain MRIs that we call Brain-Aware Replacements (BAR), which utilizes anatomically relevant regions from the Automated Anatomical Labeling Atlas (AAL) for non-linear replacements from a randomly picked MRI into an anchor MRI to produce synthetic MRIs and soft labels. Compared to CutMix, BAR produces more realistic-looking synthetic MRIs, which leads to higher local variability, thus harder-to-solve synthetic samples. On top of BAR, inspired by [11], we propose a supervised contrastive pre-training plus fine-tuning framework. However, unlike [11], our pre-training model aims to learn the relative similarity of representations, reflecting how much the mixed images have the original positives or negatives by optimizing a continuous-value-capable supervised contrastive loss [12]. This way, we do not fully exhaust the entropic capacity of our hard labels, since we only use them to create soft labels and synthetic MRI mixtures through BAR.

Our contributions are two-fold. First, we propose BAR, a novel augmentation strategy that utilizes the AAL to create realistic-looking synthetic samples and soft labels. Second, we show that training a supervised contrastive loss with the soft labels and synthetic MRIs generated through BAR leads to very powerful representation learning. We also show that the pre-trained model can be further fine-tuned utilizing the same labels that were used to create synthetic MRIs and soft labels. To the best of our knowledge, supervised mixture learning with contrastive loss has yet to be investigated, as most of the contrastive mixture learning approaches are conducted in self-supervised fashion [13,14]. Also, our work is the first application of supervised contrastive learning within AD classification research. We compare our results with a slightly modified version of CutMix by incorporating it into our framework, as well as state-of-the-art self-supervised and supervised pre-training approaches and show that our approach outperforms them on the AD-vs-cognitively-normal binary classification task. We will share our code at[1].

2 Method

Contrastive Objectives: The goal of contrastive learning [5] is to map samples to a unit hypersphere by preserving semantics, i.e., semantically similar samples are pulled together, and different ones are pushed apart. In the self-supervised approach, an anchor image X_i is augmented twice using a set of transformations, which creates two augmented views, t_1^i and t_2^i. With these two augmented views, an InfoNCE loss [15] can be calculated as follows:

$$L_{NCE} = -\sum_{i=1}^{n} log \frac{e^{\theta(t_1^i, t_2^i)}}{\frac{1}{b}\sum_{j=1}^{b} e^{\theta(t_1^i, t_2^j)}} \tag{1}$$

where θ denotes an encoder, $t_2^j|_{j \neq i}$ is a negative sample, n is the number of samples, and b is the number of samples within the batch. For each sample i, the model learns to map t_1^i, t_2^i closer while pushing the negatives further away under the assumption that they are equally different from the anchor t_1^i by optimizing the InfoNCE loss.

However, for classification problems with a small number of classes such as ours, this approach has its flaws since it is highly probable that $(t_1^i, t_2^j)_{j \neq i}$ contains false negatives, i.e., the samples that are from the same class as the anchor. To alleviate this problem, a supervised version of InfoNCE could be used as given in [8]. However, this approach requires the use of all labels during

[1] https://github.com/aldraus/BrainAwareReplacementsForAD.

pre-training, which in fact limits the model's performance during the final fine-tuning stage.

CutMix Strategy: Given a set of 3D brain MRIs $X_i|_{i=1}^n$ and their binary annotations $y_i|_{i=1}^n$ stating whether they have AD or not, it is possible to generate synthetic images $X_i^p|_{i=1}^n$ and soft labels $y_i^p|_{i=1}^n$ by transferring a 3D region from X_j into X_i, and modifying the label y_i to be a linear combination of y_i and y_j as follows [9]:

$$X_i^p = (1 - M) \odot X_i + M \odot X_j \tag{2}$$

$$y_i^p = \lambda y_i + (1 - \lambda) * y_j \tag{3}$$

Here M denotes a binary mask, and λ denotes the pixel-wise combination ratio. This process can be repeated to create a large variety of synthetic images and soft labels and is shown to be effective in natural images. For natural images, local ambiguity generally yields less optimal results, since fine-grained features are mostly localized, thus local connectivity is important. In AD detection however, the atrophy is not localized, but instead generally spread across the whole MRI, so replacing a number of smaller locally disconnected regions instead of a big patch would give more of an insight into the disease. Also, unlike natural images, our MRI data is pixel-wise aligned, thus, estimating the mixture from a big connected region is an easier problem compared to estimating the mixture when a number of smaller patches are replaced. Because the goal is to make pre-training objectives harder, it is more suitable to use a number of smaller patches when replacing parts. This way, the model is implicitly forced to have more of a global understanding.

Proposed Framework: Our framework is based on two ideas: First, to address the problems mentioned in the CutMix section, we propose Brain Aware Replacement (BAR) as an alternative augmentation strategy that non-linearly creates realistic looking mixtures within the dataset by replacing anatomically relevant 3D brain regions. BAR has some advantages over CutMix. Unlike CutMix, the generated images always look realistic, thus there is less distribution shift [16], which in turn helps network training. Also, BAR explicitly forces the model to pay attention to the relationship between medically relevant brain regions, instead of random patches provided by CutMix. Second, to alleviate the problems mentioned in the Contrastive Objectives section, we propose the use of a continuous-valued supervised contrastive objective [12] with soft labels that are produced with BAR. Inspired by [11], our framework is based on a supervised pre-training plus supervised fine-tuning approach; the overall architecture is shown in Fig. 1.

For BAR, given that $\forall i \in [1..N]$, X_i is pixel-wise aligned, in similar fashion to Eq. 2, a 3D binary mask M is generated by sampling regions from the Automated Anatomical Labeling Atlas (AAL) 2 [17], which has 62 distinct brain regions when the left and right lobes are merged. The variable λ is sampled from a left-skewed beta distribution with $\sigma = 0.2$ and $\beta = 0.8$ and is used for sampling a number of regions from the AAL to create M. A number of anatomical brain regions in proportion with λ are then carved from a randomly selected sample X_j and are replaced into the same regions of t_1^i based on Eq. 2. Equation 3 is then used to create a pseudo label y_i^p for t_1^i by linearly mixing y_i and y_j. This approach helps create a large variety of natural looking inputs, which enables the model to be further fine-tuned using the hard labels that are used to create the synthetic samples through BAR. To prevent the model from focusing on the same regions, t_1^i is further augmented by Brain Aware Masking (BAM), which fills the 20% of the anatomical brain regions that are left untouched in the swapping stage with random noise. Then, t_2^i is augmented by either in-painting [18] or out-painting [19], and local pixel shuffling [20], which are all adopted for 3D inputs. During supervised pre-training, t_1^i and t_2^i are then used to train a Siamese network. Here a continuous valued supervised contrastive loss is used as given in [12]:

$$L_{NCE}^{c} = -\sum_{k=1}^{n} \frac{\varphi(y_k^p, y_i^p)}{\sum_{j=1}^{b} \varphi(y_j^p, y_i^p)} log \frac{e^{\theta(t_1^i, t_2^k)}}{\frac{1}{b}\sum_{j=1}^{b} e^{\theta(t_1^i, t_2^j)}} \tag{4}$$

where φ denotes a distance kernel between two labels, which in our case are the soft labels of mixtures given in Eq. 3. Hence, we explicitly force our model to learn the relative similarity of augmented versions, and bring similarly mixed MRIs together by focusing on anatomically replaced brain regions. Then a 3D reconstruction (recon) objective [7] between the anchor MRI and the decoder-processed second augmented copy t_2^i (that does not contain any replacements from another MRI) is trained as follows:

$$\mathcal{L}_{recon} = \left\| t_2^i - X_i \right\|_1 \tag{5}$$

The final pre-training loss is then calculated as $\alpha_1 \mathcal{L}_{NCE}^{c} + \alpha_2 \mathcal{L}_{recon}$. We conducted a hyperparameter search based on the validation set, and found that $\alpha_1 = \alpha_2 = 1$ yields the best results.

Model Architecture. We utilize a 3D Vision Transformer (ViT) [21] as our encoder with 10 layers and 12 attention heads. Our ViT takes a 3D input volume with resolution (H, W, D) where H, W, D are each 96, and sequences it in non-overlapping flattened patches with resolutions of $16 \times 16 \times 16$. This creates $H \times W \times D/16^3 = 216$ patches for each MRI. All patches are projected into a 768-dimensional embedding space and added on the learnable positional embeddings. The learnable $[cls]$ token is added at the beginning of the sequence of embeddings, so each MRI is represented with a 217×768 dimensional matrix. Then we use multi-head self-attention and multilayer perceptron sublayers, both of which utilize layer normalization [22]. For our decoder, we use 2 layers of Convolutional Transpose layers which reconstruct the MRIs from 216×768 latent representations ($[cls]$ is not used during reconstruction). The model outputs two tensors, a reconstructed output for the recon objective and the cls token for the contrastive objective.

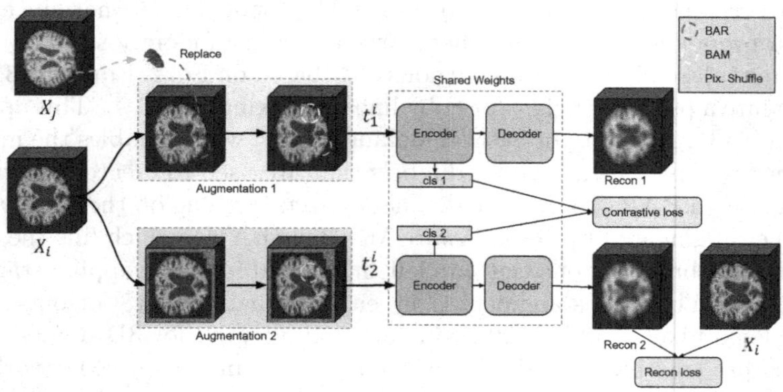

Fig. 1. Overall architecture of the proposed method. A number of anatomically-relevant 3D brain regions are taken from X_j and replaced in the same part of X_i where $j \neq i$. The replaced version, t_1^i, is followed by BAM, where another brain region is replaced with random noise. Then t_2^i undergoes inner-outer cuts and pixel shuffling, and a Siamese ViT is used to get the cls tokens for both views, which in turn are used to compute the contrastive loss. Furthermore, a decoder is used to generate an image from t_2^i (which has no replacements) and a reconstruction loss is calculated between the recon 2 and the anchor X_i.

3 Experimental Settings

Data: We used the Alzheimer's Disease Neuroimaging Initiative (ADNI) dataset[2] in this work. The ADNI was launched in 2003 as a public-private partnership, led by Principal Investigator Michael W. Weiner, MD. The primary goal of ADNI has been to test whether serial MRI, positron emission tomography, other biological markers, and clinical and neuropsychological assessment can be combined to measure the progression of mild cognitive impairment and early AD. We include all participants across the ADNI 1-2-3 cohorts that have structural T1 MRI scans divided in 246 Alzheimer's patients (AD) and 597 healthy controls (CN), which totals to 3306 MRIs. The data is first registered to the ICBM template, then skull stripping and bias field correction are conducted, and the resulting MRIs are resampled to $96 \times 96 \times 96$ voxels along the sagittal, coronal and axial dimensions. The data is split into training and testing sets that do not have overlapping subjects, which prevents data leakage [3]. The average age of subjects in the training and testing sets are roughly similar and around 77. 5-fold cross validation is conducted, and every time one of the folds is used to select the model parameters and tested on the individual test set.

Implementation Details: We used Pytorch [23] and MONAI[3] to implement our models. In all our experiments, we used the ADAM optimizer, with a learning

[2] adni.loni.usc.edu.
[3] https://monai.io/.

rate of 10^{-4} for pre training and $3 * 10^{-5}$ for fine tuning stages. $\sigma = 0.18$ is used for our RBF kernel φ. We trained our models using 4 NVIDIA RTX A4000 GPUs, having 16Gb VRAM each; a batch size of 4 is employed during pre-training due to computation limits of 3D modalities and a batch size of 12 is used during fine tuning. For fine tuning, we used 2 layers of MLP with 128 and 64 neurons that we attach on top of our $[cls]$ token and trained with binary cross entropy loss. For augmentations, we used MONAI's RandCoarseDropout and RandCoarseShuffle functions with holes = 6, spatial size = 5 for inner cut, holes = 6, spatial size = 20 for outer cut, and holes = 10 and spatial size = 5 for pixel shuffling.

Experimental Design: We compared the performance of our proposed framework against: 1) Training a model from scratch, 2) Self-Supervised pre-training + fine-tuning, 3) Modified CutMix based supervised pre-training + fine tuning. For training from scratch, we used a 3D ViT as our encoder. For the self-supervised approach, we tested with three different settings to see the individual contributions of contrastive objective, reconstruction objective and their combination. For the modified Cutmix, we replace a number of smaller 3D patches from the target MRI to the anchor MRI, instead of a single patch. For augmentations, we used inner and outer cutouts with equal probability for both augmented views, followed by pixel shuffling. Finally, we tested the performance of BAR against CutMix.

4 Experimental Results

The results for the AD vs NC task are shown in Table 1. As expected, both self-supervised and supervised pre-training outperform training from scratch. For the self-supervised approach, when trained alone, the contrastive objective yields the worst results, especially on recall. We hypothesize that this is caused by the large number of false positives due to the binary nature of the problem. Because the CN case is more abundant in the training data, that class is more affected, thus explaining the poor recall rate. In some cases, the anchor might even be pushed apart against a sample from the same subject, which is non-optimal. Interestingly, when combined with the recon objective, the contrastive objective provides a slight boost to the performance. Recon stabilizes the learning when combined with the contrastive loss (which is tricky to train as it depends on the intensity of the masking ratio from inner-outer cuts in earlier iterations, and it is quite unstable) and grants a performance boost. Finally, we compare CutMix with BAR, and see that BAR outperforms CutMix both with and without the recon objective. BAR is especially better in precision, which shows that it is better in detecting AD related atrophy. Also, in both cases, using the recon loss during pre-training yields a substantial performance boost.

Ablation Study on Brain Aware Masking in Self-supervised Case: We test the performance of BAM, (i.e., we randomly selected and filled 3D anatomical brain regions with noise) against the use of inner and outer cuts in a self-supervised manner. When fine-tuned, the performance is comparable to inner

Table 1. Fine-tuning results for AD vs. CN case; best is shown in **bold**

Framework	Method	Precision	Recall	Accuracy
No Pre Training	ViT from scratch	74.38 ± 7	85.6 ± 3.1	80.83 ± 3
Self Supervised Pre-Training + Fine Tuning	Contrastive	78.42 ± 4.5	81.18 ± 1.6	80.1 ± 1.9
	Recon	78.6 ± 5	85.57 ± 1.1	82.69 ± 2.5
	Contrastive + Recon	80.2 ± 4.1	85.77 ± 2	83.4 ± 1.7
Supervised Pre-Training +Fine Tuning	CutMIX	83.06 ± 4.8	87.08 ± 3.5	85.29 ± 2.8
	CutMIX + Recon	84.6 ± 3.8	87.9 ± 2.2	86.4 ± 1
	BAR	84.7 ± 3.3	87.6 ± 2.1	86.3 ± 1.1
	BAR + Recon	$\mathbf{86.24 \pm 3}$	$\mathbf{88.08 \pm 2.3}$	$\mathbf{87.22 \pm 0.8}$

outer cuts with an overall accuracy of 83.54 ± 1.8 when trained with Contrastive + Recon with a similar drop ratio used in inner-outer cuts.

Ablation Study on the Selection of Beta Distribution for BAR: We tried two different beta distributions for sampling brain regions in BAR, a left skewed one with parameters of beta(0.2, 0.8) and a uniform distribution with beta(1,1). We obtained an overall accuracy of 86.9 ± 1.5 with the left skewed one as opposed to 87.2 ± 1.3 with the uniform one. We argue that the replacement ratio sampled from the left skewed beta distribution makes somewhat of an easier objective with less replacements and thus is easier to solve. However, more research is needed to find the optimal replacement ratio.

Ablation Study on Further Transferability: To see how much further transferability is possible, we froze the ViT encoder in the BAR framework and trained an MLP with the cls tokens of the encoder. We obtained an accuracy of 85.2 which shows that there is further room for the same features to be used for fine tuning, as the fine-tuned model yields about 87.22. We argue that this is the case because we do not directly use our hard labels during pre-training but use them for creating soft labels and realistic looking synthetic images instead, thus their entropic capacity is not fully exhausted during the pre-training phase.

Ablation Study on Directly Using the Hard-labels During Pre-training: We also compared our soft-label supervised contrastive learning + fine-tuning approach against hard-label supervised contrastive learning + fine-tuning approach. To that end, we utilized hard-labels and no replacements during pre-training of supervised contrastive loss [8] + recon loss, using inner outer cuts and pixel shuffling for both t_1^i and t_2^i. This approach is produces lower quality embeddings compared to the soft-label approach as it yields an accuracy around 83.7% by training an MLP on top of the frozen encoder, and its fine tuning results are 84.7%.

5 Discussion and Conclusion

We proposed a new framework for AD detection that combines a novel augmentation strategy, BAR, which leverages 3D anatomical brain regions to create synthetic MRIs and labels. We showed that, when pre-trained with the synthetic samples, a continuous valued supervised contrastive loss is very effective for the AD detection task. We experimented on the public dataset ADNI and showed that our approach outperforms training from scratch as well as self-supervised approaches. Furthermore, we compared BAR with (CutMix), a popular synthetic data generation strategy into our framework, and showed that, for the AD detection task, using medically relevant brain regions is superior to replacement with arbitrary patches. For future work, We plan to expand our dataset to see how scaleable this framework is with larger datasets.

Acknowledgement. Data used in preparation of this article were obtained from the Alzheimer's Disease Neuroimaging Initiative (ADNI) database. As such, the investigators within the ADNI contributed to the design and implementation of ADNI and/or provided data but did not participate in analysis or writing of this report. A complete listing of ADNI investigators can be found at http://adni.loni.usc.edu/wp-content/uploads/how_to_apply/ADNI

References

1. Liu, S., Yadav, C., Fernandez-Granda, C., Razavian, N.: On the design of convolutional neural networks for automatic detection of Alzheimer's disease. In: Machine Learning for Health Workshop, pp. 184–201. PMLR (2020)
2. Zhao, X., Ang, C.K.E., Rajendra Acharya, U., Cheong, K.H.: Application of artificial intelligence techniques for the detection of Alzheimer's disease using structural MRI images. Biocybernet. Biomed. Eng. **41**(2), 456–473 (2021)
3. Fung, Y.R., Guan, Z., Kumar, R., Wu, J.Y., Fiterau, M.: Alzheimer's disease brain MRI classification: challenges and insights. arXiv preprint arXiv:1906.04231 (2019)
4. He, K., Fan, H., Wu, Y., Xie, S., Girshick, R.: Momentum contrast for unsupervised visual representation learning. In: Proceedings of the IEEE/CVF Conference on Computer Vision and Pattern Recognition, pp. 9729–9738 (2020)
5. Chen, T., Kornblith, S., Norouzi, M., Hinton, G.: A simple framework for contrastive learning of visual representations. In: International Conference on Machine Learning, pp. 1597–1607. PMLR (2020)
6. Zhou, Z., et al.: Models genesis: generic autodidactic models for 3D medical image analysis. In: Shen, D., et al. (eds.) MICCAI 2019. LNCS, vol. 11767, pp. 384–393. Springer, Cham (2019). https://doi.org/10.1007/978-3-030-32251-9_42
7. Tang, Y., et al.: Self-supervised pre-training of SWIN transformers for 3D medical image analysis. arXiv preprint arXiv:2111.14791 (2021)
8. Khosla, P., et al.: Supervised contrastive learning. Adv. Neural. Inf. Process. Syst. **33**, 18661–18673 (2020)
9. Yun, S., Han, D., Oh, S.J., Chun, S., Choe, J., Yoo, Y.: Cutmix: regularization strategy to train strong classifiers with localizable features. In: Proceedings of the IEEE/CVF International Conference on Computer Vision, pp. 6023–6032 (2019)
10. Zhang, X., et al.: CarveMix: a simple data augmentation method for brain lesion segmentation. In: de Bruijne, M., et al. (eds.) MICCAI 2021. LNCS, vol. 12901, pp. 196–205. Springer, Cham (2021). https://doi.org/10.1007/978-3-030-87193-2_19
11. Cao, Z., et al.: Supervised contrastive pre-training for mammographic triage screening models. In: de Bruijne, M., et al. (eds.) MICCAI 2021. LNCS, vol. 12907, pp. 129–139. Springer, Cham (2021). https://doi.org/10.1007/978-3-030-87234-2_13

12. Dufumier, B., et al.: Contrastive learning with continuous proxy meta-data for 3D MRI classification. In: de Bruijne, M., et al. (eds.) MICCAI 2021. LNCS, vol. 12902, pp. 58–68. Springer, Cham (2021). https://doi.org/10.1007/978-3-030-87196-3_6
13. Kim, S., Lee, G., Bae, S., Yun, S.-Y.: Mixco: Mix-up contrastive learning for visual representation. arXiv preprint arXiv:2010.06300 (2020)
14. Kalantidis, Y., Sariyildiz, M.B., Pion, N., Weinzaepfel, P., Larlus, D.: Hard negative mixing for contrastive learning. In: Advances in Neural Information Processing Systems, vol. 33, pp. 21798–21809 (2020)
15. Van den Oord, A., Li, Y., Vinyals, O.: Representation learning with contrastive predictive coding. arXiv e-prints, pages arXiv-1807 (2018)
16. Gontijo-Lopes, R., Smullin, S., Cubuk, E.D., Dyer, E.: Tradeoffs in data augmentation: an empirical study. In: International Conference on Learning Representations (2020)
17. Rolls, E.T., Joliot, M., Tzourio-Mazoyer, N.: Implementation of a new parcellation of the orbitofrontal cortex in the automated anatomical labeling atlas. Neuroimage **122**, 1–5 (2015)
18. DeVries, T., Taylor, G.W.: Improved, regularization of convolutional neural networks with cutout. arxiv. preprint (2017)
19. Chen, L., Bentley, P., Mori, K., Misawa, K., Fujiwara, M., Rueckert, D.: Self-supervised learning for medical image analysis using image context restoration. Med. Image Anal. **58**, 101539 (2019)
20. Kang, G., Dong, X., Zheng, L., Yang, Y.: Patchshuffle regularization. arXiv preprint arXiv:1707.07103 (2017)
21. Hatamizadeh, A., et al.: Unetr: transformers for 3d medical image segmentation. In: Proceedings of the IEEE/CVF Winter Conference on Applications of Computer Vision, pp. 574–584 (2022)
22. Ba, J.L., Kiros, J.R., Hinton, G.E.: Layer normalization. arXiv preprint arXiv:1607.06450 (2016)
23. Paszke, A., et al. Pytorch: an imperative style, high-performance deep learning library. In: Advances in Neural Information Processing Systems, vol. 32 (2019)

Heart and Lung Imaging

AANet: Artery-Aware Network for Pulmonary Embolism Detection in CTPA Images

Jia Guo[1,2], Xinglong Liu[2], Yinan Chen[2,3], Shaoting Zhang[2], Guangyu Tao[4], Hong Yu[4], Huiyuan Zhu[4], Wenhui Lei[5], Huiqi Li[1](✉), and Na Wang[2](✉)

[1] Beijing Institute of Technology, Beijing, China
huiqili@bit.edu.cn
[2] SenseTime Research, Shanghai, China
wangna2@sensetime.com
[3] West China Hospital-SenseTime Joint Lab, West China Biomedical Big Data Center, Sichuan University West China Hospital, Chengdu, China
[4] Shanghai Chest Hospital, Shanghai, China
[5] Shanghai Jiaotong University, Shanghai, China

Abstract. Pulmonary embolism (PE) is life-threatening and computed tomography pulmonary angiography (CTPA) is the best diagnostic techniques in clinics. However, PEs usually appear as dark spots among the bright regions of blood arteries in CTPA images, which can be very similar with veins that are less bright and soft tissues. Even for experienced radiologists, the evaluation of PEs in CTPA is a time-consuming and nontrivial task. In this paper, we propose an artery-aware 3D fully convolutional network (AANet) that encodes artery information as the prior knowledge to detect arteries and PEs at the same time. In our approach, the artery context fusion block (ACF) is proposed to combine the multi-scale feature maps and generate both local and global contexts of vessels as soft attentions to precisely recognize PEs from soft tissues or veins. We evaluate our methods on the CAD-PE dataset with the artery and vein vessel labels. The experimental results with the sensitivity of 78.1%, 84.2%, and 85.1% at one, two, and four false positives per scan have been achieved, which shows that our method achieves state-of-the-art performance and demonstrate promising assistance for diagnosis in clinical practice.

Keywords: Pulmonary embolism · CTPA · Pulmonary artery

1 Introduction

Pulmonary embolism (PE) is formed when a portion of a blood clot breaks off from the wall of a vein and travels through the blood stream, passes through the heart (right atrium and right ventricle), and becomes lodged in a pulmonary artery, causing a partial or complete obstruction. It was reported that the mortality rate of untreated PE is around 25% [1]. The high morbidity and mortality risk of PE shows that there are urgent needs for early and precise diagnosis. The best available diagnostic technique of PE is computed tomography pulmonary angiography (CTPA) [6], where PEs usually appear as dark

L. Wang et al. (Eds.): MICCAI 2022, LNCS 13431, pp. 473–483, 2022.
https://doi.org/10.1007/978-3-031-16431-6_45

spots among the bright regions of blood arteries. Unfortunately, PE remains high false negative among diagnoses, in part due to inconspicuous gray scale feature, physician fatigue, and diagnostic error [18]. Clinical studies suggest that by using computer-aided diagnosis system (CAD) to detect PE as a second opinion, radiologists can improve their sensitivity with a minimal decrease in specificity [5].

PE detection methods usually have two steps: generating PE candidates and eliminating false positives. As representatives, Liang et al. [12] and Bouma et al. [2] extract hand-crafted features of candidate points and use classification techniques to evaluate them. In recent years, the emerging technologies of deep learning, especially convolutional neural networks (CNN), have enabled CAD algorithms to provide fast and accurate diagnose. Huang et al. [9] proposed a 3D CNN for identifying PE, which treats the problem as a classification task but accurate location cannot be provided. Tajbakhsh et al. [20] used CNNs and a new vessel-aligned multi-planar image representation to reduce false positives in PE detection. Similarly, Lin et al. [13] also included vessel context information around the candidate point to improve detection performance; however, it was evaluated on the self-labeled test set because the official test set was not released at the time. Recently, Cano-Espinosa et al. [3] proposed a 2.5D network trained with the extended training set of CAD-PE, marking the state-of-the-art performance. Both methods in [20] and [13] take advantage of vessel information and align vessels in their second stage of false positive reduction network. However, such methods are mostly heuristic and cannot be trained end-to-end. In addition, they treat artery and vein equivalently because the vessels are segmented by intensity thresholding. Meanwhile, in CTPA images, some uneven and dark veins and lymph tissues can be similar to PE regions as they are not developed by contrast medium [15]. Therefore, most previous methods would generate false-positive detections that are not in an artery.

In this study, we propose an artery-aware 3D fully convolutional network (AANet) to segment and detect PE in CTPA images, which fully utilizes artery contexts. The network predicts artery and PE at the same time and encodes artery segmentations as an in-network prior knowledge for PE segmentation. We employ artery context fusion block (ACF) to combine the multi-scale feature maps and generate contexts of vessels as soft attentions to recognize PEs from soft tissues or veins precisely by giving more attention on artery regions. The network is trained with sampled image patches. We introduce the even-dice loss to avoid gradient exploding of the dice loss when there is no foreground (PE) in an image patch. In addition, we show that pre-training AANet on a larger medical dataset [19] can provide better performance to address the scarcity of the training set.

The contributions of this study can be summarized as:

- An artery-aware network (AANet) to segment PE in CTPA image that fully utilizes artery context is proposed.
- An artery context fusion block (ACF) generating the context of artery as in-network prior knowledge to guide PE prediction is developed.
- An even-dice loss is introduced to avoid gradient exploding when there is no foreground in an image patch during training.

We evaluate our methods on the CAD-PE dataset [7] consisting of 71 and 20 CTPA images in the training and testing set respectively. For supervising the artery segmentation of AANet, we annotate the artery and vein segmentation for all images in CAD-PE by a semi-automatic method. The vessel segmentation label is made public. Experimental result shows our proposed methods achieved state-of-the-art performance with the sensitivity of 78.1%, 84.2%, and 85.1% at one, two, and four false positives per scan at 2mm tolerance margin. We also show that the proposed AANet trained by PE and artery label can be used as a pre-trained model for transfer learning on PE datasets without artery label.

2 Methodology

The architecture of our AANet can be divided into of three parts: the U-Net shaped backbone, the artery context fusion block (ACF), and the PE segmentation head, as illustrated in Fig. 1. The U-Net backbone is used to extract image features. The ACF fuses the extracted features, generates artery predictions, and utilizes the artery prediction as soft attention to repaint the features in order to focus on artery regions. Finally, the PE segmentation head generates PE predictions using the repainted features.

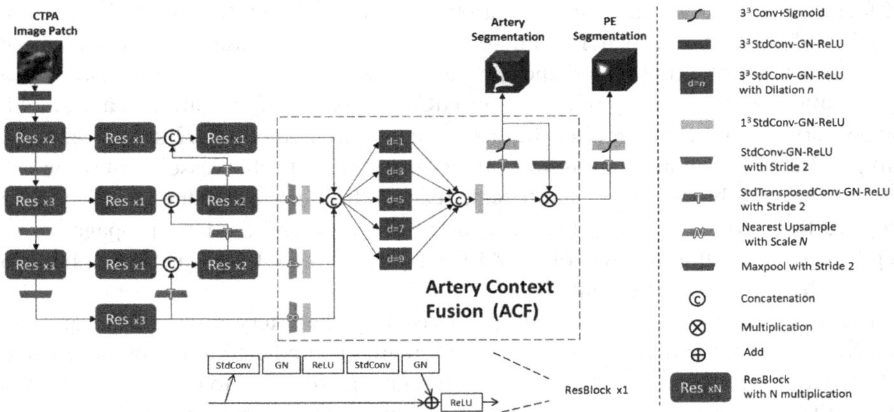

Fig. 1. Architecture of our artery-aware network (AANet).

2.1 AANet

It is non-trivial to distinguish PEs from vessels and other soft tissues, especially for those located in distal and thin artery branches. Therefore, the network must be powerful enough to capture and encode the features and difference between factual PEs and other candidates. The architecture of our network is shown in Fig. 1. The network is composed of a U-Net-like [22] backbone to capture features, a context feature fusion block to fuse the extracted features and encode vessel information, and a segmentation head to generate predictions. We adopt weight standardization [17] and group normalization [21] instead of batch normalization [10] as they are more friendly to small batch size. The backbone starts with a 3^3 Conv-GN-ReLU ($3 \times 3 \times 3$ standardized Conv followed by GroupNorm and ReLU, the same for all following) as stem layer and a down-sampling 3^3 Conv-GN-ReLU with 2 as stride. The encoder has four layers, each of which consists of two or three residual blocks [8]. A down-sampling 2^3 Conv-GN-ReLU with 2 as stride halves the spatial size at the end of each encoder layer except the last. The decoder has three layers, each of which consists of an up-sampling 2^3 TransposedConv-GN-ReLU followed by an encoder-decoder skip connection and several residual blocks. We replace the typical skip connection with a residual block as a bridge layer to alleviate the semantic gap between the encoder and decoder.

We propose ACF block to generate artery context and use the soft artery segmentation as an in-network prior knowledge for PE segmentation. In CTPA images, arteries are supposed to be brighter than veins in HU scale because of the contrast medium; however, they may look similar and should be distinguished by relative position in case of bad angiography quality. Inspired by ASPP [4], we use atrous (dilated) convolutions to combine the feature maps of the different layers to improve the receptive field needed for artery and vein identification. The feature maps with 1/4, 1/8, and 1/16 spatial size in U-Net backbone are up-sampled to 1/2 spatial size and squeezed to the same channels of the 1/2 feature maps by a 1^3 Conv-GN-ReLU. Then, they are concatenated with the 1/2 feature map and go through five paralleled dilated Conv-GN-ReLU with 1, 3, 5, 7, and 9 dilation rate. A final 1^3 Conv-GN-ReLU is used to fuse the output of these dilated convolutions. After fusion, the feature is used to predict artery segmentation by a segmentation head consisting of a TransposedConv-GN-ReLU, a Conv, and a Sigmoid. The soft artery segmentation after Sigmoid is then max-pooled to 1/2 spatial size and multiplied on the feature map as soft attention to integrate the vessel context in the feature. The ACF block utilizes multi-scale features from the backbone to enhance the receptive field, which can also facilitate the information and gradient propagating to deep layers. The output features of the ACF are passed to the PE segmentation head to generate PE segmentation output.

The artery segmentation is jointly supervised by the artery ground-truth and the backward gradient from PE segmentation. If a region is predicted to be non-artery, the multiplication operation will also force the PE segmentation head to predict non-PE. On the other hand, the PE ground-truth guides the artery segmentation head to predict arteries in PE regions. Therefore, the ACF block is robust to vessel label noise that an artery is wrongly annotated as vein. This joint learning can help the network to distinguish artery from vein and focus on detecting PE in artery region. It is noted that the artery segmentation ground-truth is only required for training, and will not be needed during inference.

2.2 Even-Dice Loss

The standard dice loss computes the loss for every sample (image patch) in the mini-batch independently and then averages the loss over the batch, which is referred to as sample-dice [16]. However, small errors in patches with no foreground can cause large gradients and dominate the parameter updates during training [11]. Image patches with no foreground are helpful when training a PE detection network because most patches of the sliding window during inference have no foreground. Batch-dice loss [11] that computes the dice value over all images in the batch (pretending they are one large image) is used to alleviate the problem of gradient exploding. However, it reduces the contribution of small PE when the same batch contains large PE as the dice value is directly computed batch-wise. In this study, we introduce even-dice loss (EDiceLoss) that combines batch-dice loss (BDiceloss) and positive-sample-dice loss (PSDiceLoss), as:

$$\text{BDiceL}oss = 1 - \frac{\sum_{i=1}^{N\times D^3} p_i g_i + \epsilon}{\sum_{i=1}^{N\times D^3} p_i^2 + g_i^2 + \epsilon}, \tag{1}$$

$$PSDiceLoss = 1 - \frac{1}{\sum_{n=1}^{N}\sum_{i=1}^{D^3} g_i > 0}\sum_{n=1}^{N}\left(\frac{\sum_{i=1}^{D^3} p_i g_i}{\sum_{i=1}^{D^3} p_i^2 + g_i^2}\cdot\sum_{i=1}^{D^3} g_i > 0\right), \tag{2}$$

$$\text{EDiceL}oss = \frac{\text{BDiceL}oss + PSDiceLoss}{2}, \tag{3}$$

where N, D, p_i, g_i and ϵ are batch size, image width, prediction, ground truth, and smoothing term, respectively. In even-dice loss, the batch-dice loss term helps to stabilize the training by reducing the harmful gradient from patches with no foreground, while positive-sample dice loss that only operates on the patches with foreground helps to focus on PEs within an image patch.

2.3 Training and Inference Details

In preprocessing, the CTPA images are isotropically resampled to [1.0, 1.0, 1.0] mm, and min-max normalized to 0–1 using the window of [–100, 500] Hounsfield Units (HU). After that, the lung area is cropped by lung segmentation.

In training, image patches with the size of 96^3 are sampled from the whole CT scan. The sampling probability for a patch to contain at least one PE is set to 75%. The training batch size is 16, consisting of 2 patches $\times$ 8 scans. We use the weighted sum of the even-dice loss of artery segmentation and PE segmentation as the total training loss. The weight for artery and PE is set to 0.25 and 0.75 respectively. AdamW [14] optimizer is employed with weight decay of 1e-4. The network is trained for 500 epochs costing about 8 h, and the network after the last epoch is used for evaluation. The initial learning rate is 1e-4 and warms up to 1e-3 in 4 epochs and cosine anneal to 1e-6 in the end. Following [13], we pre-train the network on a large-scale dataset LUNA16 [19]. The experiments are carried by PyTorch 1.1.0 on a cluster of six NVIDIA GTX 1080Ti GPUs (11GB).

During inference, we employ a sliding-window strategy with the patch size of 128^3 and overlap of 32. Weighted patch assembling is used to assemble the predictions of the patch image to a whole prediction image with the same size as the original CT. The weights of the pixels near the patch edge are lower when averaging the predictions of the overlapping area. From the assembled prediction, a threshold of 0.5 is used to generate a binary image which is then closed by a 3^3 kernel as in [3]. Finally, connected component analysis is used to generate individual PEs, each of which is assigned with a confidence score by the highest prediction score in the connected component and a center by the location of the voxel with the highest score.

3 Experiments and Results

3.1 Dataset and Metric

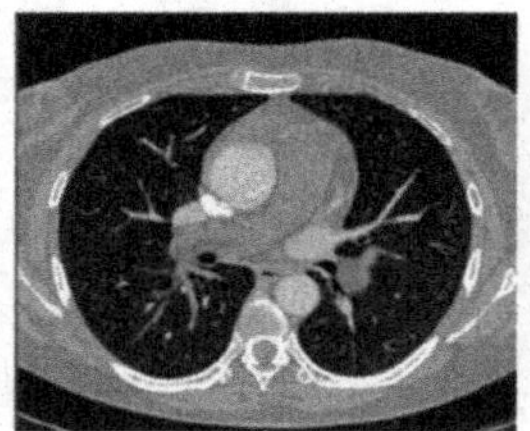
(a) Horizontal Plane

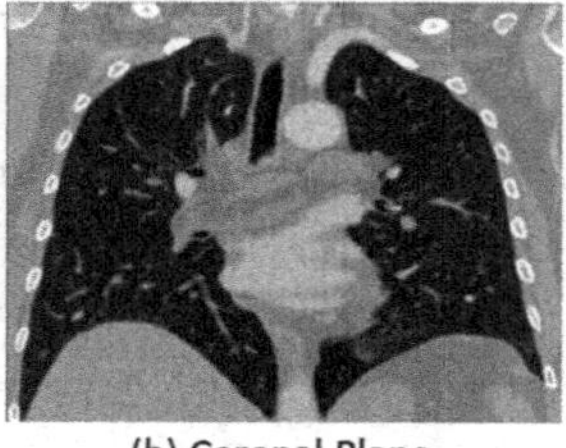
(b) Coronal Plane

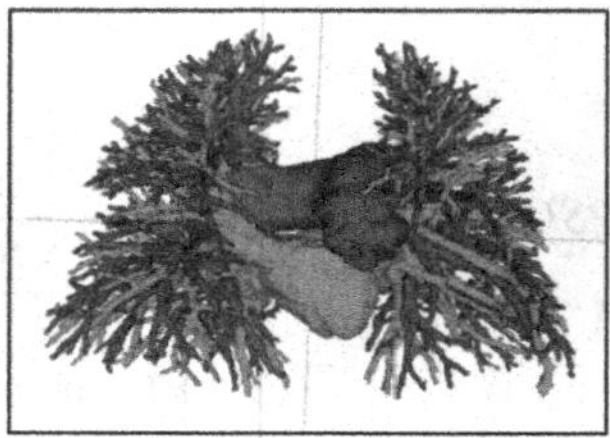
(c) Vessel Reconstruction

Fig. 2. The vessel segmentation label of 021.nrrd in CAD-PE dataset. Artery and vein are displayed in red and green respectively. The 3D reconstruction is generated by ITK-Snap. (Color figure online)

We evaluate our methods on the CAD-PE dataset [7] consisting of 71 CTPA images for training and 20 CTPA images for evaluation. Each connected component of the ground-truth segmentation is considered as an individual PE, resulting 130 PE in the evaluation set. We employ the same evaluation metric of CAD-PE challenge. Detection performance is evaluated based on Free-Response Receiver Operating Characteristic (FROC), reporting detection recall at various average false positives per scan (FPs/scan).

If a detection lies close to an embolus, it can still be useful when using CAD as a second reader [7], thus a variation in the reference standard was performed by dilating the border of each PE with an ϵ as tolerance margin ($\epsilon = 2$ mm and $\epsilon = 5$ mm). This modification affects the number of PE and their size in the reference segmentations, but is only used for evaluation purposes. $\epsilon = 0$ mm denotes using the original segmentation label with no dilation.

For supervising the artery segmentation of AANet, we annotate the artery and vein segmentation label for all images in CAD-PE. We first generate the vessel segmentation of CAD-PE dataset by a CNN that is trained by 300 in-house CT scans with vessel labels. Then, the segmentation is refined by a clinician. Due to limited resources, the final vessel label may contain a few noises such as artery wrongly annotated as vein and the contrary. An example of vessel segmentation label is shown in Fig. 2.

3.2 Results

We compare our methods with UA-2/2.5/3D [3] and two top-ranked participants in CAD-PE challenge, i.e. ASU-Mayo [20] and FUM-Mvlab. It is noted that though Lin et al. [13] made an excellent result, the measurements were performed on their own label of the CAD-PE evaluation set because the official label was not released at the time. We have looked into their unofficial label which only has 80 PEs compared to 130 PEs of the official label. We found they missed a large number of small PEs. Therefore, method in [13] is not included in the comparison study for fair comparison. ASU-Mayo and FUM-Mvlab used the original training set of the challenge consisting of 20 CTPA images, while UA-2/2.5/3D used the extended training set of CAD-PE [7] released lately, like in this paper. All compared methods did not use pre-training.

Figure 3 shows the FROC curves with three tolerance margins ϵ. Our AANet achieved state-of-the-art performance with the sensitivity of 70.8%, 76.2%, and 76.9% at 1, 2, and 4 FPs/Scan using $\epsilon = 0$ mm, and the sensitivity of 78.1%, 84.2%, and 85.1% at 1, 2, and 4 FPs/Scan using $\epsilon = 2$ mm, outperforming all previous methods by a large margin. We also evaluate our method on the unofficial labels provided by [13]. Omitting the false positives with regard to the unofficial label that hit the official label, the sensitivities are 75.1%, 78.8%, and 81.8% at 1, 2, and 4 FPs/Scan using $\epsilon = 2$ mm,. The unfavorable performance is mainly caused by the extra 12 (15% of 80) PEs non-existent in the official label, which might be falsely labeled.

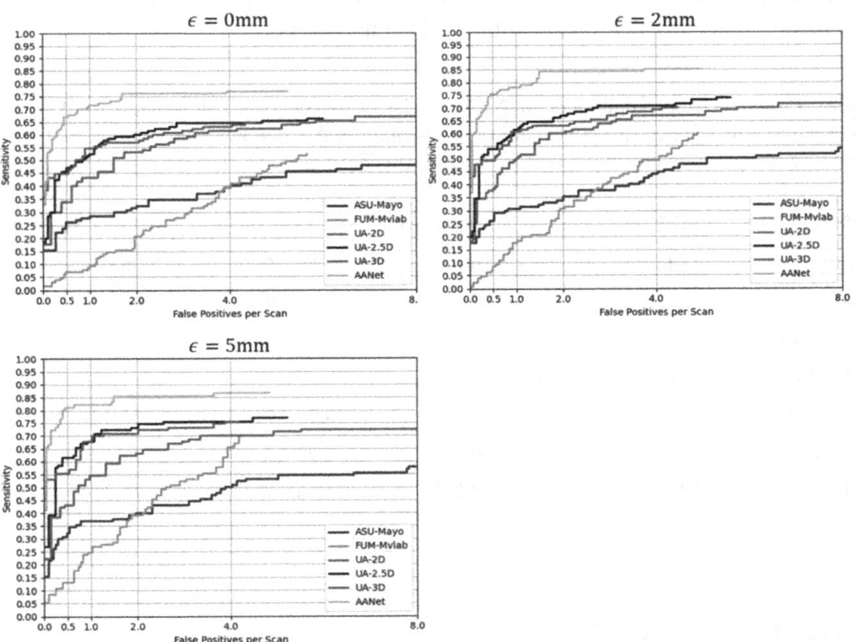

Fig. 3. FROC curves with tolerance margin $\epsilon = 0$ mm, 2 mm, and 5 mm.

3.3 Ablation Studies

We provide details on the ablation studies carried out to examine the effect of each component. We compare the results using $\epsilon = 0$ mm, reported in Table 1. First, the AANet without pre-training on LUNA16 has lower performance; however, it is still superior than the compared methods without pre-training. Because the number of CTPA scans in CAD-PE dataset is scarce, pre-training on a larger dataset with a similar domain can boost the sensitivity at 1 FPs/Scan by 5.4%. We substitute the even-dice loss by batch-dice loss (BDiceLoss) and sample-dice loss (DiceLoss). The use of batch-dice loss achieved a worse result compared to our proposed even-dice loss, while the sample-dice loss cannot converge because of the existence of image patches containing no foreground. Then, we examine the effect of the artery segmentation and the artery attention, i.e. multiplying the soft artery segmentation on the features. Removing both artery segmentation head (AS) and artery attention (AA) in ACF leads to a large decrease in sensitivity at 1 FPs/Scan by 7%. Removing only artery attention (AA) and keeping artery segmentation (AS) decrease sensitivity at 1FPs/Scan by 4.6%, which proves that introducing artery information implicitly by simply adding artery segmentation branch is beneficial but not sufficient. The result proves that the vessel information is essentially helpful to PE detection. Removing the closing operation for post-processing gives an equal or even better result, indicating that the operation [3] is unnecessary.

Table 1. Effect of each component in AANet. BDiceLoss denotes batch-dice loss. DiceLoss denotes sample-dice loss. AS denotes artery segmentation. AA denotes artery attention. Closing denotes the morphology closing operation for post processing.

Method	0.5 FPs/Scan	1 FPs/Scan	2 FPs/Scan	4 FPs/Scan
AANet	**66.9%**	**70.8%**	**76.2%**	**76.9%**
w/o LUNA16	59.2%	65.4%	73.1%	76.2%
w/ BDiceLoss	58.5%	61.5%	66.2%	70.8%
w/ DiceLoss	None convergence			
w/o AS&AA	59.2%	63.8%	71.5%	75.4%
w/o AA	65.4%	66.2%	70.8%	74.6%
w/o Closing	**67.7%**	**70.8%**	**76.9%**	**77.7%**

3.4 Fine-Tuning Without Vessel Label

The training of AANet requires extra artery segmentation label apart from the PE label. Though the improvement is promising, the annotation of vessel segmentation can be a heavy burden. Therefore, we show that our AANet trained by PE and artery label can be used as a pre-trained model that is already able to capture vessel information for transfer learning on a PE dataset without artery label. We use the AANet and AANet w/o AS&AA in previous experiments as the initialization and further fine-tune them on

FUMPE dataset [15]. In a total of 35 CTPA scans, the first 20 CTPA are selected as training set and the last 15 CTPA as evaluation set. The networks are fine-tuned for 50 epochs with the supervision of PE segmentation only. The result on the evaluation set is reported in Table 2, proving that the artery information previously learned still boosts the performance when fine-tuning without artery segmentation.

Some official PE labels of FUMPE are problematic. The 10th slice and the 100th slice of each PE label are swapped. The labels of PAT021 and PAT031 are z-axis flipped. We correct them with the confirmation of clinicians and make them public.

Table 2. The performance on the evaluation set of FUMPE after finetuning. $\epsilon = 0$ mm

Method	0.5 FPs/Scan	1 FPs/Scan	2 FPs/Scan	4 FPs/Scan
AANet	**86.1%**	**86.1%**	**88.9%**	**88.9%**
w/o AS&AA	80.6%	80.6%	86.1%	86.1%

4 Discussion and Conclusion

In this paper, we propose a novel artery-aware 3D fully convolutional network (AANet) with ACF block embedded and the even-dice loss employed. The network encodes both the local and global context as well as artery features for PE regions, which helps to recognize PEs from soft tissues or vessels precisely. The experiments on the public CAD-PE dataset demonstrate our method has achieved state-of-the-art performance, and fine-tuning on dataset without artery segmentation label shows the generality of our method. The proposed approach is applicable to clinical practice to benefit clinical diagnosis. In the future, the vessel context can be exploited by utilizing not only artery but also vein information to further reduce false positives. Besides, semi-supervised learning can be explored to train with a small number of vessel labels to alleviate the burden of vessel annotation.

Acknowledgments. This work was partially supported by the Beijing Nova Program (Z201100006820064).

References

1. Beckman, M.G., et al.: Venous thromboembolism. A public health. Concern (2010). https://doi.org/10.1016/j.amepre.2009.12.017
2. Bouma, H., et al.: Automatic detection of pulmonary embolism in CTA images. IEEE Trans. Med. Imaging. **28**, 8 (2009). https://doi.org/10.1109/TMI.2009.2013618
3. Cano-Espinosa, C., et al.: Computer aided detection of pulmonary embolism using multi-slice multi-axial segmentation. Appl. Sci. **10**, 8 (2020). https://doi.org/10.3390/APP10082945
4. Chen, L.C., et al.: Rethinking atrous convolution for semantic image segmentation liang-chieh. IEEE Trans. Pattern Anal. Mach. Intell. **40**, 4 (2018)

5. Das, M., et al.: Computer-aided detection of pulmonary embolism: Influence on radiologists' detection performance with respect to vessel segments. Eur. Radiol. **18**, 7 (2008). https://doi.org/10.1007/s00330-008-0889-x
6. Le Gal, G., Bounameaux, H.: Diagnosing pulmonary embolism: Running after the decreasing prevalence of cases among suspected patients. J. Thromb. Haemost. **2**, 8 (2004). https://doi.org/10.1111/j.1538-7836.2004.00795.x
7. González, G., et al.: Computer Aided Detection for Pulmonary Embolism Challenge (CAD-PE). arXiv Prepr. arXiv2003.13440 (2020)
8. He, K., et al.: Deep residual learning for image recognition. In: Proceedings of the IEEE Computer Society Conference on Computer Vision and Pattern Recognition (2016). https://doi.org/10.1109/CVPR.2016.90
9. Huang, S.C., et al.: PENet—a scalable deep-learning model for automated diagnosis of pulmonary embolism using volumetric CT imaging. NPJ Digit. Med. **3**(1), 1–9 (2020). https://doi.org/10.1038/s41746-020-0266-y
10. Ioffe, S., Szegedy, C.: Batch normalization: Accelerating deep network training by reducing internal covariate shift. In: Proceedings of the 32nd International Conference on Machine Learning, ICML 2015 (2015)
11. Isensee, F., Jäger, P.F., Full, P.M., Vollmuth, P., Maier-Hein, K.H.: nnU-Net for brain tumor segmentation. In: Crimi, A., Bakas, S. (eds.) BrainLes 2020. LNCS, vol. 12659, pp. 118–132. Springer, Cham (2021). https://doi.org/10.1007/978-3-030-72087-2_11
12. Liang, J., Bi, J.: Local characteristic features for computer aided detection of pulmonary embolism in CT angiography. In: Proceedings of he Pulmonary Image Analysis at Annual Conference on Medical Image Computing and Computer Assisted Intervention (2008)
13. Lin, Y., et al.: Automated pulmonary embolism detection from CTPA images using an end-to-end convolutional neural network. In: Shen, D., et al. (eds.) MICCAI 2019. LNCS, vol. 11767, pp. 280–288. Springer, Cham (2019). https://doi.org/10.1007/978-3-030-32251-9_31
14. Loshchilov, I., Hutter, F.: Decoupled weight decay regularization. In: Proceedings of the 7th International Conference on Learning Representations, ICLR 2019. (2019)
15. Masoudi, M., et al.: Data descriptor: A new dataset of computed-tomography angiography images for computer-aided detection of pulmonary embolism. Sci. Data. **5**, 1–9 (2018). https://doi.org/10.1038/sdata.2018.180
16. Milletari, F., et al.: V-Net: Fully convolutional neural networks for volumetric medical image segmentation. In: Proceedings of the 2016 4th International Conference on 3D Vision, 3DV 2016 (2016). https://doi.org/10.1109/3DV.2016.79
17. Qiao, S., et al.: Micro-batch training with batch-channel normalization and weight standardization. arXiv Prepr. arXiv1903.10520 (2019)
18. Rufener, S.L., et al.: Comparison of on-call radiology resident and faculty interpretation of 4- and 16-row multidetector CT pulmonary angiography with indirect CT venography. Acad. Radiol. **15**, 1 (2008). https://doi.org/10.1016/j.acra.2007.06.030
19. Setio, A.A.A., et al.: Validation, comparison, and combination of algorithms for automatic detection of pulmonary nodules in computed tomography images: The LUNA16 challenge. Med. Image Anal. **42**, 1–13 (2017). https://doi.org/10.1016/j.media.2017.06.015
20. Tajbakhsh, N., Shin, J.Y., Gotway, M.B., Liang, J.: Computer-aided detection and visualization of pulmonary embolism using a novel, compact, and discriminative image representation. Med. Image Anal. **58**, 101541 (2019). https://doi.org/10.1016/j.media.2019.101541
21. Wu, Y., He, K.: Group normalization. In: Ferrari, V., Hebert, M., Sminchisescu, C., Weiss, Y. (eds.) ECCV 2018. LNCS, vol. 11217, pp. 3–19. Springer, Cham (2018). https://doi.org/10.1007/978-3-030-01261-8_1
22. Zhou, Z., Rahman Siddiquee, M.M., Tajbakhsh, N., Liang, J.: UNet++: A nested U-Net architecture for medical image segmentation. In: Stoyanov, D., et al. (eds.) DLMIA/ML-CDS

-2018. LNCS, vol. 11045, pp. 3–11. Springer, Cham (2018). https://doi.org/10.1007/978-3-030-00889-5_1

Siamese Encoder-based Spatial-Temporal Mixer for Growth Trend Prediction of Lung Nodules on CT Scans

Jiansheng Fang[1,2,3], Jingwen Wang[2], Anwei Li[2], Yuguang Yan[4], Yonghe Hou[5], Chao Song[5], Hongbo Liu[2], and Jiang Liu[3(✉)]

[1] School of Computer Science and Technology, Harbin Institute of Technology, Harbin, China
[2] CVTE Research, Guangzhou, China
[3] Research Institute of Trustworthy Autonomous Systems, Southern University of Science and Technology, Shenzhen, China
liuj@sustech.edu.cn
[4] School of Computer, Guangdong University of Technology, Guangzhou, China
[5] Yibicom Health Management Center, CVTE, Guangzhou, China

Abstract. In the management of lung nodules, we are desirable to predict nodule evolution in terms of its diameter variation on Computed Tomography (CT) scans and then provide a follow-up recommendation according to the predicted result of the growing trend of the nodule. In order to improve the performance of growth trend prediction for lung nodules, it is vital to compare the changes of the same nodule in consecutive CT scans. Motivated by this, we screened out 4,666 subjects with more than two consecutive CT scans from the National Lung Screening Trial (NLST) dataset to organize a temporal dataset called NLSTt. In specific, we first detect and pair regions of interest (ROIs) covering the same nodule based on registered CT scans. After that, we predict the texture category and diameter size of the nodules through models. Last, we annotate the evolution class of each nodule according to its changes in diameter. Based on the built NLSTt dataset, we propose a siamese encoder to simultaneously exploit the discriminative features of 3D ROIs detected from consecutive CT scans. Then we novelly design a spatial-temporal mixer (STM) to leverage the interval changes of the same nodule in sequential 3D ROIs and capture spatial dependencies of nodule regions and the current 3D ROI. According to the clinical diagnosis routine, we employ hierarchical loss to pay more attention to growing nodules. The extensive experiments on our organized dataset demonstrate the advantage of our proposed method. We also conduct experiments on an in-house dataset to evaluate the clinical utility of our method by comparing it against skilled clinicians. STM code and NLSTt dataset are available at https://github.com/liaw05/STMixer.

Co-first authors: Jiansheng Fang, Jingwen Wang, Anwei Li. This work was supported in part by Guangdong Provincial Department of Education (2020ZDZX3043), and Shenzhen Natural Science Fund (JCYJ20200109140820699 and the Stable Support Plan Program 20200925174052004).

L. Wang et al. (Eds.): MICCAI 2022, LNCS 13431, pp. 484–494, 2022.
https://doi.org/10.1007/978-3-031-16431-6_46

Keywords: Lung nodule · Growth trend prediction · Siamese network · Feature fusion · Hierarchical loss · Spatial-temporal information

1 Introduction

According to current management guidelines of lung nodules [12,23], it is highly desirable to perform relatively close follow-up for lung nodules detected as suspicious malignant, then give clinical guidance according to their changes during the follow-up [4]. It is recommended to extend the follow-up time for some slow-growing nodules and perform surgery on time for some nodules that are small but speculated to grow fast. However, the imprecise follow-up recommendations may yield high clinical and financial costs of missed diagnosis, late diagnosis, or unnecessary biopsy procedures resulting from false positives [1]. Therefore, it is imperative to predict the growth trend of lung nodules to assist doctors in making more accurate follow-up decisions, thereby further reducing false positives in lung cancer diagnosis. Although deep learning approaches have been the paradigm of choice for fast and robust computer-aided diagnosis of medical images [1,27], there are few related studies on the use of data-driven approaches for assessing the growth trend of lung nodules, since it is hard to acquire large-scale sequential data and without ground-truth dynamic indicators.

Current clinical criteria for assessing lung nodule changes rely on visual comparison and diameter measurements from the axial slices of the consecutive computed tomography (CT) images [10]. The variation of nodule diameter between two consecutive CT scans reflects the most worrying interval change on CT follow-up screening. Applying the variation of nodule diameter to evaluate the growth trend has been a routine method [8,19,20,26]. Zhe *et al.* manually calculate nineteen quantitative features (including nodule diameter) of the initial CT scans to identify the growth risk of pure ground-glass nodules (GGN) [17]. Xavier *et al.* match the same nodule given the list of nodule candidates by computing the difference in diameter between them [14]. In this paper, inspired by existing medical research [17,24], we organize a new temporal dataset of CT scans to predict the evolution of lung nodules in terms of diameter variation.

To achieve this, we screened out 4,666 subjects with more than two consecutive CT scans from the NLST dataset to organize a temporal CT dataset called NLSTt. We first detect regions of interest (ROIs) covering nodules after CT registration. Then, we pair the sequential 3D ROIs with the same nodule, followed by segmentation and classification, nodule types and diameter sizes are automatically annotated. Last, we assign one of three evolution classes (dilatation, shrinkage, stability) for each nodule in terms of its changes in diameter size. Based on automatically labeling evolution classes of each nodule in the NLSTt dataset, we further perform manually double-blind annotation and experienced review to acquire reliable labels for training a deep learning network to address the growth trend prediction. Given the inputs of 3D ROI pairs, we first introduce a siamese encoder [3,14] to extract local features (lesion region) of two ROIs and global features of the current ROI. In order to jointly exploit the interval differences of lesions as well as the spatial dependencies between lesions and whole

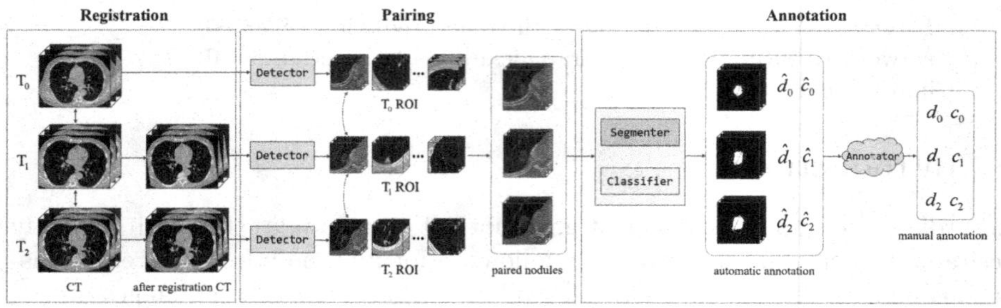

Fig. 1. The pipeline of organizing the temporal CT dataset (NLSTt), including CT scan registration, ROI pairing, and class annotation. The letter d denotes the diameter of lung nodules, and c indicates their corresponding texture types, *i.e.*, solid, part-solid (PS), ground-glass nodule (GGN).

ROIs, we design a novel spatial-temporal mixer (STM) to leverage spatial and temporal information. In addition, we employ hierarchical loss (H-loss) [25] to pay more attention to the nodule growth (dilatation class) related to possible cancerogenesis in clinical practice.

Our ***contributions*** are summarized as follows: (1) To drive the prediction of lung nodule evolution, we organize a new temporal CT dataset called NLSTt by combing automatic annotation and manual review. (2) We propose a spatial-temporal mixer (STM) to leverage both temporal and spatial information involved in the global and lesion features generated from 3D ROI pairs. (3) We conduct extensive experiments on the NLSTt dataset to evaluate the performance of our proposed method and confirm the effectiveness of our model on an in-house dataset from the perspective of clinical practice.

2 Materials and Methods

2.1 NLSTt Dataset Acquisition

Lots of subjects enrolled in the NLST dataset [21] cover scans from multiple time points (typically T0, T1, and T2 scans taken one year apart) as well as biopsy confirmations for suspicious lesions [22]. In this work, to advance the research of the growth trend prediction of lung nodules, we organize a temporal CT dataset named NLSTt by screening out 4,666 subjects from NLST, each of which has at least two CT scans up to 2 years apart from the NLST dataset.

Figure 1 illustrates the pipeline of our data organization approach, which aims to ascertain the evolution class and texture type of nodules in consecutive CT scans for the selected subjects. We first perform 3D image registration for the second (T1) and third (T2) CT scans in terms of the first scan (T0). After that, we identify 3D ROIs containing lung nodules by a detector, and then pair ROIs to match the same nodule in multiple 3D ROIs at different time points. Next, we employ a segmenter to automatically crop out the lesion of nodules in ROIs to calculate their diameters. At the same time, we apply a classifier to identify

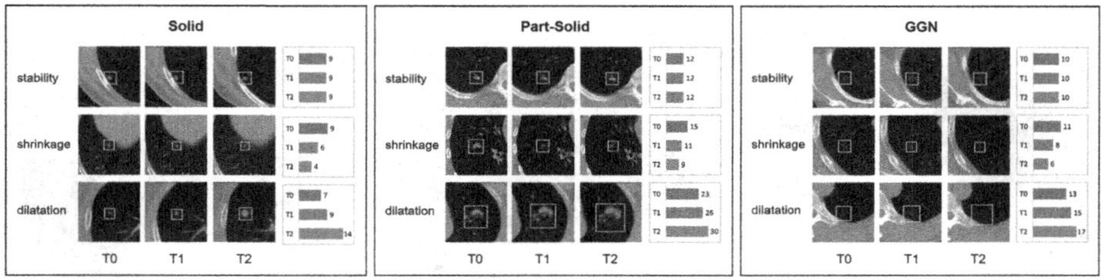

Fig. 2. Three evolution classes (dilatation, shrinkage, stability) of lung nodules in the NLSTt dataset. We depict the variation of nodule diameter (in millimeter) at three consecutive time points (T0, T1, T2) for three nodule types by horizontal histograms.

the texture types of nodules (*i.e.*, GGN, solid, part-solid (PS)). In specific, two popular CT datasets LUNA16 [16] and LNDb [13] are used to train our models, in which the detector for ROI identification is a 3D variant of the CenterNet [28], the segmenter for lesion segmentation is a multi-scale 3D UNet [9,15], and the classifier is the variant of the segmenter attaching a fully-connected layer.

In the above pipeline of organizing the NLSTt dataset, we first make registrations for the consecutive CT scans of each patient, then detect nodules. For the detected two nodules at T0 and T1, the pair criterion is that the Euclidean distance between the center points of the two nodules is less than 1.5 mm. If the nodule location changes significantly between T0 and T1 (> 1.5 millimeters), we assert that they are not the same nodule. Finally, we ask experienced clinicians to review to ensure the accuracy of paring nodules correctly. After automatically inferring the diameters and texture types, we rely on experienced clinicians to calibrate the labels by manually annotation. By combining automatic inference and manual review, we acquire the reliable label of nodules regarding the texture type and evolution class, the latter of which is determined by the diameter change of a nodule at consecutive time points. Finally, we complete the construction of the NLSTt dataset. Next, we discuss the details regarding how to construct the labels of evolution classes for nodules.

The evolution class of a nodule is based on the changes in diameter size. The diameter is the longest side of the smallest circumscribed rectangle on the maximal surface of the nodule mask generated by the segmenter. We formulate three evolution classes of lung nodules according to the diameter change of the same nodule at different time points as stability, dilatation, and shrinkage, as shown in Fig. 2. If the diameter variation of lung nodules at two consecutive CT scans is less than 1.5 mm, we refer to such trend as stability. If the diameter of lung nodules of twice CT scans varies more than 1.5 mm, we define the increment as dilatation and the reduction as shrinkage. In the following part, we present our proposed method for growth trend prediction.

2.2 Spatial-Temporal Mixer

Figure 3 overviews our proposed method for growth trend prediction, including a siamese encoder, a spatial-temporal mixer (STM), and a two-layer hierarchical

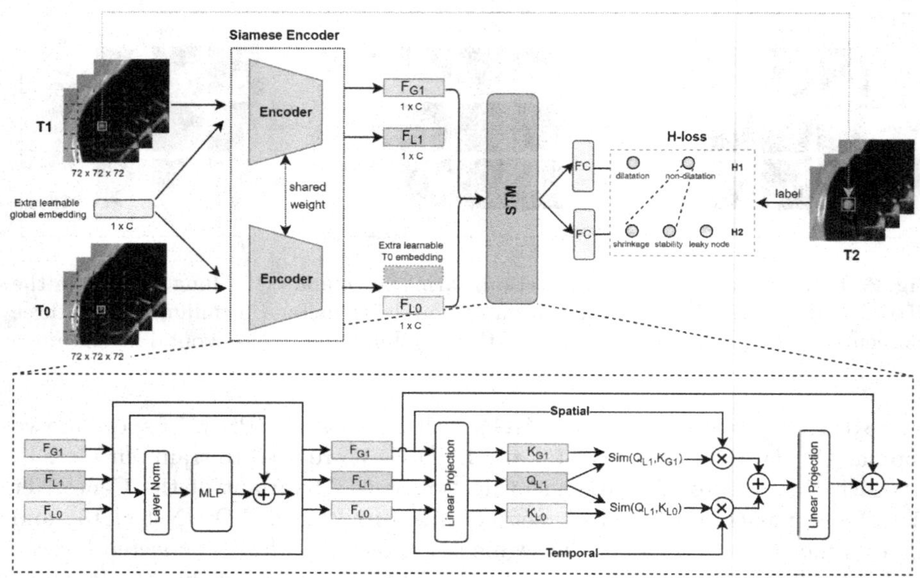

Fig. 3. Schematic of our proposed method, including a siamese encoder, a spatial-temporal mixer (STM) and a two-layer H-loss. MLP is a multilayer perception, F_{G1} denotes the global information of T1, F_{L1} indicates the local information of the lesion patch in T1, and F_{L0} is local information of the lesion patch in T0.

loss (H-loss). For a given subject, the 3D ROI pairs (T0 and T1) containing lesions are taken from CT scans at different time points, and then fed into a siamese encoder for embedding. Besides, an extra learnable global embedding is introduced to extract the global information of 3D ROIs [2]. Both encoders in the siamese encoder share the same weights, and we adopt the vision transformer (ViT) [2,6] and convolutional neural network (CNN) [7] as the backbone of the siamese encoder. We obtain three embedding vectors from sequential 3D ROIs by the siamese encoder: F_{G1} contains global information of T1, F_{L1} contains local information of the lesion patch in T1, and F_{L0} contains local information of the lesion patch in T0. We supplement a learnable embedding F_{L0} if T0 is missing, which occurs when a subject has only two CT scans.

It is worth mentioning that the global information of ROIs is changeless on T0, T1, and T2. Hence, we only learn global information from T1 without considering the global information of T0. On the contrary, the local information of the same nodule in T0, T1, and T2 are different and highly discriminative for growth trend prediction. Therefore, we learn local information from both T0 and T1 to capture the evolving local information.

Given the embeddings obtained from the siamese encoder, we propose a spatial-temporal mixer (STM) module to leverage spatial and temporal information. We firstly introduce a layer normalization, a multi-layer perception (MLP), and a residual addition operation on the three embeddings F_{L1}, F_{G1}, and F_{L0}.

After that, in order to fuse spatial and temporal information, we apply a linear projection to obtain a query vector Q_{L1} from F_{L1}, and two key vectors K_{G1} and K_{L0} from F_{G1} and F_{L0}, respectively. The spatial information is captured by $F_{G1} \cdot Sim(Q_{L1}, K_{G1})$, where $Sim(Q_{L1}, K_{G1})$ is the cosine similarity between the query-key pair of local and global embeddings of T1. Similarly, the temporal information is captured by $F_{L0} \cdot Sim(Q_{L1}, K_{L0})$, where $Sim(Q_{L1}, K_{L0})$ is the cosine similarity between the query-key pair of local embeddings of T1 and T0, which are collected at different time points. Next, we fuse the spatial and temporal information with an addition operation and a linear projection (LP). Finally, STM outputs an embedding based on the spatial-temporal information and the current local embeddings F_{L1}. In summary, the output of STM is computed as:

$$F_{L1} + LP(Sim(Q_{L1}, K_{G1}) \cdot F_{G1} + Sim(Q_{L1}, K_{L0}) \cdot F_{L0}). \tag{1}$$

2.3 Two-Layer H-Loss

After feature fusion by STM, we are ready to train a model for predicting the growth trend. We employ H-loss, which extends the weighted cross-entropy (WCE) loss function to enhance the contribution of the dilatation class by hierarchical setting. It is more cautious and attentive to the nodule growth related to possible cancerogenesis in clinical practice. Hence, it is vital to achieving high predictive accuracy for the dilatation class. To this end, we build a two-layer H-loss (H1, H2) including two separate fully-connected (FC) layer modules to analog clinical diagnosis routine. Based on three evolution classes of nodules, H1 first classifies dilatation or not for paying more attention to the dilatation class. And H2 inherits H1 to predict shrinkage or stability. The leakage node in H2 here only involves dilatation. The two-layer H-loss unifies the predictive probabilities and the ground-truth label y to train our model, as follows:

$$\mathcal{L} = \alpha \cdot WCE(P_{H1}, y) + WCE(P_{H2}, y), \tag{2}$$

where P_{H1} and P_{H2} are the probability output of H1 and H2 layers, respectively. When $\alpha = 0$, H-loss is a three-class WCE equivalent to H2. For H2, we set different weights for different classes in WCE to combat class imbalance. The weights of dilatation, shrinkage and stability in H2 are 1.0, 1.0, and 0.1, respectively. During model inference, we predict the evolution class of T2 by H1 or H2 layer.

3 Experiments

3.1 Experimental Settings

Datasets. Table 1 shows the statistical information of the datasets. Our organized dataset NLSTt with 4,666 subjects is split into a training set (3, 263 subjects), a validation set (701 subjects), and a test set (702 subjects). In addition, we collect CT scans of 199 subjects and adopt the same preprocessing approach

Table 1. Statistics of benchmark splits of the NLSTt dataset and in-house dataset*.

Types	Train set				Validation set				Test set				In-house set			
	$\Rightarrow$	$\Uparrow$	$\Downarrow$	$\sum$	$\Rightarrow$	$\Uparrow$	$\Downarrow$	$\sum$	$\Rightarrow$	$\Uparrow$	$\Downarrow$	$\sum$	$\Rightarrow$	$\Uparrow$	$\Downarrow$	$\sum$
GGN	2,496	153	34	2,683	527	28	9	564	612	35	11	658	123	6	0	129
Solid	3,804	235	82	4,121	833	40	27	900	827	47	18	892	334	12	6	352
PS	97	39	12	148	14	8	4	26	21	13	3	37	3	3	0	6
Totals	6,397	427	128	6,952	1,374	76	40	1,490	1,460	95	32	1,587	460	21	6	487

* $\Rightarrow$ denotes stability, $\Uparrow$ indicates dilatation, $\Downarrow$ represents shrinkage, and $\sum$ aggregates the number of three evolution trends for each nodule type.

Table 2. AUC (in %) of different mixers and encoders on the test and in-house sets.

Encoder	Mixer	Test Set		In-house Set	
		AUC@H1	AUC@H2	AUC@H1	AUC@H2
CNN	Concat	80.8	75.3	67.2	67.2
	MFC	81.2	75.2	69.4	66.7
	LSTM	81.8	75.0	64.0	71.0
	STM (Ours)	83.0	76.3	**73.5**	71.6
ViT	Concat	82.6	75.2	64.2	64.1
	LSTM	82.6	76.3	67.1	74.7
	STM (Ours)	**83.6**	**77.5**	72.8	**78.5**

used for NLSTt to organize an in-house dataset, which is used to evaluate the practicality of our model by comparing it against the clinicians.

Optimizer. We apply the AdamW optimizer [11] to train our model, in which CNN encoders adopt ResNet34-3D [7] and are trained from scratch, and ViT encoders are trained on the pre-trained model which uses the MAE method [6]. The batch size is set as $B = 16$ for all the conducted methods. We warm up [5] the learning rate from $10e-6$ to $lr \times B/64$ in the first 5 epochs, where $lr = 5e-4$, and then schedule the learning rate by the cosine annealing strategy [11]. The parameters of networks are optimized in 60 epochs with a weight decay of 0.05 and a momentum of 0.9.

3.2 Results and Discussions

Gain Analysis for Our STM. Table 2 reports the Area Under the Curve (AUC) of H1 (AUC@H1) and H2 (AUC@H2) layers for four feature fusion methods (Concat, MFC, LSTM, our STM) based on CNN and ViT encoders. MFC [14] combines feature maps of 3D ROI pairs and employs CNN as the encoder. Hence, we only compare MFC and our STM based on the CNN encoder. On the test set, ViT-based STM achieves the best performances. For the in-house set, CNN-based STM

Table 3. ACC and Kappa of nodule types of the test set on different methods.

Method	Test set						Extra in-house set					
	Accuracy			Kappa			Accuracy			Kappa		
	GGN	Solid	PS	GGN	Solid	PS	GGN	Solid	PS	GGN	Solid	PS
CNN+STM	90.9	88.2	56.8	27.6	25.7	26.2	87.6	91.2	58.1	14.4	7.6	27.2
ViT+STM	**92.4**	**91.6**	**59.5**	**29.1**	**33.7**	**29.2**	**93.8**	90.6	60.5	**46.9**	13.4	**29.2**
Clinician A	–	–	–	–	–	–	85.3	93.2	60.5	19.0	**19.8**	23.9
Clinician B	–	–	–	–	–	–	86.0	**94.0**	**62.8**	21.0	14.0	20.2

Table 4. AUC of two-layer H-loss with varying α on CNN encoders and our STM.

α	Test set			In-house set		
	AUC@H1	AUC@H2	AUC@H2-D	AUC@H1	AUC@H2	AUC@H2-D
0.0	–	73.2	80.3	–	62.7	63.5
0.5	82.8	78.2	82.8	71.1	71.1	66.6
1.0	83.0	76.3	**83.4**	73.5	71.6	**73.6**

obtains the best AUC@H1, and ViT-based STM gets the best AUC@H2. By further observing the high profits of ViT-based three mixers, we argue that the ViT-based siamese encoder exhibits better robustness than the CNN-based one. Furthermore, with either CNN- or ViT-based encoders, our STM brings more gains than LSTM and Concat on the test set and in-house set. Our STM also consistently outperforms MFC on two datasets in terms of two metrics. The Concat method for feature fusion only linearly combines the three embedding vectors without capturing their inter-dependencies. Hence, the predictive capability is lower than LSTM, which captures the temporal changes of two lesion features (F_{L0}, F_{L1}) extracted from T0 and T1. Besides the interval differences of lesion features at different time points. Our STM also exploits spatial dependencies of the global features F_{G1} and lesion features F_{L1} of T1, thus achieves the best performance.

Discussion of Clinical Practice. Based on the confirmation of the advantage of our STM, we further observe its performance on three nodule types and clinical applications. Since there are only six part-solid samples in the in-house set, we build an extra in-house set, which uses all the 37 part-solid samples from the test set to assemble a total of 43 part-solid samples for evaluation. As Table 3 shows, ViT-based STM obtains better Accuracy and Kappa [18] than CNN-based STM on three nodule types of the test set. By comparing with two skilled clinicians, ViT-based STM outperforms clinicians A and B on GGN while slightly weaker on solid. For part-solid, ViT-based STM achieves better Kappa but lower Accuracy than clinicians. This demonstrates that our model can carry out clinical practice in terms of GGN, solid, and part-solid. We show several cases predicted by our model and clinicians in Appendix A1.

Utility of Two-Layer H-Loss. Table 4 shows the ablation studies of the hyper-parameter α in Eq. (2). AUC@H2-D is the AUC of the dilatation class on the

H2 layer. According to AUC@H2 on the test and in-house sets, two-layer H-loss exhibits a significant advantage over the H2 layer ($\alpha = 0$) alone. Among the three evolution classes, considering clinical meaning, we preferentially ensure the predictive accuracy of the growth trend. Hence, we pay more attention on the dilatation class with the help of the two-class H1 layer. The significant difference of AUC@H2-D between $\alpha = 0$ and $\alpha = 1$ demonstrates the benefits of our strategy.

4 Conclusions

In this work, we explore how to predict the growth trend of lung nodules in terms of diameter variation. We first organize a temporal CT dataset including three evolution classes by automatic inference and manual review. Then we propose a novel spatial-temporal mixer to jointly exploit spatial dependencies and temporal changes of lesion features extracted from consecutive 3D ROIs by a siamese encoder. We also employ a two-layer H-loss to pay more attention to the dilatation class according to the clinical diagnosis routine. The experiments on two real-world datasets demonstrate the effectiveness of our proposed method.

A1: Prediction Cases

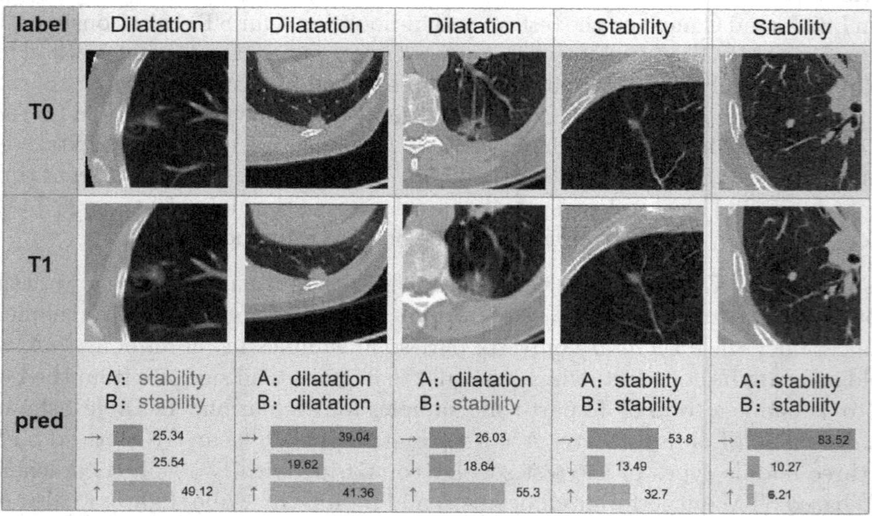

Fig. 4. Examples of predicting the growth trend by our model and clinicians A and B. The first row is the ground-truth of the evolution classes, and the predicted results in red color are the incorrect predictions. The symbols →, ↑, and ↓ denote the classes stability, dilatation and shrinkage, respectively (Color figure online)

References

1. Ardila, D., et al.: End-to-end lung cancer screening with three-dimensional deep learning on low-dose chest computed tomography. Nat. Med. **25**(6), 954–961 (2019)
2. Dosovitskiy, A., et al.: An image is worth 16 × 16 words: transformers for image recognition at scale. In: International Conference on Learning Representations (2020)
3. Fang, J., Xu, Y., Zhang, X., Hu, Y., Liu, J.: Attention-based saliency hashing for ophthalmic image retrieval. In: 2020 IEEE International Conference on Bioinformatics and Biomedicine (BIBM), pp. 990–995. IEEE (2020)
4. Gao, C., et al.: The growth trend predictions in pulmonary ground glass nodules based on radiomic CT features. Front. Oncol. **10**, 580809 (2020)
5. Goyal, P., et al.: Accurate, large minibatch SGD: training imagenet in 1 hour. arXiv preprint arXiv:1706.02677 (2017)
6. He, K., Chen, X., Xie, S., Li, Y., Dollár, P., Girshick, R.: Masked autoencoders are scalable vision learners. arXiv preprint arXiv:2111.06377 (2021)
7. He, K., Zhang, X., Ren, S., Sun, J.: Deep residual learning for image recognition. In: Proceedings of the IEEE Conference on Computer Vision and Pattern Recognition, pp. 770–778 (2016)
8. Huang, P., et al.: Prediction of lung cancer risk at follow-up screening with low-dose CT: a training and validation study of a deep learning method. Lancet Digit. Health **1**(7), e353–e362 (2019)
9. Kushnure, D.T., Talbar, S.N.: MS-UNET: a multi-scale UNET with feature recalibration approach for automatic liver and tumor segmentation in CT images. Comput. Med. Imaging Graph. **89**, 101885 (2021)
10. Larici, A.R., et al.: Lung nodules: size still matters. Eur. Respir. Rev. **26**(146) (2017)
11. Loshchilov, I., Hutter, F.: SGDR: stochastic gradient descent with warm restarts. arXiv preprint arXiv:1608.03983 (2016)
12. MacMahon, H., et al.: Guidelines for management of incidental pulmonary nodules detected on CT images: from the Fleischner Society 2017. Radiology **284**(1), 228–243 (2017)
13. Pedrosa, J., et al.: LNDb: a lung nodule database on computed tomography. arXiv preprint arXiv:1911.08434 (2019)
14. Rafael-Palou, X., et al.: Re-identification and growth detection of pulmonary nodules without image registration using 3D Siamese neural networks. Med. Image Anal. **67**, 101823 (2021)
15. Ronneberger, O., Fischer, P., Brox, T.: U-Net: convolutional networks for biomedical image segmentation. In: Navab, N., Hornegger, J., Wells, W.M., Frangi, A.F. (eds.) MICCAI 2015. LNCS, vol. 9351, pp. 234–241. Springer, Cham (2015). https://doi.org/10.1007/978-3-319-24574-4_28
16. Setio, A.A.A., et al.: Validation, comparison, and combination of algorithms for automatic detection of pulmonary nodules in computed tomography images: the LUNA16 challenge. Med. Image Anal. **42**, 1–13 (2017)
17. Shi, Z., et al.: Quantitative features can predict further growth of persistent pure ground-glass nodule. Quant. Imaging Med. Surg. **9**(2), 283 (2019)
18. Spitzer, R.L., Cohen, J., Fleiss, J.L., Endicott, J.: Quantification of agreement in psychiatric diagnosis: a new approach. Arch. Gen. Psychiatry **17**(1), 83–87 (1967)
19. Tan, M., et al.: Prediction of the growth rate of early-stage lung adenocarcinoma by radiomics. Front. Oncol. **11**, 658138 (2021)

20. Tao, G., et al.: Prediction of future imagery of lung nodule as growth modeling with follow-up computed tomography scans using deep learning: a retrospective cohort study. Transl. Lung Cancer Res. **11**(2), 250 (2022)
21. Team, N.L.S.T.R.: Reduced lung-cancer mortality with low-dose computed tomographic screening. N. Engl. J. Med. **365**(5), 395–409 (2011)
22. Veasey, B., et al.: Lung nodule malignancy classification based on NLSTX data. In: 2020 IEEE 17th International Symposium on Biomedical Imaging (ISBI), pp. 1870–1874. IEEE (2020)
23. Wood, D.E., et al.: Lung cancer screening, version 3.2018, NCCN clinical practice guidelines in oncology. J. Nat. Compr. Cancer Netw. **16**(4), 412–441 (2018)
24. Xinyue, W., et al.: Analysis of growth curve type in pulmonary nodules with different characteristics. Zhongguo Fei Ai Za Zhi **20**(5) (2017)
25. Yang, J., et al.: Hierarchical classification of pulmonary lesions: a large-scale Radio-Pathomics study. In: Martel, A.L., et al. (eds.) MICCAI 2020. LNCS, vol. 12266, pp. 497–507. Springer, Cham (2020). https://doi.org/10.1007/978-3-030-59725-2_48
26. Yoon, H.J., Park, H., Lee, H.Y., Sohn, I., Ahn, J., Lee, S.H.: Prediction of tumor doubling time of lung adenocarcinoma using radiomic margin characteristics. Thoracic cancer **11**(9), 2600–2609 (2020)
27. Zhang, H., Gu, Y., Qin, Y., Yao, F., Yang, G.-Z.: Learning with sure data for nodule-level lung cancer prediction. In: Martel, A.L., et al. (eds.) MICCAI 2020. LNCS, vol. 12266, pp. 570–578. Springer, Cham (2020). https://doi.org/10.1007/978-3-030-59725-2_55
28. Zhou, X., Wang, D., Krähenbühl, P.: Objects as points. arXiv preprint arXiv:1904.07850 (2019)

What Makes for Automatic Reconstruction of Pulmonary Segments

Kaiming Kuang[1], Li Zhang[1], Jingyu Li[3], Hongwei Li[4], Jiajun Chen[1], Bo Du[3], and Jiancheng Yang[1,2,5](✉)

[1] Dianei Technology, Shanghai, China
[2] Shanghai Jiao Tong University, Shanghai, China
jekyll4168@sjtu.edu.cn
[3] Wuhan University, Wuhan, Hubei, China
[4] Technical University of Munich, Munich, Germany
[5] EPFL, Lausanne, Switzerland

Abstract. 3D reconstruction of pulmonary segments plays an important role in surgical treatment planning of lung cancer, which facilitates preservation of pulmonary function and helps ensure low recurrence rates. However, automatic reconstruction of pulmonary segments remains unexplored in the era of deep learning. In this paper, we investigate *what makes for automatic reconstruction of pulmonary segments.* First and foremost, we formulate, clinically and geometrically, the anatomical definitions of pulmonary segments, and propose evaluation metrics adhering to these definitions. Second, we propose ImPulSe (**Im**plicit **Pul**monary **S**egment), a deep implicit surface model designed for pulmonary segment reconstruction. The automatic reconstruction of pulmonary segments by ImPulSe is accurate in metrics and visually appealing. Compared with canonical segmentation methods, ImPulSe outputs continuous predictions of arbitrary resolutions with higher training efficiency and fewer parameters. Lastly, we experiment with different network inputs to analyze what matters in the task of pulmonary segment reconstruction. Our code is available at https://github.com/M3DV/ImPulSe.

Keywords: Pulmonary segments · Surface reconstruction · Implicit fields

1 Introductions

Pulmonary segments are anatomical subunits of pulmonary lobes. There are 18 segments in total, with eight in the left lung and ten in the right [1,15,30]. Unlike pulmonary lobes, pulmonary segments are not defined by visible boundaries but bronchi, arteries and veins. Concretely, pulmonary segments should include

K. Kuang and L. Zhang—Equal contributions.

Supplementary Information The online version contains supplementary material available at https://doi.org/10.1007/978-3-031-16431-6_47.

L. Wang et al. (Eds.): MICCAI 2022, LNCS 13431, pp. 495–505, 2022.
https://doi.org/10.1007/978-3-031-16431-6_47

their segmental bronchi and arteries while establishing their boundaries along intersegmental veins [5,20]. However, boundaries between adjacent segments are ambiguous since segmenting planes are valid as long as they separate segmental bronchi and arteries while lying roughly in the area of intersegmental veins.

Automatic reconstruction of pulmonary segments helps determine the appropriate resection method in surgical treatment of lung cancer. Lobectomy (excising the affected lobe entirely) and segmentectomy (excising only the affected segment) are two major resection methods for early stage lung cancer. As the standard care of early stage lung cancer, lobectomy is challenged by segmentectomy as it preserves more pulmonary function, reduces operation time and blood loss while leading to similar recurrence rates and survival [5,9,10,27,32]. However, segmentectomy should only be considered when surgical margins can be guaranteed to ensure low recurrence rates [19,27]. Therefore, it is crucially important to reconstruct pulmonary segments before performing pulmonary surgeries.

Recent years have witnessed the great success of deep learning in medical image segmentation [13,17,26,34], as well as in segmentation of pulmonary structures such as lobes, airways and vessels [6,7,18,24]. Nonetheless, automatic reconstruction of pulmonary segments remains poorly understood. First, canonical segmentation methods such as U-Net [26] are not suitable for this task. These methods create large memory footprint with medical images of original resolutions, and output poor segmentation if inputs are downsampled. Second, it is desired that the pulmonary segment reconstruction model can generate outputs at arbitrary resolutions when given only coarse inputs. Lately, deep implicit functions show great promises in representing continuous 3D shapes [3,4,12,16,21,23]. The learned implicit function predicts occupancy at continuous locations, thus is capable of reconstructing 3D shapes at arbitrary resolutions. Moreover, implicit fields can be optimized using irregular points randomly sampled from the entire continuous space, which significantly reduces training costs. These characteristics suggest that implicit functions can be of use in the reconstruction of pulmonary segments.

In this paper, we aim to answer this question: *what makes for automatic reconstruction of pulmonary segments?* First and foremost, we give clear and concrete definitions of the anatomy of pulmonary segments (Sect. 2.1) and propose evaluation metrics that adhere to problem definitions (Sect. 2.2); Next, we present an implicit-function-based model, named ImPulSe (**Im**plicit **Pul**monary **S**egment), for the pulmonary segment reconstruction task (Sect. 2.3 and 2.4). Implicit fields render ImPulSe with high training and parameter efficiency and the ability to output reconstruction at arbitrary resolutions. Our proposed method uses only half the training time of U-Net [26] while having better accuracy. ImPulSe achieves Dice score of 84.63% in overall reconstruction, 86.18% and 87.00% in segmental bronchus and artery segmentation. Last but not least, we investigate what inputs matter for accurate reconstruction of pulmonary segments (Sect. 3.3). Our experiments indicate that bronchi and vessels play important roles in the reconstruction of pulmonary segments.

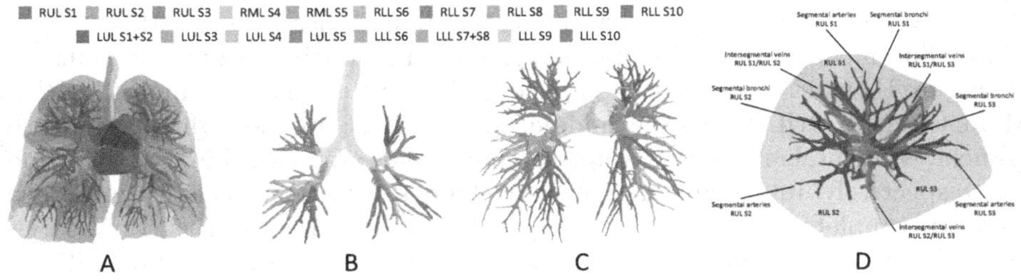

Fig. 1. Visualization of pulmonary segment anatomy. A: An overview of pulmonary segments, including bronchi, arteries and veins. **B, C**: Bronchus and artery tree are separated into segmental groups, each of which occupies a branch of the tree. **D**: A concrete example of intersegmental boundaries of RUL S1 (middle), RUL S2 (left) and RUL S3 (right). Segmental bronchi, segmental arteries and intersegmental veins are colored in gray, blue and red. Intrasegmental veins are hidden for better visualization. Each segment completely wraps its own segmental bronchi and arteries. Intersegmental boundaries lie on the branch of intersegmental veins.

2 Methods

2.1 Anatomical Definitions of Pulmonary Segments

To start the investigation of automatic pulmonary segment reconstruction, it is necessary to first set clear anatomical definitions of pulmonary segments. Pulmonary segments are subunits of pulmonary lobes. There are 5 pulmonary lobes: left upper lobe (LUL), left lower lobe (LLL), right upper lobe (RUL), right middle lobe (RML), and right lower lobe (RLL). These lobes are further divided into 18 pulmonary segments, with eight segments in the left lung and ten in the right lung [1,15,30]. These segments are numbered from S1 to S10 in either left lung and right lung, with two exceptions in the left lung (LUL S1+S2 and LLL S7+S8). Unlike other anatomical structures such as pulmonary lobes, pulmonary segments are not defined by visible boundaries but pulmonary bronchi, arteries and veins. Pulmonary bronchi and vessels (including arteries and veins) expand themselves into tree structures in lungs. Branches of bronchi and vessel trees are then divided into 18 segmental groups (*i.e.*, segmental bronchi and arteries), by which pulmonary segments are defined. Figure 1B and C show the bronchus tree and the artery tree, with segmental groups marked in different colors. Each color represents a segmental bronchus/artery. Concretely, pulmonary segments should satisfy the following three rules [5,20]:

(1) Includes its segmental bronchi. The volume of a certain segment should completely includes its corresponding segmental bronchus, as in Fig. 1B.

(2) Includes its segmental arteries. A certain segment should contain its segmental artery as in the case of segmental bronchi. See Fig. 1C.

(3) Establishes its boundaries along intersegmental veins. Pulmonary veins can be subdivided into intrasegmental or intersegmental veins. Intersegmental

veins serve as important landmarks indicating boundaries between adjacent segments. Specifically, intersegmental planes should follow intersegmental veins.

Figure 1D gives a concrete example of pulmonary segment boundaries (RUL S1 in the left, RUL S2 in the middle, and RUL S3 in the right). Segmental bronchi, segmental arteries and intersegmental veins are marked in gray, blue and red, respectively. Intrasegmental veins are hidden for better visualization. Each segment completely wraps its own segmental bronchi and arteries. Boundaries between segments lie on the branch of intersegmental veins, as per definitions. Segmental bronchi and arteries do not cross segmental boundaries.

2.2 Evaluation Metrics

Since the automatic reconstruction of pulmonary segments have not been systematically investigated, it is equally important to select appropriate evaluation metrics for this task. Dice score is a commonly-used metric in segmentation,

$$Dice_{\mathbf{s}} = \frac{2\|\mathbf{Y_s} \cap \hat{\mathbf{Y}}_\mathbf{s}\|}{\|\mathbf{Y_s}\| + \|\hat{\mathbf{Y}}_\mathbf{s}\|}, \tag{1}$$

where $\mathbf{Y_s}$ and $\hat{\mathbf{Y}}_\mathbf{s}$ are ground-truth and prediction segmentation of points in the set $\mathbf{s}$, and $\|\cdot\|$ denotes the number of elements. Definitions of pulmonary segments suggest that their boundaries are ambiguous. Therefore, it is more important to guarantee accurate reconstruction of segmental bronchi and arteries than segment reconstruction itself. Given that, we evaluate our proposed method and its counterparts not only using the overall Dice score on 18 pulmonary segments ($Dice_\mathbf{o}$) but also Dice on segmental bronchi ($Dice_\mathbf{b}$) and arteries ($Dice_\mathbf{a}$). Note that $Dice_\mathbf{b}$ and $Dice_\mathbf{a}$ are not binary Dice scores of bronchi and arteries. Instead, they are 19-class pulmonary segment Dice scores evaluated only on voxels of bronchi and arteries. These two metrics measure how accurate bronchi and arteries are divided into their corresponding pulmonary segments (as in Fig. 1 B and C). We argue that Dice scores of segmental bronchi and arteries are more important metrics than the overall Dice score. We leave veins out of this question because the segment classification of veins, especially intersegmental veins, is ambiguous. We find that Dice scores on veins are consistently lower than bronchi and arteries across all models (see supplementary materials).

2.3 The ImPulSe Network

To apply deep learning in automatic pulmonary segment reconstruction, one problem shows up: CT images come in different sizes. While high-resolution 3D inputs are not feasible due to large memory footprint, canonical segmentation methods such as U-Net [26] generate coarse segmentations of fixed sizes given low-resolution inputs (*i.e.*, low-resolution in, low-resolution out), which makes them unsuitable for this task. Unlike U-Net, implicit functions generate continuous outputs of arbitrary resolutions even when only low-resolution inputs are given. Figure 2 visualizes implicit functions and voxel-based representations. Rather

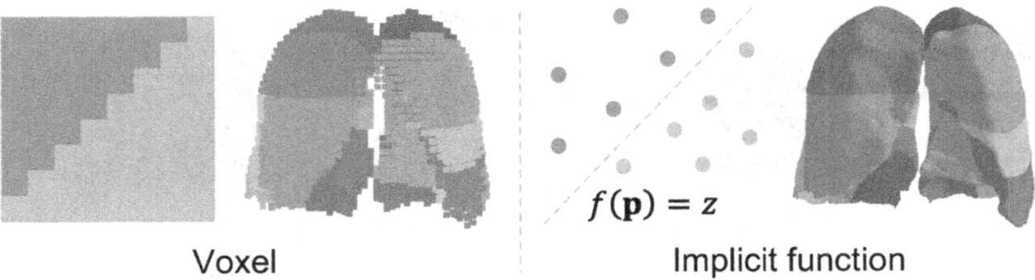

Fig. 2. Visualization of shape representations with voxel and implicit functions. Rather than classifying each voxel, implicit functions representing shapes by encoding 3D iso-surfaces with function $f(\mathbf{p}) = z$. When queried with a continuous location $\mathbf{p}$, it outputs occupancy z at the location, Thus, implicit functions can output continuous reconstruction with high resolutions and fine details. Please note that this is a conceptual visualization, not a rigorous comparison.

than predicting segmentation in a fixed-size voxel grid, the learned implicit function encodes 3D iso-surfaces with function $f(\mathbf{p}) = z$ and predicts occupancy z at a continuous location $\mathbf{p}$, thus is capable of reconstructing 3D shapes at arbitrary resolutions. To take advantage of this in reconstruction of pulmonary segments, we propose ImPulSe (**Im**plicit **Pul**monary **Se**gment), an implicit-function-based model capable of generating reconstruction of arbitrary sizes. The overall architecture of ImPulSe is shown in Fig. 3. It consists of an encoder and a decoder. The encoder f is a small CNN (3D ResNet18 [11,29] in our experiments), which takes 3D voxel grids $\mathbf{X} \in \mathbb{R}^{c \times d \times h \times w}$ as inputs and extracts a feature pyramid $\mathbf{F}_1, \mathbf{F}_2, ..., \mathbf{F}_n$. During the decoding stage, 3D continuous coordinates $\mathbf{p} \in [-1, 1]^3$ are sampled from the entire space. This is different than traditional segmentation methods, where the model runs on discrete voxel grids. These locations are queried using tri-linear interpolation to extract features at each coordinate $\mathbf{F}_1(\mathbf{p}), \mathbf{F}_2(\mathbf{p}), ..., \mathbf{F}_n(\mathbf{p})$. These features are concatenated with point coordinates $\mathbf{p}$ to form point encodings $\mathbf{F}(\mathbf{p})$, which are then fed into the decoder g, a two-layer MLP predicting segment occupancy at each location.

Compared with fully-convolutional methods [2,26,28], the ImPulSe architecture has two major advantages: it delivers continuous outputs with arbitrarily high resolutions and fine-grained details, while significantly reducing the number of parameters in its decoder. More importantly, thanks to implicit fields, ImPulSe can be trained on continuous locations instead of discrete locations on the voxel grid. These advantages render ImPulSe with high training efficiency while achieving better accuracy than traditional segmentation networks.

2.4 Training and Inference

During training and inference, ImPulSe alternates between random sampling and uniform sampling (Fig. 3). During the training phase, we train ImPulSe on randomly sampled points $\mathbf{p} \in [-1, 1]^3$. Random and continuous sampling implicitly imposes data augmentation and alleviates overfitting while ensuring sufficient

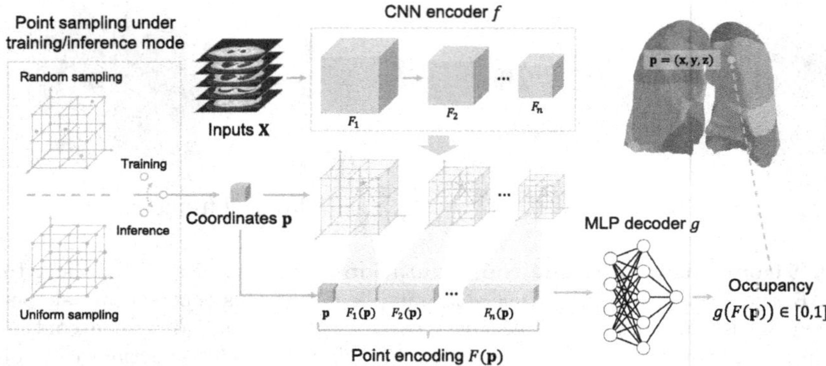

Fig. 3. The architecture of ImPulSe. ImPulSe consists of an encoder and a decoder. The encoder f takes a voxel grid $\mathbf{X}$ as input and outputs a feature pyramid $\mathbf{F}_1, \mathbf{F}_2, ..., \mathbf{F}_n$. Features at continuous coordinates $\mathbf{p}$, *i.e.*, $\mathbf{F}_1(\mathbf{p}), \mathbf{F}_2(\mathbf{p}), ..., \mathbf{F}_n(\mathbf{p})$, are sampled with tri-linear interpolation and then concatenated with $\mathbf{p}$ to form point encodings $\mathbf{F}(\mathbf{p})$. The decoder g takes $\mathbf{F}(\mathbf{p})$ and predicts occupancy at $\mathbf{p}$. Note that ImPulSe samples far fewer points during training than inference.

coverage of the entire space. Furthermore, it leads to a higher training efficiency compared against canonical segmentation methods, which trains on the entire voxel grid. With random sampling, we are able to train ImPulSe with far fewer points in each batch (*e.g.*, 16^3 for ImPulSe versus 64^3 or 128^3 for its counterparts in our experiments) while achieving better performances. Ground-truth labels on these continuous positions are queried using nearest-neighbor interpolation. The ImPulSe network is trained using a weighted combination of cross-entropy loss and Dice loss [17]. All models are trained on 4 NVIDIA RTX 3090 GPUs with PyTorch 1.10.1 [22]. During inference, we replace the random sampling with uniform sampling on the voxel grid, thus output pulmonary segment reconstruction at the original resolution of the CT image.

3 Experiments

3.1 Datasets

In this study, we compile a dataset containing 800 CT scans with annotations of pulmonary segments, pulmonary bronchi, arteries and veins. To allow elaborative analysis of reconstruction performances, pulmonary veins are further annotated as intrasegmental and intersegmental veins. CT scans are collected from multiple medical centers to improve the generalization performance of our model. Z-direction spacings of these scans range from 0.5 mm to 1.5 mm. Annotations are manually made by a junior radiologist and confirmed by a senior radiologist. All CT scans are divided into training, validation and test subsets with a 7:1:2 ratio. Results are tuned using the validation set and reported on the test set.

Table 1. Pulmonary segment reconstruction performances of ImPulSe and its counterparts. All methods are evaluated on Dice score (%) of pulmonary segments ($Dice_{\mathbf{o}}$), segmental bronchi ($Dice_{\mathbf{b}}$) and segmental arteries ($Dice_{\mathbf{a}}$). Best metrics are highlighted in bold. With fewer parameters and less training time, ImPulSe achieves better performances than its counterparts. Tr-res and Tr-time denote the input resolution during training and total training time (same number of epochs across models). DNC stands for "did not converge".

Methods	Tr-Res	$Dice_{\mathbf{o}}$	$Dice_{\mathbf{b}}$	$Dice_{\mathbf{a}}$	#Param	Tr-Time
Rikxoort et al. [25]	128^3	60.63	66.01	67.54	38.9 M	11,009 s
FCN [28]	128^3	81.70	83.67	84.92	33.4 M	10,746 s
U-Net [26]	128^3	83.25	84.71	86.07	38.9 M	11,009 s
DeepLabv3 [2]	128^3	82.44	84.48	85.64	44.4 M	11,220 s
Sliding-Window	64^3	0.11	0.05	0.02	38.9M	DNC
ImPulSe	16^3	**83.54**	**85.14**	**86.26**	**33.3 M**	**4,962 s**

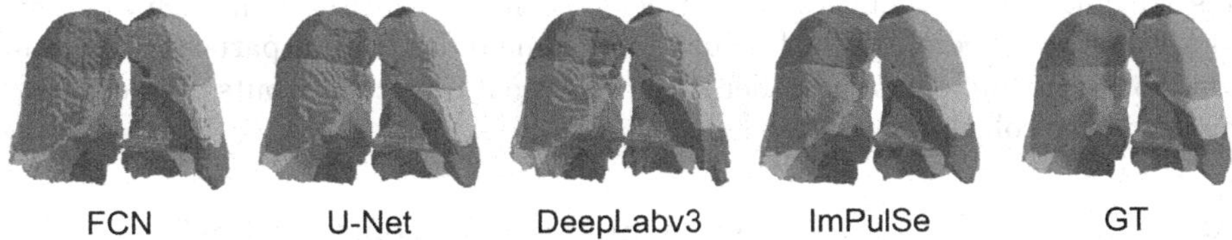

Fig. 4. Predictions of ImPulSe and some of its counterparts. Thanks to implicit functions, ImPulSe directly generates reconstruction of original resolution and therefore outputs predictions with smoother surfaces compared with its counterparts.

3.2 Reconstruction Performances

We compare ImPulse and its counterparts, including FCN [28], U-Net [26] and DeepLabv3 [2] in the pulmonary segment reconstruction task. Additionally, we roughly reproduce the pulmonary segment reconstruction method in Rikxoort et al. [25] by feeding relative coordinates into a U-Net [26]. Finally, we train an extra U-Net counterpart using the sliding-window strategy applied on raw resolution image. All models share the same encoder architecture (ResNet18 [11,29]), loss and training schedule to ensure fair comparison. Only CT images are fed as inputs in this experiment. Images are cropped around the lung area and downsampled to the size of 128^3 in pre-processing except the sliding-window experiment. During inference, ImPulSe directly outputs reconstructions of the original resolution, while predictions of all counterparts (except sliding-window) are upsampled to the original resolution with tri-linear interpolation. Table 1 shows performances of ImPulSe and its counterparts. ImPulSe achieves higher performances than its counterparts in all metrics with 14% fewer parameters and 55% less training time (compared against U-Net). The light-weighted decoder of ImPulSe (a two-layer MLP) leads to its parameter efficiency, while implicit functions enable training on

Table 2. Reconstruction performances of ImPulSe given different combinations of inputs. *I*, *L*, *B*, *A*, and *V* represent images, lobes, bronchi, arteries and veins, respectively. We evaluate ImPulSe on both ground-truth and predictions of *L*, *B*, *A* and *V*. Metrics outside/inside parentheses are evaluated on predictions/ground-truth (not applicable for *I*). Best metrics on predictions are highlighted in bold.

Inputs	*L*	*BAV*	*LBAV*	*I*	*IBAV*
$Dice_{\mathbf{o}}$	4.03 (73.77)	81.61 (81.55)	72.25 (83.21)	83.54 (n/a)	**84.63** (84.88)
$Dice_{\mathbf{b}}$	6.11 (79.01)	85.72 (85.65)	84.83 (86.77)	85.14 (n/a)	**86.18** (86.08)
$Dice_{\mathbf{a}}$	6.01 (80.62)	86.61 (86.75)	85.53 (88.02)	86.26 (n/a)	**87.00** (87.15)

a small portion of points rather than the entire voxel grid (16^3 for ImPulSe versus 64^3 or 128^3 for its counterparts) and largely save training time. Sliding-window U-Net does not even converge. This is because learning pulmonary segments requires global information, which is missing in sliding windows. Plus, sliding-window creates extreme class imbalance since only a few segments are present in each window. Figure 4 shows predictions of ImPulSe and some of its counterparts. The reconstruction of ImPulSe has smoother surfaces since it directly outputs prediction at the original resolution.

3.3 What Makes for Pulmonary Segment Reconstruction?

In this section, we evaluate effects of different inputs in reconstruction of pulmonary segments. Various combinations of the following five inputs are considered: CT images (*I*), bronchi (*B*), arteries (*A*), veins (*V*) and lobes (*L*). To obtain these inputs for inference, we train separate models for bronchus, vessel and lobe segmentation. These models are trained using the same dataset and splits as ImPulSe. For all input combinations containing *L*, *B*, *A* and *V*, ImPulSe is trained using ground-truths as input and evaluated using both ground-truths and predictions of aforementioned segmentation models. Table 2 shows performances of ImPulSe given different inputs. Metrics in parentheses are evaluated using ground-truth of *L*, *B*, *A* and *V* as input. Our findings are listed as follows:

Bronchi and Vessels Play Important Roles. Even though *I* outperforms *BAV* in overall Dice score, *BAV* surpasses *I* in bronchi and arteries. Since $Dice_{\mathbf{b}}$ and $Dice_{\mathbf{a}}$ are more important in pulmonary segment reconstruction than the overall Dice, it is clear that bronchi and vessels play crucial roles in this task. Plus, bronchi and vessels are less noisy than images since they are binary while images are encoded in HU values of wide ranges.

Images and BAV are Complementary to Each Other. IBAV achieves the highest performances in all metrics. We hypothesize that this is due to the complementary effect of *I* and *BAV*. *I* signals some basic knowledge not presented in *BAV*, *e.g.*, contours of lungs. *BAV* highlights bronchi and vessels, which have to be implicitly learned if only image inputs are given.

Lobe as Input is Prone to Overfitting. Adding lobes into inputs largely impairs performances (over 9% overall Dice drop from *BAV* to *LBAV*). We attribute this to the over-simplified structure of pulmonary lobes, which makes neural networks prone to overfitting. This is backed up by the fact that ImPulse only shows large accuracy drops transferring from ground-truth to prediction when it takes lobes as (part of) inputs. Furthermore, lobes include no information related to segmental boundaries, making it difficult to subdivide lobes into segments.

4 Conclusions

In this paper, we investigate this question: *what makes for automatic reconstruction of pulmonary segments?* We first set concrete definitions of pulmonary segment anatomy, and suggest approapraite metrics for the reconstruction task. We propose ImPulSe, an implicit-function-based model for pulmonary segment reconstruction. Using implicit fields, ImPulSe generates outputs of arbitrary resolutions with higher training and parameter efficiency compared with its counterparts. In future works, we will investigate if template deformation [8,14,31,33] and including intersegmental veins are beneficial in this task.

Acknowledgment. This work was supported in part by National Science Foundation of China (82071990, 61976238). This work was also supported in part by a Swiss National Science Foundation grant. We would like to thank Lei Liang for his generous help in proofreading, and the anonymous (meta-)reviewers for their valuable comments.

References

1. Boyden, E.A.: The intrahilar and related segmental anatomy of the lung. Surgery **18**, 706–31 (1945)
2. Chen, L.C., Papandreou, G., Schroff, F., Adam, H.: Rethinking atrous convolution for semantic image segmentation. arXiv Preprint abs/1706.05587 (2017)
3. Chen, Z., Zhang, H.: Learning implicit fields for generative shape modeling. In: Conference on Computer Vision and Pattern Recognition, pp. 5932–5941 (2019)
4. Chibane, J., Alldieck, T., Pons-Moll, G.: Implicit functions in feature space for 3d shape reconstruction and completion. In: Conference on Computer Vision and Pattern Recognition, pp. 6968–6979 (2020)
5. Frick, A.E., Raemdonck, D.V.: Segmentectomies. Shanghai. Chest **1**, 28 (2017)
6. Gerard, S.E., Patton, T.J., Christensen, G., Bayouth, J.E., Reinhardt, J.M.: Fissurenet: a deep learning approach for pulmonary fissure detection in CT images. IEEE Trans. Med. Imaging **38**, 156–166 (2019)
7. Gerard, S.E., Reinhardt, J.M.: Pulmonary lobe segmentation using a sequence of convolutional neural networks for marginal learning. In: International Symposium on Biomedical Imaging, pp. 1207–1211 (2019)
8. Groueix, T., Fisher, M., Kim, V.G., Russell, B.C., Aubry, M.: 3d-coded: 3d correspondences by deep deformation. In: European Conference on Computer Vision (2018)
9. Handa, Y., Tsutani, Y., Mimae, T., Miyata, Y., Okada, M.: Postoperative pulmonary function after complex segmentectomy. Ann. Surg. Oncol. **28**, 8347–8355 (2021)

10. Harada, H., Okada, M., Sakamoto, T., Matsuoka, H., Tsubota, N.: Functional advantage after radical segmentectomy versus lobectomy for lung cancer. Ann. Thorac. Surg. **80**, 2041–2045 (2005)
11. He, K., Zhang, X., Ren, S., Sun, J.: Deep residual learning for image recognition. In: Conference on Computer Vision and Pattern Recognition, pp. 770–778 (2016)
12. Huang, X., et al.: Representation-agnostic shape fields. In: International Conference on Learning Representations (2022)
13. Isensee, F., Jaeger, P.F., Kohl, S.A.A., Petersen, J., Maier-Hein, K.H.: nnU-Net: a self-configuring method for deep learning-based biomedical image segmentation. Nat. Methods **18**, 203–211 (2021)
14. Jack, D., et al.: Learning free-form deformations for 3d object reconstruction. In: Asian Conference on Computer Vision (2018)
15. Jackson, C.L., Huber, J.F.: Correlated applied anatomy of the bronchial tree and lungs with a system of nomenclature. Chest **9**, 319–326 (1943)
16. Mescheder, L.M., Oechsle, M., Niemeyer, M., Nowozin, S., Geiger, A.: Occupancy networks: learning 3d reconstruction in function space. In: Conference on Computer Vision and Pattern Recognition, pp. 4455–4465 (2019)
17. Milletari, F., Navab, N., Ahmadi, S.A.: V-net: fully convolutional neural networks for volumetric medical image segmentation. In: International Conference on 3D Vision, pp. 565–571 (2016)
18. Nardelli, P., et al.: Pulmonary artery-vein classification in CT images using deep learning. IEEE Trans. Med. Imaging **37**, 2428–2440 (2018)
19. Oizumi, H., et al.: Anatomic thoracoscopic pulmonary segmentectomy under 3-dimensional multidetector computed tomography simulation: a report of 52 consecutive cases. J. Thorac. Cardiovasc. Surg. **141**, 678–682 (2011)
20. Oizumi, H., Kato, H., Endoh, M., Inoue, T., Watarai, H., Sadahiro, M.: Techniques to define segmental anatomy during segmentectomy. Ann. Cardiothorac. Surg. **3**, 170–175 (2014)
21. Park, J.J., Florence, P.R., Straub, J., Newcombe, R.A., Lovegrove, S.: Deepsdf: learning continuous signed distance functions for shape representation. In: Conference on Computer Vision and Pattern Recognition, pp. 165–174 (2019)
22. Paszke, A., Gross, S., Massa, F., Lerer, A., et al.: Pytorch: an imperative style, high-performance deep learning library. In: Advances in Neural Information Processing Systems (2019)
23. Peng, S., Niemeyer, M., Mescheder, L., Pollefeys, M., Geiger, A.: Convolutional occupancy networks. In: European Conference on Computer Vision (2020)
24. Qin, Y., et al.: Learning tubule-sensitive CNNs for pulmonary airway and artery-vein segmentation in CT. IEEE Trans. Med. Imaging **40**, 1603–1617 (2021)
25. van Rikxoort, E.M., de Hoop, B., van Amelsvoort-van de Vorst, S., Prokop, M., van Ginneken, B.: Automatic segmentation of pulmonary segments from volumetric chest CT scans. IEEE Trans. Med. Imaging **28**, 621–630 (2009)
26. Ronneberger, O., Fischer, P., Brox, T.: U-net: convolutional networks for biomedical image segmentation. In: Conference on Medical Image Computing and Computer Assisted Intervention (2015)
27. Schuchert, M.J., Pettiford, B.L., Keeley, S., et al.: Anatomic segmentectomy in the treatment of stage i non-small cell lung cancer. Ann. Thorac. Surg. **84**, 926–933 (2007)
28. Shelhamer, E., Long, J., Darrell, T.: Fully convolutional networks for semantic segmentation. IEEE Trans. Pattern Anal. Mach. Intell. **39**, 640–651 (2017)

29. Tran, D., Wang, H., Torresani, L., Ray, J., LeCun, Y., Paluri, M.: A closer look at spatiotemporal convolutions for action recognition. In: Conference on Computer Vision and Pattern Recognition, pp. 6450–6459 (2018)
30. Ugalde, P., de Jesus Camargo, J., Deslauriers, J.: Lobes, fissures, and bronchopulmonary segments. Thorac. Surg. Clin. **17**(4), 587–599 (2007)
31. Wang, Y., Aigerman, N., Kim, V.G., Chaudhuri, S., Sorkine-Hornung, O.: Neural cages for detail-preserving 3d deformations. In: Conference on Computer Vision and Pattern Recognition, pp. 72–80 (2020)
32. Wisnivesky, J.P., et al.: Limited resection for the treatment of patients with stage IA lung cancer. Ann. Surg. **251**, 550–554 (2010)
33. Yang, J., Wickramasinghe, U., Ni, B., Fua, P.: Implicitatlas: learning deformable shape templates in medical imaging. In: Conference on Computer Vision and Pattern Recognition, pp. 15861–15871 (2022)
34. Zhou, Z., Siddiquee, M.M.R., Tajbakhsh, N., Liang, J.: Unet++: redesigning skip connections to exploit multiscale features in image segmentation. IEEE Trans. Med. Imaging **39**, 1856–1867 (2020)

CFDA: Collaborative Feature Disentanglement and Augmentation for Pulmonary Airway Tree Modeling of COVID-19 CTs

Minghui Zhang[1], Hanxiao Zhang[1], Guang-Zhong Yang[1], and Yun Gu[1,2](✉)

[1] Institute of Medical Robotics, Shanghai Jiao Tong University, Shanghai, China
[2] Shanghai Center for Brain Science and Brain-Inspired Technology, Shanghai, China
geron762@sjtu.edu.cn

Abstract. Detailed modeling of the airway tree from CT scan is important for 3D navigation involved in endobronchial intervention including for those patients infected with the novel coronavirus. Deep learning methods have the potential for automatic airway segmentation but require large annotated datasets for training, which is difficult for a small patient population and rare cases. Due to the unique attributes of noisy COVID-19 CTs (e.g., ground-glass opacity and consolidation), vanilla 3D Convolutional Neural Networks (CNNs) trained on clean CTs are difficult to be generalized to noisy CTs. In this work, a Collaborative Feature Disentanglement and Augmentation framework (CFDA) is proposed to harness the intrinsic topological knowledge of the airway tree from clean CTs incorporated with unique bias features extracted from the noisy CTs. Firstly, we utilize the clean CT scans and a small amount of labeled noisy CT scans to jointly acquire a bias-discriminative encoder. Feature-level augmentation is then designed to perform feature sharing and augmentation, which diversifies the training samples and increases the generalization ability. Detailed evaluation results on patient datasets demonstrated considerable improvements in the CFDA network. It has been shown that the proposed method achieves superior segmentation performance of airway in COVID-19 CTs against other state-of-the-art transfer learning methods.

Keywords: Feature Disentanglement and Augmentation · Airway Segmentation of COVID-19 CTs · Collaborative learning

1 Introduction

The novel coronavirus disease (COVID-19) has turned into a pandemic since the beginning of 2020, threatening the lives of people all over the world and causing an unprecedented social and economic crisis. Most infected patients are manifested in fever, dry cough, and malaise. Some patients progress rapidly with acute respiratory distress syndrome and multiple organ failure, which may result

L. Wang et al. (Eds.): MICCAI 2022, LNCS 13431, pp. 506–516, 2022.
https://doi.org/10.1007/978-3-031-16431-6_48

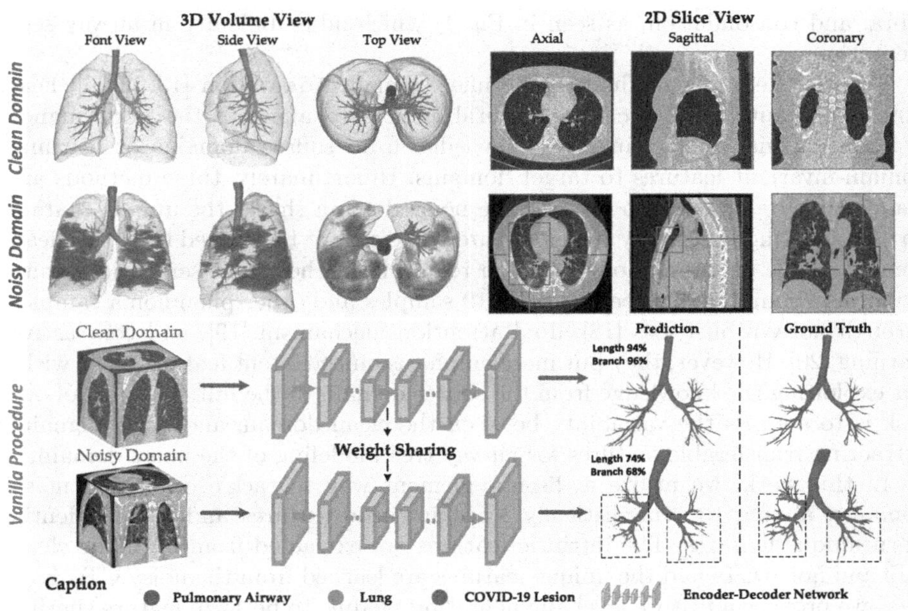

Fig. 1. Compared with the clean domain, the noisy domain introduces unique imaging patterns that affect airways. A well-trained model in the clean domain leads to low accuracy when applied in the noisy domain.

in permanent damage or even death [2,21]. Bronchoscopic-assisted intervention plays a valuable role in treating COVID-19 patients, which can help sputum suction and endobronchial intervention. The pulmonary airway tree model is essential to 3D navigated bronchoscopy. Meanwhile, the airway tree model can also provide guidance for the bronchoalveolar lavage fluid to extract cells, which can be used in analyzing the etiology and pathogenesis of COVID-19 diseases, assessing the immune responses, and evaluating the efficacy of the treatment or prognosis [16,25].

However, extraction of the detailed virtual airway tree model particularly involving distal airways remains challenging. The fine-grained pulmonary airway structure demands significant effort for manual delineation by experienced clinicians, motivating the development of automated segmentation algorithms. Bruijne *et al.* [8] organized the EXACT-09 airway segmentation challenge, and Qin *et al.* [12] published a Binary Airway Segmentation Dataset (BAS) with annotation. Thanks to these preliminary contributions, the data-driven deep learning models have been continuously proposed and promoted for airway segmentation [5,7,11,23,24,27]. However, these algorithms were developed on relative clean CT scans (termed as **clean domain**), while the COVID-19 CT scans (**noisy domain**) introduce the bias attributes, e.g., bilaterally scattered irregular patches of ground glass opacity, thickening of inter-lobular or intra-lobular septa, and consolidation, as seen in Fig. 1, which adds difficulty in airway segmentation.

Transfer Learning methods (Finetune, Domain Adaptation [13,22,26], Feature Alignment [1,17], etc.) are natural choices to alleviate the performance degradation via transferring the knowledge from source domains or learning domain-invariant features to target domains. Unfortunately, these methods are inadequate to our scenario because the noisy domain shares the unique characteristic affecting the airway tree structure while cannot be learned from the clean domain. Some other methods focus on regularizing the latent semantic feature being discriminative between COVID-19 samples and other pneumonia samples through noisy-robust loss [18], dual attention mechanism [19], and contrastive learning [20]. However, they put more emphasis on the latent feature space without exploiting the knowledge from the source domain to the full. The goal of our task is to address the variability between the clean domain and noisy domain, extracting transferable features for airway tree modeling of the noisy domain.

In this work, we utilize a disentanglement way to tackle clean and noisy domains, aiming to synergistically learn intrinsic features and independently learn unique features. The intrinsic features are extracted from both the clean CTs and noisy CTs and the unique features are learned from the noisy CTs. Further, we propose a feature level augmentation module to perform feature sharing and augmentation, which diversifies the training samples and increases the generalization ability. Besides, although the variability can be addressed via manual annotation of the new noisy domain, it is impractical for novel diseases because the voxel-wise annotation is difficult, time-consuming, the quality of annotation and the scale of a dataset is hard to guarantee. Inspired by Mahmud *et al.* [10] and Jin *et al.* [6], they firstly optimized the lesion segmentation network with a small amount of lesion labeled samples then integrated it into the following training procedure for large scale COVID-19 diagnosis and severity prediction. To enhance the effectiveness of disentanglement, we utilize a large amount of clean CTs with airway annotation (BAS dataset [12]), collaborating with the small amount of COVID-19 CTs with lesion annotation (CL dataset) [9] and airway annotation (CA dataset) [26]. This Collaborative learning method is integrated into the Feature Disentanglement and Augmentation module to construct the proposed CFDA network. Compared with other state-of-the-art methods, extensive experiments demonstrated that our method revealed the superiority in airway tree modeling under the evaluation of tree length detected rate and the branch detected rate.

2 Method

To harness the intrinsic topological knowledge from the clean domain then cooperate with unique bias features extracted from the noisy domain, we design the Collaborative Feature Disentanglement and Augmentation (CFDA) network. In this section, we detail the architecture of CFDA, which is illustrated in Fig. 2.

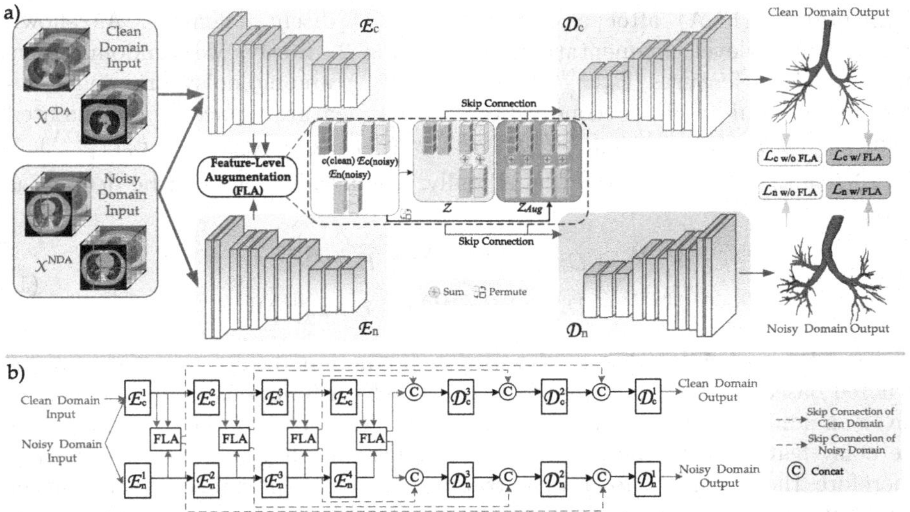

Fig. 2. a) demonstrates the overview of the proposed CFDA network. E_c extracts the sharable features from $\mathcal{X}^{\mathrm{CDA}}$ and $\mathcal{X}^{\mathrm{NDA}}$. E_n acquires the bias features of $\mathcal{X}^{\mathrm{NDA}}$. D_c and D_n reconstruct airways for clean and noisy domains respectively. As seen in b), the FLA module interacts in clean and noisy domains at each stage of the encoders, and generates augmented features for D_c and D_n respectively.

2.1 Feature Disentanglement and Augmentation

The Clean Domain contains CT scans with Airway annotation (CDA), $(\mathcal{X}^{\mathrm{CDA}}, \mathcal{Y}^{\mathrm{CDA}}) = \{(x_i^{\mathrm{CDA}}, y_i^{\mathrm{CDA}})\}_{i=1}^{n_{\mathrm{CDA}}}$, and $(\mathcal{X}^{\mathrm{NDA}}, \mathcal{Y}^{\mathrm{NDA}}) = \{(x_j^{\mathrm{NDA}}, y_j^{\mathrm{NDA}})\}_{j=1}^{n_{\mathrm{NDA}}}$ represents the samples from the Noisy Domain with Airway annotation (NDA), where $\mathcal{X}$ represents the input CT scans and $\mathcal{Y}$ represents the corresponding airway ground truth. Although the annotated samples of the noisy domain are limited, they share intrinsic topological airway structures that can be transferred from a relatively larger clean domain. The unique bias features of the noisy domain should be learned independently. Therefore, two encoders E_c and E_n with their decoders D_c and D_n are constructed. As illustrated in Fig. 2, the vanilla disentanglement takes advantage of E_c to extract the sharable features from both clean and noisy domains, and the E_n is designed to acquire the bias features. These latent features are then fed into separate decoders for airway segmentation: $\hat{y}_i^{\mathrm{CDA}} = D_c \circ E_c(x_i^{\mathrm{CDA}}), \hat{y}_j^{\mathrm{NDA}} = D_n \circ [E_c(x_j^{\mathrm{NDA}}) \oplus E_n(x_j^{\mathrm{NDA}})]$, where $\hat{y}_i^{\mathrm{CDA}}$ and $\hat{y}_j^{\mathrm{NDA}}$ represent the airway prediction of the clean domain and noisy domain, respectively. $\oplus$ denotes the layer-wise sum operation for feature interaction between clean and noisy domains.

While such a structure disentangles the intrinsic topological features and the unique bias features, the restriction for encoders is weak, D_c and D_n still tend to learn shortcuts for segmentation with given labels. To promote the performance of learning intrinsic airway tree features, we schedule the feature-level augmentation (FLA) after a certain degree of disentanglement. As shown in the feature-level augmentation part of Fig. 2, firstly, we randomly permute the bias features $E_n(x_j^{\text{NDA}})$ to $\widetilde{E_n}(x_j^{\text{NDA}})$ among noisy training samples in a mini-batch. Secondly, the latent feature space is augmented from $\mathcal{Z} = [E_c(x_i^{\text{CDA}}); E_c(x_j^{\text{NDA}}) \oplus E_n(x_j^{\text{NDA}})]$ to $\mathcal{Z}_{\text{Aug}} = [E_c(x_i^{\text{CDA}}) \oplus \widetilde{E_n}(x_j^{\text{NDA}}); E_c(x_j^{\text{NDA}}) \oplus \widetilde{E_n}(x_j^{\text{NDA}})]$. Finally, the decoder part can be defined as follows:

$$\begin{aligned} \tilde{y}_i^{\text{CDA}} &= D_c \circ [E_c(x_i^{\text{CDA}}) \oplus \widetilde{E_n}(x_j^{\text{NDA}})] \\ \tilde{y}_j^{\text{NDA}} &= D_n \circ [E_c(x_j^{\text{NDA}}) \oplus \widetilde{E_n}(x_j^{\text{NDA}})], \end{aligned} \tag{1}$$

where $\tilde{y}_i^{\text{CDA}}$ and $\tilde{y}_j^{\text{NDA}}$ denote the airway prediction results that were reconstructed based on the augmented features. The FLA functions simultaneously in the clean domain and noisy domain. As for the clean domain, the FLA integrates the noisy features into the clean features, generating hard samples for the D_c. Therefore, the E_c and D_c focus on extracting intrinsic topological features rather than learning shortcuts for airway segmentation. As for the noisy domain, the FLA diversify the noisy samples via introducing different bias features among the mini-batches, which increases the generalization ability of the E_n and D_n. The loss function of both the clean domain and the noisy domain consists of the FLA part ($\mathcal{L}_{\text{n w/ FLA}}$) and non-FLA part ($\mathcal{L}_{\text{n w/o FLA}}$):

$$\begin{aligned} \mathcal{L}(\hat{y}_i^{\text{CDA}}, y_i^{\text{CDA}}) &= \mathcal{L}_{\text{c w/o FLA}}(\hat{y}_i^{\text{CDA}}, y_i^{\text{CDA}}) + \alpha \mathcal{L}_{\text{c w/ FLA}}(\tilde{y}_i^{\text{CDA}}, y_i^{\text{CDA}}) \\ \mathcal{L}(\hat{y}_j^{\text{NDA}}, y_j^{\text{NDA}}) &= \mathcal{L}_{\text{n w/o FLA}}(\hat{y}_j^{\text{NDA}}, y_j^{\text{NDA}}) + \alpha \mathcal{L}_{\text{n w/ FLA}}(\tilde{y}_j^{\text{NDA}}, y_j^{\text{NDA}}), \end{aligned} \tag{2}$$

where $i = 1, 2, ..., n_{\text{CDA}}, j = 1, 2, ..., n_{\text{NDA}}$, $\mathcal{L}(\cdot, \cdot)$ in Eq. 2 is the Dice with Focal loss [28] and α is a balance term adjusting the weight of the feature-level augmentation.

2.2 Bias-Discriminative Encoder

$(\mathcal{X}^{\text{NDL}}, \mathcal{Y}^{\text{NDL}}) = \{(x_k^{\text{NDL}}, y_k^{\text{NDL}})\}_{k=1}^{n_{\text{NDL}}}$ represents the samples from the Noisy Domain with Lesion annotation (NDL), where $\mathcal{X}$ denotes the input CT scans and $\mathcal{Y}$ represents the lesion annotation. We adopt a small amount of CT scans from the NDL and non-lesion annotation samples from the CDA and the NDA to construct a lightweight semi-supervised-triplet (SST) network. The SST network aims to further enhance the bias-discriminative ability of the E_n via feature discrimination and supervised learning. As illustrated in Fig. 3, the triplet loss [14] is used to regularize the intra-domain cohesion and inter-domain separation of the features extracted by E_n:

$$\mathcal{L}_D(z^{\text{NDL}}, z^{\text{NDA}}, z^{\text{CDA}}) = \max\{d(z^{\text{NDL}}, z^{\text{NDA}}) - d(z^{\text{NDL}}, z^{\text{CDA}}) + d_{margin}, 0\}, \tag{3}$$

where z^{NDL}, z^{NDA}, and z^{CDA} denote the latent features extracted from the NDL, NDA, CDA respectively. $d(\cdot, \cdot)$ represents the normalized Euclidean distance and

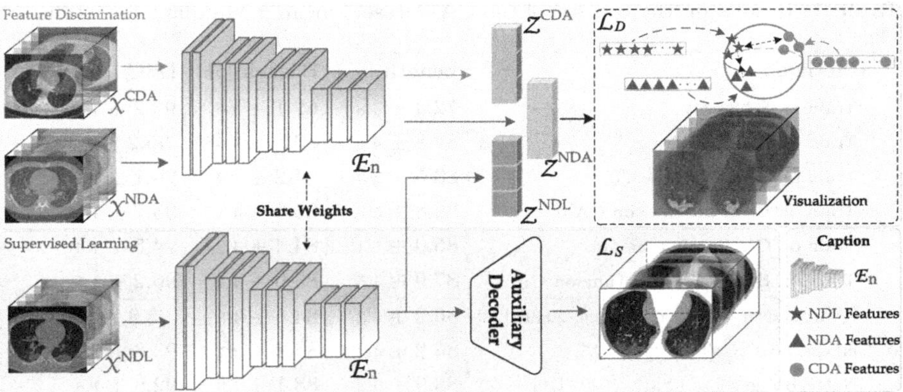

Fig. 3. The detailed structure of acquiring the BDE via the SST network. The feature discrimination part regularizes intra-domain cohesion and inter-domain separation. The supervised learning part aids E_n to be sensitive to regions of unique bias features.

d_{margin} is the margin of the distance with a value greater than 0. The auxiliary decoder aids the E_n to be more sensitive to regions of unique features. It also uses the Dice with Focal loss [28] as an objective function ($\mathcal{L}_S$) for lesion segmentation. The total loss function used for this SST network can be summarized as $\mathcal{L}_{SST} = \mathcal{L}_D + \mathcal{L}_S$. This collaborative learning of the Bias-Discriminative Encoder (BDE) could be viewed as a beneficial preliminary step for the FLA module, and it can be plugged into the proposed CFDA network in an end-to-end training fashion.

3 Experiments and Results

Dataset: Three datasets categorized into the clean domain and noisy domain were used in our work. 1) Clean Domain: Binary Airway Segmentation (BAS) dataset [12]. It contains 90 CT scans with airway annotation. The spatial resolution ranges from 0.5 to 0.82 mm and the slice thickness ranges from 0.5 to 1.0 mm. 2) Noisy Domain: 10 CT scans of the COVID-19 Lesion dataset (CL) [9] were utilized in our experiments. Their spatial resolution ranges from 0.47 to 0.86 mm and slice thickness ranges from 1.0 to 1.5 mm. COVID-19 Airway dataset (CA) [26] contains 58 COVID-19 patients with airway annotation, the spatial resolution of which ranges from 0.58 to 0.84 mm and slice thickness varies from 0.5 to 1.0 mm. We randomly split the CA dataset into 10 CT scans for training and 48 CT scans for testing.

Implementation Details: During preprocessing, the voxel values were truncated into $[-1000, 600]$ HU and normalized to $[0, 255]$. Sub-volume CT cubes with a size of $128 \times 224 \times 304$ densely cropped inside the lung field were adopted as the input. We sampled 20 patches of each scan in an epoch and chose a batch

Table 1. Results on the test set of the CA dataset (mean ± standard deviation).

Methods	Length (%)	Branch (%)	DSC (%)
Train on BAS only	72.4 ± 4.8	62.1 ± 4.5	93.2 ± 1.5
Train on BAS + CA	82.8 ± 4.8	83.8 ± 3.8	95.2 ± 1.3
Train on BAS + CA + CL	83.5 ± 5.4	84.2 ± 4.9	96.1 ± 1.7
Train on BAS, finetune on CA	86.8 ± 5.3	85.0 ± 4.1	95.7 ± 1.1
Train on CL, finetune on CA	85.0 ± 5.1	84.4 ± 4.2	94.5 ± 1.4
Train on BAS + CL, finetune on CA	87.9 ± 4.5	85.6 ± 4.4	96.2 ± 1.5
AnatomyNet (medical physics, 2019) [28]	86.2 ± 5.3	84.6 ± 4.8	95.8 ± 1.2
clDice Loss (CVPR, 2021) [15]	86.2 ± 4.9	84.1 ± 3.7	93.8 ± 1.2
Noisy-Robust loss (TMI, 2020) [18]	88.9 ± 5.2	88.1 ± 4.6	92.8 ± 1.8
Feature alignment (TMI, 2020) [1]	87.9 ± 4.9	85.5 ± 4.8	95.5 ± 1.6
Domain adaptation (TPAMI, 2018) [13]	87.0 ± 4.6	84.9 ± 4.0	96.0 ± 1.3
CDIE (JBHI, 2020) [20]	88.0 ± 3.4	86.2 ± 3.1	94.5 ± 1.2
MixStyle w/ AdaIN (ICCV, 2017) [4]	86.1 ± 4.9	84.7 ± 3.8	93.1 ± 1.5
Dual-Stream network (DART, 2021) [26]	90.2 ± 5.3	87.6 ± 4.2	**96.5 ± 1.2**
CFDA w/o BDE	91.9 ± 4.4	89.1 ± 4.2	96.1 ± 1.3
CFDA w/o FLA	91.2 ± 4.7	88.7 ± 4.4	96.2 ± 1.4
CFDA (proposed)	**92.8 ± 4.6**	**90.0 ± 4.3**	95.8 ± 1.4

size of 2 (randomly chose two clean CT scans and two noisy COVID-19 CT scans) to construct the mini-batch in the training phase. Random horizontal flipping and random rotation between $[-10°, 10°]$ were applied as the on-the-fly data augmentation. The encoder and decoder structures in Fig. 2 are based on the 3D UNet [3], where we used the pReLU and instance normalization instead of ReLU and batch normalization, respectively. All models were trained by Adam optimizer with an initial learning rate of 0.002. Ten preliminary epochs were used to acquire the pre-trained BDE via the SST network. The total epoch for the main CFDA framework was set to 30, and the learning rate was divided by 10 in the 20th epoch. The first five epochs were used as a warm-up solution for feature disentanglement, and the FLA module was introduced from the sixth epoch. The FLA module performed at each stage of the encoders. We chose $\alpha = 0.1$ and $d_{margin} = 3$ in experiments. The framework was implemented in PyTorch 1.7 with a single NVIDIA RTX A6000 GPU (48 GB graphical memory).

Quantitative Results: We adopted three metrics for evaluation: 1) tree length rate (Length, %) [8], 2) branch detected rate (Branch, %) [8], and 3) Dice score coefficient (DSC, %). Only the largest component of the prediction result is taken into account for evaluation. Experimental results in Table 1 show our proposed CFDA network achieved the highest performance of Length (92.8%), Branch (90.0%) with a compelling DSC (95.8%). Neither training on the single domain nor utilizing multi-task training across domains is ineffective for airway segmen-

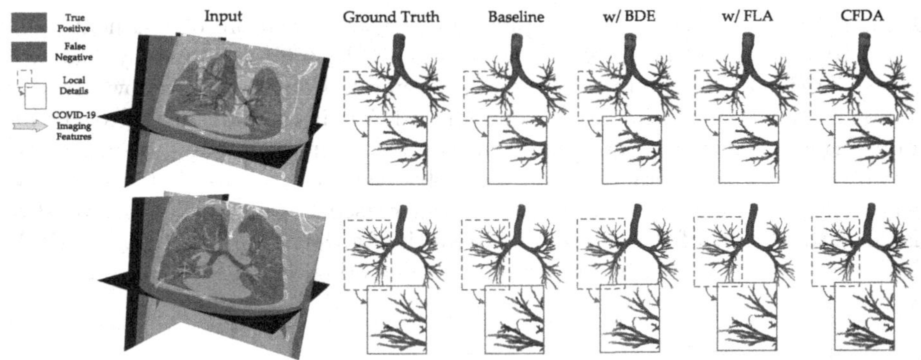

Fig. 4. Visualization of segmentation results. The true positives (TP) and false negatives (FN) are colored in red and green respectively, yellow arrows point out some COVID-19 imaging features. The blue dotted boxes indicate the local details magnified for better comparison. (Color figure online)

tation of noisy COVID-19 CTs, which implied the necessity of disentanglement and transfer learning rather than merely together different datasets. Finetune methods achieved limited improvement due to the chronic forgetting of features learned from the clean data. Under the effect of unique bias features of noisy CTs, some methods equipped with the attention mechanism [28] and robust loss function [15,18] could not generalize well. For comparison, several transfer learning methods were reimplemented to be applied in our scenario. Specifically, Feature Alignment [1] used adversarial training, Domain Adaptation [13] shared weights cross domains, and contrastive learning [20] enhanced domain-invariant features. We also introduced the noisy features as style input into the clean airway extraction procedure [4], and then finetuned on the noisy domain. However, it achieved merely 86.1% in Length, 84.7% in Branch, and 93.1% in DSC. This observation reveals that a sophisticated disentanglement should be designed for these two domains. The dual-stream network [26] tackles different domains while its decomposition ability is limited due to the weak restriction for encoders. We adopted this method as our baseline for comparison. The proposed FLA module is scheduled into the disentanglement framework and experimental results show that this can improve the Length to 91.9% and Branch to 89.1%. Further, combined with the BDE, the proposed CFDA can achieve the overall best performance. Our approach can detect more peripheral bronchi and guarantee better completeness of airway tree structure than other methods, which has clinical importance for bronchoscopy surgery. The corresponding qualitative results are demonstrated in Fig. 4. Its superior sensitivity also comes with a slight side effect of performance decline in DSC. This may be ascribed to that our model successfully detected several true tenuous airways that were too indistinct to be delineated accurately by clinicians. These real bronchi were misclassified as false positives and therefore leading to the slight decline of DSC.

Qualitative Results: Figure 4 presents the visualization of segmentation results. It is seen that the bronchi are surrounded by multifocal patchy shadowing of COVID-19 in the input CT scans, as pointed out by the yellow arrows. Compared with the baseline, the proposed CFDA gains significant improvement in detecting more bronchi and preserving better completeness under such noisy COVID-19 imaging characteristics. This observation confirms that the ability of feature disentanglement and sharing is increased by our approach. We also visualize the results of the ablation study, verifying that both the BDE and FLA components could boost the performance of detecting airways.

4 Conclusion

In this paper, we propose a Collaborative Feature Disentanglement and Augmentation (CFDA) framework to harness the intrinsic topological knowledge from the clean domain for airway segmentation of noisy COVID-19 CTs. A feature-level augmentation module is built based on the disentangled representation of clean and noisy domains. We diversify training samples via permuting and recombining the latent vectors and further utilize a lightweight semi-supervised network to enhance the bias-discriminative ability of the noisy encoder. Compared with the previous works, the highest performance of airway tree modeling achieved by the proposed approach demonstrates the potential clinical value.

Acknowledgement. This work is supported in part by the Open Funding of Zhejiang Laboratory under Grant 2021KH0AB03, in part by the Shanghai Sailing Program under Grant 20YF1420800, and in part by NSFC under Grant 62003208, and in part by Shanghai Municipal of Science and Technology Project, under Grant 20JC1419500 and Grant 20DZ2220400.

References

1. Chen, C., Dou, Q., Chen, H., Qin, J., Heng, P.A.: Unsupervised bidirectional cross-modality adaptation via deeply synergistic image and feature alignment for medical image segmentation. IEEE Trans. Med. Imaging **39**(7), 2494–2505 (2020)
2. Chen, N., et al.: Epidemiological and clinical characteristics of 99 cases of 2019 novel coronavirus pneumonia in Wuhan, china: a descriptive study. Lancet **395**(10223), 507–513 (2020)
3. Çiçek, Ö., Abdulkadir, A., Lienkamp, S.S., Brox, T., Ronneberger, O.: 3D U-Net: learning dense volumetric segmentation from sparse annotation. In: Ourselin, S., Joskowicz, L., Sabuncu, M.R., Unal, G., Wells, W. (eds.) MICCAI 2016. LNCS, vol. 9901, pp. 424–432. Springer, Cham (2016). https://doi.org/10.1007/978-3-319-46723-8_49
4. Huang, X., Belongie, S.: Arbitrary style transfer in real-time with adaptive instance normalization. In: Proceedings of the IEEE International Conference on Computer Vision, pp. 1501–1510 (2017)

5. Jin, D., Xu, Z., Harrison, A.P., George, K., Mollura, D.J.: 3D convolutional neural networks with graph refinement for airway segmentation using incomplete data labels. In: Wang, Q., Shi, Y., Suk, H.-I., Suzuki, K. (eds.) MLMI 2017. LNCS, vol. 10541, pp. 141–149. Springer, Cham (2017). https://doi.org/10.1007/978-3-319-67389-9_17
6. Jin, S., et al.: AI-assisted CT imaging analysis for COVID-19 screening: building and deploying a medical AI system in four weeks. MedRxiv (2020)
7. Garcia-Uceda Juarez, A., Tiddens, H.A.W.M., de Bruijne, M.: Automatic airway segmentation in chest CT using convolutional neural networks. In: Stoyanov, D., et al. (eds.) RAMBO/BIA/TIA -2018. LNCS, vol. 11040, pp. 238–250. Springer, Cham (2018). https://doi.org/10.1007/978-3-030-00946-5_24
8. Lo, P., et al.: Extraction of airways from CT (exact'09). IEEE Trans. Med. Imaging **31**(11), 2093–2107 (2012)
9. Ma, J., et al.: Toward data-efficient learning: a benchmark for COVID-19 CT lung and infection segmentation. Med. Phys. **48**(3), 1197–1210 (2021)
10. Mahmud, T., et al.: Covtanet: a hybrid tri-level attention-based network for lesion segmentation, diagnosis, and severity prediction of COVID-19 chest CT scans. IEEE Trans. Industr. Inf. **17**(9), 6489–6498 (2020)
11. Nadeem, S.A., et al.: A CT-based automated algorithm for airway segmentation using freeze-and-grow propagation and deep learning. IEEE Trans. Med. Imaging **40**(1), 405–418 (2020)
12. Qin, Y., et al.: Learning tubule-sensitive CNNS for pulmonary airway and artery-vein segmentation in CT. IEEE Trans. Med. Imaging **40**(6), 1603–1617 (2021)
13. Rozantsev, A., Salzmann, M., Fua, P.: Beyond sharing weights for deep domain adaptation. IEEE Trans. Pattern Anal. Mach. Intell. **41**(4), 801–814 (2018)
14. Schroff, F., Kalenichenko, D., Philbin, J.: Facenet: A unified embedding for face recognition and clustering. In: Proceedings of the IEEE Conference on Computer Vision and Pattern Recognition, pp. 815–823 (2015)
15. Shit, S., et al.: cLDice-a novel topology-preserving loss function for tubular structure segmentation. In: Proceedings of the IEEE/CVF Conference on Computer Vision and Pattern Recognition, pp. 16560–16569 (2021)
16. Sulaiman, I., et al.: Microbial signatures in the lower airways of mechanically ventilated COVID-19 patients associated with poor clinical outcome. Nat. Microbiol. **6**(10), 1245–1258 (2021)
17. Sun, B., Saenko, K.: Deep CORAL: correlation alignment for deep domain adaptation. In: Hua, G., Jégou, H. (eds.) ECCV 2016. LNCS, vol. 9915, pp. 443–450. Springer, Cham (2016). https://doi.org/10.1007/978-3-319-49409-8_35
18. Wang, G., et al.: A noise-robust framework for automatic segmentation of COVID-19 pneumonia lesions from CT images. IEEE Trans. Med. Imaging **39**(8), 2653–2663 (2020)
19. Wang, J., et al.: Prior-attention residual learning for more discriminative COVID-19 screening in CT images. IEEE Trans. Med. Imaging **39**(8), 2572–2583 (2020)
20. Wang, Z., Liu, Q., Dou, Q.: Contrastive cross-site learning with redesigned net for COVID-19 CT classification. IEEE J. Biomed. Health Inform. **24**(10), 2806–2813 (2020)
21. Wu, W., Wang, A., Liu, M., et al.: Clinical features of patients infected with 2019 novel coronavirus in Wuhan, China. Lancet **395**(10223), 497–506 (2020)
22. Xu, G.X., et al.: Cross-site severity assessment of COVID-19 from CT images via domain adaptation. IEEE Trans. Med. Imaging **41**(1), 88–102 (2021)

23. Yu, W., Zheng, H., Zhang, M., Zhang, H., Sun, J., Yang, J.: BREAK: bronchi reconstruction by geodesic transformation and skeleton embedding. arXiv preprint arXiv:2202.00002 (2022)
24. Yun, J., et al.: Improvement of fully automated airway segmentation on volumetric computed tomographic images using a 2.5 dimensional convolutional neural net. Med. Image Anal. **51**, 13–20 (2019)
25. Zhang, K., Liu, X., Shen, J., Li, Z., Sang, Y., Wu, X., Zha, Y., Liang, W., Wang, C., Wang, K., et al.: Clinically applicable AI system for accurate diagnosis, quantitative measurements, and prognosis of COVID-19 pneumonia using computed tomography. Cell **181**(6), 1423–1433 (2020)
26. Zhang, M., et al.: FDA: feature decomposition and aggregation for robust airway segmentation. In: Albarqouni, S., et al. (eds.) DART/FAIR -2021. LNCS, vol. 12968, pp. 25–34. Springer, Cham (2021). https://doi.org/10.1007/978-3-030-87722-4_3
27. Zheng, H., et al.: Alleviating class-wise gradient imbalance for pulmonary airway segmentation. IEEE Trans. Med. Imaging **40**(9), 2452–2462 (2021)
28. Zhu, W., et al.: AnatomyNet: deep learning for fast and fully automated whole-volume segmentation of head and neck anatomy. Med. Phys. **46**(2), 576–589 (2019)

Decoupling Predictions in Distributed Learning for Multi-center Left Atrial MRI Segmentation

Zheyao Gao[1], Lei Li[3], Fuping Wu[1,2], Sihan Wang[1], and Xiahai Zhuang[1(✉)]

[1] School of Data Science, Fudan University, Shanghai, China
zxh@fudan.edu.cn
[2] Department of Statistics, Fudan University, Shanghai, China
[3] Department of Engineering Science, University of Oxford, Oxford, UK
https://www.sdspeople.fudan.edu.cn/zhuangxiahai/

Abstract. Distributed learning has shown great potential in medical image analysis. It allows to use multi-center training data with privacy protection. However, data distributions in local centers can vary from each other due to different imaging vendors, and annotation protocols. Such variation degrades the performance of learning-based methods. To mitigate the influence, two groups of methods have been proposed for different aims, i.e., the global methods and the personalized methods. The former are aimed to improve the performance of a single global model for all test data from unseen centers (known as generic data); while the latter target multiple models for each center (denoted as local data). However, little has been researched to achieve both goals simultaneously. In this work, we propose a new framework of distributed learning that bridges the gap between two groups, and improves the performance for both generic and local data. Specifically, our method decouples the predictions for generic data and local data, via distribution-conditioned adaptation matrices. Results on multi-center left atrial (LA) MRI segmentation showed that our method demonstrated superior performance over existing methods on both generic and local data. Our code is available at https://github.com/key1589745/decouple_predict.

Keywords: Left atrium · Distributed learning · Segmentation · Non-IID

This work was funded by the National Natural Science Foundation of China (grant no. 61971142, 62111530195 and 62011540404) and the development fund for Shanghai talents (no. 2020015).

Supplementary Information The online version contains supplementary material available at https://doi.org/10.1007/978-3-031-16431-6_49.

L. Wang et al. (Eds.): MICCAI 2022, LNCS 13431, pp. 517–527, 2022.
https://doi.org/10.1007/978-3-031-16431-6_49

1 Introduction

Left atrial (LA) segmentation from MRI is essential in the diagnosis and treatment planning of patients suffering from atrial fibrillation, but the automated methods remain challenging [9]. Deep learning has demonstrated great potential, provided a large-scale training set from multiple centers, which nevertheless impedes the advance due to the concern of data privacy. Distributed learning has then gained great attention as it trains a model on distributed datasets without the exchange of privacy-sensitive data between centers [14]. Recently, swarm learning [22], a decentralized distributed learning method, has been proposed for medical image analysis. This new model follows the training paradigm of federated learning (FedAvg) but without a central server. The communication between centers is secured through the blockchain network [13], which further guarantees fairness in distributed learning.

Although the above methods have provided solutions for the problem of privacy and fairness, the performance could be degraded due to the non-independently identical distribution (IID)-ness of medical data [5,14]. For left atrial (LA) MRI segmentation, the heterogeneous of data distribution could be particularly worse in two aspects. One is the divergence of image distributions (also known as feature skew), originated from the difference in imaging protocols, the strength of magnetic field, or demography. The other is the inconsistency of manual labeling (referred to as label skew) due to different annotation protocols or rater variations, as manual segmentation of LA MRIs is difficult even for experienced experts [24]. Figure 1 illustrates this inconsistency from different centers.

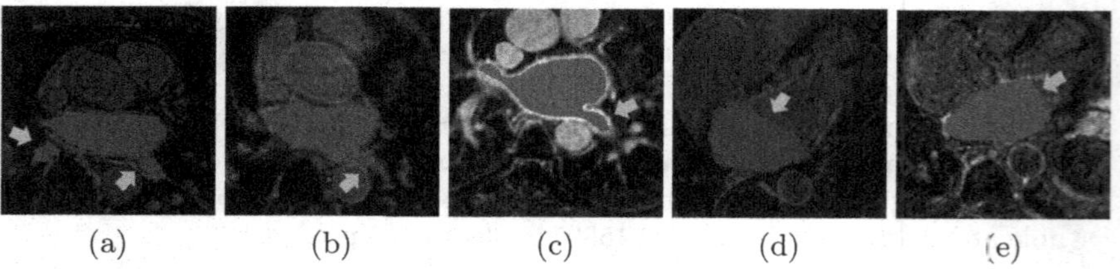

Fig. 1. Illustration of label skew in different centers. Examples are selected from public datasets [6,23]. Yellow arrows point out the segmentation bias: (a–c) show different protocols for segmenting pulmonary veins, (d-e) demonstrate inconsistency for the LA cavity. (Color figure online)

For distributed learning under data heterogeneity, solutions could be categorized as global methods or personalized methods. Global methods [7,11,17] usually regularize the local training to avoid the divergence of model updating in each center. Personalized methods [2,10,12] propose to keep distribution related parameters local or develop two sets of models through distillation. In general, global methods can perform well on generic data [5] while personalized methods have shown better results on local data. A recent work [1] has been proposed to approach the two seemingly contradictory objectives simultaneously with two predictors for image classification. Different from [1], we developed a novel distributed learning framework for LA segmentation, based on the variational Bayesian framework, which bridges the objectives of global methods and personalized methods.

In the segmentation task with non-IID data, little research has studied the influence of label skew caused by segmentation bias and variation in different centers. Therefore, we resort to the remedies for noisy label learning by considering the segmentation bias in each center as noises in the annotation. A large number of works [3,15,26] have studied the conditions for robust loss functions and proposed simple yet effective solutions for learning with label noise. Several studies [4,19] have proposed to add an adaptation layer on top of the network. A recent work [25] disentangled annotation errors from the segmentation label using an image dependent adaptation network. However, it requires annotations from multiple observers for a single case and the adaptation network could not learn the segmentation bias from labels. For our problem, we consider adapting the prediction based on the joint distribution of both image and label in each center through generative models.

In this work, our distributed learning method, tackling both problems of feature and label skew, improves the performance for both generic and local data. More specifically, our method decouples the local and global predictions through adaptation matrices conditioned on joint data distribution, thus prevents the divergence of model in local training and adapts the prediction to be consistent with local distribution during testing.

Our contribution has three folds. First, we propose a new distributed learning framework for LA MRI segmentation with non-IID data, which could improve the results on data from both unseen centers and centers involved in the training. Second, we propose a distribution-conditioned adaptation network to decouple the prediction with labels during training and adapt the prediction according

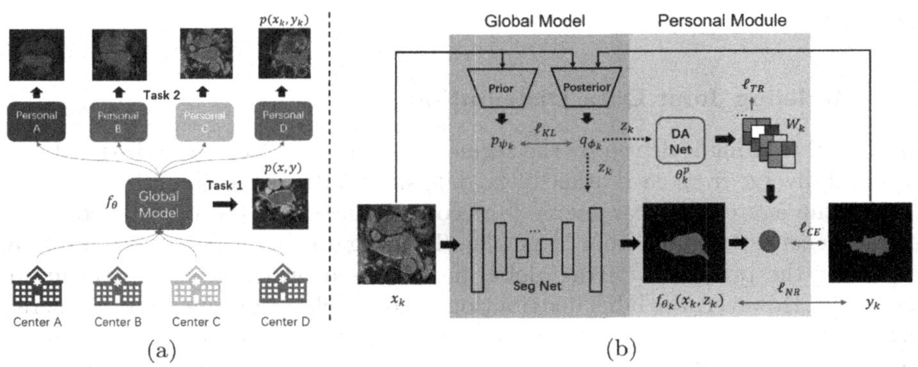

Fig. 2. Illustration of the objective (a) and framework of local training (b).

to local distribution in testing. Third, We evaluate the proposed framework on multi-center LA MRI data. The results show that our method outperforms existing methods on both generic data and local data.

2 Method

Figure 2 (a) illustrates two folds of our aim. One is to train a global model f_θ from multi-center non-IID data for Task1, and the other is to seek a personalized one for each center for Task2. The global model is expected to achieve consistent segmentation for generic data distribution [5] $p(x, y)$, *i.e.,* images from unseen centers; the personalized models, by contrast, modify predictions from the global model to favor the local distribution, e.g., $p(x_k, y_k)$ of the k-th center. The proposed method achieves the two goals simultaneously.

Based on swarm learning [22], our method involves periodically global aggregation and local training. In *global aggregation*, each center k collects locally updated parameters $\{(\theta_k)\}_{k=1}^K$ of the global model from other centers, and performs model aggregation weighted by the training size of each center (n_k), *i.e.,* $\theta = \sum_{k=1}^K \frac{n_k}{N}\theta_k$ where $N = \sum_{k=1}^K n_k$. Note that the personalized modules, denoted as $\{\theta_k^p\}$, are kept locally in this stage. In *local training*, each center then trains their own models, with the global model initialized as the aggregation result θ.

Figure 2 (b) shows the proposed framework of local training, consisting of four sub tasks respectively for segmentation network (θ_k^s), prior encoder (ψ_k), posterior encoder (ϕ_k) and the distribution adaptation (DA) network (θ_k^p). The former three maintain the information for updating the global model. The segmentation network generates predictions for generic data. Prior and posterior encoders are employed to model the joint data distribution $p(x_k, y_k)$ through a latent representation z_k. Thus, the DA network could output adaptation matrices W_k based on local distribution. The prediction for local data is obtained through the multiplication of adaptation matrices W_k and the prediction for generic data. In the following sections, we elaborate on the derivation and training details for each part of the model.

2.1 Modeling Joint Data Distribution

Since local training is driven by the segmentation risk on local data distribution, θ_k would diverge from each other if the data distributions are non-IID. This will deviate the aggregated parameter θ at convergence from the optimal solution for the generic test data from Task1 [5]. To mitigate this problem, we propose to modify the prediction of the global model based on local joint distribution $p_k(x_k, y_k)$. However, the label distribution is not available during testing. Therefore, we first model the latent representation z_k that contains the information about joint distribution following the formulation of conditional VAE [18].

Concretely, the objective is to estimate the posterior distribution $p(z_k|x_k, y_k)$. The key idea is to find a Gaussian distribution $q_{\phi_k}(z_k|x_k, y_k)$ to approximate the true posterior by Kullback-Leibler (KL) divergence. According to [18], the lower bound (LB) of the KL divergence is derived as,

$$LB = \mathbb{E}_{q_{\phi_k}(z_k|x_k,y_k)}[\log p(y_k|x_k, z_k)] - KL[q_{\phi_k}(z_k|x_k, y_k)||p_{\psi_k}(z_k|x_k)], \quad (1)$$

where the approximated prior $p_{\psi_k}(z_k|x_k)$ and posterior $q_{\phi_k}(z_k|x_k, y_k)$ are modeled by,

$$p_{\psi_k}(z_k|x_k) = \mathcal{N}\left(\mu_{prior}(x_k;\psi_k), \mathtt{diag}(\sigma_{prior}(x_k;\psi_k))\right), \quad (2)$$

$$q_{\phi_k}(z_k|x_k, y_k) = \mathcal{N}\left(\mu_{post}(x_k, y_k;\phi_k), \mathtt{diag}(\sigma_{post}(x_k, y_k;\phi_k))\right). \quad (3)$$

The parameters of the Gaussian distributions, $(\mu_{prior}, \sigma_{prior})$ and $(\mu_{post}, \sigma_{post})$ are generated by the prior and posterior network, respectively.

The posterior distribution of the latent representation z_k is approximated through maximizing Eq. (1). The KL term in the lower bound could be derived and maximized explicitly. The maximization of the first term is resolved through the training of segmentation network and DA network, which will be described in the following.

2.2 Decoupling Global and Local Predictions

To disentangle the effect of distribution shift on the prediction and avoid the parameters from drifting away during local training, we decouple the predictions for generic data with labels from local distribution. Formally, we decompose the first term in Eq. (1) as following,

$$\mathbb{E}_{q_{\phi_k}}[\log p(y_k|z_k, x_k)] = \mathbb{E}_{q_{\phi_k}}[\log \sum_{y\in\Omega} p_{\theta_k^p}(y_k|y, z_k)p_{\theta_k^s}(y|x_k, z_k)], \quad (4)$$

where $p_{\theta_k^p}(y_k|y, z_k)$ is modeled by the personalized module; $p_{\theta_k^s}(y|x_k, z_k)$ is modeled by the segmentation network; Ω is the set of all possible segmentation. Here, $p_{\theta_k^p}(y_k|y, z_k)$ is independent on x_k, as we assume that the difference between y and y_k is resulted by the shift of joint data distribution instead of x_k. In this way, the predictions for generic data and local data are decoupled. The DA network bridges the gap caused by the distribution shift.

To jointly train the segmentation network and DA network, we first apply cross-entropy loss ℓ_{CE} [20] to the multiplication of outputs from the two networks with y_k, *i.e.*, $\ell_{CE}(W_k \odot f_{\theta_k}(x_k, z_k), y_k; \theta_k)$, where $W_k \in \mathbb{R}^{C\times C\times M}$ are the adaptation matrices for all pixels. C and M represent the number of classes and pixels respectively. "$\odot$" here denotes the pixel-wise matrix multiplication. This loss ensures that the prediction for local data $W_k \odot f_{\theta_k}(x_k, z_k)$ is congruent with local labels.

However, ℓ_{CE} alone is not able to separate the effect of distribution shift from the global prediction. There are many combinations of θ_k^p and θ_k^s such that given an input image, the prediction for local data perfectly matches the labels. To avoid the analogous issues, we further introduce a regularization term for the adaptation matrices and a segmentation loss that directly compares the prediction of f_θ and the label y_k,

$$\ell_{TR}(W_k;\theta_k^p) = \frac{1}{M}\sum_{i=1}^{M} tr(W_k^i), \tag{5}$$

$$\ell_{NR}(f_{\theta_k}(x_k,z_k),y_k;\theta_k) = \sum_{i=1}^{M}\sum_{j=1}^{C} y_k^{[i,j]}(\frac{1-f_{\theta_k}^{[i,j]}(x_k,z_k)^q}{q}), \tag{6}$$

where W_k^i represent the adaptation matrix for pixel i, and ℓ_{TR} is to minimize the diagonal elements of W_k^i for each pixel. The non-diagonal elements which denote label flipping probabilities become dominant. Thus, it forces the adaptation matrices to modify the prediction as much as possible. In Eq. (6), $y_k^{[i,j]}$ and $f_{\theta_k}^{[i,j]}$ denote the jth channel of pixel i in the one-hot label and the prediction, respectively. ℓ_{NR} is the noise robust loss, which is similar to the generalized cross-entropy loss introduced in [26]. q is the hyperparameter to balance between noise tolerance and convergence rate. As $q \to 0$, it is equivalent to CE loss which emphasizes more on uncertain predictions, and it degrades to MAE loss [3], which equally penalizes on each pixel, as q approaches 1.

ℓ_{NR} guarantees that the prediction of the segmentation network $f_{\theta_k}(x_k, z_k)$ is close to y_k, while ℓ_{TR} forces the adaptation matrices to modify the uncertain predictions. As the parameters of global model are obtained in a distributed training manner, the uncertain predictions are likely to be caused by the distribution shift in local data. Therefore, the effect of distribution shift is disentangled through the combination of ℓ_{NR} and ℓ_{TR}.

2.3 Training and Testing Details

Training. The local training procedure is similar to that of probabilistic U-Net [8]. Given the input image x_k and segmentation ground truth y_k, the low-dimension representation $z_k \in \mathbb{R}^D$ is sampled from the estimated posterior q_{ϕ_k}. It is broadcast to the same size of the image as a D-channel feature to concatenate with the last feature map of the segmentation network. In our work, it is also taken as the input of DA network to yield adaptation matrices W_k. To initialize the adaptation matrices and distribution encoders, we first perform warm-up training in the first few local training epochs with,

$$\ell_{warm} = \ell_{CE}(f_{\theta_k}(x_k,z_k),y_k;\theta_k) + \beta\ell_{KL} - \ell_{TR}, \tag{7}$$

where ℓ_{KL} is derived from the KL term in Eq. (1), weighted by the hyperparameter β. ℓ_{CE} and ℓ_{KL} in Eq. (7) are to train the segmentation network and distribution encoders, respectively, such that the DA network could acquire the latent representation of the joint data distribution in the next stage. ℓ_{TR} initializes the adaptation matrices to be diagonal dominant such that it does not modify the prediction in the beginning. The overall loss applied during the main training procedure is derived as,

$$\ell = \ell_{CE} + \ell_{NR} + \alpha\ell_{TR} + \beta\ell_{KL}, \tag{8}$$

where α is applied to balance the regularization for the adaptation matrices.

Testing. During testing, the latent representation z is randomly sampled from the prior p_{ψ_k}. The prior distribution could resemble the posterior distribution in a way that reflects different modes of the segmentation results in each center due to the the KL loss [8]. Thus, it could also be applied to model the effect of distribution shift on the prediction.

3 Experiment

3.1 Dataset and Preprocessing

We evaluated our method using LA segmentation on late gadolinium enhanced (LGE) MRIs, collected from three centers, *i.e.* Beth Israel Deaconess Medical Center (Center A), Imaging Sciences at King's College London (Center B) and Utah School of Medicine (Center C, D). The datasets were obtained from MICCAI 2018 Atrial Segmentation Challenge [23] and ISBI 2012 Left Atrium Fibrosis and Scar Segmentation Challenge [6].

For Center A and B, we selected 15 cases for training and 5 cases as local test data. For the dataset from Utah, we randomly selected two sets of 35 cases as Center C and D, with a train-test split of 5:1. Following the work of [25], to simulate the annotation bias and variation, we applied morphological operations, *i.e.,* open and random erosion to training labels in Center C, and close and random dilation to training labels in Center D. Examples of results are presented in the Supplementary Materials. For the local test data in center C, D, we only applied open and close operations, since the aleatoric variation should not occur in the gold standard. Besides, we generated a test set of generic data, using 30 cases from Utah with no modification to the gold standard labels. All MRIs were resampled to the resolution of $0.6 \times 0.6 \times 1.25$ mm and cropped into the size of 256×256 centering at the heart region, with Z-score normalization. Random rotation, flip, elastic deformation and Gaussian noise were applied for data augmentation during training.

3.2 Implementation

The segmentation network was implemented using 2D UNet [16]. The structure of prior and posterior encoders was similar to the implementation in [8]. The personalized module was built with five convolution layers and SoftPlus activation (please refer to supplementary material for details). Parameters α, β and q were set to 0.01, 0.01 and 0.7 (parameter studies were presented in supplementary material), respectively. The networks were trained for 500 epochs with

50 epochs for warm-up. We used the Adam optimizer to update the parameters with a learning rate of 1e-3. The framework was implemented using Pytorch on one Nvidia RTX 3090 GPU.

3.3 Results

To validate that the proposed method could simultaneously benefit the global model and personalized models in the scenario of multiple centers and data heterogeneity, we evaluated them in two tasks, *i.e.*, one to test global model on the generic data, and the other to test personalized ones on local data. Three global models and three personalized methods were compared, as shown in Table 1.

Our method outperformed all the methods in both of the two tasks. Especially, our method improved the Swarm method by 4.2% in Dice, indicating the effectiveness of the proposed DA network in maintaining the global model from diverging during local training. For Task2, our method not only achieved better results than global methods in all centers, but also set the new state of the art when compared with the personalized methods. It is reasonable as the adaptation matrices justify the prediction according to the distribution in each center. Note that our method achieved comparable mean Dice compared with FedRep and Ditto in Center A and B, but demonstrated significant superiority in Center C and D. This was due to the fact that the personalized methods were fine-tuned

Table 1. Results (in Dice) and comparisons. For Global Methods, the global model was used for each center in Task2. For Personalized Methods, the results of Task1 were averaged from the four personalized models (for the four centers). Single: a method that trains four models solely with local data for each center. Bold text denotes the best results in each column.

Method	Task1	Task2				
	Generic	Center A	Center B	Center C	Center D	Mean
Global methods						
Swarm [22]	0.843 ± 0.044	0.749 ± 0.095	0.739 ± 0.091	0.839 ± 0.021	0.823 ± 0.031	0.795 ± 0.054
FedProx [11]	0.869 ± 0.030	0.750 ± 0.114	0.740 ± 0.086	0.851 ± 0.025	0.850 ± 0.023	0.807 ± 0.056
FedCurv [17]	0.866 ± 0.038	0.729 ± 0.218	0.758 ± 0.093	0.837 ± 0.028	0.824 ± 0.028	0.794 ± 0.081
Personalized methods						
Single	0.734 ± 0.143	0.693 ± 0.203	0.791 ± 0.054	0.708 ± 0.049	0.768 ± 0.027	0.740 ± 0.076
FedRep [2]	0.791 ± 0.054	**0.775 ± 0.218**	0.764 ± 0.062	0.769 ± 0.041	0.833 ± 0.029	0.788 ± 0.078
Ditto [10]	0.781 ± 0.068	0.762 ± 0.187	0.786 ± 0.062	0.763 ± 0.028	0.810 ± 0.033	0.781 ± 0.070
Ours	**0.885 ± 0.027**	0.772 ± 0.124	**0.793 ± 0.042**	**0.874 ± 0.020**	**0.873 ± 0.032**	**0.836 ± 0.050**

Table 2. Ablation study for the latent representation for the joint data distribution. Results were evaluated in Dice score.

Method	Task1	Task2			
	Generic	Center A	Center B	Center C	Center D
FixedAdapt	0.864 ± 0.143	0.748 ± 0.126	0.782 ± 0.067	0.808 ± 0.049	0.86 ± 0.027
ImgAdapt	0.858 ± 0.093	0.695 ± 0.188	0.735 ± 0.087	0.799 ± 0.039	0.806 ± 0.045
Ours	**0.885 ± 0.027**	**0.772 ± 0.124**	**0.793 ± 0.042**	**0.874 ± 0.02**	**0.873 ± 0.032**

on each local distribution, thus tended to be affected by aleatoric errors in the annotations which existed in Center C and D. By contrast, our method was as robust as the global methods for generic test data. This advantage was confirmed again by the results in Task2 for the personalized methods, which were affected by the segmentation bias in Center C and D and were much worse than ours, because our proposed adaptation network was able to learn this bias for local test data.

Ablation Study. To validate the importance of the latent representation z_k to the generation of adaptation matrices, we performed an ablation study, by evaluating and comparing the "FixedAdapt" and "ImgAdapt" methods. In FixedAdapt, the adaptation matrices were fixed during testing (similar to [21]); in ImgAdapt, the adaptation matrices were conditioned on the input image (analogous to [25]). Table 2 presents the results, where our proposed method was evidently better than the other two without information about the local joint distributions. This is reasonable as the adaptation network in the two ablated methods tended to overfit on the training data and could not learn the distribution shift in each center. With the latent representation z_k that contains distribution information, our method could yield more robust adaptation matrices for test data.

4 Conclusion

We have proposed a new method to bridge the gap between global and personalized distributed learning through distribution-conditioned adaptation networks. We have validated our method on LA MRI segmentation of multi-center data with both feature and label skew, and showed that our method outperformed existing global and personalized methods on both generic data and local data.

References

1. Chen, H.Y., Chao, W.L.: On bridging generic and personalized federated learning for image classification. In: International Conference on Learning Representations (2021)
2. Collins, L., Hassani, H., Mokhtari, A., Shakkottai, S.: Exploiting shared representations for personalized federated learning. arXiv preprint arXiv:2102.07078 (2021)
3. Ghosh, A., Kumar, H., Sastry, P.: Robust loss functions under label noise for deep neural networks. In: Proceedings of the AAAI Conference on Artificial Intelligence, vol. 31 (2017)
4. Goldberger, J., Ben-Reuven, E.: Training deep neural-networks using a noise adaptation layer. In: ICLR (2017)
5. Kairouz, P., et al.: Advances and open problems in federated learning. arXiv preprint arXiv:1912.04977 (2019)
6. Karim, R., et al.: Evaluation of current algorithms for segmentation of scar tissue from late gadolinium enhancement cardiovascular magnetic resonance of the left atrium: an open-access grand challenge. J. Cardiovasc. Magn. Reson. **15**(1), 1–17 (2013)

7. Karimireddy, S.P., Kale, S., Mohri, M., Reddi, S., Stich, S., Suresh, A.T.: SCAFFOLD: stochastic controlled averaging for federated learning. In: International Conference on Machine Learning, pp. 5132–5143. PMLR (2020)
8. Kohl, S., et al.: A probabilistic u-net for segmentation of ambiguous images. Adv. Neural Inf. Process. Syst. **31** (2018)
9. Li, L., Zimmer, V.A., Schnabel, J.A., Zhuang, X.: Medical image analysis on left atrial LGE MRI for atrial fibrillation studies: a review. Med. Image Anal. 102360 (2022)
10. Li, T., Hu, S., Beirami, A., Smith, V.: Ditto: fair and robust federated learning through personalization. In: International Conference on Machine Learning, pp. 6357–6368. PMLR (2021)
11. Li, T., Sahu, A.K., Zaheer, M., Sanjabi, M., Talwalkar, A., Smith, V.: Federated optimization in heterogeneous networks. Proc. Mach. Learn. Syst. **2**, 429–450 (2020)
12. Li, X., Jiang, M., Zhang, X., Kamp, M., Dou, Q.: FedBN: Federated learning on Non-IID features via local batch normalization. arXiv preprint arXiv:2102.07623 (2021)
13. Lu, Y., Huang, X., Dai, Y., Maharjan, S., Zhang, Y.: Blockchain and federated learning for privacy-preserved data sharing in industrial IoT. IEEE Trans. Industr. Inf. **16**(6), 4177–4186 (2019)
14. McMahan, B., Moore, E., Ramage, D., Hampson, S., Arcas, B.A.Y.: Communication-efficient learning of deep networks from decentralized data. In: Artificial Intelligence and Statistics, pp. 1273–1282. PMLR (2017)
15. Natarajan, N., Dhillon, I.S., Ravikumar, P.K., Tewari, A.: Learning with noisy labels. Adv. Neural Inf. Process. Syst. **26** (2013)
16. Ronneberger, O., Fischer, P., Brox, T.: U-Net: convolutional networks for biomedical image segmentation. In: Navab, N., Hornegger, J., Wells, W.M., Frangi, A.F. (eds.) MICCAI 2015. LNCS, vol. 9351, pp. 234–241. Springer, Cham (2015). https://doi.org/10.1007/978-3-319-24574-4_28
17. Shoham, N., et al.: Overcoming forgetting in federated learning on Non-IID data. arXiv preprint arXiv:1910.07796 (2019)
18. Sohn, K., Lee, H., Yan, X.: Learning structured output representation using deep conditional generative models. Adv. Neural Inf. Process. Syst. **28** (2015)
19. Sukhbaatar, S., Bruna, J., Paluri, M., Bourdev, L.D., Fergus, R.: Training convolutional networks with noisy labels. arXiv: Computer Vision and Pattern Recognition (2014)
20. Szita, I., Lörincz, A.: Learning Tetris using the noisy cross-entropy method. Neural Comput. **18**(12), 2936–2941 (2006)
21. Tanno, R., Saeedi, A., Sankaranarayanan, S., Alexander, D.C., Silberman, N.: Learning from noisy labels by regularized estimation of annotator confusion. In: Proceedings of the IEEE/CVF Conference on Computer Vision and Pattern Recognition, pp. 11244–11253 (2019)
22. Warnat-Herresthal, S., et al.: Swarm learning for decentralized and confidential clinical machine learning. Nature **594**(7862), 265–270 (2021)
23. Xiong, Z., et al.: A global benchmark of algorithms for segmenting the left atrium from late gadolinium-enhanced cardiac magnetic resonance imaging. Med. Image Anal. **67**, 101832 (2021)
24. Zhang, H., et al.: Multiple sclerosis lesion segmentation with Tiramisu and 2.5D stacked slices. In: Shen, D., et al. (eds.) MICCAI 2019. LNCS, vol. 11766, pp. 338–346. Springer, Cham (2019). https://doi.org/10.1007/978-3-030-32248-9_38

25. Zhang, L., et al.: Disentangling human error from ground truth in segmentation of medical images. Adv. Neural. Inf. Process. Syst. **33**, 15750–15762 (2020)
26. Zhang, Z., Sabuncu, M.: Generalized cross entropy loss for training deep neural networks with noisy labels. Adv. Neural Inf. Process. Syst. **31** (2018)

Scribble-Supervised Medical Image Segmentation via Dual-Branch Network and Dynamically Mixed Pseudo Labels Supervision

Xiangde Luo[1,2], Minhao Hu[3], Wenjun Liao[1], Shuwei Zhai[1], Tao Song[3], Guotai Wang[1,2](✉), and Shaoting Zhang[1,2]

[1] University of Electronic Science and Technology of China, Chengdu, China
guotai.wang@uestc.edu.cn
[2] Shanghai AI Lab, Shanghai, China
[3] SenseTime Research, Shanghai, China

Abstract. Medical image segmentation plays an irreplaceable role in computer-assisted diagnosis, treatment planning and following-up. Collecting and annotating a large-scale dataset is crucial to training a powerful segmentation model, but producing high-quality segmentation masks is an expensive and time-consuming procedure. Recently, weakly-supervised learning that uses sparse annotations (points, scribbles, bounding boxes) for network training has achieved encouraging performance and shown the potential for annotation cost reduction. However, due to the limited supervision signal of sparse annotations, it is still challenging to employ them for networks training directly. In this work, we propose a simple yet efficient scribble-supervised image segmentation method and apply it to cardiac MRI segmentation. Specifically, we employ a dual-branch network with one encoder and two slightly different decoders for image segmentation and dynamically mix the two decoders' predictions to generate pseudo labels for auxiliary supervision. By combining the scribble supervision and auxiliary pseudo labels supervision, the dual-branch network can efficiently learn from scribble annotations end-to-end. Experiments on the public ACDC dataset show that our method performs better than current scribble-supervised segmentation methods and also outperforms several semi-supervised segmentation methods. Code is available: https://github.com/HiLab-git/WSL4MIS.

Keywords: Weakly-supervised learning · Scribble annotation · Pseudo labels

1 Introduction

Recently, Convolutional Neural Networks (CNNs) and Transformers have achieved encouraging results in automatic medical image segmentation [5, 12, 26]. Most of them need large-scale images with accurate pixel-level dense annotations to train models. However, collecting a large-scale and carefully annotated medical image dataset is still an expensive and time-consuming journey, as it requires domain knowledge and clinical

L. Wang et al. (Eds.): MICCAI 2022, LNCS 13431, pp. 528–538, 2022.
https://doi.org/10.1007/978-3-031-16431-6_50

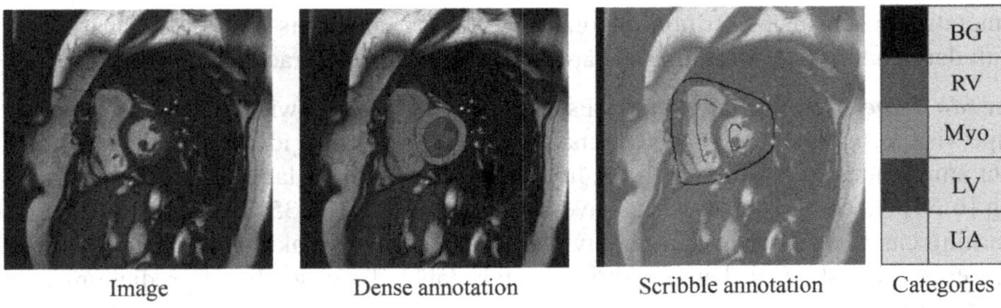

Fig. 1. Examples of dense and scribble annotations. BG, RV, Myo, LV, and UA represent the background, right ventricle, myocardium, left ventricle, and unannotated pixels respectively.

experience [19,22]. Recently, many efforts have been made to reduce the annotation cost for models training to alleviate this issue. For example, semi-supervised learning (SSL) combines a few labeled data and massive unlabeled data for network training [1,19–21]. Weakly supervised learning (WSL) uses sparse annotations to train models rather than dense annotations [7,8,30]. Considering collecting sparse annotations (points, scribbles and bounding boxes) is easier than dense annotations [16] and scribbles have better generality to annotate complex objects than bounding boxes and points [16,30] (Example in Fig. 1). This work focuses on exploring scribble annotations to train high-performance medical image segmentation networks efficiently and robustly.

Scribble-Supervised Segmentation: Using scribble annotations to segment objects has been studied for many years. Before the deep learning era, combining user-provided sparse annotations and machine learning or other algorithms was the most popular and general segmentation method, such as GraphCuts [3], GrabCut [27], Random Walker [9], GrowCut [31], ITK-SNAP [37], Slic-Seg [33], etc. Recently, deep learning with convolutional neural networks or transformers can learn to segment from dense annotations and then inference automatically. So, it is desirable to train powerful segmentation networks using scribble annotations. To achieve this goal, Lin et al. [16] proposed a graphical-based method to propagate information from scribbles to unannotated pixels and train models jointly. After that, Tang et al. [28] introduced a Conditional Random Field (CRF) regularization loss to train segmentation networks directly. For medical images, Can et al. [4] proposed an iterative framework to train models with scribbles. At first, they seeded the scribbles into the Random Walker [9] to produce the initial segmentation. Then, they used the initial segmentation to train the model and refine the model's prediction with CRF for the network retraining. Finally, they repeated the second procedure several times for powerful segmentation models. Kim et al. [13] proposed a level set-based [23] regularization function to train deep networks with weak annotations. Lee et al. [15] combined pseudo-labeling and label filtering to generate reliable labels for network training with scribble supervisions. Liu et al. [17] presented a unified weakly-supervised framework to train networks from scribble annotations, which consists of an uncertainty-aware mean teacher and a transformation-consistent strategy. More recently, Valvano et al. [30] proposed multi-scale adversarial attention gates to train models with mixed scribble and dense annotations. Although these attempts have saved the annotation cost by using scribble annotations, the performance is still lower than training with dense annotations, limiting the applicability in clinical practice.

Pseudo Labels for Segmentation: Pseudo labeling [14] is widely used to generate supervision signals for unlabeled images/pixels. The main idea is utilizing imperfect annotations to produce high-quality and reliable pseudo labels for network training [6,34]. Recently, some works have demonstrated [18,34,35] that semi-supervised learning can benefit from high-quality pseudo labels. For weakly-supervised learning, Lee et al. [15] showed that generating pseudo labels by ensembling predictions at a temporal level can boost performance. *Nevertheless, recent work* [11] *points out the inherent weakness of these methods that the model retains the prediction from itself and thus resists updates.* Recently, some works resort to perturbation-consistency strategy for semi-supervised learning [24,35], where the main branch is assisted by auxiliary branches that are typically perturbed and encouraged to produce similar predictions to the main branch. In this work, we assume that generating pseudo labels by mixing multiple predictions randomly can go against the above inherent weakness, as these auxiliary branches are added perturbations and do not enable interaction with each other.

Motivated by these observations, we present a simple yet efficient approach to learning from scribble annotations. Particularly, we employ a dual branches network (one encoder and two slightly different decoders) as the segmentation network. To learn from scribbles, the dual branches network is supervised by the partially cross-entropy loss (pCE), which only considers annotated pixels' gradient for back-propagation and ignores unlabeled pixels. At the same time, we employ the two predictions to generate hard pseudo labels for more substantial and more reliable supervision signals than scribbles. Afterward, we combine scribbles supervision and pseudo labels supervision to train the segmentation network end-to-end. Differently from threshold-based methods [15,35], we generate hard pseudo labels by dynamically mixing two branches' predictions, which can help against the inherent weakness [11]. Such a strategy imposes the segmentation network to produce high-quality pseudo labels for unannotated pixels. We evaluate our method on a public scribble-supervised benchmark ACDC [2]. Experiments results show that our proposed method outperforms existing scribble-supervised methods when using the same scribble annotations and also performs better than semi-supervised methods when taking similar annotation budgets.

The contributions of this work are two-fold. Firstly, we propose a dual-branch network and a dynamically mixed pseudo labeling strategy to train segmentation models with scribble annotations. Specifically, we generate high-quality hard pseudo labels by randomly mixing the two branches' outputs and use the generated pseudo labels to supervise the network training end-to-end. (2) Extensive experiments on the public cardiac MRI segmentation dataset (ACDC) demonstrate the effectiveness of the proposed method. Our method has achieved better performance on the ACDC dataset than existing scribble-supervised segmentation approaches and also outperformed several semi-supervised segmentation methods with similar annotation costs.

2 Method

The proposed framework for scribble-supervised medical image segmentation is depicted in Fig. 2. We firstly employ a network with one encoder and two slightly

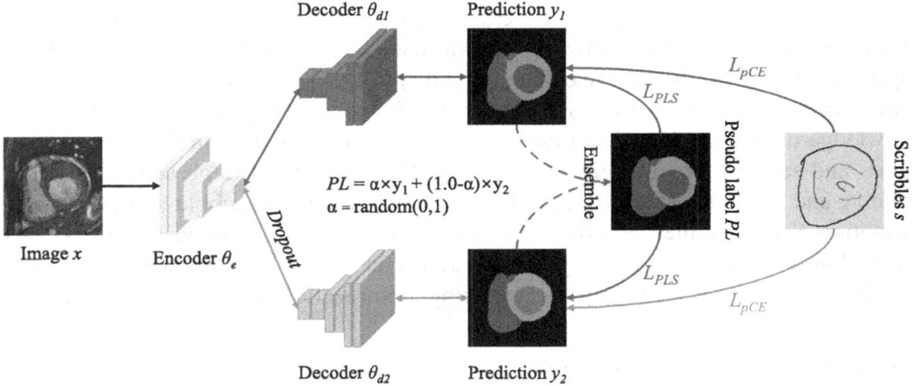

Fig. 2. Overview of the proposed method. The framework consists of an encoder (θ_e), the main decoder (θ_{d1}), and an auxiliary decoder (θ_{d2}) and is trained with scribble annotations separately (L_{pCE}). At the same time, the hard pseudo label is generated by dynamically mixing two decoders' outputs and used as the pseudo labels supervision for further network training (L_{PLS}).

different decoders to learn from scribble annotations to segment target objects. At the same time, we utilize the two branches' outputs to generate hard pseudo labels that are used to assist the network training. Note that the training procedure is in an end-to-end manner rather than the multi-stage [4] or iterative refinement strategies [16].

2.1 Learning from Scribbles

For general scribble-supervised learning, the available dataset consists of images and scribble annotations, where the scribble is a set of pixels with a category or unknown label. Previous work [4] uses interactive segmentation methods [9] to propagate annotated pixels to the whole image for a rough segmentation and then train deep networks with the segmentation in a fully-supervised manner. Recently, there are much better alternatives [15,28], e.g., using scribbles to train CNNs directly by minimizing a partial cross-entropy loss:

$$L_{pCE}(y, s) = -\sum_{c}\sum_{i\in\omega_s} \log y_i^c \tag{1}$$

where s represents the one-hot scribble annotations. y_i^c is the predicted probability of pixel i belonging class c. ω_s is the set of labeled pixels in s.

2.2 Dual-Branch Network

The proposed network ($f(\theta_e, \theta_{d1}, \theta_{d2})$) is composed of a shared encoder (θ_e) for feature extraction and two independent and different decoders (θ_{d1}, θ_{d2}) for segmentation and supplementary training (see Fig. 2). We embed a perturbed decoder into the general UNet [26], where the dropout [24] is used to introduce perturbation at the feature level. This design has two advantages: (1) It can be against the inherent weakness of pseudo-label in the single branch network [11], as the two branches' outputs are different due to the feature perturbation. (2) It can generate pseudo-label by two outputs ensemble but does not require training two networks, and the encoder benefits from the two individual supervisions to boost the feature extraction ability [24,35]. It is worthy to point out that some recent works used similar architecture for the consistency training [19,24,35] or knowledge distillation [7]. There are many significant differences in the learning scenarios and supervision strategies. Firstly, [19,24,35] concentrate on semi-supervised learning and [7] focus on knowledge distillation but we aim to scribble-supervised segmentation. Secondly, they employ consistency regularization to supervise networks, but we randomly mix two outputs to generate hard pseudo labels for fully supervised learning. These differences lead to different training, optimization strategies, and results.

2.3 Dynamically Mixed Pseudo Labels

Based on the dual-branch network, we further exploit the two decoders' outputs to boost the model training. We generate the hard pseudo labels by mixing two predictions dynamically, like mixup [38]. The dynamically mixed pseudo labels (*PL*) generation strategy is defined as:

$$PL = argmax[\alpha \times y_1 + (1.0 - \alpha) \times y_2],\ \alpha = random(0, 1) \tag{2}$$

where y_1 and y_2 are outputs of decoder 1 and 2, respectively. α is randomly generated in (0, 1) at each iteration. This strategy boosts the diversity of pseudo labels and avoids the inherent weakness of the pseudo labeling strategy (remembering itself predictions without updating) [11]. *argmax* is used to generate hard pseudo labels. Compared with consistency learning [24,35], this strategy cuts off the gradient between θ_{d1} and θ_{d2} to maintain their independence rather than enforce consistency directly. In this way, the supervision signal is enlarged from a few pixels to the whole image, as the scribbled pixels are propagated to all unlabeled pixels by the dynamically mixed pseudo labeling. Then, we further employ the generated *PL* to supervise θ_{d1} and θ_{d2} separately to assist the network training. The *P*seudo *L*abels *S*upervision (*PLS*) is defined as:

$$L_{PLS}(PL, y_1, y_2) = 0.5 \times (L_{Dice}(PL, y_1) + L_{Dice}(PL, y_2)) \tag{3}$$

where L_{Dice} is the widely-used dice loss and also can be replaced by cross-entropy loss or other segmentation loss functions. Finally, the proposed network can be trained with scribble annotations by minimizing the following joint object function:

$$L_{total} = \underbrace{0.5 \times (L_{pCE}(y_1, s) + L_{pCE}(y_2, s))}_{scribble\ supervision} + \lambda \times \underbrace{L_{PLS}(PL, y_1, y_2)}_{pseudo\ labels\ supervision} \tag{4}$$

λ is a weight factor to balance the supervision of scribbles and pseudo labels.

3 Experiment and Results

3.1 Experimental Details

Dataset: We evaluate the proposed method on the training set of ACDC [2] via five-fold cross-validation. This dataset is publicly available, with 200 short-axis cine-MRI

Table 1. Comparison with existing weakly-/semi-supervised methods on the ACDC dataset. All results are based on the *5-fold cross-validation* with same backbone (UNet). Mean and standard variance (in parentheses) values of 3D *DSC* and HD_{95} (mm) are presented in this table. * denotes p-value < 0.05 (paired t-test) when comparing with the second place method (RLoss [28]).

Type	Method	RV		Myo		LV		Mean	
		DSC	HD_{95}	*DSC*	HD_{95}	*DSC*	HD_{95}	*DSC*	HD_{95}
WSL	pCE [16]	0.625(0.16)	187.2(35.2)	0.668(0.095)	165.1(34.4)	0.766(0.156)	167.7(55.0)	0.686(0.137)	173.3(41.5)
	RW [9]	0.813(0.113)	11.1(17.3)	0.708(0.066)	9.8(8.9)	0.844(0.091)	9.2(13.0)	0.788(0.09)	10.0(13.1)
	USTM [17]	0.815(0.115)	54.7(65.7)	0.756(0.081)	112.2(54.1)	0.785(0.162)	139.6(57.7)	0.786(0.119)	102.2(59.2)
	S2L [15]	0.833(0.103)	14.6(30.9)	0.806(0.069)	37.1(49.4)	0.856(0.121)	65.2(65.1)	0.832(0.098)	38.9(48.5)
	MLoss [13]	0.809(0.093)	17.1(30.8)	0.832(0.055)	28.2(43.2)	0.876(0.093)	37.9(59.6)	0.839(0.080)	27.7(44.5)
	EM [10]	0.839(0.108)	25.7(44.5)	0.812(0.062)	47.4(50.6)	0.887(0.099)	43.8(57.6)	0.846(0.089)	39.0(50.9)
	RLoss [28]	0.856(0.101)	7.9(12.6)	0.817(0.054)	**6.0(6.9)**	0.896(0.086)	**7.0(13.5)**	0.856(0.080)	**6.9(11.0)**
	Ours	**0.861(0.096)**	**7.9(12.5)**	**0.842(0.054)***	9.7(23.2)	**0.913(0.082)***	12.1(27.2)	**0.872(0.077)***	9.9(21.0)
SSL	PS [26]	0.659(0.261)	26.8(30.4)	0.724(0.176)	16.0(21.6)	0.790(0.205)	24.5(30.4)	0.724(0.214)	22.5(27.5)
	DAN [39]	0.639(0.26)	20.6(21.4)	0.764(0.144)	9.4(12.4)	0.825(0.186)	15.9(20.8)	0.743(0.197)	15.3(18.2)
	AdvEnt [32]	0.615(0.296)	20.2(19.4)	0.760(0.151)	8.5(8.3)	0.848(0.159)	11.7(18.1)	0.741(0.202)	13.5(15.3)
	MT [29]	0.653(0.271)	18.6(22.0)	0.785(0.118)	11.4(17.0)	0.846(0.153)	19.0(26.7)	0.761(0.180)	16.3(21.9)
	UAMT [36]	0.660(0.267)	22.3(22.9)	0.773(0.129)	10.3(14.8)	0.847(0.157)	17.1(23.9)	0.760(0.185)	16.6(20.5)
FSL	FullSup [26]	0.882(0.095)	6.9(10.8)	0.883(0.042)	5.9(15.2)	0.930(0.074)	8.1(20.9)	0.898(0.070)	7.0(15.6)

scans from 100 patients, and each patient has two annotated end-diastolic (ED), and end-systolic (ES) phases scans. And each scan has three structures' dense annotation, including the right ventricle (RV), myocardium (Myo), and left ventricle (LV). Recently, Valvano et al. [30] provided the scribble annotation for each scan manually. Following previous works [1,30], we employ the 2D slice segmentation rather than 3D volume segmentation, as the thickness is too large.

Implementation Details: We employed the UNet [26] as the base segmentation network architecture, and we further extended the basic UNet to dual branches network by embedding an auxiliary decoder. We added the dropout layer (ratio = 0.5) before each conv-block of the auxiliary decoder to introduce perturbations. We implemented and ran our proposed and other comparison methods by PyTorch [25] on a cluster with 8 TiTAN 1080TI GPUs. For the network training, we first re-scaled the intensity of each slice to 0–1. Then, random rotation, random flipping, random noise were used to enlarge the training set, and the augmented image was resized to 256 × 256 as the network input. We used the SGD (weight decay = 10^{-4}, momentum = 0.9) to minimize the joint object function Eq. 4 for the model optimization. The poly learning rate strategy was used to adjust the learning rate online [20]. The batch size, total iterations, and λ are set to 12, 60*k*, and 0.5, respectively. For testing, we produced predictions slice by slice and stacked them into a 3D volume. For a fair comparison, we used the primary decoder's output as the final result during the inference stage and did not use any post-processing method. All experiments were conducted in the same experimental setting. The 3D Dice Coefficient (DSC) and 95% Hausdorff Distance (HD_{95}) are used as evaluation metrics.

Table 2. Ablation study on different supervision strategies for the dual-branch network. Single denotes the baseline UNet [26] with *pCE* only. *CR* means consistency regularization between the main and auxiliary decoders [7]. *CPS* is the cross pseudo supervision strategy in [6,35]. *Ours* is proposed *PLS*, θ_{d1} and θ_{d2} mean the prediction of main and auxiliary decoders, respectively.

Method	RV		Myo		LV		Mean	
	DSC	HD_{95}	*DSC*	HD_{95}	*DSC*	HD_{95}	*DSC*	HD_{95}
Single [16]	0.625(0.16)	187.2(35.2)	0.668(0.095)	165.1(34.4)	0.766(0.156)	167.7(55.0)	0.686(0.137)	173.3(41.5)
Dual+***CR*** [7]	0.844(0.106)	20.1(37.2)	0.798(0.07)	62.2(55.7)	0.873(0.101)	63.4(65.5)	0.838(0.092)	48.6(52.8)
Dual+***CPS*** [6,35]	0.849(0.099)	12.4(25.6)	0.833(0.056)	19.3(33.5)	0.905(0.091)	18.3(35.8)	0.863(0.082)	16.6(31.6)
Ours (α=0.5, θ_{d1})	0.855(0.101)	8.6(13.9)	0.837(0.053)	13.6(29.1)	0.908(0.086)	15.8(34.1)	0.866(0.08)	12.6(25.7)
Ours (α=random, θ_{d1})	**0.861(0.096)**	7.9(12.5)	**0.842(0.054)**	**9.7(23.2)**	**0.913(0.082)**	12.1(27.2)	**0.872(0.077)**	9.9(21.0)
Ours (α=random, θ_{d2})	0.861(0.098)	**7.3(10.3)**	0.840(0.058)	10.9(24.5)	0.911(0.086)	**11.3(26.4)**	0.871(0.08)	**9.8(20.4)**

3.2 Results

Comparison with Other Methods: Firstly, we compared our method with seven scribble-supervised segmentation methods with the same set of scribbles: 1) pCE only [16] (lower bound), 2) using pxeudo label generated by Random Walker (RW) [9], 3) Uncertainty-aware Self-ensembling and Transformation-consistent Model (USTM) [17], 4) Scribble2Label (S2L) [15], 5) Mumford-shah Loss (MLoss) [13], 6) Entropy Minimization (EM) [10], 7) Regularized Loss (RLoss) [28]. The first section of Table 1 lists the quantitative comparison of the proposed with seven existing weakly supervised learning methods. It can be found that our method achieved the best performance in terms of mean *DSC* (p-value < 0.05) and second place in the HD_{95} metric than other methods.

Afterward, we further compared our method with other popular annotation-efficient segmentation methods, e.g., semi-supervised learning methods. Following [8], we investigated the performance difference of these approaches when using very similar annotation costs. To do so, we trained networks with partially supervised and semi-supervised fashions, respectively. We used a 10% training set (8 patients) as labeled data and the remaining as unlabeled data, as the scribble annotation also takes similar annotation costs [30]. For partially supervised (*PS*) learning, we used the 10% labeled data to train networks only. For semi-supervised learning, we combined the 10% labeled data and 90% unlabeled data to train models jointly. We further employed four widely-used semi-supervised segmentation methods for comparison: 1) Deep Adversarial Network (DAN) [39], 2) Adversarial Entropy Minimization (AdvEnt) [32], 3) Mean Teacher (MT) [29], and Uncertainty Aware Mean Teacher (UAMT) [36]. The quantitative comparison is presented in the second section of Table 1. It shows that the scribbled annotation can achieve better results than pixel-wise annotation when taking a similar annotation budget. Moreover, our weakly-supervised method significantly outperforms existing semi-supervised methods in the cardiac MR segmentation. Finally, we also investigated the upper bound when using all mask annotation to train models (FullSup) in the last row of Table 1. It can be found that our method is slightly inferior compared with fully supervised learning with pixel-wise annotation. But our method requires fewer annotation costs than pixel-wise annotation. Figure 3 shows the segmentation results obtained by existing and our methods, and the corresponding ground truth

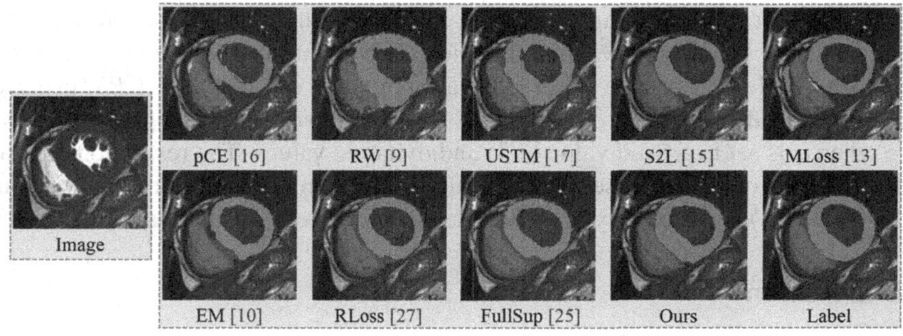

Fig. 3. Qualitative comparison of our proposed method and several existing ways.

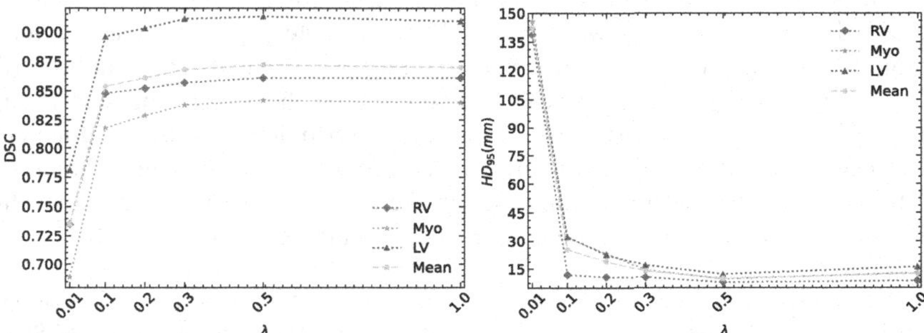

Fig. 4. Sensitivity analysis of hyper-parameter λ.

on the ACDC dataset (patient026_frame01). We can observe that the result obtained by our method is more similar to the ground truth than the others. It further shows that drawing scribble is a potential data annotation approach to reduce annotation costs.

Sensitivity Analysis of λ: The study was conducted to assess the sensitivity of λ in Eq. 4. Particularly, the *PLS* term plays a crucial role in the proposed framework, as it controls the usage of the pseudo labels during the network training. We investigated the segmentation performance of the proposed framework when the λ is set to $\{0.01, 0.1, 0.2, 0.3, 0.5, 1.0\}$. Figure 4 shows the evolution of the segmentation result of RV, Myo, LV, and their average results, all these results are based on the 5-fold cross-validation. It can be observed that increasing λ from 0.01 to 0.5 leads to better performance in terms of both *DSC* and HD_{95}. When the λ is set to 1.0, the segmentation result just decreases slightly compared with 0.5 (0.872 vs 0.870 in term of mean *DSC*). These observations show that the proposed method is not sensitive to λ.

Ablation Study: We further investigated the effect of using different supervision approaches for the dual-branch network: 1) Consistency Regularization (*CR*) [7] that encourages the two predictions to be similar, directly; 2) Cross Pseudo Supervision (*CPS*) [6,35] that uses one decoder's output as the pseudo label to supervise the other one; 3) the proposed approach dynamically mixes two outputs to generate hard pseudo labels for two decoders training separately. We trained the dual-branch network with scribbles and the above supervision strategies. The quantitative evaluation results are presented in Table 2. It can be observed that compared with *CR* and *CPS*, using our proposed *PLS* leads to the best performance. Moreover, we also investigated the performance when α is set to a fixed value (0.5) and dynamic values. The result demonstrates the effectiveness of the proposed dynamically mixing strategy. In addition, we found that the main (θ_{d1}) and auxiliary (θ_{d2}) decoders achieve very similar results.

4 Conclusion

In this paper, we presented pseudo labels supervision strategy for scribble-supervised medical image segmentation. A dual-branch network is employed to learn from scribble annotations in an end-to-end manner. Based on the dual-branch network, a dynamically mixed pseudo labeling strategy was presented to propagate the scribble annotations to the whole image and supervise the network training. Experiments on a public cardiac MR image segmentation dataset (ACDC) demonstrated the effectiveness of the proposed method, where it outperformed seven recent scribble-supervised segmentation methods using the same scribble annotations and four semi-supervised segmentation methods with very similar annotation costs. In the future, we will extend and evaluate the proposed method on other challenging medical image segmentation tasks.

Acknowledgments. This work was supported by the National Natural Science Foundations of China [81771921, 61901084] funding and key research and development project of Sichuan province, China [no. 2020YFG0084]. This work was also supported by the Beijing Nova Program [Z201100006820064].

References

1. Bai, W., et al.: Semi-supervised learning for network-based cardiac MR image segmentation. In: Descoteaux, M., Maier-Hein, L., Franz, A., Jannin, P., Collins, D.L., Duchesne, S. (eds.) MICCAI 2017. LNCS, vol. 10434, pp. 253–260. Springer, Cham (2017). https://doi.org/10.1007/978-3-319-66185-8_29
2. Bernard, O., et al.: Deep learning techniques for automatic MRI cardiac multi-structures segmentation and diagnosis: is the problem solved? TMI **37**(11), 2514–2525 (2018)
3. Boykov, Y.Y., Jolly, M.P.: Interactive graph cuts for optimal boundary & region segmentation of objects in ND images. In: ICCV, vol. 1, pp. 105–112. IEEE (2001)
4. Can, Y.B., Chaitanya, K., Mustafa, B., Koch, L.M., Konukoglu, E., Baumgartner, C.F.: Learning to segment medical images with scribble-supervision alone. In: Stoyanov, D., et al. (eds.) DLMIA/ML-CDS -2018. LNCS, vol. 11045, pp. 236–244. Springer, Cham (2018). https://doi.org/10.1007/978-3-030-00889-5_27
5. Chen, J., et al.: Transunet: Transformers make strong encoders for medical image segmentation. arXiv preprint arXiv:2102.04306 (2021)
6. Chen, X., Yuan, Y., Zeng, G., Wang, J.: Semi-supervised semantic segmentation with cross pseudo supervision. In: CVPR, pp. 2613–2622 (2021)
7. Dolz, J., Desrosiers, C., Ayed, I.B.: Teach me to segment with mixed supervision: confident students become masters. In: Feragen, A., Sommer, S., Schnabel, J., Nielsen, M. (eds.) IPMI 2021. LNCS, vol. 12729, pp. 517–529. Springer, Cham (2021). https://doi.org/10.1007/978-3-030-78191-0_40

8. Dorent, R., et al.: Inter extreme points geodesics for end-to-end weakly supervised image segmentation. In: de Bruijne, M., et al. (eds.) MICCAI 2021. LNCS, vol. 12902, pp. 615–624. Springer, Cham (2021). https://doi.org/10.1007/978-3-030-87196-3_57
9. Grady, L.: Random walks for image segmentation. TPAMI **11**, 1768–1783 (2006)
10. Grandvalet, Y., Bengio, Y., et al.: Semi-supervised learning by entropy minimization. In: NeurIPS, vol. 367, pp. 281–296 (2005)
11. Huo, X., et al.: ATSO: asynchronous teacher-student optimization for semi-supervised image segmentation. In: CVPR, pp. 1235–1244 (2021)
12. Isensee, F., Jaeger, P.F., Kohl, S.A., Petersen, J., Maier-Hein, K.H.: nnU-Net: a self-configuring method for deep learning-based biomedical image segmentation. Nat. Methods **18**(2), 203–211 (2021)
13. Kim, B., Ye, J.C.: Mumford-Shah loss functional for image segmentation with deep learning. IEEE Trans. Image Process. **29**, 1856–1866 (2019)
14. Lee, D.H., et al.: Pseudo-label: the simple and efficient semi-supervised learning method for deep neural networks. In: ICML, vol. 3, p. 896 (2013)
15. Lee, H., Jeong, W.-K.: Scribble2Label: scribble-supervised cell segmentation via self-generating pseudo-labels with consistency. In: Martel, A.L., et al. (eds.) MICCAI 2020. LNCS, vol. 12261, pp. 14–23. Springer, Cham (2020). https://doi.org/10.1007/978-3-030-59710-8_2
16. Lin, D., Dai, J., Jia, J., He, K., Sun, J.: ScribbleSup: scribble-supervised convolutional networks for semantic segmentation. In: CVPR, pp. 3159–3167 (2016)
17. Liu, X., et al.: Weakly supervised segmentation of covid19 infection with scribble annotation on CT images. PR 122, 108341 (2022)
18. Luo, W., Yang, M.: Semi-supervised semantic segmentation via strong-weak dual-branch network. In: Vedaldi, A., Bischof, H., Brox, T., Frahm, J.-M. (eds.) ECCV 2020. LNCS, vol. 12350, pp. 784–800. Springer, Cham (2020). https://doi.org/10.1007/978-3-030-58558-7_46
19. Luo, X., Chen, J., Song, T., Wang, G.: Semi-supervised medical image segmentation through dual-task consistency. AAAI **35**(10), 8801–8809 (2021)
20. Luo, X., et al.: Efficient semi-supervised gross target volume of nasopharyngeal carcinoma segmentation via uncertainty rectified pyramid consistency. In: de Bruijne, M., et al. (eds.) MICCAI 2021. LNCS, vol. 12902, pp. 318–329. Springer, Cham (2021). https://doi.org/10.1007/978-3-030-87196-3_30
21. Luo, X., et al.: Semi-supervised medical image segmentation via uncertainty rectified pyramid consistency. Media **80**, 102517 (2022)
22. Luo, X., et al.: MIDeepSeg: minimally interactive segmentation of unseen objects from medical images using deep learning. Media **72**, 102102 (2021)
23. Mumford, D.B., Shah, J.: Optimal approximations by piecewise smooth functions and associated variational problems. Commun. Pure Appl. Math. **2**(5), 577–685 (1989)
24. Ouali, Y., Hudelot, C., Tami, M.: Semi-supervised semantic segmentation with cross-consistency training. In: CVPR, pp. 12674–12684 (2020)
25. Paszke, A., et al.: Pytorch: an imperative style, high-performance deep learning library. In: NeurIPS, pp. 8026–8037 (2019)
26. Ronneberger, O., Fischer, P., Brox, T.: U-Net: convolutional networks for biomedical image segmentation. In: Navab, N., Hornegger, J., Wells, W.M., Frangi, A.F. (eds.) MICCAI 2015. LNCS, vol. 9351, pp. 234–241. Springer, Cham (2015). https://doi.org/10.1007/978-3-319-24574-4_28
27. Rother, C., Kolmogorov, V., Blake, A.: GrabCut: interactive foreground extraction using iterated graph cuts. TOG **23**(3), 309–314 (2004)

28. Tang, M., Perazzi, F., Djelouah, A., Ayed, I.B., Schroers, C., Boykov, Y.: On regularized losses for weakly-supervised CNN segmentation. In: Ferrari, V., Hebert, M., Sminchisescu, C., Weiss, Y. (eds.) ECCV 2018. LNCS, vol. 11220, pp. 524–540. Springer, Cham (2018). https://doi.org/10.1007/978-3-030-01270-0_31
29. Tarvainen, A., Valpola, H.: Mean teachers are better role models: weight-averaged consistency targets improve semi-supervised deep learning results. In: NeurIPS, pp. 1195–1204 (2017)
30. Valvano, G., Leo, A., Tsaftaris, S.A.: Learning to segment from scribbles using multi-scale adversarial attention gates. TMI (2021)
31. Vezhnevets, V., Konouchine, V.: GrowCut: interactive multi-label ND image segmentation by cellular automata. Graphicon **1**(4), 150–156 (2005)
32. Vu, T.H., Jain, H., Bucher, M., Cord, M., Pérez, P.: Advent: adversarial entropy minimization for domain adaptation in semantic segmentation. In: CVPR, pp. 2517–2526 (2019)
33. Wang, G., et al.: Slic-Seg: a minimally interactive segmentation of the placenta from sparse and motion-corrupted fetal MRI in multiple views. Media **34**, 137–147 (2016)
34. Wang, X., Gao, J., Long, M., Wang, J.: Self-tuning for data-efficient deep learning. In: ICML, pp. 10738–10748. PMLR (2021)
35. Wu, Y., Xu, M., Ge, Z., Cai, J., Zhang, L.: Semi-supervised left atrium segmentation with mutual consistency training. In: de Bruijne, M., et al. (eds.) MICCAI 2021. LNCS, vol. 12902, pp. 297–306. Springer, Cham (2021). https://doi.org/10.1007/978-3-030-87196-3_28
36. Yu, L., Wang, S., Li, X., Fu, C.-W., Heng, P.-A.: Uncertainty-aware self-ensembling model for semi-supervised 3D left atrium segmentation. In: Shen, D., et al. (eds.) MICCAI 2019. LNCS, vol. 11765, pp. 605–613. Springer, Cham (2019). https://doi.org/10.1007/978-3-030-32245-8_67
37. Yushkevich, P.A., et al.: User-guided 3D active contour segmentation of anatomical structures: significantly improved efficiency and reliability. Neuroimage **31**(3), 1116–1128 (2006)
38. Zhang, H., Cisse, M., Dauphin, Y.N., Lopez-Paz, D.: MIXUP: beyond empirical risk minimization. arXiv preprint arXiv:1710.09412 (2017)
39. Zhang, Y., Yang, L., Chen, J., Fredericksen, M., Hughes, D.P., Chen, D.Z.: Deep adversarial networks for biomedical image segmentation utilizing unannotated images. In: Descoteaux, M., Maier-Hein, L., Franz, A., Jannin, P., Collins, D.L., Duchesne, S. (eds.) MICCAI 2017. LNCS, vol. 10435, pp. 408–416. Springer, Cham (2017). https://doi.org/10.1007/978-3-319-66179-7_47

Diffusion Deformable Model for 4D Temporal Medical Image Generation

Boah Kim and Jong Chul Ye(✉)

Korea Advanced Institute of Science and Technology, Daejeon, South Korea
{boahkim,jong.ye}@kaist.ac.kr

Abstract. Temporal volume images with 3D+t (4D) information are often used in medical imaging to statistically analyze temporal dynamics or capture disease progression. Although deep-learning-based generative models for natural images have been extensively studied, approaches for temporal medical image generation such as 4D cardiac volume data are limited. In this work, we present a novel deep learning model that generates intermediate temporal volumes between source and target volumes. Specifically, we propose a diffusion deformable model (DDM) by adapting the denoising diffusion probabilistic model that has recently been widely investigated for realistic image generation. Our proposed DDM is composed of the diffusion and the deformation modules so that DDM can learn spatial deformation information between the source and target volumes and provide a latent code for generating intermediate frames along a geodesic path. Once our model is trained, the latent code estimated from the diffusion module is simply interpolated and fed into the deformation module, which enables DDM to generate temporal frames along the continuous trajectory while preserving the topology of the source image. We demonstrate the proposed method with the 4D cardiac MR image generation between the diastolic and systolic phases for each subject. Compared to the existing deformation methods, our DDM achieves high performance on temporal volume generation.

Keywords: Deep learning · Medical image generation · Image deformation · Diffusion model

1 Introduction

Exploring the progression of anatomical changes is one of the important tasks in medical imaging for disease diagnosis and therapy planning. In particular, for the case of cardiac imaging with inevitable motions, 4D imaging to monitor 3-dimensional (3D) volume changes according to the time is often required for accurate analysis [1]. Unfortunately, in contrast to CT and ultrasound, MRI requires a relatively long scan time to obtain the 4D images.

With the advances in deep learning approaches, deep learning-based medical image generation methods have been extensively developed [6,14,15,20]. However, the existing models based on generative adversarial networks (GAN) may

L. Wang et al. (Eds.): MICCAI 2022, LNCS 13431, pp. 539–548, 2022.
https://doi.org/10.1007/978-3-031-16431-6_51

generate artificial features, which should not be occurred in the medical imaging area. On the other hand, learning-based deformable image registration methods have been developed due to their capability to provide deformation fields in real-time for the source to be warped into the target [2,3,8,12]. The smooth deformation fields enable the source to deform with topology preservation. Using this property, although the generative methods employing the registration model are presented [7,9], the role of these models is to generate templates instead of generating temporal 3D volume frames along the continuous trajectory.

Recently, the denoising diffusion probabilistic model (DDPM) [10,18] has been shown impressive performance in generating realistic images by learning a Markov chain process for the transformation of the simple Gaussian distribution into the data distribution [5,17,19]. Inspired by the fact that DDPM generates images through the latent space provided by the parameterized Gaussian process, the main contribution of this work is to present a novel diffusion deformable model (DDM) composed of the diffusion and deformation modules so that the latent code has detailed spatial information of the source image toward the target. Specifically, given the source and target images, the latent code is learned by using a score function of the DDPM, which is fed into the deformation module to generate temporal frames through the deformation fields.

Once the proposed DDM is trained, intermediate frame of 4D temporal volumes can be easily generated by simply interpolating the latent code estimated by the diffusion module. More specifically, the deformation module of DDM estimates the deformation fields according to the scaled latent code and provides the continuous deformation from the source to the target images in the image registration manner, which can preserve the topology of the source image. In this sense, our score-based latent code can be interpreted to provide plausible geodesic paths between the source and target images.

We verify the proposed DDM on 4D cardiac MR image generation using ACDC dataset [4]. The experimental results show that our model generates realistic deformed volumes along the trajectory between the diastolic and systolic phases, which outperforms the methods of adjusting the registration fields. The main contributions of this work can be summarized as follows:

- We propose a diffusion deformable model for 4D medical image generation, which employs the denoising diffusion model to estimate the latent code.
- By simply scaling the latent code, our model provides non-rigid continuous deformation of the source image toward the target.
- Experimental results on 4D cardiac MRI verify that the proposed method generates realistic deformed images along the continuous trajectory.

2 Proposed Method

The overall framework of the proposed diffusion deformable model (DDM) is illustrated in Fig. 1. For the source S and the target T volumes, we design the model composed of a diffusion module and a deformation module, which is

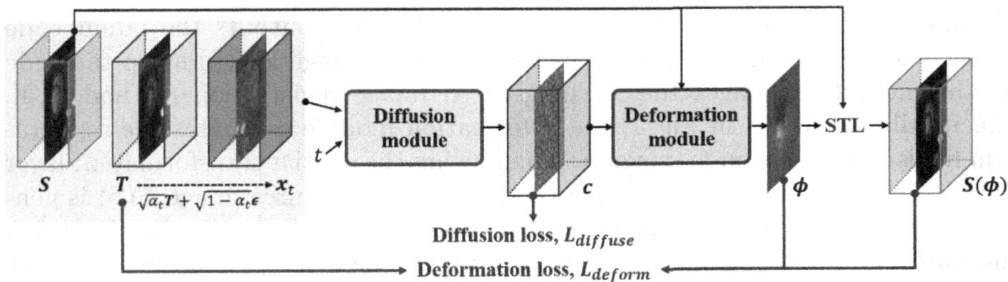

Fig. 1. The overall framework of the proposed method. The diffusion module takes a source S, a target T, and the perturbed target x_t, and estimates the latent code c. Then, the deformation module takes the code c with the source S and generates the deformed image $S(\phi)$ using the deformation fields ϕ with the spatial transformation layer (STL).

trained in an end-to-end manner. When the diffusion module estimates the latent code c that provides the spatial information of intermediate frames between the source and target, the deformation module generates deformed images along the trajectory according to the code. More details are as follows.

2.1 Diffusion Deformable Model

Diffusion Module. The diffusion module is based on the denoising diffusion probabilistic model (DDPM) [10,18]. DDPM is proposed as a generative model that learns a Markov chain process to convert the Gaussian distribution into the data distribution. The forward diffusion process adds noises to the data x_0, so that the data distribution at $t \in [0, u]$ of x can be represented as:

$$q(x_t|x_{t-1}) = \mathcal{N}(x_t; \sqrt{1-\beta_t}x_{t-1}, \beta_t\mathbf{I}), \tag{1}$$

where β_t is a noise variance in the range of (0, 1). Then, for the reverse diffusion, DDPM learns the following parameterized Gaussian transformations:

$$p_\theta(x_{t-1}|x_t) = \mathcal{N}(x_{t-1}; \mu_\theta(x_t, t), \sigma_t^2\mathbf{I}), \tag{2}$$

where $\mu_\theta(x_t, t)$ is learned mean and σ_t is a fixed variance. Accordingly, the generative process to sample the data is performed by the stochastic step: $x_{t-1} = \mu_\theta(x_t, t) + \sqrt{\beta_t}z$, where $z \sim \mathcal{N}(0, \mathbf{I})$.

By employing the property of DDPM, we design the diffusion module by considering the target as the reference data, i.e. $x_0 = T$. Specifically, given the condition of source S and target T, we first sample the perturbed target x_t by:

$$x_t = \sqrt{\alpha_t}T + \sqrt{1-\alpha_t}\epsilon, \tag{3}$$

where $\epsilon \sim \mathcal{N}(0, \mathbf{I})$, under $q(x_t|x_0) = \mathcal{N}(x_t; \sqrt{\alpha_t}x_0, (1-\alpha_t)\mathbf{I})$ where $\alpha_t = \Pi_{s=1}^t(1-\beta_s)$. Then, using the condition and the perturbed target x_t, the diffusion module learns the latent code c that compares the source and target and contains information of the score function for deformation.

Deformation Module. When the diffusion module outputs the latent code c, the deformation module generates the deformed image according to the code c. This module leverages the learning-based image registration method [2,3]. Specifically, as shown in Fig. 1, the deformation module estimates the registration fields ϕ for the source image S. Then, using the spatial transformation layer (STL) [11] with tri-linear interpolation, the deformed source image $S(\phi)$ is generated by warping the volume S with ϕ. Here, unlike [2,3] that directly takes the source and target images as a network input, the deformation module of the proposed DDM takes the latent code c and the source. As will be shown later, this enables our model to provide continuous deformation according to the code.

2.2 Model Training

To train the proposed DDM, we design the loss function as follows:

$$\min_{G_\theta} \mathcal{L}_{diffuse} + \lambda \mathcal{L}_{deform}, \tag{4}$$

where G_θ is the network of DDM with learnable parameters θ, $\mathcal{L}_{diffuse}$ is the diffusion loss, $\mathcal{L}_{deform}$ is the deformation loss, and λ is a hyper-parameter.

Specifically, the diffusion loss is to estimate the latent code c, which is originally designed for the reverse diffusion process [10]. Thus, for the source S and the target T, the diffusion loss is calculated by:

$$\mathcal{L}_{diffuse} = \mathbb{E}_{\epsilon,x,t} || G_\theta^{diffuse}(x_t, t; S, T) - \epsilon ||_2^2, \tag{5}$$

where $G_\theta^{diffuse}$ is the diffusion module of DDM whose output is c, ϵ is the Gaussian distribution data with $\mathcal{N}(0, \mathbf{I})$, and t is a uniformly sampled time step in $[0, u]$. On the other hand, the deformation loss is to generate the deformation fields to warp the source image into the target. Based on the energy function of classical image registration, we design the deformation loss as follows:

$$\mathcal{L}_{deform} = -NCC(S(\phi), T) + \lambda_R \Sigma ||\nabla \phi||^2, \tag{6}$$

where ϕ is the output of the deformation module G_θ^{deform} using the input of the latent code c and the source image S, NCC is the local normalized cross-correlation [2], and λ_R is a hyper-parameter. We set $\lambda_R = 1$ in our experiments. In (6), the first term is the dissimilarity metric between the deformed source image and the target, and the second term is the smoothness regularization for the deformation fields.

Accordingly, using the loss function (4), our model is trained by end-to-end learning in an unsupervised manner. This allows the diffusion module to learn the score function of the deformation by comparing the source and target. That is, the latent code c from the diffusion module can provide spatial information between the source and target, which enables the deformation module to generate the deformation fields of the source toward the target image.

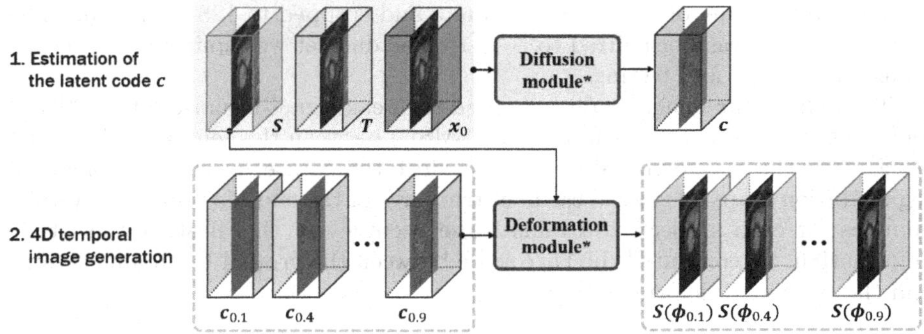

Fig. 2. The proposed inference flow for 4D temporal image generation. Once the latent code c is estimated, the continuous deformed frames between the source and the target are generated by scaling the latent code c with γ, i.e. c_γ.

2.3 Intermediate Frames for Temporal Volume Generation

Once the proposed DDM is trained, for a given condition of S and T, our model produces a deformed source image aligned with the target when x_0 is set to the target T. Here, when the latent code passed to the deformation module is set to zero, our model produces deformation fields that hardly deform the source image. Thus, the 4D temporal images from the source to the target can be obtained by adjusting the latent code c.

Specifically, as shown in Fig. 2, the latent code c is first estimated by the diffusion module with the fixed parameters θ^*:

$$c = G_{\theta^*}^{diffuse}(S, T, x_0), \tag{7}$$

where x_0 is the target T. Then, since this latent code provides the score function of deformation between the source and the target, the intermediate frame is generated by warping the source S with the deformation fields ϕ_γ:

$$\phi_\gamma = G_{\theta^*}^{deform}(S, c_\gamma), \tag{8}$$

where c_γ is the latent code adjusted by scaling c by $\gamma \in [0, 1]$, i.e. $c_\gamma = \gamma \cdot c$. Therefore, through the simple interpolation of the latent code, our model can generate temporally continuous deformed images along the trajectory from the source to the target image.

3 Experiments

Dataset and Metric. To verify the proposed method for 4D image generation, we used the publicly available ACDC dataset [4] that contains 100 4D cardiac MRI data. We trained and tested our model to provide the 4D temporal images from the end-diastolic to the end-systolic phases. All MRI scans were re-sampled with a voxel spacing of $1.5 \times 1.5 \times 3.15$ mm and cropped to $128 \times 128 \times 32$. The image intensity was normalized to $[-1, 1]$. The dataset was split into 90 and 10 scans for training and testing.

To quantify the image quality, we used the peak signal-to-noise ratio (PSNR) and the normalized mean square error (NMSE) between the generated deformed images and the real temporal data. Also, since the dataset provides the manual segmentation maps on several cardiac structures of the end-diastolic and systolic volumes for each subject, to evaluate the accuracy of the diastolic-to-systolic deformation, we computed the Dice score between the ground truth annotations and the estimated deformed maps.

Implementation Details. We built our proposed method on the 3D UNet-like structures that has the encoder and decoder with skip connections. For the diffusion module, we employed the network designed in DDPM [10] and set 8, 16, 32, and 32 channels for each stage. We set the noise level from 10^{-6} to 10^{-2}, which was linearly scheduled with $u = 2000$. For the deformation module, we used the architecture of VoxelMorph-1 [2]. We adopted Adam algorithm [13] with a learning rate 2×10^{-4}, and trained the model with $\lambda = 20$ for 800 epochs by setting the batch size as 1. The proposed method was implemented with PyTorch 1.4 [16] using an Nvidia Quadro RTX 6000 GPU. The memory usage for training and testing was about 3GB and 1GB, respectively. Also, the model training took 6 h, and the testing took an average of 0.456 s. The source code is available at https://github.com/torchDDM/DDM.

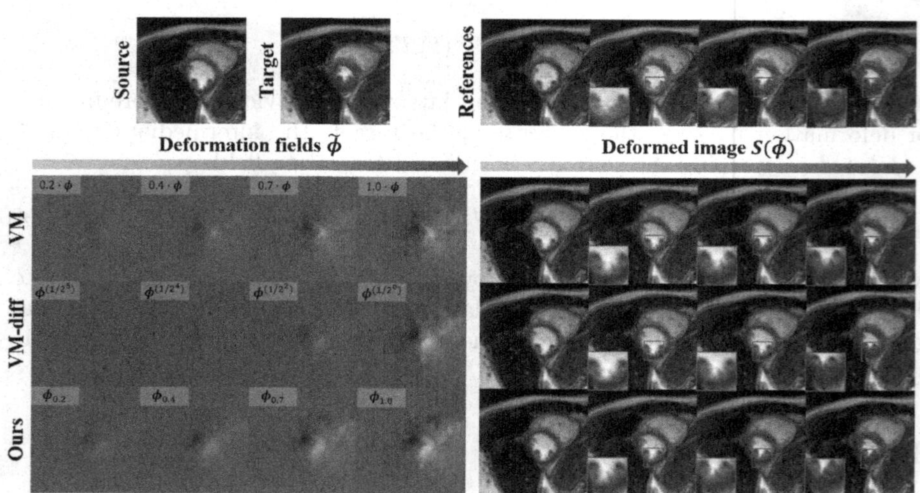

Fig. 3. Visual comparison results of the temporal cardiac image generation. Using the source and target, the deformed intermediate frames $S(\tilde{\phi})$ (right) are generated using the deformation fields $\tilde{\phi}$ (left). The references are from the ground-truth 4D images.

Table 1. Quantitative comparison results of the average PSNR, NMSE, Dice, and test runtime. Standard deviations are shown in parentheses. Asterisks denote statistical difference of the baseline methods over ours ($**$: $p < 0.005$ and $*$: $p < 0.05$).

Data	Method	PSNR (dB) ↑	NMSE ($\times 10^{-8}$) ↓	Dice ↑	Time (sec) ↓
Train	Initial	29.683 (3.116)	0.690 (0.622)	0.700 (0.185)	-
	DDM (Ours)	32.788 (2.859)	0.354 (0.422)	0.830 (0.112)	0.456
Test	Initial	28.058 (2.205)	0.790 (0.516)	0.642 (0.188)	-
	VM	30.562 (2.649) $*$	0.490 (0.467) $*$	0.784 (0.116) $*$	0.219
	VM-diff	29.481 (2.473) $**$	0.602 (0.477) $**$	0.794 (0.104)	2.902
	DDM (Ours)	**30.725** (2.579)	**0.466** (0.432)	**0.802** (0.109)	0.456

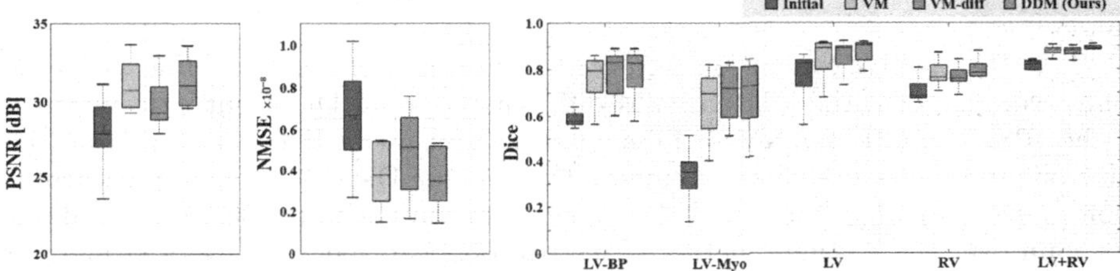

Fig. 4. Boxplots of quantitative evaluation results on test data. PSNR and NMSE are computed using the generated frames from the diastolic to the systolic phases. Dice score is computed for the segmentation maps of the cardiac structures at the end-systolic phase: left blood pool (LV-BP), myocardium (LV-Myo), epicardium of the left ventricle (LV), right ventricle (RV), and the total cardiac region (LV+RV).

Results and Discussion. We compared ours with the learning-based registration models: VM [2] and VM-diff [8]. The VM can provide the intermediate frames between the source and target by scaling the estimated deformation fields, i.e. $\gamma \cdot \phi$, while the VM-diff with diffeomorphic constraint can give the deformed images by integrating the velocity field with the timescales, i.e. $\partial\phi^{(t)}/\partial t = v(\phi^{(t)})$. We implemented these methods using the source code provided by the authors with the recommended parameters. For the VM, the weight of smooth regularization was set to 1. For the VM-diff, the number of integration steps, the noise parameter, and the prior lambda regularization for KL loss were set to 7, 0.01, and 25, respectively. For a fair comparison, we used the same network with our deformation module and trained with the learning rate 2×10^{-4} until the models converge. At the inference, we set the number of deformations according to the number of frames in each ground-truth 4D data.

Figure 3 shows visual comparisons of the temporal image generation along the trajectory between the source and the target. We can observe that the deformation fields of VM only vary in scale, but their relative spatial distributions do not change, resulting in the scaled movement of anatomical structures. Also, the VM-diff with the integration of the velocity fields hardly deforms

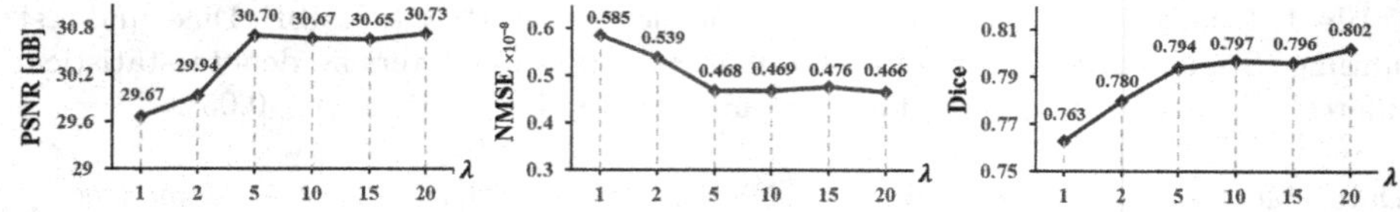

Fig. 5. Quantitative results over various hyper-parameter λ in the proposed loss function. The graphs show the average values of PSNR, NMSE, and Dice metrics.

the source in the beginning, but sharply in the end. On the other hand, in the proposed method, the deformation fields estimated from the scaled latent codes represent the dynamic changes depending on the positions. Accordingly, the generated intermediate deformed images have distinct changes from the source to the target.

The quantitative evaluation results are reported in Table 1 and Fig. 4. Specifically, compared to the VM and VM-diff, our proposed DDM outperforms with higher PSNR and lower NMSE in generating continuous deformations from the diastolic to the systolic phase volumes. The average Dice score for the segmentation maps of cardiac structures also shows that ours achieves 80.2% with about 1% gain over the baseline methods. We can observe that the proposed method tested on training data shows similar gains for all metrics when compared to the results on test data. Moreover, we evaluated the statistical significance of the results and observed that our model outperforms the VM on all metrics with p-values<0.05 under both paired t-test and Wilcoxon signed rank test. It is also worth noting that our model shows slightly higher accuracy on Dice score over the VM-diff with no significant difference but achieves significant improvement on PSNR and NMSE with p-value<0.001 under paired t-test and p-value<0.005 under Wilcoxon signed rank test. In addition, the average test time of VM-diff takes 2.902 s, while that of our DDM takes 0.456 s. These indicate that our method shows superiority in generating continuous deformations.

Furthermore, to analyze the effect of our designed loss function (4), we studied our method with various hyper-parameter λ values. Figure 5 shows the results of PSNR, NMSE, and Dice metrics. When λ increases, PSNR increases and NMSE decreases initially, and they converge when it exceeds a certain level ($\lambda = 5$), suggesting that the large value of deformation loss does not affect the generation performance more. In contrast, the Dice score improves according to the λ increases, which suggests that the deformation loss affects the registration accuracy, and also helps the diffusion module to estimate latent codes to generate temporally continuous deformed images along the trajectory. On the other hand, when we trained our model only using the deformation loss, PSNR, NMSE, and Dice scores were 30.72, 0.473×10^{-8}, and 0.799, respectively, which were lower than the optimal results of our method. This indicates that the diffusion loss is effective to learn the latent code for generating temporal images.

4 Conclusion

We propose a novel 4D image generation framework by adapting the denoising diffusion probabilistic model to the image registration model. Our model learns the distribution of the source and target and estimates the latent code to generate deformed images along the continuous trajectory. Experimental results on 4D cardiac MR image generation verify that the proposed model produces dynamic deformed images from the end-diastolic to systolic phases in real-time. Thus, our model can be a promising tool in clinical applications such as analyzing changes in anatomical structures.

Acknowledgements. This work was supported in part by the Institute of Information & communications Technology Planning & Evaluation (IITP) grant funded by the Korea government(MSIT) (No.2019-0-00075, Artificial Intelligence Graduate School Program(KAIST)), in part by the National Research Foundation of Korea under Grant NRF-2020R1A2B5B03001980, in part by the Korea Medical Device Development Fund grant funded by the Korea government (the Ministry of Science and ICT, the Ministry of Trade, Industry and Energy, the Ministry of Health & Welfare, the Ministry of Food and Drug Safety) (Project Number: 1711137899, KMDF_PR_20200901_0015), in part by the MSIT(Ministry of Science and ICT), Korea, under the ITRC(Information Technology Research Center) support program(IITP-2022-2020-0-01461) supervised by the IITP, and in part by the KAIST Key Research Institute (Interdisciplinary Research Group) Project.

References

1. Amsalu, E., et al.: Spatial-temporal analysis of cause-specific cardiovascular hospital admission in Beijing, China. Int. J. Environ. Health Res. **31**(6), 595–606 (2021)
2. Balakrishnan, G., Zhao, A., Sabuncu, M.R., Guttag, J., Dalca, A.V.: An unsupervised learning model for deformable medical image registration. In: Proceedings of the IEEE Conference on Computer Vision and Pattern Recognition, pp. 9252–9260 (2018)
3. Balakrishnan, G., Zhao, A., Sabuncu, M.R., Guttag, J., Dalca, A.V.: VoxelMorph: a learning framework for deformable medical image registration. IEEE Trans. Med. Imaging **38**(8), 1788–1800 (2019)
4. Bernard, O., et al.: Deep learning techniques for automatic MRI cardiac multi-structures segmentation and diagnosis: is the problem solved? IEEE Trans. Med. Imaging **37**(11), 2514–2525 (2018)
5. Choi, J., Kim, S., Jeong, Y., Gwon, Y., Yoon, S.: ILVR: conditioning method for denoising diffusion probabilistic models. In: Proceedings of the IEEE/CVF International Conference on Computer Vision, pp. 14367–14376 (2021)
6. Dai, X., et al.: Multimodal MRI synthesis using unified generative adversarial networks. Med. Phys. **47**(12), 6343–6354 (2020)
7. Dalca, A., Rakic, M., Guttag, J., Sabuncu, M.: Learning conditional deformable templates with convolutional networks. In: Advances in Neural Information Processing Systems 32 (2019)

8. Dalca, A.V., Balakrishnan, G., Guttag, J., Sabuncu, M.R.: Unsupervised learning for fast probabilistic diffeomorphic registration. In: Frangi, A.F., Schnabel, J.A., Davatzikos, C., Alberola-López, C., Fichtinger, G. (eds.) MICCAI 2018. LNCS, vol. 11070, pp. 729–738. Springer, Cham (2018). https://doi.org/10.1007/978-3-030-00928-1_82
9. Dey, N., Ren, M., Dalca, A.V., Gerig, G.: Generative adversarial registration for improved conditional deformable templates. In: Proceedings of the IEEE/CVF International Conference on Computer Vision, pp. 3929–3941 (2021)
10. Ho, J., Jain, A., Abbeel, P.: Denoising diffusion probabilistic models. Adv. Neural. Inf. Process. Syst. **33**, 6840–6851 (2020)
11. Jaderberg, M., et al.: Spatial transformer networks. In: Advances in Neural Information Processing Systems 28 (2015)
12. Kim, B., Kim, D.H., Park, S.H., Kim, J., Lee, J.G., Ye, J.C.: CycleMorph: cycle consistent unsupervised deformable image registration. Med. Image Anal. **71**, 102036 (2021)
13. Kingma, D.P., Ba, J.: Adam: a method for stochastic optimization. arXiv preprint arXiv:1412.6980 (2014)
14. Nie, D., et al.: Medical image synthesis with context-aware generative adversarial networks. In: Descoteaux, M., Maier-Hein, L., Franz, A., Jannin, P., Collins, D.L., Duchesne, Simon (eds.) MICCAI 2017. LNCS, vol. 10435, pp. 417–425. Springer, Cham (2017). https://doi.org/10.1007/978-3-319-66179-7_48
15. Nie, D., et al.: Medical image synthesis with deep convolutional adversarial networks. IEEE Trans. Biomed. Eng. **65**(12), 2720–2730 (2018)
16. Paszke, A., et al.: Automatic differentiation in pyTorch (2017)
17. Saharia, C., Ho, J., Chan, W., Salimans, T., Fleet, D.J., Norouzi, M.: Image super-resolution via iterative refinement. arXiv preprint arXiv:2104.07636 (2021)
18. Sohl-Dickstein, J., Weiss, E., Maheswaranathan, N., Ganguli, S.: Deep unsupervised learning using nonequilibrium thermodynamics. In: International Conference on Machine Learning, pp. 2256–2265. PMLR (2015)
19. Song, J., Meng, C., Ermon, S.: Denoising diffusion implicit models. In: International Conference on Learning Representations (2020)
20. Yang, H., et al.: Unpaired brain MR-to-CT synthesis using a structure-constrained CycleGAN. In: Stoyanov, D., et al. (eds.) DLMIA/ML-CDS -2018. LNCS, vol. 11045, pp. 174–182. Springer, Cham (2018). https://doi.org/10.1007/978-3-030-00889-5_20

SAPJNet: Sequence-Adaptive Prototype-Joint Network for Small Sample Multi-sequence MRI Diagnosis

Yuqiang Gao[1], Guanyu Yang[1,3(✉)], Xiaoming Qi[1], Yinsu Zhu[2], and Shuo Li[4]

[1] LIST, Key Laboratory of Computer Network and Information Integration, Southeast University, Ministry of Education, Nanjing 210096, China
yang.list@seu.edu.cn

[2] Department of Radiology, The First Affiliated Hospital of Nanjing Medical University, Nanjing, China

[3] Jiangsu Provincial Joint International Research Laboratory of Medical Information Processing, Southeast University, Nanjing 210096, China

[4] Department of Medical Biophysics, University of Western Ontario, London, ON, Canada

Abstract. Multi-sequence magnetic resonance imaging (MRI) images have complementary information that can greatly improve the reliability of diagnosis. However, automated diagnosis of small sample multi-sequence MR images is a challenging task due to: 1) Divergent representation. The difference between sequences and the weak correlation between contained features make the representation extracted from the network tend to diverge, which is profitless to robust classification. 2) Sparse distribution. The small sample size is reflected in the sparse distribution of the prototype, making the network only learn rough demarcation, which is inadequate for medical images with small class intervals. In this paper, we propose for the first time a network (SAPJNet) that can adapt to both multi-sequence and small sample conditions, enabling high-quality automatic diagnosis of small-sample multi-sequence MR images, which is of great help to improve clinical diagnostic efficiency. 1) The sequence-adaptive transformer (SAT) of SAPJNet generates joint representations as disease prototypes by filtering intra-sequence features and aggregating inter-sequence features. 2) The prototype optimization strategy (POS) of SAPJNet constrains the prototype distribution by approximating the intra-class prototype and alienating the inter-class prototype. The SAPJNet achieved optimal performance in three tasks: risk assessment of pulmonary arterial hypertension (PAH), classification of idiopathic inflammatory myopathies (IIM), and identification of knee abnormalities, with at least a 10%, 10%, and 6.7% improvement in accuracy of overall comparison methods.

Keywords: Multi sequence · Attention mechanism · Metric learning

---

Supplementary Information The online version contains supplementary material available at https://doi.org/10.1007/978-3-031-16431-6_52.

L. Wang et al. (Eds.): MICCAI 2022, LNCS 13431, pp. 549–559, 2022.
https://doi.org/10.1007/978-3-031-16431-6_52

1 Introduction

Multi-sequence MR images with small samples play an important role in clinical disease diagnosis. Different sequences contain diverse information at different scales and dimensions, which can greatly improve the reliability of diagnosis by complementing each other [1,8]. For example, MR images of the heart contain short axis(SAX), late gadolinium-enhanced(LGE), t1 mapping and other sequences [13]. MR images of knee joint included axial, coronal, and sagittal sequences [3]. However, the number of patient samples with complete multi-sequence MR images in the clinic is generally small. On the one hand, few patients require full sequence scans for clinical diagnosis. On the other hand, few public large datasets contain multiple sequences [2,4].

Existing studies of multi-sequence MR images mostly stack encoders to extract features, then simply concatenate all features [14] for feature fusion and classification. Rossi et al. used three MR sequences for coding to improve the diagnostic performance of prostate cancer [15]; He et al. deeply fused the two sequences of brain MR images at multiple scales, but the features obtained are still high-dimensional [10]. Guan et al. used two MR sequences to identify brain dysfunction and provided a region-level explanation through attentional maps [8]. Although the process of pooling and concatenating is relatively simple, the features learned are rough and redundant. This causes the feature dimension to explode as the number of sequences increases, and these methods are no longer applicable. Meanwhile, these methods use many training samples compared to a few test samples. Their performance under small sample size training remains questionable.

Thereby, automated diagnosis of small sample multi-sequence MR images remains challenging (Fig. 1) due to: 1) Divergent representation. The difference between sequences and the weak correlation between features make the representation extracted from the network tend to diverge, which is profitless to robust classification. 2) Sparse distribution. The small sample size is reflected in the sparse distribution of the prototype, making the network only learn rough demarcation, which is inadequate for medical images with small class intervals.

In this paper, we propose for the first time a network (SAPJNet) that can adapt to both multi-sequence and small sample conditions, enabling high-quality automatic diagnosis of small-sample multi-sequence MR images, which is of great help to improve clinical diagnostic efficiency. 1) The sequence-adaptive transformer (SAT) generates joint representations as disease prototypes(feature vectors that will eventually be used for classification) by filtering intra-sequence features and aggregating inter-sequence features. With the help of self-attention and global attention, the significance and correlation of effective features are strengthened, so the features extracted from the network are not rough and scattered. 2) The prototype optimization strategy (POS) constrains the prototype distribution by approximating the intra-class prototype and alienating the inter-class prototype. Metric learning with sparse sampling constructs positive and negative sample pairs and measures the similarity of each pair of samples so that prototypes are distributed in their respective cluster centers, which are far away from each other.

In conclusion, our main contributions are as follows:

1. We propose a sequence-adaptive prototype-joint network for small sample multi-sequence MR image classification. Its good accuracy and generalization can assist the clinical diagnosis.
2. The sequence-adaptive transformer generates joint representations as disease prototypes by filtering intra-sequence features and aggregating inter-sequence features, based on the attention mechanism.
3. The prototype optimization strategy constrains the prototype distribution by approximating the intra-class prototype and alienating the inter-class prototype, based on metric learning with sparse sampling.

2 Method

The SAPJNet first obtains global features of different sequences through differentiated encoders. They are then redivided into several groups and further entered into the self-attention mechanism to mine correlations. These correlations are finally aggregated into the first additional unit as the final prototype. Both the output of the encoder and the final prototype are monitored by the POS. It compares the prototype and pulls them closer or farther depending on the category of the prototype (Fig. 2).

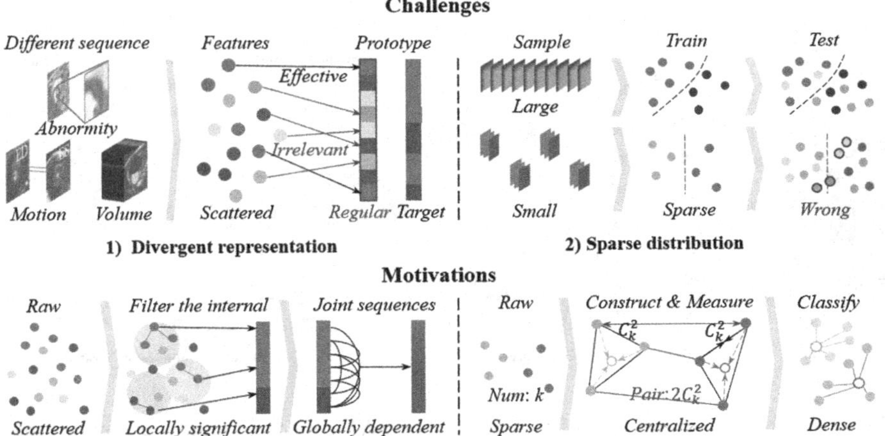

Fig. 1. Challenges and motivations. 1) Divergent representation. The difference between sequences and the weak correlation between contained features make the representation extracted from the network tend to diverge. 2) Sparse distribution. The small sample size is reflected in the sparse distribution of the prototype, making the network only learn rough demarcation. Our SAPJNet screens effective features and generates joint representations. It also constrains the distribution of prototypes by constructing sample pairs and measuring prototype similarity.

2.1 Sequence-Adaptive Transformer

The SAT extracts effective joint representations from multi-sequence MR images to solve the representation divergence caused by large sequence differences. Features contained in different MRI sequences have different meanings in terms of motion, structure, or local gray anomalies. The SAT accurately extracts these features and generates prototypes that mimic a doctor's overall assessment of a patient's disease type or risk level. Its ability to mine effective components from high-dimensional sparse features and generate low-dimensional joint representation benefits from the attention mechanism [7]. It assigns learnable keys and values to each input unit to calculate their correlation representation [16]. These representations are aggregated and translated into new semantics, which in this context are categories or degrees of disease.

To explore the correlation between features of different sequences, we design several special steps at the junction of multiple differentiated encoders and the basic transformer structure. For the output of multichannel features from encoders, the SAT firstly flattens the two-dimensional global features of each channel, and then introduces a parameter p to control the dimension and number of units input into the self-attention layer. For example, when $p = 4$, the dimension of the unit d is 4 times the size of the 2D feature graph, and the input length δ is 14 of the product of the channels number and the slice number of all three outputs of encoders. Datasets of different resolutions do not require a specially designed p value, just keeping the dimension of the prototype at a general level, 128 for example.

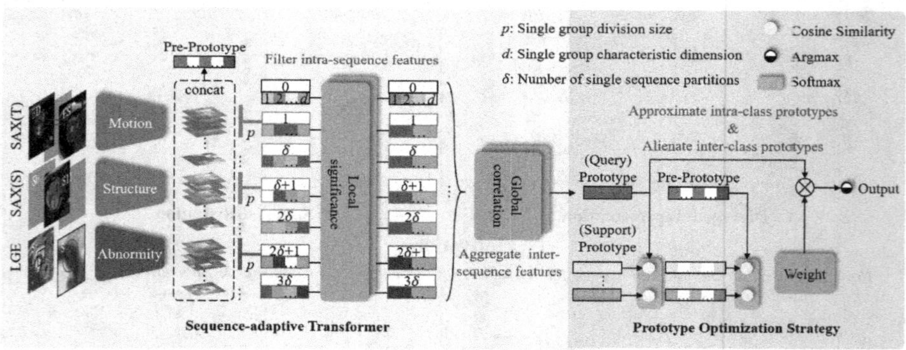

Fig. 2. The overall framework of SAPJNet. The SAT generates joint representations as disease prototypes. It first codes multiple sequences, then groups features and calculates attention weighting, and finally gathers effective information in the first feature group. The POS constrains the prototype distribution. It measures the two outputs of the SAT using different ways in two stages.

2.2 Prototype Optimization Strategy

POS constrains the prototype distribution to solve the sparse distribution problem caused by small samples. It combines two different measurement learning methods to gain two advantages at the same time. On the one hand, the pair-based way allows it to implicitly expand the training batch by constructing positive and negative sample pairs, that is, to make full use of the supervision information. During the process of measuring the similarity of prototypes, the distance of different categories is enlarged, while the distance of the same category is constantly reduced [11,12]. On the other hand, the proxy-based way makes it can use the last layer of the network to store prototypes of diseases, thus realizing the end-to-end training and testing of the network [17].

Specifically, the POS measures the two outputs of the SAT using different ways in two stages. The two outputs are respectively: the pre-prototype concatenated by features output from the encoder set and the prototype output from the SAT. The former is reserved for faster propagation of the gradient to the encoder. The first stage calculates the losses of two outputs using a pair-based way. Inside an n-way k-shot training batch, C_k^2 positive sample matches are firstly mined by in-class combination. Then for each positive match, assume that one is the query sample q_i^a and the other is the support sample s_j^b, where $a, b \in 1, 2, \ldots, k$ and $i, j \in 1, 2, \ldots, n$. Each q_i^a has $k-1$ negative matches s_j^b, where $b = 1, 2, \ldots, k, b \neq a$. The total count of all combinations will be $n^2 C_k^2$, which is much less the original number of combinations C_{nk}^2. After the sample is turned into a prototype by the SAT, the loss per batch is calculated as follows: for each $f(s_j^b)$, its cosine similarity with each $f(q_i^a)$ is fed into the cross-entropy function as the probability that it belongs to different classes, where $f(\cdot)$ represents the pre-prototype or the real prototype output from the SAT. The process can be described by Eq. 1.

$$Loss_1 = -\frac{f(s^b) \cdot f(q^a)}{\|f(s^b)\| \times \|f(q^a)\|} + \log \sum_{d \neq a} e^{\frac{f(s^b) \cdot f(q^d)}{\|f(s^b)\| \times \|f(q^d)\|}} \tag{1}$$

The second stage uses additive additive-angular-margin loss [5] to fine-tune. It maintains consistency in measuring prototypes while releasing the SAPJNet from sample pairing. It takes a full connection layer as the storage pool of category prototypes, and constantly updates parameters in training to continuously optimize the final prototype representation. The additive-angular-margin loss is described in Eq. 2.

$$Loss_3 = -\frac{1}{m} \sum_{i=1}^{m} \log \frac{e^{s(\cos(\theta_{y_i} + m))}}{e^{s(\cos(\theta_{y_i} + m))} + \sum_{j=1, j \neq y_i}^{n} e^{s(\cos \theta_{y_i})}} \tag{2}$$

The total loss of the SAPJNet is $Loss_1 + \beta Loss_2 + \gamma Loss_3$. It is adjusted at different stages with different hyperparameters. $\alpha = \beta = 1, \gamma = 0$ in the first stage, $\alpha = \beta = \gamma = 1$ in the second stage. Loss2 has the same form as Loss1, but its input variables are prototypes refined by the SAT.

3 Experiments and Results

3.1 Datasets and Setup

Two different organ datasets were used in our experiment. 1) **Local dataset**. 80 hospital MR images containing raw imaging data from 40 patients with PAH (20 at low risk and 20 at high risk) and 40 patients with IIM (20 with dermatomyositis and 20 with polymyositis). All SAX images were processed as [90, 90, 12, 12] data blocks, representing the dimensions of [width, height, slice, frame]. The LGE sequence has no time dimension and is otherwise consistent. Notice that different sequences of the same patient, although having the same voxel size after resampling, were not spatially aligned. 2) **External dataset**. A dataset used in MRNet [3] for the diagnosis of knee abnormalities, contains three 3D sequences from different angles. We follow the data format without tailoring or changing anything. Its 2D plane resolution is [256, 256], but the slice dimension is different.

ResNet-50 [9] was chosen as the backbone of the SAPJNet, the number of channels in the ResNet-50 is [12, 12, 24, 48, 96, 192, 384], For the internal dataset, the single group division size is $p = 4$. For the external dataset $p = 2$. We used the Adam optimizer to train the SAPJNet, with a learning rate of $5E-4$ for the first stage of all training and $5E-5$ for the stalemate stage. The batch size is 10 and the number of iterations in both training stages is 40. We implemented the method using PyTorch programming and carried out experiments on two GPUs (Nvidia GeForce RTX 3090).

We used five indicators commonly used in image classification tasks to evaluate the experimental results of the SAPJNet and other research methods on all datasets. These indicators include AUC, accuracy, precision, recall, and F1 scores. These indicators are also mentioned in the paper of the methods we compare. In the experiments, we preferentially select the training results with the best AUC performance in the verification set.

3.2 Comparative Study

Method comparison experiments on local data sets are designed to measure the advantages of the SAPJNet over other methods on smaller data sets. External approaches to this are: 1) **DenseNet**: a baseline that has performed well in numerous studies. 2) **RepVGG** [6]: Improved VGG network with better performance than ResNet at the same scale. 3)**FiANet** [10]: an attention-guided dual-sequence-input classifier. 4) **MMSNet** [15]: a three-sequence-input siamese network.

To be fair on data input, we made some adjustments. For DenseNet and RepVGG, we stacked three 3D encoders to classify each sequence separately and conducted soft voting to determine the final sample category. For FiANet, to maintain the original design as much as possible, we cut out an output branch from the SAPJNet, which also shows the flexibility of the SAPJNet. All methods are verified by two-fold cross-validation, and their network and training parameters maintain their default values during this process.

The comparative study shows the superiority of the SAPJNet in small sample multi-sequence MR image diagnosis. On the PAH dataset, the SAPJNet achieved 14% and 17% improvement in AUC, 2% and 10% improvement in ACC, and 7% and 2% improvement in F1 scores compared with FiANet. When three sequences were used, the performance of the SAPJNet was further improved by at least 8% on AUC and 10% on ACC, with significant improvements in accuracy and F1 scores (Table 1). Similarly, the SAPJNet achieves the best performance in the IIM dataset (Table 2). These results show that simple stacking of encoders is easy to overfit, and soft voting in the network by feature concatenation has little effect. The introduction of multiscale and the attention mechanism can improve the performance to a certain extent, but reduce the simplicity and scalability of the network. By contrast, the SAPJNet uses simple backbones with built-in feature priors and focuses on the correlation between global features, thus

Table 1. Two-fold cross-validation results of different methods on PAH datasets. B represents large structure input, S represents small structure input, and M represents motion input.

Methods	AUC	Accuracy	Recall	Precision	F1-score
DenseNet*3	66.5	50.0	50.0	50.0	50.0
RepVGG*3	68.3	50.0	100	50.0	66.7
FiANet with B & S	63.5	72.5	80.0	60.0	68.6
FiANet with M & S	57.8	65.0	100	58.8	74.1
SAPJNet with B & S	77.5	75.0	75.0	75.0	75.0
SAPJNet with M & S	74.0	75.0	80.0	72.7	76.2
MMSNet	50.0	50.0	50.0	50.0	50.0
SAPJNet with three	**85.3**	**85.0**	80.0	**88.9**	**84.2**

Table 2. Two-fold cross-validation results of different methods on IIM datasets. B represents large structure input, S represents small structure input, and M represents motion input.

Methods	AUC	Accuracy	Recall	Precision	F1-score
DenseNet*3	56.0	57.5	75.0	55.1	54.3
RepVGG*3	50.0	50.0	0.00	0.00	0.00
FiANet with B & S	79.0	77.5	80.0	76.2	78.0
FiANet with M & S	86.7	80.0	**85.0**	77.3	81.0
MMSNet	50.0	50.0	50.0	50.0	50.0
SAPJNet with three	**86.8**	**90.0**	80.0	**100.**	**88.9**

achieving better performance while reducing the number of parameters, even with fewer inputs.

The ability of the SAPJNet to find the effective components from high-dimensional features is demonstrated by the difference between the bottom features of the three encoders. In patients with PAH, abnormal ventricular septal brightness on LGE images, right heart hypertrophy, and reduced cardiac activity on SAX images are reflected in the distribution of the three output features respectively (Fig. 3). In contrast, the feature distribution of DM and PM did not show significant visual differences due to significant cardiac symptoms, but the SAPJNet was still capable of high-quality classification of prototypes.

3.3 Ablation Study

Module ablation experiments on the PAH dataset were conducted to validate the effectiveness of the three modules we designed. The baseline method we used was three ResNet encoders to extract features, a full connection layer for classification, and a cross-entropy function for training. The results presented in Table 3 show that the full version of the SAPJNet achieves optimal performance in all indicators. Specifically, the POS solves the problem of over-fitting caused by complex module structure, while the SAT satisfies the requirement of the POS for a concise and accurate prototype. They complement each other to ensure good classification performance.

The data ablation experiment on external datasets is designed to verify the generalization performance of the SAPJNet under large test samples. We deliberately used a few samples (40, 20, and 10) for training and numerous samples (1000) for testing. Compared with the baseline method (Fig. 4), it can be seen that with the reduction of training samples, the performance of the SAPJNet is better, and its performance loss is smaller in the five training sessions.

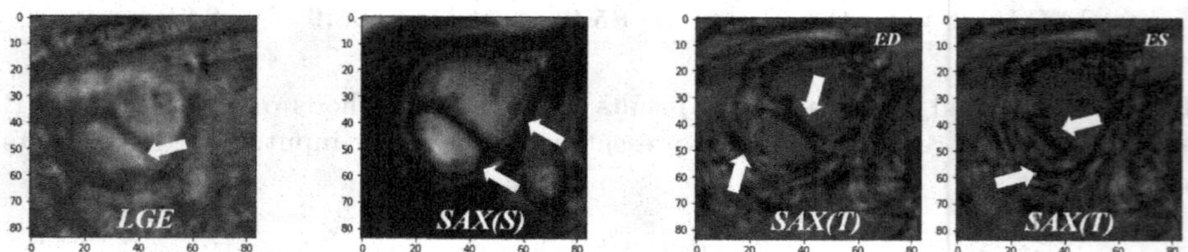

Fig. 3. The difference between the three encoders on the bottom feature map of the input layer sample. Samples with different risks or diseases showed differences in myocardial abnormalities, ventricular volume, and myocardial motor characteristics.

Table 3. The ablation trial on PAH datasets. RN represents ResNet, CE represents cross entropy loss, POS1 represents the first stage, and POS2 represents the second stage.

Methods	AUC	Accuracy	Recall	Precision	F1-score
RN + CE	63.5	60.0	80.0	57.1	66.7
RN + POS2	65.2	67.5	70.0	66.7	68.3
RN + POS1 + VOT	66.5	72.5	80.0	69.6	74.4
RN + POS1 + POS2	55.5	57.5	55.6	75.0	63.8
SAT + CE	46.8	52.5	40.0	53.3	45.7
SAT + POS2	52.5	55.0	55.0	55.0	55.0
SAT + POS1 + VOT	79.7	77.5	80.0	76.2	78.0
SAT + POS1 + POS2	**85.3**	**85.0**	**80.0**	**88.9**	**84.2**

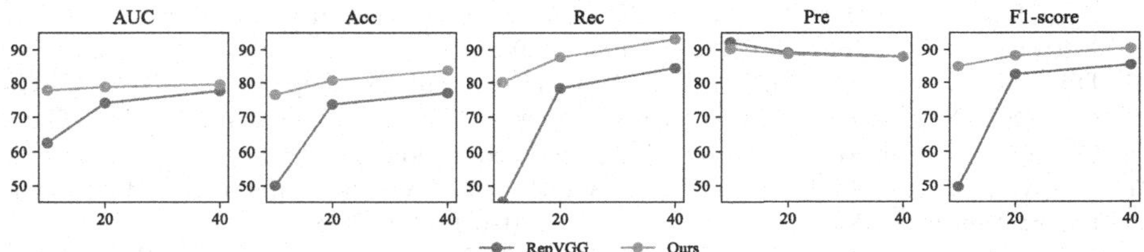

Fig. 4. The mean of 5 of the random sampling results on the external dataset. the SAPJNet is better on almost every indicator. The X-axis is the number of samples.

4 Discussion and Conclusion

For the first time, the SAPJNet enables the automated diagnosis of small sample multi-sequence MR images, addressing two major challenges and achieving optimal performance across multiple data sets. It can help doctors improve diagnostic efficiency and assist in the early study of difficult diseases. Moreover, by visualizing the network learning process, the SAPJNet demonstrates the interpretability of diagnosis, especially the different emphasis on different sequences. These two advantages make the SAPJNet more suitable for clinical application. Temporarily, we only conducted experiments in medical image data sets, or specifically multi-sequence MR image data sets. In the future, we will expand the sequence types by looking for more medical data that contain multiple signal forms to study and extend the possibility of our method. We also want to use our SAPJNet on MindSpore[1], which is a new deep learning computing framework.

Acknowledgements. This work was supported in part by the CAAI-Huawei MindSpore Open Fund, CANN(Compute Architecture for Neural Networks), Ascend AI Processor, and Big Data Computing Center of Southeast University.

[1] https://www.mindspore.cn.

References

1. Aldoj, N., Lukas, S., Dewey, M., Penzkofer, T.: Semi-automatic classification of prostate cancer on multi-parametric MR imaging using a multi-channel 3D convolutional neural network. Eur. Radiol. **30**, 1243–1253 (2020)
2. Arif, M., Schoots, I.G., Castillo T., J.M., Roobol, M.J., Niessen, W., Veenland, J.F.: Computer aided diagnosis of clinically significant prostate cancer in low-risk patients on multi-parametric MR images using deep learning. In: 2020 IEEE 17th International Symposium on Biomedical Imaging (ISBI), pp. 1482–1485 (2020)
3. Bien, N., et al.: Deep-learning-assisted diagnosis for knee magnetic resonance imaging: development and retrospective validation of MRNet. PLoS Med. **15**, e1002699 (2018)
4. Cheplygina, V., Peña, I.P., Pedersen, J.H., Lynch, D.A., Sørensen, L., de Bruijne, M.: Transfer learning for multicenter classification of chronic obstructive pulmonary disease. IEEE J. Biomed. Health Inform. **22**, 1486–1496 (2018)
5. Deng, J., Guo, J., Xue, N., Zafeiriou, S.: ArcFace: additive angular margin loss for deep face recognition. In: 2019 IEEE/CVF Conference on Computer Vision and Pattern Recognition (CVPR), pp. 4685–4694 (2019)
6. Ding, X., Zhang, X., Ma, N., Han, J., Ding, G., Sun, J.: RepVGG: Making VGG-style convNets great again. In: Proceedings of the IEEE Computer Society Conference on Computer Vision and Pattern Recognition, pp. 13728–13737 (2021)
7. Dosovitskiy, A., et al.: An image is worth 16x16 words: transformers for image recognition at scale. arXiv preprint arXiv:2010.11929 (2020)
8. Guan, H., Liu, Y., Yang, E., Yap, P.T., Shen, D., Liu, M.: Multi-site MRI harmonization via attention-guided deep domain adaptation for brain disorder identification. Med. Image Anal. **71**, 102076 (2021)
9. He, K., Zhang, X., Ren, S., Sun, J.: Deep residual learning for image recognition. In: Proceedings of the IEEE Conference on Computer Vision and Pattern Recognition, pp. 770–778 (2016)
10. He, S., et al.: Multi-channel attention-fusion neural network for brain age estimation: accuracy, generality, and interpretation with 16,705 healthy MRIS across lifespan. Med. Image Anal. **72**, 102091 (2021)
11. He, X., Zhou, Y., Zhou, Z., Bai, S., Bai, X.: Triplet-center loss for multi-view 3D object retrieval. In: Proceedings of the IEEE Conference on Computer Vision and Pattern Recognition, pp. 1945–1954 (2018)
12. Koch, G., et al.: Siamese neural networks for one-shot image recognition. In: ICML Deep Learning Workshop, vol. 2 (2015)
13. Lagan, J., Schmitt, M., Miller, C.A.: Clinical applications of multi-parametric CMR in myocarditis and systemic inflammatory diseases. Int. J. Cardiovasc. Imaging **34**(1), 35–54 (2017)
14. Mehta, P., Antonelli, M., Ahmed, H.U., Emberton, M., Punwani, S., Ourselin, S.: Computer-aided diagnosis of prostate cancer using multiparametric MRI and clinical features: a patient-level classification framework. Med. Image Anal. **73**, 102153 (2021)
15. Rossi, A., Hosseinzadeh, M., Bianchini, M., Scarselli, F., Huisman, H.: Multi-modal Siamese network for diagnostically similar lesion retrieval in prostate MRI. IEEE Trans. Med. Imaging **40**, 986–995 (2021)

16. Vaswani, A., et al.: Attention is all you need. In: Advances in Neural Information Processing Systems, pp. 5998–6008 (2017)
17. Yang, H.M., Zhang, X.Y., Yin, F., Liu, C.L.: Robust classification with convolutional prototype learning. In: Proceedings of the IEEE Conference on Computer Vision and Pattern Recognition, pp. 3474–3482 (2018)

Evolutionary Multi-objective Architecture Search Framework: Application to COVID-19 3D CT Classification

Xin He[1], Guohao Ying[2], Jiyong Zhang[3(✉)], and Xiaowen Chu[1,4(✉)]

[1] Hong Kong Baptist University, Hong Kong, China
xwchu@ust.hk
[2] University of Southern California, Los Angeles, CA, USA
[3] Hangzhou Dianzi University, Hang Zhou, China
jzhang@hdu.edu.cn
[4] The Hong Kong University of Science and Technology (Guangzhou), Guangzhou, China

Abstract. The COVID-19 pandemic has threatened global health. Many studies have applied deep convolutional neural networks (CNN) to recognize COVID-19 based on chest 3D computed tomography (CT). Recent works show that no model generalizes well across CT datasets from different countries, and manually designing models for specific datasets requires expertise; thus, neural architecture search (NAS) that aims to search models automatically has become an attractive solution. To reduce the search cost on large 3D CT datasets, most NAS-based works use the weight-sharing (WS) strategy to make all models share weights within a supernet; however, WS inevitably incurs search instability, leading to inaccurate model estimation. In this work, we propose an efficient **E**volutionary **M**ulti-objective **AR**chitecture **S**earch (**EMARS**) framework. We propose a new objective, namely **potential**, which can help exploit promising models to indirectly reduce the number of models involved in weights training, thus alleviating search instability. We demonstrate that under objectives of accuracy and potential, EMARS can balance exploitation and exploration, *i.e.,* reducing search time and finding better models. Our searched models are small and perform better than prior works on three public COVID-19 3D CT datasets.

Keywords: COVID-19 · Neural Architecture Search (NAS) · Weight-sharing · Evolutionary Algorithm (EA) · 3D Computed Tomograph (CT)

1 Introduction

The rapid spread of *coronavirus disease 2019* (COVID-19) pandemic has threatened global health. Isolating infected patients is an effective way to block the

Supplementary Information The online version contains supplementary material available at https://doi.org/10.1007/978-3-031-16431-6_53.

L. Wang et al. (Eds.): MICCAI 2022, LNCS 13431, pp. 560–570, 2022.
https://doi.org/10.1007/978-3-031-16431-6_53

transmission of the virus. Thus, fast and accurate methods to detect infected patients are crucial. Chest CT is relatively easy to perform and has been proved an important complement to nucleic acid test [7]. However, there is a serious lack of radiologists during the pandemic. Many researchers have applied deep learning (DL) techniques to assist CT diagnosis. For COVID-19 3D CT classification, there are two mainstream CNN-based methods: 1) multiview-based methods [15,22] use 2D CNN to extract features for each 2D CT slice and then fuse these features to make predictions; and 2) voxel-based methods [8,32] feed 3D CNNs with 3D CT scans to make full use of the geometric information. He *et al.* [9] benchmark a series of hand-crafted 2D and 3D CNNs and demonstrate that 3D CNNs generally outperform 2D CNNs.

Some recent works [8,11] benchmark multiple COVID-19 datasets from different countries and find that no model can maintain absolute advantages on different datasets. However, since it is difficult to design models manually for specific datasets, the neural architecture search (NAS) [6,10] has become an attractive solution to discover superior models without human assistance. Reinforcement learning [21,33], gradient descent (GD) [18], and evolutionary algorithm (EA) [23,30] are three mainstream NAS methods. The comparative results of a recent survey [10] show that the EA-based NAS can discover better networks than other types of NAS methods. However, the better performance of EA-based NAS is at the cost of more computing resources because they need to retrain all searched models to compare their performance, *e.g.,* AmoebaNet [23] took 3,150 GPU days to search. Thanks to the weight-sharing method [21,29], any model can be evaluated without retraining, and Yang et al. [30] reduced the search time of the EA-based NAS to 0.4 GPU days. NAS was originally proposed for large-scale 2D image tasks. Although some works [8,9] have extended NAS to search 3D models for COVID-19 3D datasets, they suffered from the search instability (analyzed in Sect. 3.1) incurred by weight-sharing, which leads to fluctuation in the search process and even worse results than random search in some cases. In this work, we propose an efficient **E**volutionary **M**ulti-objective **AR**chitecture **S**earch framework, dubbed as **EMARS**. We summarize our contributions below.

1. We propose a new objective, *i.e., potential*, which can help exploit promising models and indirectly reduce the number of models involved in weights training, thereby alleviating search instability.
2. We demonstrate that compared to conventional objective settings (*e.g.,* only considering accuracy), EMARS that aims at accuracy, potential, and small size objectives can trade-off between exploitation and exploration, reducing search time by 22% on average and discovering better models.
3. Our searched models are small in size and outperform prior works [8,9] on three public datasets: CC-CCII [31], MosMed [20], and Covid-CTset [22].

2 Preliminaries

In this section, we describe the common basis of weight-sharing neural architecture search (NAS) [29]. NAS is formulated as a bi-level optimization problem:

$$\begin{array}{ll} \min_\alpha L_{\text{val}}\ (w^*, \alpha) \\ \text{s.t.}\ \ w^* = \operatorname{argmin}_w L_{\text{train}}\ (w, \alpha) \end{array} \tag{1}$$

where L_{train} and L_{val} indicate the training and validation loss; w and α indicate the weights and architecture of a candidate model. The early NAS methods [23, 33] search and evaluate the networks by retraining them from scratch, resulting in huge computational cost. To reduce the burden, the weight-sharing strategy [29] was proposed, in which the SuperNet $\mathcal{N}$ contains all possible architectures (subnets) and its weights $\mathcal{W}$ are shared among these subnets. The architecture and weights of each subnet are denoted by $\mathcal{N}(\alpha)$ and $\mathcal{W}(\alpha)$, respectively, where α is the subnet architecture, encoded by one-hot sequences (described in Sect. 3.3). The loss of a subnet is expressed as $L(\alpha) = L(\mathcal{N}(\alpha), \mathcal{W}(\alpha), X, Y)$, where L, X, Y indicate the loss function, input data, and target, respectively, and the gradient of subnet weights is $\nabla_{\mathcal{W}(\alpha)} = \frac{\partial L(\alpha)}{\partial \mathcal{W}}$. Then gradients of SuperNet weights $\mathcal{W}$ can be calculated as the average gradient of all subnets, *i.e.*, $\nabla_{\mathcal{W}} = \frac{1}{N}\sum_{i=1}^{N} \nabla_{\mathcal{W}(\alpha_i)} = \frac{1}{N}\sum_{i=1}^{N} \frac{\partial L(\alpha_i)}{\partial \mathcal{W}}$, where N is the total number of subnets. Obviously, it is not practical to use all subnets to update SuperNet weights at each time. Therefore, we use a mini-batch of subnets for training, detailed as Eq. 2

$$\nabla_{\mathcal{W}} \approx \frac{1}{M}\sum_{i=1}^{M} \nabla_{\mathcal{W}(\alpha_i)} \tag{2}$$

where M is the number of subnets sampled in a mini-batch and $M << N$. In our experiments, we find that $M = 1$ works just fine, *i.e.*, we can update $\mathcal{W}$ using the gradient from any single sampled subnets for each training batch.

3 Methodology

3.1 Potential Objective: Alleviating Search Instability

By *instability*, we mean that the same subnet can produce a completely different performance at different times of the search process. The instability is caused by the weight-sharing strategy because the weights of all subnets are coupled, then an update of any subnet's weights is bound to affect (usually negatively) other subnets. Therefore, the performance of a subnet at a specific time does not necessarily represent its real performance but instead misleads the direction of evolutionary search (described in Sect. 3.2). To mitigate the search instability caused by weight-sharing, a natural idea is to reduce the number of models involved in weights training (*i.e.*, Eq. 2). For this reason, some works [2,13] directly reduce the number of models by progressively shrinking the search space based on the model performance, but this may eliminate promising models in the early stage of the search. To avoid this problem, we take an indirect approach in which we keep exploring various models in the early stage of the search and then spend more effort on training those promising models in the later stage of the search. In this way, we can indirectly reduce the number of models involved in weights training without deliberately reducing the search space. However, how do we determine whether a model is promising or not?

Here, we propose a new objective, namely *potential*, to help find promising models. Specifically, for each sampled model, we maintain and update its historical performances $Z = (E, F)$, where $E = [e_1, ..., e_m]^T$ is a column vector recording the epochs when the model is sampled, $F = [f_1, ..., f_m]^T$ is a column vector recording the corresponding validation accuracy. Note that, Z is dynamically updated with the search process, so the size of Z (*i.e.*, m) varies for models. The potential $\mathcal{P}$ of a model is calculated by ordinary least squares (OLS):

$$\mathcal{P} = (E^T E)^{-1} E^T F \tag{3}$$

To some extent, E can also reflect how promising a model is, *e.g.*, if E is densely distributed, it means this model outperforms other models in multiple rounds of search and hence wins more chances to be sampled. However, considering only E will exacerbate the Matthew effect, and the search may get trapped in a local optimum. Our proposed potential solves this problem by considering the coupling relation between sampling frequency E and validation accuracy F, *i.e.*, the growing trend of accuracy rather than the accuracy at a specific time. The larger the $\mathcal{P}$ value, the more promising the model is.

3.2 Evolutionary Search

The search algorithm (see Supplement Alg. 1) starts with a warm-up stage, followed by the evolutionary search stage. In the warm-up, the SuperNet is trained by uniformly sampling subnets, thus all candidate operations are trained equally. After the warm-up, top-P best-performing subnets form the initial population, *i.e.*, $\mathcal{A}^{(0)}$, and will be evolved for multiple generations. Each generation comprises two sequential processes: 1) *weights training*, where each individual (*i.e.*, subnet) is selected from the population and trained based on Eq 2; and 2) *architecture search*, comprising selection, crossover, and mutation (see Fig. 1).

Selection. After weights training, we record multiple objectives for all individuals in the population. We adopt NSGA-II [4] method to select Pareto-front individuals under the recorded objectives from the population. We compare different combinations of these objectives in Sect. 4.2 and find that searching with potential and accuracy can discover better models with less cost.

Crossover and Mutation. The selection produces K Pareto-front individuals, based on which we further generate $P - K$ new individuals. Each new individual is generated by randomly sampling from the SuperNet or performing crossover and mutation (CM) with certain probabilities. The basic unit of CM is the one-hot sequence, representing the candidate operation (see Fig. 1).

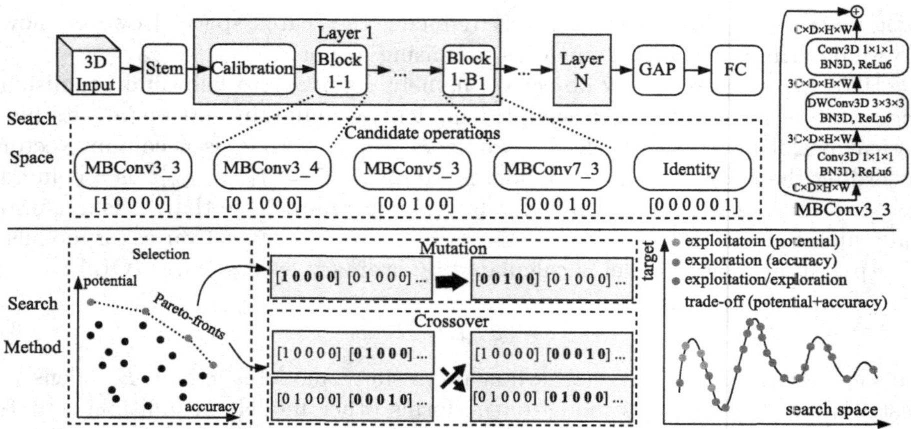

Fig. 1. Overview of search space and search method. Upper-right: MBConv3_3, where C, D, H, W indicate channels, depth, height, and width. Lower-right: An example of exploitation and exploration under different objectives. (best viewed in color) (Color figure online)

Exploitation and Exploration. Figure 1 (lower-right) shows an example of two important issues in the evolutionary algorithm (EA) based search: *exploration* and *exploitation*. Exploitation prefers the current optimal solution, which reduces search cost but may lead to a local optimum; exploration is more likely to find the optimal solution but consumes more resources. The common opinion about EA is that the steps of crossover and mutation determine the exploration, and exploitation is done by selection. However, our experiments in Sect. 4.2 show that setting different objectives in the selection step can also control the evolution direction. Specifically, accuracy and potential will make the evolution process towards exploration and exploitation, respectively, while combining accuracy and potential can balance exploration and exploitation.

3.3 Search Space

SuperNet. The search space is represented by a SuperNet $\mathcal{N}$, containing all possible subnets. SuperNet comprises two parts: 1) the searchable part, *i.e.,* $N = 6$ layers; 2) the fixed part, *i.e.,* stem block, global average pooling [17], and a fully connected layer. The stem block is a standard $3 \times 3 \times 3$ 3D convolution followed by a 3D batch normalization and a ReLu6 activation function [12].

Layer. The i-th layer comprises a calibration block and B_i searchable blocks. The calibration block is a 3D $1 \times 1 \times 1$ point-wise convolution to solve the problem of feature dimension mismatch; thus, all subsequent blocks have a stride of 1.

The number of searchable blocks and the stride of calibration block in six layers are [4,4,4,4,4,1] and [2,2,2,1,2,1], respectively. The output channels of the stem block and six layers are 32 and [24,40,80,96,192,320], respectively.

Block. Each searchable block is a candidate operation, encoded by a one-hot sequence. We adopt eight candidate operations, including a *skip-connection* operation and seven mobile inverted bottleneck convolutions [24], denoted by $MBConvk\_e$, where $k\_e \in \{3\_3, 3\_4, 3\_6, 5\_3, 5\_4, 7\_3, 7\_4\}$, k is the kernel size of the intermediate depth-wise convolution (DWConv), and e is the expansion ratio between the input channel and inner channel of MBConv.

4 Experiments

4.1 Implementation Details

Datasets. For a fair comparison, we apply the same three datasets as prior works [8,9]. CC-CCII [31] has 3,993 CT scans of three classes: novel coronavirus pneumonia (NCP), common pneumonia (CP), and normal case; MosMed [20] has 1,110 scans of NCP and normal classes; Covid-CTset [22] has 526 scans of NCP and normal classes. More details of datasets can be referred to supplement.

Search Stage. We use four Nvidia V100 GPUs to search for 100 epochs, where the warm-up stage has 10 epochs. During each search epoch, a population of models are equally trained on the training set and evaluated on the validation set. The population size is 20, where 10 Pareto-front models are selected from the population using NSGA-II [4] under multiple objectives (*e.g.,* validation accuracy, potential, and model size), and 10 new models are generated by crossover and mutation with the probabilities of 0.3 and 0.2. To improve search efficiency, we set the input size ($width \times height \times depth$) to $64 \times 64 \times 16$. We use Adam optimizer [14] with a weight decay of 3e-4 and an initial learning rate of 0.001.

Retraining Stage. After the search stage, we combine the training and validation set and retrain the Pareto-front models on the combined set for 200 epochs. We use the same Adam settings as the search stage. The 3D input sizes of CC-CCII, MosMed, and Covid-CTset datasets are $128 \times 128 \times 32$, $256 \times 256 \times 40$, and $512 \times 512 \times 32$, respectively. Our framework is based on NNI [19] and available at: https://github.com/marsggbo/MICCAI2022-EMARS.

4.2 Results and Analysis

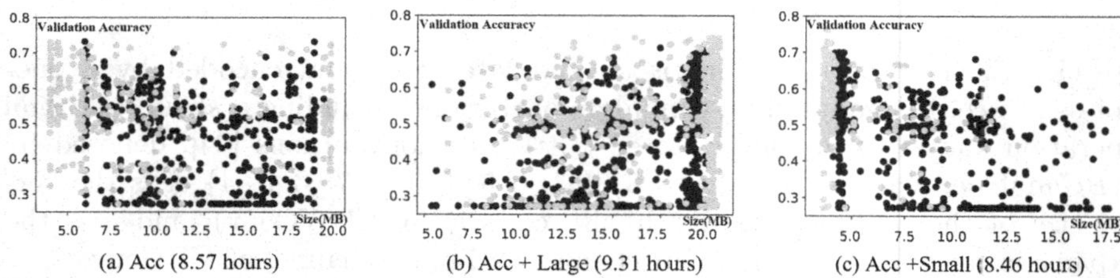

Fig. 2. The model size-aware search results. X and Y axes indicate model size and validation accuracy (Acc). The purple and yellow points indicate the sampled models in the first and last half of the search stage, respectively. (best viewed in color)

Model Size-Aware Search. Figure 2 presents model size-aware search results on CC-CCII dataset. Figure 2 (a) shows that searching under only validation accuracy (Acc) will explore both extremes of model size, but with no performance gain, while Fig. 2 (b)&(c) show that additional consideration of model size on top of Acc helps find better models in the later stage, indicating multi-objective can facilitate the search process. Besides, compared to Fig. 2 (b), searching under Acc and small model size in Fig. 2 (c) can not only reduce search time from 9.31 h to 8.46 h but also discover competitive models.

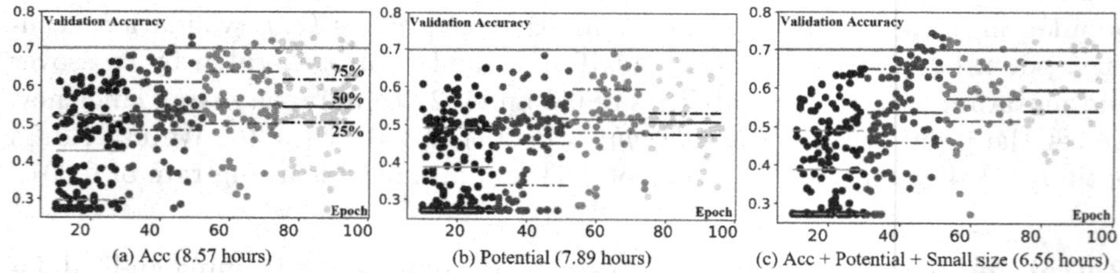

Fig. 3. The potential (P) aware search results. Different colored points indicate the models sampled in different epoch periods. The solid and dashed lines in each period indicate the average and 25/75 percentile accuracy, respectively. (best viewed in color) (Color figure online)

Potential-Aware Search. We further build three experiments on the CC-CCII dataset to validate *potential* objective. Each sub-figure of Fig. 3 divides the search process into four periods based on the search epoch. Each period is presented with different colors and marked with the accuracy of 25/50/75th percentiles. Figure 3 (a) shows that searching under Acc tends to *explore* more models, regardless of whether the model performance is good or bad, leading to wasting time on those unpromising models (lower-right points). On the contrary, in Fig. 3 (b), the difference between the 25th and 75th percentiles and the number of sampled models are gradually reduced with the search process, which implies that potential will guide the evolution process in the later stage to *exploit* promising models already discovered. Although it reduces search time, it has lower Acc due to being trapped in local optima in the early stage. Figure 3 (c) shows that searching under potential, Acc, and small size can reduce the search time by 19% on average and balance exploitation and exploration. Specifically, the first two periods are dominated by exploration, as a wide accuracy range of models is explored, and we can find models with an accuracy of more than 0.7 faster in the second period. On the other hand, the last two periods focus more on exploitation, as the number of unpromising models is significantly reduced, and the accuracy of 25/50/75th percentiles is improved steadily.

Comparison with Prior Works. Table 1 compares our searched models with prior works based on four widely used metrics: accuracy, precision, sensitivity, and f1 score. Precision and sensitivity are a pair of negatively correlated metrics, so they cannot fully describe model performance. F1 score is the harmonic mean of the precision and sensitivity; thus, it is a better metric. As can be seen, our models searched under APS (accuracy, potential, and small model size) objectives have small sizes and outperform all prior hand-crafted and NAS-based models on three datasets in terms of accuracy, precision, and f1 score. Besides, MosMed is an imbalanced dataset, and we can find that the models (*e.g.,* CovidNet3D-S/L and EMARS-A) searched without potential are overfitted on positive class (*i.e.,* NCP), as they have extremely high sensitivity but low precision. On the contrary, EMARS-P and EMARS-APS are searched with potential objective, balancing precision and sensitivity well and achieving higher accuracy and f1 scores. More results can be referred to the Supplement.

Table 1. Results on CC-CCII [31], MosMed [20], and Covid-CTset [22] datasets. A, P, and S in our model name indicate accuracy, potential, and small model size, *e.g.*, EMARS-A indicates the model searched under the accuracy objective.

Dataset	Model	Size (MB)	Type	Accuracy	Precision	Sensitivity	F1
CC- CCII [China] [31]	ResNet3D101 [26]	325.21	Manual	85.54	89.62	77.15	82.92
	DenseNet3D121 [5]	43.06		87.02	88.97	82.78	85.76
	MC3_18 [26]	43.84		86.16	87.11	82.78	84.89
	COVID-AL [28]	-		86.60	-	-	-
	VGG16-Ensemble [16]	-		88.12	84.04	89.19	86.54
	CovidNet3D-S [8]	11.48	Auto	88.55	88.78	**91.72**	90.23
	CovidNet3D-L [8]	53.26		88.69	90.48	88.08	89.26
	MNas3DNet [9]	22.91		87.14	88.44	86.09	87.25
	EMARS-A	5.93		**89.67**	89.26	89.22	89.23
	EMARS-P	5.63		88.78	88.81	88.22	88.51
	EMARS-APS	3.38		89.61	**91.48**	89.97	**90.72**
Mos- Med [Russia] [20]	ResNet3D101 [26]	325.21	Manual	81.82	81.31	97.25	88.57
	DenseNet3D121 [5]	43.06		79.55	84.23	92.16	88.01
	MC3_18 [26]	43.84		80.4	79.43	98.43	87.92
	DeCoVNet [27]	-		82.43	-	-	-
	CovidNet3D-S [8]	12.48	Auto	81.17	78.82	99.22	87.85
	CovidNet3D-L [8]	60.39		82.29	79.50	98.82	88.11
	EMARS-A	2.89		80.98	77.91	**99.61**	87.44
	EMARS-P	18.22		84.34	**93.56**	85.49	89.34
	EMARS-APS	10.69		**88.09**	93.52	90.59	**92.03**
Covid- CTset [Iran] [22]	ResNet3D101 [26]	325.21	Manual	93.87	92.34	95.54	93.92
	DenseNet3D121 [5]	43.06		91.91	92.57	92.57	92.57
	MC3_18 [26]	43.84		92.57	90.95	94.55	92.72
	CovCTx [3]	-		96.37	-	97.00	-
	Vit-32×32 [25]	-		95.36	-	83.00	-
	CovidNet3D-S [8]	8.36	Auto	94.27	92.68	90.48	91.57
	CovidNet3D-L [8]	62.82		96.88	97.50	92.86	95.12
	AutoGluon model [1]	93.00		89.00	90.00	88.00	88.00
	EMARS-A	8.36		95.16	95.77	95.16	95.46
	EMARS-P	14.41		92.87	92.73	92.74	92.74
	EMARS-APS	9.95		**97.66**	**97.61**	**97.58**	**97.59**

5 Conclusion and Future Work

In this work, we introduce an EA-based neural architecture search (EMARS) framework, which can efficiently discover superior 3D models under multiple objectives for COVID-19 3D CT classification. We demonstrate that our proposed objective, *i.e., potential*, can effectively alleviate the search instability and help exploit promising models. The models searched by EMARS under accuracy and potential objectives have small sizes and outperform the previous work on three public datasets. We believe our framework can also be extended to other types of datasets and tasks (*e.g.*, segmentation), which is also our future work.

Acknowledgement. This work was supported in part by Hong Kong Research Matching Grant RMGS2019_1_23, the Zhejiang Province Nature Science Foundation of

China under Grant LZ22F020003, and the HDU-CECDATA Joint Research Center of Big Data Technologies under Grant KYH063120009.

References

1. Anwar, T.: COVID19 diagnosis using AutoML from 3D CT scans. In: Proceedings of the IEEE/CVF International Conference on Computer Vision, pp. 503–507 (2021)
2. Chen, M., Fu, J., Ling, H.: One-shot neural ensemble architecture search by diversity-guided search space shrinking. In: Proceedings of the IEEE/CVF Conference on Computer Vision and Pattern Recognition, pp. 16530–16539 (2021)
3. Chetoui, M., Akhloufi, M.A.: Efficient deep neural network for an automated detection of COVID-19 using CT images. In: 2021 IEEE International Conference on Systems, Man, and Cybernetics (SMC), pp. 1769–1774. IEEE (2021)
4. Deb, K., Pratap, A., Agarwal, S., Meyarivan, T.: A fast and elitist multiobjective genetic algorithm: NSGA-II. IEEE Trans. Evol. Comput. **6**(2), 182–197 (2002)
5. Diba, A., et al.: Temporal 3D convnets: new architecture and transfer learning for video classification. arXiv preprint arXiv:1711.08200 (2017)
6. Elsken, T., Metzen, J.H., Hutter, F.: Neural architecture search: a survey. arXiv preprint arXiv:1808.05377 (2018)
7. Fu, Z., et al.: CT features of COVID-19 patients with two consecutive negative RT-PCR tests after treatment. Sci. Rep. **10**(1), 1–6 (2020)
8. He, X., et al.: Automated model design and benchmarking of deep learning models for COVID-19 detection with chest CT scans. In: Proceedings of the AAAI Conference on Artificial Intelligence, pp. 4821–4829 (2021)
9. He, X., et al.: Benchmarking deep learning models and automated model design for COVID-19 detection with chest CT scans. medRxiv (2020)
10. He, X., Zhao, K., Chu, X.: AutoML: A survey of the state-of-the-art. Knowl. Based Syst. **212**, 106622 (2021)
11. Horry, M.J., Chakraborty, S., Pradhan, B., Fallahpoor, M., Chegeni, H., Paul, M.: Factors determining generalization in deep learning models for scoring COVID-CT images. Math. Biosci. Eng. **18**(6), 9264–9293 (2021)
12. Howard, A.G., et al.: Efficient convolutional neural networks for mobile vision applications. arXiv preprint arXiv:1704.04861 (2017)
13. Hu, Y., et al.: Angle-based search space shrinking for neural architecture search. In: Vedaldi, A., Bischof, H., Brox, T., Frahm, J.-M. (eds.) ECCV 2020. LNCS, vol. 12364, pp. 119–134. Springer, Cham (2020). https://doi.org/10.1007/978-3-030-58529-7_8
14. Kingma, D.P., Ba, J.: Adam: a method for stochastic optimization. In: Bengio, Y., LeCun, Y. (eds.) 3rd International Conference on Learning Representations, ICLR (2015)
15. Li, L., et al.: Artificial intelligence distinguishes COVID-19 from community acquired pneumonia on chest CT. Radiology **296**(2), 65–71 (2020)
16. Li, X., Tan, W., Liu, P., Zhou, Q., Yang, J.: Classification of COVID-19 chest CT images based on ensemble deep learning. J. Healthc. Eng. **2021**, 5528441 (2021)
17. Lin, M., Chen, Q., Yan, S.: Network in network. In: Bengio, Y., LeCun, Y. (eds.) 2nd International Conference on Learning Representations, ICLR (2014)
18. Liu, H., Simonyan, K., Yang, Y.: DARTS: differentiable architecture search. In: 7th International Conference on Learning Representations, ICLR (2019)

19. Microsoftware: Neural Network Intelligence (NNI). https://github.com/microsoft/nni/tree/v1.4 (2019)
20. Morozov, S., et al.: MosMedData: Chest CT scans with COVID-19 related findings. medRxiv (2020)
21. Pham, H., Guan, M.Y., Zoph, B., Le, Q.V., Dean, J.: Efficient neural architecture search via parameter sharing. In: Dy, J.G., Krause, A. (eds.) Proceedings of the 35th International Conference on Machine Learning, ICML 2018, Stockholmsmässan, Stockholm, Sweden, 10–15 July 2018, vol. 80, pp. 4092–4101. Proceedings of Machine Learning Research, PMLR (2018)
22. Rahimzadeh, M., Attar, A., Sakhaei, S.M.: A fully automated deep learning-based network for detecting COVID-19 from a new and large lung CT scan dataset. medRxiv (2020)
23. Real, E., Aggarwal, A., Huang, Y., Le, Q.V.: Regularized evolution for image classifier architecture search. In: The Thirty-Third AAAI Conference on Artificial Intelligence, pp. 4780–4789. AAAI Press (2019)
24. Sandler, M., Howard, A.G., Zhu, M., Zhmoginov, A., Chen, L.: MobileNetV2: inverted residuals and linear bottlenecks. In: 2018 IEEE Conference on Computer Vision and Pattern Recognition (CVPR), pp. 4510–4520. IEEE Computer Society (2018)
25. Than, J.C., et al.: Preliminary study on patch sizes in vision transformers (ViT) for COVID-19 and diseased lungs classification. In: 2021 IEEE National Biomedical Engineering Conference (NBEC), pp. 146–150. IEEE (2021)
26. Tran, D., Wang, H., Torresani, L., Ray, J., LeCun, Y., Paluri, M.: A closer look at spatiotemporal convolutions for action recognition. In: 2018 IEEE Conference on Computer Vision and Pattern Recognition (CVPR), pp. 6450–6459. IEEE Computer Society (2018)
27. Wang, X., et al.: A weakly-supervised framework for COVID-19 classification and lesion localization from chest CT. IEEE Trans. Med. Imaging **39**(8), 2615–2625 (2020)
28. Wu, X., Chen, C., Zhong, M., Wang, J., Shi, J.: COVID-AL: the diagnosis of COVID-19 with deep active learning. Med. Image Anal. **68**, 101913 (2021)
29. Xie, L., et al.: Weight-sharing neural architecture search: a battle to shrink the optimization gap. ACM Comput. Surv. (CSUR) **54**(9), 1–37 (2021)
30. Yang, Z., et al.: CARS: continuous evolution for efficient neural architecture search. In: 2020 IEEE/CVF Conference on Computer Vision and Pattern Recognition (CVPR), pp. 1826–1835. IEEE (2020)
31. Zhang, K., et al.: Clinically applicable AI system for accurate diagnosis, quantitative measurements, and prognosis of COVID-19 pneumonia using computed tomography. Cell **181**(6), 1423–12433 (2020)
32. Zheng, C., et al.: Deep learning-based detection for COVID-19 from chest CT using weak label. MedRxiv (2020)
33. Zoph, B., Le, Q.V.: Neural architecture search with reinforcement learning. In: 5th International Conference on Learning Representations, ICLR 2017 (2017)

Detecting Aortic Valve Pathology from the 3-Chamber Cine Cardiac MRI View

Kavitha Vimalesvaran[1,3,4](✉), Fatmatülzehra Uslu[2], Sameer Zaman[1,4], Christoforos Galazis[1,4], James Howard[3,4], Graham Cole[3], and Anil A. Bharath[4]

[1] Artificial Intelligence for Healthcare Centre for Doctoral Training, Imperial College London, South Kensington Campus, London SW7 2BX, UK
k.vimalesvaran@imperial.ac.uk

[2] Electrical and Electronics Engineering Department, Bursa Technical University, 16310 Bursa, Türkiye
fatmatulzehra.uslu@btu.edu.tr

[3] Imperial College Healthcare NHS Trust, Du Cane Road, London W12 0HS, UK

[4] Imperial College London, Exhibition Road, London SW7 2AZ, UK

Abstract. Cardiac magnetic resonance (CMR) is the gold standard for quantification of cardiac volumes, function, and blood flow. Tailored MR pulse sequences define the contrast mechanisms, acquisition geometry and timing which can be applied during CMR to achieve unique tissue characterisation. It is impractical for each patient to have every possible acquisition option. We target the aortic valve in the three-chamber (3-CH) cine CMR view. Two major types of anomalies are possible in the aortic valve. Stenosis: the narrowing of the valve which prevents an adequate outflow of blood, and insufficiency (regurgitation): the inability to stop the back-flow of blood into the left ventricle. We develop and evaluate a deep learning system to accurately classify aortic valve abnormalities to enable further directed imaging for patients who require it. Inspired by low level image processing tasks, we propose a multi-level network that generates heat maps to locate the aortic valve leaflets' hinge points and aortic stenosis or regurgitation jets. We trained and evaluated all our models on a dataset of clinical CMR studies obtained from three NHS hospitals (n = 1,017 patients). Our results (mean accuracy = 0.93 and $F1$ score = 0.91), show that an expert-guided deep learning-based feature extraction and a classification model provide a feasible strategy for prescribing further, directed imaging, thus improving the efficiency and utility of CMR scanning.

Keywords: Cardiac MRI · Aortic valve · Explainability · Machine learning

K. Vimalesvaran and F. Uslu—contributed equally.

Supplementary Information The online version contains supplementary material available at https://doi.org/10.1007/978-3-031-16431-6_54.

L. Wang et al. (Eds.): MICCAI 2022, LNCS 13431, pp. 571–580, 2022.
https://doi.org/10.1007/978-3-031-16431-6_54

1 Introduction

Cardiac magnetic resonance (CMR) is of great utility in the assessment of valve disease, as the armamentarium of imaging techniques provides excellent evaluation [1] of anatomy, function, flow quantification, and repercussion on the left and right ventricles. In this study, we focus on aortic valve disease, the most common valvular heart disease in developed countries [2], and with increasing prevalence due to an aging population. Aortic valve disease (Fig. 1) includes: 1) aortic valve stenosis where the leaflets become stiff, causing restricted movement and reduced opening area for blood to flow through; and 2) aortic valve regurgitation, where the leaflets do not close properly, causing blood to flow backwards into the left ventricle [3].

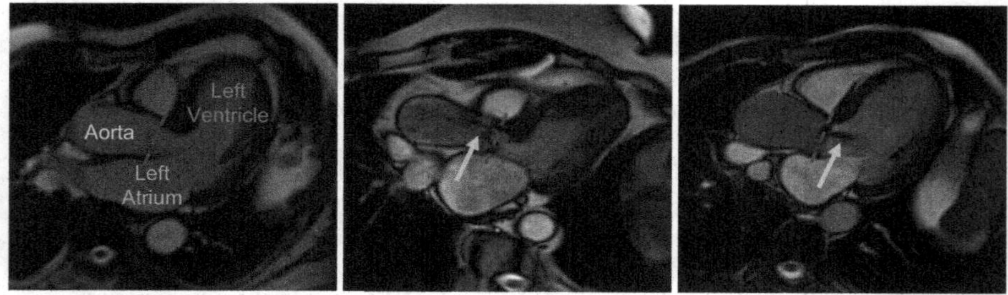

Fig. 1. Three long axis three chamber images. The red dotted line denotes the aortic valve leaflets. Left - normal aortic valve; Middle - aortic stenosis showing high velocity stenotic jet (arrow); and Right - aortic regurgitation with dark regurgitant jet due to signal drop out from turbulence (arrow). (Color figure online)

It is impractical for every patient to have every image type acquired in every image plane [4]. Extensive clinical training and experience is required to adapt the CMR protocol (pre-set list of imaging techniques) during the scan, either because of unexpected or equivocal findings. Within the first few minutes of a clinical CMR scan, all patients undergo cine imaging in the standard views [5]: four, two, three chamber and aortic valve cross section.

Image based diagnostic support systems have the potential to improve the patient journey and clinical workflow [6,7] by supporting clinical decision making [8] and enabling adaptive scanning protocols. Previous studies [9–11] have proposed the use of deep learning to classify anatomical aortic valve abnormalities from phase-contrast velocity mapping images. Phase-contrast velocity mapping is a CMR imaging technique which can be performed to accurately assess blood flow across the aortic valve [12,13]. However, this technique can be time consuming and unnecessary in patients with a normal aortic valve. To the best of our knowledge, no work has yet targeted the long-axis three chamber (3CH) cine CMR view to provide an accurate, initial classification of physiological aortic valve abnormalities based on the movement of blood.

In this study, we propose a machine learning approach to detect abnormal aortic valve activity from the 3CH cine CMR. This is enabled by the use of deep networks to localize the region containing the aortic valve. First - for localization - we utilise a small but unique, expert-annotated dataset consisting of: 1) four important aortic valve landmarks (two hinge points and two leaflets - right coronary cusp and non-coronary cusp); 2) stenotic jet; and 3) regurgitant jet. Using heat map regression, we can then detect the presence of pathological jets in this region. Secondly, we use a random forest approach [14], to classify abnormal cases by using the characteristics of estimated pathological curves obtained from the heat map regression output; this uses human- interpretable and explainable features for classification.

2 Method

Our method consists of six main steps, as shown in Fig. 2: (i) automatic localisation of hinge points in the 3CH cine CMR images, (ii) heat map estimation of aortic valve leaflets and pathological curves due to stenotic or regurgitant jets, (iii) curve tracking in estimated heat maps, (iv) quantification of detected curves in each frame, (v) feature summarising across frames, and (vi) patient classification. The first two steps are defined as regression tasks and we train copies of the same network to estimate the heat maps in these tasks.

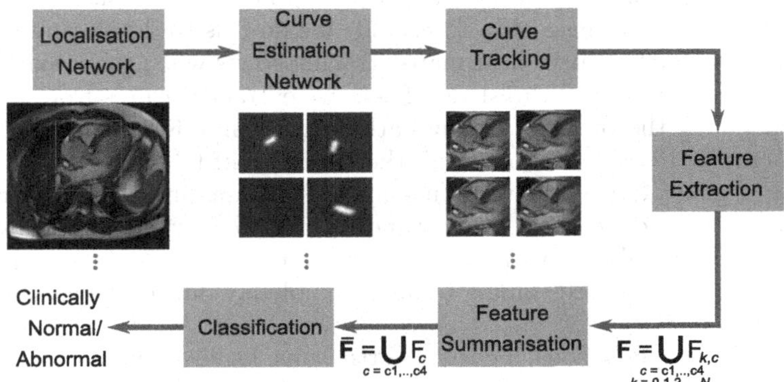

Fig. 2. Overview of our method for abnormal aortic valve classification.

2.1 The Proposed Network for Heat Map Regression

The network (Supplementary material: Fig. S1) contains three identical sub-networks refining the output of the preceding one. Each sub-network resembles U-Net [15], with its encoder, decoder and skip connections. To increase the sensitivity of sub-networks to the hinge points and valve leaflets/pathological jets, we replace the first two and the last two convolutional blocks with dense blocks

[16] which has been shown to be better at low level image processing tasks. Each dense block contains five 3×3 convolutional layers followed by a 1×1 convolutional layer. For hinge point localisation, the network generates three heat maps: one for each hinge point and one for both hinge points to constrain their position in a single frame. For curve heat map estimation, the network produces four heat maps; two for the aortic valve leaflets and two for the pathological jets. We optimise the parameter of both networks with the loss function $L = l(\tilde{H}_1, H_1) + l(\tilde{H}_2, H_2) + l(\tilde{H}_3, H_3)$, where l is the mean square error loss. H and $\tilde{H}$ respectively represent ground truth heat maps and their estimates.

Localising Hinge Points from Heat Maps: We detect the locations of the hinge points in the estimated heat maps by taking the maximum of the corresponding heat map of each hinge point for each frame of a 3CH cine CMR view. Predictions in some frames may not be accurate due to limited representation of frames in our training dataset. Therefore, we use the median of hinge points across frames to crop a CMR image at the middle of the median hinge points locations. Cropped frames are later used for heat map regression of aortic leaflets and curve-like structures that are indicative of pathology.

2.2 Pathology Classification

We use a simple tracking method (Supplementary material: Algorithm S1), which traces ridge points in heat maps starting from the location with the maximum value until reaching a stop threshold. We do this to detect any potential curves in the predicted curve heat maps. Tracing starts when the maximum of a heat map is over an initial threshold. Then, we extract features from predicted curves. Although the presence of any pathological curve is sufficient to classify the 3CH cine CMR view as abnormal, the existence of false curves - often generated by artefact during acquisition - makes this classification less accurate. Movement of the aortic valve from frame to frame can yield important information and reveal subtle abnormalities. For example, a clinician would suspect aortic stenosis in restricted (failing to open completely during the cardiac cycle) aortic valve leaflets.

Therefore, we quantify each curve - aortic valve leaflets and pathological jets - based on: (i) their proximity to hinge points: distance from the mid point of the curve to the image center, which is at the mid point of the predicted valve hinges; (ii) their orientation with respect to the line connecting the hinge points; (iii) their length; and (iv) probability of being a curve, to be able to discriminate true curves from false ones. We do this (feature (iv)) by treating generated heat maps as probability maps of being curves and take the average of probabilities sampled for traced curve locations. Given there are four curve types (two leaflets and two jets) for each frame, we analyse 16 features (see Eq. (1)) in total. These features (see Fig. 3) mimic the interpretation of the 3CH cine CMR by a clinician for aortic valve abnormalities.

$$F_f = F_{c1} \cup F_{c2} \cup F_{c3} \cup F_{c4}, \quad F_c = [L, A, D, P] \tag{1}$$

where F_f shows features for the frame f, which consists of features of four curves $c_1 \cdots c_4$ including the right coronary cusp leaflet, pathological curve of a stenotic jet, the non coronary cusp leaflet, and pathological curve of a regurgitant jet. L, A, D, P respectively represent curve length, the angle between hinge points and the curve, the distance between the middle of the curve to image patch center, and the mean value of curve probability taken along the curve.

For a patient with k frames of CMR images, there is a feature set with a size of $R^{k\times 16}$, where k depends on patient data. In abnormal cases, multiple frames in a cine will demonstrate aortic abnormalities. With increasing severity, more frames will contain pathological curves. Therefore, we summarise k frames to reduce the likelihood of missing abnormal cases, with $\bar{F} = K(F_1, F_2, \cdots F_k)$, where $\bar{F}$ represents the summarised feature set for a patient, by using a summarising technique K, explained as follows, over k frames.

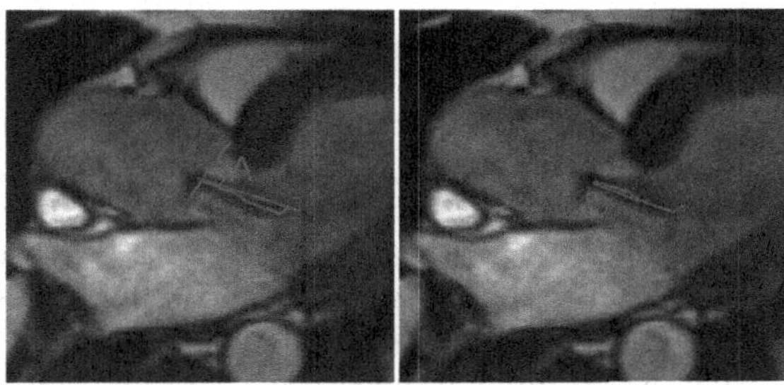

Fig. 3. Features: A: angle between hinge points and the curve; D: distance between middle of the curve to image patch center. The green curve shows the predicted AV-regurgitation curve. (Color figure online)

We present four feature summarising approaches (Supplementary Material: Figure S2): (i) calculating the median of features across frames (Median), (ii) calculating the mean of features across frames (Mean), (iii) using the features of the single frame with the maximum of all curve probabilities (SingleFrameMaxCurveP), regardless of curve types, (iv) using the features of curves showing the largest curve probabilities for each type (MaxCurveP). Apart from the third approach, all techniques use more than one frame to detect patients with aortic valve abnormalities. The first two techniques assume that features are independent, while the last two preserve the membership of features to curves.

2.3 Imaging Data and Manual Annotation

We trained and evaluated all our models on a dataset of clinical CMR studies obtained from different scanner types across three NHS hospitals (Imperial College Hospitals NHS Trust). Ethical approval was gained from the Health

Regulatory Agency (Integrated Research Application System ID 243023). See supplementary material, Table S1 for acquisition parameters. We used two separate datasets for heat map regression and pathology classification. For the former task, we utilised 1221 unique frames from eighty patients where hinge points, the aortic valve leaflets and pathological jets were manually annotated by three cardiologists. Their binary annotations were smoothed with a Gaussian-like kernel with $\sigma = 5$ pixels to generate heat maps.

900 frames were used for training, 100 for validation and 221 for performance evaluation. We also used 1000 healthy frames to increase the size of the training set for curve estimation. Data splitting was based on patient-wise selection. For the classification task, we obtained a cohort of 1017 patients with binary labels describing whether they had a normal or abnormal aortic valve. In total 496/1017 patients had an abnormal aortic valve. Of those abnormal cases, 184 patients had aortic stenosis, 222 aortic regurgitation and 90 cases had mixed valve disease. The average frame count for a patient was 31 ± 15. The resolution of images varies between 1.17×1.17 pixels to 1.56×1.56 pixels.

3 Experimental Setting and Performance Criteria

We optimised our network using the Adam optimisation algorithm with a learning rate of 0.001, for the heat map regression of hinge points, and the aortic valve leaflet and pathological curves. Training lasted for 12 epochs for the former and 80 epochs for the latter, with a batch size of 2. We use data augmentation techniques such as rotation, translation and scaling to deal with overfitting. Our localisation network takes an input image of 256×320 pixels. For curve estimation, we cropped input frames to have a size of 96×96 pixels. For pathology classification, we trained a random forest model [14] with a 10 fold cross-validation because of the limited size of our dataset. Each random forest contained 100 trees. We start curve tracking in heat-maps if their maximum is over a start threshold, where we evaluate the classification performance for a threshold of 0.1 or 0.5. Curve tracking continues while probability of the last traced point is over a stop threshold of 0.1. We assess the classification performance of our method with accuracy and $F1$ score and heat map regression performance for curves with mean absolute error (MAE), which are calculated with a Python library[1]. For hinge point detection, we calculated the distance between mid points of reference and estimated hinge points because hinge points may swap between output channels of our network.

4 Results

Localisation of the aortic valve is important for accurate classification of aortic valve abnormalities. Our network presents good localisation performance for hinge points with a MAE of 3.5 ± 4.2 mm. The performance of our network for heat map regression of aortic leaflets and pathological curves also shows small MAE scores (see Table 1).

[1] https://scikit-learn.org/stable/.

Table 1. Performance evaluation for heat map regression of aortic leaflets and pathological curves.

Estimated locations	Mean ± Std (pixels)
Right coronary cusp (RCC, Curve 1)	0.0095 ± 0.0045
Stenotic Jet (SJ, Curve 2)	0.0102 ± 0.0049
Non-coronary cusp (NCC, Curve 3)	0.0132 ± 0.0124
Regurgitant Jet (RJ, Curve 4)	0.0097 ± 0.0082

Our results for classification of abnormal aortic valves show good agreement with the ground truth labels obtained from corresponding CMR reports with a mean accuracy of up to 0.93 and mean $F1$ score of 0.91 (see Table 2) despite the small size of expert-labelled data. We observe that the start threshold for tracking has a slight effect on classification performance. We find that there is only a small performance difference depending on which feature summarising technique is used. The best performing option is selecting curves based on largest probability for each type when threshold is 0.1 while the worst one is using single frame features, selected based on largest probability of all curves.

Table 2. Performance comparison for abnormal aortic valve classification ST: Start threshold for tracking

ST	Feature summ. tech.	Accuracy	$F1$ score
0.1	Median	0.91 ± 0.04	0.88 ± 0.06
	Mean	0.92 ± 0.03	0.90 ± 0.04
	SingleFrameMaxCurveP	0.90 ± 0.03	0.88 ± 0.04
	MaxCurveP	**0.93** ± 0.03	**0.91** ± 0.04
0.5	Median	**0.92** ± 0.03	0.89 ± 0.04
	Mean	**0.92** ± 0.03	**0.90** ± 0.05
	SingleFrameMaxCurveP	0.90 ± 0.04	0.87 ± 0.05
	MaxCurveP	**0.92** ± 0.03	0.89 ± 0.05

Figure 4 shows the average feature importance for our random forest classification model with 10 fold cross-validation for proposed feature summarising techniques. All methods dominantly use the features of pathological curves as a result of stenosis (curve 2) and regurgitation (curve 4) in classification. The most important features of the curves are probabilities of being curves, curve lengths and curve distance to patch centre. Crucially, our method utilises features which are interpretable by clinicians to understand how the model predicts abnormal aortic valves.

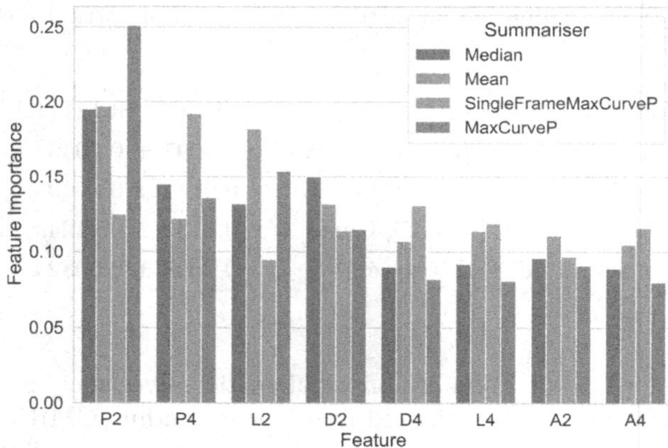

Fig. 4. Average feature importance bar plots for our feature summarising methods. See text for the meaning of L, A, D and P. Numbers 2 and 4 denote the curves for stenotic and regurgitant jets respectively.

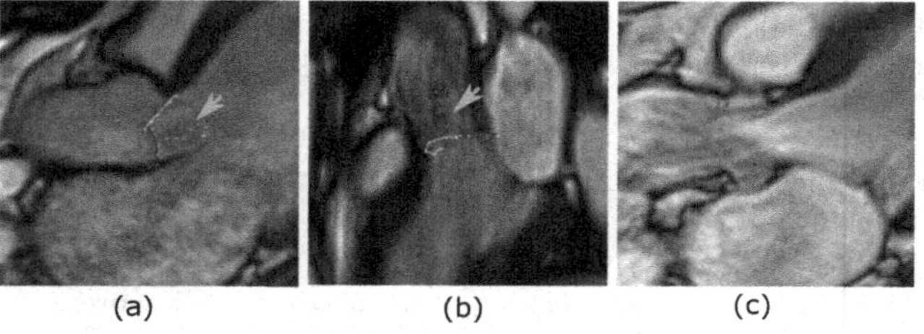

Fig. 5. Failure cases. (a) Red arrow indicates a false regurgitant jet. (b) Red arrow indicates a false stenotic jet. (c) green arrow indicates a missed stenotic jet. (Color figure online)

Table 3. Ablation Experiments. MAEs of heat-map regression of hinge points/leaflets and pathological curves. HP: Hinge point, RCC: Right coronary cusp, LCC: Left coronary cusp; SJ: stenotic jet and RJ: Regurgitant jet.

Methods	HP1	HP2	RCC	NCC	SJ	RJ
Our network	**0.0012**	0.0018	**0.0014**	**0.0017**	**0.0019**	0.0018
Network-S	0.0015	**0.0014**	0.0018	0.0019	0.0022	0.0018
Network-D	0.0012	**0.0014**	0.0016	0.0018	0.0021	**0.0016**
U-Net [15]	0.0013	0.0040	0.0037	0.0026	0.0084	0.0027

Table 3 tabulates the results of our ablation experiment, where network-S and network-D use single and double networks for heat map regression. As seen in this table, our network better approximates heat maps with smaller MAEs, compared to U-Net [15] and the other versions of our network. Note that smaller MAEs compared to Table 1 is due to the calculation of the errors in the entire image, instead of cropped images using the localisation network. Figure 5 exemplifies two failure cases.

5 Conclusion

We present a novel, multi-level machine learning framework for successful aortic valve detection from the 3CH cine CMR view. We focus on identifying abnormal aortic stenotic or regurgitant jets, characterised by a signal change, varying frame by frame in each cine, using heat map regression. We then use a binary classification model for patients with normal/abnormal aortic valves. Clinicians are often wary of machine learning models especially on medical imaging datasets due to the "black-box" nature of many models. In our study, alongside demonstrating a high predictive performance of our methods, we present a selection of features in the region of interest that corresponds to the final interpretation resulting in a normal or abnormal aortic valve classification. These complement human interpretation of identifying abnormal valve pathology.

Our results represent the first presentation of an interpretable machine learning model to accurately classify abnormal aortic valves from a single cine CMR view. Moving forward, the accuracy of our model could be improved by increasing the number of patients in our dataset. There are three major potential applications of an automatic aortic valve classification system from the long-axis 3CH cine CMR view; these are to: (i) provide indications of aortic valve abnormality; (ii) provide guidance to radiographers in suggesting the need for further directed imaging of the aortic valve; and (iii) achieve a reduction in scanning time, and improve workflow efficiency. Future work will include validation on multi-scanner, multi-institute and multi-class pathology classification.

Acknowledgements. This work was supported by the UKRI CDT in AI for Healthcare http://ai4health.io (Grant No. EP/S023283/1)

References

1. Myerson, S.G.: CMR in evaluating valvular heart disease: diagnosis, severity, and outcomes. Cardiovasc. Imaging **14**(10), 2020–2032 (2021)
2. Baumgartner, H.: What influences the outcome of valve replacement in critical aortic stenosis? Heart **91**(10), 1254 (2005)
3. Thubrikar, M.: The Aortic Valve. Routledge, Abingdon (2018)
4. Howard, J.P., et al.: Automated analysis and detection of abnormalities in transaxial anatomical cardiovascular magnetic resonance images: a proof of concept study with potential to optimize image acquisition. Int. J. Cardiovasc. Imaging **37**(3), 1033–1042 (2020). https://doi.org/10.1007/s10554-020-02050-w
5. Kramer, C.M., Barkhausen, J., Bucciarelli-Ducci, C., Flamm, S.D., Kim, R.J., Nagel, E.: Standardized cardiovascular magnetic resonance imaging (cmr) protocols: 2020 update. J. Cardiovasc. Magn. Reson. **22**(1), 1–18 (2020)

6. Lin, A., Kolossváry, M., Išgum, I., Maurovich-Horvat, P., Slomka, P.J., Dey, D.: Artificial intelligence: improving the efficiency of cardiovascular imaging. Expert Rev. Med. Dev. **17**(6), 565–577 (2020)
7. Gonzales, R.A., Lamy, J., Seemann, F., Heiberg, E., Onofrey, J.A., Peters, D.C.: TVnet: automated time-resolved tracking of the tricuspid valve plane in MRI long-axis cine images with a dual-stage deep learning pipeline. In: de Bruijne, M., et al. (eds.) MICCAI 2021. LNCS, vol. 12906, pp. 567–576. Springer, Cham (2021). https://doi.org/10.1007/978-3-030-87231-1_55
8. Petch, J., Di, S., Nelson, W.: Opening the black box: the promise and limitations of explainable machine learning in cardiology. Can. J. Cardiol. **38**, 204–213 (2021)
9. Fries, J.A., et al.: Weakly supervised classification of aortic valve malformations using unlabeled cardiac MRI sequences. Nat. Commun. **10**(1), 1–10 (2019)
10. Guala, A., et al.: Machine learning to automatically detect anatomical landmarks on phase-contrast enhanced magnetic resonance angiography. Eur. Heart J.-Cardiovasc. Imaging **22**(Supplement_2), jeab090-122 (2021)
11. Mejia Cordova, M., et al.: Reinforcement machine learning-based aortic anatomical landmarks detection from phase-contrast enhanced magnetic resonance angiography. Eur. Heart J.-Cardiovasc. Imaging **22**(Supplement_1), jeaa356-286 (2021)
12. Ebbers, T.: Flow imaging: cardiac applications of 3D cine phase-contrast MRI. Curr. Cardiovasc. Imaging Rep. **4**(2), 127–133 (2011)
13. Johnson, E.M., et al.: Detecting aortic valve-induced abnormal flow with seismocardiography and cardiac MRI. Ann. Biomed. Eng. **48**(6), 1779–1792 (2020)
14. Ho,T.K.: Random decision forests. In: Proceedings of 3rd International Conference on Document Analysis and Recognition, vol. 1, pp. 278–282. IEEE (1995)
15. Ronneberger, O., Fischer, P., Brox, T.: U-Net: convolutional networks for biomedical image segmentation. In: Navab, N., Hornegger, J., Wells, W.M., Frangi, A.F. (eds.) MICCAI 2015. LNCS, vol. 9351, pp. 234–241. Springer, Cham (2015). https://doi.org/10.1007/978-3-319-24574-4_28
16. Huang, G., Liu, Z., Van Der Maaten, L., Weinberger, K.Q.: Densely connected convolutional networks. In: Proceedings of the IEEE Conference on Computer Vision and Pattern Recognition, pp. 4700–4708 (2017)

CheXRelNet: An Anatomy-Aware Model for Tracking Longitudinal Relationships Between Chest X-Rays

Gaurang Karwande[1], Amarachi B. Mbakwe[1], Joy T. Wu[2,3], Leo A. Celi[4,5,6], Mehdi Moradi[7], and Ismini Lourentzou[1(✉)]

[1] Virginia Tech, Blacksburg, USA
{gaurangajitk,bmamarachi,ilourentzou}@vt.edu
[2] Standford Medicine, Stanford, USA
joytywu@stanford.edu
[3] IBM Research, Armonk, USA
[4] Institute for Medical Engineering and Science, Massachusetts Institute of Technology, Cambridge, USA
lceli@mit.edu
[5] Beth Israel Deaconess Medical Center, Boston, USA
[6] Harvard T.H. Chan School of Public Health, Boston, USA
[7] McMaster University, Hamilton, Ontario, Canada
moradm4@mcmaster.ca

Abstract. Despite the progress in utilizing deep learning to automate chest radiograph interpretation and disease diagnosis tasks, change between sequential Chest X-rays (CXRs) has received limited attention. Monitoring the progression of pathologies that are visualized through chest imaging poses several challenges in anatomical motion estimation and image registration, *i.e.*, spatially aligning the two images and modeling temporal dynamics in change detection. In this work, we propose `CheXRelNet`, a neural model that can track longitudinal pathology change relations between two CXRs. `CheXRelNet` incorporates local and global visual features, utilizes inter-image and intra-image anatomical information, and learns dependencies between anatomical region attributes, to accurately predict disease change for a pair of CXRs. Experimental results on the Chest ImaGenome dataset show increased downstream performance compared to baselines. Code is available at https://github.com/PLAN-Lab/ChexRelNet.

Keywords: Graph attention networks · CXR graph representations · Chest X-Ray comparison relations · Longitudinal CXR relationships

G. Karwande and A. B. Mbakwe—Equal Contribution, with authors listed in alphabetical order.

Supplementary Information The online version contains supplementary material available at https://doi.org/10.1007/978-3-031-16431-6_55.

L. Wang et al. (Eds.): MICCAI 2022, LNCS 13431, pp. 581–591, 2022.
https://doi.org/10.1007/978-3-031-16431-6_55

1 Introduction

Medical imaging research has experienced tremendous growth over the past years, spurred by continuous AI advancements [10,15,23], and in particular in the development of specialized digital devices and neural medical imaging architectures. Chest radiography is one of the most performed diagnostic examinations worldwide. The demand for chest radiography has increased the radiologists' workload. As manually interpreting Chest X-rays (CXRs) and radiology reports can be time-consuming, these challenges contribute to the delays in detecting findings and providing exemplary patient clinical management plans. Though there has been substantial progress in radiology such as disease diagnostics [4,11,18,20], medical image segmentation [8,14,19,21], *etc.*, more complex reasoning tasks remain fairly unexplored. For example, despite significant progress in the application of machine learning in chest radiograph medical diagnosis, detecting longitudinal change between CXRs has attracted limited attention. Yet, understanding whether the patient's condition has deteriorated or improved is crucial to guide the physician's decision-making and determine the patient's clinical management.

Automating this process is a challenging task. At times, differences between X-rays might go undetected, hindering early detection of disease progression that requires an immediate change in treatment plans. Previous work tackles change between longitudinal patient visits and evaluates the severity of diseases at each time point on a continuous scope on osteoarthritis in knee radiographs and retinopathy of prematurity in retinal photographs [13]. Other works target longitudinal disease tracking and outcome prediction severity for COVID-19 pulmonary diseases [12], by calculating a severity score for pulmonary X-rays via computing the Euclidean distance between each of the normal images and the image of interest. In addition, geometric correlation maps have been used to study the CXR longitudinal change detection problem [16], in which feature maps are extracted from CXR pairs and their matching scores are used to generate a geometric correlation map that can detect map-specific patterns showing lesion change. However, these works rely on global image information. To the best of our knowledge, no prior work considers capturing correlations among anatomical regions and findings when modeling change between CXRs. Yet, localizing pathologies to anatomy is critical for the radiologists' reasoning and reporting process, where correlations between image findings and anatomical regions can help narrow down potential diagnoses.

The development of imaging models that track progress or retrogression between CXR findings or diseases remains still an open issue. Therefore, in this work, we propose `CheXRelNet`, an anatomy-aware neural model that utilizes information from anatomical regions, learns their intra-image and inter-image dependencies with a graph attention network, and combines the localized region features with global image-level features to accurately capture anatomical location semantics for each finding when performing longitudinal relation comparison between CXR exams for a variety of anatomical findings.

The contributions of this work are summarized as follows: 1) we introduce `CheXRelNet`, an anatomy-aware model for tracking longitudinal relations

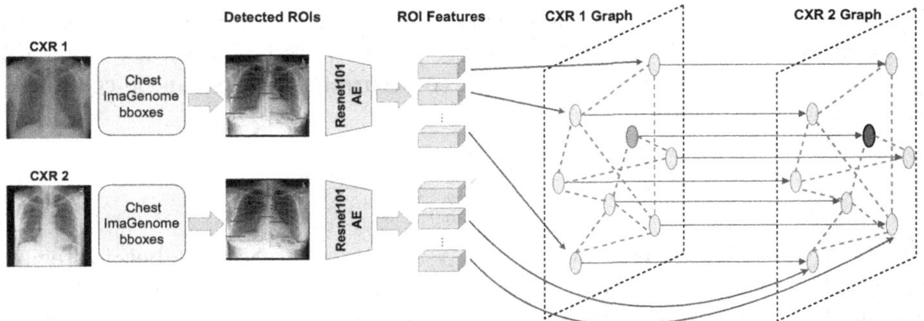

Fig. 1. Graph construction overview. Detected anatomical regions of interest (ROIs) are fed into a ResNet101 pretrained autoencoder to extract their corresponding visual features, formulating initial node representations $\mathcal{V}_i, \mathcal{V}'_i$ for CXR graphs $\mathcal{G}_i = (\mathcal{V}_i, \mathcal{E})$ and $\mathcal{G}'_i = (\mathcal{V}'_i, \mathcal{E})$. Here, $\mathcal{E}$ is constructed based on intra-image region-disease co-occurrence. Moreover, the nodes of the two graphs are connected via a set $\tilde{\mathcal{E}}$ of directed edges indicating inter-image relations.

between CXRs. The proposed model utilizes both local and global anatomical information to output accurate localized comparisons between two sequential CXR examinations, 2) we propose a graph construction to capture correlations between anatomical regions from a pair of CXRs, and 3) we conduct experimental analysis to demonstrate that our proposed `CheXRelNet` model outperforms baselines. Finally, 4) we perform transfer learning experiments to test the generalization capabilities of our model across pathologies.

2 Methodology

Let $\mathcal{C} = \{(x_i, x'_i)\}_{i=i}^N$ be the set of CXR image pairs. Each image x_i has k anatomical regions. In addition, each image is associated with a set of labels $\mathcal{Y}_i = \{y_{i,m}\}_{m=1}^M$, $y_{i,m} \in \{0,1\}$ indicating whether the label for pathology m appears in image x_i or not, and each pair (x_i, x'_i) is associated with a set of labels $\mathcal{Z} = \{z_{i,m}\}_{m=1}^M, z_{i,m} \in \{0,1\}$ indicating whether the pathology m appearing in the image pair has improved or worsened. The goal is to design a model that compares the two images and predicts their labels as accurately as possible for an unseen image pair (x, x') and a wide range of pathologies. This is achieved by utilizing (i) the correlation among anatomical region features from the images x_i, x'_i, *i.e.*, $R = f(x)$ and $R' = f(x')$, $R, R' \in \mathbb{R}^{k \times d}$, where k is the number of anatomical regions, each embedded into a row vector with dimensionality d (extracted by a pretrained feature extractor f) and (ii) the correlation among anatomical regions between the two images in the pair.

Given the initial training set of anatomical region representations $\{(R_i, R'_i)\}_{i=1}^N$, we define a normalized adjacency matrix $A \in \mathbb{R}^{2k \times 2k}$ that captures intra-image and inter-image region correlations. The intra-image correlations corresponding to the two $k \times k$ diagonal blocks of A are constructed based on the region-disease co-occurrence [1], *i.e.*, the number of times two anatomical regions co-occur with the same disease or finding in the set of images $R_i, R'_i, i = 1, \ldots, N$. Each of these $k \times k$ co-occurrence blocks can be computed via the Jaccard similarity

$$J(r_s, r_t) = \frac{1}{M} \sum_{m=1}^{M} \frac{|\mathcal{Y}_{s,m} \cap \mathcal{Y}_{t,m}|}{|\mathcal{Y}_{s,m} \cup \mathcal{Y}_{t,m}|}. \tag{1}$$

Here, r_s represents an anatomical region, $\mathcal{Y}_s^m$ is the set of disease labels for region r_s and pathology m across all images and $\cap, \cup$ denote the intersection and union over multi-sets. To overcome the shortcomings of the label co-occurrence construction tendency to overfit the training data, a filtering threshold τ is adopted, *i.e.*,

$$A_{st} = \begin{cases} 1 & \text{if } J(R_s, R_t) \geq \tau \\ 0 & \text{if } J(R_s, R_t) < \tau \end{cases}. \tag{2}$$

Here, A_{st} corresponds to an element of one of the two diagonal blocks. We further note that the two diagonal blocks are identical.

The inter-image correlations correspond to the two off-diagonal $k \times k$ blocks of the adjacency matrix A, and they are chosen to indicate a relationship between the same anatomical regions of every pair of images. More precisely, we set $A_{st} = \mathbb{1}\{t = s + k\}$ for $s = 1, \ldots, k$. The rationale of this adjacency matrix definition is that A will be associated with every pair (x_i, x'_i) and will capture useful inter-image correlations and local intra-image region-level correlations. More precisely, the upper $k \times k$ diagonal block is associated with image x_i, forming a graph $G_i = (\mathcal{V}_i, \mathcal{E})$ with nodes being the vector representations of the k anatomical regions of image x_i. Similarly, the lower $k \times k$ diagonal block is associated with image x'_i, forming a graph $G'_i = (\mathcal{V}'_i, \mathcal{E})$ as before. Finally, the upper $k \times k$ off-diagonal block indicates a set of edges $\hat{\mathcal{E}}$ between the same regions of images x_i, x'_i. This graph construction is also depicted in Fig. 1.

To capture global and local dependencies between anatomical regions, we utilize a graph attention network (GAT) [24] $Z_i = g(R_i, A) \in \mathbb{R}^{k \times d}$ to update R_i as follows:

$$R_i^{(t+1)} = \alpha_{i,i}^{(t)} W_1^{(t)} R_i^{(t)} + \sum_{j \in \mathcal{N}(i)} \alpha_{i,j}^{(t)} W_1^{(t)} R_j^{(t)}, \tag{3}$$

where $W_1 \in \mathbb{R}^{d \times d}$ is a learned weight matrix, $\mathcal{N}(i)$ denotes the neighborhood of x_i, t is the number of stacked GAT layers, and $\alpha_{i,j}$ are the attention coefficients computed as

$$\alpha_{i,j}^{(t)} = \frac{\exp\left(\text{LeakyReLU}\left(\mathbf{a}^\top \left[W_1^{(t)} R_i^{(t)}; W_1^{(t)} R_j^{(t)}\right]\right)\right)}{\sum_{k \in \mathcal{N}(i) \cup \{i\}} \exp\left(\text{LeakyReLU}\left(\mathbf{a}^\top \left[W_1^{(t)} R_i^{(t)}; W_1^{(t)} R_j^{(t)}\right]\right)\right)} \tag{4}$$

Here, $\mathbf{a}$ is a learned weight vector, and ; denotes concatenation. The final region representations are computed by a weighted combination of the neighbor vector representations, scaled by their attention scores

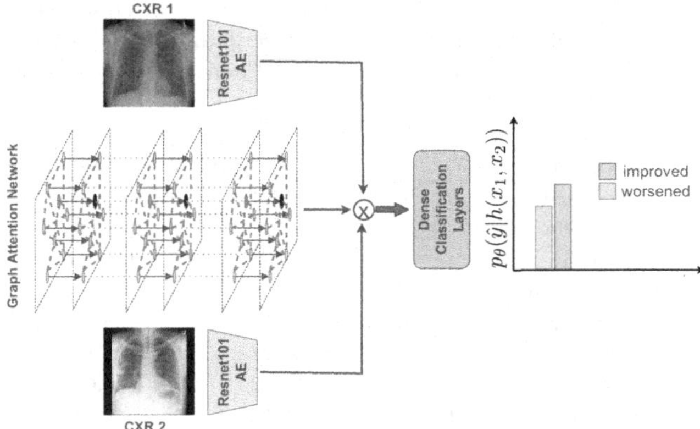

Fig. 2. Classification module. The constructed graph, *i.e.*, the anatomical regions of interest (ROIs) vector representations and the corresponding adjacency matrix, is passed through a graph attention network that learns ROI inter-dependencies, essentially capturing local ROI information. Global image-level representations extracted from a pretrained ResNet101 autoencoder model are concatenated with the ROI learned representations and are passed through a final dense classification layer. The model is trained end-to-end with a cross-entropy classification loss.

$$R_i^{(t+1)} = \phi\left(\sum_{j\in\mathcal{N}(i)} \alpha_{ij}^{(t)} R_j^{(t)}\right), \tag{5}$$

where $\phi(\cdot)$ is a non-linear transformation. Given the past history of the patient, a medical expert has enough information to direct the majority of their focus on a particular region within a CXR. In Fig. 1, the node highlighted in red corresponds to the physician-designated focus region $k^* \in [1, k]$ for the particular CXR examination. We extract the node embedding corresponding to the focus region of x'_i for each CXR image pair and forward this embedding to the final dense classification layer. Specifically, for a focus-region $k^* \in [1, k]$, the extracted node embedding $R'_i \in \mathbb{R}^d$ is given by,

$$R'_i = R'^{(t+1)}_i \mathbb{1}\{k = k^*\}. \tag{6}$$

To capture global image-level information, each image in pair (x_i, x'_i) is encoded into two d-dimensional vectors by utilizing the pretrained feature extractor f, *i.e.*, $Q_i = f(x_i)$ and $Q'_i = f(x'_i)$, $Q_i, Q'_i \in \mathbb{R}^d$. The final prediction is computed via

$$\hat{y} = [R'_i; Q_i; Q'_i]\, W_2^T, \tag{7}$$

where ; denotes the concatenation of the local region-level and global image-level features, $W_2 \in \mathbb{R}^{3d\times M}$ is a fully connected layer that obtains the label predictions. The network is trained with a multi-label cross-entropy classification loss

Table 1. Dataset characteristics. # Image pairs (number of comparison CXR pairs) and # Bboxes (number of bounding boxes) and # Training Pairs (number of training comparison CXR pairs) per pathology label. Each pathology is indexed with a pathology ID (first column).

Pathology ID	Description	# Image pairs	# Bboxes	# Training pairs
D1	Lung opacity	32,524	455,336	22,620
D2	Pleural effusion	13,122	183,708	9,192
D3	Atelectasis	9,660	135,240	6,922
D4	Enlarged cardiac silhouette	1,958	3,916	1,384
D5	Pulmonary Edema/Hazy opacity	12,090	169,260	8,424
D6	Pneumothorax	2,728	38,192	1,930
D7	Consolidation	3,332	46,648	2,310
D8	Fluid overload/Heart failure	674	9,436	132
D9	Pneumonia	3,814	53,396	2,590
All 9 pathologies	**Total**	79,902	1,095,132	55,504

$$L = \frac{1}{N}\sum_{i=1}^{N}\sum_{m=1}^{M} z_{i,m}log(\sigma(\hat{z}_{i,m})) + (1 - z_{i,m})log(1 - \sigma(\hat{z}_{i,m})), \quad (8)$$

where σ is the sigmoid function and $\{\hat{y}_i^m, y_i^m\} \in \mathbb{R}^M$ are the model prediction and the ground truth for example x_i, respectively. Figure 2 presents an overview of the model architecture.

3 Experiments

3.1 Dataset

The proposed `CheXRelNet` model is trained and evaluated on the Chest ImaGenome dataset [28]. This dataset was generated by locally labeling 242, 072 frontal MIMIC-CXRs [6] (AP or PA view) automatically through a combination of rule-based text analysis and atlas-based bounding box extraction techniques [27,29]. Chest ImaGenome represents the connections of each CXR annotation as an anatomy-centered scene graph, following a radiologist-constructed CXR ontology. The dataset contains 1, 256 combinations of relation annotations between 29 CXR anatomical locations and their attributes structured as one scene-graph per image, and about 670, 000 localized comparison relations between the anatomical locations across sequential exams. In this work, we utilize the localized comparison relations data that involves cross-image relations for the 9 pathologies. Each comparison relation in the Chest ImaGenome dataset consists of the DICOM identifiers of the two CXRs being compared, the particular pathological finding observed in those two CXRs, the anatomical region of interest on which the radiologist's comparison is focused, and the corresponding comparison label. In addition to comparison relations, the Chest ImaGenome dataset also provides bounding box information for extracting individual anatomical regions from the CXRs, viz. 'Left Lung', 'Cardiac Silhouette',

Table 2. Comparison against baselines (accuracy).

Method	D1	D2	D3	D4	D5	D6	D7	D8	D9	Al
Local	0.59	0.53	0.60	0.47	0.56	0.46	0.61	0.47	0.63	0.
Global	0.67	**0.69**	0.64	0.74	0.71	0.50	0.65	0.69	0.67	0.
`CheXRelNet`	0.67	0.68	**0.66**	**0.75**	0.71	**0.52**	**0.67**	**0.73**	0.67	**0.**

etc.. For each of the 242,072 frontal MIMIC-CXRs a list of anatomical regi (bboxes) is provided, as well as the corresponding Euclidean coordinates for eac bounding box. We utilize these coordinates to crop different anatomical regions within a CXR. There are a total of 122,444 unique comparisons in the dataset, of which 79,902 have at least one of the nine selected pathology labels, in addition to regions detected by the object detection pipeline and the overall comparison relation. For each image, except for those with the pathology label as 'Enlarged Cardiac Silhouette', 7 of the most frequently occurring anatomical regions were extracted. For the pathology label 'Enlarged Cardiac Silhouette', the dataset provides only one corresponding bounding box. Table 1 shows high-level data statistics.

3.2 Baselines

We compare the `CheXRelNet` model against the following baselines: 1) **Local** model: we utilize a previously proposed siamese network trained on cropped ROIs, encoded with a pre-trained ResNet101 autoencoder and passed through a dense layer and a final classification layer [28]. This model essentially only looks at the corresponding anatomical regions and considers neither global information nor intra-region dependencies. 2) **Global** model: we also design a siamese architecture that encodes the entire CXR as opposed to only the cropped ROIs in the Local model. Apart from the input being a full image rather than an ROI, the model architecture is the same as the Local model. Hence, this baseline incorporates the global information but does not take into consideration the anatomical region of interest nor explicitly models inter-region dependencies. These two siamese models serve as baseline methods to contrast the effectiveness of `CheXRelNet`, which not only is location-aware but can also explicitly model both inter-region and intra-image CXR dependencies.

3.3 Implementation Details

To train each model, we use the train/validation/testing splits and detected ROIs provided by Chest ImaGenome. For each image within the comparison pair, we crop the image ROIs and resize them to 224×224. Each cropped ROI is then embedded into a vector with 2048 dimensions, by utilizing a pre-trained ResNet101 autoencoder that is trained on several medical imaging datasets, e.g., NIH, CheXpert, and MIMIC datasets, *etc.* [5,6,25]. The same autoencoder is utilized for the baseline models. The co-occurrence matrix threshold is set to 0.5. Our model is a 2-layer graph neural network with 2048 and 1024 neurons per layer, in the first and second layers respectively. There are 5 and 3 multi-head-attentions in each respective layer. The output from the graph attention network is concatenated with the global information and then passed through two dense layers of sizes 768 and 128, respectively. We train the network using Adam [9] optimizer for 200 epochs, with a $0.8e^{-3}$ initial learning rate set [26] and a batch size of 32. To avoid overfitting, we utilize early stopping with 11 patience and gradient clipping that is set to 0.1. In addition, we use 0.5 Dropout [22] and a learning rate decay factor of 0.3 with the patience threshold set to 4. The model is implemented by utilizing the PyTorch [17] and pytorch_geometric [3] deep learning frameworks. The evaluation metric is accuracy and results are reported over six experimental trials.

3.4 Experimental Results

Results are summarized in Table 2. `CheXRelNet` achieves a mean accuracy of 0.683 (SD = 0.0024), while the Local model has 0.602 mean accuracy (SD = 0.0059) and the Global model has 0.672 mean accuracy (SD = 0.0046) over six trials. We observe that the Local model is generally underperforming, and it is most likely limited because it focuses on a specific anatomical region and completely neglects the global information. In contrast, radiologists often take into consideration more than one anatomical region when drawing inferences from CXRs. The Global model is a lot more effective than the Local one, and incorporating global information boosts the prediction accuracy. Yet, the Global model is also limited as it focuses on the entire image but fails to consider the relationships among anatomical regions. We additionally perform statistical significance tests, *i.e.*, an unpaired t-test (p = 0.049) and a one-tailed t-test (p = 0.018) comparing `CheXRelNet` and the Global baseline. These t-test results verify that the `CheXRelNet` and Global baseline predictions follow distinct distributions and that the improvement in accuracy is significant at $p < 0.05$. Additional experiments w.r.t. model architectures and model capacity (number of parameters) are reported in the supplementary. Overall, `CheXRelNet` improves upon the Global model's prediction accuracy by modeling the inter-image and intra-image region correlations and attending to the anatomical regions of interest.

We also perform a transfer learning experiment wherein we train `CheXRelNet` on a set of diseases and test performance on a different set of diseases. Specifically, we train `CheXRelNet` on a subset of the data with 'Pneumothorax', 'Consolidation', 'Fluid Overload/Heart Failure', 'Pneumonia' (D6-D9) pathologies, and test on the following pathology labels that are unseen during training: 'Lung Opacity', 'Pleural Effusion', 'Atelectasis', 'Enlarged Cardiac Silhouette', and 'Pulmonary Edema/Hazy Opacity' (D1-D5). Results are reported in Table 3. We perform this experiment on individual unseen pathology labels as well as on sets of multiple unseen pathology labels. We observe that our model can generalize well to unseen pathology labels. We can attribute this to the incorporation of both local and

Table 3. Transfer learning evaluation against baselines (accuracy). Models are trained on D6-D9 and tested on unseen pathologies (D1–D5). SetA consists of unseen pathologies {D1, D2}. SetB consists of unseen pathology labels, {D3, D4}. Set C consists of all unseen pathology labels {D1, D2, D3, D4, D5}.

Method	D1	D2	D3	D4	D5	SetA	SetB	SetC
Local	0.56	0.49	0.54	0.49	0.55	0.54	0.55	0.54
Global	0.61	**0.63**	0.60	0.65	0.63	0.61	0.63	0.62
`CheXRelNet` (ours)	**0.64**	0.60	**0.61**	**0.68**	**0.67**	**0.63**	**0.64**	**0.64**

global information during training. The model is learning associations between different anatomical regions and therefore can identify complex bio-markers associated with the progression of pathologies.

4 Conclusion

CXRs are commonly repeatedly requested in the clinical workflow to assess for a myriad of attributes. Diagnosis and monitoring are typically performed through comparisons of sequential CXR images, both in in-patient and outpatient settings. Given a patient with two sequential CXR exams, the goal of this work is to automatically evaluate disease change. To this end, we describe a methodology for localized relation comparisons between CXR images. The proposed `CheXRelNet` fuses global image-level information, local intra-image region-level correlations, and inter-image correlations. Experimental results show that `CheXRelNet` outperforms baselines in both traditional and transfer learning settings. As a result, our method provides necessary components for monitoring the progression of pathologies that are visualized through chest imaging. In the future, we hope to expand our work to model disease progression among several sequential CXRs, incorporate additional temporal context information and physiological data [7,28] and account for the time interval variability found in longitudinal imaging records. Finally, future work can adapt the proposed methodology to other medical imaging tasks, and include interpretability mechanisms.

References

1. Agu, N.N., et al.: AnaXNet: anatomy aware multi-label finding classification in chest x-ray. In: de Bruijne, M., et al. (eds.) MICCAI 2021. LNCS, vol. 12905, pp. 804–813. Springer, Cham (2021). https://doi.org/10.1007/978-3-030-87240-3_77
2. Chao, W.-L., Changpinyo, S., Gong, B., Sha, F.: An empirical study and analysis of generalized zero-shot learning for object recognition in the wild. In: Leibe, B., Matas, J., Sebe, N., Welling, M. (eds.) ECCV 2016. LNCS, vol. 9906, pp. 52–68. Springer, Cham (2016). https://doi.org/10.1007/978-3-319-46475-6_4
3. Fey, M., Lenssen, J.E.: Fast graph representation learning with PyTorch Geometric. In: Proceedings of the ICLR Workshop on Representation Learning on Graphs and Manifolds (2019)

4. Guo, R., Passi, K., Jain, C.K.: Tuberculosis diagnostics and localization in chest x-rays via deep learning models. Front. Artif. Intell. 74 (2020)
5. Irvin, J., et al.: CheXpert: a large chest radiograph dataset with uncertainty labels and expert comparison. In: Proceedings of the AAAI Conference on Artificial Intelligence (AAAI), vol. 33, pp. 590–597 (2019)
6. Johnson, A.E., Pollard, T.J., Berkowitz, S.J., et al.: Mimic-CXR, a de-identified publicly available database of chest radiographs with free-text reports. Sci. Data ,1–8 (2019)
7. Karargyris, A., et al.: Creation and validation of a chest x-ray dataset with eye-tracking and report dictation for AI development. Sci. Data **8**(1), 1–18 (2021)
8. Kim, M., Lee, B.D.: Automatic lung segmentation on chest x-rays using self-attention deep neural network. Sensors **21**(2), 369 (2021)
9. Kingma, D.P., Ba, J.: Adam: a method for stochastic optimization. In: Proceedings of the International Conference on Learning Representations (ICLR) (2015)
10. Kiradoo, G.: Role of application of artificial intelligence and its importance in the healthcare industry. Int. J. Adv. Res. Eng. Technol. **9**(2) (2018)
11. Kobayashi, M., et al.: Computer-aided diagnosis with a convolutional neural network algorithm for automated detection of urinary tract stones on plain x-ray. BMC Urol. **21**(1), 1–10 (2021)
12. Li, M.D., et al.: Automated assessment and tracking of COVID-19 pulmonary disease severity on chest radiographs using convolutional Siamese neural networks. Radiol. Artif. Intell. **2**(4) (2020)
13. Li, M.D., et al.: Siamese neural networks for continuous disease severity evaluation and change detection in medical imaging. NPJ Digit. Med. **3**(1), 1–9 (2020)
14. Maity, A., Nair, T.R., Mehta, S., Prakasam, P.: Automatic lung parenchyma segmentation using a deep convolutional neural network from chest x-rays. Biomed. Sig. Process. Control **73**, 103398 (2022)
15. Majkowska, A., et al.: Chest radiograph interpretation with deep learning models: assessment with radiologist-adjudicated reference standards and population-adjusted evaluation. Radiology **294**(2), 421–431 (2020)
16. Oh, D.Y., Kim, J., Lee, K.J.: Longitudinal change detection on chest x-rays using geometric correlation maps. In: Shen, D., et al. (eds.) MICCAI 2019. LNCS, vol. 11769, pp. 748–756. Springer, Cham (2019). https://doi.org/10.1007/978-3-030-32226-7_83
17. Paszke, A., et al.: PyTorch: an imperative style, high-performance deep learning library. In: Proceedings of the International Conference on Neural Information Processing Systems (NeurIPS), 8026–8037 (2019)
18. Rajpurkar, P., et al.: Deep learning for chest radiograph diagnosis: a retrospective comparison of the ChexNext algorithm to practicing radiologists. PLoS Med. **15**(11), e1002686 (2018)
19. Reamaroon, N., et al.: Robust segmentation of lung in chest x-ray: applications in analysis of acute respiratory distress syndrome. BMC Med. Imaging **20**(1), 1–13 (2020)
20. Shelke, A., et al.: Chest x-ray classification using deep learning for automated COVID-19 screening. SN Comput. Sci. **2**(4), 1–9 (2021)
21. Souza, J.C., Diniz, J.O.B., Ferreira, J.L., da Silva, G.L.F., Silva, A.C., de Paiva, A.C.: An automatic method for lung segmentation and reconstruction in chest x-ray using deep neural networks. Comput. Methods Programs Biomed. **177**, 285–296 (2019)

22. Srivastava, N., Hinton, G., Krizhevsky, A., Sutskever, I., Salakhutdinov, R.: Dropout: a simple way to prevent neural networks from overfitting. J. Mach. Learn. Res. **15**(56), 1929–1958 (2014)
23. Tang, X.: The role of artificial intelligence in medical imaging research. BJR Open **2**(1) (2019)
24. Veličković, P., Cucurull, G., Casanova, A., Romero, A., Liò, P., Bengio, Y.: Graph attention networks. In: Proceedings of the International Conference on Learning Representations (ICLR) (2018)
25. Wang, X., Peng, Y., Lu, L., Lu, Z., Bagheri, M., Summers, R.M.: Chestx-ray8: Hospital-scale chest x-ray database and benchmarks on weakly-supervised classification and localization of common thorax diseases. In: Proceedings of the IEEE Conference on Computer Vision and Pattern Recognition (CVPR), pp. 2097–2106 (2017)
26. Wilson, D.R., Martinez, T.R.: The need for small learning rates on large problems. In: Proceedings of the International Joint Conference on Neural Networks (IJCNN), vol. 1, pp. 115–119. IEEE (2001)
27. Wu, J., et al.: Automatic bounding box annotation of chest x-ray data for localization of abnormalities. In: Proceedings of the 17th International Symposium on Biomedical Imaging (ISBI), pp. 799–803. IEEE (2020)
28. Wu, J.T., et al.: Chest imagenome dataset for clinical reasoning. In: Proceedings of the International Conference on Neural Information Processing Systems (NeurIPS) Datasets and Benchmarks Track (2021)
29. Wu, J.T., Syed, A., Ahmad, H., et al.: AI accelerated human-in-the-loop structuring of radiology reports. In: Proceedings of the American Medical Informatics Association (AMIA) Annual Symposium (2020)
30. Xian, Y., Schiele, B., Akata, Z.: Zero-shot learning-the good, the bad and the ugly. In: Proceedings of the IEEE Conference on Computer Vision and Pattern Recognition (CVPR), pp. 4582–4591 (2017)

Reinforcement Learning for Active Modality Selection During Diagnosis

Gabriel Bernardino[1(✉)], Anders Jonsson[2], Filip Loncaric[3], Pablo-Miki Martí Castellote[2], Marta Sitges[4], Patrick Clarysse[1], and Nicolas Duchateau[1,5]

[1] Univ Lyon, Université Claude Bernard Lyon 1, INSA-Lyon, CNRS, Inserm, CREATIS UMR 5220, U1294, 69621 Lyon, France
gabriel.bernardino@creatis.insa-lyon.fr

[2] DTIC, Universitat Pompeu Fabra, Barcelona, Spain

[3] Department of Cardiovascular Diseases, University Hospital Centre Zagreb, Zagreb, Croatia

[4] Cardiovascular Institute, Hospital Clínic Barcelona, University of Barcelona, Institut de Investigació Biomedica Augist Pi i Sunyer (IDIBAPS). CIBERCV, Instituto de Salud Carlos III, Barcelona, Spain

[5] Institut Universitaire de France (IUF), Paris, France

Abstract. Diagnosis through imaging generally requires the combination of several modalities. Algorithms for data fusion allow merging information from different sources, mostly combining all images in a single step. In contrast, much less attention has been given to the incremental addition of new data descriptors, and the consideration of their costs (which can cover economic costs but also patient comfort and safety).

In this work, we formalise clinical diagnosis of a patient as a sequential process of decisions, each of these decisions being whether to take an additional acquisition, or, if there is enough information, to end the examination and produce a diagnosis. We formulate the goodness of a diagnosis process as a combination of the classification accuracy minus the cost of the acquired modalities. To obtain a policy, we apply reinforcement learning, which recommends the next modality to incorporate based on data acquired at previous stages and aiming at maximising the accuracy/cost trade-off. This policy therefore performs medical diagnosis and patient-wise feature selection simultaneously.

We demonstrate the relevance of this strategy on two binary classification datasets: a subset of a public heart disease database, including 531 instances with 11 scalar features, and a private echocardiographic dataset including signals from 5 standard image sequences used to assess cardiac function (2 speckle tracking, 2 flow Doppler and tissue Doppler), from 188 patients suffering hypertension, and 60 controls.

For each individual, our algorithm allows acquiring only the modalities relevant for the diagnosis, avoiding low-information acquisitions, which both resulted in higher stability of the chosen modalities and better classification performance under a limited budget.

Supplementary Information The online version contains supplementary material available at https://doi.org/10.1007/978-3-031-16431-6_56.

L. Wang et al. (Eds.): MICCAI 2022, LNCS 13431, pp. 592–601, 2022.
https://doi.org/10.1007/978-3-031-16431-6_56

Keywords: Computer aided diagnosis · Reinforcement learning · Active feature selection · Acquisition costs · Cardiac imaging

1 Introduction

Medical diagnosis is not based on a single image, but usually considers several sources of information, each with different costs and accuracy at detecting different phenomena. Given the limited amount of resources in many real-life situations, it is crucial to select the most appropriate acquisitions for each patient [6]. In clinical practice, decisions are based on guidelines and consensus recommendations [4], which are in turn based on qualitative analysis of current evidence by experts. While machine learning has shown great success for quantitative analysis and diagnosis in medical images [1,13], quantifying the appropriateness of acquisitions has been neglected, as data are often considered an immutable input of the algorithms.

Cost-aware feature selection has received substantial attention from the machine learning community: the simplest methods are based on heuristics that promote sparsity at a population level, such as L_1 regularisation [9]; or decision trees and forests that include features' costs in the split criteria, thus doing patient-specific feature selection [11]. Recent approaches are based on Markov Decision Processes (MDPs), a formalisation of a time discrete process involving decisions with uncertain outcome, [5,12]. These methods build a common space that integrates all information, treating non acquired data as missing/censored. The current state is defined as a point/probability distribution in that space, which is updated after the acquisition of a new modality. Finally, Reinforcement learning (RL) is used to discover a policy, which is the optimal modality acquisitions at each point of this space.

A downside of the previous methods is that they heavily involve sampling, for both the data imputation and RL, which can become prohibitive when the data are high dimensional objects. In addition, the handling of "missing" data assumes that not-yet-acquired data can be estimated from the present data. In [10], *Wang et al.* proposed a method that considered all possible combinations of the N features, and a single-step policy, based on cost-sensitive-learning decision trees, had to be learnt for each of these 2^N combinations. However, this approach is only tractable for a small number of features, and therefore more modern literature has focused on partially-observed data approaches.

In this work, we propose a modality- and cost-aware RL method that sequentially proposes new acquisitions until it has enough confidence to produce a diagnosis. This work uses RL to extend to multiple modalities and more complex scenarios a recent two-stage strategy that recommends when a complex modality is needed instead of a simpler one [2]. Our method is similar to the value iteration algorithm, and we use kernel methods to estimate the state-action values. To avoid sampling, we use a strategy similar to *Wang et al.* [10]. We demonstrate the relevance of our method on two clinical datasets: the publicly available Heart Disease dataset [3], involving scalar measurements, and a private echocardiography dataset of patients with arterial hypertension with disease-related changes in cardiac function (i.e. diastolic and systolic function), involving temporal signals along the whole cardiac cycle (flow and tissue motion data from speckle tracking and Doppler Imaging).

2 Methodology

2.1 Markov Decision Process and Reinforcement Learning

An MDP is a mathematical framework of sequential decisions with uncertain effects. Formally, an MDP consists of a tuple $(\mathcal{S}, \mathcal{S}_{end}, \mathcal{A}, \mathcal{R}, \mathcal{T})$, where $\mathcal{S}$ is the possible state spaces, $\mathcal{S}_{end} \subset \mathcal{S}$ is a set of ending states, $\mathcal{A}$ is the discrete action space, $\mathcal{T} : \mathcal{S} \times \mathcal{A} \times \mathcal{S} \to \mathbb{R}$ is the transition function, which gives the probability distribution over $\mathcal{S}$ of the next state, knowing the current state and the action taken. $\mathcal{R} : \mathcal{S} \times \mathcal{A} \times \mathcal{S} \to \mathbb{R}$ is the random reward function, that depends on the current and next states, as well as the action taken. An episode, starting in a given state s_1, consists of a sequence of states and actions:

$$((s_1, a_1, r_1), (s_2, a_2, r_2) \dots (s_n, a_n, r_n)), \tag{1}$$

where s_i is not a final state for $i \in [0, n-1]$ and s_n is a final state. The total reward of this episode is $\sum_i r_i$. The transitions and r_i follow the previously stated probability distributions $\mathcal{T}$ and $\mathcal{R}$. As we will show in the next section, our MDP is episodial with the number of steps being lower than the number of available modalities, so the use of a discount factor γ is not required.

A policy $\pi : \mathcal{S} \to \mathcal{A}$ is a function that chooses which action to perform at each state. We would like to find the optimal policy π^* that maximises the expected total reward over a random episode if we take actions following this policy. RL is a set of techniques to estimate such policy when the $\mathcal{R}$ and $\mathcal{T}$ distributions are unknown, but samples can be obtained. Given a policy, we define the *value* of a state s as the expected total reward over all episodes starting in s.

2.2 Problem Definition

We formalise our cost-sensitive diagnostic problem under the notation of an MDP. An episode corresponds to an examination of a single patient, where each step (action) is the acquisition of a certain modality. The state will contain information on the already acquired data. The total reward will depend on whether a correct diagnosis was reached, and the costs associated with the used modalities.

Each modality of the set of modalities M is identified by an index i, and its measurements are elements of $\mathbb{R}^{n_i}$, allowing vector-valued measurements of different dimensionalities. Therefore, to represent a combination of measurements p, we use the Cartesian product $(\times)$ of each space corresponding to a single measurement: $\times_{i \in p} \mathbb{R}^{n_i}$. To avoid data imputation, the full state space $\mathcal{S}$ is defined as the disjoint union of all possible combinations of acquired modalities $p \subset M$:

$$\mathcal{S} = \dot{\bigcup}_{p \subset M} \bigtimes_{i \in p} \mathbb{R}^{n_i} \tag{2}$$

where $\dot{\bigcup}$ denotes the disjoint union. $\mathcal{S}$ therefore consists of a connected domain for each element i of the powerset $p \subset M$. We will call each of these domains a "superstate". By allowing multidimensional measurements, we can group modalities, so that the policy is forced to either ignore or acquire them together, thus reducing the number of superstates.

The set of actions contains one action a for each modality, associated with acquiring the a-th modality, and a special action to finish the episode and produce a diagnosis. Transitions associated which each "acquire" action are to move to a new state, which includes the previously acquired data, and the newly acquired measurement of the modality a. Therefore, the next state s_{n+1} will belong to the superstate identified by $p \cup \{a\}$, where p is the superstate of the current state s_n. Note that this means that even if we cannot know the exact state resulting of applying an action, since the measurement values are unknown, the transitions between "superstates" are completely deterministic, since they only depend on which modalities were acquired, but not on the measurement that was observed.

If the new state is not a final state, the reward of each action is set to 0. For a final state s_n, belonging to the superstate p, the reward is set as follows:

$$\mathcal{R}(s_f) = \mathbb{1}(y = y_p^{pred}) - \lambda \sum_{i \in p} c_i \tag{3}$$

where $\mathbb{1}$ is the indicator function, thus reflecting the accuracy (y_p^{pred} being the label prediction of the instance at state p, y being the true label of the instance); c_i is the cost of the i-th modality and λ is the coefficient weighting the relative contribution of the cost and accuracy. The reward of acquiring a modality that has already been observed is set to $-\infty$ to discourage the algorithm to take it, therefore each action is taken at most once during an episode. The class predictions y_p^{pred} are computed statically, using a classifier learnt at each superstate p from the available modalities. In our case, we chose a Support Vector Machine with Gaussian kernel.

2.3 Policy Optimisation

We use a variation of value iteration to estimate π^*, where we estimate the state-action value $Q(s, a)$ for each state s, defined as the expected value if an action a is taken. Our method is a model-free strategy (meaning that it does not estimate the state transition function $\mathcal{T}$). We search for a solution to the recursive Bellman optimality equations:

$$Q(s, a) = E[\mathcal{R}(s, a, s') + \max_{a'} Q(s', a')], \tag{4}$$

where E refers to the expectation over the next states s', a' being the next action. For discrete spaces, this Q-function can be exactly stored in a table. However, in a continuous setting, function approximation is needed. Typical choices are neural networks, but given our limited amount of training data, and the success of kernel methods in similar applications, we used kernel ridge regression.

In a general case, Eq. 4 cannot be solved for Q directly, since it requires knowledge of $\mathcal{R}$ and $\mathcal{T}$. Therefore, value iteration repeatedly performs the following updates for all state-action pairs until convergence of Q:

$$Q^{n+1}(s,a) \leftarrow E[r_t + \max_{a'} Q^n(s',a')] \tag{5}$$

where s' is the next state.

In our case, the particular structure of the state transitions produced by the actions allows a direct solution of Eq. 5, as shown in [10]. Since our state space is disconnected, we can train an independent Q-function for each of the $2^{N_{meas}}$ superstates, noted as $Q_p(s,a)$. And, since we know which will be the next superstate of each action, Eq. 5 becomes:

$$Q_p^{n+1}(s,a) \leftarrow E[r_t + \max_{a'} Q^n_{p\cup\{a_t\}}(s',a')]. \tag{6}$$

Using the fact that subsets form a directed acyclic graph, we can visit all superstates in postorder, and train the Q_p functions on the transitions derived from the collected data. As when visiting the superstate p, all $Q_{p\cup\{a_t\}}$ have already been trained, direct optimisation is possible.

2.4 Code Availability

An implementation of the method, and the code used for the experiments, are publicly available at https://github.com/creatis-myriad/featureSelectionRL.

3 Datasets

3.1 Heart UCI Dataset

We first tested our methods on the public Heart Disease Data Set [3] available at the University of California at Irving repository, whose objective is to diagnose heart disease for patients admitted to intensive care. These data correspond to a multicentric study, and we used all the available data except the imaging information since it was only available in a single center. Data therefore consisted of 11 features and associated costs per modality, which were split in 4 different feature groups (clinical history, laboratory, vital constants and exercise testing). We removed the individuals with missing data, which would induce additional challenges out of the scope of this paper, leaving a total of 207 controls and 324 cases.

3.2 Hypertension Dataset

We also evaluated our methodology on temporal signals quantifying the cardiac function on an hypertense population. Details on the recruitment and cohort, as well as on the signal pre-processing (delineation and temporal alignment) can be found in [7]. The study protocol was approved by an internal ethical committee. We used the signals corresponding to flow (Mitral and Aortic Flow Doppler)

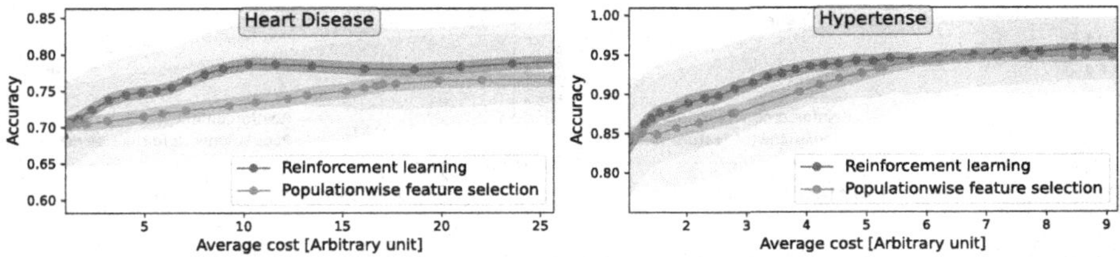

Fig. 1. Average accuracy as a function of the average acquisition costs, where each point was generated by a specific value of the cost-coefficient λ. The solid line, light shadowed and darker shadowed depict the mean value, standard deviation, and 95% confidence interval, respectively.

and myocardial motion/deformation Septal Tissue Doppler, Global Longitudinal Strain (GLS) and the local strain in the septal basal segment). These data were assigned arbitrary costs based on the frequency they are used in clinical practice (1, 1, 2.5, 5, 10 respectively). Examples of the signals of three individuals can be found in Supplementary Material. Principal Component Analysis (PCA) was applied to the signals from each modality to reduce the dimensionality of these data before using them in the RL framework.

The interest of this dataset is twofold: the hierarchical representation provided by our method allows identifying different phenotypes, and simultaneously quantifying which modality is the most appropriate to detect each of these phenotypes. In addition to the cost-effective predictive power of our algorithm, we can interpret each decision of the policy by examining which biomarkers they capture, which can be seen as a data-based clinical guideline. This usage is very relevant when current human-generated clinical guidelines are long and complex.

4 Results

4.1 Prediction Error Against Cost

For both datasets, we trained the policy for accuracy-cost coefficients λ (see definition in Eq. 3) between 10^{-3} and 10^{-1}, and reported the mean accuracy and cost on a test set, which was a class-stratified split of the full dataset. We compared our algorithm with a classical feature selection using a validation set, in which we tested all possible combination and features and kept the one with a maximal validation reward. Results can be seen in Fig. 1, where they were repeated over different train/test splits. We can observe that our method has a higher accuracy for a constrained budget, since RL allowed using expensive modalities only for a few individuals with a difficult diagnosis, while population-wise feature selection was forced to acquire the modality either for everybody or for nobody.

4.2 Stability Under Different Training Sets

We evaluated the consistency of the features selected by the algorithm, under different bootstrap samples of the training data. We separated a test subset

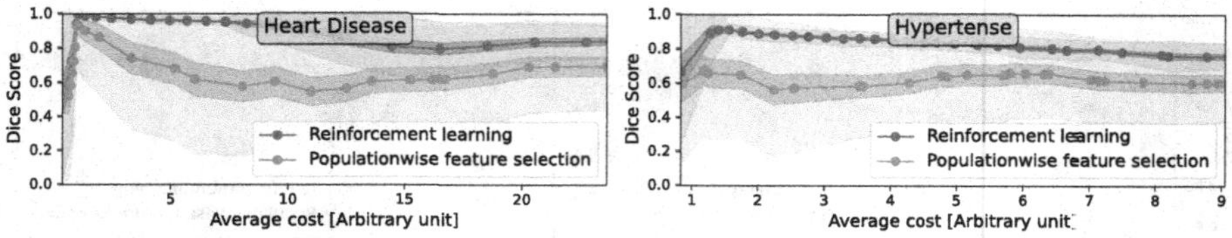

Fig. 2. Stability of the policies to changes in the training set. Different policies resulting of bootstrap samples of a train dataset are evaluated on a fixed test-set, and the overlap of the sets of acquired modalities for each test patient on different bootstrap samples is quantified using the Dice score (y-axis). This experiment is repeated for several values of the λ weighting coefficient, resulting in different average costs (x-axis). The color code is equivalent to Fig. 1.

from the dataset, and subsequently obtained several training samples of the remaining dataset, which were used to train the proposed RL model, and a classical feature selection using cross-validation. Then, we applied the trained models on the test set, and checked which were, for each subject, the selected features (i.e. the recommended acquisitions) at their final state. This set was quantitatively compared to the features' set resulting from the other training bootstraps by computing the Dice coefficient. To improve statistical stability, the procedure was tested for different values of the accuracy-cost coefficient λ and test set splits. Figure 2 shows the results of this experiment, where our method is more robust than feature selection using validation: indeed, static feature selection forces that either all or none individuals acquire a modality, while our method allows a gradual process where only a few individuals acquire an expensive modality.

4.3 Policy Interpretation on the Hypertense Dataset

We further studied the learnt policy on the hypertense dataset. We trained the RL algorithm with $\lambda = 0.05$, chosen to guarantee a low number of acquired modalities. Afterwards, we evaluated the policy on all individuals from the training set, keeping track of the superstates they visited. The full decision graph can be seen in Fig. 3. Figure 4 complements this by showing the representative signals associated to the individuals that were diagnosed at each superstate of the graph visited by more than 10 individuals, allowing us to examine and interpret the decisions of the algorithm.

The first acquisition recommended by RL was Mitral Doppler, which is consistent with clinical knowledge [8]. Hypertense individuals with a clear grade I diastolic dysfunction mitral inflow pattern (presenting E/A-wave fusion, A-peak larger than E-peak) or those clearly controls are differentiated based on the Mitral Doppler only (first row in Fig. 4). The remaining patients had still an unclear diagnosis and were referred to either Aortic Doppler or GLS depending on the findings in the Doppler.

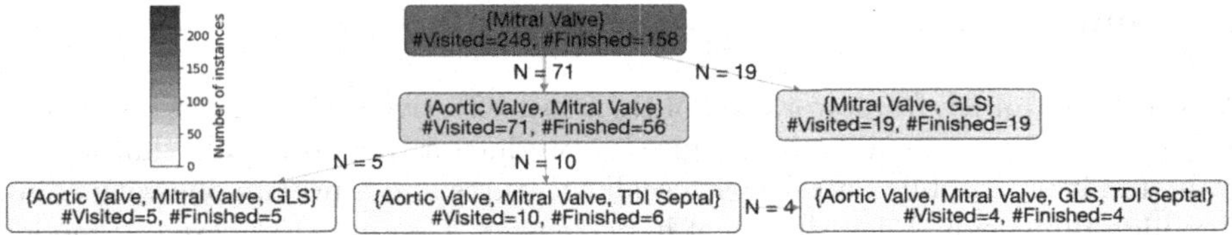

Fig. 3. Graph showing the decision paths proposed by our algorithm on the full population. Each node represents a superstate (a set of modalities), and each edge represents an action (acquiring a new modality). The number of individuals that take the "finish" action is indicated at each node.

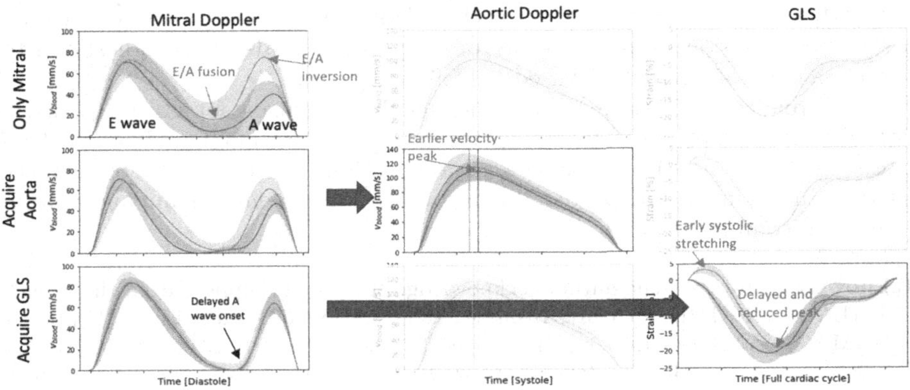

Fig. 4. Class-wise mean and standard deviation (orange hypertension, blue control) for the Mitral, Aortic and GLS signals of the individuals whose diagnosis was done using Mitral only (1st row), Mitral and Aortic (2nd row) and Mitral and GLS (3rd row). Un-used modalities are displayed with low opacity.

The Mitral Doppler of patients for which RL suggested Aortic Doppler (2nd row) presented a slightly elevated A-wave, but still lower than the E-wave; lying on the boundary between controls and cases of the first subpopulation. Although the mitral inflow pattern is not suggestive of diastolic dysfunction, an earlier, higher velocity aortic flow peak reflected increased cardiac contractility in the setting of elevated afterload due to high blood pressure, identifying patients with altered systolic function in hypertension.

The last group (3rd row) consists of individuals for which GLS was recommended. Mitral Doppler showed a later onset of atrial contraction within the cardiac cycle. This is confirmed in the GLS curves, where hypertense subjects showed prolonged left ventricular stretching during the initial part of systole - identifying a disease-related pattern in the timing of cardiac events.

5 Conclusion

We presented an RL framework to obtain a cost-effective policy that sequentially proposes the best modality to acquire to produce a diagnosis. We thoroughly evaluated it on a public dataset with scalar features and a private one with complex high-dimensional descriptors of hypertension. Compared to classical model selection using a validation set, our method improved the model performance at similar acquisition cost by making a more efficient use of expensive modalities. The proposed method also showed stability to changes in the training set.

In addition, we were able to interpret and explain the policy constructed in the diastolic dysfunction dataset, which captured physiological patterns consistent with current clinical knowledge. Our algorithm showed potential not only as a clinical decision support system for diagnosis, but also at a higher level to help clinicians to derive data-based guidelines.

On a broader perspective, the method is highly promising for assisting experts with the analysis of multiple descriptors in an efficient manner. It could lead to discovering cost-efficient policies in applications where high heterogeneity between individuals is expected, in particular for screening campaigns or in developing countries.

Acknowledgements. The authors acknowledge the partial support from the French ANR (LABEX PRIMES of Univ. Lyon [ANR-11-LABX-0063] and the JCJC project "MIC-MAC" [ANR-19-CE45-0005]) and the Spanish AEI [PID2019-108141GB-I00]. We thank Prof. B. Bijnens (IDIBAPS & ICREA, Barcelona, Spain) for fruitful discussions.

References

1. Bernard, O., et al.: Deep learning techniques for automatic mri cardiac multi-structures segmentation and diagnosis: is the problem solved? IEEE Trans. Med. Imaging **37**(11), 2514–2525 (2018)
2. Bernardino, G., et al.: Hierarchical multi-modality prediction model to assess obesity-related remodelling. In: Puyol Antón, E., Pop, M., Martín-Isla, C., Sermesant, M., Suinesiaputra, A., Camara, O., Lekadir, K., Young, A. (eds.) STACOM 2021. LNCS, vol. 13131, pp. 103–112. Springer, Cham (2022)
3. Detrano, R., et al.: International application of a new probability algorithm for the diagnosis of coronary artery disease. Am. J. Cardiol. **64**(5), 304–310 (1989)
4. Garbi, M., et al.: EACVI appropriateness criteria for the use of cardiovascular imaging in heart failure derived from European National Imaging Societies voting. Europ. Heart J. Cardiovascular Imaging **17**(7), 711–721 (2016)
5. Gong, W., et al.: Icebreaker: Element-wise Efficient Information Acquisition with a Bayesian Deep Latent Gaussian Model. In: Wallach, H., Larochelle, H., Beygelzimer, A., Alché-Buc, F., Fox, E., Garnett, R. (eds.) Advances in Neural Information Processing Systems, vol. 32. Curran Associates, Inc. (2019)
6. Hadian, M., Jabbari, A., Mazaheri, E., Norouzi, M.: What is the impact of clinical guidelines on imaging costs? J. Educ. Health Promotion **10**, 10 (2021)

7. Loncaric, F., et al.: Automated pattern recognition in whole-cardiac cycle echocardiographic data: capturing functional phenotypes with machine learning. J. Am. Soc. Echocardiography **34**(11), 1170–1183 (2021)
8. Nagueh, S.F., et al.: Recommendations for the evaluation of left ventricular diastolic function by echocardiography: an update from the American society of echocardiography and the European Association of Cardiovascular Imaging. J. Am. Soc. Echocardiography : official publication of the American Society of Echocardiography **29**(4), 277–314 (2016)
9. Ng, A.Y.: Feature selection, L1 vs. L2 regularization, and rotational invariance. In: Proceedings, Twenty-First International Conference on Machine Learning, ICML 2004 (2004)
10. Wang, J., Trapeznikov, K., Saligrama, V.: Efficient learning by directed acyclic graph for resource constrained prediction. In: Cortes, C., Lawrence, N., Lee, D., Sugiyama, M., Garnett, R. (eds.) Advances in Neural Information Processing Systems, vol. 28. Curran Associates, Inc. (2015)
11. Xu, Z.E., Kusner, M.J., Weinberger, K.Q., Chen, M., Chapelle, O.: Classifier cascades and trees for minimizing feature evaluation cost. J. Mach. Learn. Res. **15**, 2113–2144 (2014)
12. Yin, H., Li, Y., Pan, S.J., Zhang, C., Tschiatschek, S.: Reinforcement Learning with Efficient Active Feature Acquisition. arXiv, September 2020
13. Zhou, T., Ruan, S., Canu, S.: A review: deep learning for medical image segmentation using multi-modality fusion. Array 3–4, 100004 (2019)

Ensembled Prediction of Rheumatic Heart Disease from Ungated Doppler Echocardiography Acquired in Low-Resource Settings

Pooneh Roshanitabrizi[1](✉), Holger R. Roth[2], Alison Tompsett[3], Athelia Rosa Paulli[1], Kelsey Brown[3], Joselyn Rwebembera[4], Emmy Okello[4], Andrea Beaton[5], Craig Sable[3], and Marius George Linguraru[1,6]

[1] Sheikh Zayed Institute for Pediatric Surgical Innovation, Children's National Hospital, Washington, DC, USA
proshnani2@childrensnational.org
[2] NVIDIA Corporation, Santa Clara, CA, USA
[3] Division of Cardiology, Children's National Hospital, Washington, DC, USA
[4] Uganda Heart Institute, Kampala, Uganda
[5] Department of Pediatric Cardiology, Cincinnati Children's Hospital Medical Center, Cincinnati, OH, USA
[6] Departments of Radiology and Pediatrics, School of Medicine and Health Sciences, George Washington University, Washington, DC, USA

Abstract. Rheumatic heart disease (RHD) is a common medical condition in children in which acute rheumatic fever causes permanent damage to the heart valves, thus impairing the heart's ability to pump blood. Doppler echocardiography is a popular diagnostic tool used in the detection of RHD. However, the execution of this assessment requires the work of skilled physicians, which poses a problem of accessibility, especially in low-income countries with limited access to clinical experts. This paper presents a novel, automated, deep learning-based method to detect RHD using color Doppler echocardiography clips. We first homogenize the analysis of ungated echocardiograms by identifying two acquisition views (parasternal and apical), followed by extracting the left atrium regions during ventricular systole. Then, we apply a model ensemble of multi-view 3D convolutional neural networks and a multi-view Transformer to detect RHD. This model allows our analysis to benefit from the inclusion of spatiotemporal information and uses an attention mechanism to identify the relevant temporal frames for RHD detection, thus improving the ability to accurately detect RHD. The performance of this method was assessed using 2,136 color Doppler echocardiography clips acquired at the point of care of 591 children in low-resource settings, showing an average accuracy of 0.78, sensitivity of 0.81, and specificity of 0.74. These results are similar to RHD detection conducted by expert clinicians and superior to the state-of-the-art approach. Our novel model thus has the potential to improve RHD detection in patients with limited access to clinical experts.

Keywords: Classification · Color Doppler echocardiography · Deep learning · Multi-view learning · Rheumatic heart disease

L. Wang et al. (Eds.): MICCAI 2022, LNCS 13431, pp. 602–612, 2022.
https://doi.org/10.1007/978-3-031-16431-6_57

1 Introduction

Rheumatic heart disease (RHD), a consequence of heart valve impairment caused by acute rheumatic fever, is a common treatable condition in young children; however, its late detection and treatment can lead to heart failure or death, making it a big concern in low- and middle-income countries with limited access to specialized healthcare facilities [1, 2]. Color Doppler echocardiography is a popular test for RHD screening in children due to its safety, high speed, and cost-effectiveness [3]. In screening sonography, RHD often presents as mitral regurgitation (MR), which is blood backflow into the left atrium during ventricular systole/contraction [3]. While high-resource ultrasound devices are equipped with electrocardiogram (ECG) gating, which helps the determination of ventricular systole, hand-held ultrasound machines used in low-resource settings do not include this property.

The World Heart Federation (WHF) has established echocardiographic criteria to detect RHD based on the morphological and functional analysis of the heart valves. Using these criteria, RHD can be categorized into three groups: borderline, definite, and severe [3]. Whereas the WHF grading system is standard for RHD detection, it is complex and requires input from expert cardiologists [4, 5]. Although simplified grading systems have been proposed to detect RHD, these methods need further validation for clinical applications [6–8].

Imaging methods used texture analysis to assess the MR severity from echocardiograms [9, 10]. However, these methods did not detect RHD. Several other studies proposed to identify RHD from heart sound data [11, 12]. However, auscultation shows lower sensitivity than echocardiography for RHD detection [13]. Recently, convolutional neural networks (CNNs) have shown great success in several facets of automatic echocardiogram analysis [14], including view classification [15, 16], cardiac segmentation [17], and diagnosis of heart disease [18]. The state-of-the-art method for RHD detection was presented in [19]. First, a 3D CNN was used to detect RHD based on the first 16 frames of each multi-view color Doppler and B-mode echocardiogram. Then, a supervised meta-classifier was applied to aggregate the prediction results.

Recent computer vision studies have shown that fusing information from multiple views can be beneficial for improving the ability of a model to make decisions. Seeland and Mäder [20] showed that the integration of visual information through the network outperformed the fusion of classification scores by post-processing. Su et al. [21] used multiple views of a 3D object into a single deep learning model to recognize its shape. Later, Chen et al. [22] developed a multi-view vision Transformer to recognize a 3D object. Transformer, originally developed by Vaswani et al. [23], includes a self-attention mechanism in the structure of a deep learning model, which has shown great potential in many computer vision tasks [24].

We hypothesize that RHD can be accurately detected using a simplified imaging protocol based only on ungated color Doppler echocardiograms acquired at the point of care. To the best of our knowledge, the automatic detection of RHD using only color Doppler ultrasound has not been done before. In this paper, we present a complete framework for automatic RHD detection using two-view color Doppler echocardiograms (parasternal long axis [PLAXC] and apical 4-chamber [A4CC]). Our framework consists of two stages: (1) echocardiogram homogenization and (2) RHD detection using the integration of two multi-view deep learning models (3D CNNs and Transformer). Our contributions include: (1) harmonization of highly variable imaging data, (2) model ensemble of 3D CNNs and Transformer for RHD detection, (3) early fusion of visual information obtained from two views through the deep learning models, and (4) embedding spatiotemporal information with an attention mechanism to improve the accuracy of RHD detection. Early RHD detection outside of elite healthcare systems has tremendous potential to improve the quality of life and even save the lives of children in low-resource settings.

2 Materials

We acquired 2,136 color Doppler echocardiograms in video format (IRB approved) from 591 children (338 females; 253 males; mean age 12 ± 3 years; ranging from 5 to 18 years), who were examined for RHD detection in low resource settings. The data was acquired from at least two different views (PLAXC and A4CC) using a VIVID Q or VIVID IQ low-cost portable echocardiography machine (GE Milwaukee, WI) with a 5 MHz transducer. The data had an average image size of 597 × 823 pixels with a pixel resolution ranging between 0.1 and 0.4 mm. A board of expert cardiologists designated 250/591 cases as normal (no RHD) and detected RHD in 341/591 of the cases (63 definite, 260 borderline, and 18 severe RHD), as shown in Fig. 1.

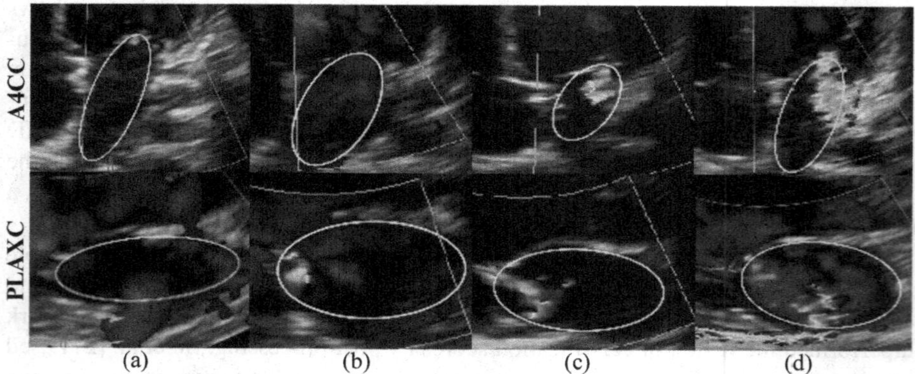

Fig. 1. Examples of the localized atrium regions (ellipsoids) on two views (A4CC and PLAXC), synchronized by the identified ventricular systole duration; (a) a normal case and patients with (b) borderline, (c) definite, and (d) severe RHD

3 Methods

Figure 2 illustrates an overview of the proposed method, including (1) echocardiogram homogenization and (2) detection of RHD.

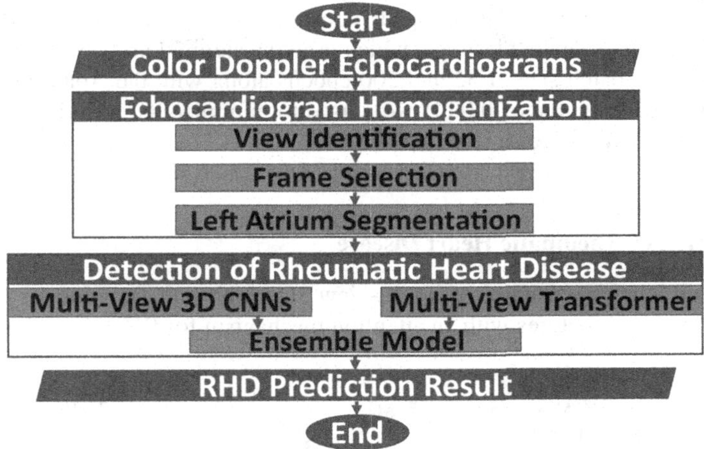

Fig. 2. Flowchart of the method proposed for RHD detection

3.1 Echocardiogram Homogenization

Echocardiogram homogenization was performed to standardize the image information that was relevant to RHD detection, including (1) view identification, (2) frame selection during ventricle contraction, and (3) left atrium segmentation.

View Identification. A4CC and PLAXC are the standard views to visualize mitral inflow/outflow for regurgitation. To retrieve these views from the multitude of collected data for each patient, we passed the first frame of each video stream (representative frame) through a deep learning classification model. The model included: (1) a ResNet-50 CNN with an input image size of 256 × 256 pixels × 3 color channels, (2) a 7 × 7 average pooling layer, (3) a fully connected layer of 512 units with rectified linear unit (ReLu) activation function, and (4) a final output layer with Softmax probability function to classify views to three categories (A4CC, PLAXC, and other). ResNet-50 CNN was pre-trained on the ImageNet datasets [25]. A size of 256 pixels was determined based on the memory required. The model was trained using the categorical cross-entropy loss function, a batch size of 32, the Adam optimization algorithm, a learning rate of 0.0001, and a total of 100 epochs.

Frame Selection During Ventricle Contraction. MR occurs during ventricle contraction/ventricular systole when the mitral valve closes. To identify frames during ventricular contraction (Fig. 1), we employed a model with the same structure as that explained in View Identification, with the following notable differences: all frames from the video were analyzed. Also, we used the binary cross-entropy loss function.

Left Atrium Segmentation. The left atrium is the region where the MR occurs. To segment the left atrium, we utilized LinkNet [26, 27] with the VGG16 encoder, pre-trained on the ImageNet datasets. In [28], LinkNet showed good accuracy for ultrasound-based kidney segmentation. The model took an input image of size 256 × 256 pixels ×

3 color channels and examined it through five resolutions. All layers employed the ReLu activation function except the last one, which used the sigmoid probability function. To train the model, a batch size of 32 and 500 epochs along with the Adam optimization algorithm and learning rate of 0.0001 were used to minimize the negative value of the Dice similarity coefficient as a loss function.

3.2 Detection of Rheumatic Heart Disease

We developed a model ensemble of multi-view 3D CNNs and Transformer to fuse the spatial and temporal features with an attention mechanism for RHD detection.

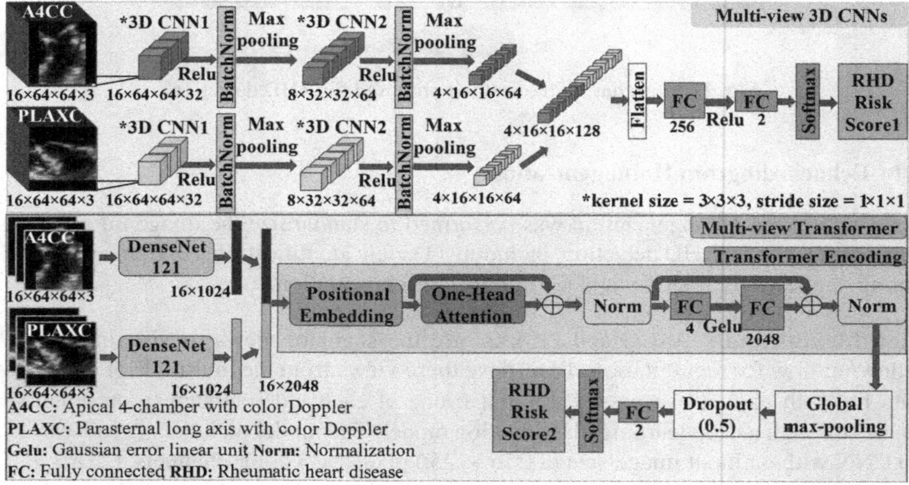

Fig. 3. Structure of the employed 3D convolutional neural networks (CNNs) and Transformer

Multi-view 3D Convolutional Neural Networks. We applied 3D CNNs to extract spatiotemporal information from two views, which fused them in a single end-to-end network (Fig. 3). Input data was created from the localized left atrium regions captured during ventricular systole from one of the two views, resampled to 64 × 64 pixels × 3 color channels × 16 frames. These sizes were determined based on memory management. Each input data was processed using two 3 × 3 × 3 convolutional filters with the ReLu activation function. Each filter was followed by a batch normalization layer and 2 × 2 × 2 max-pooling by strides of 2 in each dimension. Then, the features extracted from the two views were concatenated and processed together using two fully connected layers, including 256 units with the ReLu activation function and 2 units with the Softmax probability function. The model employed a batch size of 64 to minimize the binary cross-entropy loss function using the Adam optimization algorithm with a learning rate of 0.0001 during 350 epochs.

Multi-view Transformer. We included a Transformer to embed relevant dependencies between the frames during ventricular contraction and draw attention to the important time points (Fig. 3). First, we applied two DenseNet121 CNNs, pre-trained on the ImageNet dataset, to capture the low-level features from the two views. DenseNet121 CNN analyzed frames during ventricular systole, resampled to 16 frames with a size of 64×64 pixels $\times$ 3 color channels. The information obtained from the two views was aggregated for high-level analysis by a Transformer model, similar to the one proposed in [23]. The Transformer model first embedded the positional information into the input features to keep the sequential information. Then, the model learned relationships between frames using a self-attention module followed by a feed-forward neural network, which consisted of two layers with a Gaussian error linear unit [29]. Shortcut connections between the input and output of each block were used to directly pass the gradients through the network. After each block, a normalization layer was applied to increase the generalizability and decrease the processing time. The features encoded by the Transformer model were downsampled with a global max-pooling operation and fed to the last fully connected layer with the Softmax probability function to predict RHD. Before the last fully connected layer, we applied a dropout with a keep rate of 0.5. The model was trained using a batch size of 10 and the Adam optimization algorithm (learning rate of 0.0001) to minimize the binary cross-entropy loss function through 350 epochs.

Prediction of Rheumatic Heart Disease. The 3D CNNs analyzed all frames during ventricular systole as volume data to assess RHD, while the Transformer evaluated the data frame by frame. Since the two deep learning models analyzed the data from different perspectives, we fused their predictive scores in an ensemble model by applying the maximum voting strategy to increase the accuracy. Furthermore, when multiple acquisitions are available, as is typical in clinical practice, subjects have more than one video for each A4CC/PLAXC view, which can also include multiple ventricular systoles. We included all the ventricle contractions from the available A4CC and PLAXC views to obtain the final predictive score of RHD.

4 Experimental Results

Our method was implemented using Keras (version 2.6.0) and TensorFlow (version 2.6.2) and trained on a GeForce GTX TITAN X GPU (NVIDIA, Santa Clara, CA) with 12 GB memory. We evaluated our method using cross-validation with five folds for validation and one fold for testing and balanced the numbers of positive/negative cases. After setting aside 20% of the cases for testing, we randomly split the rest of the data into training and validation with a ratio of 80:20. We trained/validated the pre-processing steps with 95 of 591 subjects, which had expert annotated labels, and tested on the rest of 496/591 subjects. For 591 subjects, there were 8,403 videos, including 2,136 videos in A4CC or PLAXC view. For each video, the number of images-frames varied as is typical in clinical practice. The number of frames also varied with the tasks, e.g., the first frame was used to detect the view, all frames were used for frame selection, and only frames during ventricular systole were used to localize the atrium and detect RHD. The average number of training/validation images for each fold and total test images for each task were: (1) detect view: 1,578/390/6,435; (2) select frame: 9,806/2,450/84,698; (3) localize atrium: 2,512/628/29,332. RHD detection was trained on 5,108 4D data (378 cases), validated on 1,277 4D data (94 cases), and tested on 1,510 4D data (119 cases). Images from the same patient were not shared between the training, validation, and test sets. On a single GPU, the training time for view identification, frame selection, and atrium segmentation was 45 min, 100 min, and 250 min, respectively. The training time for RHD detection was 255 min, including 165 min and 90 min for the 3D CNNs and the Transformer, respectively. While not performed in this study, parallelization is feasible. We considered the number of epochs, batch size, and learning rate in the range [20:600], [10, 32, 64], and [1e-5, 1e-4, 1e-3], respectively. The best parameters were selected based on the maximum model accuracy at validation. Then, we used the best hyperparameters to retrain the model using all training datasets.

Table 1 shows a summary of quantitative results (mean and standard deviation) for each pre-processing step. Accurate performance of the identification of the image view, cardiac cycle, and atrium locations showed that our datasets were correctly homogenized, which made training easier for the RHD detection model using ungated images acquired with a manual probe. In Table 2, the quantitative RHD detection results obtained from different models and views are presented for the validation and test experiments. Integrating the information from both views increased the detection accuracy of RHD in comparison to using single views. A model ensemble of multi-view 3D CNNs and Transformer significantly improved the performance compared to each application of 3D CNNs and Transformer (p-value of 0.03 and 0.04, respectively, using the Wilcoxon signed-rank method with a significance level of 0.05).

Table 1. Quantitative results obtained for each step of ungated echocardiogram homogenization

	View identification			Frame selection			Atrium localization	
Model	ResNet-50 CNN			ResNet-50 CNN			LinkNet	
View	Accuracy	Sen	Spec	Accuracy	Sen	Spec	DC	HD (mm)
A4CC	0.99 ± 0.08	1 ± 0	0.99 ± 0.09	0.94 ± 0.22	0.96 ± 0.17	0.93 ± 0.24	0.88 ± 0.05	0.55 ± 0.43
PLAXC	0.99 ± 0.08	0.99 ± 0.07	0.99 ± 0.08	0.93 ± 0.23	0.94 ± 0.23	0.93 ± 0.23	0.9 ± 0.04	0.45 ± 0.22

Sen – Sensitivity; Spec – Specificity; DC – Dice coefficient; HD – Hausdorff distance

5 Discussion

RHD is a major concern in pediatric health, especially in low- and middle-income countries [4]. Echocardiography is an efficient exam used to diagnose RHD and trigger treatment, but it requires the input of expert cardiologists. Moreover, portable low-cost ultrasound machines used by non-experts do not use gating, which further complicates the interpretation of data. This paper proposed a fully automatic framework for the detection of RHD using only color Doppler echocardiography clips. Our approach does not require gating or ECG analysis for the detection of ventricular contraction, making it compatible with the data collected using handheld ultrasound devices. Additionally, the fully automated nature of the model will expand accessibility beyond clinical experts, allowing for RHD diagnoses to be made in areas with otherwise limited access and resources.

Table 2. Quantitative RHD detection results using different deep learning models and views

	Validation							Test
Model	3D CNNs			Transformer			Ensemble	Ensemble
View	A4CC	PLAXC	Both*	A4CC	PLAXC	Both**	Both*,**	Both
Accuracy	0.68 ± 0.37	0.71 ± 0.35	0.73 ± 0.35	0.67 ± 0.41	0.68 ± 0.36	0.7 ± 0.36	0.75 ± 0.34	0.78 ± 0.41
Sen	0.75 ± 0.35	0.82 ± 0.29	0.75 ± 0.32	0.63 ± 0.43	0.65 ± 0.36	0.75 ± 0.34	0.78 ± 0.32	0.81 ± 0.39
Spec	0.59 ± 0.37	0.56 ± 0.37	0.71 ± 0.31	0.72 ± 0.36	0.73 ± 0.35	0.63 ± 0.36	0.72 ± 0.33	0.74 ± 0.44

Sen – Sensitivity; Spec – Specificity; *p-value between the ensemble model and 3D CNNs on the validation data = 0.03; **p-value between the ensemble model and Transformer on the validation data = 0.04.

Assessment of color Doppler echocardiograms is challenging due, in part, to large variability in the pattern and timing of the MR jet, as well as the similarity between the spatial distribution of velocities at different locations. These challenges are also a reflection of the large variability between clinical ultrasound images acquired with manual probes. To overcome these challenges, our method homogenized the images by selecting consistent views and periods of the cardiac cycle. We focused on the left atrium regions during ventricular systole to provide the relevant information for RHD detection and reduce the variability in datasets. Since A4CC and PLAXC views assessed the mitral valve from different perspectives, their information was combined through the network to provide a compact structure and improve prediction accuracy. In addition, a model ensemble of multi-view 3D CNNs and Transformer was applied to assess the data using spatiotemporal information with an attention mechanism.

Previous reports showed that expert clinicians who reviewed echocardiograms based on the complex WHF criteria detected RHD with an agreement of 66–83% [4, 5]. Our method demonstrated a clinically acceptable accuracy of 0.78. It also has the potential to extend the benefits of RHD screening without requiring the input of an expert cardiologist. Additionally, our method is fully automated and reproducible. Compared to the state-of-the-art approach [19], our method detects RHD with higher accuracy (0.78 vs. 0.72) while requiring less data, i.e., only two-view color Doppler echocardiograms. The performance of our method may be affected by low-quality frames, in particular, if the left atrium is not visible. This suggests that in the absence of experts at the point of care of the patients, minimal training for data acquisition should be provided as shown in [6]. By proposing a fully automatic framework that harmonizes images without requiring ECG gating, our method can be applied for RHD detection trials in low-resource settings. In future work, we will investigate how to assess image quality before the analysis for RHD detection.

6 Conclusion

We presented an automatic deep learning-based method to detect RHD using ungated multi-view color Doppler echocardiograms. First, we homogenized the images using deep-learning approaches to reduce data variability and to focus our classifier on the image information relevant to RHD. Next, we applied multi-view deep learning models (3D CNNs and Transformer) to analyze the spatiotemporal information of frames with an attention mechanism. Finally, we employed an ensemble model to fuse the predictions from multiple ventricular contractions and to obtain the RHD risk score. Results showed that our method could detect RHD as reliably as expert clinicians and outperformed the state-of-the-art approach for detecting RHD. Our approach is compatible with low-price, handheld ultrasound devices without ECG gating, which makes it applicable for RHD screening in low-resource settings.

References

1. Marijon, E., Mirabel, M., Celermajer, D.S., Jouven, X.: Rheumatic heart disease. Lancet **379**(9819), 953–964 (2012). https://doi.org/10.1016/S0140-6736(11)61171-9
2. Liu, M., Lu, L., Sun, R., Zheng, Y., Zhang, P.: Rheumatic heart disease: Causes, symptoms, and treatments. Cell Biochem. Biophys. **72**(3), 861–863 (2015). https://doi.org/10.1007/s12013-015-0552-5
3. Reményi, B., et al.: World heart federation criteria for echocardiographic diagnosis of rheumatic heart disease–an evidence-based guideline. Nat. Rev. Cardiol. **9**(5), 297–309 (2012). https://doi.org/10.1038/nrcardio.2012.7
4. Beaton, A., et al.: The utility of handheld echocardiography for early rheumatic heart disease diagnosis: a field study. Eur. Hear. J. Cardiovasc. Imaging **16**(5), 475–482 (2015). https://doi.org/10.1093/ehjci/jeu296
5. Scheel, A., et al.: The inter-rater reliability and individual reviewer performance of the 2012 world heart federation guidelines for the echocardiographic diagnosis of latent rheumatic heart disease. Int. J. Cardiol. **328**, 146–151 (2021). https://doi.org/10.1016/j.ijcard.2020.11.013
6. Mirabel, M., et al.: Screening for rheumatic heart disease: Evaluation of a focused cardiac ultrasound approach. Circ. Cardiovasc. Imaging **8**(1), 1–8 (2015). https://doi.org/10.1161/CIRCIMAGING.114.002324
7. Nunes, M.C.P., et al.: Simplified echocardiography screening criteria for diagnosing and predicting progression of latent rheumatic heart disease. Circ. Cardiovasc. Imaging **12**(2), 1–13 (2019). https://doi.org/10.1161/CIRCIMAGING.118.007928
8. Diamantino, A., et al.: A focussed single-view hand-held echocardiography protocol for the detection of rheumatic heart disease. Cardiol. Young **28**(1), 108–117 (2018). https://doi.org/10.1017/S1047951117001676
9. Balodi, A., Anand, R.S., Dewal, M.L., Rawat, A.: Computer-aided classification of the mitral regurgitation using multiresolution local binary pattern. Neural Comput. Appl. **32**(7), 2205–2215 (2019). https://doi.org/10.1007/s00521-018-3935-x

10. Moghaddasi, H., Nourian, S.: Automatic assessment of mitral regurgitation severity based on extensive textural features on 2D echocardiography videos. Comput. Biol. Med. **73**, 47–55 (2016). https://doi.org/10.1016/j.compbiomed.2016.03.026
11. Asmare, M.H., Filtjens, B., Woldehanna, F., Janssens, L., Vanrumste, B.: Rheumatic heart disease screening based on phonocardiogram. Sensors (Basel) **21**(19), 1–17 (2021). https://doi.org/10.3390/s21196558
12. Asmare, M.H., Woldehanna, F., Janssens, L., Vanrumste, B.: Rheumatic heart disease detection using deep learning from spectro-temporal representation of un-segmented heart sounds. In: Proceedings of the Annual International Engineering in Medicine and Biology Society, pp. 168–71. IEEE, Montreal, QC, Canada (2020). https://doi.org/10.1109/EMBC44109.2020.9176544
13. Godown, J., et al.: Handheld echocardiography versus auscultation for detection of rheumatic heart disease. Pediatrics **135**(4), e939–e944 (2015). https://doi.org/10.1542/PEDS.2014-2774
14. Zamzmi, G., Hsu, L.Y., Li, W., Sachdev, V., Antani, S.: Harnessing machine intelligence in automatic echocardiogram analysis: Current status, limitations, and future directions. IEEE Rev. Biomed. Eng. **14**, 181–203 (2020). https://doi.org/10.1109/RBME.2020.2988295
15. Madani, A., Arnaout, R., Mofrad, M., Arnaout, R.: Fast and accurate view classification of echocardiograms using deep learning. NPJ Digit. Med. **1**(6), 1–8 (2018). https://doi.org/10.1038/s41746-017-0013-1
16. Kusunose, K., Haga, A., Inoue, M., Fukuda, D., Yamada, H., Sata, M.: Clinically feasible and accurate view classification of echocardiographic images using deep learning. Biomolecules **10**(5), 1–8 (2020). https://doi.org/10.3390/biom10050665
17. Chen, C., et al.: Deep learning for cardiac image segmentation: a review. Front. Cardiovasc. Med. **7**(25), 1–33 (2020). https://doi.org/10.3389/fcvm.2020.00025
18. Ghorbani, A., et al.: Deep learning interpretation of echocardiograms. NPJ Digit. Med. **3**(10), 1–10 (2020). https://doi.org/10.1038/s41746-019-0216-8
19. Martins, J.F.B.S., et al.: Towards automatic diagnosis of rheumatic heart disease on echocardiographic exams through video-based deep learning. J. Am. Med. Informatics Assoc. **28**(9), 1834–1842 (2021). https://doi.org/10.1093/JAMIA/OCAB061
20. Seeland, M., Mäder, P.: Multi-view classification with convolutional neural networks. PLoS ONE **16**(1), 1–17 (2021). https://doi.org/10.1371/journal.pone.0245230
21. Su, H., Maji, S., Kalogerakis, E., Learned-Miller, E.: Multi-view convolutional neural networks for 3D shape recognition. In: Proceedings of the International Conference on Computer Vision, pp. 1–12. IEEE, Santiago, Chile (2015). https://doi.org/10.48550/arxiv.1505.00880
22. Chen, S., Yu, T., Li, P.: MVT: Multi-view vision transformer for 3D object recognition. In: Proceedings of the 32nd British Machine Vision Conference, pp. 1–14. British Machine Vision Association, Online (2021)
23. Vaswani, A., et al.: Attention is all you need. In: Proceedings of the 31st Conference in Neural Information Processing Systems, pp. 1–11. Curran Associates Inc., Long Beach, CA, USA (2017). https://doi.org/10.48550/arxiv.1706.03762
24. Khan, S., Naseer, M., Hayat, M., Zamir, S.W., Khan, F.S., Shah, M.: Transformers in vision: A survey. ACM Comput. Surv., 1–38 (2021). https://doi.org/10.1145/3505244
25. Deng, J., Dong, W., Socher, R., Li, L.-J., Li, K., Fei-Fei, L.: ImageNet: a large-scale hierarchical image database. In: Proceedings of the Conference on Computer Vision and Pattern Recognition, pp. 248–55. IEEE, Miami, Florida, USA (2009). https://doi.org/10.1109/CVPR.2009.5206848
26. Chaurasia, A., Culurciello, E.: LinkNet: Exploiting encoder representations for efficient semantic segmentation. In: Proceedings of the Visual Communications and Image Processing, pp. 1–5. IEEE, St. Petersburg, FL, USA (2017). https://doi.org/10.1109/VCIP.2017.8305148

27. Yakubovskiy, P.: Segmentation models. GitHub repository. GitHub (2019). https://github.com/qubvel/segmentation_models. Accessed 30 Jun 2022
28. Roshanitabrizi, P., et al.: Standardized analysis of kidney ultrasound images for the prediction of pediatric hydronephrosis severity. In: Lian, C., Cao, X., Rekik, I., Xu, X., Yan, P. (eds.) MLMI 2021. LNCS, vol. 12966, pp. 366–375. Springer, Cham (2021). https://doi.org/10.1007/978-3-030-87589-3_38
29. Hendrycks, D., Gimpel, K.: Gaussian error linear units (GELUs). arXiv: Learning, 1–9 (2016)

Attention Mechanisms for Physiological Signal Deep Learning: Which Attention Should We Take?

Seong-A Park[1], Hyung-Chul Lee[1,2], Chul-Woo Jung[1,2], and Hyun-Lim Yang[1(✉)]

[1] Department of Anesthesiology and Pain Medicine, Seoul National University Hospital, Seoul, Republic of Korea
hlyang@snu.ac.kr

[2] Department of Anesthesiology and Pain Medicine, Seoul National University College of Medicine, Seoul, Republic of Korea

Abstract. Attention mechanisms are widely used to dramatically improve deep learning model performance in various fields. However, their general ability to improve the performance of physiological signal deep learning model is immature. In this study, we experimentally analyze four attention mechanisms (e.g., squeeze-and-excitation, non-local, convolutional block attention module, and multi-head self-attention) and three convolutional neural network (CNN) architectures (e.g., VGG, ResNet, and Inception) for two representative physiological signal prediction tasks: the classification for predicting hypotension and the regression for predicting cardiac output (CO). We evaluated multiple combinations for performance and convergence of physiological signal deep learning model. Accordingly, the CNN models with the *spatial* attention mechanism showed the best performance in the classification problem, whereas the *channel* attention mechanism achieved the lowest error in the regression problem. Moreover, the performance and convergence of the CNN models with attention mechanisms were better than stand-alone self-attention models in both problems. Hence, we verified that convolutional operation and attention mechanisms are complementary and provide faster convergence time, despite the stand-alone self-attention models requiring fewer parameters.

Keywords: Physiological signal · Attention · Deep learning

1 Introduction

Deep learning has dramatically improved the predictability of various phenomena based on input data of past events. For natural language processing, recurrent neural networks (RNNs) are particularly effective in analyzing time-series

Supplementary Information The online version contains supplementary material available at https://doi.org/10.1007/978-3-031-16431-6_58.

L. Wang et al. (Eds.): MICCAI 2022, LNCS 13431, pp. 613–622, 2022.
https://doi.org/10.1007/978-3-031-16431-6_58

sequences [2,6,11]. For image processing, convolutional neural networks (CNNs) that mimic human visual cognitive functions have grown in popularity [10,23,24]. However, both methods have shortcomings, such as the RNN's vanishing gradient and information loss problems [21], which limits performance, and the CNN's locality of pixel dependency [15], which make it goes deeper. To overcome these roadblocks, attention mechanisms have been used to enable neural models to pay closer attention to the most important parts of the data while ignoring irrelevant parts [7]. It gives higher weight to parts that are more relevant to produce output, and lower weights to parts that are not. Bahadnau et al. [1] introduced this idea to machine translation, resulting in superior performance over canonical RNNs. Similar concept of attention mechanism was also introduced, e.g., Luong et al. [18], and the other types of attention mechanisms were also proffered which tailored to computer vision applications [12,26,27].

In recent days, self-attention-based mechanisms had been replaced the canonical deep learning architectures and are positioned as a mainstream of AI research. Vaswani et al. [25] proposed a deep learning model that skipped the RNN and applied a self-attention mechanism by itself (so-called Transformer), achieving superior performance in machine translation and document generation. Dosoviskiy et al. [3] proposed a vision transformer, which a variant of the Transformer for image classification tasks, outperforming canonical CNNs with substantially fewer computations. Subsequently, self-attention-based deep learning was used to predict protein structures [14], compiler graph optimizers [30], and audio generation methods [13].

Consequently, the application of deep learning to physiological signal analysis has been considered [4]. For example, Hannun et al. [8] built a CNN that detects arrhythmia from electrocardiogram (ECG), showing human expert-level performance. As in other domains, attention mechanisms have been used to improve performance in physiological signal analysis. Mousavi et al. [22] proposed an attention-based CNN+RNN network to predict sleep stages from single-channel electroencephalogram. Yang et al. [28] built a CNN with attention blocks to predict stroke volume from arterial blood-pressure waveform. Unfortunately, all of these methods were tuned for specific signal types or tasks, and the best attention mechanisms for general field use for physiological signal analysis was not determined.

In this study, we experimentally determine which CNN architectures and attention mechanisms are the best for analyzing physiological signals. We focus on attention mechanisms used in computer vision, as the various features of physiological signal processes are similar, and the challenges of accurately predicting and classifying the presence of signal and object anomalies are closely related. Hence, we considered the three types of CNN models which popular for image processing and four types of attention mechanisms which suggested for computer vision tasks. Notably, a physiological signal generally has a smaller dimension than does an image, and the attention mechanism designed for computer vision may reduce efficiency by adding unnecessary calculations. Additionally, in a computer vision problem, discriminating feature detection is the main task, whereas in physiological signal analysis, not only is detecting discriminating features important, but detecting signal trends is also crucial. Therefore, for effective and efficient use of attention mechanisms, it is necessary to analyze how each attention mechanisms affects physiological signal analysis. To the best of our knowledge, this study is the first attempt to identify the most effective attention mechanism for physiological signal analysis using deep learning. We believe that our work will enable generalizable physiological signal deep learning, including the development of prototypes.

2 Methods

In this study, we analyze the efficacy of three CNN architectures (e.g., VGG-16 [23], ResNet-18 [10], and Inception-V1 [24]) with four types of attention mechanisms (e.g., squeeze-and-excitation (SE) [12], non-local (NL) [26], convolutional block attention module (CBAM) [27], and multi-head self-attention (MSA) [25]) for physiological signal deep learning. Each model uses unique feature extraction modules. VGG module includes two or three consecutive convolution layers and a pooling layer. ResNet module contains two consecutive convolution layers and a residual path. Inception module includes three convolution layers and a pooling layer in parallel. The CNN models used in this study are tailored to modality and dimension differences between image and physiological signal data. Detailed reduction criteria are described in Appendix.

The SE module is a *channel* attention mechanism. It encodes features with a squeeze part and decodes it with an excitation part to increase the quality of feature representation by considering the interdependency of channel information. The NL module is a *spatial* attention mechanism that calculates global feature information with covariance-like self-attention, which can overcome the locality of pixel dependency of CNN model, in which they fail to extract relational features between the first and last points of the input segment. CBAM is a *channel + spatial* attention mechanism. It performs channel-wise attention which is similar to SE module and performs spatial attention mechanism in that it sequentially reduces the feature size using multiple pooling and convolutional layers. The MSA module [25] is a stand-alone spatial self-attention method comprising multiple scaled dot-product attention layers in parallel, which use input data itself for queries, keys, and values. It analyzes how the given input data are self-related and helps extract enriched feature representations. The first three attention modules are harmonized to CNN models, but MSA does not use intermediate convolutional layers. A total of 13 types deep learning models (i.e., three pure CNN-based models, nine attention involved CNN-based models, and an MSA-based model) are compared.

Each model is trained to solve two representative physiological signal problems: classification for predicting intraoperative hypotension and regression for predicting intraoperative cardiac output (CO). Unexpected hypotension is a critical event that requires prompt intervention. Many risk factors have been revealed, but they do not help reduce its incidence or duration. Therefore, early prediction and prevention are crucial. Several studies have attempted to predict hypotension using deep learning [9,17]. We followed their methods of predicting hypotension events within 5 min of occurrence.

ECG, plethysmography (PPG), and demographic data were used as input variables for classification task. The output variable was binary, the positive label was defined as hypotension (mean arterial blood pressure $\leq$ 65 mmHg) lasting $>$ 1 min, and the negative label for otherwise. A pair of 20-s input segments of ECG and PPG waveforms and demographic data were extracted to predict events within 5-min. For preprocessing, we removed segments with ECG outside a range of -2 to 4.5 mV or a PPG range of zero (unitless) or less.

CO, the volume of blood being pumped by the heart per minute, is used to monitor and optimize systemic oxygen and drug delivery in critically ill or high-risk surgical patients. Especially for surgical patients, it is directly related to postoperative complications; hence, immediate treatment to keep CO levels between 4 and 8L/min during surgery may improve patient outcomes [5]. However, accurate CO monitoring requires invasive catheters, which may lead to severe complications. Some previous deep learning works attempted to predict CO using the data of invasive medical devices [19,28,29]. However, we sought a non-invasive method. Our model allows us to monitor CO for general patients by eliminating the invasiveness.

The input variables of the regression task were the same as those of the hypotension prediction model. The output variable was stroke volume index (SVI) instead of CO so that we could return a prompt result and correct the interpatient biases. Note that SVI = CO/(heart rate (HR) $\times$ body surface area). To remove outliers, only values with CO/HR between 20 and 200 mL/beat were used. The 20-s segments of input were extracted to predict immediate SVI values. Preprocessing for input segments was the same as the classification task.

3 Experiments

Training and testing datasets were obtained from VitalDB [16], an open-source physiological signal database containing perioperative physiological signs of more than 6,000 surgical patients. We extracted the required tracks for each task and conducted minimal preprocessing to determine CNN models and attention mechanisms having the best model effects using real-world physiological signal data.

To measure the effectiveness of the three attention mechanisms in each CNN model, performance variations were recorded by changing the attention fraction of the attention mechanism. Attention fraction is defined as the number of attention mechanisms divided by the number of CNN modules times 100. The 0, 50, and 100% attention fractions were considered in our experiments. Each attention mechanism was applied as the end-stage of each module. Note that for a 50% attention fraction, one attention mechanism was embedded in every two CNN modules. Notably, the MSA-based model did not include a convolutional module and do not have a standardized architecture; hence, we explored various MSA-based model types using a grid search. The search spaces for self-attention had input and output dimensionalities of 16, 32, and 64, parallel attention layers (number of heads) of two, four, six, and eight, inner-layer dimensionalities of

32, 64, and 128, and identical layers (number of layers) of one, two, and three. Through the hyperparameter search, we fixed other options as to be the best performance except number of heads and number of layers and recorded performance by changing the unit of number of heads and layers. Note that unit for number of heads increase by 2 and for number of layers increase by 1. The best setting of our MSA-based model was input and output dimensionality of 32, inner-layer dimensionality of 128. A single convolutional layer was added to the input layer of each MSA-based model to match the variable dimensionality of the self-attention models.

The input data of two tasks were two-channel (ECG and PPG) 100-Hz waveforms of 20-s. Patient demographic information was concatenated after the first fully connected layer. Detailed model architectures are illustrated in Fig. 1. It presents the final baseline model used. The green box (attention module) was replaced with the module required for each experiment.

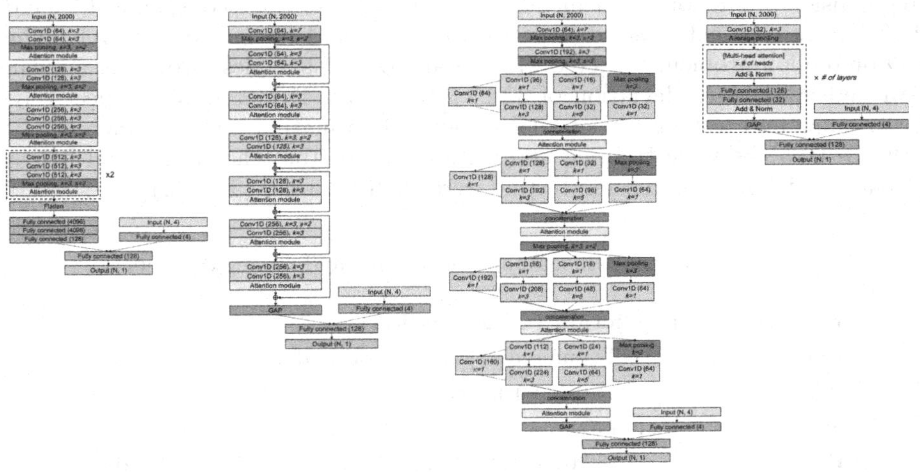

(a) VGG-based model (b) ResNet-based model (c) Inception-based model (d) MSA-based model

Fig. 1. The baseline model architectures. The models with 100% attention fraction are shown. The number in parentheses means filters or neurons. k: kernel size, s: stride.

For classification task, all models were trained with binary cross-entropy loss. The Adam optimizer was used for all models, apart from the inception-based one, which used RMSProp. The area under the receiver operating characteristics curve (AU-ROC) was used to evaluate the classification model. For the regression task, all models were trained with root mean squared error loss and the Adam optimizer. The mean absolute percentage error (MAPE) was calculated to measure model performance. Both classification and regression models were generated with a learning rate of 0.001 set to decrease by 0.1 times every

20 epochs. A batch size of 128 was used. To derive more reliable results, all models were repeated five times for training, and their performances were compared based on mean and standard deviation. We also measured the elapsed times of model convergence at given performances. The elapsed times to reach 0.7 AUROC for classification and 27.0% of MAPE for regression task were considered. All experiments, apart from those of the MSA-based models, were performed using Tensorflow 2.4.1 with Python 3.9 on a 32-core AMD EPYC 7542 processor and a single NVIDIA RTX 5000 GPU. For self-attention models, we used two NVIDIA RTX 5000 GPUs with NVLink connections to supplement GPU memory.

4 Results

Totals of 3,211 and 801 cases were extracted for hypotension and CO prediction, respectively. A randomly sampled 20% of cases were used for testing. For the hypotension prediction problem, 289,775 and 74,779 samples containing 4.74 and 4.03% positive events were collected for training and testing, respectively. The CO prediction problem collected 271,288 and 64,659 samples, providing a mean SVI and a standard deviation of 42.11 ± 13.25 and 41.71 ± 12.37, respectively, for training and testing. Patient demographic information was not different (P-value > 0.05) between training and testing, except that the weight and height of patients in the hypotension testing were slightly larger (Table 1).

Table 1. Patient demographics of training and testing datasets

Hypotension prediction (Classification)			
Characteristic	Training dataset	Testing dataset	P-value
Age, years†	61.0 (49.0–69.8)	60.0 (52.0–70.0)	0.258
Sex, # of male (%)	1409 (54.8%)	368 (57.3%)	0.278
Height, cm†	162.6 (156.3–168.7)	163.4 (157.2–170.0)	0.040
Weight, kg†	60.0 (53.4–68.6)	61.3 (53.0–68.3)	0.030
Cardiac output prediction (Regression)			
Characteristic	Training dataset	Testing dataset	P-value
Age, years†	61.0 (52.0–70.0)	62.0 (50.0–69.0)	0.660
Sex, # of male (%)	394 (61.3%)	90 (57.0%)	0.367
Height, cm†	163.8 (157.8–169.8)	162.3 (155.4–169.2)	0.178
Weight, kg†	61.5 (54.2–69.5)	61.1 (53.7–68.0)	0.478

† Data are represented as median (interquartile range).

The model performance variances of each CNN model and attention mechanism are illustrated in Fig. 2 and 3. Regarding the classification task for predicting hypotension of Fig. 2, ResNet-based model showed overall higher performance with a 50% attention fraction. ResNet-based model with NL module

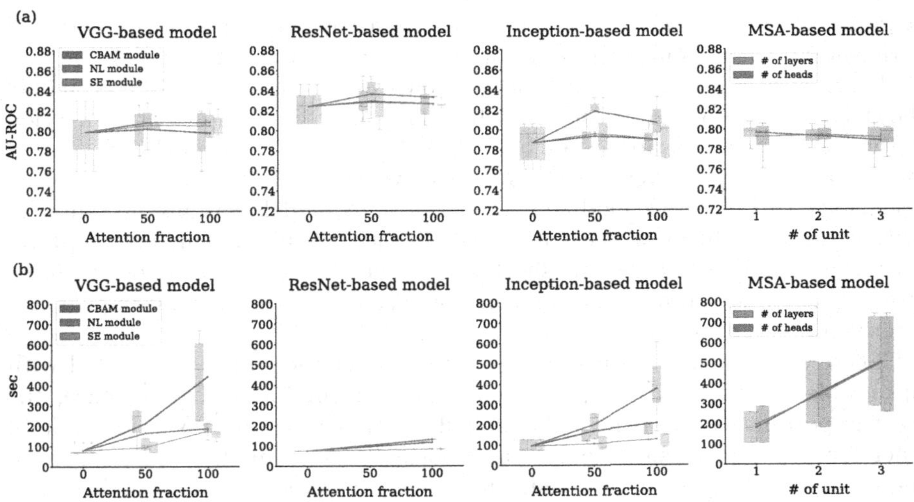

Fig. 2. Performance and convergence time in hypotension prediction problem. (a) is comparison of AU-ROC in the classification task. (b) is comparison of elapsed time to converge AU-ROC = 0.7

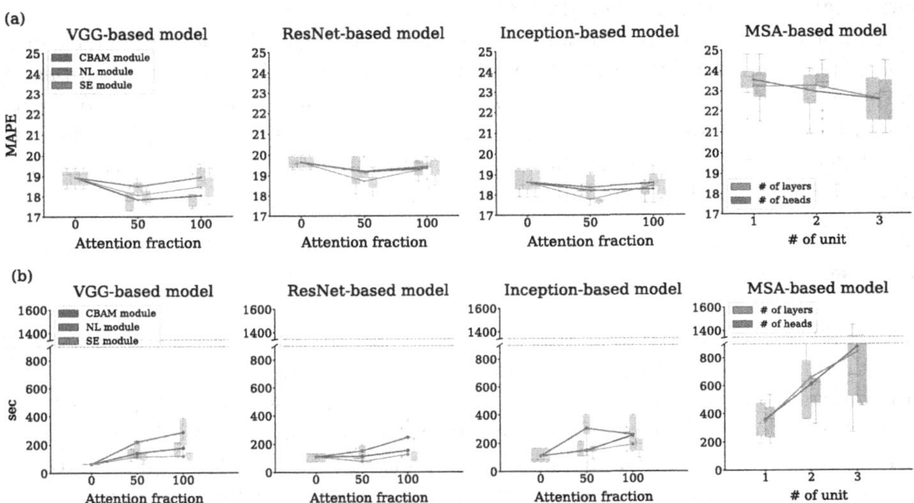

Fig. 3. Performance and convergence time in CO prediction problem. (a) is comparison of MAPE in the regression task. (b) is comparison of elapsed time to converge MAPE = 27.0%

showed the best AU-ROC of 0.854. When examining the elapsed time needed to converge 0.7 of the AU-ROC, ResNet-based model was the fastest. Additionally, the SE module added negligible additional computing overhead, but the overall CNN performance increased. There was an obvious tendency of increased performance when using *spatial* attention (i.e., NL or CBAM module).

During CO regression prediction, as shown in Fig. 3, the VGG-based model showed an overall low error. The VGG-based model with a 50% attention fraction of the CBAM module showed the best MAPE of 17.3%. However, ResNet-based model had the best convergence time to achieve 27.0% of MAPE. The computational overhead of the SE module in the three CNN models was also negligible in the regression problem, whereas it played a major role in reducing errors. Moreover, the convergence time was shortened in the ResNet-based model with SE module. There was also a clear tendency of decreasing error when using *channel* attention mechanisms (i.e., SE or CBAM module).

These experimental results can be better understood when contrasted with the problem defined. To predict hypotension within 5 min of occurrence, the most important feature is hemodynamic flow changes across 20-s of input data. Therefore, *spatial* attention plays an important role in model performance. In the prompt-CO regression problem, the waveform shape from a single beat was most important as CO is closely related to heart dynamics and the elasticity or compensation of blood vessels. Notably, each patient has a different beat pattern. Therefore, it is crucial to properly analyze the shape of the beat waveform. *Channel* attention extracts various features from the input and improves performance by helping diversify feature representations.

In both problems, the model performance was generally better when using 50% of the attention fraction rather than 100% or the fully self-attention-based model. Similar results were reported for computer vision problems [20]. We confirmed that convolution and self-attention were complementary in physiological signal deep learning, as with computer vision. Furthermore, good performance cannot be achieved by using only one building block.

5 Conclusion

In this study, we determined the best CNN and attention mechanism pairing for building deep learning models for physiological signal analysis. An attention mechanism should be selected by determining which characteristics from the raw physiological signal should be addressed to solve the problem. Convolution and attention mechanisms are complementary; therefore, there may be an ideal attention fraction for optimal performance. The ResNet-based model showed moderate performance and fast convergence in both experimental tasks. Therefore, ResNet-based model with an attention mechanism is the best candidate for prototype model. Recent studies suggest using a combined MSA with CNN for higher performance. We plan to compare physiological signal analysis performance using multiple models in a future paper.

Acknowledgement. This research was supported by a grant of the Korea Health Technology R&D Project through the Korea Health Industry Development Institute (KHIDI), funded by the Ministry of Health & Welfare, Republic of Korea (grant number

: HI21C1074); and the Korea Medical Device Development Fund grant funded by the Korea government (the Ministry of Science and ICT, the Ministry of Trade, Industry and Energy, the Ministry of Health & Welfare, Republic of Korea, and the Ministry of Food and Drug Safety) (Project Number: 202011B23)

References

1. Bahdanau, D., Cho, K., Bengio, Y.: Neural machine translation by jointly learning to align and translate. In: 3rd International Conference on Learning Representations, ICLR 2015 (2015)
2. Cho, K., et al.: Learning phrase representations using RNN encoder-decoder for statistical machine translation. In: Proceedings of the 2014 Conference on Empirical Methods in Natural Language Processing, EMNLP 2014, pp. 1724–1734 (2014)
3. Dosovitskiy, A., et al.: An image is worth 16 × 16 words: transformers for image recognition at scale. In: 9th International Conference on Learning Representations, ICLR 2021, Virtual Event, Austria, 3–7 May 2021 (2021)
4. Faust, O., Hagiwara, Y., Hong, T.J., Lih, O.S., Acharya, U.R.: Deep learning for healthcare applications based on physiological signals: a review. Comput. Methods Programs Biomed. **161**, 1–13 (2018)
5. Giglio, M., Marucci, M., Testini, M., Brienza, N.: Goal-directed haemodynamic therapy and gastrointestinal complications in major surgery: a meta-analysis of randomized controlled trials. Br. J. Anaesth. **103**(5), 637–646 (2009)
6. Graves, A., Schmidhuber, J.: Framewise phoneme classification with bidirectional LSTM networks. In: Proceedings of the 2005 IEEE International Joint Conference on Neural Networks 2005, vol. 4, pp. 2047–2052 (2005)
7. Guo, M., et al.: Attention mechanisms in computer vision: a survey. CoRR **abs/2111.07624** (2021)
8. Hannun, A.Y., Rajpurkar, P., Haghpanahi, M., Tison, G.H., Bourn, C., Turakhia, M.P., Ng, A.Y.: Cardiologist-level arrhythmia detection and classification in ambulatory electrocardiograms using a deep neural network. Nat. Med. **25**(1), 65 (2019)
9. Hatib, F., et al.: Machine-learning algorithm to predict hypotension based on high-fidelity arterial pressure waveform analysis. Anesthesiology **129**(4), 663–674 (2018)
10. He, K., Zhang, X., Ren, S., Sun, J.: Deep residual learning for image recognition. In: Proceedings of the IEEE Conference on Computer Vision and Pattern Recognition, pp. 770–778 (2016)
11. Hochreiter, S., Schmidhuber, J.: Long short-term memory. Neural Comput. **9**(8), 1735–1780 (1997)
12. Hu, J., Shen, L., Sun, G.: Squeeze-and-excitation networks. In: Proceedings of the IEEE Conference on Computer Vision and Pattern Recognition, pp. 7132–7141 (2018)
13. Huang, C.Z.A., et al.: Music transformer: generating music with long-term structure. In: International Conference on Learning Representations, ICLR 2019 (2019)
14. Jumper, J., et al.: Highly accurate protein structure prediction with AlphaFold. Nature **596**(7873), 583–589 (2021)
15. Krizhevsky, A., Sutskever, I., Hinton, G.E.: ImageNet classification with deep convolutional neural networks. In: Pereira, F., Burges, C.J., Bottou, L., Weinberger, K.Q. (eds.) Advances in Neural Information Processing Systems, vol. 25. Curran Associates, Inc. (2012). https://proceedings.neurips.cc/paper/2012/file/c399862d3b9d6b76c8436e924a68c45b-Paper.pdf

16. Lee, H.C., Jung, C.W.: Vital recorder-a free research tool for automatic recording of high-resolution time-synchronised physiological data from multiple anaesthesia devices. Sci. Rep. **8**(1), 1–8 (2018)
17. Lee, S., et al.: Deep learning models for the prediction of intraoperative hypotension. Br. J. Anasethesia **126**, 808–817 (2021)
18. Luong, T., Pham, H., Manning, C.D.: Effective approaches to attention-based neural machine translation. In: Proceedings of the 2015 Conference on Empirical Methods in Natural Language Processing, EMNLP 2015, pp. 1412–1421 (2015)
19. Moon, Y.J., et al.: Deep learning-based stroke volume estimation outperforms conventional arterial contour method in patients with hemodynamic instability. J. Clin. Med. **8**(9), 1419 (2019)
20. Park, N., Kim, S.: How do vision transformers work? In: International Conference on Learning Representations (2022)
21. Pascanu, R., Mikolov, T., Bengio, Y.: On the difficulty of training recurrent neural networks. In: International Conference on Machine Learning, pp. 1310–1318 (2013)
22. Mousavi, F.A.S., Acharya, U.R.: SleepEEGNet: automated sleep stage scoring with sequence to sequence deep learning approach. arXiv preprint arXiv:1903.02108 (2019)
23. Simonyan, K., Zisserman, A.: Very deep convolutional networks for large-scale image recognition. arXiv preprint arXiv:1409.1556 (2014)
24. Szegedy, C., et al.: Going deeper with convolutions. In: Proceedings of the IEEE Conference on Computer Vision and Pattern Recognition, pp. 1–9 (2015)
25. Vaswani, A., et al.: Attention is all you need. In: Advances in Neural Information Processing Systems, vol. 30, pp. 5998–6008 (2017)
26. Wang, X., Girshick, R., Gupta, A., He, K.: Non-local neural networks. In: Proceedings of the IEEE Conference on Computer Vision and Pattern Recognition, pp. 7794–7803 (2018)
27. Woo, S., Park, J., Lee, J.-Y., Kweon, I.S.: CBAM: convolutional block attention module. In: Ferrari, V., Hebert, M., Sminchisescu, C., Weiss, Y. (eds.) ECCV 2018. LNCS, vol. 11211, pp. 3–19. Springer, Cham (2018). https://doi.org/10.1007/978-3-030-01234-2_1
28. Yang, H.L., et al.: Development and validation of an arterial pressure-based cardiac output algorithm using a convolutional neural network: retrospective study based on prospective registry data. JMIR Med. Inform. **9**(8), e24762 (2021)
29. Yang, H.L., Lee, H.C., Jung, C.W., Kim, M.S.: A deep learning method for intraoperative age-agnostic and disease-specific cardiac output monitoring from arterial blood pressure. In: 2020 IEEE 20th International Conference on Bioinformatics and Bioengineering (BIBE), pp. 662–666 (2020)
30. Zhou, Y., et al.: Transferable graph optimizers for ML compilers. In: Advances in Neural Information Processing Systems, vol. 33, pp. 13844–13855 (2020)

Computer-Aided Tuberculosis Diagnosis with Attribute Reasoning Assistance

Chengwei Pan[1], Gangming Zhao[2,3], Junjie Fang[4,5], Baolian Qi[5], Jiaheng Liu[1], Chaowei Fang[6], Dingwen Zhang[7], Jinpeng Li[4,5(✉)], and Yizhou Yu[2,3]

[1] Institute of Artificial Intelligence, Beihang University, Beijing, China
[2] The University of Hong Kong, Hong Kong, China
[3] Deepwise AI Lab, Beijing, China
[4] HwaMei Hospital, UCAS, Ningbo, China
[5] Ningbo Institute of Life and Health Industry, UCAS, Ningbo, China
lijinpeng@ucas.ac.cn
[6] School of Artificial Intelligence, Xidian University, Xi'an, China
[7] School of Automation, Northwestern Polytechnical University, Xi'an, China

Abstract. Although deep learning algorithms have been intensively developed for computer-aided tuberculosis diagnosis (CTD), they mainly depend on carefully annotated datasets, leading to much time and resource consumption. Weakly supervised learning (WSL), which leverages coarse-grained labels to accomplish fine-grained tasks, has the potential to solve this problem. In this paper, we first propose a new large-scale tuberculosis (TB) chest X-ray dataset, namely tuberculosis chest X-ray attribute dataset (TBX-Att), and then establish an attribute-assisted weakly supervised framework to classify and localize TB by leveraging the attribute information to overcome the insufficiency of supervision in WSL scenarios. Specifically, first, the TBX-Att dataset contains 2000 X-ray images with seven kinds of attributes for TB relational reasoning, which are annotated by experienced radiologists. It also includes the public TBX11K dataset with 11200 X-ray images to facilitate weakly supervised detection. Second, we exploit a multi-scale feature interaction model for TB area classification and detection with attribute relational reasoning. The proposed model is evaluated on the TBX-Att dataset and will serve as a solid baseline for future research. The code and data will be available at https://github.com/GangmingZhao/tb-attribute-weak-localization.

Keywords: Computer-aided tuberculosis diagnosis · Weakly supervised learning · Attribute reasoning

1 Introduction

Tuberculosis (TB) is one of the most serious thoracic diseases associated with a high death rate. The early diagnosis and treatment is very important to confront the threat of TB. Computer-aided tuberculosis diagnosis (CTD) has been

C. Pan and G. Zhao—Contributed equally to this work.

L. Wang et al. (Eds.): MICCAI 2022, LNCS 13431, pp. 623–633, 2022.
https://doi.org/10.1007/978-3-031-16431-6_59

widely investigated to assist radiologists in diagnosing TB. Although deep learning algorithms have achieved stupendous success in automatic disease classification [1,24] and localization [3,11,14] for chest X-rays, their performance on CTD remains a barrier to clinical application. One obstacle lies in the absence of high quality annotated datasets and the scarcity of clinical features (attributes). In practice, such attributes have been validated to be helpful to the performance and interpretability.

In addition, the main challenges of diagnosing TB with chest X-ray images include the low visual contrast between lesions and other regions, and distortions induced by other overlapping tissues. Sometimes it is difficult for radiologists to recognize obscure diseases. For CTD, previous methods mainly concentrated on disease classification [2,19] , and several recent works have taken a step forward to detect disease regions under weak/limited supervisions. They can be grouped into two main categories: the first category [24] resorts to convolutional neural networks (CNN) trained on the classification task and output disease localization results through calculating the category activation maps [29]; the second category [11,14] uses the multiple instance learning to directly yield categorical probability maps that can be easily transformed to lesion positions. However, the performance of these methods is still far from clinical usage. Effective WSL algorithms remain the boundary to explore.

The mainstream pipeline of object detection is screening out potential proposals followed by proposal classification [21]. By stacking piles of convolutional layers, CNNs are very advantageous at extracting surrounding contextual information, however, distant relationships are still hard to capture by convolutions with small kernels. On the other hand, radiologists use attribute information for diagnosis, whereas the pure data-driven CNNs can not take this prior knowledge into account, leading to disconnections to the clinical practice.

The main contributions of this paper can be summarized as follows.

- We present a large-scale TBX-Att dataset for attribute-assisted X-ray diagnosis for TB. To the best of our knowledge, this offers the first playground to leverage the attribute information to help models detect TB areas in a weakly-supervised manner.
- We propose an effective method to fuse the attribute information and TB information. The multi-scale attribute features are extracted for each TB proposal under the guidance of a relational reasoning module and are further refined with the attribute-guided knowledge. A novel feature fusion module is devised to enhance the TB representation with its effective attribute prompt.
- The proposed method improves object detection baselines [12,21] by large margins on TBX-Att, leading to a solid benchmark for weakly supervised TB detection.

2 Related Work

2.1 Object Detection

Object detection is a widely-studied topic in both natural and medical images. It aims at localizing object instances of interest such as faces, pedestrians and disease lesions. The most famous kind of deep learning approaches for object detection is the R-CNN [7] family. The primitive R-CNN extracts proposals through selective search [23], and then predicts object bounding boxes and categories from convolution features of these proposals. Fast R-CNN [6] adopts a shared backbone network to extract proposal features via RoI pooling. Faster R-CNN [21] automatically produces object proposals from top-level features with the help of pre-defined anchors. The above methods accomplish the detection procedure through two stages, including object proposal extraction, object recognition and localization. In [12], the feature pyramid network is exploited to further improve the detection performance of Fast R-CNN and Faster R-CNN with the help of multi-scale feature maps. The other pipeline for object detection implements object localization and identification in single stage through simultaneous bounding box regression and object classification, such as YOLO [20] and SSD [15]. The RetinaNet [13] is also built upon the feature pyramid network, and uses dense box predictions during the training stage. The focal loss is proposed to cope with the class imbalance problem. The detection task has also attracted a large amount of research interest in medical images, such as lesion detection in CT scans [25] and cell detection in malaria images [9]. This paper targets at detecting diseases in chest X-ray images. Practically, we propose a *TBX-Att dataset* to exploit attribute information to enhance TB feature representations of disease proposals.

2.2 Disease Diagnosis in Chest X-ray Images

Accurately recognizing and localizing diseases in chest X-Ray images is very challenging because of low textural contrast, large anatomic variation across patients, and organ overlapping. Previous works in this field mainly focus on disease classification [1,17,24,27]. Recently, the authors in [4] propose to transfer deep models pretrained on the ImageNet dataset [5] for recognizing pneumonia in chest X-ray images. In [22], the artificial ecosystem-based optimization algorithm is used to select the most relevant features for tuberculosis recognition. Based on the category activation map [29] which can be estimated with a disease recognition network, researchers attempt to localize disease in a weakly supervised manner [24]. In [26], the triplet loss is used to facilitate the training of the disease classification model, and better performance is observed in class activation maps (CAM) estimated by the trained model. In [11,14], multiple instance learning is employed to solve the disease localization problem. In [28], a novel weakly supervised disease detection model is devised on the basis of the

DenseNet [8]. Two pooling layers including a class-wise pooling layer and a spatial pooling layer are used to transform 2-dimensional class attention maps into the final prediction scores. The performance of these methods is still far from practical usage in automatic diagnosis systems. In [16], a new benchmark is proposed to identify and search potential TB diseases, however, it lacks attribute information that has been proven to pose a great effect on the practice diagnosis of medical experts.

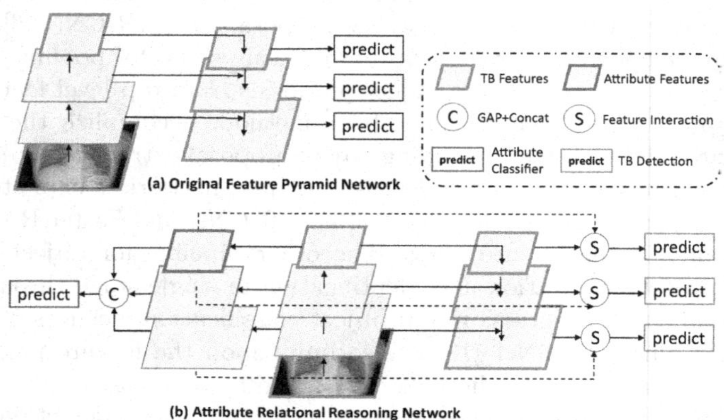

Fig. 1. Illustration of the proposed attribute relational reasoning network.

3 Methods

3.1 Overview

The method proposed in this paper mainly addresses the weakly supervised tuberculosis detection from chest X-rays. The pipeline of our proposed method is highlighted in Fig. 1, which include three parts: (1) the feature pyramid network is utilized for extracting multi-scale tuberculosis feature maps; (2) the attribute classifier, which employs the attribute supervised information to generate multi-scale attribute feature maps; (3) the feature interaction module takes the attribute feature map to establish the attribute prompt to obtain a more representative feature representation of tuberculosis. Compared with the original Feature Pyramid Network used for object detection, another branch is added to generate the corresponding multi-scale features for attribute classification, which can prompt a guidance to extract representative features for detection.

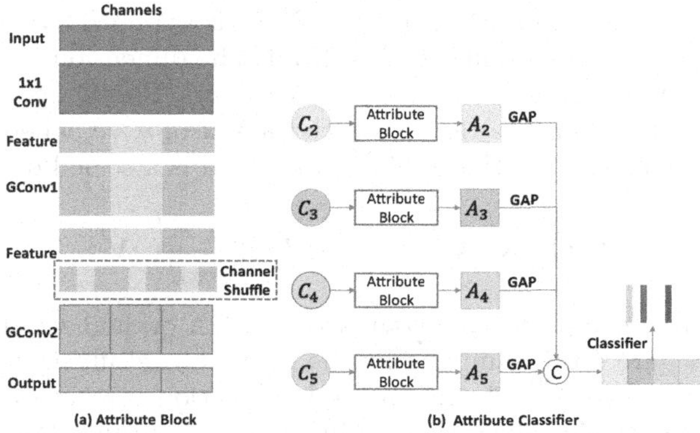

Fig. 2. Attribute feature representation. (a) Shows the process of attribute feature extraction. (b) Shows the fusion of multi-scale attribute features and the classification.

3.2 Attribute Feature Representation

The detection branch is designed to utilize multi-scale features to generate foreground detection result. The backbone network used for abstracting feature representation always generate multi-scale features. Specifically, for ResNets we use the feature activation output by each stage's last residual block. We denote the output of these blocks as $\{C_2, C_3, C_4, C_5\}$, which have strides of $\{4, 8, 16, 32\}$ pixels with respect to the input image. Correspondingly, multi-scale attribute features as shown in Fig. 2 are constructed in our method for both the task of classification and feature interaction described in the next section.

Given the input C_i from the i-th stage's last residual block, we simply attach a 1×1 convolution layer to produce a feature map having $C_a \times N_a$ channels, where N_a represents the number of kinds of attributes and C_a represents the channel dimension of each attribute feature map. Then two group convolution layers are used to generate the distinguishable feature maps for each attribute individually. For each group convolution layer, we set the number of groups as N_a and expect each group to represent a kind of attribute. For better taking advantage of information flowing across attributes' feature maps, channel shuffle operation is added in the last group convolution layer. The specially designed Attribute Block used for extracting attribute features is shown in Fig. 2(a).

After obtaining the features of each scale(A_2 to A_5 in Fig. 2(b), the Global Average Pooling operation and feature fusion by concatenation are used to get the final feature vectors used for attribute classification.

3.3 Feature Interaction

Multi-head Cross-Attention. Given two feature maps $\mathbf{X}$, $\mathbf{Y} \in R^{C\times H\times W}$, where H,W are the spatial height, width and C is the number of channels. One 1×1 convolution is used to project $\mathbf{X}$ to query embedding $\mathbf{Q}_i$, another two 1×1

convolutions are adopt similarly to project $\mathbf{Y}$ to key embedding $\mathbf{K}_i$ and value embedding $\mathbf{V}_i$, where i represents the i-th head. The dimension of embedding in each head is d, and the number of heads is $N = C/d$. The $\mathbf{Q}_i$, $\mathbf{K}_i$, $\mathbf{V}_i$ are then flattened and transposed into sequences with the size of $n \times d$, where $n = H \times W$. The output of the cross-attention (CA) in i-th head is a scaled dot-product:

$$CA_i(\mathbf{Q}_i, \mathbf{K}_i, \mathbf{V}_i) = SoftMax(\frac{\mathbf{Q}_i \mathbf{K}_i^T}{\sqrt{d}})\mathbf{V}_i. \quad (1)$$

To reduce the large computational complexity of CA on high resolution feature maps, $\mathbf{K}_i$ and $\mathbf{V}_i$ are down-sampling by the ratio of s, resulting smaller feature maps with the size of $d \times \frac{H}{s} \times \frac{W}{s}$. In our implementation, s is set to 16 in the first stage, and gradually reduce by 2 times. Finally, in a N-head attention situation, the output after multi-head cross-attention (MCA) is calculated as follows:

$$MCA(\mathbf{X}, \mathbf{Y}) = \Phi(Concat(CA_1, CA_2, ..., CA_N)) \quad (2)$$

where $\Phi(\cdot)$ is a liner projection layer that weights and aggregates the feature representation of all attention heads.

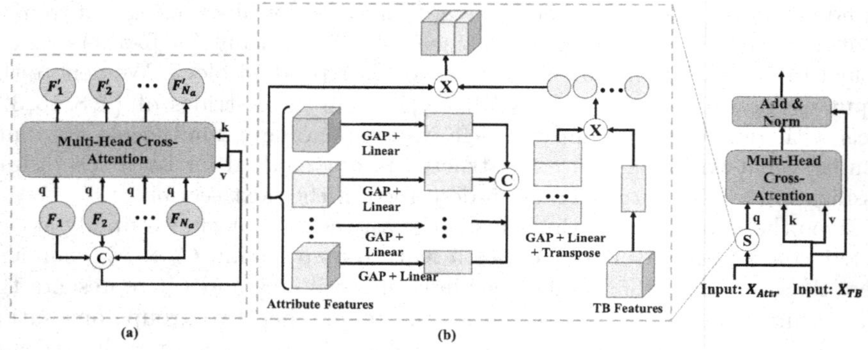

Fig. 3. Cross-attention. (a) shows the multi-head cross attention among different kinds of attribute features. (b) shows cross attention between attribute features and TB features.

Attr-Attr Attention (A^2-Attn). Given the input $\mathbf{X} = \{\mathbf{F}_1, \mathbf{F}_2, ..., \mathbf{F}_{N_a}\}$ from the i-th scale of attribute feature maps, attention modules are used to reproduce more representative features by implicitly mining the relationship among N_a kinds of attributes. To build the attention module, we firstly concatenate $\mathbf{F}_i$ channel-wisely, and then linearly projected to get $\mathbf{Y}$, which has the same channel dimension of $\mathbf{F}_i$. As shown in Fig. 3(a), the attention module takes $\{\mathbf{F}_i, i = 1, ..., N_a\}$ as queries and uses $\mathbf{Y}$ to generate key and value embeddings, so the finally aggregated feature $\mathbf{F}_i^{'}$ can be obtained by $\mathbf{F}_i^{'} = MCA(\mathbf{F}_i, \mathbf{Y})$. The operation of A^2-Attn can be applied to attribute features of each scale (A_2 to A_5 in Fig. 2(b)) to get more representative features.

Attr-TB Attention (AT-Attn). Given attribute features $\{\mathbf{F}_i, i = 1, ..., N_a\}$ and TB feature $\mathbf{F}_{tb}$, we expect to design a attribute and TB attention module (AT-Attn) to enhance the TB representation with attribute prompt. We firstly reaggregate attribute features to get $\mathbf{X}$ by taking advantage of the similarity between attribute features and TB feature. Specifically, GAP operation followed by a linear projection is performed to get attribute vectors $\{\mathbf{A}_i, i = 1, ..., N_a\}$ and TB vector $\mathbf{B}$. The similarity score s_i is calculated by $\mathbf{A}_i \cdot \mathbf{B}$, and then $\mathbf{X}$ can be obtained by $\mathbf{X} = \sum s_i \mathbf{F}_i$. As shown in Fig. 3(b), the attention module takes $\mathbf{X}$ as query and use $\mathbf{F}_{tb}$ to generate key and value embeddings. Finally, the output of the AT-attn is obtained as follows:

$$\mathbf{F}_{tb}^{'} = \mathbf{F}_{tb} + Norm(MCA(\sum s_i \mathbf{F}_i, \mathbf{F}_{tb})) \tag{3}$$

3.4 Loss Function

During the training, the classification branch and the detection branch are jointly trained, which can be seen as a type of multi-tasking learning. $Loss_{det}$ is used to optimize the detection branch, and $Loss_{cls}$ is performed to train the classification branch. For $Loss_{cls}$, we apply the most commonly used binary cross-entropy loss. For $Loss_{det}$, the same loss function is used as in [21], which contains an anchor category prediction item and a bounding box regression item. The final loss function is the sum of $Loss_{det}$ and $Loss_{cls}$, and λ is used for balancing the influence of two loss terms.

$$Loss = Loss_{det} + \lambda Loss_{cls} \tag{4}$$

3.5 Experimental Analysis

Dataset. The collection of attribute-based TB X-rays faces many difficulties [16]. The attribute information is so hard to obtain due to its unclear pathological structure. To reduce these problems, we cooperate with one of the most important hospitals in China to collect X-rays. We also combine the TBX11K dataset [16] to construct a totally new attribute-based TB dataset. The TBX11K dataset consists of 11200 X-rays, including 5000 healthy cases, 5000 sick but non-TB cases, and 1200 cases with manifestations of TB. The dataset includes 8 types of sign: Fibrotic Streaks, Pulmonary Consolidation, Diffuse Nodules, Pulmonary Cavitation, Atelectasis, Multiple Nodules, Pleural Effusion, and Pulmonary Tuberculosis. Note that all X-ray images are resized and have the resolution of about 512 × 512. Two senior radiologists participated in the annotation independently following a list of attributes. If an attribute is present, it is annotated as 1 and otherwise 0, resulting in a sparse vector of attributes. The sample was retained only if the two radiologists labeled the same. The attribute dataset consists of a total of 2000 X-ray images, including a training set of 1700 images and a validation set of 300 images. For the combination with the existing TBX11K dataset, annotations from the attribute dataset are used to train the attribute classification branch, while the TBX11K

Table 1. Performance comparison of tuberculosis diagnosis models on the TB-Xatt dataset.

Methods	Detector	Major component			Results (%) (mean ± standard deviation)		
		GroupConv	A^2-Attn	AT-Attn	F-score	Accuracy	mAP
Baseline	Two-stage Model	⊠	⊠	⊠	29.24 ± 0.76	88.08 ± 0.12	17.10 ± 0.11
SingleScale	Two-stage Model	⊠	✓	✓	33.70 ± 0.63	92.08 ± 0.71	18.20 ± 0.17
		✓	⊠	✓	32.70 ± 0.72	91.08 ± 0.12	17.20 ± 0.16
		✓	✓	⊠	32.42 ± 0.12	91.03 ± 0.64	17.09 ± 0.13
		✓	✓	✓	34.24 ± 0.63	93.15 ± 0.21	19.00 ± 0.12
MultiScale	Two-stage Model	⊠	✓	✓	33.13 ± 0.14	94.01 ± 0.21	18.79 ± 0.32
		✓	⊠	✓	34.12 ± 0.23	91.82 ± 0.33	17.92 ± 0.42
		✓	✓	⊠	34.22 ± 0.11	92.01 ± 0.24	18.01 ± 0.22
		✓	✓	✓	**39.37 ± 0.12**	**94.61 ± 0.11**	**19.20 ± 0.10**

dataset consists of TB localization information which provides the supervision for training the detection branch.

Settings. We adopt Faster RCNN as baseline, in addition, a single scale feature interaction is firstly used to explore the influence of cross attribute attention. Then a multi-scale feature interaction way is adopt to further enhance the attention function. For all experiments, each model is independently trained five times with randomly initialized weights and correspondingly 5 validations are performed for each model. The overall performance of a model is assessed with several commonly used metrics, including the mean and standard deviation of accuracy, F-score and mAP. All models are trained for 60 epoches from scratch using PyTorch [18] on NVIDIA Titan X pascal GPUs while Adam [10] being the optimizer with the initial learning rate set to 1e-3, which is reduced by a factor of 10 after every 20 epochs. The weight decay is set to 1e-4. We adopt ResNet-50 as the backbone model. The batch size is 8 on a single GPU. Different from [16], we train the two branches simultaneously rather than dividing it into two stages.

Results. On our new proposed TBX-Att dataset for weakly tuberculosis detection, we compared our attribute relational reasoning with a two-stage faster RCNN model. We focus on the impacts of three carefully designed components on both attribute classification and TB detection in experiments. As shown in Table 1, when the three components are all used, the highest performance is obtained and exceeds baseline by a large margin. Our proposed method under multi-scale setting achieves an accuracy of 94.61%, a F-score of 39.37% and a mAP of 19.20%, and outperforms the baseline by 10.13%, 6.53% and 2.1% respectively. Any pairwise combination will reduce the performance, which can demonstrate the effectiveness of each component implicitly. Some interesting phenomenon that the feature interaction can gain improvements on both tasks (not just TB detection) is found in Table 1. Moreover, the results verify that the two tasks are complementary, where the higher accuracy of attribute classification leads to higher TB detection performance. And the above findings also confirm our original intention, which takes advantage of TB attribute information to guide the effective extraction of features for TB detection.

4 Conclusion

In this paper, we present a TBX-Att dataset with attribute labels to expand the existing TBX11K dataset for weakly tuberculosis detection, considering radiologists usually use clinical features like attribute information for diagnosis. Moreover, a multi-scale feature interaction model is devised to enhance TB feature representations under the guidance of relational knowledge reasoning. We find out that with the attribute information assistance, TB classification and detection are more easily to achieve a better performance. We hope this dataset can further inspire not only researchers but also medical experts.

Acknowledgements. This work was supported by National Natural Science Foundation of China (Grants 62141605, 62106248, U21B2048) and Hong Kong Research Grants Council through General Research Fund (Grant 17207722).

References

1. Aviles-Rivero, A.I., et al.: GraphXnet chest x-ray classification under extreme minimal supervision. arXiv:1907.10085 (2019)
2. Brestel, C., Shadmi, R., Tamir, I., Cohen-Sfaty, M., Elnekave, E.: Radbot-CXR: Classification of four clinical finding categories in chest x-ray using deep learning. In: Proceedings of Medical Imaging Deep Learning, pp. 1–9 (2018)
3. Cai, J., Lu, L., Harrison, A.P., Shi, X., Chen, P., Yang, L.: Iterative attention mining for weakly supervised thoracic disease pattern localization in chest x-rays. In: Frangi, A.F., Schnabel, J.A., Davatzikos, C., Alberola-López, C., Fichtinger, G. (eds.) MICCAI 2018. LNCS, vol. 11071, pp. 589–598. Springer, Cham (2018). https://doi.org/10.1007/978-3-030-00934-2_66
4. Chouhan, V., et al.: A novel transfer learning based approach for pneumonia detection in chest x-ray images. Appl. Sci. **10**(2), 559 (2020)
5. Deng, J., Dong, W., Socher, R., Li, L.J., Li, K., Fei-Fei, L.: ImageNet: a large-scale hierarchical image database. In: 2009 IEEE Conference on Computer Vision and Pattern Recognition, pp. 248–255. IEEE (2009)
6. Girshick, R.: Fast R-CNN. In: Proceedings of IEEE International Conference on Computer Vision (ICCV), pp. 1440–1448 (2015)
7. Girshick, R., Donahue, J., Darrell, T., Malik, J.: Rich feature hierarchies for accurate object detection and semantic segmentation. In: Proceedings of the IEEE Conference on Computer Vision and Pattern Recognition, pp. 580–587 (2014)
8. Huang, G., Liu, Z., Van Der Maaten, L., Weinberger, K.Q.: Densely connected convolutional networks. In: Proceedings of IEEE Conference on Computer Vision and Pattern Recognition (CVPR), pp. 4700–4708 (2017)
9. Hung, J., Carpenter, A.: Applying faster R-CNN for object detection on Malaria images. In: Proceedings of IEEE Conference on Computer Vision and Pattern Recognition (CVPR) Workshop, pp. 56–61 (2017)
10. Kingma, D.P., Ba, J.: Adam: a method for stochastic optimization. arXiv preprint arXiv:1412.6980 (2014)

11. Li, Z., Wang, C., Han, M., Xue, Y., Wei, W., Li, L.J., Fei-Fei, L.: Thoracic disease identification and localization with limited supervision. In: Proceedings of IEEE Conference on Computer Vision and Pattern Recognition (CVPR), pp. 8290–8299 (2018)
12. Lin, T.Y., Dollár, P., Girshick, R., He, K., Hariharan, B., Belongie, S.: Feature pyramid networks for object detection. In: Proceedings of the IEEE Conference on Computer Vision and Pattern Recognition, pp. 2117–2125 (2017)
13. Lin, T.Y., Goyal, P., Girshick, R., He, K., Dollár, P.: Focal loss for dense object detection. In: Proceedings of IEEE International Conference on Computer Vision (ICCV), pp. 2980–2988 (2017)
14. Liu, J., Zhao, G., Fei, Y., Zhang, M., Wang, Y., Yu, Y.: Align, attend and locate: Chest x-ray diagnosis via contrast induced attention network with limited supervision. In: Proceedings of IEEE International Conference on Computer Vision (ICCV), pp. 10632–10641 (2019)
15. Liu, W., et al.: SSD: single shot multibox detector. In: Leibe, B., Matas, J., Sebe, N., Welling, M. (eds.) ECCV 2016. LNCS, vol. 9905, pp. 21–37. Springer, Cham (2016). https://doi.org/10.1007/978-3-319-46448-0_2
16. Liu, Y., Wu, Y.H., Ban, Y., Wang, H., Cheng, M.M.: Rethinking computer-aided tuberculosis diagnosis. In: Proceedings of the IEEE/CVF conference on computer vision and pattern recognition. pp. 2646–2655 (2020)
17. Mohd Noor, N., et al.: Texture-based statistical detection and discrimination of some respiratory diseases using chest radiograph. In: Lai, K.W., et al. (eds.) Advances in Medical Diagnostic Technology. LNB, pp. 75–97. Springer, Singapore (2014). https://doi.org/10.1007/978-981-4585-72-9_4
18. Paszke, A., et al.: PyTorch: an imperative style, high-performance deep learning library. In: Advances in Neural Information Processing Systems, pp. 8026–8037 (2019)
19. Qin, C., Yao, D., Shi, Y., Song, Z.: Computer-aided detection in chest radiography based on artificial intelligence: a survey. Biomed. Engg. OnLine **17**(1), 113 (2018)
20. Redmon, J., Divvala, S., Girshick, R., Farhadi, A.: You only look once: unified, real-time object detection. In: Proceedings of the IEEE Conference on Computer Vision and Pattern Recognition, pp. 779–788 (2016)
21. Ren, S., He, K., Girshick, R., Sun, J.: Faster R-CNN: towards real-time object detection with region proposal networks. In: Advances in Neural Information Processing Systems, pp. 91–99 (2015)
22. Sahlol, A.T., Abd Elaziz, M., Tariq Jamal, A., Damaševičius, R., Farouk Hassan, O.: A novel method for detection of tuberculosis in chest radiographs using artificial ecosystem-based optimisation of deep neural network features. Symmetry **12**(7), 1146 (2020)
23. Uijlings, J.R., Van De Sande, K.E., Gevers, T., Smeulders, A.W.: Selective search for object recognition. Int. J. Comput. Vis. **104**(2), 154–171 (2013)
24. Wang, X., Peng, Y., Lu, L., Lu, Z., Bagheri, M., Summers, R.M.: Chestx-ray8: hospital-scale chest x-ray database and benchmarks on weakly-supervised classification and localization of common thorax diseases. In: Proceedings of the IEEE Conference on Computer Vision and Pattern Recognition (CVPR), pp. 2097–2106 (2017)
25. Yan, K., Wang, X., Lu, L., Summers, R.M.: DeepLesion: automated mining of large-scale lesion annotations and universal lesion detection with deep learning. J. Med. Imaging **5**(3), 036501 (2018)
26. Zhang, C., Chen, F., Chen, Y.Y.: Thoracic disease identification and localization using distance learning and region verification. arXiv:2006.04203 (2020)

27. Zhao, G., Feng, Q., Chen, C., Zhou, Z., Yu, Y.: Diagnose like a radiologist: hybrid neuro-probabilistic reasoning for attribute-based medical image diagnosis. IEEE Trans. Pattern Anal. Mach. Intell. 1 (2021)
28. Zhou, B., Li, Y., Wang, J.: A weakly supervised adaptive densenet for classifying thoracic diseases and identifying abnormalities. arXiv:1807.01257 (2018)
29. Zhou, B., Khosla, A., Lapedriza, A., Oliva, A., Torralba, A.: Learning deep features for discriminative localization. In: Proceedings of the IEEE Conference on Computer Vision and Pattern Recognition, pp. 2921–2929 (2016)

Multimodal Contrastive Learning for Prospective Personalized Estimation of CT Organ Dose

Abdullah-Al-Zubaer Imran[1(✉)], Sen Wang[1], Debashish Pal[2], Sandeep Dutta[2], Evan Zucker[1], and Adam Wang[1]

[1] Stanford University, Stanford, CA 94305, USA
aimran@stanford.edu
[2] GE Healthcare, Waukesha, WI 53188, USA

Abstract. The increasing frequency of computed tomography (CT) examinations has sparked development of dose reduction techniques to reduce the radiation dose to patients. Optimal dose while maintaining image quality can be achieved through accurate and realistic dose estimates. Unfortunately, existing dosimetric measures are either prohibitively slow or heavily reliant on absorbed dose within a cylindrical phantom, thereby ignoring the impact of patient anatomy and organ radiosensitivity on effective dose. We propose a novel deep learning-based patient-specific CT organ dose estimation method namely, multimodal contrastive learning with Scout images (Scout-MCL). Our proposed Scout-MCL gives accurate and realistic dose estimates in real-time and prospectively, by learning from multi-modal information leveraging image (lateral and frontal scouts) and profile (patient body size). Additionally, the incorporation of an accurately modeled tube current modulation (TCM) enables Scout-MCL to learn realistic dose variations. We evaluate our proposed method on a scout-CT paired scan dataset and show its effectiveness on predicting diverse TCM doses.

Keywords: Computed tomography · Dosimetry · Scout · Tube current modulation · Contrastive learning

1 Introduction

Computed tomography (CT) has been one of the most frequent and useful imaging modalities to image-based disease diagnosis and prognosis as well as to treatment and surgery planning [23]. The clinical utility has considerably increased especially with numerous advances—improvements in rotation time and detector size, helical scanning, advanced reconstruction, energy-integrating detector to photon-counting detector [15,22]. Despite all the successes, the exposure of patients to ionizing radiation remains a major concern. Although radiation dose can be reduced to comply with the so called as low as reasonably achievable (ALARA) principle, it causes problems, such as image noise and artifacts. Therefore, it is extremely important to minimize dose while maintaining diagnostic

L. Wang et al. (Eds.): MICCAI 2022, LNCS 13431, pp. 634–643, 2022.
https://doi.org/10.1007/978-3-031-16431-6_60

value (maximized image quality). Considering the dose reduction objective, optimal CT acquisition requires an accurate dose estimate. Dose quantities, such as CT dose index (CTDIvol) do not represent patient dose, ignoring the anatomical variability from patient to patient. Patient water-equivalent diameter D_w provides information about the patient body size and anatomical variability. Size-specific dose estimate such as [11,16] still do not consider the radio-sensitivity of different organs and imaging tasks (organs-of-interest). Instead, organ-level dose optimization is highly desirable. Moreover, tube current modulation (TCM) in a CT scan enables automatic exposure control (AEC), lowering tube current (mA) without compromising image quality. While organ doses are directly related to TCM, previous work on dose calculations was performed for constant-current rotational exposures [12]. Considering the relationship between TCM and organ dose, realistic dose estimates should take TCM into account.

CT Dose Estimation: Despite many different efforts, CT dose estimation has remained an open and interesting problem to solve. Monte Carlo (MC) simulation is the gold standard for patient dose estimation, which calculates per voxel dose deposition by simulating particle tracking of the voxels [7]. MC-GPU is a GPU implementation that provides accelerated patient-specific dose estimation for a CT scan [2]. Several studies have adopted deep learning or more specifically convolutional neural networks (CNNs) for the direct estimation of personalized CT doses [6,8,9,19]. However, while such estimates can give direct and fast dose predictions, they are retrospective and are limited to the requirement of the patient CT data itself. If we want to optimize the CT dose, then the dose prediction must be made prospectively, before doing the actual scans. The recent scout-based method [12] predicts organ doses prospectively, but it does not represent realistic variation of doses at varying tube currents.

Contrastive Learning: Self-supervised learning (SSL) is a form of unsupervised learning that enables learning visual representations without requiring human supervision [10,18]. SSL uses pretext tasks for pre-training based on the raw data and then fine-tuning on downstream tasks is performed for the final evaluation. Most recently, contrastive learning based SSL pretexts have been gaining attention. Contrastive learning learns invariant representations by organizing training samples as positives and negatives in order to maximize and minimize respectively [1,3]. A simple contrastive learning framework has been proposed which works under large batch setting [4]. Existing contrastive learning methods are heavily reliant on the availability of large scale datasets [14]. Especially in medical imaging, training data are limited making the contrastive learning methods less effective and under-explored [1]. A more fruitful approach would be to perform task-specific contrastive learning for more effectiveness and context-awareness. We leverage multi-modal data to better align the visual features in a more task-specific manner.

Contributions: In CT scans, patient water-equivalent diameter (D_w) is used as a descriptor of the patient body size and is computed from the scout images.

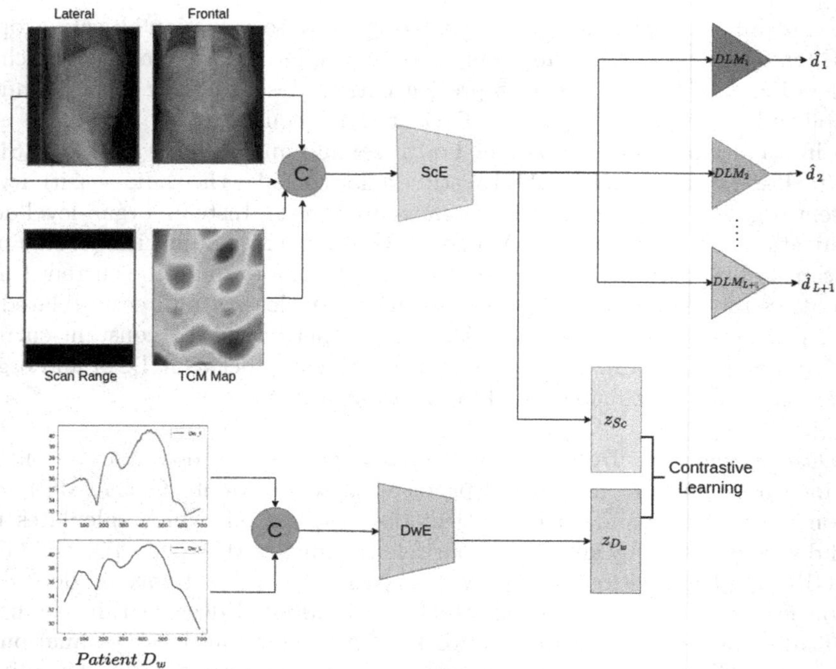

Fig. 1. Schematic of the proposed *Scout-MCL* model: Contrastive learning is performed based on the representations learned from the image modality containing scout views, scan range, and TCM map via Scout-based embedding (ScE) and the patient (body) size profile containing lateral and frontal D_w's via D_w-based embedding (DwE). A contrastive loss is calculated on projection heads of scout views (z_{Sc}) and D_w views (z_{Dw}) In the fine-tuning phase, organ-specific dose learning modules (DLMs) are added and the dose prediction losses are calculated on predicting the organ doses ($\hat{d}_1, \hat{d}_2, \ldots, \hat{d}_L$) as well as patient body dose ($\hat{d}_{L+1}$).

Since the task is to estimate patient-specific dose, D_w can bring more personalization into it. Our multi-modal contrastive learning optimizes a contrastive loss as pretext, on the representations learned from scout images and D_w inputs. This yields representation learning better aligned to the patient size and thereby improved estimation of CT organ doses. Our specific contributions in the present paper can be summarized as:

- A novel contrastive learning technique leveraging multimodal information from image space (scouts) and profile space (patient sizes)
- Real-time, patient-specific, and realistic CT organ dose estimation prospectively from scout images with TCM
- Generation of TCM map from tube current profile to calculate realistic CT organ doses

2 The Scout-MCL Model

We propose a novel multimodal contrastive learning model leveraging CT scout images and patient water-equivalent diameter (D_w). As seen in Fig. 1, Scout-MCL takes the lateral and frontal scout view images as input in order to predict the organ-specific doses in a prospective manner. In addition to the scout views, the model is provided the scan range to make it better informed about the actual scan length of the subsequent CT exam to be performed. Tube current modulation provides the control in dose distribution by changing the tube current during the CT exam. The generated TCM map via our TCM generator (see Sect. 3), is also passed to the model in order to learn the realistic variation of organ doses as well as their relationship.

While the model is trained on estimating personalized organ doses, a novel contrastive learning is employed between image (scouts, scan range, TCM map) based embedding and the embedding of the patient size profile (D_w). Since our goal is to predict the patient-specific doses, the patient body size information better informs the model, leading to more personalized doses by aligning the learned representations. Moreover, the scout (lateral and frontal) images and the patient D_w (lateral and frontal) of the same patient should learn similar representations. Therefore, these two modalities should be close to each other in the embedding space.

Contrastive Loss: Usually, a contrastive learning is formulated based on two augmented versions of the same input image [4]. Since the augmentations (positive samples) are from the same image, the model is trained on minimizing the distance between them. On the other hand, the distance is maximized for the augmented versions of other images (negative samples). In the case of multimodal contrastive learning, a similar objective could be set between two different modalities [24]. Our Scout-MCL model treats the two modality information (images and patient size profile) from the same patient as the positive examples and the modality information from other patients as the negative examples. In the Scout-MCL model, two different embedding networks are used for the scout images and D_w's. First, the representation learned from the scouts via ScE (Scout-based Embedding) is projected to an embedding from a projection head. The pretrained ResNet-18 is used as the ScE, removing the last FC layer and then using a small MLP (FC(256)-LeakyReLU-FC(64)) as the projection head, the projection embedding (z_{S_c}) is obtained. Similarly, the model uses DwE (D_w-based embedding) to project the patient body size profile to the embedding space (z_{D_w}). The DwE initialized randomly, takes the 2-channel D_w input and applies a 1D convolution with 512 filters of 1×1 followed by the same small MLP (FC(256)-LeakyReLU-FC(64)). D_w information is precomputed and stored in the scout DICOM metadata, thus it is freely available for self-supervision. The contrastive loss is calculated for pairs of positive examples (the projections (z_{S_c}) and (z_{D_w}) of the same patient) as

$$C(i,j) = -\log \frac{\exp(sim(z_{Sc_i}, z_{Dw_j})/\tau)}{\sum_{k=1}^{2M} \mathbb{1}_{[k \neq i]} \exp(sim(z_{Sc_i}, z_{Dw_k})/\tau)}, \quad (1)$$

where $sim(a, b)$ denotes cosine similarity between a and b ($a^T b/||a||||b||$), M is the minibatch size, $\mathbb{1}_{[k \neq i]} \in \{0, 1\}$ and the temperature parameter τ is set to $1/M$. Then the final contrastive loss is calculated by taking an arithmetic mean of the losses for all positive pairs in the batch based on (1)

$$\mathcal{L}_C = \frac{1}{2M} \sum_{k=1}^{M} [C(2k-1, 2k) + C(2k, 2k-1)]. \tag{2}$$

Dose Loss: Dropping the DwE and the MLPs from the contrastive learning, the pretrained ScE is used as the feature extractor. The learned representations from the scouts are passed to organ-specific dose learning modules (DLMs). Each DLM consists of two FC layers (256 and 1 neurons respectively). A LeakyReLU activation with slope 0.2 is applied after the first FC and finally by performing sigmoid, dose $\hat{d}_l$ at organ l is obtained in the normalized scale. From the reference doses d_l and the model predicted doses $\hat{d}_l$ at an organ l ($l \in \{1, 2, \ldots, L+1\}$) including the patient body dose ($d_{(L+1)}$), the individual dose losses is therefore calculated as

$$\mathcal{L}_D = \frac{1}{M} \sum_{i}^{M} ||d_l(i) - \hat{d}_l(i)||_2. \tag{3}$$

3 TCM Generation and CT Organ Dose Calculation

TCM is a feature in modern CT scanners to balance the image quality and radiation dose to the patient by modulating the tube current during a scan, such as GE's Auto mA, Philips' Dose-Right Dose Modulation, Canon's Real E.C., and Siemens' CARE Dose 4D [13]. Although exact TCM strategies are generally proprietary for various CT manufacturers, we know that the TCM realization is driven by tube current profiles, which define the tube current values at specific gantry angles and z locations [17].

In Fig. 2, we illustrate a representative tube current profile with the corresponding CT coronal view. For the adaptation to Scout-MCL that also utilizes 2D images as the input, we generated synthetic 2D TCM maps by fitting profiles from DCT (discrete cosine transform) basis images to TCM profiles as is shown in Fig. 2. The horizontal axis and vertical axis denote rotation angles and z locations of source positions, respectively. The 2D map defines all possible source positions, and the orange dotted lines refer to the source positions for a specific helical scan. An image can be expressed as the linear combination of DCT basis images. To generate the synthetic TCM map, we need to determine the weights or coefficients for corresponding DCT basis images. We sampled the profiles at the orange dotted lines from each DCT basis image and fit these profiles to the tube current profile in the left figure. This fitting gives us the coefficients for each DCT basis image. Thus, we can get the synthetic TCM map based on the coefficients and DCT basis images.

With the intention to model a realistic CT system that includes a non-isotropic source due to the anode heel effect and bowtie filtration, we enhanced

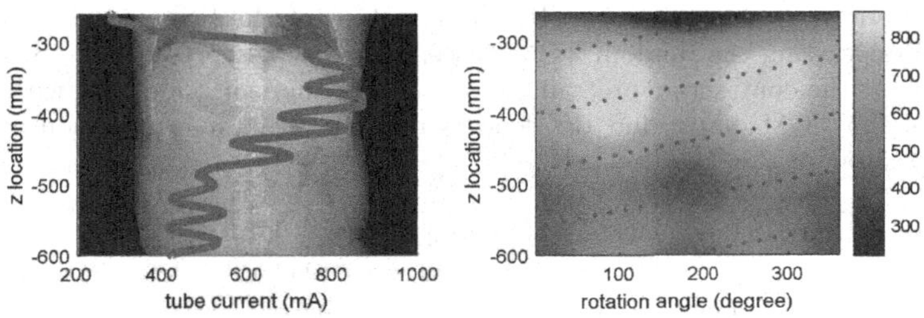

Fig. 2. (left) Tube current modulation profile overlaid on scout image, showing variation in tube current. (right) TCM map showing tube current for each z location and gantry angle. The TCM profile of a helical scan is shown with dotted orange lines. (Color figure online)

the capabilities of MC-GPU [2] for calculating individualized dose in near real time [21]. The direct product from the MC-GPU tool measures the dose contribution of each photon. Keeping in mind that the total flux of photons in one exposure at a specific projection view is proportional to the tube current (assuming the exposure time is equal for all views, as is typical in CT), organ doses under TCM can be viewed as the weighted summation of such single-view doses, where the weights are the product of a conversion factor from empirical calibration and the tube currents at these scanning positions that are defined in the tube current profile.

Hence, we further modified the MC-GPU tool for calculating patient doses at each projection view. The dose under TCM can be calculated as,

$$D = \sum_i w_i s_i, \tag{4}$$

where D denotes the dose under TCM from a CT scan, s_i is the single-view dose of i-th view, and w_i is the corresponding weight at the same view. Organ-specific reference doses d_l ($l \in L$) are then calculated by applying organ masks from the 3D segmentation tool [5] and taking the mean.

4 Experiments and Results

4.1 Data

We use a set of 175 adult body scans, primarily contrast-enhanced, acquired from Revolution CT scanners (GE Healthcare), paired with (frontal and lateral) scout views. The CT scans were acquired with a standard body protocol of 120 kVp, body bowtie, and helical pitch 0.992. Note that the scout images do not contain contrast, and there are small inconsistencies between the scouts and CT scan due to patient motion, as is realistic in clinical use. All the data were collected with IRB approval from the participating institutions. The dataset has representative coverage of patient populations, with water equivalent diameter

Table 1. Performance comparison (RMSE in mGy) of the Scout-MCL against the baseline in estimating the random TCM doses. Scout-NCL represents the Scout-MCL model without the contrastive pretraining. R denotes augmentation only with random selection of the TCM maps, and RS denotes random TCM map selection as well as on-the-fly scaling of the TCM maps)

Model	Augm.	Body	Organs-of-interest						Avg.
			Lungs	Kidneys	Liver	Bladder	Spleen	Pancreas	
Scout-Net [12]	R	1.667	3.048	4.773	3.049	6.024	3.003	4.119	3.669
Scout-NCL		0.963	**2.126**	4.933	3.222	**4.346**	2.429	3.266	3.0 41
Scout-MCL		**0.726**	2.371	**4.084**	**1.906**	4.781	**2.255**	**2.331**	**2.636**
Scout-Net [12]	RS	2.237	2.768	5.424	2.928	5.716	3.250	4.056	3.768
Scout-NCL		1.163	2.088	**3.223**	2.016	3.507	2.350	**2.548**	2.414
Scout-MCL		**1.084**	**1.746**	3.364	**1.734**	**3.096**	**2.062**	2.787	**2.267**

(D_w) [20] ranging from 25 cm to 36 cm. We randomly split the dataset into train (130) and test (45) scans.

4.2 Implementation

The input scout pairs (frontal and lateral) were registered to the scan range at 690×530 pixels and resized to 224×224 before feeding to the model. Similarly, the lateral and frontal D_w were first aligned to the scan range at 690 and resized to 224 to be fed into the model. Following Sect. 3, we generated 200 random TCM maps and split them into two equal subsets. We used the Adam optimizer with adaptive learning rate starting at $2e-3$ and L2 regularization with a weight decay of $1e-6$. We trained the model for 100 epochs with a minibatch size of 64 and then fine-tuned for 20 epochs. Our experiments include two scenarios. First, the train and test samples were augmented by randomly selecting 10 TCM maps. Then with on-the-fly 4 random scaling (40%–99%) of the TCM maps, the training data were further augmented by 4 times. The corresponding reference doses were also scaled as per the scaling percentages. As baselines, we used the Scout-Net model as well as the retrospective CT-Net [12]. We report the mean percentage relative error $\left(100\% \cdot \frac{|d_l - \hat{d}_l|}{d_l}\right)$, RMSE score in mGy $\left(\sum_{i=1}^{N} \frac{(d_l - \hat{d}_l)^2}{N}\right)$, and R^2 goodness of fit $\left(1 - \frac{\sum(d_l - \hat{d}_l)^2}{\sum(d_l - \bar{d}_l)^2}\right)$ at organ-specific ($l \in L$) as well as patient body doses ($l = L+1$). The model is implemented in Python with PyTorch and run on an *Intel Core i7 64GiB* machine across two *Titan RTX* GPUs adopting data parallelism.

4.3 Results

We evaluated the model performance on predicting the patient body dose as well as the individual organ doses. Table 1 reports the RMSE scores of the baselines Scout-Net and the Scout-MCL model without the contrastive pretraining

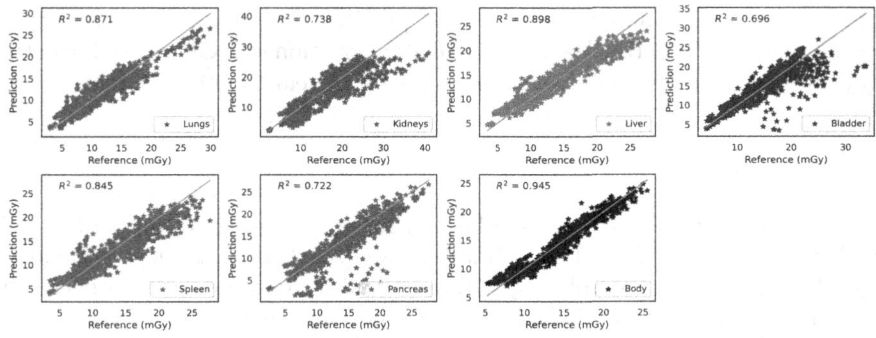

Fig. 3. Good agreement between the Scout-MCL predicted TCM doses with the reference in each of the organs as well as patient body dose.

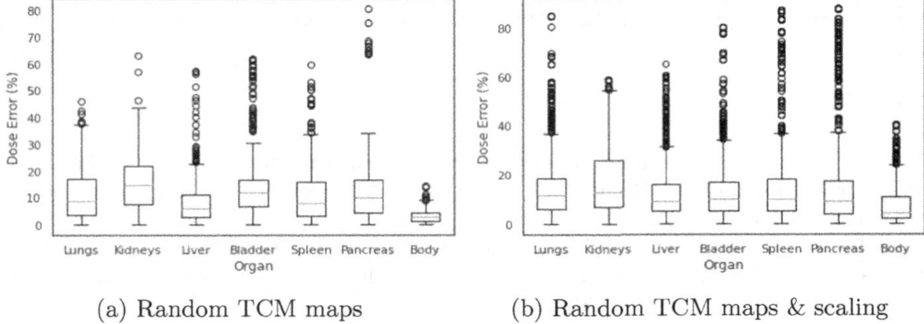

(a) Random TCM maps (b) Random TCM maps & scaling

Fig. 4. Box plots showing effectiveness (percentage relative error) of the proposed Scout-MCL model in predicting the prospective CT organ doses.

(NCL: non contrastive learning) as well as the proposed Scout-MCL model, when trained on augmented samples with 10 random TCM maps. As seen, our proposed Scout-MCL model consistently outperforms the baseline Scout-Net model in each of the dose predictions with an overall dose prediction RMSE of 2.636 mGy—approximately 28% improvement over Scout-Net and 13% improvement over the non-contrastive baseline. Except the lungs and bladder, Scout-MCL predicted patient body dose and other organs are in better agreement with the reference doses.

Similarly, the superiority of the proposed Scout-MCL model is observed in predicting the doses, when trained on augmented samples with 10 random TCM maps and on-the-fly scaling by 4 times per scan. The Scout-MCL model achieves an overall RMSE of 2.267 mGy (5.012% error). The further improvement of the model also shows the effectiveness in predicting the patient body dose and organ doses from varying TCM maps. The model predicted body dose and the organ doses are in good agreement with the reference doses as evident from the R^2 scores (Fig. 3) and the percentage errors (Fig. 4). The less agreed cases are primarily in the bladder and pancreas, possibly due to segmentation errors. Overall, Scout-MCL better learns the realistic variation of TCM doses for better personalized dose prediction from scout images, in real time (7 ms/scan).

5 Conclusions

We have proposed a novel multimodal contrastive learning technique for end-to-end, fully-automated, and real-time personalized CT organ dose estimation (Scout-MCL) which is capable of predicting realistic organ doses prospectively. Comparing against the state-of-the-art scout-based dose prediction method, our Scout-MCL has demonstrated the correct prediction of doses at six different organs, with improved generalization across varying tube current modulation. In our future work, we envision performing organ dose estimation for additional organs as well as at other CT scan locations on much larger dataset, potentially guiding the automatic exposure control to balance dose and image quality.

References

1. Azizi, S., et al.: Big self-supervised models advance medical image classification. In: Proceedings of the IEEE/CVF International Conference on Computer Vision, pp. 3478–3488 (2021)
2. Badal, A., Badano, A.: Accelerating Monte Carlo simulations of photon transport in a voxelized geometry using a massively parallel graphics processing unit. Med. Phys. **36**(11), 4878–4880 (2009)
3. Chaitanya, K., Erdil, E., Karani, N., Konukoglu, E.: Contrastive learning of global and local features for medical image segmentation with limited annotations. Adv. Neural Inf. Process. Syst. **33**, 12546–12558 (2020)
4. Chen, T., Kornblith, S., Norouzi, M., Hinton, G.: A simple framework for contrastive learning of visual representations. In: International Conference on Machine Learning, pp. 1597–1607. PMLR (2020)
5. Dutta, S., Das, B., Kaushik, S.: Assessment of optimal deep learning configuration for vertebrae segmentation from CT images. In: Medical Imaging 2019: Imaging Informatics for Healthcare, Research, and Applications, vol. 10954, pp. 298–305. SPIE (2019)
6. Fan, J., Xing, L., Dong, P., Wang, J., Hu, W., Yang, Y.: Data-driven dose calculation algorithm based on deep U-Net. Phys. Med. Biol. **65**(24), 245035 (2020)
7. Furhang, E.E., Chui, C.S., Sgouros, G.: A Monte Carlo approach to patient-specific dosimetry. Med. Phys. **23**(9), 1523–1529 (1996)
8. Götz, T.I., Schmidkonz, C., Chen, S., Al-Baddai, S., Kuwert, T., Lang, E.: A deep learning approach to radiation dose estimation. Phys. Med. Biol. **65**(3), 035007 (2020)
9. Guerreiro, F.: Deep learning prediction of proton and photon dose distributions for paediatric abdominal tumours. Radiotherapy Oncol. **156**, 36–42 (2021)
10. Hadsell, R., Chopra, S., LeCun, Y.: Dimensionality reduction by learning an invariant mapping. In: 2006 IEEE Computer Society Conference on Computer Vision and Pattern Recognition (CVPR 2006), vol. 2, pp. 1735–1742. IEEE (2006)

11. Hardy, A.J., Bostani, M., Kim, G.H.J., Cagnon, C.H., Zankl, M., McNitt-Gray, M.: Evaluating size-specific dose estimate (SSDE) as an estimate of organ doses from routine CT exams derived from monte carlo simulations. Med. Phys. **48**, 6160–6173 (2021)
12. Imran, A.-A.-Z., Wang, S., Pal, D., Dutta, S., Patel, B., Zucker, E., Wang, A.: Personalized CT organ dose estimation from scout images. In: de Bruijne, M., et al. (eds.) MICCAI 2021. LNCS, vol. 12904, pp. 488–498. Springer, Cham (2021). https://doi.org/10.1007/978-3-030-87202-1_47
13. Jadick, G., Abadi, E., Harrawood, B., Sharma, S., Segars, W.P., Samei, E.: A scanner-specific framework for simulating CT images with tube current modulation. Phys. Med. Biol. **66**(18), 185010 (2021)
14. Kinakh, V., Taran, O., Voloshynovskiy, S.: Scatsimclr: self-supervised contrastive learning with pretext task regularization for small-scale datasets. In: Proceedings of the IEEE/CVF International Conference on Computer Vision, pp. 1098–1106 (2021)
15. Lell, M.M., Kachelrieß, M.: Recent and upcoming technological developments in computed tomography: high speed, low dose, deep learning, multienergy. Invest. Radiol. **55**(1), 8–19 (2020)
16. Leng, S., Shiung, M., Duan, X., Yu, L., Zhang, Y., McCollough, C.H.: Size-specific dose estimates for chest, abdominal, and pelvic CT: effect of intrapatient variability in water-equivalent diameter. Radiology **276**(1), 184–190 (2015)
17. Li, X., Segars, W.P., Samei, E.: The impact on CT dose of the variability in tube current modulation technology: a theoretical investigation. Phys. Med. Biol. **59**(16), 4525 (2014)
18. Liu, X., et al.: Self-supervised learning: generative or contrastive. IEEE Trans. Knowl. Data Eng. (2021)
19. Maier, J., Eulig, E., Dorn, S., Sawall, S., Kachelrieß, M.: Real-time patient-specific CT dose estimation using a deep convolutional neural network. In: 2018 IEEE Nuclear Science Symposium and Medical Imaging Conference Proceedings (NSS/MIC), pp. 1–3. IEEE (2018)
20. McCollough, C., et al.: Use of water equivalent diameter for calculating patient size and size-specific dose estimates (SSDE) in CT: the report of AAPM task group 220. AAPM Rep. **2014**, 6 (2014)
21. Wang, S., Imran, A., Pal, D., Zucker, E., Wang, A.: Fast monte carlo simulation of non-isotropic x-ray source for CT dose calculation. In: Medical Physics, vol. 48. Wiley, Hoboken (2021)
22. Willemink, M.J., Persson, M., Pourmorteza, A., Pelc, N.J., Fleischmann, D.: Photon-counting CT: technical principles and clinical prospects. Radiology **289**(2), 293–312 (2018)
23. Withers, P.J., et al.: X-ray computed tomography. Nat. Rev. Methods Primers **1**(1), 1–21 (2021)
24. Yuan, X., et al.: Multimodal contrastive training for visual representation learning. In: Proceedings of the IEEE/CVF Conference on Computer Vision and Pattern Recognition, pp. 6995–7004 (2021)

RTN: Reinforced Transformer Network for Coronary CT Angiography Vessel-level Image Quality Assessment

Yiting Lu[1], Jun Fu[1], Xin Li[1], Wei Zhou[1], Sen Liu[1], Xinxin Zhang[2], Wei Wu[2], Congfu Jia[2], Ying Liu[2], and Zhibo Chen[1](✉)

[1] University of Science and Technology of China, Hefei, Anhui, China
{luyt31415,fujun,lixin666,weichou}@mail.ustc.edu.cn,
elsen@iat.ustc.edu.cn, chenzhibo@ustc.edu.cn

[2] The First Affiliated Hospital of Dalian Medical University, Dalian, Liaoning, China

Abstract. Coronary CT Angiography (CCTA) is susceptible to various distortions (e.g., artifacts and noise), which severely compromise the exact diagnosis of cardiovascular diseases. The appropriate CCTA Vessel-level Image Quality Assessment (CCTA VIQA) algorithm can be used to reduce the risk of error diagnosis. The primary challenges of CCTA VIQA are that the local part of coronary that determines final quality is hard to locate. To tackle the challenge, we formulate CCTA VIQA as a multiple-instance learning (MIL) problem, and exploit **T**ransformer-based **MIL** module (termed as T-MIL) to aggregate the multiple instances along the coronary centerline into the final quality. However, not all instances are informative for final quality. There are some quality-irrelevant/negative instances intervening the exact quality assessment(*e.g.*, instances covering only background or the coronary in instances is not identifiable). Therefore, we propose a **P**rogressive **R**einforcement learning based **I**nstance **D**iscarding module (termed as PRID) to progressively remove quality-irrelevant/negative instances for CCTA VIQA. Based on the above two modules, we propose a **R**einforced **T**ransformer **N**etwork (RTN) for automatic CCTA VIQA based on end-to-end optimization. The experimental results demonstrate that our proposed method achieves the state-of-the-art performance on the real-world CCTA dataset, exceeding previous MIL methods by a large margin.

Keywords: Image quality assessment · CCTA · Reinforced learning · Transformer

1 Introduction

Coronary Computed Tomography Angiography (CCTA) plays an important role in the diagnosis of cardiovascular diseases for providing vital visual clues.

Y. Lu and J. Fu—The Authors contribute equally to this work.

Supplementary Information The online version contains supplementary material available at https://doi.org/10.1007/978-3-031-16431-6_61.

L. Wang et al. (Eds.): MICCAI 2022, LNCS 13431, pp. 644–653, 2022.
https://doi.org/10.1007/978-3-031-16431-6_61

However, the CCTA images are easily degraded by various factors (*i.e.*, breathing motion artifacts and insufficient contrast agent dose) and contain hybrid distortions [11], which inevitably affects the subsequent analysis of doctors [5]. When artifacts appear in the coronary artery stenosis, it is difficult for doctors to diagnose stenosis [8]. To ensure accurate diagnosis, it is necessary to provide doctors with high-quality CCTA images. Therefore, there is an urgent need to develop CCTA Vessel-level Image Quality Assessment (CCTA VIQA) algorithms.

With the rapid development of machine learning, the seminal work [18] maps hand-crafted global and local features (*i.e.*, noise, contrast, misregistration scores, and un-interpretability index) of coronary arteries onto quality scores through machine learning algorithms. However, its input features are not rich since they only include four types of image characteristics, which always causes the sub-optimal performance and lacks of enough flexibility. Also, quality metric [13,14] designed for natural image are not suitable for medical image. During the dataset annotation process, the professional doctors only provide the vessel-level label when browsing the complete CT. So no position labels are provided for quality relevant regions and the key local parts that determine the vessel-level quality are hard to locate, which shows CCTA VIQA is an obvious weakly-supervised problem [24]. So the quality relationship between various local parts of coronary arteries in CCTA image can be excavated by modeling CCTA VIQA as a multiple-instance learning (MIL) problem. Therefore, we propose Transformer-based MIL backbone (T-MIL) in CCTA VIQA. Specifically, since the quality of CCTA images is only associated with the coronary arteries, we utilize the centerline tracking algorithm [21] to detect the regions of coronary arteries. Then we define 3D cubes cropped along the vessel centerline as instances. Finally, the discriminative features from multiple instances extracted by 3D convolutional neural networks are aggregated into the quality space through the latest network architecture, *i.e.*, transformer. Recently, there are various *instance* aggregators in MIL methods, like attention [6,10,15], RNN [2], sparse convolution [9], and graph [23]. Specially, transformer-based MIL frameworks [7,17,19,22] have achieved remarkable success in a broad of medical tasks, such as whole slide image classification.

Although the instances (*i.e.*, cubes) have covered all possible quality-associated contents, the quality-irrelevant contents also infiltrate the instances severely, which is detrimental for the estimation of overall quality. For instance, the quality-related cubes only take a small proportion of all cubes. According to our observation, there are three typical cases of quality-irrelevant instances *i.e.*, the instance that does not match the vessel-level label, the coronary in instances is not identifiable, and the instance contains only background. To remove these negative instances while mining the most informative instances, we propose a **P**rogressive **R**einforcement learning based **I**nstance **D**iscarding module (termed as PRID) to preserve informative instances as the inputs of the transformer. The reinforcement learning (RL) agent from PRID accepts the output feature embedding of transformer as states, and selects one instance to discard. Then we input the new instance set into T-MIL to obtain the states (both in training and testing) for the next iteration and the reward (just for training) to refine current action.

We call the T-MIL together with PRID as **R**einforced **T**ransformer **N**etwork, which is denoted as RTN. We summarize our contributions as follows.

- To our knowledge, we propose the first fully automatic CCTA VIQA algorithm RTN based on end to end optimization. We formulate the CCTA VIQA as the typical MIL problem, and introduce transformer to aggregate multiple instances and map them to final quality.
- To elide the intervention from quality-irrelevant/negative isntances, we propose a progressive reinforced learning based instance discarding strategy (*i.e.*, PRID) to mine the most informative instances for transformer network.
- Extensive experimental results reveal that our proposed RTN achieves the state-of-the-art (SOTA) performance on hospital-built CCTA dataset, exceeding previous MIL methods by a large margin.

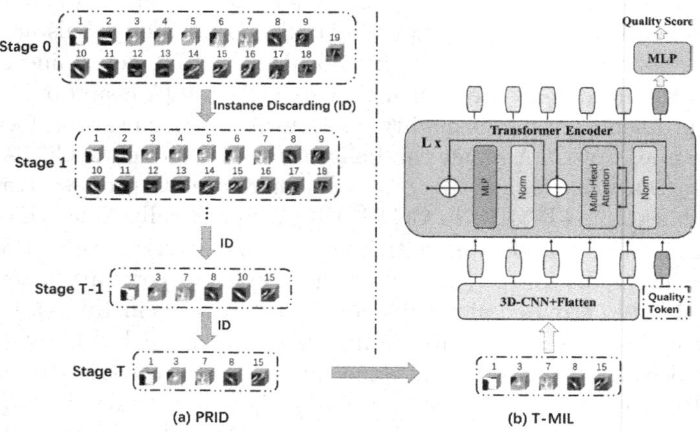

Fig. 1. Our RTN includes two modules: (a) Progressive Reinforcement Learning based Instance Discarding (PRID), (b) Transformer-based MIL Backbone (T-MIL).

2 Methods

Figure 1 depicts the overall framework of RTN for the CCTA VIQA task, which is composed of two basic modules *i.e.*, Progressive Reinforcement Learning based Instance Discarding (PRID) and Transformer-based MIL Backbone (T-MIL). Given one CCTA image, we first collect the cubes cropped along the coronary centerline as instances. Then PRID module employs a reinforcement learning (RL) agent to determine which instance should be discarded progressively. After obtaining the most informative instances, T-MIL is devoted to classifying the final vessel-level quality grade. In the following sections, we will clarify the T-MIL and PRID of our RTN from both implementation and principal perspectives.

2.1 Transformer-Based MIL

Multiple-instance learning (MIL) is a strong tool to solve weakly-supervised problem. In the definition of MIL, a set of multiple instances can be regarded as a bag and only bag-level label is provided. In our method, we define the ith 3D cube sampled on the coronary artery centerline as an instance x_i, and the whole coronary artery region is taken as a bag $\mathcal{B} = \{\boldsymbol{x_i} | 1 \leq i \leq n\}$. Then the perceptual quality y of whole coronary $\mathcal{B}$ is:

$$y(\mathcal{B}) = h(f(\boldsymbol{x_1}), f(\boldsymbol{x_2}), ..., f(\boldsymbol{x_i}), ..., f(\boldsymbol{x_n})),\ 1 \leq i \leq n \tag{1}$$

where, $x_i \in \mathbb{R}^{C_1 \times D \times H \times W}$ is the ith instance in the bag $\mathcal{B}$. T-MIL contains $f(.)$ and $h(.)$, which are separately as instance feature extractor and transformer-based aggregator. In this paper, the instance feature extractor f is composed of several 3D convolution based residual blocks [3] and flatten operation.

Transformer-Based Aggregator. To capture the long-range dependency between different instances, we employ the transformer architecture in ViT [4] as the aggregator of MIL. As shown in Fig. 1 each transformer encoder layer is consist of multi-head self-attention (MHSA) layer and feed-forward (FF) layer. We follow the ViT [4] and add the quality token $\boldsymbol{c_0}(\mathcal{B})$ to the instance token groups. The input token embeddings can be written as:

$$\boldsymbol{z_0} = [\boldsymbol{c_0}(\mathcal{B}), f(\boldsymbol{x_1}), f(\boldsymbol{x_2}), ..., f(\boldsymbol{x_i}), ..., f(\boldsymbol{x_n})],\ 1 \leq i \leq n. \tag{2}$$

In MHSA, we firstly transform instance embedding to key K, query Q and value V, and then calculate the similarity of key and query as attention weight matrix. The matrix's each item means dependencies between any pair of instances. The output of MHSA contains aggregation information, especially quality token embedding that aggregates the contribution of each instance to final vessel-level quality prediction.The full process of the lth transformer layer is as follows, in which LN is layernorm and MLP includes two fully-connected layers with a GELU non-linearity:

$$\begin{aligned} \boldsymbol{z}_l' &= MHSA(LN(\boldsymbol{z}_{l-1})), \quad l = 1, 2, ...L \\ \boldsymbol{z}_l &= MLP(LN(\boldsymbol{z}_l')) + \boldsymbol{z}_l', \quad l = 1, 2...L \end{aligned} \tag{3}$$

After feeding input token embedding into L transformer layers, we can obtain the output token embeddings $\boldsymbol{z_L} \in \mathbb{R}^{(n+1) \times D}$, in which D is the dimension of the token embedding. The first quality token embedding $\boldsymbol{c_L}(\mathcal{B}) = \boldsymbol{z_L}[0]$ is used to quality classification and following instance embedding $\boldsymbol{b_L} = \boldsymbol{z_L}[1, 2, ..., n]$ can be used as the states of PRID, in which the instance embeddings $\boldsymbol{b_L}$ can be understood as features of n instances in one vessel extracted by T-MIL.

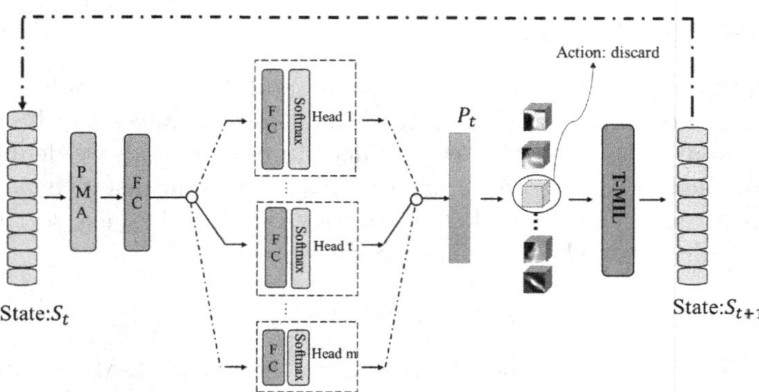

Fig. 2. The agent network of PRID, contains common Pooling by Multi-Head Attention(PMA) module and various MLP layers.

2.2 PRID

To reduce the intervention of negative instances (*e.g.*, the instance that do not match vessel-level labels or the instance contains only background), we propose to utilize reinforcement learning (RL) agent to adaptively identify them and discard them progressively [20]. Specifically, we model the process of progressively instance discarding as a Markov Decision Process (MDP) [1,12] and introduce a RL agent to obtain the optimal solution for it. The state, action, reward and agent in RL are clarified clearly as follows.

States. As shown in Fig. 2, in the tth iteration, the state $\boldsymbol{s}_t$ is defined as the output instance embedding $\boldsymbol{b}_L(t-1) \in \mathbb{R}^{(n-t+1)\times D}$ of the $(t-1)^{th}$ iteration's transformer layer, since the features captured by transformer are more representative for quality prediction.

Action. Action $\boldsymbol{a}_t$ is the instance index that is discarded within the scope of the instance set. In the t-th iteration, the action search space $\mathcal{A} = \{1, 2, ..k, ..., n-t+1\}$ is the current instances' index list. The agent's output probability vector $\boldsymbol{p}_t \in \mathbb{R}^{n-t+1}$can be regarded as the selected distribution of current instance set. Thus we can encode action as multinomial distribution sampling when training and top one sampling when testing, the selected k is equal to action: $k = sample(\boldsymbol{p}_t)$. The state $\boldsymbol{s}_t$ transforms to $\boldsymbol{s}_{t+1}$ through the action because of changes in the instance set: $\{\boldsymbol{x}_i\}_{i=1}^{n-t+1} \longrightarrow \{\boldsymbol{x}_i\}_{i=1,i\neq k}^{n-t+1}$.

Reward. The reward $\boldsymbol{r}_t$ need to reflect the effect of transforming from the state $\boldsymbol{s}_t$ to $\boldsymbol{s}_{t+1}$ due to the action. After discarding one instance, we feed new instance set into the pre-trained T-MIL and compare the new predicted result with the label to calculate the reward $(t > 1)$:

$$r_t = \begin{cases} 2, & if\ y_t = label \quad and \quad y_{t-1} = label \\ 1, & if\ y_t = label \quad and \quad y_{t-1} \neq label \\ -1, & if\ y_t \neq label \quad and \quad y_{t-1} = label \\ -2, & if\ y_t \neq label \quad and \quad y_{t-1} \neq label \end{cases} \tag{4}$$

In the first selection, the predict result y_1 need to compare with label. If the prediction is correct, give a positive reward $(+1)$, otherwise give a negative reward (-1). In the next choice, as the Eq. 4 shown, the value of reward is not only related to the accuracy of the current selection's prediction result, but also to the last selection's result. This is because in the MDP problem, the current selection (iteration) is related to the last selection (iteration).

Agent. As shown in Fig. 2, the agent in PRID receives the states from T-MIL. We first aggregate the n tokens of states into one token through PMA module $PMA(.)$ [7]. In the tth iteration, this module sets a learnable embedding $I \in \mathbb{R}^{1\times D}$ as a query, and directly regards the instance embedding $\boldsymbol{b}_L(t-1) \in \mathbb{R}^{(n-t+1)\times D}$ as key and value to calculate a attention matrix of $1 \times (n-t+1)$ dimension to gather these feature embedding. Similarly, the cross attention here is also implemented in the form of multi-head. Then we feed the fused token into the MLP head $g_t(.)$ to obtain the probability vector $\boldsymbol{p_t} \in \mathbb{R}^{n-t+1}$. Note that in t^{th} iteration, we will use t^{th} MLP head $g_t(.)$:

$$\boldsymbol{p_t} = g_t(PMA(\boldsymbol{b}_L(t-1))) \tag{5}$$

Instance Discarding Strategy. The implementation requires above two modules: PRID and T-MIL. In the first stage, we need to pre-train the T-MIL by randomly selecting $n-m$ instances from n instances. Secondly, fix the parameters of T-MIL and update the agent's parameter through m progressive selections through interaction with T-MIL. At each iteration, we can obtain the selected index (k) probability from distribution $\boldsymbol{p_t}$ and reward $\boldsymbol{r_t}$, so the training loss is

$$loss = -\sum\nolimits_{t=1}^{m} log(\boldsymbol{p_t}[k]) \times \boldsymbol{r_t} \tag{6}$$

3 Experiment

3.1 Implementation Details

Our CCTA VIQA dataset is collected with the help of a partner hospital, where the vessel-level quality labels of each CCTA image are provided by experienced imaging doctors and the resolution (*i.e.*, 512×512) of CCTA slices along axis is commonly used in the hospital. There are two quality levels in our dataset *i.e.*, "1" and "0". "1" means the CCTA image is high-quality and accepted by doctors, while "0" represents the CCTA image is low-quality and cannot be used

Table 1. Performance comparisons with state-of-the-arts on the CCTA dataset.

MIL methods	Accuracy	AUC
AttentionMIL [6]	0.7574	0.7576
MIL-RNN [2]	0.7322	0.6842
CLAM [15]	0.7761	0.7161
DSMIL [10]	0.6917	0.5378
T-MIL (ours)	**0.8036**	**0.7658**
RTN(PRID+T-MIL) (ours)	**0.8546**	**0.8461**

for diagnosis. The dataset consists of 80 CCTA scans from 40 patients in both systole and diastole, which can generate 210 coronary branches by the centerline tracking algorithm [21]. The centerline algorithm realizes the rough detection of coronary region, and the rough detection accuracy can reach 94%, which has little influence on the subsequent VIQA. Therefore, our dataset contains 210 pairs of coronary branches and vessel-level quality labels, where the ratio of label "1" and label "0" is 114/96. And we possibly plan to make this CCTA VIQA dataset public later.

We adopt the numbers of instances (*i.e.*, cubes) n in MIL as 19, which are uniformly cropped along the vessel centerline. All cubes are with the size of $20 \times 20 \times 20$ and cover the whole coronary artery branch. We also augment the data by moving the cube's center point randomly to three voxels in any direction along 6 neighborhoods as in [16]. We follow the 5-fold cross validation setting with 80% of data for training and 20% for testing in each split. Both T-MIL and PRID are implemented with Pytorch and trained on one NIVDIA 1080Ti GPU. In the training process, we first train T-MIL for 200 epoches with the batchsize 2. Then, we optimize the PRID module for 400 epoches with batchsize 2. We utilize two metrics of quality classification at vessel level to measure the effectiveness of the proposed framework: Accuracy and Area Under the Curve (AUC) scores. Moreover, the best selection of instance discarding number m is based on experiments. And m pick 14 as baseline.

Table 2. Performance comparison with different discarding numbers and discarding strategies in RTN on the CCTA dataset.

Discarding number	PRID(Accuracy/AUC)	Random(Accuracy/AUC)
4	0.7964/0.7777	0.7682/0.7409
9	0.8253/0.7674	0.8007/0.7567
14	0.8546/0.8461	0.8107/0.7994

Table 3. Performance comparison with different pooling module in agent network of RTN on the CCTA dataset and different cube size on one vessel.

Pooling module	Accuracy	AUC	Crop size	Accuracy	AUC
PMA	**0.8546**	**0.8461**	15	0.8042	0.7459
Avg Pooling	0.8443	0.8257	20	**0.8546**	0.8461
Max Pooling	0.8273	0.8198	30	0.8510	**0.8668**

3.2 Comparisons with State-of-the-arts

We compare our methods with the state-of-the-art MIL methods on our dataset, including attention-based MIL [6], RNN-based MIL [2], attention-based and cluster-based MIL [15], non-local attention based MIL [10]. In order to ensure fairness, the feature extraction process of the above methods shares the same two layers of 3D residential blocks. As shown in Table 1, transformer-based MIL exceeds the second best method CLAM [15] by 2.75%, thanks to its better long-range relationship modeling capability. Furthermore, our proposed RTN achieves the best performance, outperforming previous MIL-based methods by 7.85%, which reveals the effectiveness of our PRID. In other words, discarding quality-irrelevant instances is vital for CCTA VIQA. See supplementary material, the visualization of index distribution of discarded instance and remained instance shows that only limited instances will play a role in CCTA VIQA.

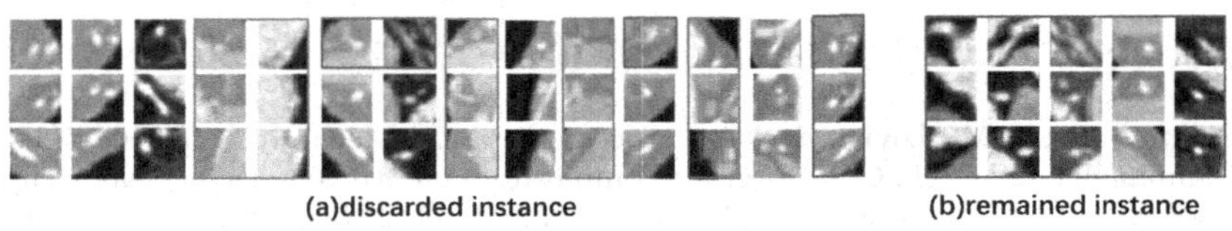

Fig. 3. Example of instance discarding with label “1”. Three rows represent views in axial, sagittal, and coronal orientations from all 3D cubes on the coronary artery, among them, the blue box is the case where the coronary in instances is not identifiable, and the red box is the case with obvious distortion. (Color figure online)

3.3 Ablation Study

In this section, we verify the effectiveness of our proposed PRID from four aspects: the number of discarding instances, discarding strategy, pooling operations and cube size. Table 2 shows the comparison results of different discarding numbers and different discarding strategies. According the results, the discarding number $m = 14$ is the best solution. This is because after iterative discarding, the five instances with the most information are retained at last, which will make it easier for network to classify, as shown in Fig. 3. We also compare the PRID with random discarding strategy in Table 2. Our PRID exceeds random discarding strategy by a large margin regardless of the discarding number, which reveals the effectiveness of our PRID on instance selection. In Table 3, we compare the different pooling operations for RL agent in PRID. We can draw a conclusion that PMA has a stronger aggregation ability to input instance embedding. This also shows that it is more explanatory to aggregate tokens through cross attention [7]. The comparison of different cube size in Table 3 shows that the cubes with small size cannot cover the whole vessel and the cubes with larger size will contain a little more quality-unrelated content.

4 Conclusion

In this paper, we present a novel Reinforced Transformer Network(RTN) model for CCTA VIQA, which contains two modules: Transformer-based MIL backbone (T-MIL) and Progressive Reinforcement learning based Instance Discarding module (PRID). T-MIL can solve the challenge that local part of coronary that determines final quality is hard to locate. Moreover, PRID can overcome the intervention from quality-irrelevant/negative instances. Compared with previous MIL methods, our RTN has achieved great improvement. In the future, we plan to adaptively select the number of discarded instances, which will continue to be improved in the later work and put into clinical use.

Acknowledgement. This work was supported in part by NSFC under Grant U1908209, 62021001 and the National Key Research and Development Program of China 2018AAA0101400.

References

1. Bellman, R.: A Markovian decision process. J. Math. Mech. **6**(5), 679–684 (1957)
2. Campanella, G., et al.: Clinical-grade computational pathology using weakly supervised deep learning on whole slide images. Nat. Med. **25**(8), 1301–1309 (2019)
3. Chen, S., Ma, K., Zheng, Y.: Med3D: transfer learning for 3D medical image analysis. arXiv preprint arXiv:1904.00625 (2019)
4. Dosovitskiy, A., et al.: An image is worth 16 x 16 words: transformers for image recognition at scale. arXiv preprint arXiv:2010.11929 (2020)
5. Ghekiere, O., et al.: Image quality in coronary CT angiography: challenges and technical solutions. Br. J. Radiol. **90**(1072), 20160567 (2017)
6. Ilse, M., Tomczak, J., Welling, M.: Attention-based deep multiple instance learning. In: International conference on machine learning, pp. 2127–2136. PMLR (2018)
7. Lee, J., Lee, Y., Kim, J., Kosiorek, A., Choi, S., Teh, Y.W.: Set transformer: a framework for attention-based permutation-invariant neural networks. In: International Conference on Machine Learning, pp. 3744–3753. PMLR (2019)
8. Leipsic, J., et al.: Adaptive statistical iterative reconstruction: assessment of image noise and image quality in coronary CT angiography. Am. J. Roentgenol. **195**(3), 649–654 (2010)
9. Lerousseau, M., Vakalopoulou, M., Deutsch, E., Paragios, N.: SparseConvMIL: sparse convolutional context-aware multiple instance learning for whole slide image classification. In: MICCAI Workshop on Computational Pathology, pp. 129–139. PMLR (2021)

10. Li, B., Li, Y., Eliceiri, K.W.: Dual-stream multiple instance learning network for whole slide image classification with self-supervised contrastive learning. In: Proceedings of the IEEE/CVF Conference on Computer Vision and Pattern Recognition, pp. 14318–14328 (2021)
11. Li, X.: Learning disentangled feature representation for hybrid-distorted image restoration. In: Vedaldi, A., Bischof, H., Brox, T., Frahm, J.-M. (eds.) ECCV 2020. LNCS, vol. 12374, pp. 313–329. Springer, Cham (2020). https://doi.org/10.1007/978-3-030-58526-6_19
12. Littman, M.L.: Reinforcement learning improves behaviour from evaluative feedback. Nature **521**(7553), 445–451 (2015)
13. Liu, J., Li, X., Peng, Y., Yu, T., Chen, Z.: SwinIQA: learned swin distance for compressed image quality assessment. In: Proceedings of the IEEE/CVF Conference on Computer Vision and Pattern Recognition, pp. 1795–1799 (2022)
14. Liu, J., Zhou, W., Xu, J., Li, X., An, S., Chen, Z.: LIQA: lifelong blind image quality assessment. arXiv preprint arXiv:2104.14115 (2021)
15. Lu, M.Y., Williamson, D.F., Chen, T.Y., Chen, R.J., Barbieri, M., Mahmood, F.: Data-efficient and weakly supervised computational pathology on whole-slide images. Nature Biomed. Eng **5**(6), 555–570 (2021)
16. Ma, X., Luo, G., Wang, W., Wang, Kuanquan: Transformer network for significant stenosis detection in CCTA of coronary arteries. In: de Bruijne, M., et al. (eds.) MICCAI 2021. LNCS, vol. 12906, pp. 516–525. Springer, Cham (2021). https://doi.org/10.1007/978-3-030-87231-1_50
17. Myronenko, A., Xu, Z., Yang, D., Roth, H.R., Xu, D.: Accounting for dependencies in deep learning based multiple instance learning for whole slide imaging. In: de Bruijne, M., et al. (eds.) MICCAI 2021. LNCS, vol. 12908, pp. 329–338. Springer, Cham (2021). https://doi.org/10.1007/978-3-030-87237-3_32
18. Nakanishi, R., et al.: Automated estimation of image quality for coronary computed tomographic angiography using machine learning. Eur. Radiol. **28**(9), 4018–4026 (2018). https://doi.org/10.1007/s00330-018-5348-8
19. Shao, Z., et al.: TransMIL: transformer based correlated multiple instance learning for whole slide image classification. In: Advances in Neural Information Processing Systems 34 (2021)
20. Tang, Y., Tian, Y., Lu, J., Li, P., Zhou, J.: Deep progressive reinforcement learning for skeleton-based action recognition. In: Proceedings of the IEEE conference on computer vision and pattern recognition, pp. 5323–5332 (2018)
21. Wolterink, J.M., van Hamersvelt, R.W., Viergever, M.A., Leiner, T., Išgum, I.: Coronary artery centerline extraction in cardiac CT angiography using a CNN-based orientation classifier. Med. Image Anal. **51**, 46–60 (2019)
22. Yu, S., et al.: MIL-VT: multiple instance learning enhanced vision transformer for fundus image classification. In: de Bruijne, M., et al. (eds.) MICCAI 2021. LNCS, vol. 12908, pp. 45–54. Springer, Cham (2021). https://doi.org/10.1007/978-3-030-87237-3_5
23. Zhao, Y., et al.: Predicting lymph node metastasis using histopathological images based on multiple instance learning with deep graph convolution. In: Proceedings of the IEEE/CVF Conference on Computer Vision and Pattern Recognition, pp. 4837–4846 (2020)
24. Zhou, Z.H.: A brief introduction to weakly supervised learning. Natl. Sci. Rev. **5**(1), 44–53 (2018)

A Comprehensive Study of Modern Architectures and Regularization Approaches on CheXpert5000

Sontje Ihler[1(✉)], Felix Kuhnke[2], and Svenja Spindeldreier[1]

[1] Institute of Mechatronic Systems, Leibniz Universität Hannover, Hanover, Germany
sontje.ihler@imes.uni-hannover.de

[2] Institute for Information Processing, Leibniz Universität Hannover, Hanover, Germany

Abstract. Computer aided diagnosis (CAD) has gained an increased amount of attention in the general research community over the last years as an example of a typical limited data application - with experiments on labeled 100k–200k datasets. Although these datasets are still small compared to natural image datasets like ImageNet1k, ImageNet21k and JFT, they are large for annotated medical datasets, where 1k–10k labeled samples are much more common. There is no baseline on which methods to build on in the low data regime. In this work we bridge this gap by providing an extensive study on medical image classification with limited annotations (5k). We present a study of modern architectures applied to a fixed low data regime of 5000 images on the CheXpert dataset. Conclusively we find that models pretrained on ImageNet21k achieve a higher AUC and larger models require less training steps. All models are quite well calibrated even though we only fine-tuned on 5000 training samples. All 'modern' architectures have higher AUC than ResNet50. Regularization of Big Transfer Models with MixUp or Mean Teacher improves calibration, MixUp also improves accuracy. Vision Transformer achieve comparable or on par results to Big Transfer Models.

Keywords: Medical image classification · Limited data · Transfer learning

1 Introduction

Automated analysis from radiology images like X-ray, CT, MRI, and ultrasound is becoming a valuable support tool for diagnosing and treating injuries and diseases. With the success of deep learning for image classification more and more applications come within reach of becoming standard tools for radiologists. The tremendous increase in research on medical image classification is usually fueled

Supplementary Information The online version contains supplementary material available at https://doi.org/10.1007/978-3-031-16431-6_62.

L. Wang et al. (Eds.): MICCAI 2022, LNCS 13431, pp. 654–663, 2022.
https://doi.org/10.1007/978-3-031-16431-6_62

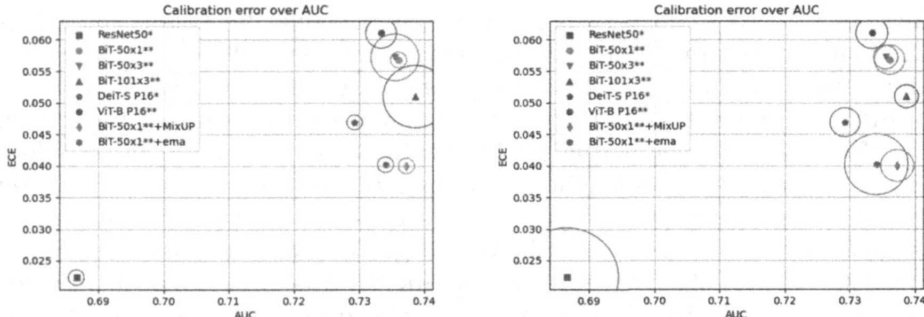

Fig. 1. Uncalibrated ECE vs. AUC on CheXpert5000. Metrics are computed on resplit test set. (left) circle size corresponds to number of model parameters, (right) circle size corresponds to our required training steps until convergence. Models pretrained on ImageNet21k achieve a higher AUC. All 'modern' architectures have higher AUC than ResNet50. Larger Models require less training steps. All models are quite well calibrated even though we only fine-tuned on 5000 training samples. Regularization with MixUp and Mean Teacher (ema) reduces calibration error, MixUp simultaneously increases accuracy for BiT-50×1. *) pretrained on ImageNet **) pretrained on ImageNet21k

by strategies from the natural image domain (computer vision). At the same time computer aided diagnosis (CAD) has gained an increased amount of attention in the general research community over the last years as an example of a typical limited data application - with experiments on labeled 100k–200k datasets. A common approach to handle limited data regimes is transfer learning. Models are pretrained on large natural image datasets and then afterwards fine-tuned to the medical application. However, radiology images are quite different from natural (photographic) images and computer vision (CV) methods need to be adapted to the medical image domain. To complicate matters, large labeled datasets are widely available in the CV community and associated methods build on this fact. In the medical domain, annotations are much harder to obtain. These burdens are a major bottleneck in the development and transfer of CV model to image classification applications in radiology. These connections also generate a phenomenon in the medical image community. While in CV new methods are usually compared in large studies and promoted as general problem solvers, in the medical community specific adaptations of a solution are presented. For newcomers, an overview of the available baselines is often missing.

In the recent past, great improvements have been achieved in transfer learning on natural images. An elementary component for successful transfer learning is the scaling of model capacity and pretraining datasets. Bigger is better. So-called Big Transfer Models (BiT) [8] have reached the state of the art for many visual benchmarks in the natural image domain. BiT models are tweaked ResNet [6] variants which are optimized for transfer performance. Recently, [12] applied these models to the CheXpert dataset [7] (among others) to study the transfer effect to the medical image domain. The authors found these models to outperform a ResNet50 baseline on all studied datasets. The overall findings in this work are very promising, but it should be noted that CheXpert is comparably large for a medical dataset with over 200k X-ray images. Is it not uncommon for medical dataset to be drastically smaller (four-digit range). The authors do indeed study the effect of fewer labels, by reducing the amount of transfer labels and comparing to the full ResNet50 baseline. This top-down approach naturally provides information on robustness to data reduction, but it still misses a very important point. For safety-critical applications, the ultimate goal should always be to achieve the greatest accuracy on the greatest amount of data available. **An important question is therefore, if only few examples are available, how can we get the most out of them**. This just mentioned top-down approach has limitations for directly comparing methods specifically in a low data regime. We believe that a direct comparison of methods is essential to improve performance on low medical data regimes. One might argue that performance on small data regimes will become obsolete in the near future due to federated learning [13] but we believe that large datasets will still be correlated to economic interest. High quality labels from specialized domain experts are expensive and are subject to chance/privilege for individual patients (like selection of relevant diseases and patient groups). We therefore provide a comprehensive study of modern architectures and regularization approaches at a fixed low data regime to fill the gap of method comparisons to improve performance specifically on limited medical data. We not only test Big Transfer models in a very small data regime but also Vision Transformers [3], another model class pretrained on huge datasets which have recently outperformed CNNs.

Our contribution is a study of these modern architectures/methods applied to a fixed low data regime of 5000 images on the CheXpert dataset, which to our knowledge does not exist in this form. We include:

1. Big Transfer models [8] at varying sizes,
2. ConvNets vs. Vision Transformer [3,19],
3. gains of standard regularization: Mean Teacher [17] and MixUp [21],
4. public data splits to enable direct method comparison in the future.

2 Related Work

This work is mainly inspired by 'Supervised Transfer Learning at Scale for Medical Imaging' [12]. This work compares accuracy on the CheXpert dataset for Big Transfer (BiT) models [8] of different sizes pretrained on datasets of different sizes. BiT models are updated/improved (wide) ResNet-v2 models [6] which are optimized for transfer learning. Batch normalization is replaced with group normalization. Studied BiT models are BiT-50×1, BiT-101×1, BiT-50×3, and BiT-101×3, the baseline is a standard Resnet50 [5]. The BiT models are pretrained on three datasets of increasing size: 1. *ImageNet* [14]: 1.3M image, 1000 classes, multi-class labels, 2. *ImageNet-21k* [2]: 14M images, 21000 classes, multi-class labels and 3. non-public *JFT-300M* [16]: 300M images, 18k classes, average of 1.29 labels per image, with pretrained weights for the former two publicly available. In [12], all BiT models outperform the ResNet50 baseline (all pretrained on ImageNet) on all medical datasets.

Generally accuracy increases for increase in model size and increased number of pretraining samples. There are more improvements from ImageNet to ImageNet21k pretraining than from ImageNet21k to JFT. Larger models are more data efficient and models pretrained on larger datasets are more data efficient. A BiT-50×1 model pretrained on ImageNet only requires 75% of the data to achieve the same accuracy as a standard Resnet50. A BiT-101×3 model pretrained on Imagenet21K [2] only requires 49%. They find that larger models converge faster (taking only about half of the time, sometimes even less) and larger models show better generalization performance to other datasets in the same domain (small domain shift). The authors find improved calibration for the larger models on a dermatology dataset [9] (derived from camera images which are presumably closer to natural images than X-ray imaging) but not for the CheXpert dataset.

For applications in safety-critical applications like medical image analysis it is essential that models do not only provide highly accurate predictions on the test domain (as well as under domain shift) but they should also be well calibrated. For a well predicted model, the predicted numerical output (referred to as confidence) should correlate to the real accuracy of the model. The predicted confidence then provides a reliable measure of certainty/uncertainty of a prediction. In the last years the best performing convnet models like ResNet, DenseNet and EfficientNet unfortunately showed poor calibration properties [4]. Recently Minderer et al. [11] have shown that newer model architectures MLP-Mixer [18] and Vision Transformer (ViT) [3] not only achieve higher accuracy on ImageNet but also improved calibration. This also applies to BiT models to some extent. These findings apply also to performance under distribution shift. These findings have been established in natural images at large scale. It is not clear how these results transfer to very limited medical data.

ViT models been successfully applied to medical images with results on par with a ResNet50 baseline [10]. We hope to transfer their successes to our case as well.

3 Experiments

We study BiT and ViT models in our experiments, as well as established regularization methods Mean Teacher and MixUp. For comparison to older works we compare all results to ResNet50, a very common backbone in medical image classification. All experiments are based on pretrained models which are then fine-tuned to CheXpert5000. Our baseline ResNet50 and DeiT are pretrained on ImageNet, while all other models are pretrained on Imagenet21k. We use models and pretrained weights from the timm library [20] (PyTorch). For all experiments we use the weak image augmentation described in 3.1. The best model has highest AUC on the validation set. We trained all models on five different train sets with 5000 images each and provide mean and standard deviation of the five best models. All 5k models are trained on the same five train sets. For comparison we also trained ResNet50 on the full CheXpert dataset.

3.1 CheXpert5000

We perform our method study on CheXpert [7], a publicly available dataset of chest X-ray images. It contains 224,316 radiographs from 65,240 patients. The validation set was created from 234 manually annotated X-rays. The train set was automatically extracted from patients' reports. The labels on the train set were created using natural language processing on these reports. To handle incomplete diagnosis in the patients' reports there are 4 categories for 14 observations (diseases). For each observation the categories are (1) certain disease observation, (u) uncertain disease observation, (-) disease not mentioned, and (0) disease ruled out. The dataset also provides age, gender and view point. Most images are frontal view and PA.

Preprocessing of CheXpert: Mustafa et al. [12] have found that comparing models on the provided validation set is unreliable due to it's small size (234 images). We follow their protocol and resplit the original train set into a new train (75%: 124,664 images), validation (10%: 16,989 images) and test set (15%: 25,205) based on patient id.

We further follow the common practice to reduce the number of classes to 5 (multi-label classification)and map uncertain labels (u) to 1 and missing observations (-) to 0. To study the effect of limited annotations we create five subsets of the train set of 5000 labeled samples each which are all frontal view and PA, half male, half female. We therefore also reduce the validation and test set to frontal view and PA (validation set: 12,115 images and test set: 18,363 images). We provide lists of all mentioned splits and subsets[1]. All subsets are created by random sampling (uniform distribution) so that all datasets show the same label distribution as the original dataset.

Mustafa et al. studied varying image resolution as model input and found no improvement in accuracy on the CheXpert dataset with resolution higher than 224×224. We therefore use the 'small' version of the dataset with image size 320×320 which is a lot more accessible due to its smaller memory consumption. All images are normalized with mean 0.5 and standard deviation 0.5.

Large validation sets are obviously not representative for a true small data regime, however the focus of this work is an analysis of the potential of the varying model architectures which requires a robust stop criterion. The effect of small validation sets and resulting fluctuation in convergence of the validation loss are subject of future work.

Image augmentations in the medical image domain is a risky task, we therefore only employ limited augmentations again following [12]. Spatial Transformation: We scale images to 248×248, perform random rotation by angle $\alpha \sim Uniform(-20, 20)$, random crops to 224×224, random horizontal flips with probability 50%. Chromatic Transformation: We employ random brightness and contrast jitter with factor 0.2.

[1] https://gitlab.uni-hannover.de/sontje.ihler/chexpert5000.

3.2 Finetuning: Models and Augmentation

Our **ResNet50 baselines** were fine-tuned using SGD with learning rate 0.003, momentum 0.9, weight decay 2e–5, 10 warm up epochs, and plateau scheduler (patience: 10 epochs, decay rate : 0.1). Training was terminated when the learning rate dropped below 1e–6. We used batchsize 512 [8] and batch size 32 which we found to improve results.

We study the following **BiT models** [8]: BiT-50×1, BiT-50×3 and BiT-101×3 which can also be found in [12]. For fine-tuning our BiT models we follow the BiT-HyperRule which proposes SGD with an initial learning rate of 0.003, momentum 0.9, and batch size 512. During fine-tuning the learning rate should be decayed by a factor of 10 at 30%, 60% and 90% of the training steps. For datasets with less than 20k samples, the authors propose 500 fine-tuning steps with a batch size of 512 and no MixUp [8]. We therefore planned on training all 5k models for 60 epochs but we found that BiT-50×1 did not universally converge in that time. We therefore trained BiT-50×1 for 100 epochs ($\approx$ 900 steps). We also found that finetuning the BiT models with a smaller batch size yielded better results. We therefore also fine-tuned all BiT models with our baseline protocol with a batch size of 32.

We employ two configurations of **ViT models**: ViT-B and DeiT. (ViT-B) In accordance to the finding of [15] that more pretraining data generally outperforms any data augmentation or regularization, we use models pretrained on ImageNet21k . We therefore do not use any regularization or advanced data augmentation (apart from the augmentation described in Sect. 3.1). Following their fine-tuning protocol for their smalles datasets, we use SGD with a momentum of 0.9, use a learning rate of 0.003 and a batchsize of 512. We train for 500 steps with cosine decay learning rate and 3 warm-up epochs, as well as gradient clipping at norm 1. (DeiT-S) For the DeiT model we follow the training protocol of [10]. We use DeiT-S with a patch size 16. We use Adam optimizer with learning rate 1e-4, momentum 0.9, and weight decay 1e-5, and a plateau scheduler with minimal learning rate 1e-6 (warm up for 10 epochs). Batch size is 64.

Mean Teacher and MixUp: Low data regime strategies like contrastive and consistency learning are based on (strong) image augmentations. This is always challenging for medical image diagnosis as perturbations can alter images in a way that eliminate disease relevant landmarks and features. We therefore study the effect of two established low data regime approaches Mean Teacher and MixUp that do not rely on strong data augmentation. We apply both regularizations to BiT-50×1. We follow our baseline protocol with batch size 32 and set α to 0.5 for MixUp regularization with a MixUp probability of 50%.

4 Results

To evaluate the prediction quality of all models fine-tuned models we provide the common metric AUC (also: AUROC), area under the curve for the ROC curve

Table 1. BiT models vs. ResNet50 on 5k train set. The BiT models are trained according to the BiT-HyperRule. We used the same batch size for ResNet50. We provide the amount of images the model has seen during training (image iter.) before convergence of validation loss. To enable a comparison on a larger scale we also provide results from training BiT-50×1 on the full train set.

Model	Params.	Pretr.	Training samples	Batch size	Image iter.	AUC (resplit)
ResNet50	23.51M	**in1k**	5000	512	438k	0.5615±.0028
BiT-50×1	23.51M	in21k	5000	512	256k	0.7206±.0021
BiT-50×3	211.19M	in21k	5000	512	112k	0.7218±.0009
BiT-101×3	381.81M	in21k	5000	512	99k	**0.7317±.0015**
BiT-50×1	23.51M	in21k	89944*	512	2330k	0.7718
BiT-50×1	23.51M	in21k	124663**	512	2070k	0.7714

*) all Frontal/AP views from full resplit train set **) full resplit train set

Table 2. BiT models vs. ResNet50 on 5k train set trained according to baseline protocol with a batch size of 32. Reducing the batch size (in comparison to BiT-HyperRule) decreases training time drastically while also improving the accuracy for all models.

Model	Params.	Pretr.	Batch size	Image iter.	AUC (resplit)	AUC (official)
ResNet50	23.51M	**in1k**	32	487k	0.6866±.0030	0.7949±.0106
BiT-50×1	23.51M	in21k	32	42k	0.7361±.0018	**0.8380±.0095**
BiT-50×3	211.19M	in21k	32	27k	0.7355±.0029	0.8359±.0169
BiT-101×3	381.81M	in21k	32	**26k**	**0.7388±.0008**	0.8342±+.0212

(true positive rate vs. false positive rate), and additionally we also provide class-wise AUPRC, the area under the PR curve (precision vs. recall) in the appendix. To quantify the calibration we compute ECE i.e. the expected calibration error. The calibration error is computed based on histogram binning with equal number of samples per bin (31 bins). We only compute the calibration on our test split, as the official validation set is too small to obtain representative results.

For the highest possible comparisons with other works on CheXpert we provide accuracy measures on our new test split of automatically labeled samples as well as the provided validation set of manual annotations. We provide the numbers of parameters for the used models as well as the amount of images the best models has seen during training until convergence of validation AUC. The computation steps can be computed by dividing image iterations by the batch size.

In Table 1 we show BiT models trained according to BiT-Hyperrule. Equivalent to the results on the full CheXpert datset, larger models show higher accuracy. The BiT-Hyperrule recommends a high batch size of 512 - but we saw great improvement regarding accuracy and training time when reducing the batchsize for CheXpert, see Table 2. Fine-tuned with a batch size of 32 the BiT models

Table 3. ResNet50 variations vs. Vision Transformer. The ViT-B P16* model achieves comparable results to Bit-50×1.

Model	Params.	Pretr.	Batch size	Image iter.	AUC (resplit)	AUC (official)
ResNet50	23.51M	**in1k**	32	487k	0.6866±.0030	0.7949±.0106
BiT-50×1	23.51M	in21k	32	42k	**0.7361±.0018**	**0.8380±.0095**
DeiT-S P16	21.67M	in1k	64	41k	0.7293±.0035	0.8161±.0173
ViT-B P16	87.46M	in21k	512	352k	0.6394±.0070	0.7704±.0054
ViT-B P16*	87.46M	in21k	64	46k	0.7334±.0032	0.8299±.0139

*)ViT-B trained on DeiT-S fine-tuning protocol

Table 4. Vanilla BiT-50×1 (none) vs. regularized BiT-50×1 using MixUp and Mean Teacher. We provide results for the teacher (ema) model as the teacher generally achieves higher accuracies.

Regularization	Params	Pretr.	Batch size	Image iter.	AUC (resplit)	AUC (official)
none	23.51M	in21k	32	42k	0.7361±.0018	**0.8380±.0095**
MixUP	23.51M	in21k	32	50k	**0.7373±.0005**	0.8313±.0116
Mean Teacher	23.51M	in21k	32	190k	0.7341±.0141	0.8267±.0103

arrive at their optimum in significantly less training steps (less than 28 times) with simultaneously higher accuracy. The BiT models based on ResNet-v2 architecture outperform standard ResNet50 not only on large but also on our very small data regime. ResNet50 and BiT-50×1 are almost identical in architecture and ResNet50 can be interchanged easily by BiT-50×1 in existing frameworks to probably gain a performance boost and shorter training times.

In Table 3 we compare results for ResNet variants ResNet50 and BiT-50×1 (ResNet-v2) to ViT models. Even though ViT models are infamous to require large amounts of data, they show great performance in this very small data regime outperform the ResNet50 baseline. The DeiT model is not yet quite on par with the BiT model, however the bigger ViT model shows comparable performance (at least when trained with the DeiT-S protocol). These findings are similar to [10].

We finally show the effect of regularization in the small data regime in Table 4. To our surprise regularization has little effect on accuracy. However, both regularizations improve calibration, see Fig. 1.

Note: Accuracy on the automatic labels is considerably lower than on the manual labels. This is on par with the works of [1,12]. We provide the class-wise AUC und AUPRC in the appendix for all five classes on the official validation set.

5 Conclusion and Discussion

In this work we presented a method study of modern architectures applied to a fixed low data regime of 5000 images on the CheXpert dataset and provide subsets and data splits for reproducibility. Conclusively we find/verify that model pretrained on ImageNet21k achieve a higher AUC and larger models require less training steps. All models are quite well calibrated even though we only fine-tuned on 5000 training samples. All 'modern' architectures have higher AUC than ResNet50. Regularization of BiT-50×1 with MixUp or Mean Teacher improves calibration and accuracy. Vision Transformer achieve comparable or on par results to BiT-50×1. While BiT-50×1 is one of many updates of the ResNet variants, ViTs are still in their infancy for small data regime szenarios. As ViTs can outperform CNNs in a large data regime, it may therefore only be a question of time until they outgrow CNNs for low data regimes.

References

1. Azizi, S., et al.: Big self-supervised models advance medical image classification. In: Proceedings of the IEEE/CVF International Conference on Computer Vision, pp. 3478–3488 (2021)
2. Deng, J., Dong, W., Socher, R., Li, L.J., Li, K., Fei-Fei, L.: ImageNet: a large-scale hierarchical image database. In: 2009 IEEE Conference on Computer Vision and Pattern Recognition, pp. 248–255. IEEE (2009)
3. Dosovitskiy, A., et al.: An image is worth 16x16 words: transformers for image recognition at scale. arXiv preprint arXiv:2010.11929 (2020)
4. Guo, C., Pleiss, G., Sun, Y., Weinberger, K.Q.: On calibration of modern neural networks. In: International Conference on Machine Learning, pp. 1321–1330. PMLR (2017)
5. He, K., Zhang, X., Ren, S., Sun, J.: Deep residual learning for image recognition. In: Proceedings of the IEEE Conference on Computer Vision and Pattern Recognition, pp. 770–778 (2016)
6. He, K., Zhang, X., Ren, S., Sun, J.: Identity mappings in deep residual networks. In: Leibe, B., Matas, J., Sebe, N., Welling, M. (eds.) ECCV 2016. LNCS, vol. 9908, pp. 630–645. Springer, Cham (2016). https://doi.org/10.1007/978-3-319-46493-0_38
7. Irvin, J., et al.: CheXpert: a large chest radiograph dataset with uncertainty labels and expert comparison. In: Proceedings of the AAAI Conference on Artificial Intelligence, vol. 33, pp. 590–597 (2019)
8. Kolesnikov, A., et al.: Big transfer (BiT): general visual representation learning. In: Vedaldi, A., Bischof, H., Brox, T., Frahm, J.-M. (eds.) ECCV 2020. LNCS, vol. 12350, pp. 491–507. Springer, Cham (2020). https://doi.org/10.1007/978-3-030-58558-7_29
9. Liu, Y., et al.: A deep learning system for differential diagnosis of skin diseases. Nat. Med. **26**(6), 900–908 (2020)
10. Matsoukas, C., Haslum, J.F., Söderberg, M., Smith, K.: Is it time to replace cnns with transformers for medical images? arXiv preprint arXiv:2108.09038 (2021)
11. Minderer, M., et al.: Revisiting the calibration of modern neural networks. In: Advances in Neural Information Processing Systems 34 (2021)

12. Mustafa, B., et al.: Supervised transfer learning at scale for medical imaging. arXiv preprint arXiv:2101.05913 (2021)
13. Rieke, N., et al.: The future of digital health with federated learning. NPJ Digit. Med. **3**(1), 1–7 (2020)
14. Russakovsky, O., et al.: ImageNet large scale visual recognition challenge. Int. J. Comput. Vision **115**(3), 211–252 (2015)
15. Steiner, A., Kolesnikov, A., Zhai, X., Wightman, R., Uszkoreit, J., Beyer, L.: How to train your ViT? data, augmentation, and regularization in vision transformers. arXiv preprint arXiv:2106.10270 (2021)
16. Sun, C., Shrivastava, A., Singh, S., Gupta, A.: Revisiting unreasonable effectiveness of data in deep learning era. In: Proceedings of the IEEE International Conference on Computer Vision, pp. 843–852 (2017)
17. Tarvainen, A., Valpola, H.: Mean teachers are better role models: weight-averaged consistency targets improve semi-supervised deep learning results. In: Advances in neural information processing systems 30 (2017)
18. Tolstikhin, I.O., et al.: MLP-Mixer: An all-MLP architecture for vision. In: Advances in Neural Information Processing Systems 34 (2021)
19. Touvron, H., Cord, M., Douze, M., Massa, F., Sablayrolles, A., Jégou, H.: Training data-efficient image transformers & distillation through attention. In: International Conference on Machine Learning, pp. 10347–10357. PMLR (2021)
20. Wightman, R.: PyTorch image models. https://github.com/rwightman/pytorch-image-models (2019). https://doi.org/10.5281/zenodo.4414861
21. Zhang, H., Cisse, M., Dauphin, Y.N., Lopez-Paz, D.: mixup: Beyond empirical risk minimization. arXiv preprint arXiv:1710.09412 (2017)

LSSANet: A Long Short Slice-Aware Network for Pulmonary Nodule Detection

Rui Xu[1,2], Yong Luo[1,2(✉)], Bo Du[1,2(✉)], Kaiming Kuang[4], and Jiancheng Yang[3,4,5]

[1] National Engineering Research Center for Multimedia Software, School of Computer Science, Institute of Artifical Intelligence, and Hubei Key Laboratory of Multimedia and Network Communication Engineering, Wuhan University, Wuhan, China
{luoyong,dubo}@whu.edu.cn
[2] Hubei Luojia Laboratory, Wuhan, China
[3] Shanghai Jiao Tong University, Shanghai, China
[4] Dianei Technology, Shanghai, China
[5] EPFL, Lausanne, Switzerland

Abstract. Convolutional neural networks (CNNs) have been demonstrated to be highly effective in the field of pulmonary nodule detection. However, existing CNN based pulmonary nodule detection methods lack the ability to capture long-range dependencies, which is vital for global information extraction. In computer vision tasks, non-local operations have been widely utilized, but the computational cost could be very high for 3D computed tomography (CT) images. To address this issue, we propose a long short slice-aware network (LSSANet) for the detection of pulmonary nodules. In particular, we develop a new non-local mechanism termed long short slice grouping (LSSG), which splits the compact non-local embeddings into a short-distance slice grouped one and a long-distance slice grouped counterpart. This not only reduces the computational burden, but also keeps long-range dependencies among any elements across slices and in the whole feature map. The proposed LSSG is easy-to-use and can be plugged into many pulmonary nodule detection networks. To verify the performance of LSSANet, we compare with several recently proposed and competitive detection approaches based on 2D/3D CNN. Promising evaluation results on the large-scale PN9 dataset demonstrate the effectiveness of our method. Code is at https://github.com/Ruixxxx/LSSANet.

Keywords: Pulmonary nodule detection · Long short slice grouping

Supplementary Information The online version contains supplementary material available at https://doi.org/10.1007/978-3-031-16431-6_63.

L. Wang et al. (Eds.): MICCAI 2022, LNCS 13431, pp. 664–674, 2022.
https://doi.org/10.1007/978-3-031-16431-6_63

1 Introduction

Lung cancer is the leading cause of cancer-related death worldwide. Prompt diagnosis and timely treatments of the pulmonary nodules are critical solutions for lung cancer, which can significantly improve the prospects of survival. In order to realize the pulmonary nodules' early diagnosis, thoracic computed tomography (CT) is widely adopted and demonstrated to be an effective tool. Nonetheless, manually identifying nodules in CT images is a labor-intensive task, since interpreting CT data requires doctors to analyze hundreds of slices every time. Hence, it is desirable to introduce machine learning to automatically and accurately identify and diagnose the pulmonary nodule.

In recent years, with the prosperity of deep learning, convolutional neural networks (CNNs) have been introduced to assist doctors in the field of pulmonary nodule detection [2–4,7–9,12,13,16–18,24]. Owing to the 3D nature of CT images, for approaches based on 2D CNNs [2,16], post-processing is often required to integrate the 2D regions detected into 3D, but this may affect the efficiency and also accuracy of the pulmonary nodule detection. Therefore, 3D CNN based methods [3,4,7–9,12,13,17,18,24] are dominant in this field, where most state-of-the-art frameworks consist of three components: U-Net-like backbone [15], 3D region proposal network, and the false positive reduction module. However, the aforementioned methods lack the ability to capture long-range dependencies, which is vital for global information extraction in vision tasks. Compact non-local operation [20,23] is able to alleviate this issue and improve the discrimination ability of fine-grained objects. However, when we directly apply such operations to pulmonary nodule detection, the computational or memory cost may be tremendous resulting from the 3D nature of CT images.

A possible solution to reduce the computational cost is to employ the idea of grouping, which is studied by many existing works, such as Xception [1], MobileNet [6], ResNeXt [22], Group normalization [21], and CrossFormer [19]. Since the computational complexity is usually nonlinear w.r.t. the input size, dividing the input into groups has been demonstrated to be effective in reducing the computational cost. Considering the fact that one nodule usually exists in several adjacent slices, we follow the way of grouping in [12], which groups the neighboring embeddings by depth. Yet, the adjacent depth grouping operations will lose the correlations among elements across long-distance slices, and hence may undermine the elimination of continuous pipe-like structures and the localization of pulmonary nodules.

Therefore, we propose a substitute of the vanilla compact non-local mechanism termed long short slice grouping (LSSG). In particular, we split the compact non-local operation into short-distance slice grouping (SSG) and long-distance slice grouping (LSG). SSG captures the long-range dependencies among any elements of adjacent slices, while LSG takes charge of the dependencies among any elements of slices far away from each other. More importantly, the proposed LSSG can further improve the performance by focusing on the small-scale features of nodules existing in consecutive slices, and realizing global feature extraction from the sparse slices.

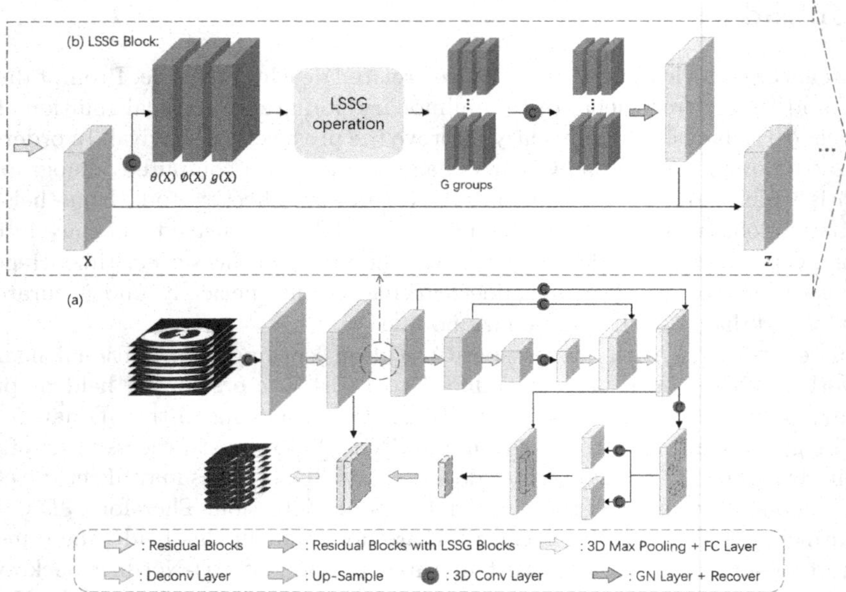

Fig. 1. (a) An illustration of our long short slice-aware network (LSSANet), where a designed long short slice grouping (LSSG) block is integrated into the encoder of U-shaped backbone. (b) The LSSG block. Embeddings resulting from three $1 \times 1 \times 1$ convolutional layers are grouped into G slice groups using our proposed LSSG operation. Then 3D convolution and GN are applied for different groups. After depth recovery and integrating with the original representation, we obtain an improved representation that explores the long-range dependencies among the elements across long/short-distance slices.

Based on the LSSG introduced above, we construct a new end-to-end framework termed LSSANet, for the detection of pulmonary nodules. Following [12], 3D ResNet50 [5] followed by two deconvolutional layers is adopted as a U-shaped encoder-decoder backbone [15]. Our proposed LSSG is integrated into the encoder to improve the feature extraction. Then, a 3D region proposal network is employed on the output of the backbone to generate pulmonary nodule candidates. Finally, multi-scale regions of interest are fed into one false positive reduction network to further decrease false positives.

To summarize, the main contributions of this paper are:

- We develop a long short slice-aware network (LSSANet) aiming at pulmonary nodule detection, where a novel long short slice grouping (LSSG) module is designed to simultaneously reduce the computational cost and capture the long-distance dependencies.
- Multiple variants of the designed LSSG by placing different numbers of LSG and SSG operations in different orders in the encoder are compared.

We conduct experiments on the large-scale PN9 [12] dataset, and the results demonstrate that our proposed LSSANet can significantly outperform the state-of-the-art 2D CNN based and 3D CNN based pulmonary nodule detection approaches.

2 Methods

Inspired by the tremendous success of non-local mechanism [20] for modeling long-range dependencies in computer vision, and to take full advantage of the CT images' 3D nature, we develop a long short slice-aware network (LSSANet). The overall architecture of LSSANet is illustrated in Fig. 1, which comprises an encoder-decoder backbone, a 3D region proposal component, and a multi-scale false positive reduction component. Particularly, we design a long short slice-grouping (LSSG) module to explore the long-range dependencies among any elements of one long/short-distance sliced group in the feature map and is integrated into the encoder.

2.1 Long Short Slice Grouping

In the thoracic CT images, pulmonary nodules usually show up as isolated spherical shapes, while other tissues like vessels and bronchus appear as the continuous pipe-like structures. Thus, looking at a few successive scans and examining the correlation among them is enough for doctors to differentiate nodules from other tissues. Meanwhile, viewing multiple slices also helps with the elimination of continuous pipe-like structures and the localization of nodules. Considering these two aspects, we propose a long short slice grouping (LSSG) operation based on the non-local operation in [20,23], which is illustrated in Fig. 2.

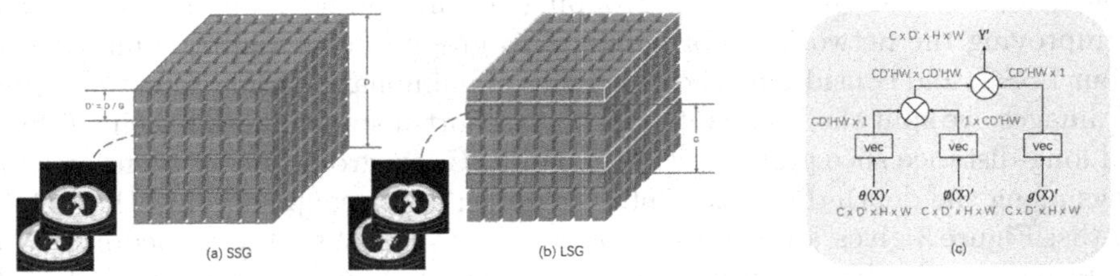

Fig. 2. Long short slice grouping (LSSG) operation. (a) Short-distance slice grouping (SSG). Embeddings (blue cubes) are grouped by red boxes along the depth dimension. (b) Long-distance slice grouping (LSG). Embeddings with the same color borders belong to the same group. (c) The detailed LSSG operation in one group. (Color figure online)

We first begin by reviewing the original non-local operation [20]. Suppose $\mathbf{X} \in \mathbb{R}^{C \times D \times H \times W}$ is the input feature map, where C, and D, H, W denote the number of channels, and its depth, height, width. The definition of the non-local operation in [20] is:

$$\mathbf{Y} = f(\theta(\mathbf{X}), \phi(\mathbf{X}))g(\mathbf{X}), \tag{1}$$

where $\mathbf{Y} \in \mathbb{R}^{C\times D\times H\times W}$. $\theta(\cdot), \phi(\cdot), g(\cdot) \in \mathbb{R}^{C\times DHW}$ come from $1\times 1\times 1$ convolutional transformations, and they are defined as:

$$\theta(\mathbf{X}) = \mathbf{W}_\theta \mathbf{X}, \phi(\mathbf{X}) = \mathbf{W}_\phi \mathbf{X}, g(\mathbf{X}) = \mathbf{W}_g \mathbf{X}, \tag{2}$$

where $\mathbf{W}_\theta$, $\mathbf{W}_\phi$, and $\mathbf{W}_g$ are learnable weight matrices. The function $f(\cdot,\cdot)$ implements the computation between all positions in the feature embeddings. There are several choices for f in [20]. In this paper, we adopt the simple dot-product version:

$$f(\theta(\mathbf{X}), \phi(\mathbf{X})) = \theta(\mathbf{X})^T \phi(\mathbf{X}). \tag{3}$$

Through this original non-local operation [20], the long-range dependencies among any locations in the feature map can be captured. Nonetheless, the affinity between any channels is not included, which is demonstrated to be vital for the fine-grained object differentiation [23]. Thus, we merge channel into position to reshape the output of Eq. (2), and obtain vectors $vec(\theta(\cdot)), vec(\phi(\cdot)), vec(g(\cdot)) \in \mathbb{R}^{CDHW}$ for the next similarity computation. By doing so, the compact non-local operation can absorb richer long-range dependencies among any elements. Our LSSG operation calculates the response $\mathbf{Y}$ as:

$$\mathbf{Y} = f(vec(\theta(\mathbf{X})), vec(\phi(\mathbf{X})))vec(g(\mathbf{X})), \tag{4}$$

where vec denotes vector reshape operation. Yet, there is a $CDHW \times CDHW$ pairwise matrix; therefore the computational or memory cost of LSSG is vast. Thus, it is infeasible to directly implement the LSSG operation.

Some recent works adopt the idea of grouping, e.g. Xception [1], MobileNet [6], ResNeXt [22], Group normalization [21], and CrossFormer [19]. It has been demonstrated that channel grouping or embedding grouping is effective in improving the network performance. This group idea can also be introduced in our LSSG, and considering the properties of pulmonary nodule detection using CT images, we split the LSSG into two parts: short-distance slice grouping (SSG) and long-distance slice grouping (LSG). For SSG, we group the adjacent embeddings along the depth dimension into G groups; each group contains $D^{'} = D/G$ depths. Figure 2 gives an example where $G = 4$. For LSG, the embeddings are sampled along the depth direction with a fixed interval G. For instance, in Fig. 2, all embeddings with a red border belong to a group, and those with a yellow border belong to another. The resulting number of groups is also G, and each group contains $D^{'} = D/G$ depths as well. After grouping embeddings, both SSG and LSG employ the LSSG operation within each group:

$$\mathbf{Y}^{'} = f(vec(\theta(\mathbf{X})^{'}), vec(\phi(\mathbf{X})^{'}))vec(g(\mathbf{X})^{'}), \tag{5}$$

where $\mathbf{Y}^{'} \in \mathbb{R}^{C\times D'\times H\times W}$, and $vec(\theta(\mathbf{X})^{'}), vec(\phi(\mathbf{X})^{'}), vec(g(\mathbf{X})^{'}) \in \mathbb{R}^{CD^{'}HW}$. The nodule usually appears in a few successive CT scans, making it unnecessary to utilize all depths for detecting the nodule, whereas depth sparse slices help with the elimination of tissues other than nodules and the localization of nodules. These two situations are nicely covered by two slice groupings of our LSSG, where the similarity between any positions and any channels across slices in long or short distance can be absorbed; thus it facilitates the distinguishment of nodules with various sizes.

The LSSG is then wrapped into the LSSG block, whose definition is:

$$\mathbf{Z} = recover(GN(\mathbf{W}_z\mathbf{Y}')) + \mathbf{X}, \tag{6}$$

where $\mathbf{W}_z$ denotes the weight matrx to be learned for a $1 \times 1 \times 1$ convolution, and GN is a Group Normalization [21]. *recover* denotes that positions of the slices in all groups are recovered in the depth direction. To make the LSSG block compatible with the existing neural network, the residual connection is adopted.

With regard to the LSSG block configuration, SSG and LSG blocks are added into the pulmonary nodule detection backbone alternately to construct our proposed LSSANet, as shown in Fig. 1. It is noteworthy that SANet is a special case of our LSSANet when all the blocks added into the backbone are SSG blocks.

2.2 Long Short Slice-Aware Network

Encoder-Decoder Backbone. Following prior work [12], we adopt 3D ResNet50 [5] as the encoder considering its outstanding feature extraction performance. The decoder comprises two $2 \times 2 \times 2$ deconvolutional layers; thereby the feature map can be upsampled to proper sizes. Furthermore, two output feature maps from the encoder layers are individually modulated by a $1 \times 1 \times 1$ convolutional layer, and then concatenated with the corresponding output of the decoder layers. Our proposed LSSG blocks are integrated into the second and third stages of the ResNet50 encoder (2 to the second and 3 to the third), with SSG and LSG appearing alternately. The U-shaped encoder-decoder architecture [15] is as shown in Fig. 1.

3D Region Proposal Network. In order to generate highly sensitive pulmonary nodule candidates, a $3 \times 3 \times 3$ convolutional layer is first applied to the output of the backbone. It is subsequently followed by two parallel $1 \times 1 \times 1$ convolutional layers for predicting the classification probability associated with each anchor at each voxel and regressing the 3D bounding box. Each anchor needs to specify six parameters: central point coordinates, width, height, and depth. In this work, five cube anchors with sizes 5, 10, 20, 30, and 50 are chosen. We use the multi-task loss objective in [12] for this stage.

False Positive Reduction. The nodule candidates generated in the last stage usually contain many false positives. As shown in Fig. 1, the feature maps produced by a shallow block in ResNet50 [5] and the last block of the backbone are cropped using nodule candidates; then the latter feature map is upsampled and concatenated with the former one to obtain the final multi-scale region of interest (RoI). A 3D max pooling is then applied to this RoI. Eventually, two fully connected layers are utilized for generating classification probability and 3D bounding box regression terms. Here we utilize the same loss function as the above 3D region proposal network's. It can not only decrease the number of false positives but also further refine the bounding box regression.

3 Empirical Study

In this section, we evaluate our LSSANet's pulmonary nodule detection performance on the large-scale pulmonary nodule dataset PN9 [12] using the Free-Response Receiver Operating Characteristic (FROC). Several state-of-the-art approaches are compared, including algorithms based on 2D CNN Faster RCNN [14], RetinaNet [10], SSD512 [11], and algorithms based on 3D CNN Leaky Noisy-OR [9], 3D Faster RCNN [24], DeepLung [24], NoduleNet [18], I3DR-Net [4], DeepSEED [8], SANet [12].

Dataset. We use PN9 [12] for evaluating the performance of LSSANet. This dataset is so far the largest and most diverse one for the detection of pulmonary nodules to the best of our knowledge, which comprises 8,796 CT images and 40,436 annotated nodules of 9 different classes. To entail a fair comparison, the 8,796 CT images are split into 6,037 for training, 670 for validation, and 2,089 for testing as in [12]. By the way, the PN9 is a preprocessed version, so there are no more data preprocessing steps needed.

Implementation Details. We use the model in [12] pretrained on PN9 [12] to initialize the weights of our LSSANet. The Stochastic Gradient Descent (SGD) optimizer is applied for training, with the batch size of 16, the learning rate of 0.001, the momentum of 0.9, and the weight decay of 1×10^{-4}. The LSSANet is trained for 9 epochs before RCNN gets involved. All the other hyperparameters and experimental settings are the same with [12]. Besides, we use a PyTorch implementation of our method. The experiments are conducted on 3 NVIDIA GeForce RTX 3090 GPU with 24GB memory.

Table 1. Comparison of pulmonary nodule detection performance on PN9 dataset in terms of FROC. Values in each column represent the pulmonary nodule detection sensitivities (unit: %) under different average numbers of false positives per CT image.

Method	0.125	0.25	0.5	1.0	2.0	4.0	8.0	Avg
2D CNN Based Algorithms:								
Faster RCNN [14]	10.79	15.78	23.22	32.88	46.57	61.94	75.52	38.10
RetinaNet [10]	8.42	13.01	20.13	29.06	40.41	52.52	65.42	32.71
SSD512 [11]	12.26	18.78	28.00	40.32	56.89	73.18	86.48	45.13
3D CNN Based Algorithms:								
Leaky Noisy-OR [9]	28.08	36.42	46.99	56.72	66.08	73.77	81.71	55.68
3D Faster RCNN [24]	27.57	36.59	46.76	58.00	70.00	80.02	88.32	58.18
DeepLung [24]	28.59	39.08	50.17	62.28	72.60	82.00	88.64	60.48
NoduleNet [18]	27.33	38.25	49.40	61.09	73.11	83.28	89.93	60.33
I3DR-Net [4]	23.99	34.37	46.80	60.04	72.88	83.60	89.57	58.75
DeepSEED [8]	29.21	40.64	51.15	62.20	73.82	83.24	89.70	61.42
SANet [12]	38.08	45.05	54.46	64.50	75.33	83.86	**89.96**	64.46
LSSANet	**51.59**	**51.59**	**58.18**	**66.88**	**77.33**	**85.35**	89.87	**68.69**

Nodule Detection Performance. The experimental results are listed in Table 1. As we can see, LSSANet obtains the highest average FROC score over other methods, leading to an improvement of 4.23% over the second-best SANet, which is actually a special case of our LSSANet. It is noted that our LSSANet surpasses the compared detection approaches to a large extent at all the average numbers of false positives per CT image except for the 8.

Table 2. Effect of implementing with different configurations for the proposed LSSG module (%). The SSG/LSG denotes the number of SSG vesus LSG.

#SSG/LSG	FROC
5/0	64.46
3/2	68.24
2/3	**68.69**
0/5	68.05

Table 3. Effect of using different number of Groups G for the proposed LSSG module (%).

#Groups	FROC
2	68.27
4	**68.69**
8	66.29

Table 4. Ablation study for compact non-local operation (%).

Module	FROC
NL-LSSG	68.30
CNL-LSSG	**68.69**

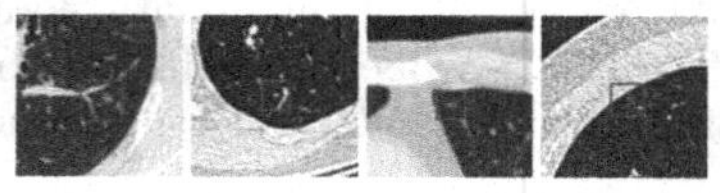

Fig. 3. Visualization of some pipe-like false positives that SANet detects.

Different Configurations of LSSG. Table 2 gives the results of our proposed LSSG module using different configurations. We conduct the experiments by making the LSSG blocks in the LSSANet encoder be all SSG blocks, all LSG blocks, or alternate back and forth. The results of LSSANet with 3 LSG and 2 SSG blocks appearing alternately in the encoder are the best, which improves the LSSANet with all SSG blocks and the one with all LSG blocks by 4.23% and approximately 1% respectively, in terms of average FROC score. The increase in overall performance could be attributed to the ability to focus on the small-scale features of nodules existing in consecutive slices, and extract the global features from the sparse slices.

We also analyze the sensitivity of our proposed LSSG w.r.t. different numbers of groups G, as listed in Table 3. The number of groups 4's FROC score is the best, which improves the setting $G = 8$ by over 2%.

Effectiveness of CNL. We add an ablation study to compare the network equipped with the compact non-local operation (CNL-LSSG) with the one utilizing the original non-local operation (NL-LSSG). From Table 4, we can see that the compact non-local operation is beneficial for the pulmonary nodule detection.

Visualization Analysis. Here we provide visualization of some pipe-like structures in Fig. 3, which have high probabilities to be detected as false positives by the original SANet. In contrast, our LSSANet is able to discriminate the nodules from continuous pipe-like structures (such as blood vessels) beyond a few neighboring slices.

4 Conclusion

In this paper, we develop a long short slice-aware network (LSSANet) for detecting the pulmonary nodules. A novel long short slice grouping (LSSG) module is proposed to explore explicit correlations among any positions and any channels of long-distance or short-distance slice groups. Experimental results on the large-scale dataset PN9 [12] demonstrate that LSSANet can achieve the state-of-the-art performance in pulmonary nodule detection, and both the long-distance slice grouping (LSG) and short-distance slice grouping (SSG) are critical in improving the performance. Besides, our LSSG is flexible and can be plugged into many nodule detection models. In the future, we intend to incorporate the proposed LSSG in more nodule detection frameworks to improve their performance.

Acknowledgements. This work was partially supported by National Natural Science Foundation of China under Grant 62141112, and the Special Fund of Hubei Luojia Laboratory under Grant 220100014. This work was also supported in part by National Science Foundation of China (82071990, 61976238).

References

1. Chollet, F.: Xception: deep learning with depthwise separable convolutions. In: CVPR, pp. 1800–1807 (2017)
2. Ding, J., Li, A., Hu, Z., Wang, L.: Accurate pulmonary nodule detection in computed tomography images using deep convolutional neural networks. In: Descoteaux, M., Maier-Hein, L., Franz, A., Jannin, P., Collins, D.L., Duchesne, S. (eds.) MICCAI 2017. LNCS, vol. 10435, pp. 559–567. Springer, Cham (2017). https://doi.org/10.1007/978-3-319-66179-7_64
3. Dou, Q., Chen, H., Yu, L., Qin, J., Heng, P.: Multilevel contextual 3-D CNNs for false positive reduction in pulmonary nodule detection. IEEE Trans. Biomed. Eng. **64**(7), 1558–1567 (2017)
4. Harsono, I.W., Liawatimena, S., Cenggoro, T.W.: Lung nodule detection and classification from thorax CT-scan using retinanet with transfer learning. J. King Saud Univ. Comput. Inf. Sci. (2020)
5. He, K., Zhang, X., Ren, S., Sun, J.: Deep residual learning for image recognition. In: CVPR, pp. 770–778 (2016)
6. Howard, A.G., et al.: Mobilenets: efficient convolutional neural networks for mobile vision applications. arXiv preprint (2017). http://arxiv.org/abs/1704.04861
7. Kim, B., Yoon, J.S., Choi, J., Suk, H.: Multi-scale gradual integration CNN for false positive reduction in pulmonary nodule detection. Neural Netw. **115**, 1–10 (2019)
8. Li, Y., Fan, Y.: DeepSEED: 3D squeeze-and-excitation encoder-decoder convolutional neural networks for pulmonary nodule detection. In: ISBI, pp. 1866–1869 (2020)
9. Liao, F., Liang, M., Li, Z., Hu, X., Song, S.: Evaluate the malignancy of pulmonary nodules using the 3-D deep leaky noisy-or network. IEEE Trans. Neural Networks Learn. Syst. **30**(11), 3484–3495 (2019)
10. Lin, T., Goyal, P., Girshick, R.B., He, K., Dollár, P.: Focal loss for dense object detection. In: ICCV, pp. 2999–3007 (2017)
11. Liu, W., et al.: SSD: single shot MultiBox detector. In: Leibe, B., Matas, J., Sebe, N., Welling, M. (eds.) ECCV 2016. LNCS, vol. 9905, pp. 21–37. Springer, Cham (2016). https://doi.org/10.1007/978-3-319-46448-0_2
12. Mei, J., Cheng, M.M., Xu, G., Wan, L.R., Zhang, H.: SANet: a slice-aware network for pulmonary nodule detection. IEEE Trans. Pattern Anal. Mach. Intell. (2021). https://doi.org/10.1109/TPAMI.2021.3065086
13. Ozdemir, O., Russell, R.L., Berlin, A.A.: A 3D probabilistic deep learning system for detection and diagnosis of lung cancer using low-dose CT scans. IEEE Trans. Medical Imaging **39**(5), 1419–1429 (2020)
14. Ren, S., He, K., Girshick, R.B., Sun, J.: Faster R-CNN: towards real-time object detection with region proposal networks. In: NIPS, pp. 91–99 (2015)
15. Ronneberger, O., Fischer, P., Brox, T.: U-Net: convolutional networks for biomedical image segmentation. In: Navab, N., Hornegger, J., Wells, W.M., Frangi, A.F. (eds.) MICCAI 2015. LNCS, vol. 9351, pp. 234–241. Springer, Cham (2015). https://doi.org/10.1007/978-3-319-24574-4_28

16. Setio, A.A.A., et al.: Pulmonary nodule detection in CT images: false positive reduction using multi-view convolutional networks. IEEE Trans. Med. Imaging **35**(5), 1160–1169 (2016)
17. Song, T., et al.: CPM-Net: a 3D center-points matching network for pulmonary nodule detection in CT scans. In: Martel, A.L., et al. (eds.) MICCAI 2020. LNCS, vol. 12266, pp. 550–559. Springer, Cham (2020). https://doi.org/10.1007/978-3-030-59725-2_53
18. Tang, H., Zhang, C., Xie, X.: NoduleNet: decoupled false positive reduction for pulmonary nodule detection and segmentation. In: Shen, D., et al. (eds.) MICCAI 2019. LNCS, vol. 11769, pp. 266–274. Springer, Cham (2019). https://doi.org/10.1007/978-3-030-32226-7_30
19. Wang, W., Yao, L., Chen, L., Cai, D., He, X., Liu, W.: Crossformer: a versatile vision transformer based on cross-scale attention. arXiv preprint (2021). https://arxiv.org/abs/2108.00154
20. Wang, X., Girshick, R.B., Gupta, A., He, K.: Non-local neural networks. In: CVPR, pp. 7794–7803 (2018)
21. Wu, Y., He, K.: Group normalization. In: Ferrari, V., Hebert, M., Sminchisescu, C., Weiss, Y. (eds.) ECCV 2018. LNCS, vol. 11217, pp. 3–19. Springer, Cham (2018). https://doi.org/10.1007/978-3-030-01261-8_1
22. Xie, S., Girshick, R.B., Dollár, P., Tu, Z., He, K.: Aggregated residual transformations for deep neural networks. In: CVPR, pp. 5987–5995 (2017)
23. Yue, K., Sun, M., Yuan, Y., Zhou, F., Ding, E., Xu, F.: Compact generalized non-local network. In: NIPS, pp. 6511–6520 (2018)
24. Zhu, W., Liu, C., Fan, W., Xie, X.: DeepLung: deep 3D dual path nets for automated pulmonary nodule detection and classification. In: WACV, pp. 673–681 (2018)

Consistency-Based Semi-supervised Evidential Active Learning for Diagnostic Radiograph Classification

Shafa Balaram[1,2], Cuong M. Nguyen[2], Ashraf Kassim[3], and Pavitra Krishnaswamy[2(✉)]

[1] National University of Singapore, Singapore, Singapore
shafa.balaram@u.nus.edu

[2] Institute for Infocomm Research, A*STAR, Singapore, Singapore
pavitrak@i2r.a-star.edu.sg

[3] Singapore University of Technology and Design, Singapore, Singapore

Abstract. Deep learning approaches achieve state-of-the-art performance for classifying radiology images, but rely on large labelled datasets that require resource-intensive annotation by specialists. Both semi-supervised learning and active learning can be utilised to mitigate this annotation burden. However, there is limited work on combining the advantages of semi-supervised and active learning approaches for multi-label medical image classification. Here, we introduce a novel Consistency-based Semi-supervised Evidential Active Learning framework (CSEAL). Specifically, we leverage predictive uncertainty based on theories of evidence and subjective logic to develop an end-to-end integrated approach that combines consistency-based semi-supervised learning with uncertainty-based active learning. We apply our approach to enhance four leading consistency-based semi-supervised learning methods: *Pseudo-labelling*, *Virtual Adversarial Training*, *Mean Teacher* and *NoTeacher*. Extensive evaluations on multi-label Chest X-Ray classification tasks demonstrate that CSEAL achieves substantive performance improvements over two leading semi-supervised active learning baselines. Further, a class-wise breakdown of results shows that our approach can substantially improve accuracy on rarer abnormalities with fewer labelled samples.

Keywords: Semi-supervised learning · Active learning · Theory of evidence · Subjective logic · Multi-label classification

---

A. Kassim and P. Krishnaswamy—Equal Contribution.

Supplementary Information The online version contains supplementary material available at https://doi.org/10.1007/978-3-031-16431-6_64.

L. Wang et al. (Eds.): MICCAI 2022, LNCS 13431, pp. 675–685, 2022.
https://doi.org/10.1007/978-3-031-16431-6_64

1 Introduction

Deep learning approaches offer leading-edge performance for automated image classification applications in the domain of radiology [5,14,17]. However, training deep learning models requires large datasets that are carefully labelled by clinical specialists. In practice, expert annotation of large-scale medical image databases is highly laborious, resource-intensive, and often infeasible.

To address the annotation challenge, it is desirable to (a) prioritise labelling of the most informative images and (b) simultaneously leverage large unlabelled image collections that are readily available within routine clinical databases for model training. However, each of these objectives requires distinct approaches: (a) active learning to select images for labelling, and (b) semi-supervised learning to leverage limited labelled datasets alongside larger unlabelled datasets. Semi-supervised active learning strategies combine advantages of the two approaches to generate better feature representations, even when starting with small randomly sampled labelled sets [2,7]. Although these approaches have been explored for many applications involving natural scene images, there are limited demonstrations of semi-supervised active learning for diverse medical image classification tasks [3,25,28].

Here, we propose a novel Consistency-based Semi-supervised Evidential Active Learning (CSEAL) framework that learns and leverages predictive uncertainty alongside the classification objective in an end-to-end manner. Our approach builds upon the Dempster-Shafer theory of evidence [4] and the principles of subjective logic [11], concurrently estimates the label uncertainty and prediction consistency to facilitate active learning and semi-supervised learning respectively, and is well-suited to handle the diverse challenges for radiology image classification. Our main contributions are as follows:

1. We propose a new end-to-end integrated approach to combine consistency-based semi-supervised learning with uncertainty-based active learning, by learning and utilising the class-wise evidences for both parameter optimisation and uncertainty estimation.
2. We apply CSEAL to develop evidential analogues of four leading consistency-based semi-supervised methods: Pseudo-labelling (PSU) [13], Virtual Adversarial Training (VAT) [15], Mean Teacher (MT) [20], and NoTeacher (NoT) [21].
3. In realistic experiments on the NIH-14 Chest X-Ray dataset [24], we demonstrate that CSEAL outperforms leading semi-supervised active learning baselines with very low labelling budgets of under 5%. In particular, for rarer abnormalities with $< 5\%$ prevalence, we observe that CSEAL enables substantial performance gains of up to 17% in AUROC over evidential supervised learning with random sampling.

2 Related Works

We briefly review two recent state-of-the-art semi-supervised active learning approaches characterised in this work. First, [10] proposed an unlabelled sample loss estimation method which applies Temporal Output Discrepancy (TOD) for semi-supervised learning and Cyclic Output Discrepancy (COD) for active learning. This TOD+COD method compared the change in model predictions between optimisation steps based on MT and active learning cycles. Second, Virtual Adversarial Training with Augmentation Variance (VAT+AugVar) [7], an augmentation-based active learning approach, quantifies informativeness using the variance across augmented sample predictions as demonstrated with VAT.

Other salient semi-supervised active learning approaches in the computer vision literature include Cost-Effective Active Learning (CEAL), and MixMatch-based methods. CEAL incorporated PSU in multiple uncertainty-based active learning heuristics [23], but this approach can propagate label noise during model training and also exhibit uncertainty in network predictions even when the probability output after softmax is high [6]. MixMatch, a dominant semi-supervised learning method, was combined separately with various active learning techniques such as data summarisation [2], label propagation [9], and k-means and cosine similarity distances [19]. However, the MixUp augmentation technique used in MixMatch limits its applicability to medical image classification tasks.

Some studies have also proposed semi-supervised active learning for medical image applications. For instance, CEAL [28] was adapted with representativeness sampling for computerised tomography (CT) in lung nodule segmentation. Further, Co-training active learning (COAL) [25] used a deep co-training with a hybrid active learning acquisition function for mammography image classification. Recently, PSU was combined with Batch Active learning by Diverse Gradient Embeddings (BADGE) [1], and demonstrated on three biomedical datasets [3]. Yet, there remains a need for a unified and customisable framework that can address diverse challenges in radiology image classification. For example, COAL's choice of active learning criteria is highly data and task specific, hindering its generalisability. Additionally, BADGE is not easily scalable to multi-label settings, due to its computationally expensive k-means clustering.

3 CSEAL Framework

In this section, we provide an overview of our proposed CSEAL framework. We consider the semi-supervised active learning setup with held-out validation and test sets. The training labelled set and validation set are initialised by random sampling until the annotation budget is met or there is complete class coverage. The notations $\{\mathbf{x}_i^{\mathrm{L}}, \mathbf{y}_i\}_{i=1}^{L_T}$ and $\{\mathbf{x}_i^{\mathrm{U}}\}_{i=1}^{L_U}$ are used to denote the labelled and remaining unlabelled training samples respectively. The validation set is denoted as $\{\mathbf{x}_i^{\mathrm{L}}, \mathbf{y}_i\}_{i=1}^{L_V}$. For a realistic process, we assume that $L_V \ll L_T$.

Figure 1 illustrates the Consistency-based Semi-supervised Active Learning (CSEAL) framework for binary classification. A multi-label classifier can be achieved by having multiple binary classification heads. The full version of CSEAL (with two networks) involves two major steps: (1) evidential-based semi-supervised learning and (2) evidential-based active learning.

(1) Evidential-based Semi-supervised Learning: Given a training image x, the transformation functions η_1 and η_2 are applied to the training images to generate the augmented samples $\{\mathbf{x}_1^{\mathrm{L}}, \mathbf{x}_1^{\mathrm{U}}\}$ and $\{\mathbf{x}_2^{\mathrm{L}}, \mathbf{x}_2^{\mathrm{U}}\}$ respectively. We apply two neural networks F_1 and F_2 to generate outputs from the corresponding augmented inputs. Since we are dealing with binary classification, the class predictors $\mathbf{p}_1 = [p_1^+, p_1^-]^\top$ and $\mathbf{p}_2 = [p_2^+, p_2^-]^\top$ can be obtained by applying a sigmoid function on the output logits $\mathbf{f_1}$ and $\mathbf{f_2}$. However, they are just point estimates which do not carry uncertainty information. Inspired from the evidential-based uncertainty estimation works [8,18], we assume that the Bernoulli variables $\mathbf{p}_1$ and $\mathbf{p}_2$ have priors in the form of Beta distributions parameterised by $\boldsymbol{\tau}_1 = [\alpha_1, \beta_1]$ and $\boldsymbol{\tau}_2 = [\alpha_2, \beta_2]$, respectively. We use the output logits from the networks to compute evidences and estimate $\boldsymbol{\tau}_1$ and $\boldsymbol{\tau}_2$ using $\boldsymbol{\tau} = \exp(\mathbf{f}) + \mathbf{1}$ with $\mathbf{f}$ clamped to $[-10, 10]$. Unlike a standard neural network classifier where the output is squashed into a probability assignment (or class predictor), i.e., $P(y = +) = p^+$, CSEAL uses network outputs to parameterise a Beta prior instead, which represents the density of each and every possible probability assignment. Hence, CSEAL models the second-order probabilities and uncertainty [11]. The Beta prior also enables the estimation of aleatoric uncertainty, which will be discussed in the next step.

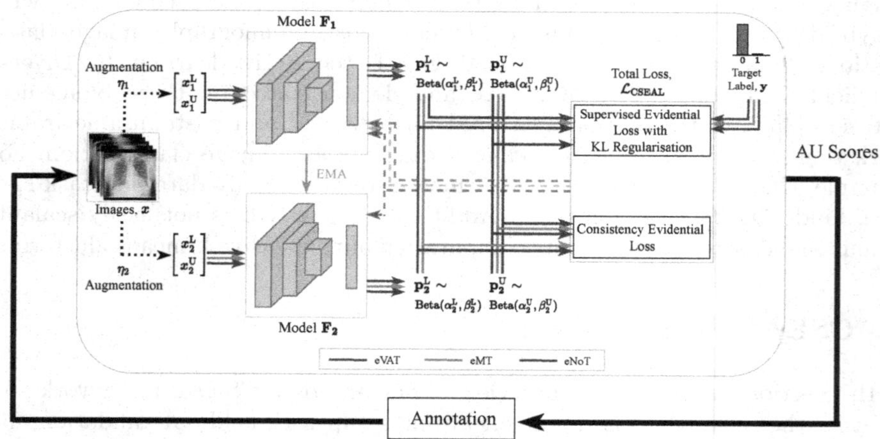

Fig. 1. Overview of the Consistency-based Semi-supervised Evidential Active Learning (CSEAL) framework. Augmentations η_1 and η_2 are applied to the input image x to generate x_1 and x_2. These are fed into the parametrised networks F_1 and F_2 before computing the supervised evidential loss and the consistency evidential loss. The forward and backward propagation are denoted by solid and dashed arrows respectively. The aleatoric uncertainty (AU) score is calculated for each unlabelled image to prioritise them for annotation. CSEAL is shown here with the evidential analogues of the 3 most recent consistency-based semi-supervised learning methods, namely evidential Virtual Adversarial Training (eVAT), Mean Teacher (eMT) and NoTeacher (eNoT). In the case of ePSU, the pseudo-labels inferred by the network F_1 are used in the supervised evidential loss of the unlabelled samples.

At inference, the prediction probabilities of each class are computed as the mean of the Beta distribution, i.e. $\hat{\mathbf{p}}_1 = [\hat{p}_1^+, \hat{p}_1^-]^\top = [\alpha_1/E_1, \beta_1/E_1]^\top$ where the total evidence is computed as $E = \alpha + \beta$. Here, the term *evidence* refers to the amount of support obtained from a data sample in favor of positive or negative predictions. Subsequently, we define a semi-supervised loss as a function of the class predictors and compute its Bayes risk w.r.t. Beta distribution priors. Additionally, a Kullback-Leibler (KL) divergence term is also included to penalise predictions with high uncertainty. Specifically, this KL term computes divergence between the Beta prior with adjusted parameters $\tilde{\boldsymbol{\tau}} = \mathbf{y} + (1 - \mathbf{y}) \odot \boldsymbol{\tau}$ and the uniform Beta distribution which represents the state of complete uncertainty. The general loss of CSEAL is:

$$\begin{aligned}\mathfrak{L}_{\text{CSEAL}}(\mathbf{x}, \mathbf{y}) = \lambda_{sup} \left[\mathfrak{L}_{err}(\mathbf{y}, \hat{\mathbf{p}}) + \mathfrak{L}_{var}(\hat{\mathbf{p}}, \boldsymbol{\tau}) + \lambda_t \; \mathfrak{L}_{reg}(\boldsymbol{\tau}, \mathbf{y})\right] \\ + \lambda_{cons} \mathfrak{L}_{cons}(\hat{\mathbf{p}_1}, \hat{\mathbf{p}_2})\end{aligned} \tag{1}$$

where $\lambda_t = \min(1.0, t/10)$ is the adaptive regularisation coefficient over the first t epochs. The supervised loss terms $\mathfrak{L}_{err}(\mathbf{y}, \hat{\mathbf{p}})$ and $\mathfrak{L}_{var}(\hat{\mathbf{p}}, \boldsymbol{\tau})$ originate from the Bayes risk of the squared error between $\mathbf{y}$ and $\mathbf{p}$. The term $\mathfrak{L}_{reg}(\boldsymbol{\tau}, \mathbf{y})$ is a result of regularisation using KL divergence. The (optional) consistency term $\mathfrak{L}_{cons}(\hat{\mathbf{p}_1}, \hat{\mathbf{p}_2})$ is only computed between the outputs of two separate networks.

When CSEAL is applied to SUP (a supervised learning baseline), PSU and VAT, we adopt a single-network architecture and drop the second network F_2. When applied to MT and NoT, CSEAL takes the full form with two separate networks. We now describe how $\mathfrak{L}_{\text{CSEAL}}(\mathbf{x}, \mathbf{y})$ can be adapted for different parametrised consistency-based semi-supervised learning models:

- **ePSU:** We infer the pseudo-labels from the mean of the Beta distribution.
- **eVAT:** We optimise the Bayes risk of the squared error between the class predictor on an unlabelled sample and its adversarial counterpart.
- **eMT:** We take the Bayes risk of the squared error between the class predictors of the student and the EMA-updated teacher networks w.r.t. their respective Beta distribution priors.
- **eNoT:** We optimise the Bayes risk of the squared error terms in the log likelihood w.r.t. their Beta distribution priors.

We provide additional details about these loss functions in the Supplement.

(2) Evidential-based Active Learning: To facilitate active learning, we compute the aleatoric uncertainty (AU) as previous work has shown that AU is more effective than epistemic uncertainty (or model uncertainty) in evidential graphical semi-supervised learning [27]. We estimate AU for each class as the expected entropy of the class predictor $\mathbf{p}$ given its Beta distribution prior as follows:

$$\text{AU} = \mathbb{E}_{p \sim \text{Beta}(\alpha,\beta)} \left\{\mathcal{H}[p]\right\} = \frac{1}{\ln 2} \sum_{\gamma \in \{\alpha,\beta\}} \frac{\gamma}{E} \left(\psi(E+1) - \psi(\gamma+1)\right) \tag{2}$$

where $\psi(\cdot)$ is the *digamma* function. The label-level AU scores are aggregated to obtain the image-level uncertainty score.

We annotate the unlabelled images with the highest scores and use these to augment the labelled training set. We also randomly select and label additional validation samples such that the ratio $L_T : L_V$ is maintained throughout the active learning process. The networks F_1 and F_2 continue to be trained with backpropagation and the above steps are repeated as part of an iterative process until the final labelling budget is met.

4 Experiment Setup

Dataset: We demonstrate our method on the NIH-14 Chest X-Ray dataset [24]. This dataset contains 112,120 high-dimensional frontal radiographs that are labelled for presence or absence of one or more abnormalities from 14 pathologies. The dataset exhibits high class imbalance. Out of the 46.1% of images containing at least one abnormality, 40.1% are multi-labelled. We use publicly available training (70%), validation (10%) and test (20%) splits generated without patient overlaps [26].

Realistic Active Sampling Process and Labelling Regimes: We ensured our active sampling process is reflective of practical clinical annotation workflows. First, as we used a separate validation pool, the labelled validation sets are representative of the held-out test set throughout the experiment, and sized to be much smaller than the training set. Second, our CSEAL-based annotation process is realistic as it does not require stratified or class-balanced initial labelled training sets or aligned training and validation class distributions. We performed experiments in two labelling regimes, namely: (a) the low-range regime from 2% to 5% in steps of 0.5% and (b) the mid-range regime from 5% to 10% in steps of 1%.

Experiments and Baselines: We evaluated the performance of ePSU, eVAT, eMT, and eNoT, and benchmarked against two competitive semi-supervised active learning methods, TOD+COD [10] and VAT+AugVar [7]. We also included an evidential supervised baseline (eSUP) to assess gains arising from inclusion of the unlabelled data. All evidential methods are evaluated using both Random and AU sampling. We use the same held-out test set across all experiments for fair comparisons.

Implementation Details: We describe the evidential supervised learning setup and hyperparameter tuning process. Additional details are in the Supplement.

Model Training: For fair comparisons across all the methods evaluated, we utilise the same DenseNet121 classifier backbone followed by a dropout layer and fully-connected layer as per [8]. The logits of each class are mapped to the parameters $[\alpha, \beta]$ of the Beta distribution using an exponential activation function for the evidential classifiers, and to the sigmoidal prediction probabilities for the standard classifiers. After each annotation round, we reset the network to its pre-trained ImageNet weights for retraining. The input images are augmented using random affine transformations with rotation of $\pm 10°$, translation of up to

10% and scaling between 0.9 and 1.1 followed by random horizontal flipping, resizing and centre cropping during training. The transformed images are then normalised using ImageNet mean and standard deviation. We use an Adam optimiser ($\beta = [0.9, 0.999], \epsilon = 1 \times 10^{-8}$) with learning rate of 1×10^{-4}, linear decay scheduler of 0.1 based on the validation loss, and a weight decay of 1×10^{-5}. All methods are implemented in PyTorch v1.4.0 [16].

Hyperparameter Tuning: Since a complete grid search to find the optimal hyperparameters for each semi-supervised active learning method is computationally infeasible, we use optimal hyperparameter configurations employed in previous works on non-evidential semi-supervised learning [21,22] and tune the dropout rate for each method. For TOD+COD, we utilise the original implementation with EMA decay rate of 0.999 and consistency weight $\lambda = 0.05$.

EMA Averaging: We retain an exponential moving average (EMA) copy of the model weights for all methods to ensure a fairer comparison against eMT and TOD+COD. This enables us to attribute the performance improvements observed to parameter averaging, consistency mechanism and/or active sampling. The active learning scores are computed from the model or its EMA copy based on the validation AUROC. For all evaluations, we report the best test AUROC either from the model or from its EMA copy.

5 Results

Here, we report performance of CSEAL in relation to the baselines in the low-range labelling regime. The Supplement includes results of CSEAL and the baselines in the mid-range labelling regime, as well as the performance comparison of the evidential semi-supervised learning approaches against their non-evidential counterparts.

Average AUC vs. Labelling Budget: Figure 2 shows the average test AUROC obtained as a function of the labelling budget in the low-range labelling regime. In most cases, the AU-based active learning methods outperform their random sampling based counterparts. In particular, the performance gains under the best CSEAL method are higher in early annotation rounds, wherein the evidential active supervised method (eSUP+AU) could be biased on account of starting with very small randomly sampled labelled sets. This suggests that the integration of semi-supervised and active learning under CSEAL offers increased robustness to the cold-start problem [12]. Amongst the 4 semi-supervised evidential active learning methods, eNoT outperforms the others, possibly on account of its better consistency enforcement mechanism. Finally, we find that the best performing CSEAL method eNoT+AU outperforms the competing semi-supervised active learning baselines, TOD+COD and VAT+AugVar, by up to 5.4% in AUROC and up to 2.6% in AUPRC on average at the final labelling budget of 5%.

Analysis of Class-Wise Performance Gains: At the end of the active learning process in the low-range labelling regime, i.e. at a labelling budget of 5%, we compute the AUROC gain of all methods evaluated over eSUP+Random and plot the class-wise breakdowns of the AUROC gain in Fig. 3. The classes are ordered by their prevalence in the test set. We observe that eMT+AU and eNoT+AU are comparable to or better than the competing baselines VAT+AugVar and TOD+COD. In particular, they have substantially larger gains over eSUP+Random for rarer classes. For example, eNoT+AU gains 17.05% for Hernia and 3.52% for Pleural Thickening which have a prevalence

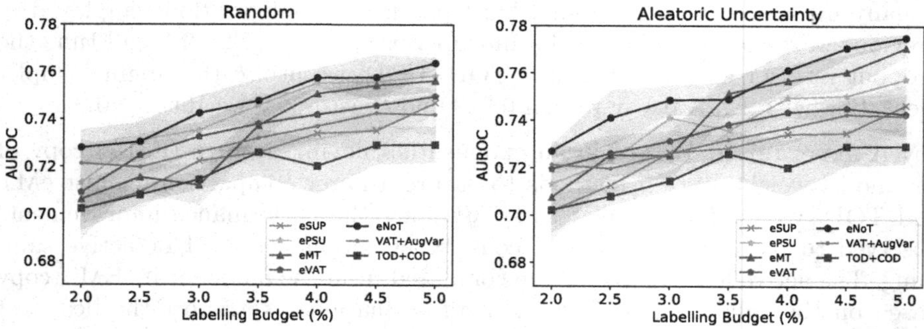

Fig. 2. Average test AUROC (mean ± std across 5 runs) on the NIH-14 Chest X-Ray dataset with different labelling budgets in the low-range labelling regime for random sampling (*left*) and aleatoric uncertainty (*right*). The proposed evidential counterparts of Pseudo-labelling (ePSU), Virtual Adversarial Training (eVAT), Mean Teacher (eMT) and NoTeacher (eNoT) are compared against the semi-supervised active learning baselines, VAT+AugVar and TOD+COD.

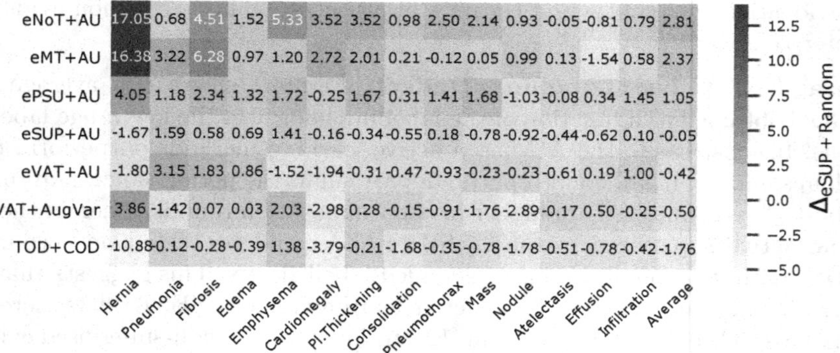

Fig. 3. Class-wise breakdown of the AUROC gains (in %) against the eSUP+Random baseline at the 5% labelling budget, i.e., the end of the active learning process for the low-range labelling regime. All numbers in the heatmap are averaged over 5 runs. The proposed evidential counterparts of Pseudo-labelling (ePSU), Virtual Adversarial Training (eVAT), Mean Teacher (eMT) and NoTeacher (eNoT) with aleatoric uncertainty sampling (AU) are compared against the semi-supervised active learning baselines, VAT+AugVar and TOD+COD.

of $< 0.2\%$ and $< 3.3\%$ respectively. We posit that the enhanced performance of these CSEAL methods could be attributed to their consistency enforcement mechanism. Also, CSEAL estimates the aleatoric uncertainty as the informativeness criterion for active learning more effectively, especially for classes with few labelled samples.

6 Conclusions

We introduced CSEAL, a novel end-to-end semi-supervised active learning framework for multi-label radiology image classification. We applied our framework to propose evidential analogues of 4 leading consistency-based semi-supervised learning methods, namely: evidential Pseudo-labelling (ePSU), evidential Virtual Adversarial Training (eVAT), evidential Mean Teacher (eMT) and evidential NoTeacher (eNoT). During active learning within a realistic annotation process, we demonstrate that CSEAL methods improve upon two leading semi-supervised active learning baselines. Amongst the CSEAL methods, we find that eNoT+AU provides the best performance across classes. Further, a class-wise breakdown shows that our best performing methods can gain up to 17% in AUROC over the evidential supervised learning approaches for rarer classes (with $< 5\%$ prevalence). Although our approach has currently been demonstrated on a single albeit challenging task of multi-label Chest X-Ray image classification with a specific convolutional backbone, it is amenable for future extensions to additional radiology modalities, such as CT and MRI, and networks. Further, future work could focus on investigating the different loss components empirically and theoretically to provide insight into the specific factors contributing to the effectiveness of CSEAL.

Acknowledgements. Research efforts were supported by the Singapore International Graduate Award (SINGA Award to Shafa Balaram), Agency for Science, Technology and Research (A*STAR), Singapore; the A*STAR AME Programmatic Funds (Grant No. A20H6b0151); as well as funding and infrastructure for deep learning and medical imaging R&D from the Institute for Infocomm Research, Science and Engineering Research Council, A*STAR. We also thank Mike Shou from the National University of Singapore for his support and insightful discussions.

References

1. Ash, J.T., Zhang, C., Krishnamurthy, A., Langford, J., Agarwal, A.: Deep batch active learning by diverse, uncertain gradient lower bounds. In: International Conference on Learning Representations (2019)
2. Borsos, Z., Tagliasacchi, M., Krause, A.: Semi-supervised batch active learning via bilevel optimization. In: ICASSP 2021–2021 IEEE International Conference on Acoustics, Speech and Signal Processing (ICASSP), pp. 3495–3499. IEEE (2021)
3. Boushehri, S.S., Qasim, A., Waibel, D., Schmich, F., Marr, C.: Systematic comparison of incomplete-supervision approaches for biomedical imaging classification (2021)

4. Dempster, A.P.: A generalization of Bayesian inference. J. Roy. Stat. Soc. Ser. B (Methodol.) **30**(2), 205–232 (1968)
5. Flanders, A.E., et al.: Construction of a machine learning dataset through collaboration: the RSNA 2019 brain CT hemorrhage challenge. Radiol. Artif. Intell. **2**(3), e190211 (2020)
6. Gal, Y.: Uncertainty in deep learning. University of Cambridge. Ph.D. thesis (2016).http://mlg.eng.cam.ac.uk/yarin/thesis. Accessed 07 Dec 2020
7. Gao, M., Zhang, Z., Yu, G., Arık, S.Ö., Davis, L.S., Pfister, T.: Consistency-based semi-supervised active learning: towards minimizing labeling cost. In: Vedaldi, A., Bischof, H., Brox, T., Frahm, J.-M. (eds.) ECCV 2020. LNCS, vol. 12355, pp. 510–526. Springer, Cham (2020). https://doi.org/10.1007/978-3-030-58607-2_30
8. Ghesu, F.C., et al.: Quantifying and leveraging predictive uncertainty for medical image assessment. Med. Image Anal. **68**, 101855 (2021)
9. Guo, J., et al.: Semi-supervised active learning for semi-supervised models: exploit adversarial examples with graph-based virtual labels. In: Proceedings of the IEEE/CVF International Conference on Computer Vision, pp. 2896–2905 (2021)
10. Huang, S., Wang, T., Xiong, H., Huan, J., Dou, D.: Semi-supervised active learning with temporal output discrepancy. In: Proceedings of the IEEE/CVF International Conference on Computer Vision, pp. 3447–3456 (2021)
11. Jsang, A.: Subjective Logic: A Formalism for Reasoning Under Uncertainty. Springer, Cham (2016). https://doi.org/10.1007/978-3-319-42337-1
12. Konyushkova, K., Sznitman, R., Fua, P.: Learning active learning from data. In: Advances in Neural Information Processing Systems, vol. 30 (2017)
13. Lee, D.H.: Pseudo-label: the simple and efficient semi-supervised learning method for deep neural networks. In: ICML Workshop on Challenges in Representation Learning, vol. 3 (2013)
14. McKinney, S.M., et al.: International evaluation of an AI system for breast cancer screening. Nature **577**(7788), 89–94 (2020)
15. Miyato, T., Maeda, S.I., Koyama, M., Ishii, S.: Virtual adversarial training: a regularization method for supervised and semi-supervised learning. IEEE Trans. Pattern Anal. Mach. Intell. **41**(8), 1979–1993 (2018)
16. Paszke, A., et al.: Automatic differentiation in PyTorch (2017)
17. Rajpurkar, P., et al.: Deep learning for chest radiograph diagnosis: a retrospective comparison of the CheXNeXt algorithm to practicing radiologists. PLoS Med. **15**(11), e1002686 (2018)
18. Sensoy, M., Kaplan, L., Kandemir, M.: Evidential deep learning to quantify classification uncertainty. In: Advances in Neural Information Processing Systems, vol. 31 (2018)
19. Song, S., Berthelot, D., Rostamizadeh, A.: Combining mixmatch and active learning for better accuracy with fewer labels. arXiv preprint arXiv:1912.00594 (2019)
20. Tarvainen, A., Valpola, H.: Mean teachers are better role models: weight-averaged consistency targets improve semi-supervised deep learning results. In: Advances in Neural Information Processing Systems, pp. 1195–1204 (2017)
21. Unnikrishnan, B., Nguyen, C., Balaram, S., Li, C., Foo, C.S., Krishnaswamy, P.: Semi-supervised classification of radiology images with NoTeacher: a teacher that is not mean. Med. Image Anal. **73**, 102148 (2021)
22. Unnikrishnan, B., Nguyen, C.M., Balaram, S., Foo, C.S., Krishnaswamy, P.: Semi-supervised classification of diagnostic radiographs with NoTeacher: a teacher that is not mean. In: Martel, A.L., et al. (eds.) MICCAI 2020. LNCS, vol. 12261, pp. 624–634. Springer, Cham (2020). https://doi.org/10.1007/978-3-030-59710-8_61

23. Wang, K., Zhang, D., Li, Y., Zhang, R., Lin, L.: Cost-effective active learning for deep image classification. IEEE Trans. Circuits Syst. Video Technol. **27**(12), 2591–2600 (2016)
24. Wang, X., Peng, Y., Lu, L., Lu, Z., Bagheri, M., Summers, R.: Hospital-scale chest X-ray database and benchmarks on weakly-supervised classification and localization of common thorax diseases. In: IEEE Conference on Computer Vision and Pattern Recognition, pp. 3462–3471 (2017)
25. Yang, Z., Wu, W., Zhang, J., Zhao, Y., Gu, L.: Deep co-training active learning for mammographic images classification. In: 2020 Chinese Automation Congress (CAC), pp. 1059–1062. IEEE (2020)
26. Zech, J.: reproduce-chexnet. GitHub repository (2018). https://github.com/jrzech/reproduce-chexnet
27. Zhao, X., Chen, F., Hu, S., Cho, J.H.: Uncertainty aware semi-supervised learning on graph data. Adv. Neural. Inf. Process. Syst. **33**, 12827–12836 (2020)
28. Zotova, D., Lisowska, A., Anderson, O., Dilys, V., O'Neil, A.: Comparison of active learning strategies applied to lung nodule segmentation in CT scans. In: Zhou, L., et al. (eds.) LABELS/HAL-MICCAI/CuRIOUS -2019. LNCS, vol. 11851, pp. 3–12. Springer, Cham (2019). https://doi.org/10.1007/978-3-030-33642-4_1

Self-Rating Curriculum Learning for Localization and Segmentation of Tuberculosis on Chest Radiograph

Kunlei Hong[1], Lin Guo[1], and Yuan-ming Fleming Lure[1,2](✉)

[1] Shenzhen Zhiying Medical Imaging Co., Ltd, Shenzhen, China
f.lure@hotmail.com
[2] MS Technologies Corporation, Rockville, MD, USA

Abstract. Tuberculosis (TB) is the second leading cause of infectious disease death, and chest X-ray is one of the most commonly used methods to detect TB. In this work, we bring forward a Self-Rating Curriculum Learning (SRCL) method to exploit the task of localization and segmentation of tuberculosis on chest radiographs. A total number of 12,000 CXR images of healthy subjects and bacteriologically-confirmed TB patients, retrospectively collected from multi-center local hospitals, are used in the study. A classical instance localization and segmentation framework (Mask-RCNN with backbone Resnet-50) is presented to compare traditional one-step training method and our proposed SRCL method in metrics and efficiency. First, a teacher model with self-rating function without human participation is developed to output the rating score of each sample, and all the samples are classified into three categories, namely easy set, moderate set and hard set, by using kernel density estimate (KDE) plot. After grouping the cases images in order of difficulty, the SRCL training is conducted on progressively harder images in three stages. We evaluate the proposed SRCL method in metrics and efficiency. Results indicate that the proposed SRCL method is able to boost the performance of the compared traditional method.

Keywords: Curriculum learning · Deep learning · Ranking function · Tuberculosis · Chest X-ray

1 Introduction

The chest X-ray (radiograph) is commonly performed to diagnose various abnormalities, such as pulmonary tuberculosis (TB) and nodules [1]. It is a fast and painless screening test, and the chest radiograph could provide a wide range of visual diagnostic information with minimal radiation exposure. However, it's still a challenging and time-consuming

K. Hong and L. Guo — contribute equally to this work.

Supplementary Information The online version contains supplementary material available at https://doi.org/10.1007/978-3-031-16431-6_65.

L. Wang et al. (Eds.): MICCAI 2022, LNCS 13431, pp. 686–695, 2022.
https://doi.org/10.1007/978-3-031-16431-6_65

task to identify and distinguish the various chest abnormalities for radiologists. Therefore, developing computer-aided diagnosis (CAD) method is urgent to assist radiologists in reality.

In 2019, an estimated 10 million new TB cases emerged worldwide, and more than 1.4 million people died of TB, making it the leading single infectious disease cause of death that year [2]. Artificial intelligence (AI) has played an increasingly important role in medical imaging, and researchers have realized that AI-based chest X-ray is a very promising tool for diagnosing TB, especially in resource-limited rural areas [3]. Reviews of different CAD methods for TB can be found in the published studies [4, 5]. These CAD systems can automatically score the TB likelihood of each chest image. However, most of previous reports relied on publicly available CXR datasets, which may bring bias for AI model development due to demographic imbalances [6], poor-quality or even without radiologists' annotation [7], etc. As creating a large annotated medical dataset is not easy, it may be a limitation in applying the systems in real-life situation.

Bengio et al. firstly proposed the original concept of curriculum learning, which means "training from easier data to harder data" [8]. The basic idea is to "start small" to train the learning model with easier data subsets and then gradually increase the difficulty level of data until the whole training datasets. Different methods and routes for curriculum learning, such as self-paced learning and transfer teacher method, have been developed with extensive experiments from early work to recent efforts [9, 10]. Based on previous experiments, the extent of contribution of introducing curriculum into machine learning depends on how we design the curriculum for specific applications and datasets [11]. The key issue for curriculum learning is how to build effective difficulty measurer and training scheduler to divide learning curriculum. Predefined curriculum learning use human expert's prior knowledge to design difficulty and training scheduler, and automatic curriculum learning needs at least one of difficulty measurer and training scheduler designed by data-driven method automatically.

In this paper, we propose a self-rating curriculum learning (SRCL) method for the task of localization and segmentation tuberculosis on retrospectively collected chest radiograph from multi-center local hospitals. The difficulty measurer is established by a well-designed teacher model. By utilizing the teacher model predicting, self-rating to every sample in training set could be achieved automatically by self-designed ranking function, and the curriculum course could be thus obtained. The training scheduler is conducted by multi-stage training strategy following curriculum by transfer learning. Human prior knowledge is not involved in the established of both difficulty measurer and training scheduler to reduce or eliminate the subjective biases. The teacher model can judge the training set and give rating score based on each performance of algorithm predicting, and multi-stage transfer training can ensure knowledge delivering to next generation during curriculum learning period. Furthermore, we build up a universal curriculum learning route map to apply on other datasets, not just limited in some field.

To evaluate our method, we test our proposed SRCL method on tuberculosis dataset constructed from multi-center local hospitals. Results show that our proposed SRCL method outperforms the compared traditional one-step method on TB detection.

2 Methods

2.1 Dataset

The dataset consists of 6,000 cases of tuberculosis (TB) chest radiograph and 6,000 cases of normal chest radiograph, which is collected from multi-center local hospitals with local Institutional Review Board approval. All patient identification has been removed. These cases are randomly divided into three datasets for training, validating, and testing at a ratio of 8:1:1. All the training, validation and test sets are the same, and the only difference is that training set is fed for one-step training process in the baseline (called teacher model training), whereas same training set is used by gradually adding more difficult images into training process in SRCL.

2.2 Deep Learning-Based Classification and Localization Framework

An open-source project Detectron2, which is Facebook AI Research's next generation library that provides state-of-the-art detection and segmentation algorithms, is used in this work. We choose Resnet50-FPN with Mask R-CNN as deep learning framework.

2.3 Training Platform and Image Pre-processing

All experiments are conducted on single NVIDIA 2080-Ti. Initial learning rate is set to 0.01 with SGD optimizer. Batch size is set to 8 images per batch. Max iteration epoch is 60, warm up epoch is 3, and learning rate decreasing strategy is cosine down. Input Image size is resized to 512 * 512 pixels. Data Augmentation is then performed, including random crop with relative ratio of 0.95*0.95, random horizontal flip with probability of 0.5, random brightness and contrast adjustment with ratio from 0.8 to 1.2. The initialized parameter of network is transferred from model in Detecton2 model zoo.

2.4 Training Platform and Image Pre-processing

Procedure of SRCL. As displayed in Fig. 1, SRCL is developed as follows. First, total training set is utilized to develop the teacher model with preset learning frame and learning settings, and the teacher model could judge every sample in this dataset by directly predicting. Then, according to AI predicted result and ground truth information, the rating score of each sample is obtained. By statistical analysis, the training set can be divided into three independent parts, including easy set, moderate set and hard set. After grouping the cases from easy to hard, the SRCL method is conducted following curriculum learning method. For our SRCL training schedule, we train our network on progressively harder images in three stages:

Stage 1: easy set only.

Stage 2: easy set + moderate set.

Stage 3: easy set + moderate set + hard set.

Specifically, easy set is first used to develop stage 1 model. The optimized stage 1 model with highest AP50 is introduced as the initial parameters in the next step by using transfer learning. Second, moderate set combined with easy set are used to develop the stage 2 model, and similarly, the optimized stage 2 model with highest AP50 is introduced in the next step. Finally, the combination of easy, moderate and hard set is incorporated to develop the stage 3 model, and the optimized stage 3 model with highest AP50 is set as the final SRCL output model.

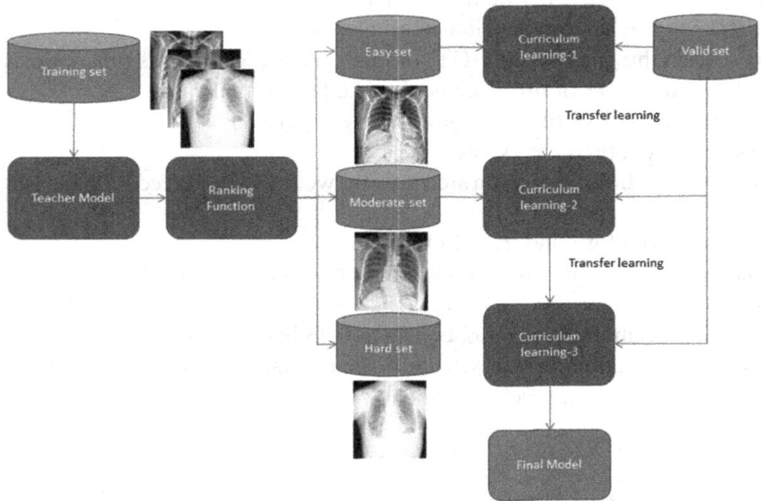

Fig. 1. Aerial view of SRCL process.

Ranking Function Design. In curriculum learning, scoring images by difficulty is often the core problem as addressed in many previous studies [12]. In most cases, an ideal ranking function is rarely available. The most commonly used method for ranking is to involve certified radiologists by giving quantitative scores based on their experience to group the difficulty levels [11, 13], which is time-consuming and may cause subjective biases. In order to solve this problem, an effective ranking function is designed by developing a teacher model in our study based on self-rating scores instead of human participation. The initial teacher model is trained on total training set, and the optimized teacher model with highest AP50 as the final teacher model to output the sample learnable difficulty score. For each image of Chest x-ray, the ranking function can be expressed as follows:

$$score = score_{predicted} + score_{missed} \tag{1}$$

where score means one chest x-ray graph score, including two parts:

$$score_{predicted} = y_{predicted} * \sum_{\text{i}=1}^{\text{N}} \lambda 1 * (1 - \text{Pi}) + \lambda 2 * (1 - \max(\{IOU_{i-l} \in A : 1 \leq l \leq T\})) \tag{2}$$

$$score_{missed} = y_{gt} * \sum_{j=1}^{M} \lambda_3 * 1 + \lambda_4 * 1 \quad (3)$$

The prediction threshold score of teacher model is 0.1.

$y_{predicted} = 1$ when teacher model predicts this slice as tuberculosis case.

$ypredicted = 0$ when predicts it as normal case.

$y_{gt} = 1$ when the ground truth (GT) of this slice is identified as tuberculosis case.

$y_{gt} = 0$ when the GT of this slice is identified as normal case.

N: total predicted quantity of Region of Interest (ROI).

P_i: teacher model predicted score.

IOU_{i-l}: Intersection over Union area ratio between i-th predicted ROI and l-th ground truth ROI.

T: total quantity of ground truth ROIs.

M: total missed GT quantity, which means this GT ROI's IOU equals zero with all predicted ROI.

λ_1 λ_2 λ_3 λ_4: normalization ratio, equals to 0.25 in this case.

Based on the assumption that similar difficulty sample's score will follow normal distribution under teacher model judgement.

3 Experiments and Results

3.1 Teacher Model Training

First, the teacher model training is conducted under the condition described in Sect. 2.3. All 9,600 samples are fed as input images for the one-step training process, and the highest mAP50 value is achieved at the total steps of 32,400 (Supplementary F1). Thus, the optimized teacher model could be decided under such circumstance.

3.2 Self-Rating Process

Each sample of the training set is predicted by using the teacher model developed in Sect. 3.1. After outputting the prediction results each sample is labeled with a self-rating score by using ranking function. The descriptive statistics are listed in Table 1.

Table 1. Descriptive statistics of self-rating results.

Item	Value
Sample number	9,600
Mean	0.376
Standard deviation	0.373
Minimum value	0
25% Quartile value	0

(*continued*)

Table 1. (*continued*)

Item	Value
50% Quartile value	0.289
75% Quartile value	0.572
Maximum value	2.146

In order to group the images into three categories with easy, moderate and hard, a statistical method is involved to build the curriculum. Kernel density estimate (KDE) plot is a standard method for visualizing the distribution of observations in a dataset. The density curve and cumulative density curve of our study dataset are showed as Fig. 2.

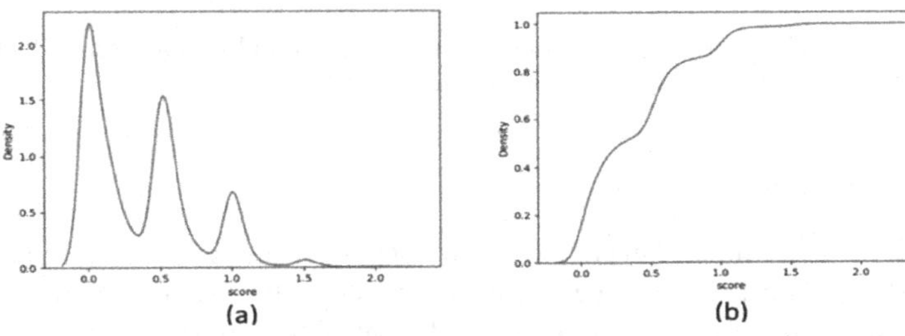

Fig. 2. (a) Density curve and (b) cumulative density curve of training set.

With analysis on the density and cumulative density curves, two specified partition points could be found at 0.34 and 0.84, which is the changing inflectional point as displayed in Fig. 2(b). Therefore, samples with self-rated scores ranged from 0 to 0.34 is regarded as easy set, samples with scores ranged from 0.34 to 0.84 is regarded as moderate set and samples whose score is higher than 0.84 is regarded as hard set. All the easy, moderate and hard sets follow Normal distribution. Finally, the 9,600 training cases are divided into 4,949 (52%) easy samples, 3,287 (34%) moderate samples and 1,364 (14%) hard samples. The CXR image samples of normal and TB after separation in easy, moderate and hard sets are shown in Fig. 3.

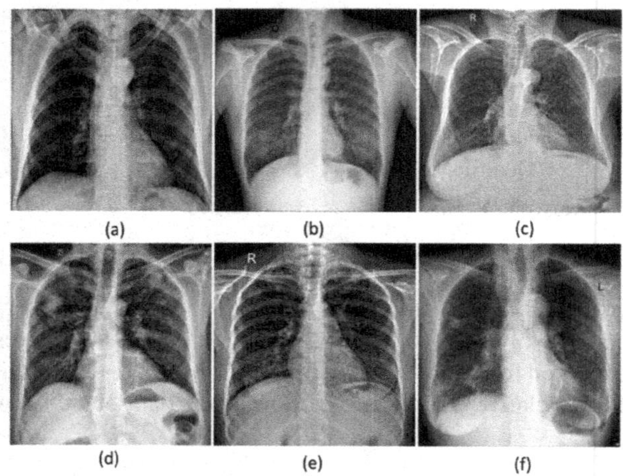

Fig. 3. Normal samples from (a) simple set, (b) moderate set and (c) hard set and TB samples from (d) simple set, (e) moderate set and (f) hard set.

3.3 SRCL Training

Stage 1. All the CXR images in easy set with 4,949 CXR images are chosen as training set at this stage. The learning rate during every step, the loss downstream, and the mAP and mAP50 curves of validation set are displayed in Supplementary F2. When the training step numbers achieve 4,943, the highest mAP50 value is obtained with 78.7.

Stage 2. Training set at this stage contains both easy set and moderate set. The total number of samples in this stage is 8,236. According to the training metrics, the best model's mAP50 is reached to 82.88 at step of 13,380 (Supplementary F3).

Stage 3. Training set contains the easy, moderate and hard sets with all the 9,600 samples. The initial parameters of model at this stage are transferred from stage 2 model at step of 13,380. According to the training metrics, mAP50 value of the best model is reached to 83.16 at step 1,199. Therefore, this best model under such circumstance is chosen as the final SRCL model (Supplementary F4).

3.4 Performance Evaluation

As shown in Table 2, by using SRCL method, the box-mAP, box-AP50, mask-mAP and mAP50 values on validation sets are all improved with margins of 1.38, 0.62, 1.08 and 0.33, respectively. In accordance with the validation results, all the testing performance on testing set 1,200 of CXR images also possess better performance of SRCL model than teacher model. Furthermore, the optimized steps are reduced from 32,400 to 19,522, revealing a 39.75% efficiency and energy-saving improvement by using SRCL Model.

Table 2. Performance of teacher model and SRCL model on validation and testing sets.

	Methods	Evaluation index					
		box-mAP	box-AP50	mask-mAP	mask-AP50	Optimized steps	AUC
Validation	Teacher model	40.86	82.54	41.75	81.76	32,400	0.97
	SRCL model	42.24	83.16	42.83	82.09	19,522	0.97
Testing	Teacher model	37.59	79.25	38.30	79.16	-	0.96
	SRCL model	39.30	81.59	40.12	80.91	-	0.96

Notation: mAP = mean average precision; AP50 = average precision at IoU threshold = 0.5.

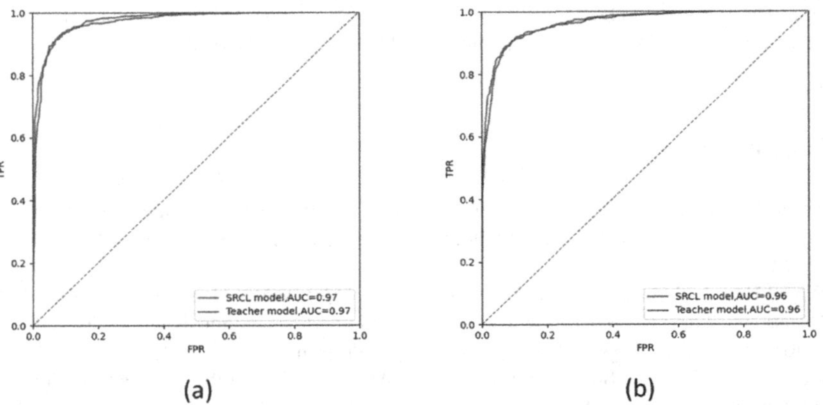

Fig. 4. ROC curves of (a) validation and (b) testing results by using teacher model and SRCL model. The green lines represent teacher model and the red lines represent SRCL model.

It needs to emphasize that both the AUC values on validation and testing sets are very high, with 0.97 and 0.96 respectively (Fig. 4). One of the great contributions may be our established high-quality multi-center dataset. The generalizability of AI model is important, and the acquisition protocols in each hospital may be different to impact the model performance. We did observe that AI model performed differently on different sources of datasets, such as Ho et al. reported AUC values of 0.77–0.95 on public ChestXray14 as training dataset and Montgomery and Shenzhen as two external testing datasets [14], and Lee reported AUC values of 0.79–0.83 on retrospectively gathered 6,654 TB and 3,182 normal chest radiographs [15], for which unstandardized image acquisition protocols played an important role. However, it should be noted that unstandardized image acquisition protocols were a common issue and it remained discussed in the field of medical imaging [16]. Therefore, in the current study, in order to improve the generalizability of the AI model, only those datasets met the image quality criteria of each hospital, and with gender and disease balance are included.

4 Conclusion

In this work, we propose a novel self-rating curriculum learning method for tuberculosis detection on CCR images. We create a large annotated tuberculosis dataset from multi-center hospitals, and establish a teacher model with ranking function based on self-rating scores to group cases in order of difficulty for curriculum learning training. Results show that the proposed SRCL method outperforms the compared traditional one-step method with improved image localization and segmentation performance and efficiency. As future work, we plan to validate our proposed model with prospective study and further transfer the method on other pulmonary detection including nodule and pneumonia.

Acknowledgements. This work was supported by the National Key Research and Development Program of China [Grant No.: 2019YFE0121400]; the Shenzhen Science and Technology Program [Grant No.: KQTD2017033110081833; JSGG20201102162802008]; and the Shenzhen Fundamental Research Program [Grant No.: JCYJ20190813153413160].

References

1. Zhou, W., et al.: Deep learning-based pulmonary tuberculosis automated detection on chest radiography: Large-scale independent testing. Quant. Imaging Med. Surg. **12**(4), 2344–2355 (2022)
2. World Health Organization: Global tuberculosis report. World Health Organization (2020)
3. Nijiati, M., et al.: Deep learning assistance for tuberculosis diagnosis with chest radiography in low-resource settings. J. X-ray Sci. Technol. (Preprint), 1–12 (2021)
4. Hooda, R., Mittal, A., Sofat, S.: A survey of CAD methods for tuberculosis detection in chest radiographs. In: Ray, K., Sharma, T.K., Rawat, S., Saini, R.K., Bandyopadhyay, A. (eds.) Soft Computing: Theories and Applications. AISC, vol. 742, pp. 273–282. Springer, Singapore (2019). https://doi.org/10.1007/978-981-13-0589-4_25
5. Jaeger, S., et al.: Automatic screening for tuberculosis in chest radiographs: a survey. Quant. Imaging Med. Surg. **3**(2), 89–99 (2013)
6. Paul, H.Y., Kim, T.K., Siegel, E., Yahyavi-Firouz-Abadi, N.: Demographic reporting in publicly available chest radiograph data sets: Opportunities for mitigating sex and racial disparities in deep learning models. J. Am. Coll. Radiol. **19**(1), 192–200 (2022)
7. Irvin, J., Rajpurkar, P., Ko, M., Yu, Y., Ciurea-Ilcus, S., Chute, C., Ng, A.Y.: Chexpert: A large chest radiograph dataset with uncertainty labels and expert comparison. In: Proceedings of the AAAI Conference on Artificial Intelligence, pp. 590–597. AAAI Press, Hawaii, USA (2019)
8. Bengiom, Y., Louradour, J., Collobert, R., Weston, J.: Curriculum learning. In: Proceedings of the 26th Annual International Conference on Machine Learning, pp. 41–48. Quebec, Australia (2009)
9. Kumar, M., Packer, B., Koller, D.: Self-paced learning for latent variable models. In: 24th Annual Conference on Neural Information Processing Systems, pp. 1189–1197. Vancouver, Canada (2010)

10. Weinshall, D., Gad, C., Dan, A.: Curriculum learning by transfer learning: Theory and experiments with deep networks. In: Proceedings of the 35th International Conference on Machine, p. 80. Stockholm, Sweden (2018)
11. Wei, J., Suriawinata, A., Ren, B., Liu, X., Lisovsky, M., Vaickus, L., Hassanpour, S.: Learn like a pathologist: curriculum learning by annotator agreement for histopathology image classification. In: Proceedings of the IEEE/CVF Winter Conference on Applications of Computer Vision, pp. 2473–2483. IEEE, Virtual/online, United States (2021)
12. Wang, X., Chen, Y., Zhu, W.: A survey on curriculum learning. IEEE Trans. Pattern Anal. Mach. Intell. **14**(8) (2021)
13. Luo, J., Kitamura, G., Arefan, D., Doganay, E., Panigrahy, A., Wu, S.: Knowledge-guided multiview deep curriculum learning for elbow fracture classification. In: Lian, C., Cao, X., Rekik, I., Xu, X., Yan, P. (eds.) MLMI 2021. LNCS, vol. 12966, pp. 555–564. Springer, Cham (2021). https://doi.org/10.1007/978-3-030-87589-3_57
14. Ho, T.K.K., Gwak, J., Prakash, O., Song, J.-I., Park, C.M.: Utilizing pretrained deep learning models for automated pulmonary tuberculosis detection using chest radiography. In: Nguyen, N.T., Gaol, F.L., Hong, T.-P., Trawiński, B. (eds.) ACIIDS 2019. LNCS (LNAI), vol. 11432, pp. 395–403. Springer, Cham (2019). https://doi.org/10.1007/978-3-030-14802-7_34
15. Lee, S., et al.: Deep learning to determine the activity of pulmonary tuberculosis on chest radiographs. Radiology **301**(2), 435–442 (2021)
16. Snaith, B., Field, L., Lewis, E.F., Flintham, K.: Variation in pelvic radiography practice: Why can we not standardise image acquisition techniques? Radiography **25**(4), 374–377 (2019)

Rib Suppression in Digital Chest Tomosynthesis

Yihua Sun[1], Qingsong Yao[3], Yuanyuan Lyu[4], Jianji Wang[5], Yi Xiao[6], Hongen Liao[1], and S. Kevin Zhou[2,3](✉)

[1] Department of Biomedical Engineering, School of Medicine, Tsinghua University, Beijing, China

[2] Center for Medical Imaging, Robotics, Analytic Computing & Learning (MIRACLE), School of Biomedical Engineering, Suzhou Institute for Advanced Research, University of Science and Technology of China, Suzhou, China

skevinzhou@ustc.edu.cn

[3] Key Laboratory of Intelligent Information Processing of Chinese Academy of Sciences (CAS), Institute of Computing Technology, CAS, Beijing, China

[4] Z2Sky Technologies Inc., Suzhou, China

[5] Affiliated Hospital of Guizhou Medical University, Guizhou, China

[6] Department of Radiology, Changzheng Hospital, Naval Medical University, Shanghai, China

Abstract. Digital chest tomosynthesis (DCT) is a technique to produce sectional 3D images of a human chest for pulmonary disease screening, with 2D X-ray projections taken within an extremely limited range of angles. However, under the limited angle scenario, DCT contains strong artifacts caused by the presence of ribs, jamming the imaging quality of the lung area. Recently, great progress has been achieved for rib suppression in a single X-ray image, to reveal a clearer lung texture. We firstly extend the rib suppression problem to the 3D case at the software level. We propose a **T**omosynthesis **RI**b Su**P**pression and **L**ung **E**nhancement **Net**work (TRIPLE-Net) to model the 3D rib component and provide a rib-free DCT. TRIPLE-Net takes the advantages from both 2D and 3D domains, which models the ribs in DCT with the exact FBP procedure and 3D depth information, respectively. The experiments on simulated datasets and clinical data have shown the effectiveness of TRIPLE-Net to preserve lung details as well as improve the imaging quality of pulmonary diseases. Finally, an expert user study confirms our findings. Our code is available at https://github.com/sunyh1/Rib-Suppression-in-Digital-Chest-Tomosynthesis.

Keywords: Digital chest tomosynthesis · Rib suppression · Limited angle artifacts

1 Introduction

Digital chest tomosynthesis (DCT) is a relatively novel imaging modality using limited angle tomography to provide the benefits of 3D imaging [22,30], which

Supplementary Information The online version contains supplementary material available at https://doi.org/10.1007/978-3-031-16431-6_66.

L. Wang et al. (Eds.): MICCAI 2022, LNCS 13431, pp. 696–706, 2022.
https://doi.org/10.1007/978-3-031-16431-6_66

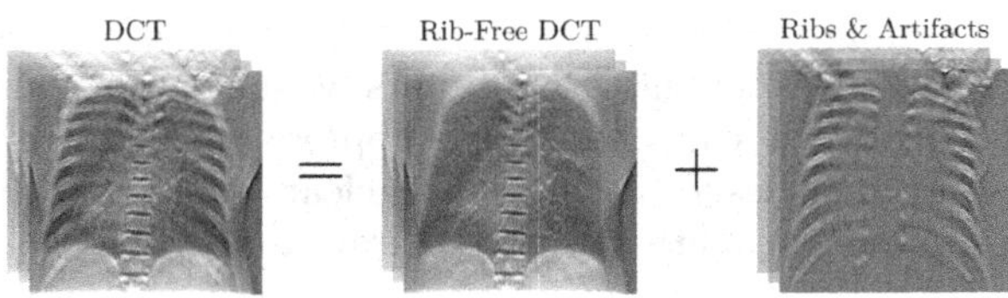

Fig. 1. Visualizations of the DCT, rib-free DCT, and ribs with artifact.

is reconstructed from a series of X-ray projections acquired within an extremely limited angle range [9] using filtered back projection (FBP) [17]. DCT shares some tomographic benefits with computed tomography (CT) as an adjunct to a conventional chest radiography exam for diagnosing pulmonary disease, and carries promise in clinical decision making. Compared with conventional chest radiography, DCT greatly improves the performance in lung nodule detection [14,20]. While compared with low-dose CT, DCT achieves competitive performance with lower cost and less radiation dosage [9] on the diagnosis of lung cancer [24]. DCT has advantages in the detection of the early stages of COVID-19 [21,28], too.

However, the limited angle scenario of DCT leaves strong artifacts of ribs overlapped with lung textures, making it difficult for doctors to identify some lung disease contexts close to the rib artifacts. Recently, great progress has been achieved for rib suppression in a single 2D X-ray radiograph [29]. At the software level, DecGAN [18,19] proposes a CycleGAN-based network that translates chest X-ray images to the simulated images. Furthermore, RSGAN [11] improves the performance by using a disentanglement network. For rib suppression in DCT at the device level, dual-energy chest tomosynthesis (DE-DCT) [23] can provide rib-free DCT by irradiating the tissues with two different energy levels of radiation, but exposes the patient to extra radiation doses. Therefore, rib suppression for DCT at the software level remains an important and unsolved problem.

We first visualize the difference between DCT and rib-free DCT in Fig. 1. The difference consists of the rib voxel itself and the artifacts generated by limited angle reconstruction using FBP, which should be removed simultaneously. In this paper, we propose a **T**omosynthesis **RI**b Su**P**pression and **L**ung **E**nhancement **Net**work (TRIPLE-Net) to suppress the ribs and their artifacts in DCT without additive radiation dose. TRIPLE-Net leverages knowledge in both 2D X-ray projection images and 3D DCT reconstructed volumes with **three** convolutional neural networks: projection-net, volume-net, and aggregation-net.

The projection-net suppresses rib component in the 2D projection, resulting in rib-free volume reconstructed by FBP, which is better at modeling artifacts based on the knowledge of FBP. However, decomposing rib components with high accuracy from the overlapped 2D projective textures might be difficult. Conversely, the volume-net directly extracts the artifacts from lung tissues in 3D DCT volumes, which is expert in accurate and fine modeling. While our proposed **TRIPLE-Net** reaps the advantages of two sub-modules by merging their outputs with an aggregation-net.

To the best of our knowledge, we are the first to suppress ribs along with their artifacts in DCT. Extensive quantitative results, visualizations, and user studies validate that TRIPLE-Net can effectively suppress rib artifacts in DCT and keep the textures of lung tissues accurate and clear, which greatly outperforms competitive rib suppression methods in 2D X-ray radiograph.

2 TRIPLE-Net

2.1 Problem Formulation

The acquisition of DCT can be viewed as taking X-ray projection from multiple viewing angles within a limited range. The detector is stationary while the X-ray radiator moves vertically, which forms a cone-beam geometry. The collected X-ray projection logarithm can be expressed as,

$$I^\theta = -ln \int \eta(E) e^{-\mathcal{R}_\theta(f(E))} dE, \tag{1}$$

where $f(E)$ is the 3D object's attenuation coefficient at energy level E, $\mathcal{R}_\theta$ is the 3D Radon transformation of DCT geometry at angle θ, and η is the energy distribution function of the X-ray spectrum.

Denote I^θ and I^θ_{rs} as the 2D X-ray projection at angle $\theta \in \Theta_\alpha$ and its corresponding rib-free projection, where $\Theta_\alpha = \{-\frac{\alpha}{2}, \cdots, \frac{\alpha}{2}\}$ is a set of angles in range of α. The difference image I^θ_Δ represents the rib component and artifacts:

$$I^\theta_\Delta = I^\theta - I^\theta_{rs}. \tag{2}$$

The 3D DCT volume V and rib-free volume V_{rs} reconstructed by a FBP operator $\mathcal{B}$ can be expressed as:

$$V_\alpha = \sum_{\theta \in \Theta_\alpha} \mathcal{B}(I^\theta), \; V_{\alpha,rs} = \sum_{\theta \in \Theta_\alpha} \mathcal{B}(I^\theta_{rs}). \tag{3}$$

Similarly, we denote the rib artifacts in 3D DCT as $V_{\alpha,\Delta}$. Given the linearity of FBP operator [17], we have

$$V_{\alpha,\Delta} = V_\alpha - V_{\alpha,rs} = \sum_{\theta \in \Theta_\alpha} \mathcal{B}(I^\theta) - \sum_{\theta \in \Theta_\alpha} \mathcal{B}(I^\theta_{rs}) = \sum_{\theta \in \Theta_\alpha} \mathcal{B}(I^\theta_\Delta). \tag{4}$$

Accordingly, there are two approaches to suppress the rib artifacts in DCT: (i) remove rib component I^θ_Δ from I^θ in 2D projection at each projection angle $\theta \in \Theta_\alpha$; (ii) or estimate $V_{\alpha,\Delta}$ directly from V_α in 3D.

2.2 Method

Figure 2 shows the framework of TRIPLE-Net, consisting of three convolutional neural networks: projection-net $\mathcal{M}_{2D}$, volume-net $\mathcal{M}_{3D}$, and aggregation-net $\mathcal{F}$.

2D Rib Component Modeling. A 2D Residual U-Net (ResUNet) [15] $\mathcal{M}_{2D}$ is utilized to extract $I^{\theta,P}_\Delta = \mathcal{M}_{2D}(I^\theta)$ from I^θ. $\mathcal{M}_{2D}$ shares weights for each projection I^θ across all $\theta \in \Theta_\alpha$. The FBP operator $\mathcal{B}$ is employed to reconstruct 3D rib component in DCT with the 2D predictions $I^{\theta,P}_\Delta$ from all viewing angles.

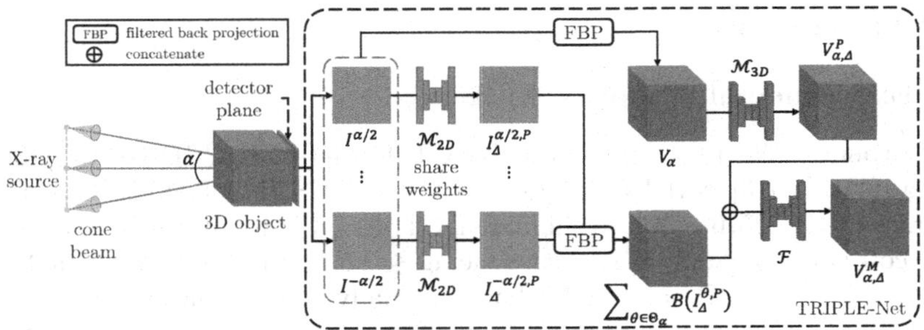

Fig. 2. The overall framework of TRIPLE-Net. DCT has a cone-beam limited angle geometry. $\mathcal{M}_{2D}$, which shares weights across different views, models rib components in the 2D domain before FBP. $\mathcal{M}_{3D}$ models rib components in the 3D domain after FBP. $\mathcal{F}$ takes the advantage of both worlds.

3D Rib Component Modeling. In the 2D projection domain, the rib component and other textures in lung tissues are overlapped. Besides, there is a lack of 3D consistency of the rib structures across different θ. To more accurately model rib artifacts, a 3D ResUNet $\mathcal{M}_{3D}$ is employed to directly predict $V_{\alpha,\Delta}^{P} = \mathcal{M}_{3D}(V_\alpha)$, modeling rib artifacts in 3D reconstructed volume domain.

Merging the Results. $\mathcal{M}_{2D}$ and $\mathcal{M}_{3D}$ are able to suppress ribs in DCT based on the knowledge in 2D X-ray projections and 3D DCT volumes respectively. Here is the instinctive difference: $\mathcal{M}_{2D}$ is good at modeling artifacts, while $\mathcal{M}_{3D}$ is expert in modeling rib components with high accuracy. Therefore, $\mathcal{F}$ is trained to learn the difference and leverages both the advantages of $\mathcal{M}_{2D}$ and $\mathcal{M}_{3D}$. A 3D ResUNet $\mathcal{F}$ is employed to merge the output of $\mathcal{M}_{2D}$ and $\mathcal{M}_{3D}$, producing the final prediction $V_{\alpha,\Delta}^{M}$ of ribs and their artifacts in DCT:

$$V_{\alpha,\Delta}^{M} = \mathcal{F}\left(\sum_{\theta\in\Theta_\alpha} \mathcal{B}\left(I_{\Delta}^{\theta,P}\right), V_{\alpha,\Delta}^{P}\right) \tag{5}$$

Loss Functions. With the ground truth I_{Δ}^{θ} and $V_{\alpha,\Delta}$, we perform supervised learning with the loss function defined as follows,

$$\mathcal{L}_{2D} = \lambda_{2D} \cdot ||I_{\Delta}^{\theta,P} - I_{\Delta}^{\theta}||_1,\ \mathcal{L}_{3D} = \lambda_{3D} \cdot ||V_{\alpha,\Delta}^{P} - V_{\alpha,\Delta}||_1. \tag{6}$$

Since we have ground truth for both 2D and 3D domain, $\mathcal{M}_{2D}$ and $\mathcal{M}_{3D}$ are trained separately and the parameters are fixed before training $\mathcal{F}$. With the pretrained $\mathcal{M}_{2D}$ and $\mathcal{M}_{3D}$, the loss function for training $\mathcal{F}$ is defined as,

$$\mathcal{L}_{M} = \lambda_{M} \cdot ||V_{\alpha,\Delta}^{M} - V_{\alpha,\Delta}||_1 \tag{7}$$

We experimentally set the hyperparameters λ_{2D}, λ_{3D} and λ_M to 20, 50, and 50.

3 Experiment

3.1 Experimental Setup

CT Datasets. We use 4 online available CT datasets, LIDC-IDRI("LI") [3,4,8], RibFrac ("RF") [13], MIDRC-RICORD-1a ("MR") [8,25,26] and NSCLC Radiogenomics ("NR") [5,6,8,10], to simulate DCT. As DCT requires a high resolution in the coronal plane, only CTs with spacing $\leq$ 2.5 mm in the longitudinal direction are selected. "MR" and "NR" have manually labeled masks for COVID-19 and lung cancer (mostly in nodule manifestation), which are utilized for testing the DCT image quality on lung disease. We use 90% of "LI" and 90% of "RF" for training, and 10% of "LI", 10% of "RF", "MR", and "NR" for testing. In total, there are 1353 training and 402 test data.

Simulating DCT from CT. With CT, we have the 3D attenuation coefficient distribution of an object. We can simulate 2D X-ray projection images with our DCT geometry from CT using (1) by deriving $f(E)$ from HU values [27]. This procedure is known as the digitally reconstructed radiography (DRR) technique. We segment ribs in CT with a 3D U-Net [7] and inpaint rib mask with surrounding tissues, deriving the 3D attenuation coefficient distribution function denoted as $f_{rs}(E)$. Then we can obtain rib suppressed 2D X-ray projections I^{θ}_{rs} with DRR similarly. The volumes V_{α} and $V_{\alpha,rs}$ are accordingly reconstructed by FBP with the simulated projections as described in (3). In this paper, we do not focus on reconstruction algorithms for DCT, but on rib suppression in DCT; so we refer to $V_{\alpha,\Delta}$ and $V_{\alpha,rs}$ as "ground truth" for ribs on DCT and rib-free DCT.

Implementation Details. The 3D field of view (FOV) for DCT is set to 409.6 mm $\times$ 300 mm $\times$ 409.6 mm with a shape of 256 $\times$ 128 $\times$ 256, for the desired resolution of DCT in the anterior-posterior direction is lower. Before DRR, the lung area of CT is placed at the center of FOV. The DRR and FBP procedures are implemented with ODL [2]. The shape of I^{θ} and I^{θ}_{rs} are 256 $\times$ 256. We train and evaluate models separately for $\alpha = 30°$ and $\alpha = 15°$, with 59 and 29 projections taken equiangularly in the range of α. $\mathcal{M}_{2D}$, $\mathcal{M}_{3D}$ and $\mathcal{F}$ are implemented with MONAI [1] in PyTorch framework and trained with Adam [16] optimizer with a learning rate of 1×10^{-4} for 100 epochs.

Performance Metrics. We segment the lung area LA in CT with a lung mask [12], and calculate L_1 and L_2 criteria within LA of DCT (L_1^{LA}, L_2^{LA}) and on the whole DCT volume (L_1, L_2) for evaluation over the whole test dataset. Besides, we use peak signal-to-noise ratio (PSNR) for evaluation within disease masks of "MR" and "NR" images pre-normalized to $[0,1]$. We use paired t-test deriving p-values to compare the metrics of other methods with TRIPLE-Net's.

Clinical Study. We collect 4 clinical DCTs with $\alpha = 15°$ and 1 clinical DCT with $\alpha = 30°$ for evaluation, referred as the clinical dataset. We randomly select 30 cases from the test dataset and simulate DCTs with both $\alpha = 15°$ and 30°, resulting in 60 simulated DCTs. We invited 2 clinical doctors to give rankings

Table 1. Quantitative comparison of different methods. The p-value is summarized in the square bracket and asterisks indicate p-values <0.001. TRIPLE-Net achieves the best performance in different regions and in terms of all metrics.

α	Method	$L_1(\times 10^{-2})$	$L_2(\times 10^{-4})$	$L_1^{LA}(\times 10^{-2})$	$L_2^{LA}(\times 10^{-4})$	$PSNR(dB)$
30°	RSGAN	0.853*	3.272*	1.891*	6.997*	25.09*
	$\mathcal{M}_{2D}$	0.753*	2.406*	1.793*	6.344*	25.07*
	$\mathcal{M}_{3D}$	0.662*	1.590*	1.033*	2.280*	33.37 [<0.027]
	TRIPLE-Net	**0.471**	**0.948**	**0.869**	**1.658**	**33.56**
15°	RSGAN	1.164*	5.957*	2.641*	13.45*	25.11*
	$\mathcal{M}_{2D}$	1.013*	4.323*	2.446*	11.51*	25.46*
	$\mathcal{M}_{3D}$	0.873*	2.756*	1.640*	5.356*	30.52*
	TRIPLE-Net	**0.687**	**2.058**	**1.397**	**4.200**	**30.85**

from 1–5 (higher is better) for the DCTs by paying attention to the rib suppression performance and lung details. The DCT processed by different methods are randomly shuffled before presenting to the doctors. Doctor A is a proficient radiologist for chest imaging with over 20 years of reading experience. Doctor B is an orthopedist but with the knowledge of DCT. We use paired Wilcoxon signed-rank test deriving p-values to compare scores of other methods with TRIPLE-Net's.

3.2 Comparison on Simulated Dataset

Quantitative Metrics. RSGAN [11] is a disentanglement method with generative adversarial networks designed for better performance on clinical chest X-ray images, which may have deteriorated performances on the simulated dataset. Therefore, we train $\mathcal{M}_{2D}$ solely on the DRR domain to make a fair comparison of 2D and 3D methods on our simulated dataset. Table 1 shows the quantitative comparison of different methods, where TRIPLE-Net achieves the best performance. TRIPLE-Net has the best rib suppression ability on the whole DCT volume, and within the lung area which is the major concern of DCT. Moreover, TRIPLE-Net has better image quality where the pulmonary disease lies.

Visualization on Rib Suppression Performance. Figure 3 shows the rib suppression performance of different methods on DCT with $\alpha = 30°$. The lung details are magnified in the green rectangles with a unified window level, whose corresponding difference maps are in the red rectangles. We can observe that with DCT rib suppression techniques, the lung details (green rectangles) intersecting with rib artifacts can be revealed more clearly.

Since 2D methods extract rib components in the 2D domain and model $V_{\alpha,\Delta}$ with the exact FBP procedure, they only affect the "rib-FBP" area and no more, as pointed by red arrows in Fig. 3. On the contrary, $\mathcal{M}_{3D}$ simply approximates the volumetric ground truth with a 3D neural network without the knowledge of the FBP mechanism, leaving error widely spread on the whole volume. Moreover, bone components are sometimes complex (bones around shoulders) or with

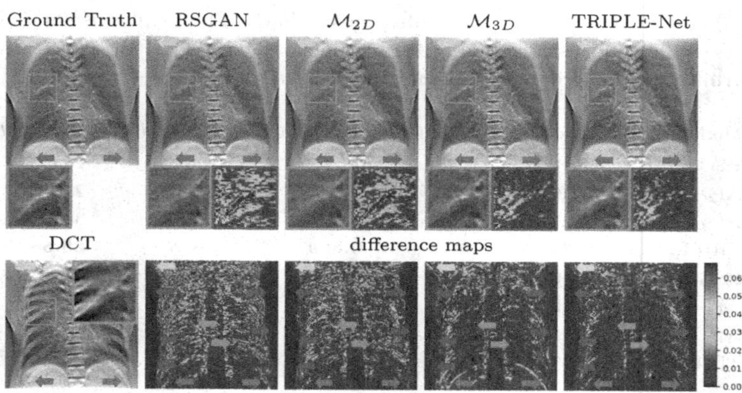

Fig. 3. Visualization of rib suppression results on DCT with $\alpha = 30°$. The methods with 3D information can better preserve lung details while removing rib components (green arrows/rectangles). The 2D methods only affect the "rib-FBP" area and no more, having better accuracy in "non-rib" and "tricky-artifacts" area outside lung area (red/yellow/pink arrows). TRIPLE-Net leverages the advantages of both 2D and 3D methods. (Color figure online)

lower contrast (ribs in tissue), as pointed by yellow and pink arrows in Fig. 3. So, 3D models have difficulty extracting features and modeling those tricky artifacts. 2D models are easier to model those tricky artifacts in DCT for they are caused by FBP and not that tricky in the 2D domain. With both 2D and 3D information, TRIPLE-Net effectively learns the modality difference by leveraging the advantages of both 2D and 3D domains and achieves the best visual quality with high accuracy within and beyond the lung area.

Visualization on Disease Imaging Quality. In Fig. 4, images with lung nodules are visualized, where the affected area is magnified in the green rectangles with a unified window level for $\alpha = 15°$ and 30°, respectively. The corresponding original CT, from which the DCTs are simulated, is also presented. With DCT rib suppression techniques, lung nodules buried by rib artifacts can be identified more easily with clearer boundaries. In the result of 2D models, the lung nodule and lung texture are smoothed. The lung nodule is more distinguishable with better contrast in the result of 3D models. Besides, TRIPLE-Net has a more accurate intensity compared with $\mathcal{M}_{3D}$, within and around the diseased area.

Reducing the DCT's Acquisition Angle Range. Because of a more limited range of angles, DCT with $\alpha = 15°$ contains stronger artifacts than $\alpha = 30°$. It is harder to identify lung nodules when $\alpha = 15°$ than $\alpha = 30°$ as shown in Fig. 4. But with TRIPLE-Net, visually the DCT with $\alpha = 15°$ has a comparative imaging quality as $\alpha = 30°$ for lung textures, especially lung nodules. This shows that TRIPLE-Net has the potential to further reduce the DCT acquired angle

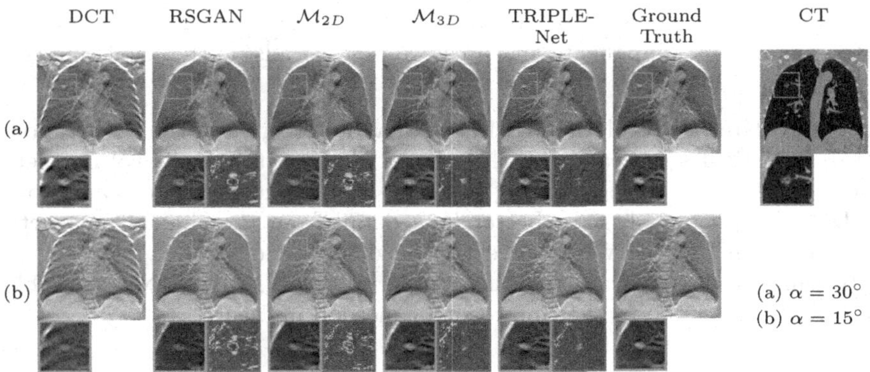

Fig. 4. Lung nodules imaging performance of different methods. With DCT rib suppression techniques, lung details buried by rib artifacts are revealed more clearly. The 2D methods provide smoothed results while the 3D methods present the lung nodule with better contrast and sharper boundary. TRIPLE-Net has better accuracy within and around the lung nodule.

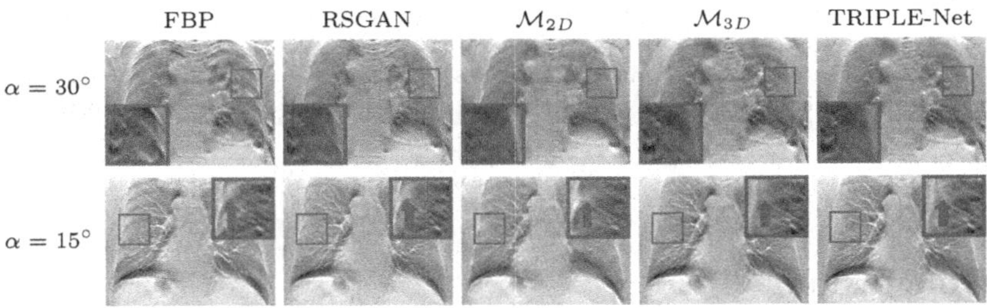

Fig. 5. Visualization of 2 different patients from clinical data, with $\alpha = 15°$ and $\alpha = 30°$. $\mathcal{M}_{2D}$ provided smoother lung textures. 3D models can better suppress rib artifacts in DCT compared with RSGAN, pointed by red arrows. (Color figure online)

range. It is beneficial to scenarios that do not allow for a large source movement, *e.g.*, intraoperative imaging. Furthermore, it also carries a promise of reducing the radioactive dose, which is harmful to patients.

3.3 Clinical Study

Visualization. We directly run our trained model on clinical data from two patients and visualize the results in Fig. 5. To better generalize TRIPLE-Net to clinical data, RSGAN is substituted for the 2D sub-module $\mathcal{M}_{2D}$. The noticeable difference is contained in the red rectangle, being magnified and showed in a unified window level respectively for $\alpha = 15°$ and $\alpha = 30°$. RSGAN preserves more lung detail than $\mathcal{M}_{2D}$, for its stronger generalization ability in the clinical data. However, without cross-view or 3D information, there is still some rib component leftover in the result of RSGAN (red arrows). Lung details in the red rectangle in the result of $\mathcal{M}_{2D}$ are smoothed. With 3D information, models can better remove rib components and preserve lung details on the whole image.

Expert Rating. In Table 2, TRIPLE-Net has a score greater than other models with a statistical significance on the simulated dataset. On clinical dataset, it is demonstrated that deep learning based rib suppression methods can improve the DCT imaging quality for better lung detail from the clinical perspective, compared with FBP. However, given the insufficiency of clinical data, the user study can not demonstrate the sheer superiority of TRIPLE-Net in Table 2, which motivates us to further improve our method in future.

Table 2. Ratings of clinical and simulated DCT images processed by different methods. For comparison of TRIPLE-Net with other methods, p-values are also summarized.

Method	Clinical dataset				Simulated dataset			
	Doctor A		Doctor B		Doctor A		Doctor B	
	Rating	p-value	Rating	p-value	Rating	p-value	Rating	p-value
FBP	1.00 ± 0.00	0.031	1.00 ± 0.00	0.031	1.00 ± 0.00	<0.001	1.02 ± 0.13	<0.001
RSGAN	2.20 ± 0.40	0.031	2.00 ± 0.00	0.031	2.38 ± 0.49	<0.001	2.03 ± 0.26	<0.001
$\mathcal{M}_{2D}$	3.00 ± 0.63	0.031	3.00 ± 0.00	0.031	2.92 ± 0.78	<0.001	3.02 ± 0.34	<0.001
$\mathcal{M}_{3D}$	4.40 ± 0.80	0.500	$\mathbf{4.60 \pm 0.49}$	0.688	4.20 ± 0.81	0.057	4.18 ± 0.53	<0.001
TRIPLE-Net	$\mathbf{4.40 \pm 0.49}$	n.a	4.40 ± 0.49	n.a	$\mathbf{4.50 \pm 0.65}$	n.a	$\mathbf{4.75 \pm 0.43}$	n.a

4 Conclusion

In this paper, we have proposed TRIPLE-Net to model rib artifacts in DCT caused by limited angle FBP, leveraging information in both 2D and 3D domains and reaps the benefits from both worlds. TRIPLE-Net can suppress rib artifacts in DCT to obtain a clearer lung texture and better visualization of pulmonary disease areas, which has the potential for better diagnosis of lung nodules and COVID-19 in clinics. In future, research could be furthered for higher resolution in rib-suppressed DCT and better performance in clinical data.

References

1. Medical open network for artificial intelligence (monai). https://monai.io/. Accessed 27 Feb 2022
2. Adler, J., Kohr, H., Oktem, O.: Operator discretization library (odl) (2017). Software available from https://github.com/odlgroup/odl
3. Armato, S.G., III., et al.: The lung image database consortium (lidc) and image database resource initiative (idri): a completed reference database of lung nodules on ct scans. Med. Phys. **38**(2), 915–931 (2011)
4. Armato III, S.G., et al.: Data from lidc-idri [data set]. Cancer Imaging Arch. (2015)

5. Bakr, S., et al.: A radiogenomic dataset of non-small cell lung cancer. Sci. Data **5**(1), 1–9 (2018)
6. Bakr, S., et al.: Data for nsclc radiogenomics collection. Cancer Imaging Arch. (2017)
7. Çiçek, Ö., Abdulkadir, A., Lienkamp, S.S., Brox, T., Ronneberger, O.: 3D U-Net: learning dense volumetric segmentation from sparse annotation. In: Ourselin, S., Joskowicz, L., Sabuncu, M.R., Unal, G., Wells, W. (eds.) MICCAI 2016. LNCS, vol. 9901, pp. 424–432. Springer, Cham (2016). https://doi.org/10.1007/978-3-319-46723-8_49
8. Clark, K.: The cancer imaging archive (tcia): maintaining and operating a public information repository. J. Dig. Imaging **26**(6), 1045–1057 (2013)
9. Dobbins, J.T., III., McAdams, H.P.: Chest tomosynthesis: technical principles and clinical update. Eur. J. Radiol. **72**(2), 244–251 (2009)
10. Gevaert, O., et al.: Non-small cell lung cancer: identifying prognostic imaging biomarkers by leveraging public gene expression microarray data-methods and preliminary results. Radiology **264**(2), 387–396 (2012)
11. Han, L., Lyu, Y., Peng, C., Zhou, S.K.: Gan-based disentanglement learning for chest x-ray rib suppression. Med. Image Anal. **77**, 102369 (2022)
12. Hofmanninger, J., Prayer, F., Pan, J., Röhrich, S., Prosch, H., Langs, G.: Automatic lung segmentation in routine imaging is primarily a data diversity problem, not a methodology problem. Eur. Radiol. Exp. **4**(1), 1–13 (2020). https://doi.org/10.1186/s41747-020-00173-2
13. Jin, L., et al.: Deep-learning-assisted detection and segmentation of rib fractures from ct scans: development and validation of fracnet. EBioMedicine (2020)
14. Jung, H., Chung, M., Koo, J., Kim, H., Lee, K.: Digital tomosynthesis of the chest: utility for detection of lung metastasis in patients with colorectal cancer. Clin. Radiol. **67**(3), 232–238 (2012)
15. Kerfoot, E., Clough, J., Oksuz, I., Lee, J., King, A.P., Schnabel, J.A.: Left-ventricle quantification using residual u-net. In: Pop, M., et al. (eds.) STACOM 2018. LNCS, vol. 11395, pp. 371–380. Springer, Cham (2019). https://doi.org/10.1007/978-3-030-12029-0_40
16. Kingma, D.P., Ba, J.: Adam: a method for stochastic optimization. In: 3rd International Conference on Learning Representations (2015)
17. Lauritsch, G., Härer, W.H.: Theoretical framework for filtered back projection in tomosynthesis. In: Medical Imaging 1998: Image Processing, vol. 3338, pp. 1127–1137. International Society for Optics and Photonics (1998)
18. Li, H., et al.: High-resolution chest x-ray bone suppression using unpaired CT structural priors. IEEE Trans. Med. Imaging **39**(10), 3053–3063 (2020)
19. Li, Z., Li, H., Han, H., Shi, G., Wang, J., Zhou, S.K.: Encoding CT anatomy knowledge for unpaired chest X-ray image decomposition. In: Shen, D., et al. (eds.) MICCAI 2019. LNCS, vol. 11769, pp. 275–283. Springer, Cham (2019). https://doi.org/10.1007/978-3-030-32226-7_31
20. Machida, H., et al.: Whole-body clinical applications of digital tomosynthesis. Radiographics **36**(3), 735–750 (2016)
21. Miroshnychenko, O., Miroshnychenko, S., Nevgasymyi, A., Khobta, Y.: Contrasts comparison of same cases of chest pathologies for radiography and tomosynthesis. In: 2020 International Symposium on Electronics and Telecommunications (ISETC), pp. 1–4. IEEE (2020)
22. Molk, N., Seeram, E.: Digital tomosynthesis of the chest: a literature review. Radiography **21**(2), 197–202 (2015)

23. Sone, S.: Chest imaging with dual-energy subtraction digital tomosynthesis. Acta Radiologica **34**(4), 346–350 (1993)
24. Terzi, A., et al.: Lung cancer detection with digital chest tomosynthesis: baseline results from the observational study sos. J. Thoracic Oncol. **8**(6), 685–692 (2013)
25. Tsai, E.B., et al.: The RSNA international covid-19 open radiology database (ricord). Radiology **299**(1), E204–E213 (2021)
26. Tsai, E.B., et al.: Data from the medical imaging data resource center - RSNA international covid radiology database release 1a - chest ct covid+ (midrc-ricord-1a). Data Cancer Imaging Arch. (2022)
27. Unberath, M., et al.: DeepDRR – a catalyst for machine learning in fluoroscopy-guided procedures. In: Frangi, A.F., Schnabel, J.A., Davatzikos, C., Alberola-López, C., Fichtinger, G. (eds.) MICCAI 2018. LNCS, vol. 11073, pp. 98–106. Springer, Cham (2018). https://doi.org/10.1007/978-3-030-00937-3_12
28. Yao, Q., Xiao, L., Liu, P., Zhou, S.K.: Label-free segmentation of Covid-19 lesions in lung CT. IEEE Trans. Med. Imaging **40**(10), 2808–2819 (2021)
29. Zhou, S.K., et al.: A review of deep learning in medical imaging: imaging traits, technology trends, case studies with progress highlights, and future promises. In: Proceedings of the IEEE (2021)
30. Zhou, S.K., Rueckert, D., Fichtinger, G.: Handbook of Medical Image Computing and Computer Assisted Intervention. Academic Press, London (2019)

Multi-task Lung Nodule Detection in Chest Radiographs with a Dual Head Network

Chen-Han Tsai(✉) and Yu-Shao Peng

HTC DeepQ, Taipei, Taiwan
{maxwell_tsai,ys_peng}@htc.com

Abstract. Lung nodules can be an alarming precursor to potential lung cancer. Missed nodule detections during chest radiograph analysis remains a common challenge among thoracic radiologists. In this work, we present a multi-task lung nodule detection algorithm for chest radiograph analysis. Unlike past approaches, our algorithm predicts a global-level label indicating nodule presence along with local-level labels predicting nodule locations using a Dual Head Network (DHN). We demonstrate the favorable nodule detection performance that our multi-task formulation yields in comparison to conventional methods. In addition, we introduce a novel Dual Head Augmentation (DHA) strategy tailored for DHN, and we demonstrate its significance in further enhancing global and local nodule predictions.

Keywords: Nodule detection · Chest radiograph · Dual Head Network · Dual Head Augmentation · Faster R-CNN

1 Introduction

Lung cancer ranks among the top causes of cancer-related deaths worldwide. Pulmonary nodule findings, though typically benign, are an alarming sign for potential lung cancer. Given its simple and inexpensive operating cost, chest radiography (i.e., x-rays) is the most widely adopted chest imaging solution available. However, one concern during radiograph analysis is the proportion of nodules thoracic radiologists often miss due to the nature of the imaging modality [5,26]. A chest radiograph is a 2D projection of a patients' chest. Thus, nodules appear less visible when occluded by other organs (e.g., rib cages) or foreign bodies (e.g., CVADs). With the rising workload already posing a challenge for thoracic radiologists, assistive tools to reduce missed nodules during chest radiograph analysis are gaining significant clinical relevance [4,10,19].

To identify pulmonary nodules on chest radiographs, several works propose potential solutions. Some [1,21,28] focus on image-level prediction indicating

Supplementary Information The online version contains supplementary material available at https://doi.org/10.1007/978-3-031-16431-6_67.

L. Wang et al. (Eds.): MICCAI 2022, LNCS 13431, pp. 707–717, 2022.
https://doi.org/10.1007/978-3-031-16431-6_67

nodule presence per scan (we refer to as *global* methods). Others [11,13,18,24] study patch-level prediction exploring nodule detection with local bounding box information for each nodule (we refer to as *local* methods). Although both local and global methods offer information regarding nodule presence in a given chest radiograph, adopting just a single method alone can be undesirable. Local methods offer the benefit of pinpointing each nodule, but the adopted labeling criterion can be highly subjective and prone to inconsistency [2,12]. Global methods alleviate this issue by predicting a single label indicating nodule presence, but further effort is required by the examiner to locate these nodules.

Li et al. [14] and Pesce et al. [20] attempted to address this disparity. The former formulated a multiple instance learning (MIL) model to classify nodule presence using a grid of patches across the input scan. The image-level label is then computed using the joint probability across patch predictions. Despite its attempt in combining local and global predictions, such MIL model is not translation equivariant by design, causing the predicted global label to be highly dependent on the nodule location innate to the scan. The latter proposed CONAF, a network composed of a shared backbone between its classification and localization head. The classification head outputs a global label indicating nodule presence, and its localization head outputs a downsized score map. Considering that most nodule sizes are relatively small with respect to the scan size, the low resolution scoremap limits the localizer head from explicit localization purposes.

In this work, we present a novel multi-task lung nodule detection algorithm using a Dual Head Network (DHN) and an accompanying Dual Head Augmentation (DHA) training strategy. Our multi-task objective is similar to [3], but instead of using a RetinaNet [16], we adopt a modified Faster-RCNN [22] architecture customized for nodule detection. Considering the properties of pulmonary nodule, we propose to use deformable convolutions [6] and the gIOU loss [23]. In addition, we incorporate our novel DHA strategy during DHN training, and we demonstrate its importance in further enhancing global and local nodule predictions. The remaining of the paper is organized as follows. We first illustrate the preliminaries in Sect. 2. The proposed methods will be detailed in Sect. 3, followed by experimental results in Sect. 4. Conclusion will be given in Sect. 5.

2 Preliminaries

The Faster-RCNN [22] is a two-stage network originally designed for object detection. The first stage of the network is a feature extractor (i.e. VGG-16 [25]), and the second stage consists of an RPN, ROIPool, and ROIHead module. For a given input image, the feature extractor outputs a feature representation of that image, and this representation is fed to the second stage. The RPN generates bounding box proposals around regions that contain potential non-background objects. Crops for each proposed region are then taken from the extracted feature representation, and they are resized to a fixed dimension using ROIPool. The fixed-size feature crops are then independently classified using the ROIHead, and an updated bounding-box is predicted for each crop.

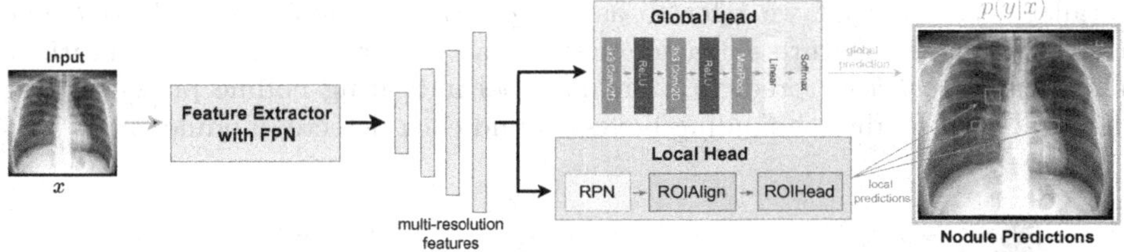

Fig. 1. An illustration of the DHN architecture during inference. The input x is fed through a feature extractor with FPN to obtain a set of multi-resolution features. The *global head* predicts the global label indicating nodule presence in the scan, and the *local head* predicts local bounding boxes indicating nodule locations.

Modern implementations of the Faster-RCNN often include two modifications to the original design that enhances detection performance. The first is the attachment of the Feature Pyramid Network (FPN) [15] behind the first stage feature extractor. FPN allows cross-level information flow between multi-resolution representations which is beneficial for detecting small objects. The second is the replacement of ROIPool with Multi-Scale ROIAlign [7,15] to avoid quantization while cropping the multi-resolution features.

During training, the RPN generates an objectness loss ℓ_{obj} from the background foreground classification and a regression loss ℓ_{reg} from the distance computed between the proposed regions with their matched ground truth bounding boxes. The ROIHead generates a classification loss ℓ_{cls} during feature crop classification and a bounding box loss ℓ_{bbox} from the distance computed between the updated bounding boxes and their matched ground truth bounding boxes. The default setup utilizs a Smooth L1 Loss to train ℓ_{reg}, ℓ_{bbox} and the Cross Entropy Loss to train ℓ_{obj}, ℓ_{cls}. The final loss function L_{local} is formulated in Eq. 1

$$L_{local} = \ell_{obj} + \ell_{reg} + \ell_{cls} + \ell_{bbox}. \tag{1}$$

3 Methods

In this section, we present our multi-task Dual Head Network (DHN) architecture and the accompanying Dual Head Augmentation (DHA) strategy. The DHN takes advantage of the two-stage structure seen in the Faster-RCNN by adding an additional network in parallel to the original second stage. The DHA strategy is applied during training, and it utilizes the DHN's dual head design in improving nodule detection performance. The specifics are detailed in the following sections.

3.1 Dual Head Network

The DHN architecture is designed in a two-stage approach (see Fig. 1). The first stage is a feature extractor with FPN, and the second stage consists of two parallel networks that we refer to as the *global head* and the *local head.* For a given scan, the feature extractor first extracts its respective representation. Then, the *global head* predicts a binary label indicating nodule presence, and the *local head* predicts bounding boxes around each detected nodule.

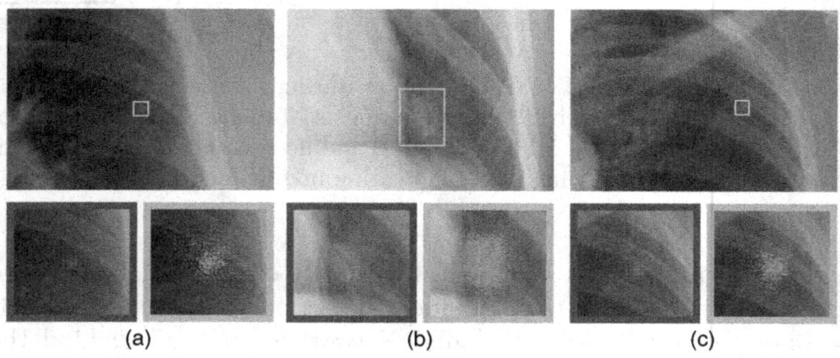

Fig. 2. Comparison between deformable (turquoise) versus standard convolution (dark blue) applied on three sample cases (a–c) using the 3rd layer of a trained ResNet-18 feature extractor. The receptive fields are fixed if regular convolution is applied, whereas deformable convolutions allow dynamic focus on the regions of interest. (Color figure online)

The feature extractor of our DHN is a modified ResNet-18 [8]. We conduct a series of experiments comparing various CNN architectures from the ResNet family, and we observe slightly better performance as model size increased. However, the training time increases significantly for larger models. Hence, we select the smallest model from the ResNet family, the ResNet-18, to serve as a lower bound for DHN performance throughout our experiments. We modify the ResNet-18 by replacing standard convolutions with deformable convolutions [6] in the final three layers. Applying deformable convolutions allows more dynamic focus on particular regions in the image where the nodules size might be small (see Fig. 2). For a given scan x, we take the intermediate representations the ResNet-18 extracts, and we pass them to the FPN to obtain set of multi-resolution representations. This set of multi-resolution representations is fed to the *global* and *local heads* for further processing.

The *global head*'s primary purpose is to classify whether nodules are present in a given scan. Thus, we select the representation with the largest receptive field [17] as input. As shown in Fig. 3, this representation is passed through two consecutive 2D convolutions and ReLUs before being max-pooled into a single vector. Then, a linear layer and a softmax layer are applied to obtain the probability indicating nodule presence in the scan. For a set of N labeled scans $\{(x_i, y_i)\}_{i=1}^{N}$ where $x_i \in \mathbf{R}^{h \times w}$ is the ith scan with resolution $h \times w$ (we set $h, w = 512$) and $y_i \in \{0, 1\}$ is the corresponding global label, we formulate the *global head loss* L_g (see Eq. 2) using a weighted cross-entropy:

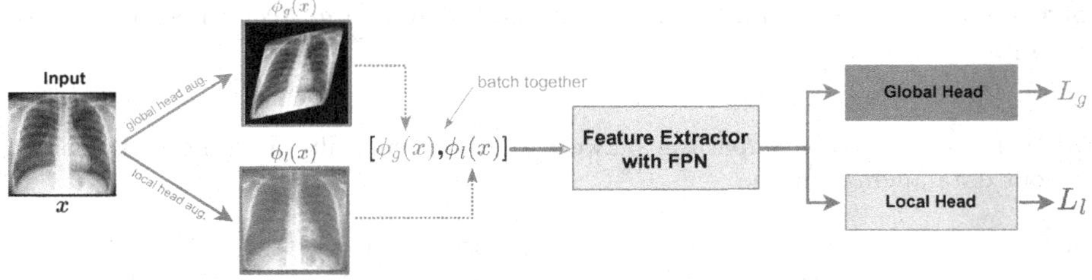

Fig. 3. An illustration of DHA. Two augmented images $\phi_g(x)$ and $\phi_l(x)$ of x are generated. They are batched together and passed into the feature extractor with FPN. The output features are then split according to their corresponding head. Individual head losses L_g and L_l are summed during training. Notice that $\phi_g(x)$ only updates the *global head*, and $\phi_l(x)$ only updates the *local head* during back-propagation.

$$L_g = -\sum_{i}^{N} \alpha_1 y_i \log\left(p(y_i|x_i)\right) + \alpha_2(1-y_i)\log\left(1-p(y_i|x_i)\right), \tag{2}$$

where α_1 and α_2 are two hyperparameters.

The *local head* serves as a nodule detector, and we adopt the second-stage design as specified in Sect. 2. During training, however, we replace the Smooth L1 loss in the RPN and the ROIHead with the gIOU loss [23] since the gIOU's scale-invariant property is beneficial during small scale bounding box optimization. The *local head loss* L_l follows the formulation in Eq. 1. We propose an end-to-end optimization method to consider the *local head loss* and the *global head loss* simultaneously. The final multi-task loss L (see Eq. 3) is a weighted sum of the two losses, i.e.,

$$L = \lambda_1 L_g + \lambda_2 L_l. \tag{3}$$

3.2 Dual Head Augmentation

Data augmentation is a well-known technique that increases diversity in the training data to improve model generalization. In classical image classification or object detection tasks, one augmentation strategy (i.e., a pre-defined set of stochastic transforms) is applied per training image during the forward-pass. However, training with just one augmentation strategy per image for a dual head architecture can easily lead to one head being particularly optimized while the other head performs mediocrely.

To fully optimize each head to their specific objectives, we propose a novel data augmentation strategy called the Dual Head Augmentation (DHA). DHA takes advantage of the DHN's dual head structure by applying an augmentation strategy for each head. As shown in Fig. 3, we designate an augmentation function ϕ_g for the *global head* and ϕ_l for the *local head.* Given an input scan x, the augmented images $\phi_g(x)$ and $\phi_l(x)$ are generated and batched together. This batch is fed into the feature extractor with FPN, and we obtain the multi-resolution representations. We split the batch, and we feed the representations corresponding to $\phi_g(x)$ and $\phi_l(x)$ to the *global* and *local heads* respectively to optimize each head.

Table 1. Performance comparison between the DHN and alternate methods trained without data augmentation.

Model	Classification metrics		Localization metrics		
	ROC-AUC	PR-AUC	FROC-AUC	AFROC-AUC	mAP
FCOS [27]	–	–	0.540 ± 0.054	0.564 ± 0.024	0.166 ± 0.009
RetinaNet [16]	–	–	0.576 ± 0.009	0.507 ± 0.023	0.166 ± 0.003
DenseNet-121 [9]	0.856 ± 0.006	0.612 ± 0.006	–	–	–
CONAF [20]	0.800 ± 0.004	0.450 ± 0.007	–	–	–
Global head only	0.809 ± 0.010	0.524 ± 0.014	–	–	–
Local head only	–	–	0.618 ± 0.006	0.606 ± 0.029	0.184 ± 0.008
DHN	$\mathbf{0.873 \pm 0.004}$	$\mathbf{0.654 \pm 0.016}$	$\mathbf{0.628 \pm 0.011}$	$\mathbf{0.626 \pm 0.033}$	$\mathbf{0.188 \pm 0.009}$

We select a set of image transformations[1] that involves illumination transforms (e.g., histogram equalization, random brightness, etc.) and geometric transforms (e.g., horizontal flip, rotation, etc.). In this work, we consider two types of augmentation strategies that exhibit favorable single-head performance, and we define the augmentation strategies by their sampling method. The first sampling method is binomial sampling. Formally, for a given probability p and a set of transforms $\Theta = \{\theta_j\}_{j=1}^{M}$, binomial sampling selects transform θ_j for $j \in \{1, ..., M\}$ with probability p, and we refer to the corresponding augmentation strategy as $\phi^{\mathtt{bin}}(x; p, \Theta)$. The second sampling method is uniform sampling. Specifically, uniform sampling selects a transform from Θ with probability $1/M$, and we refer to this augmentation strategy as $\phi^{\mathtt{uni}}(x; \Theta)$. During each training iteration, the augmentation strategy samples from Θ using one of the above sampling methods and applies the selected transforms.

4 Experiments

In this section, we evaluate the nodule detection performance of a DHN in comparison to other notable approaches. In addition, we perform a series of experiments to compare the influence different augmentation strategies play towards enhancing global and local predictions. Classification performance is evaluated using the *global head* prediction, and localization performance[2] is evaluated with *local head* predictions. We report on the test set mean and standard deviation of the top 8 performing validation checkpoints throughout each experiment. Each model is trained with SGD and momentum for 70 epochs (~3 days on an NVIDIA

[1] The complete list of transformations are detailed in the supplementary materials.
[2] FROC and AFROC are computed with an Intersection over Union (IOU) threshold of 0.4, and the FROC-AUC is computed with a False Positive Per-Image up to 1.

P100 16 GB GPU) with a step size of 5e–5 and momentum of 0.975 at a batch size of 2. Loss weights α_1, α_2 are set to 0.69 and 1.76 while λ_1, λ_2 are set to 0.35 and 2.5.

4.1 Dataset

The dataset we evaluate in this study is a subset of the National Institute of Health (NIH) Chest X-Ray dataset [28]. Each case in our dataset was labeled with bounding box annotations for each nodule by three thoracic radiologists. Prior to labeling, each radiologist had to pass an assessment test requiring them to correctly identify over 80% of the cases from a held-out test set with potential lung nodules. The radiologists were then asked to independently label each case in our dataset. Labels were aggregated upon completion, and the radiologists reviewed each case to reach a consensus on which labels to keep. A final senior radiologist then reviewed and control the quality of annotations. In total, we randomly sample 26000 scans from NIH dataset and 21,189 qualified scans are added to our dataset. These scans are split into training, validation, and test sets following the ratio of 80 : 10 : 10 based on patient identifiers.

4.2 Dual Head Network Analysis

For our first set of experiments, we evaluate the performance of a DHN in comparison to several *global* and *local* methods [9,16,27]. We also compare our DHN with a notable dual-head approach CONAF [20]. Although CONAF's localization head generates a heatmap-like mask for nodule localization, the mask is too coarse to derive precise bounding box coordinates from. Hence, we only evaluate the classifier head's prediction in our comparison. For fair comparison, we do not apply data augmentation to the training strategy. We also analyze the single head performance of the DHN. Specifically, we consider the case where we train only the *global head* and the case where we train only the *local head*.

As shown in Table 1, the DHN architecture yields favorable improvements in both classification and localization metrics. We believe this improvement is that our DHN extracts more informative representations due to cross-task supervision. Specifically, training with local labels using the *local head* encourages the feature extractor to extract more meaningful representations that highlight local findings. Since these representations are shared with the *global head*, classification can benefit from the available local information the representations possess. Conversely, global labels are trained with high level features that have the widest receptive field. This encourages the feature extractor to leverage potential ancillary information beneficial for global prediction which may also assist local predictions. As our experiments demonstrate, we observe complementing nodule detection performance from the two heads when they are jointly trained.

4.3 Dual Head Augmentation Evaluation

In this subsection, we are interested in analyzing the effects different augmentation strategies impose on the DHN's global and local predictions. As mentioned in Sect. 3.2, we consider both binomial and uniform augmentation strategies ϕ^{bin} and ϕ^{uni}. We also include an identity transform (no augmentation) strategy denoted ϕ^{id} as a baseline reference.

Table 2. Comparison of single head DHA strategies (ϕ^{id} on one of the heads). First five rows are the results of joint training with single head DHA. The bottom four rows are the results when only a single head is trained with the specified augmentation strategy.

DHA		Classification metrics		Localization metrics		
Global	Local	ROC-AUC	PR-AUC	FROC-AUC	AFROC-AUC	mAP
ϕ^{id}	ϕ^{id}	0.873 ± 0.004	0.654 ± 0.016	0.628 ± 0.011	0.626 ± 0.033	0.188 ± 0.009
ϕ^{bin}	ϕ^{id}	$\mathbf{0.879 \pm 0.007}$	$\mathbf{0.675 \pm 0.017}$	0.624 ± 0.012	0.575 ± 0.037	0.184 ± 0.006
ϕ^{uni}	ϕ^{id}	0.877 ± 0.004	0.673 ± 0.012	0.627 ± 0.009	0.597 ± 0.057	0.182 ± 0.06
ϕ^{id}	ϕ^{bin}	0.858 ± 0.007	0.631 ± 0.011	0.656 ± 0.009	$\mathbf{0.687 \pm 0.014}$	0.182 ± 0.007
ϕ^{id}	ϕ^{uni}	0.877 ± 0.003	0.643 ± 0.009	$\mathbf{0.658 \pm 0.017}$	0.663 ± 0.015	$\mathbf{0.200 \pm 0.005}$
ϕ^{bin}	–	$\mathbf{0.849 \pm 0.004}$	$\mathbf{0.628 \pm 0.007}$	–	–	–
ϕ^{uni}	–	0.834 ± 0.008	0.585 ± 0.019	–	–	–
–	ϕ^{bin}	–	–	0.637 ± 0.032	$\mathbf{0.686 \pm 0.028}$	0.177 ± 0.024
–	ϕ^{uni}	–	–	$\mathbf{0.651 \pm 0.021}$	0.662 ± 0.015	$\mathbf{0.204 \pm 0.008}$

We first analyze the performance of a DHN when only one augmentation strategy is applied on one of the two heads. From the results shown in Table 2, we can see that better classification performance is observed when the *global head* uses the strategy ϕ^{bin}. Similarly, we observe slightly better localization performance when the *local head* adopts the strategy ϕ^{uni}. We can also observe noticeable improvements with dual head training versus single head training. When only one augmentation strategy is employed on a single DHN head, its performance still surpasses that of a single head network that utilizes the same augmentation strategy. This observation reflects the results we perceive in Sect. 4.2. Thus, even when just one augmentation strategy is applied on a single DHN head, we can expect favorable performance from DHN's over single head networks.

Table 3. Comparison of dual head DHA strategies (i.e., ϕ^{bin} and ϕ^{uni} applied on the global and local heads).

DHA		Classification metrics		Localization metrics		
Global	Local	ROC-AUC	PR-AUC	FROC-AUC	AFROC-AUC	mAP
ϕ^{bin}	ϕ^{bin}	0.882 ± 0.003	0.674 ± 0.008	0.668 ± 0.008	$\mathbf{0.707 \pm 0.013}$	0.187 ± 0.006
ϕ^{uni}	ϕ^{uni}	0.881 ± 0.009	0.675 ± 0.015	0.664 ± 0.008	0.702 ± 0.009	0.159 ± 0.012
ϕ^{bin}	ϕ^{uni}	$\mathbf{0.903 \pm 0.003}$	$\mathbf{0.702 \pm 0.003}$	$\mathbf{0.708 \pm 0.005}$	0.705 ± 0.008	$\mathbf{0.245 \pm 0.005}$
ϕ^{uni}	ϕ^{bin}	0.878 ± 0.003	0.667 ± 0.010	0.658 ± 0.011	0.676 ± 0.011	0.202 ± 0.005

With the observed characteristics for the *global* and *local head*, we propose a hypothesis that heavier augmentations $\phi^{\mathtt{bin}}$ are more suitable for the *global head* but not necessarily the *local head*. We believe that the *global head* favors heavier augmentations to increase its robustness against image-level distortions. In contrary, the *local head* seems to prefer lighter augmentations as heavy augmentations might have imposed an overwhelming amount of distortions to region specific features. We verify our hypothesis with an experiment shown in Table 3. The results verify our hypothesis that the DHN obtains optimal performance when the DHA strategy applies $\phi^{\mathtt{bin}}$ on the *global head* and $\phi^{\mathtt{uni}}$ on the *local head*.

5 Conclusion

In this work, we present a multi-task lung nodule detection algorithm using a DHN. Our DHN architecture features a *global head* and a *local head* that performs lung nodule detection on the global and local level simultaneously. In addition, we introduce a novel DHA strategy that leverages the dual head design of the DHN to enhance global and local nodule detection performance while training. Throughout our experiments, we demonstrate the performance gain our DHN yields in comparison to conventional single head networks in both classification and localization abilities. Furthermore, we identified the DHA strategy that applies the appropriate augmentations for each head. Together with the optimal DHA strategy, our DHN obtained a performance otherwise not attainable if regular single head augmentation strategies are employed.

Acknowledgements. We would like to thank Che-Han Chang and the anonymous reviewers for their valuable suggestions. We also thank the members: Chun-Nan Chou, Fu-Chieh Chang, Yu-Quan Zhang, and Hao-Jen Wang for their support in collecting annotated data, and Yi-Hsiang Chin for his efforts in conducting experiments.

References

1. Ausawalaithong, W., Thirach, A., Marukatat, S., Wilaiprasitporn, T.: Automatic lung cancer prediction from chest x-ray images using the deep learning approach. In: BMEiCON 2018–11th Biomedical Engineering International Conference, vol. 1 (2019). https://doi.org/10.1109/BMEICON.2018.8609997
2. Busby, L.P., Courtier, J.L., Glastonbury, C.M.: Bias in radiology: the how and why of misses and misinterpretations. Radiographics **38**, 236–247 (2018), https://pubs.rsna.org/doi/abs/10.1148/rg.2018170107
3. Sainz de Cea, M.V., Diedrich, K., Bakalo, R., Ness, L., Richmond, D.: Multi-task learning for detection and classification of cancer in screening mammography. In: Martel, A.L., et al. (eds.) MICCAI 2020. LNCS, vol. 12266, pp. 241–250. Springer, Cham (2020). https://doi.org/10.1007/978-3-030-59725-2_24
4. Cha, M.J., Chung, M.J., Lee, J.H., Lee, K.S.: Performance of deep learning model in detecting operable lung cancer with chest radiographs. J. Thorac. Imaging **34**, 86–91 (2019). https://doi.org/10.1097/RTI.0000000000000388

5. del Ciello, A., Franchi, P., Contegiacomo, A., Cicchetti, G., Bonomo, L., Larici, A.R.: Missed lung cancer: when, where, and why? Diagn. Intervent. Radiol. **23**, 118 (2017). https://doi.org/10.5152/DIR.2016.16187
6. Dai, J., et al.: Deformable convolutional networks. In: 2017 IEEE International Conference on Computer Vision (ICCV), pp. 764–773 (2017). https://doi.org/10.1109/ICCV.2017.89
7. He, K., Gkioxari, G., Dollár, P., Girshick, R.B.: Mask R-CNN. In: 2017 IEEE International Conference on Computer Vision (ICCV), pp. 2980–2988 (2017)
8. He, K., Zhang, X., Ren, S., Sun, J.: Deep residual learning for image recognition. In: 2016 IEEE Conference on Computer Vision and Pattern Recognition (CVPR), pp. 770–778 (2016)
9. Huang, G., Liu, Z., Maaten, L.V.D., Weinberger, K.Q.: Densely connected convolutional networks. In: Proceedings - 30th IEEE Conference on Computer Vision and Pattern Recognition, CVPR 2017 2017-January, pp. 2261–2269, November 2017. https://doi.org/10.1109/CVPR.2017.243
10. Hwang, E.J., Park, C.M.: Clinical implementation of deep learning in thoracic radiology: potential applications and challenges. Korean J. Radiol. **21**, 511–525 (2020). https://doi.org/10.3348/KJR.2019.0821
11. Kim, Y.G., et al.: Short-term reproducibility of pulmonary nodule and mass detection in chest radiographs: comparison among radiologists and four different computer-aided detections with convolutional neural net. Sci. Rep. **2019** 9:1 9, 1–9 (2019). https://doi.org/10.1038/s41598-019-55373-7
12. Larici, A.R., et al.: Lung nodules: size still matters. Eur. Respir. Rev. official J. Eur. Respir. Soc. **26** (2017). https://doi.org/10.1183/16000617.0025-2017
13. Li, X., Shen, L., Luo, S.: A solitary feature-based lung nodule detection approach for chest x-ray radiographs. IEEE J. Biomed. Health Inf. **22**, 516–524 (2018). https://doi.org/10.1109/JBHI.2017.2661805
14. Li, Z., et al.: Thoracic disease identification and localization with limited supervision. In: 2018 IEEE/CVF Conference on Computer Vision and Pattern Recognition, pp. 8290–8299 (2018). https://doi.org/10.1109/CVPR.2018.00865
15. Lin, T.Y., Dollár, P., Girshick, R., He, K., Hariharan, B., Belongie, S.: Feature pyramid networks for object detection. In: 2017 IEEE Conference on Computer Vision and Pattern Recognition (CVPR), pp. 936–944 (2017). https://doi.org/10.1109/CVPR.2017.106
16. Lin, T.Y., Goyal, P., Girshick, R., He, K., Dollár, P.: Focal loss for dense object detection. In: 2017 IEEE International Conference on Computer Vision (ICCV), pp. 2999–3007 (2017). https://doi.org/10.1109/ICCV.2017.324
17. Luo, W., Li, Y., Urtasun, R., Zemel, R.: Understanding the effective receptive field in deep convolutional neural networks. In: Proceedings of the 30th International Conference on Neural Information Processing Systems, NIPS 2016, pp. 4905–4913. Curran Associates Inc. (2016)
18. Mendoza, J., Pedrini, H.: Detection and classification of lung nodules in chest x-ray images using deep convolutional neural networks. Comput. Intell. **36**, 370–401 (2020). https://doi.org/10.1111/COIN.12241
19. Nam, J.G., et al.: Development and validation of deep learning-based automatic detection algorithm for malignant pulmonary nodules on chest radiographs. Radiology **290**, 218–228 (2019). https://pubs.rsna.org/doi/abs/10.1148/radiol.2018180237

20. Pesce, E., Withey, S.J., Ypsilantis, P.P., Bakewell, R., Goh, V., Montana, G.: Learning to detect chest radiographs containing pulmonary lesions using visual attention networks. Med. Image Anal. **53**, 26–38 (2019). https://doi.org/10.1016/J.MEDIA.2018.12.007
21. Rajpurkar, P., et al.: Chexnet: radiologist-level pneumonia detection on chest x-rays with deep learning. ArXiv abs/1711.05225 (2017)
22. Ren, S., He, K., Girshick, R., Sun, J.: Faster R-CNN: towards real-time object detection with region proposal networks. IEEE Trans. Pattern Anal. Mach. Intell. **39**(6), 1137–1149 (2017). https://doi.org/10.1109/TPAMI.2016.2577031
23. Rezatofighi, H., Tsoi, N., Gwak, J., Sadeghian, A., Reid, I., Savarese, S.: Generalized intersection over union: a metric and a loss for bounding box regression. In: 2019 IEEE/CVF Conference on Computer Vision and Pattern Recognition (CVPR), pp. 658–666 (2019). https://doi.org/10.1109/CVPR.2019.00075
24. Schultheiss, M., et al.: A robust convolutional neural network for lung nodule detection in the presence of foreign bodies. Sci. Rep. **2020** 10:1 10, 1–9 (2020). https://doi.org/10.1038/s41598-020-69789-z
25. Simonyan, K., Zisserman, A.: Very deep convolutional networks for large-scale image recognition. In: International Conference on Learning Representations (2015)
26. Tack, D., Howarth, N.: Missed lung lesions: side-by-side comparison of chest radiography with MDCT. In: Hodler, J., Kubik-Huch, R.A., von Schulthess, G.K. (eds.) Diseases of the Chest, Breast, Heart and Vessels 2019-2022. ISS, pp. 17–26. Springer, Cham (2019). https://doi.org/10.1007/978-3-030-11149-6_2
27. Tian, Z., Shen, C., Chen, H., He, T.: Fcos: fully convolutional one-stage object detection. In: 2019 IEEE/CVF International Conference on Computer Vision (ICCV), pp. 9626–9635 (2019). https://doi.org/10.1109/ICCV.2019.00972
28. Wang, X., Peng, Y., Lu, L., Lu, Z., Bagheri, M., Summers, R.M.: Chestx-ray8: hospital-scale chest x-ray database and benchmarks on weakly-supervised classification and localization of common thorax diseases. In: 2017 IEEE Conference on Computer Vision and Pattern Recognition (CVPR) 2017-January, pp. 3462–3471, July 2017. https://doi.org/10.1109/CVPR.2017.369

Dermatology

Data-Driven Deep Supervision for Skin Lesion Classification

Suraj Mishra[1(✉)], Yizhe Zhang[2], Li Zhang[3], Tianyu Zhang[4], X. Sharon Hu[1], and Danny Z. Chen[1]

[1] University of Notre Dame, Notre Dame, USA
{smishra3,shu,dchen}@nd.edu
[2] Nanjing University of Science and Technology, Nanjing, China
[3] Qingdao Women and Children's Hospital of Qingdao University, Qingdao, China
[4] University of Connecticut, Storrs, USA

Abstract. Automatic classification of pigmented, non-pigmented, and depigmented non-melanocytic skin lesions have garnered lots of attention in recent years. However, imaging variations in skin texture, lesion shape, depigmentation contrast, lighting condition, etc. hinder robust feature extraction, affecting classification accuracy. In this paper, we propose a new deep neural network that exploits input data for robust feature extraction. Specifically, we analyze the convolutional network's behavior (field-of-view) to find the location of deep supervision for improved feature extraction. To achieve this, first we perform activation mapping to generate an object mask, highlighting the input regions most critical for classification output generation. Then the network layer whose layer-wise effective receptive field matches the approximated object shape in the object mask is selected as our focus for deep supervision. Utilizing different *types* of convolutional feature extractors and classifiers on three melanoma detection datasets and two vitiligo detection datasets, we verify the effectiveness of our new method.

1 Introduction

Automatic classification of skin lesions such as pigmented/non-pigmented lesions concerning skin cancers [1,3,4,6,23,26] and depigmented and hypomigmented lesions concerning vitiligo (vitiligo vs vitiligo-type, e.g., pityriasis alba, nevus depigmentous [13,29]) in RGB images has garnered lots of research attention in recent years. Convolutional neural network (CNN) based models have shown promising results in accurate classification of skin lesions. However, imaging variations in skin texture, lesion shape, depigmentation contrast, lighting condition, etc. hinder robust feature extraction, thus affecting classification accuracy.

For input data specific robust feature extraction, ensuring the discriminativeness of the learned features in the hidden layers of a CNN is critical [15]. In [22],

Supplementary Information The online version contains supplementary material available at https://doi.org/10.1007/978-3-031-16431-6_68.

L. Wang et al. (Eds.): MICCAI 2022, LNCS 13431, pp. 721–731, 2022.
https://doi.org/10.1007/978-3-031-16431-6_68

deep supervision was explored as an effective way to encourage feature maps at different layers to be directly predictive of the final output. By providing a companion objective using auxiliary classifiers [12], deep supervision generates feedback for a local output (at a hidden layer). The combined error (i.e., global and local) is back-propagated, influencing the hidden layer update process and ensuring highly discriminative feature maps. But, where and how to apply deep supervision still remains an active research topic. Recently, Mishra et al. [17] studied data-driven deep supervision for medical image segmentation utilizing target object labels (or masks) as guidance for deep supervision. Yet, for image classification problems in general, the key issue of deep supervision location selection in a CNN becomes challenging due to the absence of object-level labels, since in classification problems, only image-level labels are normally available.

Previous work [15,17] exploited the observation that if the receptive field (RF) or field-of-view of a segmentation CNN matches in size the target object size in the input images, then the segmentation accuracy can be improved. Relative to the CNN's RF, smaller objects are lost in the network's sub-sampling operations while robust global features of larger objects are not well captured due to a smaller RF [17]. We extend the observation/technique of deep supervision to tackle various skin lesion classification tasks, proposing data-driven deep supervision. Specifically, for deep supervision location selection, we utilize the input domain information (target object sizes) to identify the layer with preeminent contribution to feature extraction, and select this layer as the target location for deep supervision as auxiliary classification. Hence, as in [17], we employ an RF based approach to determine the preeminent layer in a classification CNN. But, note that for image classification problems, object-level labels are usually not available (only image-level labels, e.g., {yes, no}, are given). Thus, a difficulty to our approach is how to devise an effective method to estimate the sizes of objects in input images that possibly influence the classification results.

Each convolutional (conv) layer of a CNN extracts local features from a region of an input image to generate its output. The RF of the CNN is defined as the input region that directly affects the CNN's output [11]. By analyzing the near-*Gaussian* distribution of RF, Luo et al. [14] identified the unresponsive boundary regions of RF as non-output-affecting. The actual output-affecting region of the input image, which is smaller than the theoretical RF, is defined as the effective receptive field (ERF) of the network [14]. Intuitively, convolutional filters extract improved features when the target object size (called morphological size) matches the filters' ERF and the target lies at the center of the ERF [17].

From the viewpoint of ERF matching, for skin lesion classification using a deep CNN, we need to address two challenges: (1) determining the characteristic size of objects in the input images that the CNN is applied to; (2) determining the CNN's ERF, whose size should match the characteristic object size. For the first challenge, we explore activation mapping to determine the morphological size of the input objects (approximate regions of interest in input images which have significant contribution to output generation). Since such *regions of interest* can vary largely in size, for the second challenge, instead of using the fixed (or default) network ERF of a deep CNN, we utilize the layer-wise ERF (LERF) [15]. LERF represents the ERF of each conv layer in the CNN with respect to the image

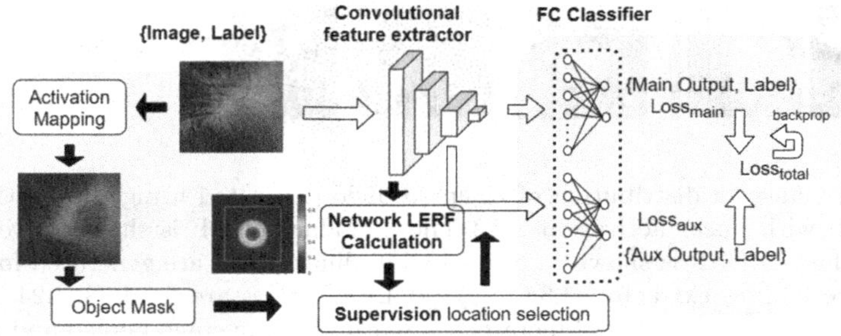

Fig. 1. Our proposed framework for data-driven deep supervision for skin lesion classification. The major components of our framework are (1) LERF determination, (2) morphological object size approximation, and (3) deep supervision employment, shown in blue, orange, and red background, respectively. Specifically, utilizing activation maps, a morphological object mask is generated. The network layer whose LERF size well matches the morphological object size is chosen for deep supervision. (Color figure online)

input. The conv layer with an LERF similar to the dataset-specific approximate object size is used as the target layer for deep supervision.

In data-driven deep supervision for skin lesion classification, we first approximate the morphological input object size from image-level labels using activation mapping. Then we introduce deep supervision from the layer of the network whose LERF size matches (as well as possible) the morphological object size. We validate our approach on three pigmented/non-pigmented lesion classification datasets concerning skin cancer (ISIC 2016 [6], ISIC 2017 [1], and ISIC 2018 [2]) and two depigmented/hypopigmented lesion classification datasets concerning vitiligo (vitiligo vs vitiligo-type, e.g., pityriasis alba, nevus depigmentous). For the public ISIC 2016, ISIC 2017, and ISIC 2018 datasets, we achieve comparable scores with the challenge winners. For both the vitiligo datasets [29], we outperform existing methods.

2 Method

The overall framework of our proposed data-driven deep supervision for skin lesion classification is shown in Fig. 1. The major components of our framework are (1) LERF determination, (2) morphological object size approximation, and (3) deep-supervision employment. Section 2.1 and Sect. 2.2 highlight LERF determination and morphological object size approximation, respectively. We then show how to use the information thus obtained for data-driven deep supervision in Sect. 2.3, along with some details of deep supervision implementation.

2.1 Layer-Wise Effective Receptive Field (LERF)

We take a partial derivative based approach to measure the influence on an input image region by an output node (a pixel) of a conv layer as in [14,17].

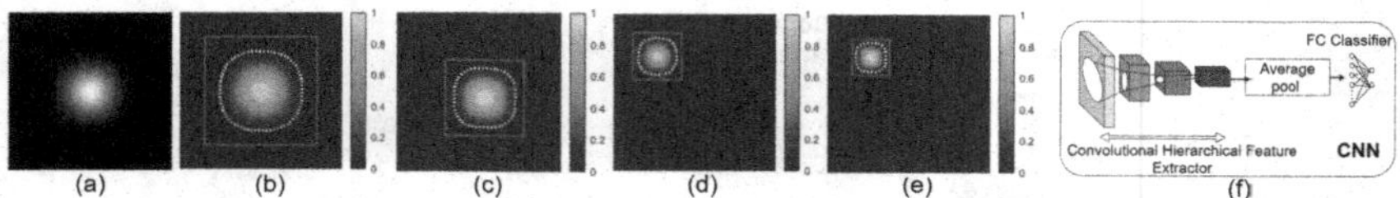

Fig. 2. (a) Gaussian distribution of receptive field generated using the VGG13 architecture [21] with linear activations. (b) The boundary of RF is shown in red square. The boundary of ERF is shown in dashed white lines. Plots are generated for the last layer of the feature extractor (L34). (c)–(e) LERF plots are for L28, L24, and L21. (f) The extracted convolutional hierarchical features are average-pooled and classified. (Color figure online)

For a classification CNN with N conv layers, $LERF_h$ is associated to layer h ($h = 1, 2, \ldots, N$). Our objective is to determine $\frac{\partial q_{i_t,j_t}}{\partial p_{i,j}}$, where $p_{i,j}$ is an input node (pixel) and q_{i_t,j_t} is the target node at the output of the conv layer h for which the influence measurement is being performed. Following [17], the region of influence (i.e., $\frac{\partial q_{i_t,j_t}}{\partial p_{i,j}}$) is determined using backpropagation (as all non-zero gradients on the input image). In Fig. 2, LERFs associated to different layers of a representative CNN network are shown. Any node q_{i_t,j_t} of a target layer can be selected for backpropagation since different nodes only generate translational shifts in the output (see Fig. 2).

To extract the $LERF$ size from $\frac{\partial q_{i_t,j_t}}{\partial p_{i,j}}$, first we threshold (1 if $\frac{\partial q_{i_t,j_t}}{\partial p_{i,j}}$ ¿ th; else 0) to remove the unresponsive boundary regions. Then, the pixel count of the active region is square-rooted to approximate the size of the $LERF$. Such a size approximation is inspired by the near-Gaussian distribution of ERF [17]. Following [14], a $th = 0.0455$ (i.e., a range of two standard deviations of the *Gaussian* distribution, 4.55% of the maximum value, i.e., 1 at the center) is used in $LERF$ calculation. In Fig. 2, the convex hull of the $LERF$ pixels (the smallest convex area covering all the pixels) is marked with a closed white dotted curve. With non-linear activations (i.e., ReLU), this convex hull has a shape of a deformed square or an evolved ellipse. In the absence of network non-linearity, $LERF$ has a circular shape (see Fig. 2). Hence, a square root based operation is used to approximate such a deformed shape of $LERF$ with network non-linearity.

2.2 Object Size Approximation Using Activation Mapping

We explore an activation mapping based morphological object size approximation. Consider a classification CNN (generating an output label for an input image), which contains a stack of conv, activation, and sub-sampling layers for feature extraction. CNN-extracted features are average-pooled and fed to a fully connected (FC) classifier for final output generation (see Fig. 2(f)).

For an input image, propagating through the CNN, let $f_k(i, j)$ represent the activation of a unit/channel/kernel k in the last conv layer at spatial location (i, j). Then performing average pooling for unit k results in $F_k = \sum_{i,j} f_k(i, j)$. Ignoring the bias term, the FC classifier layer with weights w_k^c associated to

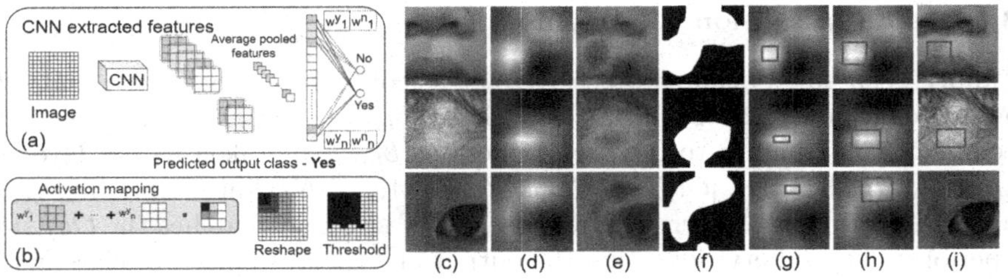

Fig. 3. (a) A CNN with extracted conv features being pooled and classified for final output generation. (b) An activation map generated for a predicted class *yes*. The generated activation map is thresholded with $\delta = 0.5$ to generate a summary-mask. (c)–(d)–(e)–(f) Example images, activation maps in gray, activation maps in color, and summary-masks, respectively. (g) Bounding boxes for the central regions. (h)–(i) Object masks shown on activation maps and input images, respectively. (Color figure online)

a class c generates output $\sum_k w_k^c F_k$ for class c. This output is used for loss calculation after a softmax-*type* operation. By plugging the value of F_k into the class score, we obtain $\sum_k w_k^c \sum_{i,j} f_k(i,j) = \sum_k \sum_{i,j} w_k^c f_k(i,j)$. In [31], the term $\sum_k w_k^c f_k(i,j)$ is defined as the activation map (A_c) for class c (see Fig. 3(b)), indicating the importance of the activation leading to the classification of an image for class c. As activation mapping indicates the importance of activations on an input image, we utilize it for object mask generation.

For object mask generation, we utilize the activation map A_c and reshape it to the input image size using cubic-interpolation to obtain A'_c. The reshaped map A'_c is then thresholded with a threhhold δ' (i.e., $A'_c(i,j) = 0$ if $A'_c(i,j) < \delta'$; else $A'_c(i,j) = 1$). We use $\delta' = 0.9$ to determine the *central region*. The smallest bounding box containing the central region is computed (see Fig. 3(g)). Then we perform a region-growing type process to increase the size of the bounding box (equally in four directions) till we reach a pixel with a 50% probability value ($\delta = 0.5$) in the activation map. A box in Fig. 3(g) is increased in size till it touches the boundary of the summary mask (see Fig. 3(f)), and the result is shown in Fig. 3(h). Such processes are based on the special and stable properties of skin images (similar appearance, captured in similar conditions, using specific devices, etc.). A common target lesion in skin images is a single connected region lying around the center of an image. With our approach, the most critical regions associated with the classification of the images in Fig. 3(c) are shown in Fig. 3(i). In Fig. 3(h)–(i), example object masks are shown for some vitiligo images. It is interesting to note that the transition regions have the highest impact on final output generation. As in Sect. 2.1, square root of the pixel count of an object mask (red bounding box regions in Fig. 3(i)) is used as an approximate size of the object mask. The average size of all such object masks over sampled training set images is used as the morphological object size (Obj) of a specific dataset.

2.3 Deep Supervision Employment

To apply deep supervision, we use $LERF_h$'s (computed in Sect. 2.1) and Obj (computed in Sect. 2.2) to find a target layer L_{target}, i.e., the conv layer with preeminent contribution to feature extraction. L_{target} is taken as the layer h^* with $|Obj - LERF_{h^*}| \leq |Obj - LERF_h|, \forall\ h \in \{1, \ldots, N\}$. Intuitively, L_{target} is the conv layer in an N-layer CNN whose $LERF$ size best matches the Obj size. To perform deep supervision using the output of L_{target}, we need to classify the features associated with L_{target}. The classification scores are then used for the auxiliary loss calculation. Our total loss is computed as $Loss_{total} = Loss_{main} + Loss_{aux}$, where $Loss_{main}$ is the main loss and $Loss_{aux}$ is the auxiliary loss.

Remarks on Implementation. Here, we discuss two aspects related to deep supervision implementation: (1) auxiliary classifier design; (2) transfer learning.

For deep supervision employment, conv features are processed by a classifier. The classified output is then used to compute the auxiliary loss. In this work, we explore two types of classifier architectures for auxiliary classifier design. (i) In a *VGG*-type classifier [21], multiple FC layers are used along with activation and dropout layers. In contrast, (ii) a *ResNet*-type classifier [7] is light weight and uses a single FC layer as the classifier (details in Supplemental Material).

Transfer learning plays a critical role (better/quicker optimization) in image classification by initiating the training process with a previously trained model [20], and has shown to be effective with improved classification accuracy. In data-driven deep supervision, an auxiliary loss is introduced from the preeminent layer of the network. Such model modifications encourage us to adopt a super-model based encapsulation approach, in which the base model is initialized by transfer learning and the auxiliary branch is initialized randomly (He-initialization [8]). Benefits of transfer learning are shown in Sect. 3.

3 Experiments and Results

Datasets. Our experiments use the following datasets. **ISIC 2016** [6]: It contains 900 training and 379 test images. We randomly divide the original training set into 649 images for training and 251 images for validation. **ISIC 2017** [1]: It contains 2000 training, 150 validation, and 600 test images. Experiments are performed for task-3A, i.e., melanoma detection. **ISIC 2018** [2]: It contains 10,015 training, 193 validation, and 1512 test images with external evaluation. Considering the smaller size of the validation set, we randomly select 999 images ($\approx$ 10%) from the training set and add them to the validation set (9016 train and 1192 validation images). **Vitiligo (in-house)**: It contains 2188 images (1227 training, 308 validation, and 653 test images). The in-house dataset consists of images from retrospective consecutive outpatients obtained by the dermatology department of Qingdao Women and Children's Hospital (QWCH) in China. For each patient with suspected vitiligo (e.g., pityriasis alba, hypopigmented nevus), clinical photographs of the affected skin areas were taken by medical assistants using a point-and-shoot camera (as described in [29]). **Vitiligo (public)** [29]: It contains 1341 images (672 training, 268 validation, and 401 test images).

Table 1. Lesion classification results on skin image datasets. Proposed*/Proposed—without/with deep supervision.

ISIC 2016 Dataset					ISIC 2017 Dataset				
Method	AUC	Acc	Sen	Spe	Method	AUC	Acc	Sen	Spe
VGG-16 [21]	0.826	0.826	0.413	0.928	Galdran *et al.* [3]	0.765	0.480	**0.906**	0.377
GoogleNet [22]	0.801	0.847	0.507	0.931	Vasconcelos *et al.* [23]	0.791	0.830	0.171	0.990
DRN-50 [27]	0.783	0.855	0.547	0.931	Yang *et al.* [26]	0.830	0.830	0.436	0.925
Gutman *et al.* [6]	0.804	0.855	0.507	0.941	Diaz *et al.* [4]	0.856	0.823	0.103	**0.998**
ARDT-DenseNet [25]	0.837	0.857	**0.816**	0.756	ARDT-DenseNet [25]	**0.879**	**0.868**	0.668	0.896
SDL [28]	0.829	**0.857**	-	-	SDL [28]	0.830	0.830	-	-
Proposed* - VGG13	0.830	0.852	0.334	0.980	Proposed* - VGG13	0.808	0.845	0.325	0.971
Proposed - VGG13	0.837	0.871	0.387	**0.990**	Proposed - VGG13	0.831	0.860	0.590	0.926
Proposed* - RES18	0.817	0.850	0.400	0.961	Proposed* - RES18	0.786	0.825	0.282	0.956
Proposed - RES18	**0.850**	0.842	0.440	0.951	Proposed - RES18	0.791	0.848	0.470	0.940
Vitiligo (in-house) Dataset					Vitiligo (public) Dataset				
Method	AUC	Acc	Sen	Spe	Method	AUC	Acc	Sen	Spe
VGG-13 [21]	0.851	-	0.791	0.913	VGG-13 [21]	0.995	-	0.972	0.963
ResNet [7]	0.840	-	0.775	0.902	ResNet [7]	0.958	-	0.952	0.957
DenseNet [9]	0.847	-	0.784	0.906	DenseNet [9]	0.982	-	0.962	0.961
Dermatologists [29]	-	-	**0.811**	**0.999**	Dermatologists [29]	-	-	0.964	0.803
Proposed* - VGG13	0.852	0.882	0.797	0.907	Proposed* - VGG13	0.970	0.973	0.992	0.951
Proposed - VGG13	**0.931**	**0.904**	0.777	0.941	Proposed - VGG13	**0.998**	**0.988**	**0.996**	**0.975**
Proposed* - RES18	0.897	0.865	0.662	0.925	Proposed* - RES18	0.987	0.950	0.962	0.932
Proposed - RES18	0.920	0.893	0.757	0.933	Proposed - RES18	0.990	0.968	0.962	0.975

Experimental Setup. We resize all the images of each dataset to 224×224. To reduce overfitting on a small training set, data augmentation is performed using random flipping and cropping. In cropping, a random portion of an image is extracted and resized to the target size (224×224). We use a standard back-propagation implementing *Adam* [10] with a fixed learning rate of 0.00002 ($\beta_1 = 0.9, \beta_2 = 0.999$, and $\epsilon = 1\text{e}{-8}$). Experiments are performed on NVIDIA-TITAN and Tesla P100 GPUs, using PyTorch for $1k$ epochs. The batch size for each case is selected as the maximum size allowed by the GPU. We use a network initialization as discussed in Sect. 2.3. For all the datasets, cross-entropy loss is used as $Loss_{main}$ and $Loss_{aux}$. The best performing model version on the validation set is saved as the model checkpoint.

VGG13 [21] and ResNet18 [7] are used as the backbone models. Untrained networks with *He*-initialization [8] are used for the *LERF* experiments. To neutralize variations in *LERF* computation caused by network non-linearity, the mean over 20 iterations is used. We should note that the LERF computing time is much less than the CNN training time (less than a minute on an NVIDIA P100 machine). Following the processes presented in Sect. 2.2, *Obj* for resized images is determined. Using the *LERF* values of the network layers, deep supervision locations are decided. ResNet-type classifiers are used as the default architecture for the auxiliary classifier (more details in Supplemental Material).

Results and Discussions. Table 1 shows the results. For each dataset, the *Proposed** or *Proposed* row gives our results obtained without or with deep

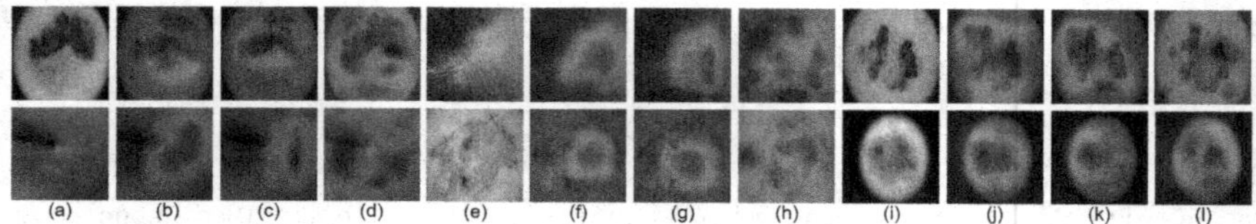

Fig. 4. (a)–(e)–(i) Example images. (b)–(f)–(j) Activation maps generated without deep supervision. (c)–(g)–(k) Activation maps by deep-supervision (main branch). (d)–(h)–(l) Activation maps by deep-supervision (aux branch).

Table 2. Ablation study analysis. C: Classifier type (R(*ResNet-type*)/V(*VGG-type*)); DS: Deep supervision location; TL: Transfer learning (with(✓)/without(✗)).

Method				ISIC 2016 Dataset				Vitiligo (in-house) Dataset				Vitiligo (public) Dataset			
Backbone	C	DS	TL	AUC	Acc	Sen	Spe	AUC	Acc	Sen	Spe	AUC	Acc	Sen	Spe
VGG13	R	✗	✓	0.830	0.852	0.334	0.980	0.852	0.882	0.797	0.907	0.970	0.973	0.992	0.951
VGG13	R	L28	✓	0.837	**0.871**	0.387	**0.990**	0.931	0.904	0.777	0.941	**0.998**	**0.988**	**0.996**	**0.975**
VGG13	R	L24	✓	0.830	0.839	0.293	0.974	0.924	**0.907**	**0.791**	0.941	-	-	-	-
VGG13	R	L24&L28	✓	0.834	0.847	0.320	0.977	0.931	0.897	0.716	**0.950**	-	-	-	-
VGG13	R	L31	✓	0.838	0.852	0.387	0.967	**0.936**	0.891	0.811	0.915	0.997	0.973	0.983	0.957
VGG13	R	✗	✗	0.715	0.810	0.227	0.954	0.809	0.860	0.703	0.915	0.892	0.905	0.962	0.821
VGG13	R	L28	✗	0.744	0.829	0.213	0.980	0.842	0.862	0.804	0.879	0.921	0.928	0.954	0.889
VGG13	V	L28	✓	**0.840**	0.847	**0.400**	0.957	0.917	0.891	0.703	0.947	**0.998**	0.980	0.992	0.963

supervision. The backbone CNN networks used in each experiments (i.e., VGG13 and RES18) are shown. Applying data-driven deep supervision indeed improves classification accuracy. On the ISIC 2016 dataset, our method exceeds the results of the challenge winner [6]. On ISIC 2017, our results are comparable to the best performing models in the leader-board with published results [3,4,23,26].

On the **ISIC 2018** dataset, external evaluation is performed and we receive the results as an "overall score" (aggregated over all the 7 classes), i.e., no Acc, Sen, Spe, and AUC results for it. Our *Proposed**-VGG13 achieves an overall score of 0.664 (comparable to the challenge rank 50 [18]); our *Proposed*-VGG13 yields an overall score of 0.701 (comparable to the challenge rank 10 [5] among the published results without using external data, with the best reported score being 0.845 [32]). For our *Proposed**-RES18, an overall score of 0.626 is received. With supervision, our *Proposed*-RES18 achieves an overall score of 0.661. These highlight the benefits of our deep supervision. To further examine the impact of data-driven deep supervision on skin lesion classification, we compare changes in activation maps with deep supervision. In Fig. 4, changes in activations due to deep supervision for example images are shown.

Ablation Study. Using ISIC 2016, Vitiligo (in-house), and Vitiligo (public) as example cases, we conduct an ablation study to examine the contributions of deep supervision location selection, classifier architecture, and transfer learning. Table 2 shows the results. **Deep Supervision Location Selection:** Contribution of our deep supervision is shown by the DS column in Table 2. In addition to data-driven deep supervision (at L28 in VGG13), experiments are conducted by applying deep supervision at different layers (L24, L24 & L28, and L31). On the Vitiligo (in-house) dataset, better AUC and specificity are observed for deep supervision with the L31, and L24 & L28 case, respectively. **Transfer Learning:** Contribution of transfer learning is shown in the TL column of Table 2. In the absence of transfer learning, there is a significant drop in all accuracy metrics. **Classifier Architecture:** Experiments using VGG-type classifier architectures are discussed in Sect. 2.3. Both the classifier architectures generate comparative results (in column C of Table 2).

4 Conclusions

In this paper, we proposed a new data-driven deep supervision approach for skin lesion classification. Utilizing CNNs' layer-wise ERF information and input object size approximated by activation mapping, deep supervision location is selected. Experiments on various datasets verify the effectiveness of our approach.

In future work, we plan to improve the robustness of our proposed method by substituting CAM [31] with improved frameworks such as grad-CAM [19] and score-CAM [24]. Further, the single target lesion assumption is a limitation of the proposed method. We plan to extend our method to multiple target objects of different sizes based on the data-driven deep supervision segmentation method for multiple object sizes as in [16,17,30].

Acknowledgement. This work was supported in part by NSF grant CCF-1617735.

References

1. Codella, N.C.F., et al.: Skin lesion analysis toward melanoma detection: a challenge at the 2017 international symposium on biomedical imaging (ISBI), hosted by the international skin imaging collaboration (ISIC). In: ISBI 2018, pp. 168–172 (2018). https://doi.org/10.1109/ISBI.2018.8363547
2. Codella, N.C.F., et al.: Skin lesion analysis toward melanoma detection 2018: a challenge hosted by the international skin imaging collaboration (ISIC). CoRR abs/1902.03368 (2019). http://arxiv.org/abs/1902.03368
3. Galdran, A., et al.: Data-driven color augmentation techniques for deep skin image analysis. CoRR abs/1703.03702 (2017). http://arxiv.org/abs/1703.03702
4. González-Díaz, I.: Incorporating the knowledge of dermatologists to convolutional neural networks for the diagnosis of skin lesions. CoRR abs/1703.01976 (2017). http://arxiv.org/abs/1703.01976
5. Goyal, M., Rajapakse, J.C.: Deep neural network ensemble by data augmentation and bagging for skin lesion classification. CoRR abs/1807.05496 (2018). http://arxiv.org/abs/1807.05496
6. Gutman, D.A., et al.: Skin lesion analysis toward melanoma detection: a challenge at the international symposium on biomedical imaging (ISBI) 2016, hosted by the international skin imaging collaboration (ISIC). CoRR abs/1605.01397 (2016). http://arxiv.org/abs/1605.01397

7. He, K., Zhang, X., Ren, S., Sun, J.: Deep residual learning for image recognition. CoRR abs/1512.03385 (2015). http://arxiv.org/abs/1512.03385
8. He, K., Zhang, X., Ren, S., Sun, J.: Delving deep into rectifiers: surpassing human-level performance on imagenet classification. CoRR abs/1502.01852 (2015). http://arxiv.org/abs/1502.01852
9. Huang, G., Liu, Z., Van Der Maaten, L., Weinberger, K.Q.: Densely connected convolutional networks. In: 2017 IEEE Conference on Computer Vision and Pattern Recognition (CVPR), pp. 2261–2269 (2017). https://doi.org/10.1109/CVPR.2017.243
10. Kingma, D.P., Ba, J.: Adam: a method for stochastic optimization. In: Bengio, Y., LeCun, Y. (eds.) ICLR (2015). http://arxiv.org/abs/1412.6980
11. LeCun, Y., et al.: Backpropagation applied to handwritten zip code recognition. Neural Comput. **1**(4), 541–551 (1989)
12. Lee, C.Y., Xie, S., Gallagher, P., Zhang, Z., Tu, Z.: Deeply-supervised nets. In: International Conference on Artificial Intelligence and Statistics, pp. 562–570 (2015)
13. Liu, J., Yan, J., Chen, J., Sun, G., Luo, W.: Classification of vitiligo based on convolutional neural network. In: Sun, X., Pan, Z., Bertino, E. (eds.) Artificial Intelligence and Security, pp. 214–223 (2019)
14. Luo, W., Li, Y., Urtasun, R., Zemel, R.: Understanding the effective receptive field in deep convolutional neural networks. In: NeurIPS, pp. 4905–4913 (2016)
15. Mishra, S., Chen, D.Z., Hu, X.S.: A data-aware deep supervised method for retinal vessel segmentation. In: ISBI, pp. 1254–1257 (2020)
16. Mishra, S., Chen, D.Z., Hu, X.S.: Objective-dependent uncertainty driven retinal vessel segmentation. In: ISBI, pp. 453–457 (2021). https://doi.org/10.1109/ISBI48211.2021.9433774
17. Mishra, S., Zhang, Y., Chen, D.Z., Sharon Hu, X.: Data-driven deep supervision for medical image segmentation. IEEE Trans. Med. Imaging (2022). https://doi.org/10.1109/TMI.2022.3143371
18. Pal, A., Ray, S., Garain, U.: Skin disease identification from dermoscopy images using deep convolutional neural network. CoRR abs/1807.09163 (2018). http://arxiv.org/abs/1807.09163
19. Selvaraju, R.R., Cogswell, M., Das, A., Vedantam, R., Parikh, D., Batra, D.: Grad-CAM: visual explanations from deep networks via gradient-based localization. In: ICCV, pp. 618–626 (2017)
20. Shin, H.C., et al.: Deep convolutional neural networks for computer-aided detection: CNN architectures, dataset characteristics and transfer learning. IEEE Trans. Med. Imaging **35**(5), 1285–1298 (2016)
21. Simonyan, K., Zisserman, A.: Very deep convolutional networks for large-scale image recognition. In: ICLR (2015)
22. Szegedy, C., et al.: Going deeper with convolutions. In: CVPR, pp. 1–9 (2015)
23. Vasconcelos, C.N., Vasconcelos, B.N.: Increasing deep learning melanoma classification by classical and expert knowledge based image transforms. CoRR abs/1702.07025 (2017). http://arxiv.org/abs/1702.07025
24. Wang, H., Du, M., Yang, F., Zhang, Z.: Score-CAM: improved visual explanations via score-weighted class activation mapping. CoRR abs/1910.01279 (2019). http://arxiv.org/abs/1910.01279
25. Wu, J., Hu, W., Wen, Y., Tu, W., Liu, X.: Skin lesion classification using densely connected convolutional networks with attention residual learning. Sensors **20**(24), 7080 (2020). https://doi.org/10.3390/s20247080. https://www.mdpi.com/1424-8220/20/24/7080

26. Yang, X., Zeng, Z., Yeo, S.Y., Tan, C., Tey, H.L., Su, Y.: A novel multi-task deep learning model for skin lesion segmentation and classification. CoRR abs/1703.01025 (2017). http://arxiv.org/abs/1703.01025
27. Yu, L., Chen, H., Dou, Q., Qin, J., Heng, P.A.: Automated melanoma recognition in dermoscopy images via very deep residual networks. IEEE Trans. Med. Imaging **36**(4), 994–1004 (2017)
28. Zhang, J., Xie, Y., Wu, Q., Xia, Y.: Medical image classification using synergic deep learning. Med. Image Anal. **54**, 10–19 (2019)
29. Zhang, L., Mishra, S., et al.: Design and assessment of convolutional neural network based methods for vitiligo diagnosis. Front. Med. 8 (2021). https://doi.org/10.3389/fmed.2021.754202
30. Zhang, Y., Ying, M., Chen, D.: Decompose-and-integrate learning for multi-class segmentation in medical images. In: MICCAI, pp. 641–650 (2019)
31. Zhou, B., Khosla, A., Lapedriza, A., Oliva, A., Torralba, A.: Learning deep features for discriminative localization. In: CVPR (2016)
32. Zhuang, J., et al.: Skin lesion analysis towards melanoma detection using deep neural network ensemble. ISIC Challenge **2018**(2), 1–6 (2018)

Out-of-Distribution Detection for Long-Tailed and Fine-Grained Skin Lesion Images

Deval Mehta[1,2(✉)], Yaniv Gal[3], Adrian Bowling[3], Paul Bonnington[2], and Zongyuan Ge[1,2,4]

[1] Monash Medical AI, Monash University, Melbourne, Australia
deval.mehta@monash.edu
[2] eResearch Centre, Monash University, Melbourne, Australia
[3] Kahu AI, Auckland, New Zealand
[4] Monash Airdoc Research, Monash eResearch Centre, Melbourne, Australia
https://www.monash.edu/mmai-group

Abstract. Recent years have witnessed a rapid development of automated methods for skin lesion diagnosis and classification. Due to an increasing deployment of such systems in clinics, it has become important to develop a more robust system towards various Out-of-Distribution (OOD) samples (unknown skin lesions and conditions). However, the current deep learning models trained for skin lesion classification tend to classify these OOD samples incorrectly into one of their learned skin lesion categories. To address this issue, we propose a simple yet strategic approach that improves the OOD detection performance while maintaining the multi-class classification accuracy for the known categories of skin lesion. To specify, this approach is built upon a realistic scenario of a long-tailed and fine-grained OOD detection task for skin lesion images. Through this approach, 1) First, we target the mixup amongst middle and tail classes to address the long-tail problem. 2) Later, we combine the above mixup strategy with prototype learning to address the fine-grained nature of the dataset. The unique contribution of this paper is two-fold, justified by extensive experiments. First, we present a realistic problem setting of OOD task for skin lesion. Second, we propose an approach to target the long-tailed and fine-grained aspects of the problem setting simultaneously to increase the OOD performance.

Keywords: Skin lesion · Out-of-distribution · Openset · Mixup · Prototype

1 Introduction

Early detection and diagnosis of skin cancer remains a global challenge. Following the advent and success of deep learning for various computer vision tasks,

Supplementary Information The online version contains supplementary material available at https://doi.org/10.1007/978-3-031-16431-6_69.

L. Wang et al. (Eds.): MICCAI 2022, LNCS 13431, pp. 732–742, 2022.
https://doi.org/10.1007/978-3-031-16431-6_69

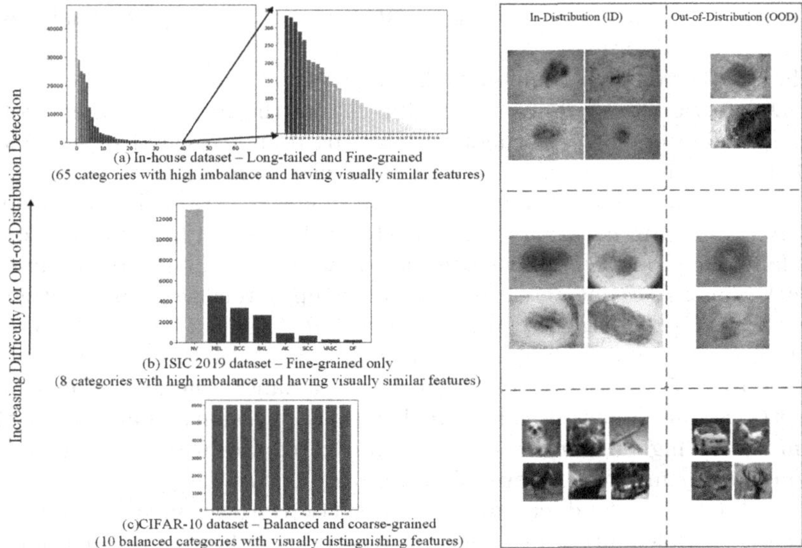

Fig. 1. Difficulty of Out-of-Distribution (OOD) task for different datasets (a) Our in-house long-tailed and fine-grained dataset, (b) ISIC 2019 fine-grained dataset, (c) CIFAR-10 dataset - Balanced and coarse-grained. (x: Classes; y: Number of samples).

medical research community has seen many deep learning models being developed for detection and classification of skin cancer [7]. While many models [11,21] have been developed to achieve a high performance, their evaluation and validation is limited to small datasets [5,28] which have limited shifts and variations in input data. Particularly, if such models are used in real practice, they are bound to encounter many Out-of-Distribution (OOD) samples which may represent unknown skin conditions, hardware device variations, and different clinical settings. Since these OOD samples are not in their training set, these models may assign a high confidence score and classify them as one of the known categories of their training set. Thus, it becomes vital to build the model's capability of rejecting the unknown OOD samples and make them robust for practical deployment.

The OOD detection task and openset recognition (OSR) problem was formalized by [24]. Since then, it has received much needed attention from the research community in general computer vision [10]. There have been many methods proposed with diverse strategies ranging from scaling the output softmax scores [1] (OpenMax), energy-based separation [19] and perturbed input pre-processing [13] (MSP). Techniques such as ODIN [18] have become standard baselines for separating the OOD samples based on the scaled softmax score. Another category of work concentrated on supplying synthetic samples generated from GANs for making the classification model aware of the OOD boundaries [4,9,17]. Other directions include using reconstruction based errors [25,31] for separating closed set and OOD samples, and using self-supervised learning [14,26] for better learning of closed set feature space. There have also been some recent works [2,6,15,22,23] based on the above ideas for developing OOD detection techniques for skin lesions utilizing the ISIC dataset.

Although the research community has made good contributions recently for the OOD detection task, we want to highlight that the commonly used problem settings and datasets for performance evaluation do not resemble a scenario for clinical deployment purposes. There are two shortcomings in this aspect - 1) The closed set dataset used for training the model is extremely well balanced and coarse-grained in nature for e.g. CIFAR-10 [16], which have visually distinguishable features between their categories (e.g. dog, car, plane etc. as shown in Fig. 1(c)). This makes the detection of OOD samples relatively easy as they are some of the reserved categories from the dataset. 2) Even if a fine-grained dataset is selected, its distribution is usually not long-tailed in nature. An example for that is the ISIC [5] dataset which only contains a handful number of categories. These settings thus do not resemble a real-world application scenario where there are significantly higher number of fine-grained categories with a long-tailed distribution like our in-house dataset shown in Fig. 1(a).

In this work, we conduct our OOD detection study on an in-house collected dataset (Molemap) from a clinical environment following tele-dermatology labelling standards. The dataset includes 208,287 tele-dermatology verified dermoscopic images categorized into 65 different skin conditions making it a long-tailed and fine-grained skin lesion dataset as shown in Fig. 1(a). The long-tailed and fine-grained challenges of our dataset inspires us to develop a method for targeting both these aspects to achieve a better OOD performance compared to the existing techniques. Specifically, we develop an inter-subset mixup strategy for targeting the middle and tail classes and combine it with prototype learning to tackle the fine-grained aspect. We conduct extensive experiments on both our in-house dataset and ISIC2019 dataset for benchmarking and validation.

2 Proposed Method

Learning better decision boundaries between closed set categories helps the classifier for detecting OOD samples more accurately [29]. This learning is dependent on the nature of a dataset. For a long-tailed dataset, head classes dominate the learning process as they contain a large number of samples. Although random oversampling/undersampling and techniques like SMOTE [3] can be used to tackle this problem, repeating/removing samples of classes does not help the classifier learn any better decision boundaries. Instead, our proposed approach employs a combination of data augmentation using mixup and better feature space learning using prototype loss specifically targeted to middle and tail classes. This enables us to improve the classification performance for those middle and tail categories which also increases the OOD detection performance. Our simple yet strategic method combines two established techniques to target the long-tailed and fine-grained aspects of our challenging dataset.

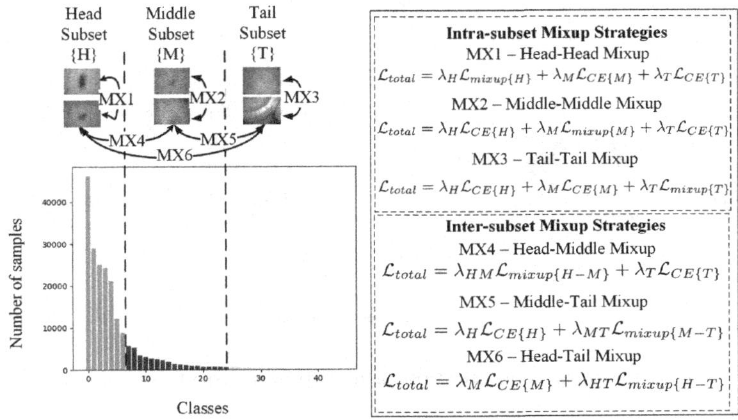

Fig. 2. Proposed mixup strategies - intra-subset and inter-subset

2.1 Intra-subset and Inter-subset Mixup Strategies

The approach of Mixup [32] has become quite popular in data augmentation space where it combines a pair of samples and their labels. This simple technique has been shown to increase robustness towards adversarial samples, a better estimate of uncertainty [27], and better generalization of the trained model [33]. The Mixup sample helps the model training to differentiate between the two categories in its sample even more and thus creates better decision boundaries between the two categories. We believe that Mixup strategy for skin lesion images will provide a better latent space representation between the different categories. However, simply adopting the Mixup strategy for a long-tailed dataset will still be heavily influenced by the head classes thus limiting its advantages only to a small part of the dataset.

To mitigate this problem, we devise different strategies for adopting mixup targeted to a certain part of the dataset only. We partition the total categories set (C) into three subsets based on the number of samples - Head ($H \subset C$), Middle ($M \subset C$), and Tail ($T \subset C$) as shown in Fig. 2. With this partition, we devise three intra-subset and three inter-subset mixup strategies. As the name suggests, in an intra-subset strategy, both the independent category samples belong to the same subset (for e.g. in MX1 strategy, both the independent samples will come from the categories in the $\{H\}$ subset). Whereas, in an inter-subset strategy, one sample comes from one of the categories of that subset and the other sample likewise (for e.g. in MX4 strategy, one sample comes from subset $\{H\}$ and the other sample comes from subset $\{M\}$). Thus, we can have six such strategies MX1 to MX6 as depicted in Fig. 2. The network training strategy is changed accordingly based on the mixup strategy adopted. It is updated with the combined loss of cross-entropy for the conventional subset learning and the mixup loss (given by Eq. 1) for the specific subset selected for mixup. In our experiments, we find

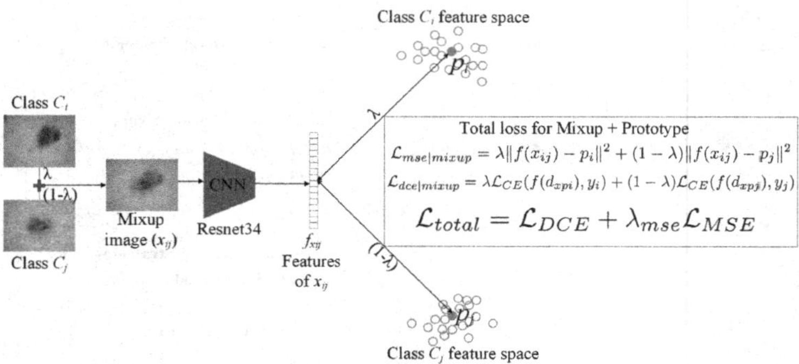

Fig. 3. Proposed combination of mixup and prototype learning

that targeting the mixup only amongst the middle and tail classes is effective for achieving better OOD detection performance.

$$\mathcal{L}_{mixup} = \lambda\mathcal{L}_{CE}(f(x_i), y_i) + (1-\lambda)\mathcal{L}_{CE}(f(x_j), y_j) \quad (1)$$

where $\{x_i,\ x_j\}$ are the input samples of class $\{C_i, C_j\}$ having labels $\{y_i, y_j\}$ respectively, λ is the weightage parameter for each sample between $\{0{,}1\}$.

2.2 Integration of Mixup with Prototype Learning

[30] introduced prototype learning for increasing the robustness of classifiers by proposing the prototype loss for reducing the intra-class distance variation and increasing the inter-class distance, which has proved extremely effective for learning fine-grained features. A standard prototype loss consists of the mean squared error loss and distance based cross entropy loss between the latent features from an encoder and the corresponding prototypes of all the categories. In our proposed approach, we further *enhance* the capability of prototype learning by combining it with mixup. The framework for this integration is shown in Fig. 3. Firstly, a mixup sample is created from two independent samples belonging to different class categories $\{C_i, C_j\}$ with weights λ and $(1-\lambda)$ respectively. This is then fed to a deep neural network (CNN) to generate its feature embeddings f_{xij}. As this input is a combination of a mixup of two different classes with a specific weightage, we propose to calculate the distance of the features f_{xij} from the class specific prototypes $\{p_i, p_j\}$ based on the weightage parameter λ and also do the same for the distance-based cross-entropy loss. With this, we design a custom loss function for the mixup sample combined with prototype learning given by Eqs. 2 and 3, which is adopted for the mixup samples only. For the normal samples, we adopt the standard prototype loss. Finally, we combine our best performing mixup strategy with prototype learning to address both the long-tailed and fine-grained aspects of our dataset.

$$\mathcal{L}_{mse|mixup} = \lambda\|f(x_{ij}) - p_i\|^2 + (1-\lambda)\|f(x_{ij}) - p_j\|^2 \quad (2)$$

$$\mathcal{L}_{dce|mixup} = \lambda \mathcal{L}_{CE}(d_{xpi}, y_i) + (1 - \lambda)\mathcal{L}_{CE}(d_{xpj}, y_j) \quad (3)$$

where $\{d_{xpi}, d_{xpj}\}$ is the square of the distance between the feature f_{xij} from the class specific prototypes $\{p_i, p_j\}$. The total loss is depicted in Fig. 3.

3 Experimental Results

3.1 Dataset Settings and Evaluation Metrics

In-house and ISIC 2019 Dataset. For evaluation purposes, we do extensive experiments on our in-house dataset and ISIC 2019 dataset. For our in-house dataset, we partition the 65 categories into 6 Head (more than 10,000 samples), 17 Middle (500 to 10,000 samples), 22 Tail (less than 500), and the rest 20 are reserved as OOD categories. As we wish to replicate a realistic clinical setting, we select this split where most samples are concentrated in the head categories, with a moderate number of samples in the middle and the least number of samples in the tail categories. Thus, we have 45 ID categories and 20 OOD categories. For the 45 ID categories, we split the data into 85%-15% train-test split. The train set is further divided into 80%-20% for training and validation. We utilize our in-house dataset for selecting all the hyperparameters via grid search (supplementary material) and use those for ISIC 2019 dataset as well. ISIC 2019 dataset has a class distribution as - MEL-4522, NV-12875, BCC-3323, AK-867, BKL-2624, DF-239, VASC-253, SCC-628. Thus, we select DF and VASC as the OOD categories. The 6 ID categories are separated as Head (NV & MEL), Middle (BCC & BKL), and Tail (AK & SCC) as we wish to have at least two categories in a subset. We adopt the five-fold cross-validation for the ID part of the ISIC 2019 dataset.

CIFAR-10 and Unusual OOD Samples. For a comprehensive study on different types of OOD samples, we add 1000 more OOD samples distributed equally amongst the 10 categories from the CIFAR-10 test set. We further enhance our OOD testing set with some commonly encountered **unusual** images in a clinic such as blurred images of skin lesions and ones that are completely covered by hair, ear, etc. A good model should give a low confidence score for these samples.

Training Implementation and Evaluation Metrics. We use Resnet34 [12] as our backbone architecture and train all the strategies using Adam optimizer with a batch size of 32, an initial learning rate of 1e-4 with exponential decay for 45 epochs. We resize the input image to a size of 224×224 and adopt the standard data augmentation of random crop and horizontal flip. Our proposed approach is trained only on the ID categories and we **do not** utilize OOD samples for fine-tuning our hyperparameters. We use precision (**pre**), recall (**rec**), and f1-score (**f1**) as the evaluation parameters for the closed set performance and Area Under Receiver Operator Characteristic (**AUROC**) as the OOD detection metric, which are the standard metrics for measuring the performance of a model for OOD detection task [10]. Our implementation code is available here[1].

[1] https://github.com/DevD1092/ood-skin-lesion.

Table 1. Performance evaluation of proposed mixup strategies on our in-house dataset

Mixup strategy	Closed set (ID) (Acc%)				OOD (AUROC%)
	Head	Middle	Tail	Total	
Baseline	66.67	38.26	36.98	60.56	65.67
Standard Mixup	67.23	45.18	33.89	**63.90**	66.35
H-H Intrasubset (MX1)	**70.11**	34.74	25.14	60.84	64.18
M-M Intrasubset (MX2)	63.12	**55.36**	31.54	61.80	66.47
T-T Intrasubset (MX3)	64.29	47.96	**39.49**	61.06	66.25
H-M Intersubset (MX4)	66.92	44.97	22.21	62.31	64.33
M-T Intersubset (MX5)	63.67	55.14	38.76	60.97	**68.78**
H-T Intersubset (MX6)	66.95	36.32	36.67	59.24	64.45

3.2 Ablation Study of Mixup Strategies

Table 1 shows the experimental results of the six proposed mixup strategies for our in-house dataset. We show the closed set accuracy for the head, middle and tail subsets separately. We also depict the corresponding OOD performance on the 20 OOD classes of our in-house dataset. Firstly, it can be noted that compared to the standard baseline, simply adopting the Mixup strategy increases the overall closed set performance as well as OOD performance by 3.3% and 0.7%, however, we want to highlight that the closed set accuracy of tail classes reduces. Secondly, compared to the standard mixup, intra-subset mixup strategies - MX1, MX2, MX3 help to increase the corresponding subset closed set accuracies. From the intra-subset experiments, it should be specifically noted that MX2 and MX3 help to increase the OOD performance suggesting that targeting the middle and tail classes is important for a better OOD detection. For the inter-subset mixup strategies - MX4, MX5, M6, we note that MX5 significantly increases the OOD performance by 3%, which suggests inter-subset mixup is more effective compared to the intra-subset strategies for middle and tail categories. It should also be noted that although the OOD performance increases for MX5, the overall accuracy only increases slightly when compared to the baseline. This is due to the reduced influence of the head subset of the dataset. Thus, we use the setting of MX5 for integration with the prototype learning as our final framework strategy.

3.3 Benchmarking with Other Methods

In Table 2 we show the performance of all the existing OOD techniques and compare it to our proposed strategies both on our in-house dataset and ISIC2019 dataset. For our in-house dataset, we also have some unusual OOD samples depicted by OOD(unk). We would firstly like to highlight two important observations from this evaluation - 1) It can be seen that all the OOD techniques perform significantly better in detecting CIFAR-10 OOD samples with OOD

Table 2. Benchmarking of OOD techniques on both In-house and ISIC2019 dataset. ID metrics -{Precision (pre), Recall (rec), and f1-score(f1)}; OOD metrics - {AUROC(%)} (best viewed in zoom).

Method	In-house dataset						ISIC2019				
	ID(pre)	ID(rec)	ID(f1)	OOD(20cl)	OOD(unk)	OOD(Cifar)	ID(pre)	ID(rec)	ID(f1)	OOD(2cl)	OOD(Cifar)
Baseline	0.58	0.59	0.585	65.67	52.90	73.24	0.86 ± 0.03	0.86 ± 0.02	0.86 ± 0.02	68.15 ± 0.9	76.43 ± 0.4
Baseline+LS+RandAug+LRS [29]	0.62	0.63	0.625	66.19	63.13	96.34	0.87 ± 0.015	0.86 ± 0.017	0.865 ± 0.015	69.41 ± 0.5	94.87 ± 0.6
ODIN [18]	0.61	0.59	0.60	64.92	62.79	96.48	0.83 ± 0.03	0.81 ± 0.02	0.82 ± 0.025	66.21 ± 1.3	95.60 ± 1.2
OLTR [20]	0.63	0.62	0.625	67.42	70.72	98.00	0.85 ± 0.01	0.86 ± 0.02	0.855 ± 0.015	71.66 ± 0.6	98.45 ± 0.5
MC-Dropout [8]	0.59	0.58	0.585	66.07	68.83	97.57	0.84 ± 0.023	0.84 ± 0.02	0.84 ± 0.02	72.18 ± 0.3	96.41 ± 0.3
ARPL [4]	**0.64**	**0.63**	**0.635**	68.55	80.61	99.42	0.85 ± 0.01	0.86 ± 0.016	0.855 ± 0.012	74.16 ± 0.7	97.20 ± 0.4
Mixup [32]	0.63	0.62	0.625	66.35	66.70	97.10	$\mathbf{0.87} \pm 0.02$	$\mathbf{0.88} \pm 0.01$	$\mathbf{0.875} \pm 0.013$	71.72 ± 0.7	96.65 ± 0.6
Prototype [30]	0.63	0.62	0.625	68.82	74.54	98.04	0.85 ± 0.02	0.86 ± 0.02	0.855 ± 0.02	72.84 ± 0.6	97.02 ± 0.5
M-T Mixup (Ours)	0.61	0.60	0.605	68.78	70.81	99.29	0.85 ± 0.03	0.85 ± 0.02	0.85 ± 0.022	73.86 ± 0.6	97.10 ± 0.6
M-T Mixup + Prototype (Ours)	0.62	0.61	0.615	**71.10**	**82.71**	**99.59**	0.85 ± 0.01	0.86 ± 0.02	0.855 ± 0.015	$\mathbf{76.37} \pm 0.5$	$\mathbf{98.46} \pm 0.4$

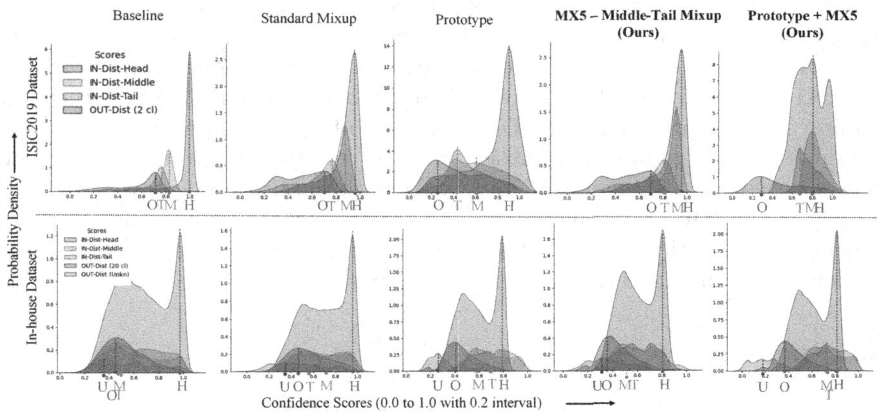

Fig. 4. Confidence Scores visualization for different methods on our In-house dataset and ISIC dataset. {H,M,T} refer to Head, Middle, and Tail subsets. {O} refers to OOD(2cl) and OOD(20cl) for ISIC and In-house dataset. {U} refers to the OOD(unk) for In-house dataset. (best viewed in zoom).

performance>95%. This shows that the existing OOD techniques are capable of detecting relatively easy OOD samples coming from a completely different domain. 2) The performance of all the techniques drops drastically when the OOD samples are from the same domain. Specifically, this is more evident for a long-tailed nature dataset such as our in-house dataset where the performance degradation is more severe compared to that of the ISIC2019. Moreover, for our in-house dataset, the OOD performance is slightly better for the unusual OOD(unk) images compared to those reserved from the extreme tail part OOD(20cl). It can be further noted that our proposed approach of M-T mixup (MX5) strategy combined with prototype learning performs the best for OOD detection while maintaining the overall ID performance compared to the baseline on both datasets.

3.4 Confidence Scores Visualization

In Fig. 4, we analyse the performance results in more detail by showing the probability density of the confidence scores for different subsets. The larger the separation between the distribution of OOD$\{O, U\}$ from ID$\{H, M, T\}$, the better the technique. It can be seen that our proposed approach achieves the largest separation for both the datasets. Specifically, it is to be noted that this is achieved by making the {M,T} subsets more confident which justifies our targeted strategy.

4 Conclusion

In this work, we present a detailed study and a strategic method for out-of-distribution detection for skin lesion images. Several vital conclusions can be made from our extensive experimentation. Firstly, we highlight that the current OOD techniques are still far away from clinical deployment where they will encounter many similar domain OOD images. To fill this gap, we propose a simple combination of middle-tail subset targeted mixup with prototype learning which is necessary for increasing the OOD performance for a long-tailed and fine-grained dataset. We believe that the experimental settings and proposed approach shown in this paper will guide the community to develop OOD detection techniques for a practical deployment application.

References

1. Bendale, A., Boult, T.E.: Towards open set deep networks. In: Proceedings of the IEEE Conference on Computer Vision and Pattern Recognition, pp. 1563–1572 (2016)
2. Budhwant, P., Shinde, S., Ingalhalikar, M.: Open-set recognition for skin lesions using dermoscopic images. In: Liu, M., Yan, P., Lian, C., Cao, X. (eds.) MLMI 2020. LNCS, vol. 12436, pp. 614–623. Springer, Cham (2020). https://doi.org/10.1007/978-3-030-59861-7_62
3. Chawla, N.V., Bowyer, K.W., Hall, L.O., Kegelmeyer, W.P.: Smote: synthetic minority over-sampling technique. J. Artif. Intell. Res. **16**, 321–357 (2002)
4. Chen, G., Peng, P., Wang, X., Tian, Y.: Adversarial reciprocal points learning for open set recognition. arXiv preprint arXiv:2103.00953 (2021)
5. Codella, N.C., et al.: Skin lesion analysis toward melanoma detection: a challenge at the 2017 international symposium on biomedical imaging (ISBI), hosted by the international skin imaging collaboration (ISIC). In: 2018 IEEE 15th International Symposium on Biomedical Imaging (ISBI 2018), pp. 168–172. IEEE (2018)
6. Combalia, M., Hueto, F., Puig, S., Malvehy, J., Vilaplana, V.: Uncertainty estimation in deep neural networks for dermoscopic image classification. In: Proceedings of the IEEE/CVF Conference on Computer Vision and Pattern Recognition Workshops, pp. 744–745 (2020)
7. Esteva, A., et al.: Dermatologist-level classification of skin cancer with deep neural networks. Nature **542**(7639), 115–118 (2017)
8. Gal, Y., Ghahramani, Z.: Dropout as a Bayesian approximation: representing model uncertainty in deep learning. In: International Conference on Machine Learning, pp. 1050–1059. PMLR (2016)

9. Ge, Z., Demyanov, S., Chen, Z., Garnavi, R.: Generative openmax for multi-class open set classification. arXiv preprint arXiv:1707.07418 (2017)
10. Geng, C., Huang, S.J., Chen, S.: Recent advances in open set recognition: a survey. IEEE Trans. Pattern Anal. Mach. Intell. **43**(10), 3614–3631 (2020)
11. Gessert, N., Nielsen, M., Shaikh, M., Werner, R., Schlaefer, A.: Skin lesion classification using ensembles of multi-resolution EfficientNets with meta data. MethodsX **7**, 100864 (2020)
12. He, K., Zhang, X., Ren, S., Sun, J.: Deep residual learning for image recognition. In: Proceedings of the IEEE Conference on Computer Vision and Pattern Recognition, pp. 770–778 (2016)
13. Hendrycks, D., Mazeika, M., Dietterich, T.: Deep anomaly detection with outlier exposure. arXiv preprint arXiv:1812.04606 (2018)
14. Hendrycks, D., Mazeika, M., Kadavath, S., Song, D.: Using self-supervised learning can improve model robustness and uncertainty. In: Advances in Neural Information Processing Systems, vol. 32 (2019)
15. Kim, H., Tadesse, G.A., Cintas, C., Speakman, S., Varshney, K.: Out-of-distribution detection in dermatology using input perturbation and subset scanning. arXiv preprint arXiv:2105.11160 (2021)
16. Krizhevsky, A., Hinton, G., et al.: Learning multiple layers of features from tiny images (2009)
17. Lee, K., Lee, H., Lee, K., Shin, J.: Training confidence-calibrated classifiers for detecting out-of-distribution samples. arXiv preprint arXiv:1711.09325 (2017)
18. Liang, S., Li, Y., Srikant, R.: Enhancing the reliability of out-of-distribution image detection in neural networks. arXiv preprint arXiv:1706.02690 (2017)
19. Liu, W., Wang, X., Owens, J., Li, Y.: Energy-based out-of-distribution detection. Adv. Neural. Inf. Process. Syst. **33**, 21464–21475 (2020)
20. Liu, Z., Miao, Z., Zhan, X., Wang, J., Gong, B., Yu, S.X.: Large-scale long-tailed recognition in an open world. In: Proceedings of the IEEE/CVF Conference on Computer Vision and Pattern Recognition, pp. 2537–2546 (2019)
21. Mahbod, A., Schaefer, G., Wang, C., Dorffner, G., Ecker, R., Ellinger, I.: Transfer learning using a multi-scale and multi-network ensemble for skin lesion classification. Comput. Methods Programs Biomed. **193**, 105475 (2020)
22. Pacheco, A.G., Sastry, C.S., Trappenberg, T., Oore, S., Krohling, R.A.: On out-of-distribution detection algorithms with deep neural skin cancer classifiers. In: Proceedings of the IEEE/CVF Conference on Computer Vision and Pattern Recognition Workshops, pp. 732–733 (2020)
23. Roy, A.G., et al.: Does your dermatology classifier know what it doesn't know? Detecting the long-tail of unseen conditions. Med. Image Anal. **75**, 102274 (2022)
24. Scheirer, W.J., de Rezende Rocha, A., Sapkota, A., Boult, T.E.: Toward open set recognition. IEEE Trans. Pattern Anal. Mach. Intell. **35**(7), 1757–1772 (2013). https://doi.org/10.1109/TPAMI.2012.256
25. Sun, X., Yang, Z., Zhang, C., Ling, K.V., Peng, G.: Conditional gaussian distribution learning for open set recognition. In: Proceedings of the IEEE/CVF Conference on Computer Vision and Pattern Recognition, pp. 13480–13489 (2020)
26. Tack, J., Mo, S., Jeong, J., Shin, J.: CSI: novelty detection via contrastive learning on distributionally shifted instances. Adv. Neural. Inf. Process. Syst. **33**, 11839–11852 (2020)
27. Thulasidasan, S., Chennupati, G., Bilmes, J.A., Bhattacharya, T., Michalak, S.: On mixup training: improved calibration and predictive uncertainty for deep neural networks. In: Advances in Neural Information Processing Systems, vol. 32 (2019)

28. Tschandl, P., Rosendahl, C., Kittler, H.: The ham10000 dataset, a large collection of multi-source dermatoscopic images of common pigmented skin lesions. Scientific data **5**(1), 1–9 (2018)
29. Vaze, S., Han, K., Vedaldi, A., Zisserman, A.: Open-set recognition: a good closed-set classifier is all you need. arXiv preprint arXiv:2110.06207 (2021)
30. Yang, H.M., Zhang, X.Y., Yin, F., Liu, C.L.: Robust classification with convolutional prototype learning. In: Proceedings of the IEEE Conference on Computer Vision and Pattern Recognition, pp. 3474–3482 (2018)
31. Yoshihashi, R., Shao, W., Kawakami, R., You, S., Iida, M., Naemura, T.: Classification-reconstruction learning for open-set recognition. In: Proceedings of the IEEE/CVF Conference on Computer Vision and Pattern Recognition, pp. 4016–4025 (2019)
32. Zhang, H., Cisse, M., Dauphin, Y.N., Lopez-Paz, D.: mixup: beyond empirical risk minimization. arXiv preprint arXiv:1710.09412 (2017)
33. Zhang, L., Deng, Z., Kawaguchi, K., Ghorbani, A., Zou, J.: How does mixup help with robustness and generalization? In: International Conference on Learning Representations (2020)

FairPrune: Achieving Fairness Through Pruning for Dermatological Disease Diagnosis

Yawen Wu[1], Dewen Zeng[2], Xiaowei Xu[3], Yiyu Shi[2(✉)], and Jingtong Hu[1(✉)]

[1] University of Pittsburgh, Pittsburgh, PA, USA
{yawen.wu,jthu}@pitt.edu
[2] University of Notre Dame, Notre Dame, IN, USA
{dzeng2,yshi4}@nd.edu
[3] Guangdong Provincial People's Hospital, Guangzhou, China
xxu8@nd.edu

Abstract. Many works have shown that deep learning-based medical image classification models can exhibit bias toward certain demographic attributes like race, gender, and age. Existing bias mitigation methods primarily focus on learning debiased models, which may not necessarily guarantee all sensitive information can be removed and usually comes with considerable accuracy degradation on both privileged and unprivileged groups. To tackle this issue, we propose a method, FairPrune, that achieves fairness by pruning. Conventionally, pruning is used to reduce the model size for efficient inference. However, we show that pruning can also be a powerful tool to achieve fairness. Our observation is that during pruning, each parameter in the model has different importance for different groups' accuracy. By pruning the parameters based on this importance difference, we can reduce the accuracy difference between the privileged group and the unprivileged group to improve fairness without a large accuracy drop. To this end, we use the second derivative of the parameters of a pre-trained model to quantify the importance of each parameter with respect to the model accuracy for each group. Experiments on two skin lesion diagnosis datasets over multiple sensitive attributes demonstrate that our method can greatly improve fairness while keeping the average accuracy of both groups as high as possible.

1 Introduction

In AI-assisted medical image analysis, deep neural networks (DNNs) tend to capture relevant statistical information such as colors and textures from the training data. This data-driven paradigm can help the network learn task-specific features for high accuracy on the target task. However, to maximize the accuracy, the network may use the information present in some data but not in other data, and thus show discrimination towards certain demographics (i.e. skin tone

Y. Wu and D. Zeng—Equal contributions. Listing order determined by coin flipping.

L. Wang et al. (Eds.): MICCAI 2022, LNCS 13431, pp. 743–753, 2022.
https://doi.org/10.1007/978-3-031-16431-6_70

or gender) [21]. For example, [5,28] demonstrate that the network learned on the CelebA dataset [17] performs better on the female group when the task is predicting facial attributes such as wavy hair and smiling. Dermatological disease classification networks trained on two public dermatology datasets (Fitzpatrick-17k and ISIC 2019 Challenge) have been reported to be biased across different skin tones [8,13]. However, no solution is proposed to mitigate the bias in these works. An X-ray computer-aided diagnosis (CAD) system was found to exhibit disparities across genders [14]. Once these biased models are deployed in the real-world system, they could be harmful to both individuals and society. For example, AI algorithms could misdiagnose people from different demographic groups, leading to increased health care disparities. This leads to a variety of research techniques that aim to alleviate the bias in DNNs.

One of the most widely used bias mitigation methods is adversarial training [2,7,12,26,29]. Normally, an adversarial network is added to the tail of an encoder or a classifier to predict the protected attributes and form a minimax game: maximize the network's ability to predict the class while minimizing the adversarial network's ability to predict the protected attributes. In this way, the model will be able to learn fair features that are irrelevant to the protected attributes. However, as mentioned in [27], the main drawback of this method is that even if the target protected attribute has been removed, the combination of other features may still be a proxy for this protected attribute. In addition, forcing the model to ignore the protected attribute relevant features may harm its classification accuracy as those features may contain important information for the final prediction [27]. Fairness through explanation is another bias mitigation technique [10,20,22], this technique requires fine-grained feature-level annotation as the domain knowledge to train the model to only focus on bias-unrelated features in the original input. Such suppression of sensitive information can also potentially remove useful information and thus greatly degrade the classification performance. Therefore, to achieve fairness, these state-of-the-art (SOTA) methods usually need to sacrifice considerable accuracy for both groups.

To avoid this, we introduce FairPrune, a technique to achieve fairness via pruning. Conventionally, pruning is used to reduce the model size for efficient inference. However, one interesting thing we found is that pruning can be a powerful tool for fairness. Our work is motivated by the observation that during pruning, the parameter that is important for one demographic group may be unimportant for another. By controlling the parameters to prune, we can reduce the accuracy difference between the privileged group and the unprivileged group to improve fairness while keeping their overall accuracy as high as possible. To this end, we utilize the saliency of each parameter (computed based on the second derivative [15]) to quantify its importance regarding the model accuracy. Specifically, we compute the saliencies of all parameters for each demographic group, which will then be used to prune the parameters that show large importance differences for these two groups to mitigate biases. In addition, the trade-off between fairness improvement and accuracy drop can be adjusted to satisfy different user requirements. As a byproduct, FairPrune can also reduce the network size for efficient deployment. We evaluate FairPrune on two skin lesion analysis datasets over two sensitive attributes and show improved fairness with a lower accuracy drop over SOTA methods for fairness.

2 Related Work

Existing bias mitigation methods can be generally categorized into three groups: pre-processing, in-processing, and post-processing.

For pre-processing methods, a straightforward solution is to remove the sensitive information from the training data. One could also assign different weights to different data samples to suppress the sensitive information during training [11]. These pre-processing techniques are not suitable for dermatological data because the sensitive information exists in the target (i.e., diseased area), which is necessary for diagnosis purposes and cannot be removed.

In-processing bias mitigation methods usually involve modifying the training loss function to regularize the model for fairness. For example, [16] added a task-specific prior to implicitly regularize the model not to pay attention to the sensitive attribute related information for its prediction. Adversarial training [1,12,25,26,29] achieve fairness by removing sensitive information with an adversarial learner. Fair meta-learning [19] trains a classifier to classify the sensitive attribute and jointly optimized with the main prediction objective. However, these methods can not explicitly protect the unprivileged group when enforcing the fairness constraints, and the accuracy of both groups will drop. Instead, our method can protect the accuracy of one group when achieving fairness goals, resulting in high overall accuracy and fairness simultaneously.

As for post-processing techniques, during the inference-time, calibration is performed by taking the model's prediction and the sensitive attribute as input [6,9,30]. The goal is to enforce the prediction distribution to match the training distribution or a specific fairness metric. While these methods show effectiveness for bias mitigation, they need access to the sensitive attribute for inference, which may not be available for disease diagnosis when taking images as input. Different from these methods, our method modifies the pre-trained model only based on the training set. During inference, the sensitive attribute is not needed. Recent works [18,23] analyze the impact of pruning on fairness. Different from these works, we achieve fairness by pruning.

3 Method

3.1 Problem Definition

Given a dataset $D = \{x_i, y_i, c_i\}$, $i \in 1, ..., N$ where x_i is the input image, y_i is the class label, c_i is the sensitive attribute (e.g., skin tone, gender, age), and a pre-trained classification model $f_\theta(\cdot)$ with parameters θ that maps the input x_i to the final prediction $\hat{y_i} = f_\theta(x_i)$. Our goal is to reduce the discrimination in $f_\theta(\cdot)$ with respect to the sensitive attribute c by only modifying some of the parameters

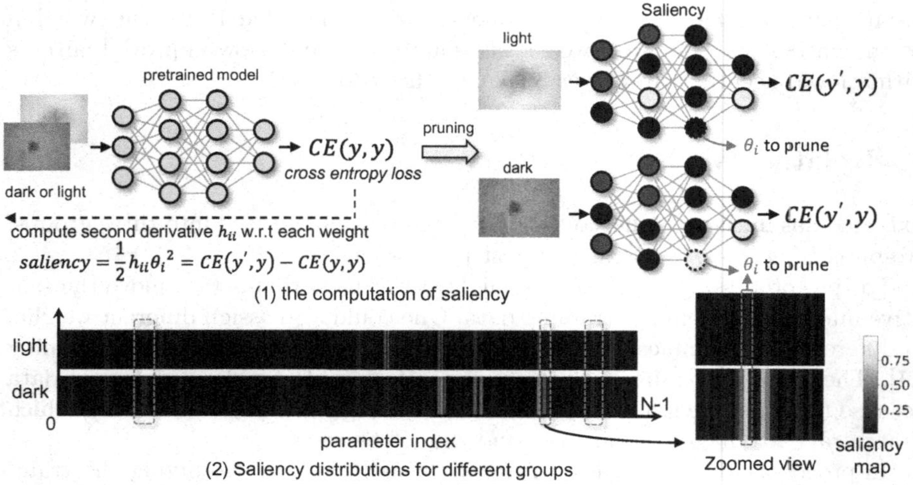

Fig. 1. Illustration of (1) saliency computation and (2) the parameters' saliency distribution for dark and light skin tone groups on the Fitzpatrick-17k dataset. The x-axis is the index of parameters in the first layer of the VGG-11, the color represents the normalized saliency for each demographic group. The blue boxes highlight those parameters that have relatively low saliency for the light (unprivileged group) but high saliency for the dark (privileged group). (Color figure online)

in $f_\theta(\cdot)$ without further finetuning. In this paper, we only consider the binary sensitive attribute (i.e., $c_i \in \{0,1\}$). $c_i = 0$ represents unprivileged samples (the model shows discrimination against), while $c_i = 1$ represents privileged samples.

3.2 Pruning for Fairness

Saliency Reflects the Accuracy Drop After Pruning. Given the pre-trained model f_θ and its objective function E, the change of the objective function after pruning parameters Θ can be approximated by a Taylor series [15]:

$$\begin{aligned} \Delta E &= E(D|\Theta = 0) - E(D) \\ &= -\sum_i g_i\theta_i + \frac{1}{2}\sum_i h_{ii}\theta_i^2 + \frac{1}{2}\sum_{i \neq j} h_{ij}\theta_i\theta_j + O(||\Theta||^3) = \frac{1}{2}\sum_i h_{ii}\theta_i^2. \end{aligned} \quad (1)$$

where $g_i = \frac{\partial E}{\partial \theta_i}$ is the gradient of E with respect to θ_i, which is close to 0 because we assume the pre-trained model has converged and the objective function is at its local minimum. $h_{ii} = \frac{\partial^2 E}{\partial^2 \theta_i}$ is the element at row i and column i of the second derivative Hessian matrix $\mathbf{H}$. The approximation assumes that ΔE caused by pruning several parameters is the sum of ΔE caused by pruning each parameter individually, so the third term is neglected. $\frac{1}{2}h_{ii}\theta_i^2$ is called the saliency of θ_i, which represents the increase of error after pruning this parameter.

Using Saliency to Achieve Group Fairness. Our idea is based on our empirical observation that in a pre-trained network, the importance (saliency) of some parameters can be totally different for different demographic groups. That is, some parameters may have a small saliency for one group but a large saliency for another group. For example, Fig. 1 shows a dermatological disease diagnosis model pre-trained by vanilla training and is biased against patients with light skin tone, which is the unprivileged group. In the coordinate, we show the saliency distribution of the parameters in the first layer of VGG-11 for both dark and light groups. The x-axis represents parameter indices, where one coordinate shows the same parameter for both the light and the dark groups. The color denotes normalized saliency for both groups. It can be seen that there exist differences between the distribution of saliency for these two groups. In the blue boxes, we highlight those parameters with low saliency for the light group but high saliency for the dark group. By pruning these parameters, the accuracy difference of the privileged group (dark) and the unprivileged group (light) can be reduced. In this way, the model is pruned to be fair for the unprivileged group.

To identify those parameters to prune, we propose a fairness-aware saliency computation method to identify the parameters which are unimportant for the unprivileged group but important for the privileged group. The importance of each parameter for fairness is quantified by integrating the saliency of the unprivileged group and the saliency of the privileged group weighted by a negative scalar. To be specific, we formulate a multi-objective optimization problem:

$$\min \Delta E_{c=0}(\Theta), \quad \max \Delta E_{c=1}(\Theta). \tag{2}$$

where $\Delta E_{c=0}(\Theta)$ and $\Delta E_{c=0}(\Theta)$ are the error changes of unprivileged group $c = 0$ and privileged group $c = 1$ after pruning parameters Θ, respectively.

To solve this problem, Eq.(2) can be transformed into a single objective as:

$$\min_{\Theta} J = \Delta E_{c=0}(\Theta) - \beta \Delta E_{c=1}(\Theta) = \sum_i s_i, \tag{3}$$

$$s_i = (\frac{1}{2} h_{ii}^0 \theta_i^2) - \beta \cdot (\frac{1}{2} h_{ii}^1 \theta_i^2) = \frac{1}{2} \theta_m^2 (h_{ii}^0 - \beta \cdot h_{ii}^1). \tag{4}$$

where β is a hyper-parameter controlling the trade-off between minimizing $\Delta E_{c=0}$ and maximizing $\Delta E_{c=1}$. θ_i is the parameter of the model and can be treated as a constant. s_i is the saliency of parameter θ_i for achieving the fairness objective, where a smaller value represents a larger benefit when we prune it since a smaller s_i contributes to minimizing the objective J. $h_{ii}^c, c \in \{0, 1\}$ is the Hessian element of θ_i for demographic group c, which is the juncture to inject the information of each group into the pruning process to achieve fairness.

3.3 Pruning Recipe

FairPrune prunes a pre-trained model by the following iterative steps.

1. Sample mini-batches $\{B^0\}$ and $\{B^1\}$ from the unprivileged group and privileged group, respectively.

2. For each pair of mini-batches (B^0, B^1), compute the second derivatives h_{ii} for each parameter, and compute parameter salience s_i.
3. After several mini-batches, average the saliency of each parameter over the mini-batches. Remove $p\%$ parameters with the smallest saliency.
4. Repeat steps 1–3 until the target fairness metric is achieved.

4 Experiments and Results

Fairness Metrics. We use multi-class equalized opportunity (Eopp) and equalized Odds (Eodd) [9] to evaluate the fairness of the model. The Eopp0 is the True Negative Rate difference between two groups, the Eopp1 is the True Positive Rate difference between two groups, while the Eodd is the summation of the True Positive Rate difference and False Positive Rate difference. Suppose TP_k^c, FN_k^c, TN_k^c, and FP_k^c are the True Positive, False Negative, True Negative, and False Positive of class k and group c, then the True Positive Rate, True Negative Rate, and False Positive Rate of class k and group c can be computed by $TPR_k^c = \frac{TP_k^c}{TP_k^c + FN_k^c}$, $TNR_k^c = \frac{TN_k^c}{TN_k^c + FP_k^c}$, and $FPR_k^c = \frac{FP_k^c}{TN_k^c + FP_k^c}$, respectively. The EOpp and Eodd can be computed by the following equations:

$$EOpp0 = \sum_{k=1}^{K} |TNR_k^1 - TNR_k^0|, \quad EOpp1 = \sum_{k=1}^{K} |TPR_k^1 - TPR_k^0|. \tag{5}$$

$$EOdd = \sum_{k=1}^{K} |TPR_k^1 - TPR_k^0 + FPR_k^1 - FPR_k^0|. \tag{6}$$

Dataset and Preprocessing. The proposed methods are evaluated on two dermatology datasets for disease classification, including the Fitzpatrick-17k [8] and ISIC 2019 challenge [3,24] datasets. The Fitzpatrick-17k contains 16,003 images in 114 skin conditions. The skin tones are categorized in six levels from 1 to 6, where a smaller value represents lighter skin and a larger value means darker skin. We categorize the skin tones into two groups, where types 1 to 3 are light skin with 11,057 images and types 4 to 6 are dark skin with 4,946 images. While light skin has more images than dark skins, the vanilla trained model still has a higher accuracy on the dark skins. The ISIC 2019 dataset contains 25,331 dermoscopic images in 9 diagnostic categories. Since skin tones are not provided by this dataset, we use the gender label to group the data as female and male. We exclude 1,373 images without gender labels and keep the remaining 23,958 images, with 11,600 female images and 12,358 male images. The vanilla trained model has a higher accuracy on the female images.

Pre-training Details. We use VGG-11 as the backbone. On both datasets, we resize all the images to 128×128. Data augmentation includes random horizontal flipping, vertical flipping, rotation, scaling, and autoaugment [4]. We randomly split the dataset into training (60%), validation (20%), and test (20%) partitions. The model is pre-trained for 200 epochs with the Adam optimizer. The batch size is 256 and the learning rate is 1e-4 decayed by a factor of 10 at epoch 160.

Table 1. Results of accuracy and fairness of different methods on Fitzpatrick-17k dataset, using skin tone as the sensitive attribute. The dark skin is the privileged group with higher accuracy by vanilla training. (pr is the pruning ratio).

Method	Skin tone	Accuracy			Fairness		
		Precision	Recall	F1-score	Eopp0 ($\times 10^{-3}$) ↓	Eopp1 ↓	Eodd ↓
Vanilla	Dark	0.563	0.581	0.546	1.331	0.361	0.182
	Light	0.482	0.495	0.473			
	Avg. ↑	0.523	0.538	0.510			
	Diff. ↓	0.081	0.086	0.073			
AdvConf [29]	Dark	0.506	0.562	0.506	1.106	0.339	0.169
	Light	0.427	0.464	0.426			
	Avg. ↑	0.467	0.513	0.466			
	Diff. ↓	0.079	0.098	0.080			
AdvRev [25]	Dark	0.514	0.545	0.503	1.127	0.334	0.166
	Light	0.489	0.469	0.457			
	Avg. ↑	0.502	0.507	0.480			
	Diff. ↓	0.025	0.076	0.046			
DomainIndep [27]	Dark	0.547	0.567	0.532	1.210	0.344	0.172
	Light	0.455	0.480	0.451			
	Avg. ↑	0.501	0.523	0.492			
	Diff. ↓	0.092	0.087	0.081			
OBD [15] (pr=35%)	Dark	0.557	0.570	0.536	1.244	0.360	0.180
	Light	0.488	0.494	0.475			
	Avg. ↑	0.523	0.532	0.506			
	Diff. ↓	0.069	0.076	0.061			
FairPrune (pr=35%, β=0.33)	Dark	0.567	0.519	0.507	**0.846**	**0.330**	**0.165**
	Light	0.496	0.477	0.459			
	Avg. ↑	0.531	0.498	0.483			
	Diff. ↓	0.071	**0.042**	0.048			

Pruning Details. On the Fitzpatrick-17k dataset, the batch size for computing the saliency is 2. The saliency of each parameter is averaged every 500 mini-batches to prune 5% of the parameters in each pruning iteration. On the ISIC 2019 dataset, the batch size for computing the saliency is 64. The saliency of each parameter is averaged every 200 mini-batches to prune 10% of the parameters. Grid search is used to find the best hyper-parameter β and pruning ratio with the optimal accuracy and fairness tradeoff on the validation set for each dataset. The pre-training and pruning are performed on one Nvidia V100 GPU.

Baselines. We compare FairPrune with multiple baselines. *Vanilla* is standard training without fairness constraints. *AdvConf* and *AdvRev* are two adversarial training based de-biasing methods. *AdvConf* [2,25] employs a uniform confusion loss to minimize the classifier's ability for predicting the sensitive attribute. *AdvRev* [29] de-biases the model by maximizing the loss of predicting the sensitive attribute with loss reversal and gradient projection. *DomainIndep* [27] trains multiple classifiers, one classifier for each group to explicitly encode separate group information. *OBD* [15] is the pruning-based method, which uses all training data to compute the saliency without fairness constraints.

Table 2. Results of accuracy and fairness of different methods on ISIC 2019 dataset, using gender as the sensitive attribute. The female group is the privileged group with higher accuracy by vanilla training. (pr is the pruning ratio).

Method	Gender	Accuracy			Fairness		
		Precision	Recall	F1-score	Eopp0($\times 10^{-3}$)↓	Eopp1($\times 10^{-3}$)↓	Eodd($\times 10^{-3}$)↓
Vanilla	Female	0.758	0.733	0.744	6.1	49.7	55.8
	Male	0.766	0.684	0.716			
	Avg ↑	0.762	0.709	0.730			
	Diff ↓	0.008	0.049	0.028			
AdvConf [29]	Female	0.691	0.688	0.686	**4.0**	75.1	79.1
	Male	0.681	0.656	0.665			
	Avg ↑	0.686	0.672	0.675			
	Diff ↓	0.010	0.032	0.021			
AdvRev [25]	Female	0.638	0.714	0.670	5.0	59.2	64.2
	Male	0.642	0.666	0.650			
	Avg ↑	0.640	0.690	0.660			
	Diff ↓	0.004	0.048	0.020			
DomainIndep [27]	Female	0.782	0.693	0.729	5.0	74.7	79.7
	Male	0.783	0.653	0.697			
	Avg ↑	0.782	0.673	0.713			
	Diff ↓	**0.001**	0.040	0.032			
OBD [15] (pr=50%)	Female	0.771	0.734	0.749	6.1	55.5	61.6
	Male	0.762	0.678	0.711			
	Avg ↑	0.767	0.706	0.730			
	Diff ↓	0.009	0.056	0.038			
FairPrune (pr=50%, β=0.2)	Female	0.754	0.674	0.707	7.8	**21.0**	**28.8**
	Male	0.762	0.675	0.710			
	Avg ↑	0.758	0.675	0.709			
	Diff ↓	0.008	**0.001**	**0.003**			

Results on Fitzpatrick-17k Dataset. Table 1 shows the accuracy and fairness results of all methods. For accuracy metrics, we report the precision, recall, and F1-score. From the table, we can see that when achieving a similar level of fairness, FairPrune shows better mean accuracy and a smaller accuracy difference in terms of F1-score compared with two adversarial training baselines *AdvConf* and *AdvRev*. This is because Fairprune can almost keep the accuracy of the unprivileged group unchanged when balancing the accuracy of two groups. We can also observe that although vanilla OBD has better overall accuracy than FairPrune, it does not show significant fairness improvements.

Results on ISIC 2019 Dataset. Table 2 shows the comparison of accuracy and fairness results of all methods. FairPrune achieves significantly better fairness while preserving accuracy. First, our FairPrune method achieves significantly better fairness than the baselines. Our FairPrune method achieves 21.0×10^{-3} Eopp1 and 28.8×10^{-3} Eodd, which are 57.7% and 48.3% lower than 49.7×10^{-3} and 55.8×10^{-3} achieved by the best baseline. Second, the accuracy of FairPrune is comparable to the Vanilla method without fairness constraints, while the F1-score difference between two groups of our method is only 0.003, an order of magnitude lower than the Vanilla training and other methods.

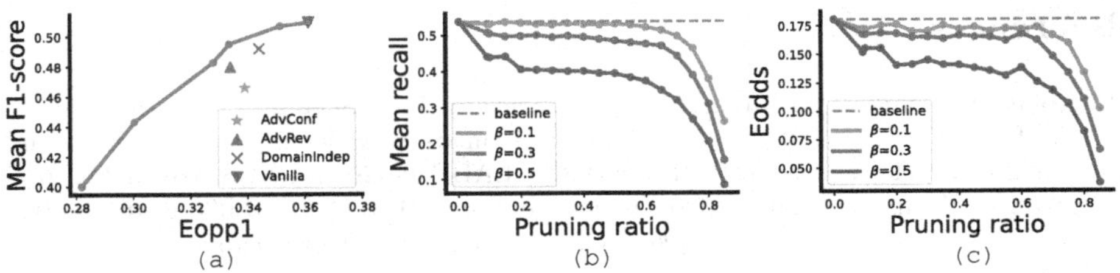

Fig. 2. Ablation study on hyper-parameter β and pruning ratio for skin tone fairness on Fitzpatrick-17k dataset.

Ablation Study. In this section, we discuss how the hyper-parameter β and pruning rate will affect the performance of FairPrune. We apply FairPrune to the model pre-trained on the Fitzpatrick-17k dataset and change β defined in Eq.(4) to see how the accuracy and fairness performance change. After that, we fix β and change the pruning ratio from 0 to 80% and repeat the evaluation.

In Fig. 2(a), we show the accuracy (F1-Score) and fairness (Eopp1) trade-off by varying β under a fixed pruning ratio of 35%. Points on the upper left of this figure represent better fairness (smaller Eopp1) and higher accuracy (larger F1-score). With varying β, our methods achieve consistently better fairness and accuracy trade-off. We can see that when achieving the same fairness, FairPrune has higher accuracy than the baselines. At the same accuracy, FairPrune can also show better fairness than the baselines. Therefore, by tuning β, the trade-off between fairness and accuracy can be adjusted to satisfy the user requirement.

Figure 2(b) and Fig. 2(c) show the effect of the pruning ratio on accuracy and fairness, respectively. It can be seen that the mean recall drops slowly and then quickly, as well as the fairness metrics Eopp1. Although the fairness improvement can be larger at a larger pruning ratio (e.g., $pruning\ ratio = 0.8$), the accuracy may also be lower. Therefore, we suggest that the optimal pruning ratio usually exists at a lower pruning ratio where we do not see a significant accuracy drop on both groups but better fairness.

5 Conclusion

In this paper, we propose FairPrune, a method to achieve fairness by pruning. Based on our observation that each model parameter has different importance for different groups' accuracy, by pruning the parameters based on this importance difference, we can reduce the accuracy difference between the privileged group and the unprivileged group to improve fairness without a large accuracy drop. We evaluate our method on two skin lesion analysis datasets over multiple sensitive attributes. The experiment results show that FairPrune can greatly improve fairness while keeping the average accuracy of both groups as high as possible.

References

1. Abbasi-Sureshjani, S., Raumanns, R., Michels, B.E.J., Schouten, G., Cheplygina, V.: Risk of training diagnostic algorithms on data with demographic bias. In: Cardoso, J., et al. (eds.) IMIMIC/MIL3ID/LABELS -2020. LNCS, vol. 12446, pp. 183–192. Springer, Cham (2020). https://doi.org/10.1007/978-3-030-61166-8_20
2. Alvi, M., Zisserman, A., Nellåker, C.: Turning a blind eye: explicit removal of biases and variation from deep neural network embeddings. In: Proceedings of the European Conference on Computer Vision (ECCV) Workshops (2018)
3. Combalia, M., et al.: Bcn20000: dermoscopic lesions in the wild. arXiv preprint arXiv:1908.02288 (2019)
4. Cubuk, E.D., Zoph, B., Mane, D., Vasudevan, V., Le, Q.V.: Autoaugment: learning augmentation policies from data. arXiv preprint arXiv:1805.09501 (2018)
5. Du, M., Mukherjee, S., Wang, G., Tang, R., Awadallah, A., Hu, X.: Fairness via representation neutralization. Adv. Neural Inf. Process. Syst. **34**, 12091–12103 (2021)
6. Du, M., Yang, F., Zou, N., Hu, X.: Fairness in deep learning: a computational perspective. IEEE Intell. Syst. **36**(4), 25–34 (2020)
7. Elazar, Y., Goldberg, Y.: Adversarial removal of demographic attributes from text data. In: Proceedings of the 2018 Conference on Empirical Methods in Natural Language Processing, pp. 11–21 (2018)
8. Groh, M., et al.: Evaluating deep neural networks trained on clinical images in dermatology with the fitzpatrick 17k dataset. In: Proceedings of the IEEE/CVF Conference on Computer Vision and Pattern Recognition, pp. 1820–1828 (2021)
9. Hardt, M., Price, E., Srebro, N.: Equality of opportunity in supervised learning. Adv. Neural Inf. Process. Syst. **29** (2016)
10. Hendricks, L.A., Burns, K., Saenko, K., Darrell, T., Rohrbach, A.: Women also snowboard: overcoming bias in captioning models. In: Proceedings of the European Conference on Computer Vision (ECCV), pp. 771–787 (2018)
11. Kamiran, F., Calders, T.: Data preprocessing techniques for classification without discrimination. Knowl. Inf. Syst. **33**(1), 1–33 (2012)
12. Kim, B., Kim, H., Kim, K., Kim, S., Kim, J.: Learning not to learn: Training deep neural networks with biased data. In: Proceedings of the IEEE/CVF Conference on Computer Vision and Pattern Recognition, pp. 9012–9020 (2019)
13. Kinyanjui, N.M., et al.: Fairness of classifiers across skin tones in dermatology. In: Martel, A.L., et al. (eds.) MICCAI 2020. LNCS, vol. 12266, pp. 320–329. Springer, Cham (2020). https://doi.org/10.1007/978-3-030-59725-2_31
14. Larrazabal, A.J., Nieto, N., Peterson, V., Milone, D.H., Ferrante, E.: Gender imbalance in medical imaging datasets produces biased classifiers for computer-aided diagnosis. Proc. Nat. Acad. Sci. **117**(23), 12592–12594 (2020)
15. LeCun, Y., Denker, J., Solla, S.: Optimal brain damage. Adv. Neural Inf. Process. Syst. **2** (1989)
16. Liu, F., Avci, B.: Incorporating priors with feature attribution on text classification. In: Proceedings of the 57th Annual Meeting of the Association for Computational Linguistics, pp. 6274–6283 (2019)
17. Liu, Z., Luo, P., Wang, X., Tang, X.: Deep learning face attributes in the wild. In: Proceedings of International Conference on Computer Vision (ICCV), December 2015
18. Paganini, M.: Prune responsibly. arXiv preprint arXiv:2009.09936 (2020)

19. Puyol-Antón, E., et al.: Fairness in cardiac MR image analysis: an investigation of bias due to data imbalance in deep learning based segmentation. In: de Bruijne, M., et al. (eds.) MICCAI 2021. LNCS, vol. 12903, pp. 413–423. Springer, Cham (2021). https://doi.org/10.1007/978-3-030-87199-4_39
20. Rieger, L., Singh, C., Murdoch, W., Yu, B.: Interpretations are useful: penalizing explanations to align neural networks with prior knowledge. In: International Conference on Machine Learning, pp. 8116–8126. PMLR (2020)
21. Seyyed-Kalantari, L., Zhang, H., McDermott, M., Chen, I.Y., Ghassemi, M.: Underdiagnosis bias of artificial intelligence algorithms applied to chest radiographs in under-served patient populations. Nat. Med. **27**(12), 2176–2182 (2021)
22. Singh, K.K., Mahajan, D., Grauman, K., Lee, Y.J., Feiszli, M., Ghadiyaram, D.: Don't judge an object by its context: learning to overcome contextual bias. In: Proceedings of the IEEE/CVF Conference on Computer Vision and Pattern Recognition, pp. 11070–11078 (2020)
23. Stoychev, S., Gunes, H.: The effect of model compression on fairness in facial expression recognition. arXiv preprint arXiv:2201.01709 (2022)
24. Tschandl, P., Rosendahl, C., Kittler, H.: The ham10000 dataset, a large collection of multi-source dermatoscopic images of common pigmented skin lesions. Sci. Data **5**(1), 1–9 (2018)
25. Tzeng, E., Hoffman, J., Darrell, T., Saenko, K.: Simultaneous deep transfer across domains and tasks. In: Proceedings of the IEEE International Conference on Computer Vision, pp. 4068–4076 (2015)
26. Wang, T., Zhao, J., Yatskar, M., Chang, K.W., Ordonez, V.: Balanced datasets are not enough: estimating and mitigating gender bias in deep image representations. In: Proceedings of the IEEE/CVF International Conference on Computer Vision, pp. 5310–5319 (2019)
27. Wang, Z., et al.: Towards fairness in visual recognition: effective strategies for bias mitigation. In: Proceedings of the IEEE/CVF Conference on Computer Vision and Pattern Recognition, pp. 8919–8928 (2020)
28. Xu, T., White, J., Kalkan, S., Gunes, H.: Investigating bias and fairness in facial expression recognition. In: Bartoli, A., Fusiello, A. (eds.) ECCV 2020. LNCS, vol. 12540, pp. 506–523. Springer, Cham (2020). https://doi.org/10.1007/978-3-030-65414-6_35
29. Zhang, B.H., Lemoine, B., Mitchell, M.: Mitigating unwanted biases with adversarial learning. In: Proceedings of the 2018 AAAI/ACM Conference on AI, Ethics, and Society, pp. 335–340 (2018)
30. Zhao, J., Wang, T., Yatskar, M., Ordonez, V., Chang, K.W.: Men also like shopping: reducing gender bias amplification using corpus-level constraints. arXiv preprint arXiv:1707.09457 (2017)

Reliability-Aware Contrastive Self-ensembling for Semi-supervised Medical Image Classification

Wenlong Hang[1], Yecheng Huang[1], Shuang Liang[2,3(✉)], Baiying Lei[4], Kup-Sze Choi[5], and Jing Qin[5]

[1] School of Computer Science and Technology, Nanjing Tech University, Nanjing, China
[2] Smart Health Big Data Analysis and Location Services Engineering Lab of Jiangsu Province, Nanjing University of Posts and Telecommunications, Nanjing, China
[3] State Key Laboratory for Novel Software Technology, Nanjing University, Nanjing, China
Shuang.liang@njupt.edu.cn
[4] School of Biomedical Engineering, Shenzhen University, Shenzhen, China
[5] School of Nursing, Hong Kong Polytechnic University, Hung Hom, Hong Kong

Abstract. Self-ensembling framework has proven to be a powerful paradigm for semi-supervised medical image classification by leveraging abundant unlabeled data. However, the unlabeled data used in most of self-ensembling methods are equally weighted, which adversely affects the classification performance of models when difference exists among unlabeled data acquired from different populations, equipment and environments. To address this issue, we propose a novel reliability-aware contrastive self-ensembling framework, which can leverage the reliable unlabeled data selectively. Concretely, we introduce a weight function to the mean teacher paradigm for mapping the probability predictions of unlabeled data to corresponding weights that reflect their reliability. Hence, we can safely leverage the predictions of related unlabeled data under different perturbations to construct a reliable consistency loss. Besides, we further design a novel reliable contrastive loss to achieve better intra-class compactness and inter-class separability for the normalized embeddings derived from related unlabeled data. As a result, our reliability-aware scheme enables the contrastive self-ensembling framework concurrently capture both the reliable data-level and data-structure-level information, thereby improving the robustness and generalization power of the model. Experiments on two publicly available medical image datasets demonstrate the superiority of the proposed method. Our model is available at https://github.com/Mwnic/RAC-MT.

Keywords: Self-ensembling · Weight function · Reliability · Medical image classification

L. Wang et al. (Eds.): MICCAI 2022, LNCS 13431, pp. 754–763, 2022.
https://doi.org/10.1007/978-3-031-16431-6_71

1 Introduction

With large amounts of labeled training data, deep learning methods have greatly advanced the classification performance of medical images [15,18]. However, collecting abundant labeled medical images is inevitably laborious and requires specialized medical knowledge, which hinders the application of deep learning methods in clinical practice. Since unlabeled medical images is generally much easier to obtain, an alternative strategy is to develop deep semi-supervised learning methods [3,20] that can exploit abundant unlabeled data.

Self-ensembling framework has proven to be a powerful paradigm for semi-supervised medical image classification by leveraging abundant unlabeled data to mitigate the reliance on large labeled datasets [16]. For example, Gyawali *et al.*. [3] presented a global smoothness regularized semi-supervised learning method via linearly interpolation between the labeled and unlabeled data in both the input and latent spaces. Wang *et al.*. [20] developed a neighbor matching-based mean teacher (MT) model [17], in which the neighbor matching strategy, *i.e.*, a pseudo-label estimator, generated the pseudo-labels for the unlabeled data that kept the same with its neighbors. Although effective, these methods only consider the data-level information. To further utilize the data-structure-level information, Liu *et al.*. [9] developed a semi-supervised classification method, which pre-trained MT model using the self-supervised contrastive learning [6] and then fine-tuned the MT model using labeled data. Liu *et al.*. [11] introduced the relation consistency paradigm to MT model to capture the relationship information between different samples. They further developed a federated semi-supervised method [10], which incorporated the consistency regularization with the inter-client relation matching scheme to train network using the data distributed in different institutions. Despite promising progress, current self-ensembling methods do not consider the reliability of unlabeled data. In clinical practice, difference inevitably exists among unlabeled medical images acquired from different populations, equipment and environments. Therefore, leveraging the unlabeled data in a flat way, *i.e.*, assigning an identical weight for all unlabeled data, may adversely affect the classification performance of self-ensembling methods.

Herein, we propose a novel reliability-aware contrastive self-ensembling framework for semi-supervised medical image classification by leveraging the unlabeled data selectively. Our framework is based on MT model, which encourages consensus among ensemble predictions of unlabeled data under small input perturbations. To leverage reliable unlabeled data, we propose a novel reliability-aware contrastive MT (RAC-MT) model, where a weight function is introduced to map the probability prediction of unlabeled data to their corresponding weight. We can then design a reliability-aware consistency loss that only enforces the prediction consistency of reliable unlabeled data under different perturbations. To exploit the data-structure-level information (*e.g.*, cluster) of the images, we further design a reliability-aware contrastive loss that only encourages the normalized embeddings derived from reliable unlabeled data to attain better intra-class compactness and inter-class separability. Our RAC-MT model can concurrently capture both

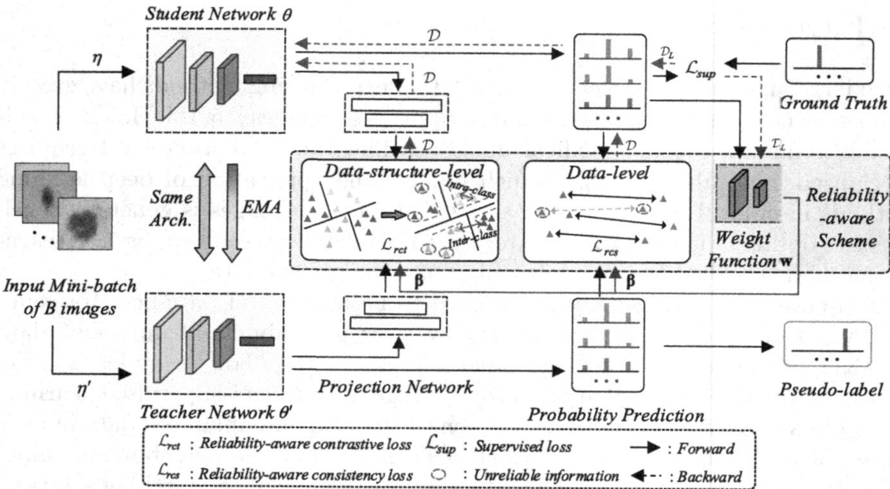

Fig. 1. The pipeline of the proposed reliability-aware framework for semi-supervised classification. The reliability-aware consistency loss and reliability-aware contrastive loss exploit reliable data-level and data-structure-level information on the training set $\mathcal{D}$. The supervised loss is optimized over the labeled data $\mathcal{D}_L$. β denotes the weight of unlabeled data. The parameter $\mathbf{w}$ of the weight function can be updated based on the gradient of the supervised loss. The framework iteratively alternates between the update of the parameter $\mathbf{w}$ and the network parameter θ.

the reliable data-level and data-structure-level information of the images, thereby improving the robustness and generalization power of the model. Experiments on the skin lesion dataset and the nucleus dataset demonstrate the superiority of our method in semi-supervised medical classification.

2 Methodology

Figure 1 illustrates the pipeline of the proposed RAC-MT model for semi-supervised medical image classification. The weight function maps the probability prediction of each unlabeled data to a unique weight that estimates its reliability. The reliability-aware consistency loss and reliability-aware contrastive loss exploit reliable data-level and data-structure-level information respectively, thus improving the robustness and generalization power of the model.

2.1 Reliability-Aware Contrastive Mean Teacher

Suppose that the training set $\mathcal{D}$ consists of L labeled data and U unlabeled data. We denote the label dataset as $\mathcal{D}_L = \{(\mathbf{x}_i, \mathbf{y}_i)\}_{i=1}^{L}$ and the unlabeled dataset as $\mathcal{D}_U = \{\mathbf{x}_i\}_{i=L+1}^{L+U}$. Here, $\mathbf{x}_i \in \mathbb{R}^{h\times w}$ denotes the ith input data and $\mathbf{y}_i \in \mathbb{R}^c$

is the corresponding one hot ground-truth label. We aim to learn a reliability-aware classification model f using limited labeled data and abundant reliable unlabeled data, where the model is expected to perform well on the test data.

The proposed reliability-aware scheme achieves semi-supervised learning based on MT model. The objective function of the MT [23] is formulated as:

$$\min_{\theta} \sum_{i=1}^{L} \mathcal{L}_{sup}\left(f\left(\mathbf{x}_i;\theta\right),\mathbf{y}_i\right) + \lambda_c \sum_{i=1}^{L+U} \mathcal{L}_{cs}\left(f\left(\mathbf{x}_i;\theta',\eta'\right),f\left(\mathbf{x}_i;\theta,\eta\right)\right). \tag{1}$$

Here, $\mathcal{L}_{sup}$ denotes the supervised loss. $\mathcal{L}_{cs}$ is the consistency loss. θ and θ' are the parameters of student model and teacher model respectively. The teacher network parameter θ' is updated as the exponential moving average (EMA) of the student network parameter θ. η and η' are different input perturbations. λ_c is the weight parameter that balances the supervised and unsupervised loss. Note that current semi-supervised learning methods use a single scalar to equally weight all the unlabeled data. As mentioned above, it may adversely affect the performance of models when difference exists among the unlabeled data.

Weight Function. To prevent potential performance degradation caused by treating all unlabeled data equally, we design a weight function which can automatically assign each unlabeled data with a unique weight. The weight function $g : \mathbb{R}^c \rightarrow \mathbb{R}$ parameterized by $\mathbf{w} \in \mathbb{R}^c$, maps the probability prediction of unlabeled data $\mathbf{x}_i$ to the corresponding weight β_i which can be expressed as:

$$\beta_i = g\left(f\left(\mathbf{x}_i;\theta,\eta\right),\mathbf{w}\right). \tag{2}$$

To find the proper weights of unlabeled data, we enable the model to keep tracking the classification performance of the labeled data to prevent performance degradation when using the weighted unlabeled data.

Reliability-Aware Consistency Loss. Self-ensembling methods are based on the smoothness assumption that data points close to each other in data space are more likely to share the same label [8,19]. Following this assumption, the MT model encourages the consensus among ensemble predictions of unlabeled data under small input perturbations. Although effective, current methods do not consider the reliability of unlabeled data and use them in a flat way, which may adversely affect the classification performance of self-ensembling methods.

To tackle this issue, we propose a novel reliability-aware consistency loss, incorporating the weight function to prevent performance degradation caused by leveraging the unreliable unlabeled data, which is defined as:

$$\mathcal{L}_{rcs} = g\left(f\left(\mathbf{x}_i;\theta,\eta\right),\mathbf{w}\right) \cdot \mathcal{L}_C = \beta_i \mathcal{L}_C, \tag{3}$$

where the weight parameter β_i is automatically identified using Eq. (2). $\mathcal{L}_C$ is a the distance metric, such as the Kullback-Leibler (KL) divergence [22] and the minimum mean-square error (MSE). It can be seen that the proposed reliability-aware consistency loss only penalizes inconsistent predictions of reliable unlabeled data under different input perturbations.

Reliability-Aware Contrastive Loss. Current self-ensembling methods only consider the perturbations around individual data point, while ignoring the connections between them, leaving the data-structure-level information unexploited. To this end, Liu *et al.*. [11] proposed to address this issue by using data relationship. However, when the unlabeled data turn out to be unreliable, the model may suffer from more severe performance degradation.

To mitigate performance degradation due to misleading data-structure-level information caused by unreliable unlabeled data, we propose a novel reliability-aware contrastive loss, which can be expressed as:

$$\mathcal{L}_{rct} = \sum_{i \in \mathcal{Z}^s \cup \mathcal{Z}^t} \frac{-1}{|P(i)|} \sum_{p \in P(i)} \log \frac{\exp\left[(\beta_i \mathbf{z}_i) \cdot (\beta_p \mathbf{z}_p)/\tau\right]}{\sum_{a \in A(i)} \exp\left[(\beta_i \mathbf{z}_i) \cdot (\beta_a \mathbf{z}_a)/\tau\right]}. \tag{4}$$

Here, $\mathcal{Z}^s \equiv \{\mathbf{z}_k\}_{k=1}^{B}$ denotes the normalized embeddings output by the student model derived from a batch of B unlabeled data. Similarly, $\mathcal{Z}^t$ is the normalized embeddings output by the teacher model. $\beta_i, i = 1, 2, \cdots, 2B$ is the weight parameter of $\mathbf{z}_i$. For the ith unlabeled data $\mathbf{x}_i$, the hard target pseudo-label obtained by the teacher model is expressed as $\hat{\mathbf{y}}_i$. $A(i) \equiv \mathcal{Z}^s \cup \mathcal{Z}^t \backslash \{i\}$, and $P(i) \equiv \{p \in A(i) : \hat{\mathbf{y}}_p = \hat{\mathbf{y}}_i\}$ denotes the set of indices of all positives in $A(i)$. τ is a positive scalar temperature parameter.

Our reliability-aware contrastive loss focuses on the connections between the reliable unlabeled data, leading to better intra-class compactness and inter-class separability for the normalized embeddings of them.

Objective Function. The objective of our RAC-MT can be formulated as the following bi-level optimization problem [1,12]:

$$\begin{aligned} &\min_{\mathbf{w}} \mathcal{L}_s\left(\mathcal{D}_L; \hat{\theta}(\mathbf{w})\right) \\ &s.t.\ \hat{\theta}(\mathbf{w}) = \operatorname*{argmin}_{\theta} \mathcal{L}_s(\mathcal{D}_L; \theta) + \lambda_1 \sum_{\mathbf{x} \in \mathcal{D}} \mathcal{L}_{rcs}(\mathbf{x}, \boldsymbol{\beta}; \theta) + \lambda_2 \sum_{\mathbf{x} \in \mathcal{D}} \mathcal{L}_{rct}(\mathbf{x}, \boldsymbol{\beta}; \theta), \end{aligned} \tag{5}$$

where $\boldsymbol{\beta}$ subsumes $\boldsymbol{\beta}_L \in \mathbb{R}_{=1}^{|\mathcal{D}_L|}$ and $\boldsymbol{\beta}_U \in \mathbb{R}_{\geq 0}^{|\mathcal{D}_U|}$ is automatically identified using Eq. (2). $\mathcal{L}_s(\cdot; \theta)$ denotes the supervised loss. Intuitively, Eq. (5) minimizes $\mathcal{L}_s$ with respect to the parameter $\mathbf{w}$ of the weight function, while being given the model parameters $\hat{\theta}(\mathbf{w})$ which minimize the overall training loss.

2.2 Optimization

To solve the bi-level optimization problem in Eq. (5), we denote the outer-level objective as $\mathcal{L}_{outer}(\theta)$ and the inner-level objective as $\mathcal{L}_{inner}(\theta, \mathbf{w})$. Then, we propose to iteratively alternate between the update of the parameter $\mathbf{w}$ of the weight function and the network parameter θ.

Updating θ. By fixing the parameter $\mathbf{w}_t$ of the weight function g, the parameter θ_{t+1} can be easily optimized as the following single-level optimization:

$$\theta_{t+1} = \theta_t - \eta_\theta \cdot \nabla_\theta \mathcal{L}_{inner}(\theta_t, \mathbf{w}_t), \tag{6}$$

where η_θ is the step size.

Updating $\mathbf{w}$. With the network parameter θ_{t+1}, the parameter $\mathbf{w}$ of the weight function g can be updated based on the gradient of the supervised loss:

$$\begin{aligned}\mathbf{w}_{t+1} &= \mathbf{w}_t - \eta_{\mathbf{w}} \cdot \nabla_{\mathbf{w}} \mathcal{L}_{outer}\left(\theta_{t+1}\right) \\ &= \mathbf{w}_t - \eta_{\mathbf{w}} \cdot \nabla_{\mathbf{w}} \mathcal{L}_{outer}\left(\theta_t - \eta_\theta \cdot \nabla_\theta \mathcal{L}_{inner}\left(\theta_t, \mathbf{w}_t\right)\right),\end{aligned} \tag{7}$$

where $\eta_{\mathbf{w}}$ is the step size. The gradient of $\mathcal{L}_{outer}\left(\theta_{t+1}\right)$ with respect to $\mathbf{w}_t$ can be easily computed through the automatic differentiation techniques in existing deep learning frameworks.

3 Experiments and Results

The proposed RAC-MT was extensively evaluated on two medical image datasets, *i.e.*, ISIC 2018 skin dataset[1] and CRCHistoPhenotypes nucleus dataset[2].

3.1 Implementation Details

The ISIC 2018 skin dataset provides 10015 skin lesion dermoscopy images, including 7 kinds of skin lesions. All the images were resized into the size of 224×224. The CRCHistoPhenotypes dataset contains a total of 22444 nuclei patches, which were labeled into 4 types of nucleus. The size of all input patches was set to $32 \times 32 \times 3$. To utilize the pre-trained model, we normalized each image of the two datasets with the statistics collected from the ImageNet dataset [13]. The entire dataset was randomly divided, 70% of which was for training, 10% for validation and 20% for testing.

Our framework was implemented with PyTorch and trained on a RTX 3090 GPU. Referring to [11], we used two types of perturbations in the model training. The smoothing parameter of EMA was set to 0.99. In the first T training epochs, the time-dependent Gaussian warming up function $\lambda(t) = \exp\left(-5(1-t/T)^2\right)$ [4] was adopted to ramp up the parameters λ_1 from 0 to 1. λ_2 was ramped up from 0 to 0.05 and 0.1 respectively for the two datasets. The temperature parameter τ was set to 0.5. The batch size was set to 64 with 16 labeled samples and 48 unlabeled samples. A single linear layer with 128 hidden neurons was used as the projection network for RAC-MT. Besides, MSE was used for $\mathcal{L}_{rcs}$. We adopted Adam optimizer for model training. $\eta_{\mathbf{w}}$ was fixed as $3e^{-5}$. The value of η_θ for the two datasets was initialized as $1e^{-4}$ and $2e^{-4}$ respectively, which was decayed with cosine annealing [21]. The number of training iterations was set to 180.

Our RAC-MT was compared with the baseline DenseNet 121 [5], and several state-of-the-art semi-supervised methods, including Π model [14], MT [17], SRC-MT [11], SelfMatch [7] and DS^{3}L [2]. DenseNet was used as the backbone for all the comparison methods. Following [11], we adopted AUC, Accuracy, Sensitivity, Specificity and F1 as the metrics to evaluate classification performance.

[1] https://challenge2018.isic-archive.com.

[2] https://warwick.ac.uk/fac/cross_fac/tia/data/crchistolabelednucleihe.

Table 1. Classification results of all comparison methods on skin dataset

Methods	Percentage		Metrics				
	Labeled	Unlabeled	AUC	Sensitivity	Specificity	Accuracy	F1
Upper Bound	100%	0	96.43	85.20	95.21	95.45	74.96
Baseline [5]	20%	0	90.90	69.37	91.77	91.42	51.89
Π model [14]	20%	80%	91.54	69.84	92.21	91.74	58.15
MT [17]	20%	80%	93.34	70.15	92.45	92.39	60.36
SRC-MT [11]	20%	80%	93.45	70.42	92.31	92.59	58.18
SelfMatch [7]	20%	80%	92.92	72.09	92.44	91.96	59.67
DS^3L [2]	20%	80%	93.85	70.33	92.29	92.53	61.08
RAC-MT	20%	80%	**94.42**	**73.41**	**92.68**	**93.27**	**63.95**

Table 2. Classification results with different percentages of labeled data on skin dataset

Methods	Percentage		Metrics				
	Labeled	Unlabeled	AUC	Sensitivity	Specificity	Accuracy	F1
Upper Bound	100%	0	96.43	85.20	95.21	95.45	74.96
Baseline	5%	0	83.57	54.89	89.45	88.70	41.65
DS^3L	5%	95%	85.08	58.82	89.52	89.27	44.19
RAC-MT	5%	95%	**87.92**	**59.34**	**90.51**	**91.11**	**48.54**
Baseline	10%	0	86.00	61.82	90.62	90.02	44.49
DS^3L	10%	90%	88.73	62.62	90.79	91.72	52.63
RAC-MT	10%	90%	**90.98**	**64.67**	**91.01**	**91.75**	**54.23**
Baseline	20%	0	90.90	69.37	91.77	91.42	51.89
DS^3L	20%	80%	93.85	70.33	92.29	92.53	61.08
RAC-MT	20%	80%	**94.42**	**73.41**	**92.68**	**93.27**	**63.95**

3.2 Evaluation of Medical Image Classification

Results on ISIC 2018 Skin Dataset. The classification results of comparison methods on the ISIC 2018 Skin dataset with 20% labeled data setting are listed in Table 1. The classification performance of the fully supervised DenseNet with 100% labeled data is considered as the upper-bound, and that with 20% labeled data as the baseline. In Table 1, our RAC-MT achieved better performance than Π model and MT, both of which utilized the data-level information. In addition, our RAC-MT outperformed SRC-MT and SelfMatch, which jointly considered the data-level and data-structure-level information. Compared to the most competitive method DS^3L, RAC-MT shows an absolute increase in AUC, Sensitivity, Specificity, Accuracy and F1 by 0.57%, 3.08%, 0.39%, 0.74% and 2.87% respectively. The consistent improvements on all metrics over the comparison methods demonstrates the effectiveness of our reliability-aware scheme which jointly exploits the reliable data-level and data-structure-level information.

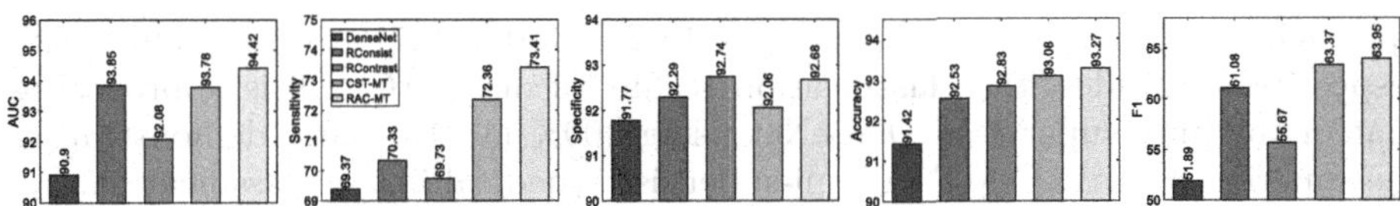

Fig. 2. Ablation study on different components.

Table 3. Classification results of all comparison methods on nucleus dataset.

Methods	Percentage		Metrics				
	Labeled	Unlabeled	AUC	Sensitivity	Specificity	Accuracy	F1
Upper Bound	100%	0	94.26	82.56	93.24	91.16	79.75
Baseline [5]	20%	0	87.23	73.75	87.67	85.34	68.43
Π model [14]	20%	80%	90.79	73.94	87.68	86.67	69.90
MT [17]	20%	80%	90.89	75.90	88.65	87.20	72.11
SRC-MT [11]	20%	80%	91.18	75.95	88.66	87.14	71.90
SelfMatch [7]	20%	80%	91.97	78.24	88.07	87.60	72.83
DS^3L [2]	20%	80%	91.61	77.59	86.72	86.61	71.52
RAC-MT	20%	80%	**92.99**	**82.15**	**88.78**	**88.57**	**74.90**

To verify the performance of RAC-MT using different percentages of labeled data, we compared it with the baseline and DS^3L on skin dataset. As shown in Table 2, RAC-MT is able to make improvements over the baseline and DS^3L in all cases. Notably, the RAC-MT trained with 5% labeled data shows comparable classification results with that of DS^3L with 10% labeled data. The promising results further demonstrate the efficiency of the proposed reliability-aware scheme.

We further conducted an ablation study to investigate the role of each component in RAC-MT, as illustrated in Fig. 2. The ablation study was performed from five aspects: 1) baseline DenseNet ($\lambda_1 = \lambda_2 = 0$), 2) DenseNet with reliability-aware consistency loss (RConsist) ($\lambda_2 = 0$), 3) DenseNet with reliability-aware contrastive loss (RContrast) ($\lambda_1 = 0$), 4) DenseNet with consistency loss and contrastive loss (CST-MT) ($\beta = 1$), and 5) RAC-MT. As shown in Fig. 2, both RConsist and RContrast achieve better classification results than DenseNet, showing the efficiency of two reliability-aware losses. The outperformance of RAC-MT reveals that the significance of the complementary roles of reliable data-level and data-structure-level information. RAC-MT outperforms CST-MT, which further verifies the reliability of the proposed method.

Results on CRCHistoPhenotypes Dataset. We further evaluated our RAC-MT on the nucleus dataset. The classification results of all methods with 20% labeled data setting are given in Table 3. Similarly, our RAC-MT obtains the best classification results among all comparison methods. Compared to the most competitive method SelfMatch, RAC-MT achieves absolute AUC, Sensitivity,

Specificity, Accuracy and F1 increase by 1.02%, 3.91%, 0.71%, 0.97% and 2.07% respectively. Besides, the classification results of our RAC-MT is approaching that of the fully supervised DenseNet using 100% labeled data, demonstrating the superiority of RAC-MT in semi-supervised medical image classification.

4 Conclusion

We propose a novel reliability-aware contrastive MT model for semi-supervised medical image classification. The designed reliability-aware scheme automatically assign each unlabeled data a unique weight that can reflect its reliability. Based on this, we further exploited a self-ensembling framework to concurrently capture the reliable data-level and data-structure-level information. Extensive experiments demonstrate the superiority of our model in medical image classification. Future work includes investigating the effect of different weight functions and applying our framework to other medical image classification tasks.

Acknowledgments. This work is supported in part by the National Natural Science Foundation of China (61902197, 61802177), the Open Project of State Key Laboratory for Novel Software Technology at Nanjing University (KFKT2020B11), and Hong Kong Research Grants Council under General Research Fund (15218521).

References

1. Bard, J.F.: Practical Bilevel Optimization: Algorithms and Applications, vol. 30. Springer, Dordrecht (2013). https://doi.org/10.1007/978-1-4757-2836-1
2. Guo, L.Z., Zhang, Z.Y., Jiang, Y., Li, Y.F., Zhou, Z.H.: Safe deep semi-supervised learning for unseen-class unlabeled data. In: International Conference on Machine Learning, pp. 3897–3906. PMLR (2020)
3. Gyawali, P.K., Ghimire, S., Bajracharya, P., Li, Z., Wang, L.: Semi-supervised medical image classification with global latent mixing. In: Martel, A.L., et al. (eds.) MICCAI 2020. LNCS, vol. 12261, pp. 604–613. Springer, Cham (2020). https://doi.org/10.1007/978-3-030-59710-8_59
4. Hang, W., et al.: Local and global structure-aware entropy regularized mean teacher model for 3D left atrium segmentation. In: Martel, A.L., et al. (eds.) MICCAI 2020. LNCS, vol. 12261, pp. 562–571. Springer, Cham (2020). https://doi.org/10.1007/978-3-030-59710-8_55
5. Huang, G., Liu, Z., Van Der Maaten, L., Weinberger, K.Q.: Densely connected convolutional networks. In: Proceedings of the IEEE Conference on Computer Vision and Pattern Recognition, pp. 4700–4708 (2017)
6. Khosla, P., et al.: Supervised contrastive learning. Adv. Neural Inf. Process. Syst. **33**, 18661–18673 (2020)
7. Kim, B., Choo, J., Kwon, Y.D., Joe, S., Min, S., Gwon, Y.: Selfmatch: combining contrastive self-supervision and consistency for semi-supervised learning. arXiv preprint arXiv:2101.06480 (2021)
8. Li, X., Yu, L., Chen, H., Fu, C.W., Xing, L., Heng, P.A.: Transformation-consistent self-ensembling model for semisupervised medical image segmentation. IEEE Trans. Neural Netw. Learn. Syst. **32**(2), 523–534 (2020)

9. Liu, F., Tian, Yu., Cordeiro, F.R., Belagiannis, V., Reid, I., Carneiro, G.: Self-supervised mean teacher for semi-supervised chest X-Ray classification. In: Lian, C., Cao, X., Rekik, I., Xu, X., Yan, P. (eds.) MLMI 2021. LNCS, vol. 12966, pp. 426–436. Springer, Cham (2021). https://doi.org/10.1007/978-3-030-87589-3_44
10. Liu, Q., Yang, H., Dou, Q., Heng, P.-A.: Federated semi-supervised medical image classification via inter-client relation matching. In: de Bruijne, M., et al. (eds.) MICCAI 2021. LNCS, vol. 12903, pp. 325–335. Springer, Cham (2021). https://doi.org/10.1007/978-3-030-87199-4_31
11. Liu, Q., Yu, L., Luo, L., Dou, Q., Heng, P.A.: Semi-supervised medical image classification with relation-driven self-ensembling model. IEEE Trans. Med. Imaging **39**(11), 3429–3440 (2020)
12. Liu, R., Gao, J., Zhang, J., Meng, D., Lin, Z.: Investigating bi-level optimization for learning and vision from a unified perspective: a survey and beyond. IEEE Trans. Pattern Anal. Mach. Intell. (2021). https://doi.org/10.1109/TPAMI.2021.3132674
13. Russakovsky, O., et al.: Imagenet large scale visual recognition challenge. Int. J. Comput. Vis. **115**(3), 211–252 (2015)
14. Samuli, L., Timo, A.: Temporal ensembling for semi-supervised learning. In: International Conference on Learning Representations, vol. 4 (2017)
15. Sirinukunwattana, K., Raza, S.E.A., Tsang, Y.W., Snead, D.R., Cree, I.A., Rajpoot, N.M.: Locality sensitive deep learning for detection and classification of nuclei in routine colon cancer histology images. IEEE Trans. Med. Imaging **35**(5), 1196–1206 (2016)
16. Su, H., Shi, X., Cai, J., Yang, L.: Local and global consistency regularized mean teacher for semi-supervised nuclei classification. In: Shen, D., et al. (eds.) MICCAI 2019. LNCS, vol. 11764, pp. 559–567. Springer, Cham (2019). https://doi.org/10.1007/978-3-030-32239-7_62
17. Tarvainen, A., Valpola, H.: Mean teachers are better role models: weight-averaged consistency targets improve semi-supervised deep learning results. In: Advances in Neural Information Processing Systems, pp. 1195–1204 (2017)
18. Tschandl, P., Rosendahl, C., Kittler, H.: The ham10000 dataset, a large collection of multi-source dermatoscopic images of common pigmented skin lesions. Sci. Data **5**(1), 1–9 (2018)
19. van Engelen, J.E., Hoos, H.H.: A survey on semi-supervised learning. Mach. Learn. **109**(2), 373–440 (2019). https://doi.org/10.1007/s10994-019-05855-6
20. Wang, R., Wu, Y., Chen, H., Wang, L., Meng, D.: Neighbor matching for semi-supervised learning. In: de Bruijne, M., et al. (eds.) MICCAI 2021. LNCS, vol. 12902, pp. 439–449. Springer, Cham (2021). https://doi.org/10.1007/978-3-030-87196-3_41
21. Wang, Z., Liu, Q., Dou, Q.: Contrastive cross-site learning with redesigned net for covid-19 CT classification. IEEE J. Biomed. Health Inf. **24**(10), 2806–2813 (2020)
22. Yang, P., Chen, B.: Robust kullback-leibler divergence and universal hypothesis testing for continuous distributions. IEEE Trans. Inf. Theor. **65**(4), 2360–2373 (2018)
23. Yu, L., Wang, S., Li, X., Fu, C.-W., Heng, P.-A.: Uncertainty-aware self-ensembling model for semi-supervised 3D left atrium segmentation. In: Shen, D., et al. (eds.) MICCAI 2019. LNCS, vol. 11765, pp. 605–613. Springer, Cham (2019). https://doi.org/10.1007/978-3-030-32245-8_67

Author Index

Printed in the United States
by Baker & Taylor Publisher Services